twelfth edition

Environmental Geology

Carla W. Montgomery

Professor Emerita Northern Illinois University

Gina Seegers Szablewski

University of Wisconsin- Milwaukee

Mc
Graw
Hill

ENVIRONMENTAL GEOLOGY

Cover Image: *Gina S. Szablewski*

In Dedication

This edition of *Environmental Geology* is
dedicated in memory of Carla for leading
the way, forging on, and then graciously
handing this work to me.

–GSS–

Preface

About the Course

Environmental Geology Is Geology Applied to Humans

The *environment* is the sum of all the features and conditions surrounding an organism that may influence it. An individual's physical environment encompasses rocks and soil, air and water, such factors as light and temperature, and other organisms. Geology is the study of Earth. Because Earth provides the basic physical environment in which we live, all of geology might in one sense be regarded as environmental geology. However, the term *environmental geology* is usually restricted to refer particularly to geology as it relates directly to human activities, and that is the focus of this book. Environmental geology is how geology affects us and how we affect the planet. We will examine the relationship between human activities and various geologic processes and hazards, the geologic aspects of pollution and waste-disposal problems, and several other ways in which geology directly and indirectly affects our lives.

Why Study Environmental Geology?

One reason for studying environmental geology is simply curiosity about the way Earth works, about the *how* and *why* of natural phenomena. Another reason is that we are increasingly faced with environmental problems to be solved and decisions to be made, particularly about how our actions are affecting climate, which in turn are changing the environment in which we live. It is essential that we understand geologic processes so we can make informed choices and find appropriate solutions to the challenges we are facing.

Many environmental problems cannot be fully assessed and solved using geologic data alone. The problems vary widely in size and in complexity. In a specific instance, we need to take into account data from other branches of science as well as economics, politics, and social priorities. Because a variety of considerations may influence the choice of a solution, we frequently disagree about which solution is best. Our personal choices and assessment of risk will often depend strongly on our group identity and beliefs about which considerations are most important.

About the Book

An introductory text cannot explore all aspects of environmental concerns. Here, the emphasis is on the physical constraints imposed on human activities by the geologic processes that have shaped and are still shaping our natural environment. In a real sense, these are the most basic, inescapable constraints; we cannot, for instance, use a resource that is not there, or build a secure home or a safe dam on land that is fundamentally unstable. Geology is a logical place to start in developing an understanding of many environmental issues. The principal aim of this book is to present the reader with a broad overview of environmental geology. Because geology doesn't exist in a vacuum, the text introduces related considerations from outside geology to clarify other ramifications of the subjects discussed. Likewise, the present doesn't exist in isolation from the past and future; occasionally, the text looks at both how Earth developed into its present condition and where matters seem to be moving for the future. One of the primary goals of this book is that the knowledge contained within it will provide the reader with a useful foundation for discussing and evaluating specific environmental issues, as well as for developing ideas about how the problems should be solved.

Features Designed for the Student

This text is intended for an introductory-level college course. It doesn't assume any prior exposure to geology or college-level mathematics or science courses. The metric system is used throughout, except where other units are conventional within a discipline. For the convenience of students not yet fluent in metric units, a conversion table is included in Appendix C, and in some cases, metric equivalents in English units are included within the text.

Each chapter opens with an introduction and an outline that set the stage for the material to follow. Each chapter section has a list of **Learning Objectives,** those ideas that are central to that section, and can be used as a study guide for the student. In the course of the chapter, important terms and concepts are identified by boldface type, and these terms are collected as **Key Terms and Concepts** at the end of the chapter for quick review. The **Glossary** includes both these boldface terms and the additional, italicized terms that many chapters contain. Each chapter includes one or more **Case Studies.** Some involve a situation, problem, or application that might be encountered in everyday life. Others offer additional case histories or examples relevant to chapter contents. Every chapter concludes with review exercises, which allow students to test their comprehension and apply their knowledge. The **Exploring Further** section of each chapter includes a number of activities in which students can engage, some involving online data, and some, quantitative analysis. For example, they may be directed to examine real-time stream-gauging or landslide-monitoring data, or information on current or recent earthquake activity; they can manipulate historic climate data from NASA to examine trends by region or time period; they may calculate how big a wind farm or photovoltaic array would be required to replace a conventional power plant; they can even learn how to reduce sulfate pollution by buying SO_2 allowances.

Any text of this kind must necessarily be a snapshot in time: Earth keeps evolving and presenting us with new geologic challenges; our understanding of our world advances; our responses to our environment change. And of course, there is vastly more relevant material that might be included than will fit in one volume. To address both of these issues, at least in part, online resources have been developed (and will be periodically updated) for each chapter, and will be available to instructors to share with students as needed. These will include references, suggested readings, and other relevant content.

New and Updated Content

Environmental geology is, by its very nature, a dynamic field in which new issues continue to arise and old ones to evolve. Every chapter has been updated with regard to data, examples, events, photos, and figures with the most recent information available at the time of authoring.

Images Geology is a visual subject, and photographs, satellite imagery, diagrams, and graphs all enhance students' learning. Accordingly, this edition includes more than 100 new or improved photographs/images and nearly sixty new figures, and revisions have been made to dozens more. Image descriptions are now directly below each individual image rather than being grouped underneath a collection of images.

New Author Approach This edition contains more of the active voice, with the intention that students understand that there are real people behind all the data and actions contained in the book. The language used is also less formal, attempting to make science a bit more approachable. Each chapter opening now has a content list for that chapter, and each chapter section has a list of learning objectives. Chapter subsections also have clearer introductions about what content is covered. Some titles have changed or been added in order to clarify the content contained within. All case studies have been moved to the end of their pertinent chapter, and the images within them have unique figure numbers. For the digital edition, a few short videos are included, and many more direct links to data used and referenced have been inserted. Common themes addressed in multiple chapters and brought more to the forefront in this edition include climate, environmental justice, and the availability of data collected by remote sensing and summarized in GIS. Overall, this edition was the first step in transitioning the content to a new author, and for it to be a digital product first, since that is the most common way that students and instructors will approach the material.

The existing format and order or appearance has been maintained. Many chapters have new or altered subtitles for existing content. Significant organizational and content changes by chapter include:

Chapter 1 The four outer planets are separated into gas giants and ice giants; more information about the planets has been added in section 1.1.

Chapter 2 More information about classifying rocks has been added in section 2.4; reference information has been added to table 2.1; Figure 2.2 Periodic Table columns are now numbered 1–18.

Chapter 4 Historic earthquakes (mentioned in this chapter) are summarized in map form in figure 4.1; the subsection "Induced Seismicity" has been moved to section 4.5 for clearer organization.

Chapter 5 Subsections "Volcano Location" and "Hot Spot Volcanoes" have been added to section 5.2; terminology has been updated so that *nuee ardente* is replaced with *pyroclastic flow*.

Chapter 6 Updated terminology is used to describe flood hazard designation and flooding probability.

Chapter 7 Section 7.5 has added information about managed retreat.

Chapter 8 The Caraballeda, Thistle landslide, and Pakistan stories have been moved to a new **Examples** subsection in 8.2; subsections in 8.4 have been reorganized.

Chapter 9 Section 9.1 rearranges possible causes and adds more plate tectonics information; section 9.3 rearranges and supplements causes of natural deserts; Case Study 9 discusses more about the Himalayas and Andes.

Chapter 11 Content has been streamlined so that less relevant content (such as international water disputes) has been removed.

Chapter 13 Figure 13.2 removes gold, tin, and lead and adds graphite, lithium, and REE; tables 13.1, 13.2, and 13.3 also have changes to the resources included.

Chapter 14 Removes Figure 14.1 (energy consumption vs GDP) and replaces it with a table showing GDP and CO_2 emissions data; removes Figure 14.7 (peak oil) and its discussion; adds a new subsection in 14.2 titled "Shale Gas and Shale Oil"; overall discussion changes from the United States being oil-poor and using much coal to the United States being oil-rich and using much less coal.

Chapter 15 New information about batteries and storage of energy has been added to section 15.3; subsection "Wood" has been added to section 15.8.

Chapter 16 "Open Dumps" subsection has been added in 16.2; section 16.6 now summarizes former content under new subsection "Possibilities for High-Level Waste Disposal."

Chapter 17 Adds information about the Flint, Michigan, water crisis; section 17.4 removes the MTBE paragraph and replaces it with information about PCBs, PFAs, and pharmaceuticals; the content of Case Study 17.1 (minimata disease) has been moved to section 17.4;

Case Study 17.2 is now 17.1; the content of section 17.8 (Gulch Superfund Site) is now Case Study 17.2.

Chapter 19 Information about ecosystem services and calculation of their worth has been added in section 19.5.

Organization

This book is informally divided into **six sections.** The **first** contains background information: a brief outline of Earth's development to the present, and a look at the rapidly growing human population as one of the primary reasons why environmental problems today are so pressing. This is followed by a short discussion of the basic materials of geology—rocks and minerals—and some of their physical properties, which introduces a number of basic terms and concepts that are used in later chapters.

The **second** section contains several chapters that cover internal geologic processes in detail. These are relatively large-scale processes, which can involve motions and forces in Earth hundreds of kilometers below the surface, and may lead to dramatic, often catastrophic events like earthquakes and volcanic eruptions. The **third** section involves processes that occur near Earth's surface, altering the landscape and occasionally causing their own special problems. This section concludes with a chapter on climate, which connects a number of the processes described earlier.

The **fourth** section covers the availability of various resources, the rates at which they are being consumed, probable amounts remaining; and projections of future availability and use. The three chapters of the **fifth** section examine the interrelated problems of air and water pollution and the strategies available for the disposal of various kinds of wastes.

The **sixth** section looks briefly at a sampling of laws, policies, and international agreements related to geologic matters discussed earlier in the book; and examines geologic constraints on construction schemes and the broader issue of trying to determine the optimum use(s) for particular parcels of land.

Supplementary material is contained in the appendices. Appendix A explores the concept of geologic time and its measurement. Appendix B provides short reference keys to aid in rock and mineral identification, and Appendix C includes units of measurement and conversion factors.

The complex interrelationships among geologic processes and features mean that any subdivision into chapter-sized pieces is somewhat arbitrary, and different instructors may prefer different sequences or groupings. An effort has been made to design chapters so that they can be resequenced in such ways without great difficulty.

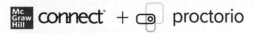

The 12th edition of *Environmental Geology* is available online with Connect, McGraw Hill's integrated assignment and assessment platform. Connect also offers SmartBook for the new edition, which is the first adaptive reading experience proven to improve grades and help students study more effectively. All of the title's website and ancillary content is also available through Connect, including a Question Bank, Test Bank, Lecture PowerPoints, and Instructor's Manual with answers to review questions.

Proctorio: Remote Proctoring and Browser-Locking Capabilities

Remote proctoring and browser-locking capabilities, hosted by Proctorio within Connect, provide control of the assessment environment by enabling security options, verifying the identity of the student, and restricting browser activity.

Instant and detailed reporting gives instructors an at-a-glance view of potential academic integrity concerns, thereby avoiding personal bias and supporting evidence-based claims.

OLC-Aligned Courses: Implementing High-Quality Instruction and Assessment Through Pre-configured Courseware

In consultation with the Online Learning Consortium (OLC) and our certified Faculty Consultants, McGraw Hill has created pre-configured courseware using OLC's quality scorecard to align with best practices in online course delivery. This turnkey courseware contains a combination of formative assessments, summative assessments, homework, and application activities, and can easily be customized to meet an individual instructor's needs and desired course outcomes. For more information, visit https://www.mheducation.com/highered/olc.

Test Builder in Connect

Available within Connect, Test Builder is a cloud-based tool that enables instructors to format tests that can be printed, administered within a Learning Management System, or exported as a Word document. Test Builder offers a modern, secure interface for easy content configuration that matches course needs, without requiring a download. It allows for just-in-time updates to flow directly into assessments.

Application-Based Activities in Connect

Application-Based Activities in Connect are highly interactive, assignable exercises that provide students a safe space to apply the concepts they have learned to real-world, course-specific problems. Each Application-Based Activity involves the application of multiple concepts, allowing students to synthesize information and use critical thinking skills to solve realistic scenarios.

About the New Author

Gina Seegers Szablewski has taught large introductory geology classes at the University of Wisconsin-Milwaukee for 20 years with a total of over 18,000 students. Having received her BA in geology from Lawrence University and MS in geology (sedimentology) from the University of Wisconsin-Milwaukee, she first worked as a geologist in environmental consulting for 7 years. Currently she teaches physical and environmental geology and earth science classes in person and online,

Courtesy of Joshua Szablewski

while also working with McGraw Hill in a variety of roles, not just as the author of *Environmental Geology* but also as a digital faculty consultant and lead digital author. She enjoys sharing with other teachers her experiences using digital content in and out of the classroom to improve both teaching and learning. She is an active member of the National Association of Geoscience Teachers and the Geological Society of America, and she frequently attends short courses and workshops associated with geoscience education. When she is not learning, teaching, and thinking about science and science education, she enjoys hiking, traveling, yoga, walking her standard poodle, reading, cooking, and crochet. She and her husband have two children—one in college (an engineer) and one in grad school (an earth scientist).

Acknowledgments

There are many people to thank, both for supporting me so I could get to the point of tackling a project such as this, and then those who did actually help me with the project. For the former, my parents Carol and Dennis Seegers; my husband Josh; and Erin and Evan for being my first and always students, grudgingly or not.

There are a few people who fall between these two categories: namely, the authors whose books I have used, created digital content for, and learned much from—Julia Johnson, David McConnell, Jim Reichard, Keith Sverdrup, and Steve Reynolds; Jodi Rhomberg and Joan Weber, for giving me the work that inspired me to make quality science content for other instructors to use; John Isbell—mentor, colleague, friend—for taking a chance on me twice; and Trent McDowell, on and off collaborator in all things geology. I would also like to thank the nearly 20,000 students I have had the honor to share my love of science and geology with.

In the latter category are those who helped with subject matter expertise: Dyanna Czeck, Jean Creighton, Jonathan Godt, Rob Graziano, John Isbell, Brett Ketter, Charlie Paradis, Steve Reynolds, Bradd Seegers, and Dylan Wilmeth; and those who helped with the creation and production of the text: Krystal Faust, Jane Mohr, Melissa Homer, Ramya Thirumavalavan, Adina Lonn, and Kathryn Pauls. (My sincerest apologies for leaving out anyone.)

Gina Seegers Szablewski

Students
Get Learning that Fits You

Effective tools for efficient studying

Connect is designed to help you be more productive with simple, flexible, intuitive tools that maximize your study time and meet your individual learning needs. Get learning that works for you with Connect.

Study anytime, anywhere

Download the free ReadAnywhere® app and access your online eBook, SmartBook® 2.0, or Adaptive Learning Assignments when it's convenient, even if you're offline. And since the app automatically syncs with your Connect account, all of your work is available every time you open it. Find out more at
mheducation.com/readanywhere

"I really liked this app—it made it easy to study when you don't have your text-book in front of you."

- Jordan Cunningham, Eastern Washington University

iPhone: Getty Images

Everything you need in one place

Your Connect course has everything you need—whether reading your digital eBook or completing assignments for class—Connect makes it easy to get your work done.

Learning for everyone

McGraw Hill works directly with Accessibility Services Departments and faculty to meet the learning needs of all students. Please contact your Accessibility Services Office and ask them to email accessibility@mheducation.com, or visit **mheducation.com/about/accessibility** for more information.

Contents

SECTION II INTERNAL PROCESSES

SECTION III **SURFACE PROCESSES**

CHAPTER **6**

Streams and Flooding   132

CHAPTER **7**

Coastal Zones and Processes   162

SECTION IV **RESOURCES**

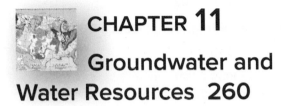

CHAPTER **11**

Groundwater and Water Resources   260

CHAPTER **12**

Weathering, Erosion, and Soil Resources   290

SECTION V WASTE DISPOSAL, POLLUTION, AND HEALTH

CHAPTER 18

Air Pollution   471

SECTION VI OTHER RELATED TOPICS

CHAPTER 19

Environmental Law and Policy   500

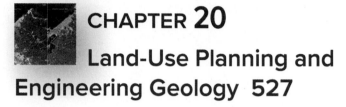

CHAPTER 20

Land-Use Planning and Engineering Geology 527

Chapter-opening photo credits: Chapter 1: Stewart Innes/Alamy Stock Photo; Chapters 2, 9, 15: ©Gina S. Szablewski; Chapter 3: Image Source ; Chapter 4: NASA Earth Observatory images by Jesse Allen; Chapter 5: Daniel Freyr Jónsson/Alamy Stock Photo; Chapter 6: NASA Earth Observatory images by Joshua Stevens, using Landsat data from the U.S. Geological Survey; Chapter 7: NASA Earth Observatory images by Lauren Dauphin; Chapter 8: U.S. Geological Survey/Photograph by Bob Van Wagenen; Chapters 10, 14, 18: NASA Earth Observatory images by Joshua Stevens; Chapter 11: U.S. Geological Survey; Chapter 12: USDA Natural Resources Conservation Service; Chapters 13, 20: Source: NASA Earth Observatory images by Lauren Dauphin; Chapter 16: NASA Earth Observatory image by Joshua Stevens, using Landsat data provided by the U.S. Geological Survey.; Chapter 17: Source: Jim West/Alamy Stock Photo; Chapter 19: MIHAI ANDRITOIU/Alamy Stock Photo.

CHAPTER 1

Planet and Population

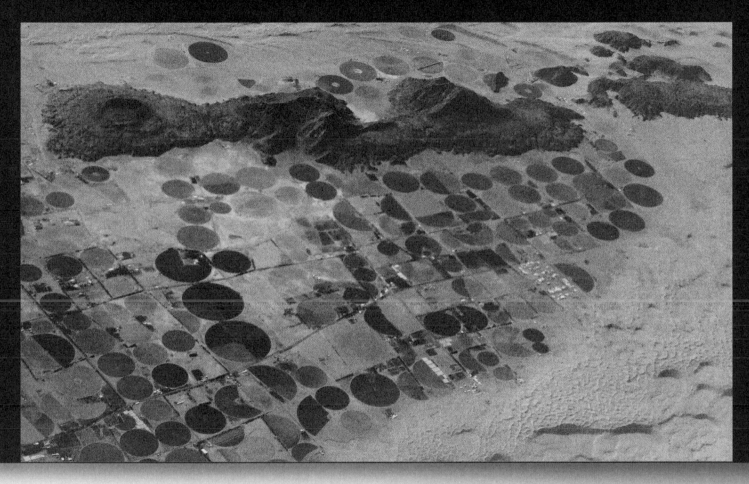

Our solar system started to evolve out of a swirling mass of gas and dust approximately 4.6 billion years ago. Fueled by a yellow-dwarf, nuclear-powered star at its center, a system of varied planets developed. Earth evolved and changed over geologic time, cooling off yet retaining enough of its interior heat to fuel changes at its surface with volcanic activity and shifting crust. The changes in Earth's geology both affected and were influenced by the development and evolution of an atmosphere and life-forms, with the abundance of liquid water playing a major role in all Earth processes.

The diversity of life on Earth has included modern humans for approximately the last 200,000 years and their close ancestors for about 2 million years. Human population

Chapter Outline

1.1 Earth in Space and Time

1.2 Geology, Past and Present

1.3 Nature and Rate of Population Growth

1.4 Impacts of the Human Population

Geology provides the ground on which we live, the soil in which we grow our crops, mineral and energy resources that support our modern lifestyles, and the water necessary for life. These irrigation fields in the Saudi Arabian desert demonstrate our often challenging dependence on Earth resources.

Stewart Innes/Alamy Stock Photo

has increased dramatically in the last 220 years, causing humans to live in large numbers in places earlier determined to be inhabitable or unsafe, increasingly competing for space and survival with Earth's other inhabitants. With the combination of intelligence and manual dexterity, we have learned to use plant, animal, mineral, fuel, and other natural resources. In some respects, humans have learned to modify natural processes to our advantage but causing disruptions in natural systems that we do not yet fully understand. One of the most important aspects of environmental geology is to deepen our understanding of the natural processes and the complex nature of how Earth systems are intertwined. We need to know how Earth works before we can fully understand how our actions, both individually and as species, can cause or solve problems posed by our geological environment. **Environmental geology** explores these many and varied interactions.

As the human population grows, these interactions expand in number and frequency. It becomes increasingly difficult for us to survive on the resources and land remaining, to avoid those hazards that cannot be controlled, and to refrain from making irreversible and undesirable changes in environmental systems. The urgency of deepening our understanding, not only of natural processes but also of our impact on the planet, is becoming more and more apparent worldwide, and has motivated increased international cooperation and dialogue on environmental issues. In 1992, nearly 200 nations came together in Rio de Janeiro for the United Nations Conference on Environment and Development to address such issues as global climate change, sustainable development, and environmental protection. The resultant UN Framework Convention on Climate Change marked the start of a series of meetings and agreements on environmental issues that continues to this day; the most recent such agreement, adopted in Glasgow in 2021 advances previous commitments to limit carbon emissions and global warming. We will explore these and other environmental accords further in chapters 17 and 19. For now, we can note that even when nations agree on what the problematic issues are, agreement on solutions is commonly more difficult to achieve, and implementation of those solutions is frequently both complex and slow. Meanwhile, the global population continues to grow.

1.1 Earth in Space and Time

Learning Objectives

- Describe the formation of the solar system
- Compare the inner and outer planets
- Differentiate Earth's internal layers
- Summarize the development of Earth's atmosphere
- Summarize the development of life on Earth through geologic time

This section provides a brief introduction to Earth history beginning with the creation of the solar system, helping us to understand the similarities between the planets. It then moves on to summarize the development of Earth's internal layers, atmosphere, water, and life, supporting the unique status of Earth as defined by the interrelated systems that occur within and on the planet's surface.

The Early Solar System

Scientists continue to construct an ever-clearer conception for the origin of the solar system that begins with the formation of the universe. The current explanation starts with all matter and energy compressed into an enormously dense, hot volume a few millimeters across approximately 14 billion years ago. This nearly instantaneous expansion of everything is commonly referred to as the Big Bang, as it was flung violently apart across an ever-larger volume of space. We can estimate when the Big Bang occurred in several ways; the most direct is the back-calculation of the universe's expansion to its apparent beginning. Other methods depend on astrophysical models of creation of the elements or the rate of evolution of different types of stars.

Stars formed from the debris of the Big Bang as locally high concentrations of mass were collected together by gravity, and some became large and dense enough that energy-releasing atomic reactions were set off deep within them. Stars are not permanent objects. They are constantly losing energy and mass as they burn their nuclear fuel. The mass of material that initially formed a star determines how rapidly the star burns; some stars burned out billions of years ago, while others are probably forming now from the original matter of the universe mixed with the debris of older stars.

Our star, the Sun, and its system of circling planets formed from a rotating cloud of gas and dust (small bits of rock and metal) as explained by the nebular hypothesis (**figure 1.1**). Most of the mass of the cloud coalesced to form the Sun, which became a star and began to release light energy when its interior became so dense and hot from the crushing effects of its own gravity that nuclear reactions were triggered inside it. Meanwhile, dust condensed from the gases remaining in the flattened cloud disk rotating around the young Sun. The dust clumped into protoplanets and then planets, the formation of which was essentially complete over 4½ billion years ago.

The Planets

The composition of each planet depended largely on its proximity to the hot Sun. The planets that formed nearest to the Sun contained mainly metallic iron and a few minerals with very high melting temperatures, with little water or gas. Somewhat farther out, where temperatures were lower, the developing

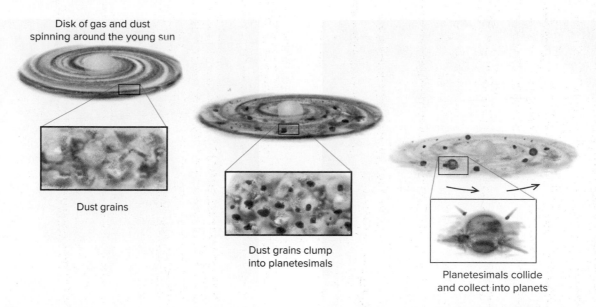

Disk of gas and dust spinning around the young sun

Dust grains

Dust grains clump into planetesimals

Planetesimals collide and collect into planets

Figure 1.1

Our solar system formed as dust condensed from the gaseous nebula, then clumped together to make protoplanets (or planetesimals) and then planets.

planets had more lower-temperature minerals including some that contain water locked within their crystal structures. Still farther from the Sun, temperatures were so low that nearly all of the materials in the original gas cloud condensed, even compounds such as methane and ammonia which are gases at normal Earth surface temperatures and pressures. The result was a series of planets with a variety of compositions, most quite different from that of Earth. Notice that the planet densities listed in **table 1.1** are consistent with higher metal and rock content in the four inner planets and a much larger proportion of gas and ice in the outer planets. With interest in mining other planetary bodies for needed mineral resources, we will want to keep in mind that both the chemistry and resource-forming processes may not be the same as those that have occurred on Earth. In addition, our principal current energy sources are fossil fuels

that required living organisms to form, and so far no such life-forms have been found on other planets or moons.

We can briefly compare Earth to its nearest neighbors to better understand how planets differ from one another. Venus is close to Earth in space and similar in size and density, but still shows marked differences. Venus's dense, cloudy atmosphere is so thick with carbon dioxide that it produces planetary surface temperatures hot enough to melt lead; Venus (**figure 1.2A**) is our best example of how the greenhouse effect works to warm an atmosphere. Mercury (**figure 1.2B**) is closer to the Sun than Venus, but its lack of an atmosphere results in a lower mean temperature. The inner planet farthest from the Sun is Mars (**figure 1.2C**). It does have an atmosphere that is 95% carbon dioxide but it is very thin, so it does not result in high surface temperatures, leaving Mars to be quite cold. We are currently conducting

Table 1.1	Select Physical Properties of the Planets			
Planet	Mean Distance from the Sun (millions of km)	Mean Temperature (°C)	Equatorial Diameter, Relative to Earth	Density* (g/cu. cm)
Mercury	58	167	0.38	5.4 ⎤
Venus	108	464	0.95	5.2 ⎬ Inner rocky planets
Earth	150	15	1.00	5.5 ⎪
Mars	228	−65	0.53	3.9 ⎦
Jupiter	778	−110	11.19	1.3 ⎤
Saturn	1427	−140	9.41	0.7 ⎬ Outer gas giants and ice giants
Uranus	2870	−195	4.06	1.3 ⎪
Neptune	4498	−200	3.88	1.6 ⎦

Source: Data from NASA.

*No other planets have been extensively sampled to determine their compositions directly, though we have some data on their surfaces. Their approximate bulk compositions are inferred from the assumed starting composition of the solar nebula and the planets' densities. For example, the higher densities of the inner planets reflect a significant iron content and relatively little gas.

A

Figure 1.2A

This 1974 image taken by the Mariner 10 probe has been recently reprocessed, showing Venus covered in sulfuric acid clouds.

JPL-Caltech/NASA

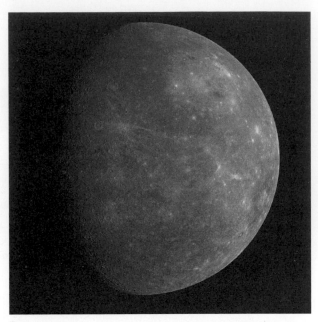

B

Figure 1.2B

Mercury, as shown in this image from a 2008 *Messenger* spacecraft flyby, is rocky, iron-rich, dry, and pockmarked with craters.

NASA Science Operations center at Johns Hopkins University Applied Physics laboratory

C

Figure 1.2C

Mars shares many surface features with Earth (volcanoes, canyons, dunes, slumps, stream channels, and more) but the surface is now dry and barren.

NASA/USGS

D

Figure 1.2D

This image of layered rock taken by the Curiosity Mars Rover represents sedimentary lake bed deposits.

JPL-Caltech/MSSS/NASA

intense scientific research on Mars, searching for signs of water and life by studying Mars's geology. The indications for a formerly warmer, watery Mars are abundant (**figure 1.2D**); there is no liquid water now but the planet does have some ice. We will continue our geologic investigations of Mars while looking for evidence of past life and simultaneously planning for the first human astronaut missions to the planet.

The comparisons to Earth get fairly drastic when we look at the planets of the outer solar system (**figure 1.3**). The

Figure 1.3

The planets in order and to scale according to size, including the dwarf planet Pluto. Note that the distances between the planets are not to scale.

John J Lutheran/Shutterstock

massive gas giant Jupiter is covered in clouds of ammonia and water, has an atmosphere of hydrogen and helium, and has no solid surface. Its gaseous companion is Saturn, whose apparent rings made of small, dusty chunks of ice and rock make it a favorite. Scientists are very interested in some of the numerous moons of these two gas giants because they appear to have liquid oceans (covered by an icy surface) that make them a prudent choice in searching for other life in the solar system. Uranus and Neptune are compositionally similar to the gas giants but contain more heavy elements; a large amount of these elements is compressed in a slushy mix, so we distinguish these two planets as ice giants. Pluto is one of the five known dwarf planets in the outer reaches of the solar system. Scientists are still analyzing the data collected from the New Horizon's flyby in 2015 in order to understand the uniqueness of Pluto.

Earth's Internal Structure and Atmosphere

Earth has changed continuously since its formation, undergoing some particularly profound transformations in its early history. The early Earth was very different from what it is today, lacking the modern oceans and atmosphere and having a much different surface from its present one. Like other planets, Earth was formed by accretion, as gravity collected together the solid bits that had condensed from the solar nebula. Some water may have been contributed by gravitational capture of icy comets, but scientists are still debating the amount. The planet was heated by the impact of the colliding dust particles and meteorites as they came together to form Earth, and by the energy released from decay of several naturally radioactive elements. These heat sources combined to raise the planet's internal temperature enough that parts of it melted, although it was probably never molten all at once. Dense materials, such as metallic iron, would have tended to sink toward Earth's center. As cooling progressed, lighter, low-density minerals crystallized and floated out toward the surface. The eventual result was an Earth differentiated into several major compositional zones: the central **core,** the surrounding **mantle,** and a thin **crust** at the surface

(see **figure 1.4**). The process was complete well before 4 billion years ago.

We have only directly sampled and analyzed the crust and a few bits of uppermost mantle that have been carried up into the crust by volcanic activity. Our understanding of the composition of Earth's deeper interior comes from various indirect sources. First, scientists can estimate from analyses of stars the starting composition of the cloud from which the solar system formed. Geologists can also infer aspects of Earth's bulk composition from analyses of certain meteorites likely formed at the same time as, and under conditions similar to, Earth. Geophysical data demonstrate that Earth's interior is zoned and also provides information on the densities of the different layers within Earth, which further limits their possible compositions. Put together, this evidence indicates that the core is made up mostly of iron, with some nickel and a few minor elements, with solid inner and molten outer sections. The mantle consists mainly of iron, magnesium, silicon, and oxygen combined in varying proportions in several different minerals. The crust is much more varied in composition and is very different chemically from the average composition of Earth (see **table 1.2**). Notice that many of the metals we have come to prize as resources are relatively uncommon elements in the crust. The crust and uppermost mantle together form a somewhat brittle shell we refer to as the **lithosphere** (**figure 1.4**).

The heating and subsequent differentiation of the early Earth led to another important result: formation of the atmosphere and oceans. Many minerals that had contained water or gases in their crystals released them during the heating and melting, and as the surface cooled, the water condensed to form the oceans. Without this abundant surface water, most life as we know it could not exist. The oceans filled basins, while the buoyant continents made of lower-density rocks and minerals stood above the sea surface. Life on Earth likely began in the ocean and eventually moved onto the land but this modification required a monumental change in Earth's atmosphere.

The early atmosphere was quite different from the modern one, aside from the effects of anthropogenic pollution. The

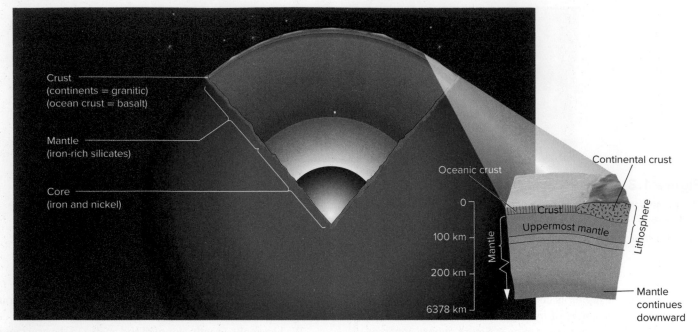

Figure 1.4

Earth is chemically differentiated into layers: the solid inner and liquid outer core, rich in iron; the silicon- and oxygen-rich crust at the top; and the mantle in-between in both location and composition. The crust (vertically exaggerated) is very thin and has two types: oceanic crust is thinner, denser, and has a basaltic composition; continental crust is thicker, more buoyant, and granitic. Both types of crust and the uppermost mantle make the brittle layer of the lithosphere.

Table 1.2	Most Common Chemical Elements in the Earth		
WHOLE EARTH		**CRUST**	
Element	Weight Percent	Element	Weight Percent
Iron	32.4	Oxygen	46.6
Oxygen	29.9	Silicon	27.7
Silicon	15.5	Aluminum	8.1
Magnesium	14.5	Iron	5.0
Sulfur	2.1	Calcium	3.6
Nickel	2.0	Sodium	2.8
Calcium	1.6	Potassium	2.6
Aluminum	1.3	Magnesium	2.1
(All others, total)	0.7	(All others, total)	1.5

(Compositions cited are averages of several independent estimates.)

first atmosphere had little or no free oxygen in it. It probably consisted dominantly of nitrogen and carbon dioxide (the gas most commonly released from volcanoes, aside from water) with minor amounts of gases such as methane, ammonia, and various sulfur compounds. Oxygen-breathing life of any kind could not exist before cyanobacteria appeared in large numbers to modify the atmosphere. Their microfossils and structures they built are found in rocks as old as 2 billion years; there are others that could be as old as 3.5 billion years. Cyanobacteria manufacture food through photosynthesis using sunlight for energy, consuming carbon dioxide, and releasing oxygen as a by-product. In time, the atmosphere developed into a thick enough layer with sufficient oxygen to support oxygen-breathing organisms and protect them from cell-damaging solar radiation.

Life on Earth

The rock record shows when different plant and animal groups appeared. Some are represented schematically in **figure 1.5**. The earliest creatures left very few remains because they had no hard skeletons, teeth, shells, or other hard parts that could be preserved in rocks. The first multi-celled oxygen-breathing creatures probably developed about 1 billion years ago, after oxygen in the atmosphere was well established. By about 541 million years ago (mya), marine animals with shells had become widespread.

The development of organisms with hard parts greatly increased the number of preserved animal remains in the rock record. Consequently, we have a better understanding of the biological developments since that time. Dry land was still barren of large plants or animals half a billion years ago. The first evidence of vertebrates comes from a jawless fish fossil that exists in rocks about 530 million years old. Plants colonized land soon thereafter starting about 465 mya. Insects show up around 400 mya and trees 385 mya. Dinosaurs appeared about 230 mya and the mammals about the same time

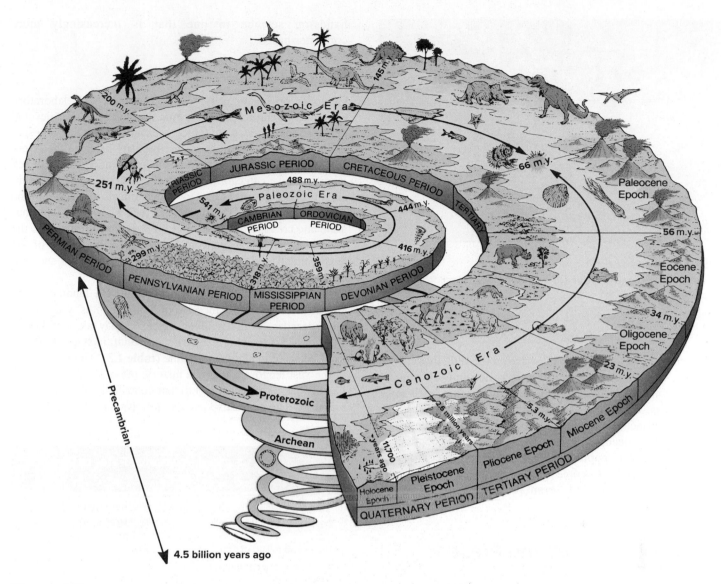

Figure 1.5

The geological spiral of time shows when important plant and animal groups first appear in significant numbers. Note that an alternative for the end of the Tertiary is the Neogene Period which contains the Pliocene and Miocene Epochs. For another way to look at these data, see table A.1 in appendix A.

Source: Modified after U.S. Geological Survey publication Geologic Time.

around 210 mya. The first true birds, evolved from dinosaurs, are represented by the infamous *Archaeopteryx* fossil dated at 150 mya (**figure 1.6**). Both birds and mammals are well established by 100 mya.

How is the understanding of how life developed on Earth applicable to our own modern lives? One of the most apparent answers is that the rapid development of human societies has been based on easily accessible, concentrated energy sources in the form of fossil fuels. Oil, natural gas, and coal are all created by the remains of particular creatures that lived in distinct environments and climates. By understanding information preserved in the rock record, we know the times when the particular groups of organisms appeared and

flourished, which has been integral in searching for the rocks of appropriate ages and in assessing the amount of the energy sources available.

On a timescale of billions of years, human beings have just arrived. The most primitive *Homo* genus remains are between 1.5 and 2.5 million years old, and anatomically modern humans *(Homo sapiens)* developed only about 300,000 years ago. Hundreds of thousands of years may seem like a long time compared to a single human lifetime, but in a geologic sense, it is a very short time. If we equate the whole of Earth's history to a 365-day year, with us currently sitting on the verge of New Year's Eve, then shelled organisms appeared in mid-November, fish on November 20, and land plants on November

Figure 1.6

Cast of the 150-million year old Archaeopteryx fossil from southern Germany.

Wicki58/E+/Getty Images

22. December sees the mammals arrive on the 13th, the birds on the 18th, and flowers on the 20th. It is not until 36 minutes before midnight do *Homo Sapiens* show up for this Earth party. Nevertheless, we have had an enormous impact on Earth, at least at its surface, an impact far out of proportion to the length of time we have occupied it. Our impact is likely to continue to increase rapidly as population and consumption continue to increase. The need for people who understand Earth's history, development over time, and current processes is more apparent now than it ever has been.

1.2 Geology, Past and Present

Learning Objectives

- List the major challenges of studying geology
- Describe the role of the scientific method to geology
- Exemplify current issues that geologists play a key role in addressing
- List cyclic processes that occur on Earth

Two centuries ago, geology was mainly a descriptive field science involving careful observation of natural processes and their products. The study of Earth has become both more quantitative and more interdisciplinary through time. Modern geoscientists draw on the principles of chemistry to interpret the compositions of geologic materials, apply the laws of physics to explain these materials' physical properties and behavior, use the biological sciences to develop an understanding of ancient life-forms, and rely on engineering principles to design safe structures in the presence of geologic hazards. Aided by new remote-sensing technologies, we are currently seeing a transition in this traditionally hands-on science to one that is increasingly more computer-based.

The Geologic Perspective

Geologic observations now are combined with laboratory experiments, careful measurements, and calculations to develop explanations of how natural processes operate. Geology is especially challenging because of the disparity between the scientist's laboratory and the natural environment. In the research laboratory, we can carefully control conditions of temperature and pressure, as well as the flow of chemicals into or out of the system under study. The researcher then knows just what has gone into creating the product of the experiment. In nature, the geoscientist is often confronted only with the results of the "experiment" and must deduce the starting materials and processes involved.

Another complicating factor is time. The laboratory scientist must work on a timescale of hours, months, years, or, at most, decades. Natural geologic processes may take a million or a billion years to achieve a particular result, in stages too slow to be detected in a human lifetime (**table 1.3**). The recognition of the vast length of geologic history, often described as *deep time,* is one of the most significant contributions of early geoscientists. The qualitative and quantitative tools for sorting out

Table 1.3	Representative Rates of Geologic Processes
Process	**Approximate Timeframe**
Rising and falling of tides	1 day
Movement of a continent by 2–3 centimeters	1 year
Accumulation of energy between large earthquakes on a major fault zone	10–100 years
Rebound (rising) by 1 meter of a continent depressed by ice sheets during the Ice Age	100 years
Flow of heat through 1 meter of rock	1000 years
Deposition of 1 centimeter of fine sediment on the deep-sea floor	1000–10,000 years
Ice sheet advance and retreat during an ice age	10,000–100,000 years
Life span of a small volcano	100,000 years
Life span of a large volcanic center	1–10 million years
Creation of an ocean basin such as the Atlantic	100 million years
Duration of a major mountain-building episode	100 million years
History of life on Earth	Over 3 billion years

geologic events and putting dates on them are outlined in appendix A. For now, bear in mind that the immensity of geologic time can make it difficult for us to arrive at a full understanding of how geologic processes operated in the past because our observations are made on a human timescale. A few hundred years of human activity may disrupt natural systems in the short term, but might make modeling possible future changes problematic given that many natural processes operate over much, much longer time frames.

Geology and the Scientific Method

The **scientific method** is a means of discovering basic scientific principles. One begins with a set of observations and/or a body of data, based on measurements of natural phenomena or on experiments. One or more *hypotheses* are formulated to explain the observations or data. A **hypothesis** can take many forms, ranging from a general conceptual framework or model describing the functioning of a natural system, to a very precise mathematical formula relating several kinds of numerical data. What all hypotheses have in common is that they must all be susceptible to testing and particularly, to *falsification*. The idea is not simply to look for evidence to support a hypothesis but to examine relevant evidence with the understanding that it may show the hypothesis to be wrong.

In the classical conception of the scientific method, one uses a hypothesis to make a set of predictions. Then one devises and conducts experiments to test each hypothesis, to determine whether experimental results agree with predictions based on the hypothesis. If they do, the hypothesis gains credibility. If not, if the results are unexpected, the hypothesis must be modified to account for the new data as well as the old or perhaps discarded altogether. Several cycles of modifying and retesting may be required before a hypothesis that is consistent with all the observations and experiments is achieved. A hypothesis that is repeatedly supported by new experiments advances to the status of a **theory,** a generally accepted explanation for a set of data or observations.

The term *theory* is used much differently by the general public, generally as a synonym for a scientific hypothesis or just an educated guess. People may misunderstand a scientist describing a theory as simply telling a plausible story to explain some data, when in fact the scientist is describing a well-tested model with a very substantial and convincing body of evidence that supports it. A hypothesis may be advanced by just one individual whereas a theory has survived the challenge of extensive testing to merit acceptance by many, often most, experts in a field. The Big Bang theory is not just a creative idea. It accounts for our observations that all the objects we can observe in the universe seem to be moving apart. If it is correct, the universe's origin was very hot; scientists have detected the cosmic microwave background radiation consistent with this. And astrophysicists' calculations predict that the predominant elements that the Big Bang would produce would be hydrogen and helium, which indeed overwhelmingly dominate the observed composition of our universe.

The classical scientific method is not strictly applicable to many geologic phenomena because of the difficulty of experimenting with natural systems, given the time and scale considerations noted earlier. For example, one may be able to conduct experiments on a single igneous rock but not to construct a whole volcano in the laboratory, nor to replicate a large meteorite impact to study its effects (**figure 1.7**). In such cases, hypotheses are tested entirely through further observations or theoretical calculations and modified as necessary until they accommodate all the relevant observations, or are discarded when they cannot be reconciled with new data. This broader conception of the scientific method is well illustrated by the development of the theory of plate tectonics, discussed in chapter 3. Continental drift was once considered a wildly implausible idea advanced by a few eccentrics, but in the latter half of the twentieth century scientists found new evidence that was well and consistently explained by the movement of plates over Earth's surface. Those data were made available by the development of new observational and analytical techniques that did not exist at the time of the original explanation. In this way, an unpopular hypothesis may become a well-established theory and/or a well-established theory may be rejected. The formation of Meteor Crater, shown in **figure 1.7**, is a good example of multiple hypotheses being tested through time by various people. The current explanation of its creation took 70 years of proposing, testing, analyzing, dismissing, and synthesizing in a process that ultimately moved forward in understanding Earth processes but was neither a straight nor a one-way path.

It would be very rare for a geologist to start a study by saying to themselves "Let's use the scientific method!" and make a strict set of rules to follow in a distinct order. We do need to incorporate these best practices in order for our research to be acceptable in the wider scientific community but there is no single correct order or way to follow these standards.

Figure 1.7

Meteor Crater, Arizona, was initially explained as resulting from volcanic activity. Further research, lasting decades, later confirmed it was formed from a relatively recent meteorite impact.

David J. Roddy, USGS, Branch of Astrogeology

The Motivation to Find Answers

In spite of the difficulties inherent in trying to explain geologic phenomena, the search for explanations goes on, spurred not only by the basic quest for knowledge but also by the practical problems posed by geologic hazards, the need for resources, and concerns about global-scale human impacts such as climate change.

Geologic hazards create the most dramatic scenes and headlines and the most abrupt consequences: 230,000 dead and 1 million homeless from the 2010 Haiti earthquake (**figure 1.8**); $300 billion in damage from the 2011 Honshu, Japan, earthquake; $5 billion in damage from the 2010 eruption in Iceland (**figure 1.9**); $130 billion in damage from the 2017 flooding in Texas associated with Hurricane Harvey (**figure 1.10**). The eruption of the Eyjafjallajökull Volcano in Iceland itself did not surprise scientists. However, the unpredictable movement of the eruption underneath a glacier after two months activity caused significant flooding and an unexpected, although not unprecedented, movement of ash toward Europe that shutdown airline travel for a week.

Scientists and engineers continue to improve early warning systems for hazards such as earthquakes, volcanic eruptions, and landslides in order to save lives and to limit disrupting infrastructure damage. Flood damage in the United States has increased significantly, from an average of $3.2 billion/year from 1980–2010 to $5.6 billion from 2010–2021 while yearly deaths in that entire timeframe remain less than 20. The science of stream dynamics is already advanced and improvements in flood control and damage will need to come from both technological advances and strict zoning policies and regulations surrounding floodplains, wetlands, and coastal areas. Landslides and other slope and ground failures (**figure 1.11**) are widespread and costly and can be reduced by engineering practices that improve slope stability, but also by the

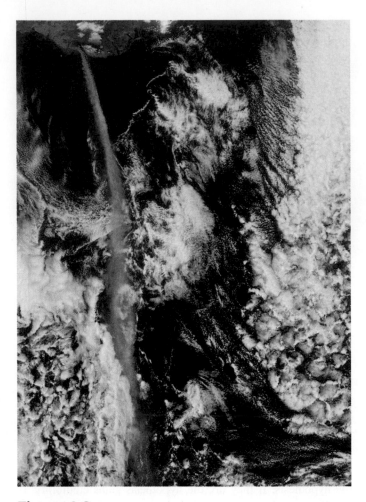

Figure 1.9

Ash from Eyjafjallajökull Volcano in May 2010 moves south at an altitude of 4,300–5,200 meters.

NASA image by Jeff Schmaltz, MODIS Rapid Response Team at NASA GSFC

Figure 1.8

Vehicles damaged by the second-story collapse of a hotel in Port-au-Prince, Haiti, in January 2010.

Walter D. Mooney/USGS

Figure 1.10

Floating cars in Houston demonstrate the flood damage created by Hurricane Harvey in 2017.

FEMA News Photo

Figure 1.11

Damage from the 2014 Oso, Washington, landslide.

Mark Reid/U.S. Geological Survey

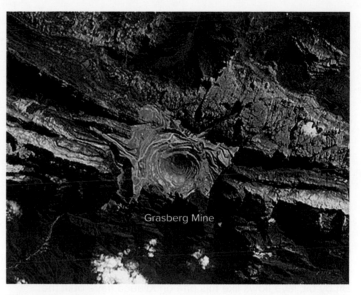

Grasberg Mine

Figure 1.12

The Grasberg Mine in Irian Jaya, Indonesia, is one of the world's largest gold- and copper-mining operations. The surface pit is nearly 4 km (2½ miles) across; note the sharp contrast with surrounding topography. Slopes oversteepened by mining have produced deadly landslides, and local residents worry about copper and acid contamination in runoff water.

Earth Sciences and Image Analysis Laboratory, NASA Johnson Space Center.

same civil decisions just described. It is not only the more dramatic hazards that are costly: on average, the cost of structural damage from unstable soils each year approximately equals the combined costs of landslides, earthquakes, and flood damages in this country.

As scientists become better able to predict events such as volcanic eruptions and floods, new challenges arise: How certain should they be before a prediction is issued? How best to educate the public and public officials about the science behind the predictions and its limitations, so that they can prepare/respond appropriately? What if a prediction is wrong? Such issues will be examined in later chapters.

Our demand for resources of all kinds continues to grow and so do the consequences of resource use. In the United States, average per-capita water use is 1000 gallons per day. In many places, groundwater supplies upon which we have come to rely heavily are being measurably depleted. Worldwide, water-resource disputes between nations are increasing in number. As we mine more extensively for mineral resources, we face the issues of degrading land that may be reserved or suited for other purposes and polluting water supplies (**figure 1.12**). The grounding of the *Exxon Valdez* in 1989, dumping 11 million gallons of oil into Prince William Sound, Alaska, and the massive spill from the 2010 explosion of the *Deepwater Horizon* drilling platform in the Gulf of Mexico are reminders of the negative consequences of petroleum exploration.

As we consume more resources, we create more waste. In the United States, total municipal waste generation is 292 million tons per year, nearly five pounds a day for each person. Careless waste disposal, in turn, leads to pollution. The Environmental Protection Agency continues to identify toxic-waste disposal sites in urgent need of cleanup; by 2021, over 1800 so-called priority sites had been identified. Cleanup costs per site range from $25 to $30 million, and the projected total costs to remediate these sites is over $1 trillion. As fossil fuels are burned, carbon dioxide in the atmosphere rises, and

modelers of global climate strive to understand what that may do to global temperatures, weather, and agriculture decades in the future.

These are just a few of the kinds of issues that geologists play a key role in addressing.

Earth Cycles and Systems

Earth is a dynamic, constantly changing planet with its crust shifting to build mountains; lava spewing out of its warm interior; and moving ice, water, and sand, along with gravity, reshaping its surface, over and over. Some changes proceed in one direction only—Earth has been cooling progressively since its formation and will continue to cool, with the released heat providing energy for other Earth's systems, many of which are cyclic in nature.

Consider, for example, such basic materials as water or rocks. Streams drain into oceans and would soon run dry if not replenished; but water evaporates from oceans, to make the rain and snow that feed the streams to keep them flowing. This describes just a part of the *hydrologic* (water) *cycle,* explored more fully in chapters 6 and 11. Rocks, despite their appearance of permanence in the short term of a human life, participate in the *rock cycle* (chapters 2 and 3). The kinds of evolutionary paths rocks may follow through this cycle are many, but consider this illustration: A volcano's lava (**figure 1.13**) hardens into rock; the rock is weathered into sand and dissolved chemicals; the debris, transported to and deposited in an ocean basin,

Figure 1.13

Bit by bit, lava flows like this one on Kilauea have built the Hawaiian Islands.

©Carla Montgomery

Figure 1.14

Rocks tell a story of constant change. The folds of the Valley and Ridge Province in Pennsylvania formed deep in the crust when Africa and North America converged hundreds of millions of years ago; now they eroded and crosscut by rivers.

NASA image created by Jesse Allen & Robert Simmon, using data provided courtesy of NASA/GSFC/METI/ERSDAC/JAROS, and U.S./Japan ASTER Science Team

is solidified into a new rock of quite a different type; and some of that new rock may be carried into the mantle via plate tectonics to be melted into a new lava. The time frame over which this process occurs is generally much longer than that over which water cycles through atmosphere and oceans, but the principle is similar. The Appalachian Mountains (and their rocks) as we see them today are not as they formed tens or hundreds of millions of years ago in water-filled basins and deserts from material eroded from even more-ancient mountains. They were buried, heated, squeezed, and uplifted by enormous forces, and then eroded from their maximum height by water and ice (**figure 1.14**).

Chemicals also cycle through the environment. The carbon dioxide that we exhale into the atmosphere is taken up by plants through photosynthesis, and when we eat those plants for food energy, we release CO_2 again. The same exhaled CO_2 may also dissolve in rainwater to make carbonic acid that dissolves continental rock. The weathering products may wash into the ocean, where dissolved carbonate then contributes to the formation of carbonate shells and carbonate rocks in the ocean basins. Those rocks may later be exposed and weathered by rain, releasing CO_2 back into the atmosphere or dissolved carbonate into streams that carry it back to the ocean. The cycling of chemicals and materials in the environment is often complex, as we will see in later chapters.

These processes and cycles are often interrelated and seemingly local actions can have distant consequences. We dam a river to create a reservoir as a source of irrigation water and hydroelectric power, inadvertently trapping stream-borne sediment at the same time; downstream, patterns of erosion and deposition in the stream channel change, and at the coast, where the stream pours into the ocean, coastal erosion of beaches increases because a part of their sediment supply, from the stream, has been cut off. The volcano that erupts the

lava to make the volcanic rock also releases water vapor into the atmosphere, and sulfur-rich gases and dust that influence the amount of sunlight reaching Earth's surface to heat it, which, in turn, can alter the extent of evaporation and resultant rainfall, which will affect the intensity of landscape erosion and weathering of rocks by water. . . . So although we divide the great variety and complexity of geologic processes and phenomena into more manageable, chapter-sized units for purposes of discussion, it is important to keep such interrelationships in mind. And superimposed on, influenced by, and subject to all these natural processes are humans and human activities.

1.3 Nature and Rate of Population Growth

Learning Objectives

- Recognize historic and projected human population growth
- Compare human population growth in different regions
- Explain exponential growth and doubling time in terms of human populations
- Exemplify regions with different human population growth issues

Table 1.4 World and Regional Population Growth and Projections (in millions)

Year	World	Africa	Asia	Europe	Latin America and Caribbean	North America	Oceania
1950	2526	229	1394	549	169	172	13
2021	7838	1373	4651	744	656	371	43
2050 (projected)	9688	2529	5192	731	762	412	62
Growth rate (%/year)	1.0	2.5	0.9	−0.3	0.9	0.1	0.9
Doubling time (yrs)	70	28	78	NA	78	700	78

Sources: Data for 1950 from United Nations World Population Estimates and Projections; all other data from 2021 World Population Data Sheet, Population Reference Bureau, 2021. Population projections to 2050 involve longer-term estimates of future growth rates, which in most areas are expected to decrease from the present levels reported here.

Animal populations, as well as primitive human populations, are generally quite limited both in the areas that they can occupy and in the extent to which they can grow. They must live near food and water sources. The climate must be one to which they can adapt. Predators, accidents, and disease take a toll. If the population grows too large, disease and competition for food are particularly likely to cut it back to sustainable levels.

The human population grew relatively slowly for hundreds of thousands of years. Not until the middle of the nineteenth century did the world population reach 1 billion. However, by then, a number of factors had combined to accelerate the rate of population increase. The second, third, and fourth billion were reached far more quickly; the world population is now over 7.8 billion and is expected to rise to 9.7 billion by 2050 (**table 1.4**).

Humans are no longer constrained to live only where conditions are ideal. We build habitable quarters even in extreme climates, using heaters or air conditioners to bring the temperature to a level we can tolerate. In addition, people need not live where food can be grown or harvested or where there is abundant fresh water. The food and water is transported to where the people choose to live.

In this section, we will review human population growth in terms of its historic and current standing, how it varies by region, and recent issues that may affect it. We will also introduce some important terms used when discussing population growth.

Growth Rates: Causes and Consequences

Population growth occurs when new individuals are added to the population faster than existing individuals are removed from it. On a global scale, the population increases when its birthrate exceeds its death rate. In assessing an individual nation's or region's rate of population change, immigration and emigration must also be taken into account. Improvements in nutrition and health care typically increase life expectancies, decrease mortality rates, and thus increase the rate of population growth. Increased use of birth-control or family-planning methods reduces birthrates and, therefore, also reduces the rate of population growth; in fact, a population can begin to decrease if birthrates are severely restricted.

The sharply rising rate of population growth over the past few centuries can be viewed another way. It took until about 1803 for the world's population to reach 1 billion. The population climbed to 2 billion in the next 125 years, and to 3 billion in just 30 more years, as ever more people contributed to the population growth and individuals lived longer. The next 4 billion people were added from 1960 to 2011, when we reached the 7 billion population mark.

There are wide differences in growth rates among regions (table 1.4; **figure 1.15**), and there are many reasons for this variability. Religious or social values may cause larger or smaller families to be regarded as desirable in particular regions or cultures. High levels of economic development are commonly associated with reduced rates of population growth; conversely, low levels of development are often associated with rapid population growth. The impact of improved education, especially for women, clearly decreases growth rates through direct and indirect factors such as delayed fertility and confidence that offspring will survive childhood.

A few governments of nations with large and rapidly growing populations have encouraged or mandated family planning. India and the People's Republic of China have taken active measures, with varying results: China's population growth rate is just 0.1% per year; its population is expected to peak at 1.5 billion in the early 2030s and then decline. However, China's One-Child-Policy is not solely responsible for the falling population, and has been recently amended to allow couples to have two, and now three, children. India's growth rate remains relatively high at 1.4% per year; its population will soon surpass that of China and is projected to be over 1.6 billion by 2050.

New developments in medical treatments have lessened the concerns with how AIDS might affect population growth, especially in African countries. In Eswatini, where over 26% of the population aged 15 to 49 is afflicted, nearly all of these people are being treated. Life expectancy, while still below the world average of 73 years, has risen from 49 to 58 years over the past 5 years. Will the recent pandemic affect population growth? Current estimates have deaths from COVID-19 at 5 to 10 million, which is still only 0.1% of global population; compare that to the 1918 flu pandemic which eliminated

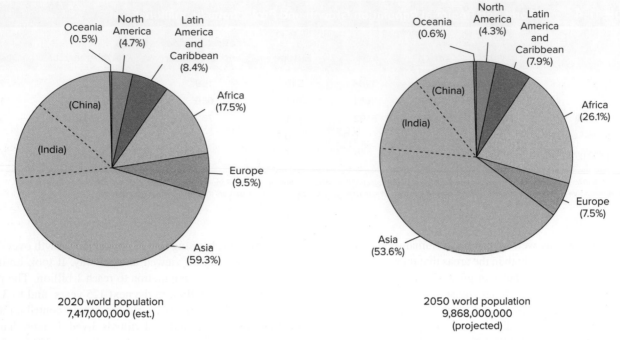

Figure 1.15

Population distribution by region in mid-2020 with projections to the year 2050. Size of circle reflects relative total population. The most dramatic changes in proportion are the relative growth of the population of Africa and decline of population in Europe. Data from table 1.4.

1% to 5% of the population yet did very little to affect long-term growth.

Even when the population growth rate is constant, the number of individuals added per unit of time increases over time. This is called **exponential growth,** a concept we will return to when discussing resources later on in this section. The effect of exponential growth is similar to interest compounding. If one invests $100 at 10% per year and withdraws the interest each year, one earns $10/year, and after 10 years, one has collected $100 in interest. If one invests $100 at 10% per year, compounded annually, then, after one year, $10 interest is credited, for a new balance of $110. But if the interest is not withdrawn, then at the end of the second year, the interest is $11 (10% of $110), and the new balance is $121. By the end of the tenth year (assuming no withdrawals), the interest for the year is $23.58, but the interest rate has not increased. And the balance is $259.37 so, subtracting the original investment of $100, this means total interest of $159.37 rather than $100. Similarly with a population of 1 million growing at 5% per year: in the first year, 50,000 persons are added; in the tenth year, the population grows by 77,566 persons. The result is that a graph of population versus time steepens over time, even at a constant growth rate. If the growth rate itself also increases, the curve rises still more sharply.

For many mineral and fuel resources, consumption has been growing very rapidly, even more rapidly than the population. The effects of exponential increases in resource demand are like the effects of exponential population growth (**figure 1.16**). If demand increases by 2% per year, it will double not in 50 years, but in 35. A demand increase of 5% per year leads to a doubling in demand in 14 years and a tenfold increase in demand in 47 years. In other words, a prediction of how soon mineral or energy supplies will be exhausted is very sensitive to the assumed rate of change in demand. Even if the population is no longer growing exponentially, consumption of many resources is.

Growth Rate and Doubling Time

Another way to look at the rapidity of world population growth is to consider the expected **doubling time,** the length of time required for a population to double in size. Doubling time (*D*) in years may be estimated from growth rate (*G*), expressed in percent per year, using the simple relationship $D = 70/G$, which is derived from the equation for exponential growth (see "Exploring Further" question 2 at the chapter's end). The higher the growth rate, the shorter the doubling time of the population (see again table 1.4). By far the most rapidly growing segment of the population today is that of Africa. Its population, estimated at 1.37 billion in 2021, is growing at about 2.5% per year. The largest segment of the population, that of Asia, is increasing at 0.9% per year, and since the over 4.5 billion people there represent well over half of the world's total population, this leads to a relatively high global average growth rate. Europe's population has begun to

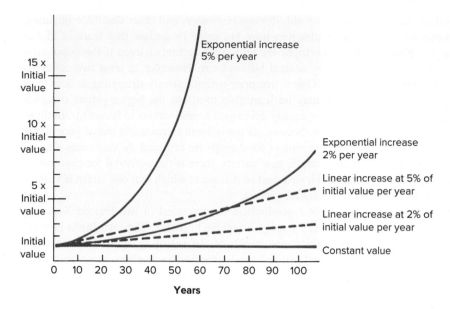

Figure 1.16

Graphical comparison of the effects of linear and exponential growth, whether on consumption of minerals, fuels, water, and other consumable commodities, or population. With linear growth, one adds a fixed percentage of the *initial* value each year (dashed lines). With exponential growth, the same percentage increase is computed each year, but year by year the value on which that percentage is calculated increases, so the annual increment keeps getting larger.

decline slightly, but Europe contains less than 10% of the world's population.

The average worldwide population growth rate is about 1.0% per year, which corresponds to a doubling time of about 70 years. At that, the present population growth rate actually represents a substantial decline from nearly 2% per year in the mid-1960s, and that decline is expected to continue. However, a decreasing growth rate is not at all the same as a decreasing population. Depending upon projected fertility rates, estimates of world population in the year 2050 can vary by several billion. Using a medium fertility rate, the Population Reference Bureau projects a 2050 world population of almost 9.7 billion. **Figure 1.15** illustrates how those people will be distributed by region, considering differential growth rates from place to place.

Even breaking the world down into regions of continental scale masks a number of dramatic individual-country cases. Discussion of these, and of their political, economic, and cultural implications, is well beyond the scope of this chapter, but consider the following: while the population of Europe is nearly stable, in many parts of northern and eastern Europe, sharp declines are occurring. By 2050, the populations of Poland, Ukraine, and Bulgaria are expected to drop by 8%, 26%, and 16%, respectively. Conversely, parts of the Middle East are experiencing explosive population growth, with projected increases by 2050 of 68% in Israel, 38% in Saudi Arabia, 41% in the Palestinian Territory, and 59% in Iraq. The demographics differ widely between countries, too. Globally, 26% of the population is under age 15, and 10% above age 65. But in Japan, only 12% of the population is below age 15, with 29% age 65 or older; in Afghanistan, 47% of people are under age 15 and only 3% age 65 or over. Thus, different nations face different challenges. Where rapid population growth meets scarcity of resources, problems arise.

1.4 Impacts of the Human Population

Learning Objectives

- Recognize current issues with food production
- Recognize the geologic factors associated with carrying capacity
- Generalize the relationship between human populations and locations of both resources and industrialized areas
- Relate population size and disruption of natural systems

The problems posed by a rapidly growing world population have often been discussed in the context of food: that is, how to produce sufficient food and distribute it effectively to prevent the starvation of millions in overcrowded countries or in countries with minimal agricultural development. So far, the production of food has kept up with that needed by the current population. The uneven distribution of that food across the globe, as well as other resources, is an apparent problem that will become even more prevalent as the population grows both in size and in wealth. Geologists will increasingly be involved in work associated with the acquisition, use, and distribution of soil, mineral, energy, and water resources.

Farmland and Food Supply

Whether or not Earth can sustainably support 8 to 10 billion people, or more, is uncertain. In part, it depends on the quality of life, the level of technological development, and other standards that societies wish to maintain. Yet even when considering the most basic factors, such as food, it is unclear just how many people Earth can sustain. Projections about the adequacy of food production, for example, require far more information than just the number of people to be fed and the amount of available land. The total arable land (land suitable for cultivation) in the world is an estimated 3.4 billion acres, or just less than half an acre per person of the present population. The major limitation on this figure is availability of water, either as rainfall or through irrigation. Further considerations relating to the nature of the soil include the soil's fertility, water-holding capacity, and general suitability for farming. Soil character and productivity varies tremendously. Moreover, farmland can deteriorate through loss of nutrients and by wholesale erosion of topsoil. Soil degradation and loss is occurring at rates that are at 10 times faster than natural (**figure 1.17**).

There is also the question of what crops can or should be grown. Today, this is often a matter of preference or personal taste, particularly in nations rich in farmland, energy, and water. The world's people are not currently being fed in the most resource-efficient ways. To produce one ton of corn requires about 250,000 gallons of water; a ton of wheat, 375,000 gallons; a ton of rice, 1,000,000 gallons; a ton of beef, 7,500,000 gallons. Some new high-yield crop varieties may require irrigation whereas native varieties did not. The total irrigated acreage in the world continues to increase each year. In some countries such as the United States, water use for irrigation continues to decrease with improved technologies and conservation techniques yet water resources are dwindling in many places. The cost of water is integral in discussing food production.

Genetic engineering is now making important contributions to food production, as varieties are selectively developed for high yield, disease resistance, and other desirable qualities. These advances have led some to declare that fears of global food shortages are no longer warranted, even if the population grows by several billion more. However, at least two concerns remain. One is that poor nations already struggling to feed their people may be least able to afford the higher-priced designer seed or specially developed animal strains to benefit from these advances. Second, as many small farms using many, genetically diverse strains of food crops are replaced by vast areas planted with a single, new variety, there is the potential for devastating losses if a new pest or disease to which that one strain is vulnerable enters the picture.

Food production as practiced in the United States is also a very energy-intensive business. The farming is heavily mechanized and much of the resulting food is extensively processed, stored, and prepared in various ways requiring considerable energy. The products are elaborately packaged and often transported long distances to the consumer. Exporting the same production methods to poor, heavily populated countries short on energy and the capital to buy fuel, as well as food, may be neither possible nor practical. Even if possible, it would substantially increase the world's energy demands.

Carrying Capacity

Food is at least a renewable resource. Within a human life span, many crops can be planted and harvested from the same land and many generations of food animals raised. By contrast, the supplies of many of the resources considered in later chapters—minerals, fuels, even land itself—are finite. There is only so much oil to burn, rich ore to exploit, and suitable land on which to live and grow food. When these resources are exhausted, we will need to find alternatives or people will have to do without.

Earth's supply of many such materials is severely limited, especially considering the rates at which these resources are presently being used. Many could be effectively exhausted within decades, yet most people in the world are consuming very little in the way of minerals or energy. Current consumption is strongly concentrated in a few highly industrialized societies. Per-capita consumption of most mineral and energy resources is higher in the United States than in any other nation. For the world population to maintain a standard of living comparable to that of the United States, mineral production would have to increase about fourfold, on the average. There are neither the recognized resources nor the production capability to maintain that level of consumption for long, and the problem becomes more acute as the population grows.

Some scholars believe that we are already on the verge of exceeding Earth's **carrying capacity**, its ability to sustain its population at a basic, healthy, moderately comfortable standard of living. Estimates of sustainable world population made over the last few decades range from well under 7 billion—and remember, we are already past that—to over 100 billion persons. The wide range is attributable to considerable variations in

Figure 1.17

Soil erosion on Canadian agricultural land.

canadabrian/Alamy Stock Photo

Figure 1.18

The landslide hazard to these structures sitting at the foot of steep slopes in Rio de Janeiro is obvious, but space for building here is limited.

Will & Deni McIntyre/Getty Images

Figure 1.19

Even where safer land is abundant, people may choose to settle in hazardous—but scenic—places, such as barrier islands. Miami Beach, Florida.

Getty Images/Tetra images

model assumptions, including standard of living and achievable productivity of farmland. Certainly, given global resource availability and technology, even the present world population could not enjoy the kind of high-consumption lifestyle to which the average resident of the United States has become accustomed.

It is true that, up to a point, the increased demand for minerals, fuels, and other materials associated with an increase in population tends to raise prices and promote exploration for these materials. The short-term result can be an apparent increase in the resources' availability, as more exploration leads to discoveries of more oil fields, ore bodies, and so on. However, the quantity of each of these resources is finite. The larger and more rapidly growing the world population, and the faster its level of development and standard of living rise, the more rapidly limited resources are consumed, and the sooner those resources will be exhausted.

Land is clearly a basic resource. Seven, 10, or 100 billion people must be put somewhere. Already the global average population density is about 60 persons per square kilometer of land surface (155 persons per square mile), and that is counting all the land surface including jungles, deserts, and mountain ranges, leaving out only the Antarctic continent. The ratio of people to readily inhabitable land is plainly much higher (565 per square kilometer). Land is also needed for agriculture, manufacturing, energy production, transportation, and a variety of other uses. Large numbers of people consuming vast quantities of materials generate vast quantities of wastes. Some of these wastes can be recovered and recycled, but others cannot. It is essential to find places to put the latter, and ways to isolate the harmful materials from contact with the growing population. This effort claims still more land and, often, resources. All of this is why land-use planning—making the most of every bit of land available—is becoming increasingly important. At present, it is too often true that the ever-growing population settles in areas that are demonstrably unsafe or in which the possible problems are imperfectly known (**figures 1.18** and **1.19**).

Uneven Distribution of People and Resources

Even if global carrying capacity were ample in principle, that of an individual region may not be. None of the resources—livable land, arable land, energy, minerals, or water—is uniformly distributed over Earth. Neither is the population (see **figure 1.20**). In 2021, the population density in persons per square kilometer of arable land was 211 in the United States, 99 in Canada, 84 in Australia, and a staggering 1,012,000 in Singapore. Such data clearly have implications for a nation's ability to feed itself, whatever the availability of farmland globally may be.

Many of the most densely populated countries are resource-poor. In some cases, a few countries control the major share of one resource. Oil is a well-known example, but there are many others. Economic and political complications enter into the question of resource adequacy. Just because one nation

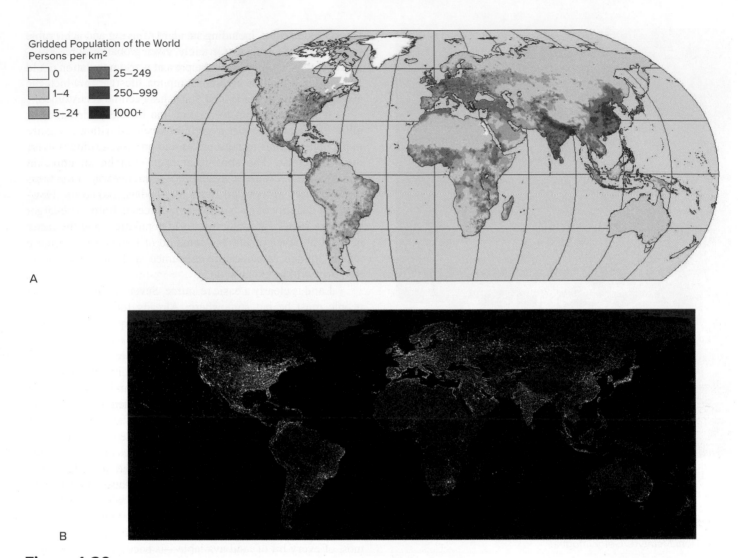

A

B

Figure 1.20

(A) Global population density, 2020; the darker the shading, the higher the population density. Comparison with the distribution of lights at night in 2016. The image (B) shows that overall population density and urbanization/industrialization are often, but not always, closely correlated.

Sources: (A) Center for International Earth Science Information Network (CIESIN), Columbia University; and Centro Internacional de Agricultura Tropical (CIAT), Gridded Population of the World (GPW), Version 3. Palisades, NY: CIESIN, Columbia University. Available at: http://sedac.ciesin.columbia.edu/data/collection/gpw-v4. (B) Source: C. Mayhew & R. Simmon (NASA/GSFC), NOAA/NGDC, DMSP

controls enough of some commodity to supply all the world's needs does not necessarily mean that the country will choose to share its resource wealth or to distribute it at modest cost to other nations. Some resources, like land, are simply not transportable and therefore cannot readily be shared. Some of the complexities of global resource distribution will be highlighted in subsequent chapters.

Disruption of Natural Systems

Natural systems tend toward a balance or equilibrium among opposing factors or forces. When one factor changes, compensating changes occur in response. If the disruption of the system is relatively small and temporary, the system may in time return to its original condition, and evidence of the disturbance will disappear. For example, a coastal storm may wash away beach vegetation and destroy colonies of marine organisms living in a tidal flat, but when the storm has passed, new organisms will start to move back into the area and new grasses will take root in the dunes. The violent eruption of a volcano like Mount Pinatubo (**figure 1.21**) may spew ash and gases high into the atmosphere partially blocking sunlight and causing Earth to cool, but within a few years the ash will have settled back to the ground and normal temperatures will be restored. Dead leaves falling into a lake provide food for the microorganisms that within weeks or months will break the leaves down and eliminate them.

Figure 1.21

The 1991 eruption of Mount Pinatubo in the Philippines lowered global temperatures by 0.5°C from 1991 to 1993.

Karin Jackson/U.S. Air Force

This is not to say that permanent changes never occur in natural systems. The size of a river channel reflects the maximum amount of water it normally carries. If long-term climatic or other conditions change so that the volume of water regularly reaching the stream increases, the larger quantity of water will, in time, carve out a correspondingly larger channel to accommodate it. The soil carried downhill by a landslide certainly does not begin moving back upslope after the landslide is over; the face of the land is irreversibly changed. Even so, a hillside forest uprooted and destroyed by the slide may, within decades, be replaced by new growth in the soil deposited at the bottom of the hill.

Human activities can cause or accelerate permanent changes in natural systems. The impact of humans on the global environment is broadly proportional to the size of the population, as well as to the level of technological development achieved. This can be illustrated especially easily in the context of pollution. The smoke from one campfire pollutes only the air in the immediate vicinity; by the time that smoke is dispersed through the atmosphere, its global impact is negligible. The collective smoke from a century and a half of increasingly industrialized society, on the other hand, has caused measurable increases in several atmospheric pollutants worldwide, and these pollutants continue to pour into the air from many sources. It was once assumed that the seemingly vast oceans would be an inexhaustible sink for any excess CO_2 that we might generate by burning fossil fuels, but decades of steadily climbing atmospheric CO_2 levels have proven that in this sense, at least, the oceans are not as large as we thought. Likewise, eight people carelessly dumping wastes into the ocean would not appreciably pollute that huge volume of water. The prospect of 8 billion people doing the same thing, however, is quite another matter. And every hour, now, world population increases by nearly 10,130 people.

Earth's Moon

Scientists have long strived to explain the origin of Earth's large and prominent satellite. Through much of the twentieth century, several different models competed for acceptance, and within the last few decades a new theory of lunar origin has been developed. While a complete discussion of the merits and shortcomings of these is beyond the scope of this book, they provide good examples of how objective evidence can provide support for, or indicate weaknesses in, hypotheses and theories.

Any acceptable theory of lunar origin has to explain a number of facts. The Moon orbits Earth in an unusual orientation (**figure CS 1.1.1**), neither circling around Earth's equator nor staying in the plane in which the planets' orbits around the Sun lie (the ecliptic plane). Its density is much lower than that of Earth, meaning that it contains a much lower proportion of iron. Otherwise, it is broadly similar in composition to Earth's mantle. However, analysis of samples from the *Apollo* missions revealed that relative to Earth, the Moon is depleted not only in most gases, but also in volatile metals such as lead and rubidium, indicating a hot history for lunar material.

The older accretion model proposed that the Earth and Moon accreted close together during solar system formation, and that is how the Moon comes to be orbiting Earth. But in that case, why is the Moon not orbiting in the ecliptic plane, and why the significant chemical differences between the two bodies?

The fission hypothesis postulated that the Moon was spun off from a rapidly rotating early Earth after Earth's core had been differentiated, so the Moon formed mainly from Earth's mantle. That would account for the Moon's lower density and relatively lower iron content. But analyses of the lunar samples revealed many additional chemical differences between the Moon and

Earth's mantle. Furthermore, calculations show that any moon formed in this way should be orbiting in the equatorial plane, and that far more angular momentum would be required to make it happen than is present in the Earth–Moon system.

A third suggestion was that the Moon formed elsewhere in the solar system and then passed close enough to Earth to be captured into orbit by gravity. A major flaw in this idea involves the dynamics necessary for capture. The Moon is a relatively large satellite for a body the size of Earth. For Earth's gravity to snare it, the rate at which the Moon came by would have to be very, very slow. But Earth is hurtling around the Sun at about 107,000 km/hr (66,700 mph), so the probability of the Moon happening by at just the right distance and velocity for capture to occur is extremely low. Nor does capture account for a hot lunar origin.

So how to explain the Moon? The generally accepted theory for the past two decades involves a collision, or giant impact, between Earth and a protoplanet named Theia, whose orbit in the early solar system put it on course to intercept Earth. The tremendous energy of the collision would have destroyed the impactor and caused extensive heating and melting on Earth, ejecting quantities of vaporized minerals into space around Earth. If this impact occurred after core differentiation, that material would have come mainly from the mantles of Earth and Theia. The orbiting material would have condensed and settled into a rotating disk of dust that later accreted to form the Moon.

This theory, exotic as it sounds, does a better job of accounting for all the necessary facts. It provides for a (very) hot origin for the material that became the Moon, explaining the loss of volatiles. With mantle material primarily involved, a resulting lunar composition similar to that of Earth's mantle is reasonable, and

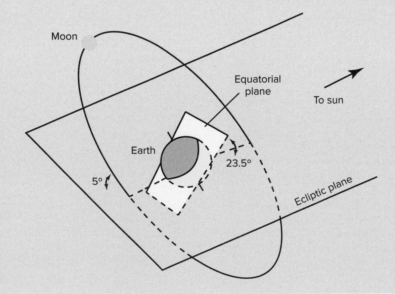

Figure CS 1.1.1

The Moon's unusual orbit.

(Continued)

contributions from the impactor would introduce some differences as well. An off-center hit by Theia could easily produce a dust disk (and eventual lunar orbit) oriented at an angle to both the ecliptic plane and Earth's equatorial plane. We know that the Moon was extensively cratered early in its history, and accretion of debris from the collision could account for this. And computer models designed to test the mechanical feasibility of this theory have shown that indeed, it is physically plausible. So the giant-impact theory is likely to remain the prevailing explanation of lunar origin—until and unless new evidence is found that does not fit.

Though humans have not been back to the Moon for four decades, new information about it can still be generated. Since *Apollo*, computers have become much more powerful, and chemical-analysis techniques more sensitive and sophisticated. Reprocessing of *Apollo*-era seismic data has indicated that the Moon has an iron-rich core like Earth's, though proportionately much smaller. Reanalysis of lunar samples has provided more-detailed information on the Moon's composition, including evidence for water in the lunar mantle. Such data will provide further tests of the giant impact theory.

New chemical and mathematical analyses are also being applied to questions such as how much of the Moon was derived from Earth material and how much from Theia, or whether, indeed, there were multiple impactors over an extended period of time.

Humans will soon return to the Moon through NASA's Artemis mission, one of its primary goals being to establish a sustainable environment capable of supporting human exploration. The conditions these new astronauts will encounter is a daunting one (**figure CS 1.1.2**). The Moon has essentially no atmosphere, to breathe or to trap heat to moderate surface temperatures, so in sunlight the surface soars to about 120°C (250°F) and during the lunar night plunges to about −175°C (−280°F). There is no vegetation or other life. They will need to access the water ice that does exist at the poles, or the water molecules present in the shallow soil on the sunlit side of the

Moon, or bring very expensive bottled water with them. Some mineral resources may be mined, while other raw materials will have to be shipped from Earth. The surface blanket of abrasive rock dust that can abrade and foul equipment and, if breathed, injure lungs will complicate many procedures. This new phase of lunar exploration will again result in technologies that we find useful, if not necessary, on Earth too.

Figure CS 1.1.2

The lunar surface is not an environment in which humans could live outside protective structures, with high energy and resource costs. Geologist/astronaut Harrison Schmitt, lunar module commander of the *Apollo* 17 mission.

NASA

Summary

The solar system formed over 4½ billion years ago. Earth is unique among the planets in its chemical composition, abundant surface water, and oxygen-rich atmosphere. Earth passed through a major period of internal differentiation early in its history, which led to the formation of the atmosphere and the oceans. Earth's surface features have continued to change throughout the last 4+ billion years through a series of processes that are often cyclical in nature, and commonly interrelated. The oldest rocks in which remains of simple organisms are recognized are at least 2 billion, and may be up to 3.5 billion years old. The earliest photosynthetic organisms were responsible for the development of free oxygen in the atmosphere, which in turn, made it possible for oxygen-breathing animals to survive. Human-genus remains are unknown in rocks over 2.5 million years of age. In a geologic sense, therefore, human beings are quite a new addition to Earth's cast of characters, but they have had a very large impact. Geologic processes, in turn, can have an equally large impact on us.

The world population, now over 7.8 billion, might well be close to 9.6 billion by the year 2050. Even our present population cannot entirely be supported at the level customary in more economically developed countries, given the limitations of land and resources. Scientifically based and technically improved use of our remaining resources will be essential.

Key Terms and Concepts

carrying capacity	doubling time	exponential growth	scientific method
core	environmental	hypothesis	theory
crust	geology	mantle	

Test Your Learning

1. Describe the process by which the solar system formed, and explain why it led to planets of different compositions, even though the planets formed simultaneously.

2. Compare the age of Earth with the length of time humans have both inhabited the planet and come to influence their geologic environment.

3. Explain how the newly formed Earth differed from the Earth we know today.

4. List the kinds of information that are used to determine Earth's internal composition.

5. Outline the processes by which Earth's atmosphere and oceans formed, and how the atmosphere has changed over Earth's history.

6. Differentiate facts, scientific hypotheses, and scientific theories.

7. Many Earth materials are transformed through processes that are cyclical in nature. Describe one example.

8. The size of Earth's human population directly affects the severity of many environmental problems. Illustrate this idea in the context of (a) resources and (b) pollution.

9. If Earth's population has already exceeded the planet's carrying capacity, explain the implications for achieving a sustainable worldwide standard of living.

10. Recall the world's present population. Explain how population growth rates over the last few centuries compare with those of earlier times, and why. Describe how global growth rates have changed over the last half-century.

11. Explain the concept of doubling time and how population doubling time has generally been changing through history. Recall the approximate doubling time of the world's population at present.

12. Indicate what regions of the world currently have the fastest and slowest rates of population growth.

13. Describe any one of the less-favored hypotheses of lunar origin, and note at least one fact about the Moon that it fails to explain.

Exploring Further

1. The urgency of population problems can be emphasized by calculating such "population absurdities" as the time at which there will be one person per square meter or per square foot of land or the time at which the weight of people will exceed the weight of Earth. Try calculating these population absurdities by using the world population projections from table 1.4 and the following data concerning Earth:

   Mass 5976×10^{21} kg*

   Land surface area (approx.) 149,000,000 sq. km

   Average weight of human body (approx.) 75 kg

2. Derive the relation between doubling time and growth rate, starting from the exponential-growth relation

$$N = N_0\, e^{(G/100)\cdot t}$$

where N = the growing quantity (number of people, tons of iron ore consumed annually, dollars in a bank account, or whatever); N_0 is the initial quantity at the start of the time period of interest; G is growth rate expressed in percent per year, and "percent" means "parts per hundred"; and t is the time in years over which growth occurs. Keep in mind that doubling time, D, is the length of time required for N to double.

3. Select a single country or small set of related countries; examine recent and projected population growth rates in detail, including the factors contributing to the growth rates and trends in those rates. Compare with similar information for the United States or Canada. A useful starting point may be the latest World Population Data Sheet from the Population Reference Bureau (**www.prb.org**).

---

*This number is so large that it has been expressed in scientific notation, in terms of powers of 10. It is equal to 5976 with twenty-one zeroes after it. For comparison, the land surface area could also have been written as 149×10^6 sq. km, or $149,000 \times 10^3$ sq. km, and so on.

CHAPTER 2

Rocks and Minerals

The study of environmental geology begins with the rocks and minerals that are on and under Earth's surface. Considering that the most common rocks and minerals are composed of a small subset of chemical elements, they are remarkably diverse in color, texture, and other physical properties. Some minerals we value for their conductivity and malleability, while others we prize as gemstones. The different physical properties of rocks and soils determine how we use them: as sources of water, metals, or petroleum; locations for waste disposal; or sources of materials needed for construction, in manufacturing, and for agriculture.

Each rock or rock unit contains clues that indicate the processes involved in its formation and its associated geologic setting. The nature of a volcano's rocks indicate what hazards it presents to us. Our search for new sources of ores or fuels is guided by an understanding of the specialized geologic environments in which they occur.

For all these reasons, and so many more not yet mentioned, it is helpful for us to understand the nature of geological materials. We will explore the roles and utilities of specific minerals and rocks in later chapters but for now, we will take a brief overview into elements, minerals, and rocks and how they are connected.

Chapter Outline

2.1 Chemistry Basics
2.2 Fundamentals of Minerals
2.3 Types of Minerals
2.4 Rocks

Much of the Grand Teton range is composed of gneiss and schist, both metamorphic rocks that have been part of the rock cycle for at least 2.7 billion years. These rocks originated as sedimentary and igneous rocks, were buried, heated, and deformed, and were eventually uplifted into mountains.

©Gina S. Szablewski

2.1 Chemistry Basics

Learning Objectives

- Recognize the parts of an atom
- Describe the importance of isotopes to geology
- Explain the layout of the periodic table
- Compare ionic and covalent bonding

Before we start discussing rocks and minerals, let us briefly describe some chemistry basics. This review will provide a solid understanding that rocks are made of minerals and minerals are made of elements, and that the nature of the elements and how they are put together in turn affect the nature of minerals and rocks.

Atomic Structure

Rocks and minerals are composed of the ninety-two naturally occurring chemical elements on Earth. The fundamental unit of an element is the **atom,** the smallest particle into which an element can be divided while still retaining the chemical characteristics of that element (see **figure 2.1**). At the center of the atom is the **nucleus** which contains one or more **protons** and usually some **neutrons.** Circling the nucleus are the negatively charged **electrons.** Protons and neutrons are similar in mass, and together they account for most of the mass of an atom. Electrons are much lighter and are difficult to portray in an image; the most popular models show the electrons around the nucleus either as a cloud, as in **figure 2.1**, or as particles in concentric shells, as in **figure 2.3**. The −1 charge of one electron exactly balances the +1 electrical charge of a single proton.

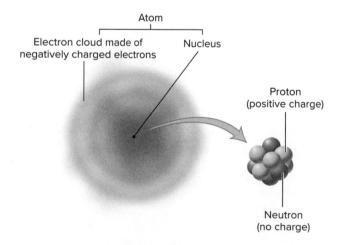

Figure 2.1

Cloud model of general atomic structure; in reality the nucleus is about 1/1000th the overall size of the atom.

The number of protons in the nucleus of an atom determines the chemical element. For example, every hydrogen atom contains one proton in its nucleus; every oxygen atom contains eight protons; every carbon atom, six; every iron atom, twenty-six. The characteristic number of protons is the **atomic number** of the element.

Elements and Isotopes

With the exception of the simplest hydrogen atoms, all nuclei contain neutrons, and the number of neutrons is the same as or somewhat greater than the number of protons. The sum of the number of protons and the number of neutrons in a nucleus is the atom's **atomic mass number.** The number of neutrons in atoms of one element can vary. Atoms of a given element with different atomic mass numbers—in other words, atoms with the same number of protons but different numbers of neutrons—are distinct **isotopes** of that element. Some elements have only a single isotope, while others may have ten or more.

For most geologic applications, we are concerned only with the elements involved, not with specific isotopes. When a particular isotope is designated, it is done by naming the element (which, by definition, specifies the atomic number or number of protons) and the atomic mass number (protons plus neutrons). Carbon, for example, has three natural isotopes; the most abundant is carbon-12, which has six neutrons in the nucleus in addition to the six protons common to all carbon atoms. The two rarer isotopes are carbon-13 (seven neutrons) and carbon-14 (eight neutrons). Chemically, they all behave alike. The human body cannot, for instance, distinguish between sugar containing carbon-12 and sugar containing carbon-13.

Other differences between isotopes may make a particular isotope useful. Some isotopes are *radioactive,* meaning that over time their nuclei will change into nuclei of other elements, releasing energy. Each such radioactive isotope will decay at its own specific rate, which allows us to use these isotopes to date geologic materials and events, as described in appendix A. A familiar example is carbon-14 which we use to date young materials containing carbon including archeological remains such as cloth, charcoal, and bones. Differences in the properties of two uranium isotopes are important in understanding nuclear power options: U-235 is suitable for use as reactor fuel but it is much less common than U-238. The fact that radioactive elements will inevitably decay and release energy at their own fixed, constant rates is part of what makes radioactive-waste disposal such a challenging problem, because no chemical or physical treatment can make those waste isotopes nonradioactive and inert.

Ions

In an electrically neutral atom, the number of protons and the number of electrons are the same; the negative charge of one electron equals the positive charge of one proton. Most atoms, however, can lose or gain some electrons. When this happens, the atom has a positive or negative electrical charge and is called an **ion.** If it loses electrons, it becomes positively charged, since the number of protons then exceeds the number of electrons. If the atom gains electrons, the ion has a negative electrical charge. Positively and negatively charged ions are called, respectively, **cations** and **anions.** Both solids and liquids are electrically neutral overall, with the total positive and negative charges of cations and anions balanced. Also, free ions don't exist in solids; cations and anions are bonded together. In a solution, however, individual ions may exist and move independently. Many minerals break down into ions as they dissolve in water. Individual ions may then be taken up by plants as nutrients or react with other materials. For example, the acidic nature of mine tailings runoff, which is determined by the concentration of hydrogen ions (pH) in solution, can cause serious contamination of surface water bodies.

The Periodic Table

Knowledge of the **periodic table** can help us understand the probable chemical behavior of elements (**figure 2.2**). Russian scientist Dmitri Mendeleev observed that certain groups of elements showed similar chemical properties that seemed to be related in a regular way to their atomic numbers. Mendeleev published his periodic table in 1869 and arranged the elements to reflect these similarities of behavior; at the time, not all elements had even been discovered, so there were some gaps in the table. The addition of elements identified later confirmed the concepts underlying their arrangement. In fact, some of the missing elements were found more easily because their properties could to some extent be anticipated from their expected positions in the periodic table.

We can relate the periodicity of chemical behavior to the electronic structures of the elements. Electrons around an atom occur in levels (or shells) of different energies, each of which can hold a fixed maximum number of electrons. Those elements in the Column 1 of the periodic table, the alkali metals, have one electron in the outermost shell of the neutral atom. Thus, they all tend to form cations of +1 charge by losing that odd electron. Outermost electron shells become increasingly full from left to right across a row. Column 17 has the halogens; these elements are lacking only one electron in the outermost shell, and they tend to gain one electron to form anions of charge −1. The inert gases, whose neutral atoms contain fully filled electron shells, are in Column 18.

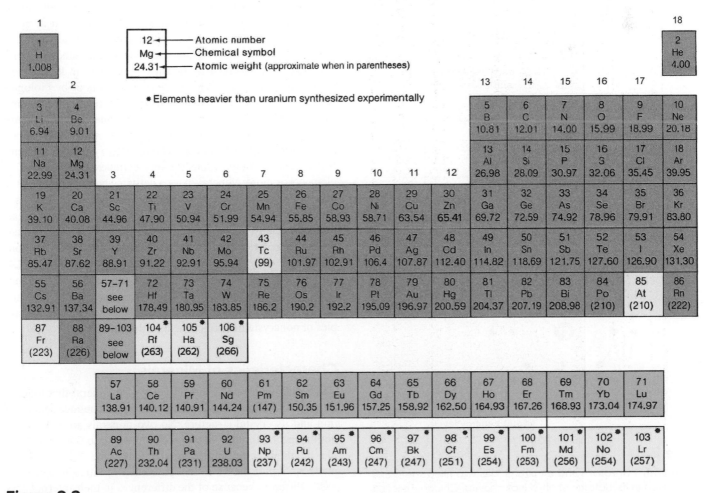

Figure 2.2

The periodic table.

Source/reference: https://iupac.org/what-we-do/periodic-table-of-elements/

Scientists discovered the last of the naturally occurring elements on Earth in the mid-1900s. Additional new elements were created, not simply discovered, because these very heavy elements with atomic numbers above 92 (uranium) are too unstable to have lasted from the early days of the solar system to the present. Some, such as plutonium, are by-products of nuclear-reactor operation; others are created by nuclear physicists who, in the process, learn more about atomic structure and stability.

Compounds and Bonding

A **compound** is a chemical combination of two or more chemical elements, bonded together in particular proportions, with a distinct set of physical properties that are often very different from those of any of the individual elements in it. In minerals, most bonds are *ionic* or *covalent,* or a mix of the two. In **ionic bonding,** the bond is based on the electrical attraction between oppositely charged ions. When atoms share electrons, **covalent bonding** occurs. Table salt (sodium chloride) is a common example of ionic bonding (**figure 2.3**). Sodium, an alkali metal, loses its outermost electron to chlorine, a halogen. The two elements now have filled electron shells, but sodium is left with a +1 net charge, chlorine −1. The ions bond ionically to form sodium chloride. Sodium is a silver metal and chlorine is a

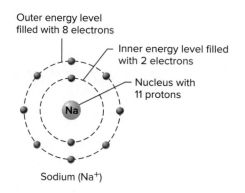

Sodium (Na⁺)

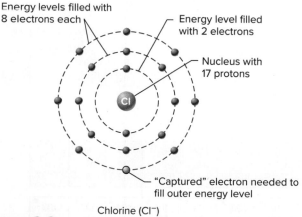

Chlorine (Cl⁻)

Figure 2.3

Sodium, with 11 protons and electrons, has two filled shells and one electron in its outermost shell (not shown). Chlorine accepts that odd electron, filling its own outermost shell exactly. The resulting oppositely charged ions attract and bond to form sodium chloride.

greenish gas that is poisonous in large doses. When equal numbers of sodium and chlorine atoms combine to make sodium chloride, the resulting compound forms colorless crystals that don't resemble either of the component elements.

2.2 Fundamentals of Minerals

Learning Objectives

- List the defining requirements of a mineral
- Describe how minerals are identified by their chemical composition and internal crystal structure
- Exemplify properties of minerals

In this section, we discuss what a mineral is, how we distinguish one mineral from another, and what properties we use to characterize minerals.

Minerals Defined

A **mineral** is a naturally occurring, inorganic, solid element or compound with a definite chemical composition and a regular internal crystal structure. *Naturally occurring,* rather than synthetic, means that minerals do not include the thousands of chemical substances invented by humans. *Inorganic,* in this context, means not produced solely by living organisms or by biological processes. *Solid* means just that—it is not a liquid or a gas at normal Earth temperatures and pressures. Chemically, minerals may consist either of one element—such as diamonds, which are pure carbon—or they may be compounds of two or more elements. Some mineral compositions are very complex, consisting of ten elements or more. Minerals have a definite chemical composition or a compositional range. The presence of certain elements in certain proportions is one of the identifying characteristics of each mineral. Finally, minerals are *crystalline,* at least on the microscopic scale. **Crystalline** materials are solids in which the atoms or ions are arranged in regular, repeating patterns (**figure 2.4**). These patterns may not be apparent to the naked eye, but most solid compounds are crystalline, and we can recognize and study their crystal structures using X rays and other techniques. Examples of noncrystalline solids include glass and plastic.

Characteristics of Minerals

The two fundamental characteristics of a mineral that together distinguish it from all other minerals are its chemical composition and its crystal structure. No two minerals are identical in both respects, though they may be the same in one. For example, diamond and graphite are chemically the same—both are made up of pure carbon. Their physical properties, however, are vastly different because of the differences in their internal crystalline structures (**figure 2.5**). In a diamond, each carbon atom is firmly bonded to every adjacent carbon atom in every direction by covalent bonds. In graphite, the carbon atoms are bonded

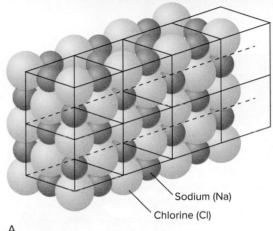

Sodium (Na)
Chlorine (Cl)

A

Figure 2.4A

Sodium and chloride ions are arranged alternately in the mineral halite. Lines model the repeating cubic crystal structure. The external cubic structure of this sample of halite reflects its internal structure.

McGraw Hill

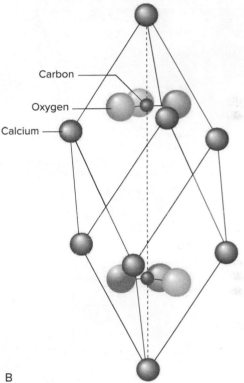

Carbon
Oxygen
Calcium

B

Figure 2.4B

The crystal structure of calcite is more complex. Here, the atoms are spread apart to make the structure more apparent. Lines model the shape of the repeating crystal structure. The external structure of this sample of calcite reflects its internal structure.

©*Carla Montgomery*

strongly in two dimensions into sheets, but the sheets are only weakly held together in the third dimension. Diamond is clear, colorless, and very hard, making it desirable both as a cutting implement and as an ornament. Graphite is black, opaque, and soft, and its sheets of carbon atoms tend to slide apart as the weak bonds between them are broken; it is very useful as a lubricant and in pencils.

We can usually only determine the distinct composition and internal crystal structure of a mineral by using sophisticated laboratory equipment. Identifying a hand sample of a mineral is difficult for a few reasons: recognizable, well-developed crystal shapes are relatively rare; different minerals share similar external forms; and the same mineral may appear with different external forms (**figure 2.6**). No one can just look at a mineral and know its chemical composition without first recognizing what mineral it is. When scientific instruments are not at hand, we base mineral identification on a variety of other physical properties that in some way reflect the mineral's composition and structure. These other properties are often what make the mineral commercially valuable. However, they are rarely unique

to one mineral and often are deceptive. A few examples of such properties follow.

Other Physical Properties of Minerals

Color is usually the most apparent characteristics of a mineral and often the least reliable in terms of identification. While

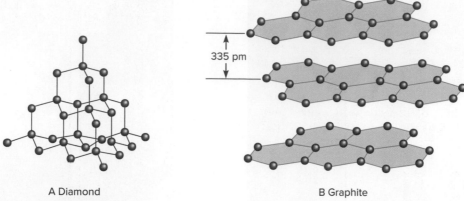

A Diamond   B Graphite

Figure 2.5

Crystalline structure of diamond versus graphite.

McGraw-Hill Education

A

B

C

D

Figure 2.6

Minerals may share the same external crystal form: (A) Galena and (B) fluorite form cubes, as do halite (figure 2.4A) and pyrite (C). However, these minerals may show other forms; (D), for example, this distinctive form of pyrite is called a pyritohedron.

(A, C) Doug Sherman/Geofile; (B, D) ©Carla Montgomery

some minerals always appear the same color, many vary from specimen to specimen. Variation in color is usually due to the presence of small amounts of chemical impurities in the mineral that have nothing to do with the mineral's characteristic composition, and such variation is especially common when the pure mineral is light-colored or colorless. The very common mineral quartz, for instance, is colorless in its pure form. However, quartz also occurs in rose pink, golden yellow, smoky brown, purple (amethyst), and milky white varieties. Clearly, quartz cannot always be recognized by its color or lack of it (**figure 2.7A**).

Corundum, a simple compound of aluminum and oxygen, is another example of how mineral color can vary. It is colorless in its pure form and quite hard, which makes it a good abrasive. It is often used for the grit on sandpaper. Yet the tiny introduction of an impurity not only changes its color, but it also transforms this utilitarian material into highly prized gems, traces of chromium produce the deep bluish-red gem we call ruby, while sapphire is just corundum tinted blue by iron and titanium. The color of a mineral can even vary within a single crystal (**figure 2.7B**). Even when the color of a mineral sample is the true color of the pure mineral, it is probably not unique. There are over 5000 known minerals, so there are many of any one particular color. *Streak,* the color of the powdered mineral as revealed when the mineral is scraped across a piece of unglazed tile, may be quite different from the color of the bulk sample, and more consistent for a single mineral. However, a tile is not always handy and some samples are too valuable to treat this way.

Hardness, the ability to resist scratching, is another easily measured physical property that can help to identify a mineral, although it usually doesn't uniquely identify the mineral. We measure relative hardness using the Mohs hardness scale (**table 2.1**), in which ten common minerals are

Table 2.1	The Mohs Hardness Scale	
Mineral	**Assigned Hardness**	**Referenced Common Objects**
Talc	1	
Gypsum	2	Fingernail (2.5)
Calcite	3	Penny (3)
Fluorite	4	
Apatite	5	Glass (5-6)
Orthoclase	6	
Quartz	7	
Topaz	8	
Corundum	9	
Diamond	10	

arranged in order of hardness. Unknown minerals are assigned a hardness on the basis of which minerals they can scratch and which minerals scratch them. A mineral that scratches gypsum and is scratched by calcite is assigned a hardness of 2½ (the hardness of an average fingernail). Because a diamond is the hardest natural substance known on Earth, and corundum the second-hardest mineral, these minerals might be identifiable from their hardnesses. Among the thousands of softer or more readily scratched minerals, there are many of any particular hardness, just as there are many of any particular color.

Cleavage is another characteristic of minerals that is related to internal structure. Cleavage occurs when minerals break cleanly in preferred directions that correspond to planes of weak bonding in the crystal, producing planar cleavage faces. There are many different types of cleavage: mica

A

B

Figure 2.7

(A) The many color of these quartz sample illustrate why color alone is a poor guide for mineral identification. (B) The varying colors in this tourmaline sample indicate that chemical conditions changed during its formation.

(A, B) ©Carla Montgomery

A

B

C

Figure 2.8

Because of their internal crystalline structures, many minerals break apart preferentially in certain directions. (A) The cleavage of halite reflects its cubic structure as shown in figure 2.4 and breaks into cubic or rectangular pieces, cleaving parallel to the crystal faces. (B) Fluorite also forms cubic crystals, but cleaves into octahedral fragments, breaking along different planes of its internal structure. (C) Quartz shows one type of fracture— conchoidal.

(A) Bob Coyle/McGraw Hill; (B) Charles D. Winters/Timeframe Photography/McGraw Hill; (C) Studio-Annika/iStock/Getty Images

minerals show one distinct direction of cleavage whereas other minerals may have two or three directions of good cleavage. Cleavage surfaces are characteristically shiny (**figure 2.8**). Not all minerals have cleavage. Some minerals such as quartz *fracture,* simply crumbling or shattering into irregular fragments, because there are no preferential planes of weakness in the crystal structure. Quartz actually has uniformly strong bonding as reflected in its hardness; it just doesn't have one set of bonds that is distinctly weaker than another so it fractures rather than cleaves.

There are other physical properties that may be common to many minerals. *Luster* describes the appearance of the surfaces—glassy, metallic, pearly, etc. Some minerals are noticeably denser than most. Some minerals are magnetic. Usually, it is only by considering a whole set of such non-unique properties as color, hardness, cleavage, density, and others together that we can identify a mineral without complex instruments. For instance, there are many colorless minerals; but if a sample of such a mineral has a hardness of 3, cleaves into rhombohedral shapes, and fizzes when weak acid is dripped on it, it is probably calcite.

Unique or not, the physical properties arising from a mineral's composition and crystal structure are often what give a mineral value from a human perspective—the slickness of talc (the main ingredient of talcum powder), the malleability and conductivity of copper, the durability of diamond, and the rich colors of tourmaline gemstones are all examples. Some minerals have several useful properties: halite is a necessary nutrient that also imparts a taste we find pleasant, dissolves readily to flavor liquids, serves as a food preservative in high concentrations, and lowers the freezing temperature of water, helping to keep roadways ice-free in winter. Recognizing the nature of a mineral will help us both identify it and understand its role in geologic processes that affect us.

2.3 Types of Minerals

Learning Objectives

- Recognize the silicate tetrahedron
- Exemplify silicate minerals
- Exemplify nonsilicate minerals and mineral groups

We group or subdivide minerals on the basis of their two fundamental characteristics—composition and crystal structure. Classification by composition is typically dependent on common ions or ion groups that a set of minerals has. There are well over 5000 different Earth minerals, but most of them are very rare. In this section, we will briefly review some of the most common on the basis of mineral categories. A summary of physical properties of selected minerals is in appendix B.

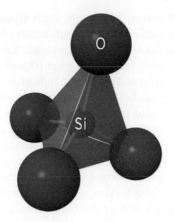

Figure 2.9

The basic silica tetrahedron is the building block of all silicate minerals.

Silicates

The two most common elements in Earth's crust are oxygen and silicon. Unsurprisingly, the largest compositional group of minerals is the **silicate** group made of compounds containing silicon and oxygen, and most of which contain other elements as well. Because this group of minerals is so large, it is subdivided on the basis of crystal structure, by the ways in which the silicon and oxygen atoms are linked together. The basic building block of all silicates is a tetrahedral arrangement of four oxygen atoms (anions) around the much smaller silicon cation (**figure 2.9**). In different silicate minerals, these *silica tetrahedra* may be linked into chains, sheets, or three-dimensional frameworks by the sharing of oxygen atoms (**figure 2.10**). The different crystal structures are closely related to the physical properties of the silicates. Next, we will briefly review a few of the more common, geologically important silicate minerals.

While not the most common, *quartz* is probably the best-known silicate. Compositionally, it is the simplest, containing only silicon and oxygen. It is a framework silicate, with silica tetrahedra linked in three dimensions, which helps make it relatively hard and weathering-resistant. Quartz occurs in a variety of rocks and soils and is a regular component of sand. Commercially, the most common use of pure quartz is in the manufacture of glass. Quartz-rich sand and gravel are used in very large quantities in construction. Examples of quartz are shown in **figures 2.7A** and **2.8C**.

The most abundant group of minerals in the crust are the chemically similar minerals known collectively as the *feldspars*. They are composed of silicon, oxygen, aluminum, and either sodium, potassium, or calcium, or some combination of these three. Again, these common minerals are made from elements abundant in the crust. We use them extensively in the manufacture of ceramics. Potassium feldspar is shown in **figure 2.11A**.

Iron and magnesium are also among the more common elements in the crust and therefore occur in many silicate minerals. **Ferromagnesian** is the general term used to describe these

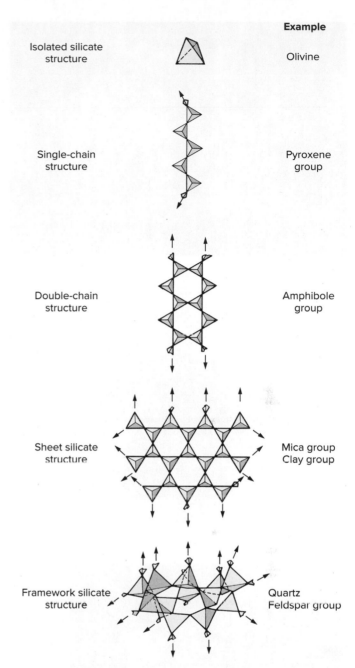

		Example
Isolated silicate structure		Olivine
Single-chain structure		Pyroxene group
Double-chain structure		Amphibole group
Sheet silicate structure		Mica group Clay group
Framework silicate structure		Quartz Feldspar group

Figure 2.10

Silica tetrahedra link together by sharing oxygen atoms, where the corners of the tetrahedra meet, to form a variety of structures. Some of these examples are described in the text.

usually dark-colored silicates (black, brown, or green) that contain iron and/or magnesium with or without additional elements. Most ferromagnesian minerals weather relatively readily. Rocks containing a high proportion of ferromagnesian minerals, then, also tend to weather easily, which is an important consideration in construction. The minerals olivine (**figure 2.11B**), pyroxene, and amphibole are examples that belong to this group.

Like the feldspars, the *micas* are another group of several silicate minerals with similar physical properties, compositions,

A

B

C

Figure 2.11

A collection of silicates. (A) Potassium feldspar; (B) olivine;
(C) mica.

(A) Aleksandr Pobedimskiy/Shutterstock; (B) RF Company/
Alamy Stock Photo; (C) ©Carla Montgomery

and crystal structures. Micas are sheet silicates, built on an atomic scale of stacked-up sheets of linked silicon and oxygen atoms. Because the bonds between sheets are relatively weak, the sheets can easily be broken apart (**figure 2.11C**). Mica is used in the electrical and electronic industries as an insulator.

Clays are another family within the sheet silicates; in clays, the sheets tend to slide past each other, a characteristic that contributes to the slippery feel of many clays and related minerals. Clays are somewhat unusual among the silicates in that their structures can absorb or lose water, depending on how wet conditions are. Absorbed water may increase the slippery tendencies of the clays. Also, some clays expand as they soak up water and shrink as they dry out. A soil rich in these expansive clays is a very unstable base for a building. On the other hand, clays also have important uses such as in making ceramics and building materials, for use in drilling operations, and as an impermeable layer to retard the movement of liquids.

Nonsilicates

Each nonsilicate mineral group is defined by a chemical constituent or characteristic that all members of the group have in common. Most often, that component is the same negatively charged ion or group of atoms. Some of the nonsilicate mineral groups with examples of common or familiar members are included in **table 2.2** and discussed in the following section.

The **carbonates** all contain carbon and oxygen combined in the proportions of one atom of carbon to three atoms of oxygen, written CO_3. The carbonate minerals dissolve relatively easily, particularly in acids, and the oceans contain a great deal of dissolved carbonate. Geologically, the most important and abundant carbonate mineral is calcite (calcium carbonate). Many marine rocks are formed through the precipitation of calcium carbonate from seawater. Another common carbonate mineral is dolomite, which contains both calcium and magnesium in approximately equal proportions. Carbonates may contain many other elements—iron, manganese, or lead, for example. The limestone and marble we use extensively for building and sculpture consist mainly of carbonates, generally calcite. Calcite is also an important ingredient in cement and is shown in **figure 12.12A**.

The **sulfates** all contain sulfur and oxygen in the ratio of 1:4, written SO_4. Gypsum is a calcium sulfate and is the most important because it is both relatively abundant and commercially useful, particularly as a major constituent in plaster and drywall. Sulfates of many other elements also exist.

When sulfur is present without oxygen, the resultant minerals are **sulfides**. A common and well-known sulfide mineral is the iron sulfide *pyrite*. We commonly refer to pyrite (**figure 2.6C, D**) as *fool's gold* because its metallic golden color is both deceiving and alluring. Pyrite is not a commercial source of iron because there are richer ores of this metal. Sulfides comprise some of the most economically important metallic ore minerals and include compounds of lead, copper, and zinc. A familiar example is the lead sulfide mineral *galena,* which often forms silver-colored cubes (**figure 2.6A**). Sulfides cause

Table 2.2   Some Nonsilicate Mineral Groups*

Group	Compositional Characteristic	Examples
Carbonates	Metal(s) plus carbonate (1 carbon + 3 oxygen ions, CO_3)	Calcite (calcium carbonate, $CaCO_3$)
		Dolomite (calcium-magnesium carbonate, $CaMg(CO_3)_2$)
Sulfates	Metal(s) plus sulfate (1 sulfur + 4 oxygen ions, SO_4)	Gypsum (calcium sulfate, with water, $CaSO_4 \cdot 2H_2O$)
		Barite (barium sulfate, $BaSO_4$)
Sulfides	Metal(s) plus sulfur, without oxygen	Pyrite (iron disulfide, FeS_2)
		Galena (lead sulfide, PbS)
		Cinnabar (mercury sulfide, HgS)
Oxides	Metal(s) plus oxygen	Magnetite (iron oxide, Fe_3O_4)
		Hematite (ferric iron oxide, Fe_2O_3)
		Corundum (aluminum oxide, Al_2O_3)
		Spinel (magnesium-aluminum oxide, $MgAl_2O_4$)
Halides	Metal(s) plus halogen element (fluorine, chlorine, bromine, or iodine)	Halite (sodium chloride, NaCl) fluorite (calcium fluoride, CaF_2)
Native elements	Mineral consists of a single chemical element	Gold (Au), silver (Ag), copper (Cu), sulfur (S), graphite (carbon, C)

*Other groups exist, and some complex minerals contain components of several groups (carbonate and hydroxyl groups, for example).

A

B

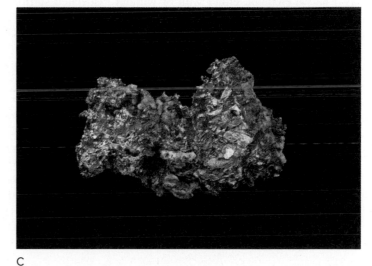

C

Figure 2.12

A collection of nonsilicates: (A) Calcite; (B) hematite; (C) native copper.

(A) P. Leveille/Shutterstock; (B) Doug Sherman/Geofile; (C) Scot-tOrr/E+/Getty Images

problems when they are exposed at Earth's surface and are chemically weathered, forming runoff rich in sulfuric acid that pollutes surface and groundwater supplies.

The **oxides** are minerals containing one or more metal combined with oxygen and lacking the other elements necessary to classify them as belonging to another mineral group. Iron combines with oxygen in different proportions to form more than one oxide mineral. One of these, *magnetite,* has magnetic properties as its name suggests, which is relatively unusual among minerals. Scientists use rocks rich in magnetite to interpret the strength and changing nature of Earth's magnetic field over geologic time. Another iron oxide, *hematite,* may sometimes be silvery black but often has a red color and gives a reddish tint to many soils (**figure 2.12B**). Iron oxides on Mars's surface are responsible for that planet's orange hue. Many other oxide minerals exist, including corundum, the very hard and useful aluminum oxide mineral mentioned earlier.

Native elements are minerals that consist of a single chemical element and the minerals' names are usually the same as the corresponding elements. Most elements do not occur in this pure form but rather form compounds with other elements. Some of our most highly prized materials such as gold, silver, and platinum, often occur as native elements. Diamond and graphite are both examples of native carbon. Sulfur may occur as a native element, either with or without associated sulfide minerals. Some of the richest copper ores contain native copper (**figure 2.12C**). Other metals that may occur as native elements include tin, iron, and antimony.

Several of the factors that make Earth unique in the solar system also increase its mineralogical diversity: plate tectonics, abundance of liquid water, and life. Earth is just the right size to retain sufficient internal heat to keep churning and reprocessing crust and mantle. Abundant surface water allows for hydrous minerals and supports most life on Earth, and those organisms, in turn, modify the chemistry of the atmosphere and oceans, which creates additional mineral possibilities. In comparison, remote sensing by orbiters around Mars have identified about 40 different minerals on the planet's surface. That estimate will likely increase with further data and analyses but will likely not approach the thousands of minerals on Earth, given the different histories of the two planets.

2.4 Rocks

Learning Objectives

- Contrast plutonic and volcanic igneous rocks
- Use igneous rock color and texture to determine composition and cooling history
- Describe lithification
- Contrast clastic and chemical sedimentary rocks
- List changes to a rock resulting from metamorphism
- Compare regional to contact metamorphism
- Exemplify the path a rock might take through the rock cycle

A **rock** is a solid, cohesive aggregate made of one or more types of Earth materials. Those materials can be minerals, pieces of other rocks, glass, and fossils. Sand is not a rock because the many mineral grains of beach sand fall apart when handled, although in time, sand grains may become cemented together to form the rock sandstone. The properties of rocks are important in determining their suitability for particular applications, such as for construction materials or for the carving of a sculpture. Each rock contains within it a record of at least a part of its history in the nature of its minerals and in the way the mineral grains fit together. The three broad categories of rocks—*igneous, sedimentary,* and *metamorphic*—are distinguished by the processes of their formation.

Rocks are linked by the **rock cycle** which summarizes the possible paths rock materials can take over geologic time. The essence of the rock cycle is that rocks are far from being the permanent objects we may imagine them to be and are continually being changed by geological processes. Over the billions of years of geologic time, a given bit of material may have been subject to many, many changes and may have been part of many different rocks. Earth is a constantly changing body. Mountains come and go; seas advance and retreat over the faces of continents; surface processes and deep crustal and mantle processes are constantly altering the planet. One aspect of this continual change is that rocks, too, are always subject to change. We do not have a single sample of rock that has remained unchanged since Earth formed. Therefore, when we describe a rock as being of a particular type, or give it a specific rock name, it is important to realize that we are really describing the form it has most recently taken, the results of the most recent processes acting on it. The following sections examine in more detail the various types of rock-forming processes involved and some of the rocks they form.

Igneous Rocks

At high enough temperatures, rocks and minerals melt. **Magma** is naturally occurring hot, molten rock material. Magmas contain dissolved water and gases and generally have some solid crystals suspended in the melt. An **igneous** rock is a rock formed by the solidification and crystallization of a cooling magma. One way in which magmas, and the igneous rocks that are formed from them, are classified is according to their chemical composition. Magmas and igneous rocks that are rich in silicate minerals are referred to as *silicic* (or *felsic*) and are generally light in color; those that are poor in silicates are *mafic* and are generally dark in color.

Most magmas form through the melting of the uppermost mantle or in deep levels of the crust where temperatures are high enough to melt rocks. Magma may or may not reach the surface before it cools enough to crystallize and solidify. The depth at which a magma crystallizes will affect how rapidly it cools and the subsequent sizes of the mineral grains in the new igneous rock. The size of the mineral grains and the way in which they fit together are called *texture,* and is the second parameter we use to classify igneous rocks.

If a magma remains well below the surface during cooling, it cools relatively slowly, insulated by overlying rock and soil. It may take hundreds of thousands of years or more to

A

B

Figure 2.13

Compare these two igneous rocks: (A) granite is felsic in composition and plutonic whereas (B) basalt is mafic and volcanic. Note the basalt is full of *vesicles,* indicating it formed from a gassy lava. The granite sample is roughly 4 cm across.

(A) Doug Sherman/Geofile; (B) Stephen Reynolds

crystallize completely. Under these conditions, the crystals have ample time to form and grow and the resulting rock has mineral grains large enough to be seen individually with the naked eye; we call this a *coarse-grained* texture. We also refer to a rock formed in this way as a **plutonic** igneous rock, derived from Pluto the Greek god of the underworld. *Granite* is the most widely used example of a plutonic rock (**figure 2.13A**). A typical granite consists principally of quartz and feldspars, and it usually contains some ferromagnesian minerals or other silicates. The proportions and compositions of these constituent minerals may vary, but all igneous rocks classified as granite show the coarse-grained, interlocking crystals characteristic of a plutonic rock. Much of the mass of the continents consists of granite or rocks of granitic composition.

A magma that flows out on Earth's surface is called **lava.** Lava is a common product of volcanic eruptions, and the term **volcanic** is given to an igneous rock formed at or close to Earth's

A

B

Figure 2.14

(A) Obsidian is a natural glass made of approximately 70% SiO_2; its dark color is based on impurities rather than a mafic composition. The textural term *porphyry* is used for a rock that has two distinct grain sizes (B), and is typically paired with a compositional descriptor, that is, granite porphyry. The obsidian sample is roughly 10 cm across.

(A, B) Bob Coyle/McGraw-Hill Education

surface. Magmas that crystallize very near the surface cool more rapidly. There is less time during crystallization for large crystals to form from the magma, so volcanic rocks are typically *fine-grained,* with most crystals too small to be distinguished with the naked eye. The most common volcanic rock is *basalt,* a dark, fine-grained rock rich in ferromagnesian minerals and feldspar (**figure 2.13B**). The ocean floor consists largely of basalt. In some cases, where cooling occurs extremely fast, even tiny crystals may not form before the magma solidifies and its atoms are frozen in a disordered state. The resulting clear, noncrystalline solid is a natural **glass,** *obsidian* (**figure 2.14A**). Occasionally, magma begins to

Figure 2.15

A sample of granite as seen through a petrographic microscopic. Geologists refer to such a sample as a *thin section.*

Melba Photo Agency/Alamy Stock Photo

crystallize slowly at depth, growing some large crystals, and then is subjected to rapid cooling (following a volcanic eruption, for instance). This results in coarse crystals in a fine-grained groundmass, a *porphyry* (**figure 2.14B**).

Regardless of the details of their compositions or cooling histories, all igneous rocks have some textural characteristics in common. Large or small, crystals in igneous rocks are tightly interlocking or intergrown (**figure 2.15**), with some exceptions such as rock made from volcanic ash. Individual crystals tend to be angular in shape. There is usually little pore space, little empty volume that could be occupied by such fluids as water. Structurally, most plutonic rocks are relatively strong unless they have been fractured, broken, or weathered.

As we have just discussed, igneous rocks are classified according to their composition and texture. We can look at a hand sample of an igneous rock and fairly easily be able to tell if it was formed from a magma rich in silica (light-colored) or poor in silica (dark-colored) and whether or not it cooled slowly inside Earth (coarse-grained) or cooled quickly at Earth's surface (fine-grained). There is a relationship between the composition of magma and the way in which (and where) it was created and its associated eruptive style, which we will explore in chapter 5.

Sediments and Sedimentary Rocks

Sedimentary rocks form at relatively low temperatures in a large variety of settings on Earth's surface. **Sediments** are loose, unconsolidated accumulations of mineral or rock particles that are transported by wind, water, or ice, or shifted under the influence of gravity, and deposited. Beach sand is a kind of sediment and so is the mud on a river bottom; they differ in grain size but may also differ in composition depending on their source. Soil is a mixture of Earth materials that has a sediment component of up to 45%. Some sediments originate through the weathering of preexisting rocks by physical breakup into finer

and finer fragments. Other sediments precipitate as crystals out of solutions full of dissolved materials. The physical properties of sediments and soils are important for a broad range of environmental issues including the stability of slopes and building foundations, the selection of optimal waste-disposal sites, and how readily water drains into the subsurface after a rainstorm.

When sediments are compacted or cemented together into a solid, cohesive mass, they become sedimentary rocks. The set of processes by which sediments are transformed into rock is collectively described as **lithification.** The resulting rock is generally denser, as well as more cohesive, than the original sediment. Sedimentary rocks are subdivided into two groups—clastic and chemical.

Clastic sedimentary rocks are formed from the products of mechanical weathering of other rocks. Natural processes continually attack rocks exposed at the surface: rain and waves pound them, windblown dust scrapes them, frost and tree roots crack them. In consequence, rocks are broken up into smaller and smaller pieces and ultimately into individual mineral grains. The resulting rock and mineral fragments may be transported by wind, water, or ice, and accumulate as sediments in streams, lakes, oceans, deserts, or soils. Later geologic processes can cause these sediments to become lithified. Burial under the weight of more sediments may pack the loose particles so tightly that they hold firmly together in a cohesive mass. Except with very fine-grained sediments, however, compaction alone is rarely enough to transform the sediment into rock. Water seeping slowly through rocks and sediments underground also carries dissolved minerals, which may precipitate out of solution to bind the sediment particles together with a natural mineral cement.

Clastic sedimentary rocks are classified according to grain size. *Sandstone,* for instance, is a rock composed of sand-sized sediment particles, 1/16 to 2 millimeters (0.002 to 0.08 inch) in diameter. *Shale* is made up of the smallest clasts—silt and clay—which we refer to as fine-grained because we cannot see them without an aide. *Conglomerate* is a relatively coarse-grained rock made of rounded clasts larger than 2 millimeters (0.08 inches) in diameter held together by finer-grained sediments (**figure 2.16**). Regardless of grain size, clastic sedimentary rocks tend to have a relatively large amount of pore space between grains. The shape and size range of the sediments affect how well the particles pack together. For example, well-rounded sand particles that reflect a large amount of transportation do not fit together very well. Many clastic sedimentary rocks are not particularly strong structurally unless they have been extensively cemented. The porosity of sedimentary rocks is important for the storage of natural resources ranging from groundwater to petroleum.

Chemical sedimentary rocks are composed of crystals formed by precipitation or growth from solution. A common example is *limestone,* composed mostly of mineral calcite (**figure 2.17A**). The chemical sediment that makes limestone may come from fresh or salt water; under favorable chemical conditions, thick limestone beds hundreds of meters thick may form. Another example of a chemical sedimentary rock is *gypsum,* which we refer to as both a mineral and a rock. Gypsum layers commonly

A

B

C

Figure 2.16

Clastic sedimentary rocks. (A) Sandstone; (B) shale; (C) conglomerate. The staff next to the shale outcrop is marked in 10-cm increments.

(A, B) ©Gina S. Szablewski; (C) ©Carla Montgomery

A

B

Figure 2.17

Chemical sedimentary rocks. (A) A limestone outcrop and (B) gypsum rock. The gypsum sample is roughly 10 cm across.

(A) I.J. Witkind, U. S. Geological Survey; (B) Doug Sherman/Geofile

form when a body of salt water is isolated and evaporates, leaving behind a deposit (**figure 2.17B**). Gypsum is used in the production of fertilizers, cement, wallboard, and plaster.

Some chemical sedimentary rocks have a large biological contribution and we refer to them as *biochemical*. Limestone is commonly biochemical in nature, as it is formed in shallow, warm, calm marine settings where creatures with hard parts are abundant. Many organisms living in water have shells or skeletons made of calcium carbonate or of silica (SiO_2, chemically equivalent to quartz). The materials of these shells or skeletons are drawn from the water in which the organisms grow. In environments where great concentrations of such creatures live and die, the hard parts—the shells or skeletons—may pile up on the bottom, eventually to be buried and lithified to form limestone (**figure 2.18**). *Chalk* is another biochemical sedimentary rock and is made almost entirely of microscopic marine creature skeletons. A sequence of sedimentary rocks may include layers of lithified **organic sediments,** carbon-rich remains of living

Figure 2.18

Fossiliferous limestone.

hsvrs/iStock/Getty Images

Figure 2.19

Horizontal bedding in sedimentary rocks along the Green River, Utah.

©*Gina S. Szablewski*

organisms; *coal* is an important example, derived from the remains of land plants that flourished and died in swamps. Coal beds throughout the world indicate that some locations and time periods were much warmer than they are today.

Gravity plays a role in the formation of all sedimentary rocks. Mechanically broken-up sediments accumulate when the wind or water is moving too weakly to overcome gravity; repeated cycles of transport and deposition can pile up, layer by layer, great thicknesses of sediment. Minerals crystallized from solution, or the shells of dead organisms, tend to settle out of the water under the force of gravity, and again, in time, layers of sediment build up. Layers, referred to as beds or *bedding,* are a common feature of sedimentary rocks and frequently a way in which we can identify their sedimentary origins (**figure 2.19**).

Sedimentary rocks preserve information about the settings in which the sediments were deposited. For example, a sandstone that has coarse-grained, well-rounded sand particles, and symmetrical ripple marks on the tops of the layers indicate the presence of an ancient beach. Sequences of sedimentary rocks indicate not only past individual environments, but also what environments were adjacent or came after, and how quickly or dramatically they changed from one to another. Distribution of glacially deposited sedimentary rocks and sediments contributes to our understanding not only of global climate change but also of plate tectonics.

Metamorphic Rocks

Metamorphic rocks are formed from other rocks that are subjected to high pressure and/or high temperature, but not hot enough to melt the rocks. Because the new conditions are different from which the rock originally formed, changes in texture and composition occur; these are the two categories we use to classify metamorphic rocks. Heat and pressure commonly cause the minerals in the rock to *recrystallize* meaning they change in size, shape, or in their arrangement. Also, some existing minerals break down completely while other new minerals form under the higher temperature and pressure conditions. The stresses that often accompany metamorphism can deform the rock through compression, stretching, folding, and compaction. Rocks remain solid through the process of metamorphism and deformation, even though individual minerals in the rock may start to melt, depending on the particular set of conditions.

There are many sources of the elevated temperatures and pressures needed for metamorphism. One source of heat is Earth itself. In general, crustal temperatures increase at the rate of about 30°C per kilometer of depth (close to 60°F per mile). Another source of heat is a cooling magma. When hot magma formed at depth rises to shallower levels in the crust, it heats and bakes the adjacent, cooler rocks, and they may metamorphose; this is **contact metamorphism.** An important source of pressure is simply burial under many kilometers of overlying rock. The weight of the overlying rock can put great pressure on the rocks below. Rocks buried deep in the crust are subjected to enough heat and pressure to show the deformation and recrystallization characteristic of metamorphism. Metamorphism results from the stresses and heating that accompany mountain-building processes and plate-tectonic movements. Such metamorphism on a large scale, not localized around a magma body, is **regional metamorphism.**

Any kind of preexisting rock can be metamorphosed. Some names of metamorphic rocks suggest what the earlier rock may have been. *Metaconglomerate* and *metavolcanic* describe, respectively, a metamorphosed conglomerate and a metamorphosed volcanic rock. *Quartzite* is a quartz-rich metamorphic rock, often formed from a very quartz-rich sandstone. The quartz crystals are more tightly interlocked in the quartzite, and the quartzite is a denser and stronger rock than the original sandstone. *Marble* is metamorphosed limestone in which the individual calcite grains have recrystallized and become tightly interlocking (**figure 2.20**). The remaining

A

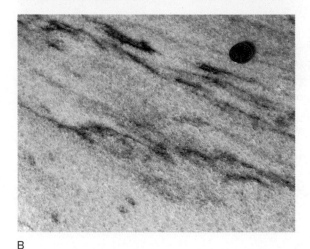

B

Figure 2.20

Metamorphic rocks. (A) Quartzite and (B) marble. The quartzite sample is roughly 50 cm across. Recognize the similar texture of these two rocks made of different types of interlocking crystals.

(A) Doug Sherman/Geofile; (B) ©Carla Montgomery

A

B

C

Figure 2.21

Metamorphic rocks. (A) Slate; (B) schist; (C) gneiss. The bands of light and dark color in the gneiss, and the irregular nature of the layers, indicate this rock was close to melting—we would refer to this as a high degree of metamorphism.

(A, B, C) ©Carla Montgomery

sedimentary layering that the limestone once showed may be folded and deformed in the process, if not completely obliterated by the recrystallization.

Some metamorphic-rock names indicate only the rock's current composition, with no particular implication of what it was before. A common example is *amphibolite,* which can be used for any metamorphic rock rich in the mineral amphibole. It might have been derived from a sedimentary, metamorphic, or igneous rock of appropriate chemical composition. The presence of abundant metamorphic amphibole indicates moderately intense metamorphism and the fact that the rock is rich in iron and magnesium, but not the previous rock type.

Other metamorphic rock names describe the characteristic texture of the rock, regardless of its composition. When the pressure of metamorphism is not uniform in all directions, rocks are subject to *differential stress* and are compressed or stretched in a particular direction. When you stamp on an aluminum can before tossing it in a recycling bin, you are technically subjecting it to differential stress—vertical compression—and it flattens in that direction in response. In a rock subjected to differential stress, minerals that form elongated or platy crystals may line up parallel to each other. The resultant texture is described as **foliation.** *Slate* is a

metamorphosed shale that has developed foliation under stress (**figure 2.21A**). The resulting rock tends to break along the foliation planes, parallel to the alignment of those minerals, and this characteristic makes it easy to break up into slabs for flagstones. The same characteristic is observed in *schist,* a coarsergrained, mica-rich metamorphic rock in which the mica flakes are similarly oriented, given this rock a layered and shiny appearance (**figure 2.21B**). The presence of foliation can cause planes of structural weakness in the rock, affecting how it weathers and whether it is prone to slope failure or landslides.

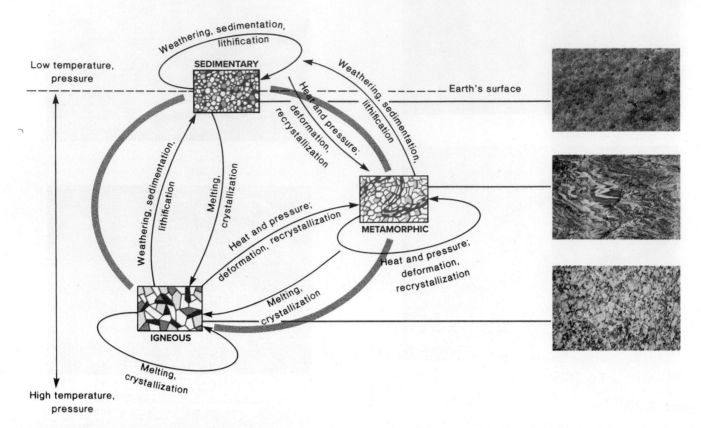

Low temperature, pressure

Earth's surface

High temperature, pressure

Figure 2.22

This schematic view of the rock cycle attempts to show that the geologic environment is not static. A variety of processes can transform any rock into a new igneous, metamorphic, or sedimentary rock.

©Carla Montgomery

In other metamorphic rocks, different minerals may be concentrated in irregular bands, often with darker, ferromagnesian-rich bands alternating with light bands rich in feldspar and quartz. Such a rock is called a *gneiss* (**figure 2.21C**). Because such terms as *schist* and *gneiss* are purely textural, the rock name can be modified by adding key features of the rock composition: biotite-garnet schist or granitic gneiss are two examples

The Rock Cycle

How do the rocks and rock-forming processes that we just reviewed fit together? They are all part of the **rock cycle** which summarizes the ways in which earth materials move and are recycled into different types of rocks throughout geologic time (**figure 2.22**). There is no one set path a rock can take through

the rock cycle, so it can be confusing to discuss this concept without having distinct starting and stopping places. Rather, we can provide possible routes that materials and rocks may take through the cycle. A sandstone formed in a desert may be weathered into pieces at Earth's surface; its fragments are then transported by a river and redeposited in a delta, and eventually lithified to form another sedimentary rock. That sedimentary rock might be deeply buried, heated, and compressed, transforming it into the metamorphic rock quartzite. Or it could be buried so deeply that it melts to form a magma that cools and crystallizes within the crust forming an intrusive igneous rock.

The processes of the rock cycle are more easily understood by recalling the vast amount of time both necessary and available for them to occur, but also in the context of plate tectonics, the subject of chapter 3.

Asbestos—A Tangled Topic

Billions of dollars are spent each year on asbestos abatement—removing and/or isolating asbestos in the materials of schools and other buildings—much of it mandated by law. The laws are intended to protect public health. But are they well-considered?

Asbestos is not a single mineral. To a geologist, the term describes any of a number of ferromagnesian chain silicates when they occur in needlelike or fibrous crystal forms. To the Occupational Safety and Health Administration (OSHA), asbestos means one of six very specific minerals—five of them amphiboles—and only when they occur in small, elongated crystals of specific dimensions. Above certain very low concentration limits in air, abatement is mandatory.

The relevant regulations, first formulated in 1972 and modified since, were enacted after it was realized that certain lung diseases occurred much more frequently in workers in the asbestos industry than in the population at large. Occupational exposure to asbestos had been associated with later increased incidence of mesothelioma (a type of lung cancer) and other respiratory diseases, including a scarring of the lungs called asbestosis. The risks sometimes extended to families of asbestos workers, exposed secondhand to fibers carried home on clothes, and to residents of communities where asbestos processing occurred.

However, the precise level of risk, and appropriate exposure limits, cannot be determined exactly. Most of the diseases caused by asbestos can be caused, or aggravated, by other agents and activities (notably smoking) too, and they generally develop long after exposure. Even for workers in the asbestos industry, the link between asbestos exposure and lung disease is far from direct. For setting exposure limits for the general public, it has been assumed that the risk from low, incidental exposure to asbestos can be extrapolated linearly from the identified serious effects of high occupational exposures, while there is actually no evidence that this yields an accurate estimate of the risks from low exposures. Some advocate what is called the "one-fiber theory" (really only a hypothesis!)—that is, if inhaling a lot of asbestos fibers is bad, inhaling even one does some harm. Yet we are all exposed to some airborne asbestos fibers from naturally occurring asbestos in the environment. In fact, measurements have indicated that an adult breathing typical outdoor air would inhale *nearly 4000 asbestos fibers a day*. The problem is very much like that of setting exposure limits for radiation, discussed in chapter 16, for we are all exposed daily to radiation from our environment, too.

Moreover, studies have clearly shown that different asbestos minerals present very different degrees of risk. In general, the risks from amphibole asbestos seem to be greatest. Mesothelioma is associated particularly with occupational exposure to the amphibole *crocidolite,* "blue asbestos," and asbestosis with another amphibole, *amosite,* "grey asbestos,"

mined almost exclusively in South Africa. On the other hand, the non-amphibole asbestos *chrysotile,* "white asbestos" (**figure CS 2.1.1**), which represents about 95% of the asbestos used in the United States, appears to be by far the least hazardous asbestos, and to pose *no* significant health threat from incidental exposure in the general public, even in buildings with damaged, exposed asbestos-containing materials. So, arguably, chrysotile could be exempted from the abatement regulations applied outside the asbestos industry proper, with a corresponding huge reduction in national abatement expenditures.

The case of Libby, Montana, is also interesting. Beginning in 1919, *vermiculite* (a ferromagnesian sheet silicate; **figure CS 2.1.2**), used in insulation and as a soil conditioner, was mined from a deposit near Libby. Unfortunately, it was found that the vermiculite in that deposit is associated with asbestos, Including *tremolite,* one of the OSHA-regulated amphibole-asbestos varieties. Decades ago, workers appealed to the state board of health for safer working conditions, but little was done for years. Meanwhile, Libby vermiculite was used in attic insulation in local homes, and waste rock from the processing was tilled into local soil, so exposures went beyond mining operations. Over time, at least 2400 people in Libby—workers and town residents—became ill, and an estimated 400 have died, from lung diseases attributed to asbestos. The mine closed in 1990, EPA started investigations in 1999, mitigation followed, and the cleanup was completed in 2019. Geologic studies have revealed other pertinent facts: (1) Only about 6% of the asbestos in the Libby vermiculite deposit is tremolite. Most consists of chemically similar varieties of amphibole, which might

Figure CS 2.1.1

Chrysotile asbestos, or "white asbestos," accounts for about 95% of asbestos mined and used in the United States.

Bob Coyle/McGraw-Hill Education

(Continued)

Figure CS 2.1.2

Vermiculite. This sample has been processed for commercial use; it has been heated to expand the grains, "popping" them apart between silicate sheets.

©Carla Montgomery

also cause lung disease but which are *not* OSHA-regulated. (2) The soils of the town contain other asbestos that is chemically and mineralogically distinct from the asbestos of the deposit. (3) The deposit weathers easily, and lies in hills upstream from Libby. So, not surprisingly, there is deposit-related asbestos in town soils that clearly predate any mining activity, meaning that residents could easily have been exposed to deposit-related asbestos that was unrelated to mining activity.

The evidence indicates that significant risk is associated with Libby amphibole species that are not OSHA-regulated. Perhaps they should be—and given the health effects observed in Libby, cleanup certainly seems desirable. On the other hand, given that little of the asbestos at Libby is actually one of the OSHA six, and considering the other facts noted, how much liability should the mining company bear. The project lasted 20 years; cleanup involved inspection of 8100 properties, the replacement of contaminated soil with clean soil, and the removal of asbestos-bearing building materials. Of the estimated total costs of $600 million, W.R. Grace was ordered to pay over $54 million. On the other hand, in 2009, the company was acquitted of criminal wrongdoing. And in early 2022, just one of the hundreds of cases awaiting trial against the company's insurer awarded $36.5 million to a single laborer for failing to warn workers of the health dangers. The cleanup may be complete, but the repercussions are not.

With asbestos abatement generally, should the regulations be modified to take better account of what we now know about the relative risks of different asbestos varieties? That might well be appropriate. Should geologists or lawmakers define what asbestos is? If left to non-scientists, asbestos might simply be "elongated mineral particles" which could be used to define common minerals that pose no risk to the general population and would divert time and funds from the minerals posing real dangers. This question of weighing the benefits of regulations against the costs will arise again in chapter 19.

Summary

The smallest possible unit of a chemical element is an atom. Isotopes are atoms of the same element that differ in atomic mass number; chemically, they are indistinguishable. Atoms may become electrically charged ions through the gain or loss of electrons. When two or more elements combine chemically in fixed proportions, they form a compound.

Minerals are naturally occurring inorganic solids, each of which is characterized by a particular composition and internal crystalline structure. They may be compounds or single elements. By far the most abundant minerals in Earth's crust and mantle are the silicates. These can be subdivided into groups on the basis of their crystal structures, by the ways in which the silicon and oxygen atoms are arranged. The nonsilicate minerals are generally grouped on the basis of common chemical characteristics.

Rocks are cohesive solids formed from rock or mineral grains or glass. The way in which rocks form determines how they are classified into one of three major groups: igneous rocks, formed from magma; sedimentary rocks, formed from low-temperature accumulations of particles or by precipitation from solution; and metamorphic rocks, formed from preexisting rocks through the application of heat and pressure. Through time, geologic processes acting on older rocks change them into new and different ones so that, in a sense, all kinds of rocks are interrelated. This concept is the essence of the rock cycle.

Key Terms and Concepts

anion	cation	contact metamorphism	foliation
atom	chemical sedimentary rock	covalent bonding	glass
atomic mass number	clastic sedimentary rock	crystalline	igneous
atomic number	cleavage	electron	ion
carbonate	compound	ferromagnesian	ionic bonding

isotope native element plutonic sedimentary
lava neutron proton silicate
lithification nucleus regional metamorphism sulfate
magma organic sediments rock sulfide
metamorphic oxide rock cycle volcanic
mineral periodic table sediment

Test Your Learning

1. Briefly define the following terms: *ion, isotope, compound, mineral,* and *rock.*

2. Recall the two properties that uniquely define a specific mineral.

3. Name the distinctive chemical characteristics of each of the following mineral groups: silicates, carbonates, sulfides, oxides, and native elements.

4. Explain how and why volcanic and plutonic rocks differ in texture.

5. Compare the two principal classes of sedimentary rocks.

6. Name several possible sources of the heat or pressure that can cause metamorphism. Describe the kinds of physical changes that may occur in a rock as a result.

7. Explain how a granite might be transformed into a sedimentary rock.

8. Briefly outline the rock cycle, indicating the kinds of processes that can lead to transformation of one rock into another.

Exploring Further

1. The variety of silicate formulas arises partly from the different ways the silica tetrahedra are linked. Consider a single tetrahedron: If the silicon cation has a +4 charge, and the oxygen anion −2, there must be a net negative charge of −4 on the tetrahedron as a unit. Yet minerals must be electrically neutral. In quartz, this is accomplished by the sharing of all four oxygen atoms between tetrahedra, so only half the negative charge of each must be balanced by Si^{+4}, and there is a net of two oxygen atoms per Si (SiO_2). But in olivine, the tetrahedra share no oxygen atoms. If the magnesium (Mg) cation has a +2 charge, explain why the formula for a Mg-rich olivine is Mg_2SiO_4. Now consider a pyroxene, in which tetrahedra are linked in one dimension into chains, by the sharing of two of the four oxygen atoms. Again using Mg^{+2} for charge-balancing, show that the formula for a Mg-rich pyroxene would be $MgSiO_3$. Finally, consider the

feldspars. They are framework silicates like quartz, but Al^{+3} substitutes for Si^{+4} in some of their tetrahedra. Explain how this allows feldspars to contain some sodium, potassium, or calcium cations and still remain electrically neutral.

2. What kinds of rocks underlie your region of the country, and what do they indicate about the geologic history of the area? (Your local geological survey could assist in providing the information.) Many states have a state mineral and/or rock. Does yours? If so, what is it, and why was it chosen?

3. California was the first state to name a state rock, in 1965; that rock is serpentine. In 2010, some California legislators moved to remove serpentine as the state rock. Investigate what prompted this, the public reaction, and the outcome.

Plate Tectonics

Over 50 years ago, multiple hypotheses explaining Earth's features and processes ultimately came together in the theory of plate tectonics, which was hailed as a revolution in the earth sciences. The concept of drifting continents that had once fit together had been published much earlier but had been generally rejected by the scientific community. Not until we began to explore the sea floor with new technologies did we acquire the needed missing evidence for this widely encompassing explanation of how Earth works. **Tectonics** is the study of large-scale movement and deformation of Earth's outer layers. **Plate tectonics** relates such deformation to the existence of rigid plates of rock that move over a weaker, more plastic layer in Earth's upper mantle. It has given geoscientists a powerful tool for understanding such phenomena as mountain-building and

Chapter Outline

The Himalayas owe their impressive elevation to the collision between the Indian and Eurasian continents in a process that continues today.

Image Source

the occurrence and distribution of earthquakes and volcanic activity.

In this chapter, we will summarize the development of plate tectonic theory starting with continental drift, then layer on the supporting data gathered in the twentieth century.

Before explaining just what the theory entails, we will briefly review how rocks respond to the huge stresses associated with moving tectonic plates. The chapter ends with a description of how we measure plate movements, and then how the theory enhances our understanding of the rock cycle.

3.1 Continental Drift

Learning Objectives

- Summarize the continental drift hypothesis
- List early evidence that suggested the continents were once joined
- Recall reasons why continental drift was not accepted by the scientific community

From the time the first reasonably accurate global maps were produced, observers noticed the similarity in outline of the continents on either side of the Atlantic Ocean (**figure 3.1**). In 1855, Antonio Snider-Pellegrini published a sketch showing how the Americas could fit together with Europe and Africa, basing his idea on plant fossils. He was not the first to suggest that the separate continents had once been part of a single landmass that had later broken up, and he was not the last. In the early 1900s, the most well-known person associated with drifting continents took up this idea.

Alfred Wegener was an accomplished and multi-subject scientist, researcher, and polar explorer. After receiving the gift of an atlas, he was struck by the matching coastlines and began to assemble other data to support the idea that continents were

not stationary. Wegener studied multiple lines of evidence that included fossil, glacial, and geologic data collected by other scientists primarily in the Southern Hemisphere. Wegener first published his ideas in 1912 in "The Origin of Continents and Oceans" and continued to refine them for nearly two decades until his death in 1930. In this seminal work, he proposed that all the continental landmasses had once formed a single super-continent, *Pangaea,* which had then split apart via a process called **continental drift.** Let's look at some of the evidence Wegener used to support his hypothesis.

Evidence of Drift

Fossils are remains of ancient life primarily preserved in sedimentary rocks that provide us with information not only about the size and structure of the creatures, but also about the environment and climate they lived in. Just as now, some long extinct plants and animals lived in very restricted environments. We have found fossils of the plant *Glossopteris* in limited areas of five widely separated continents that stretch from Australia to South America (**figure 3.2**). Could this distinctive land plant have developed simultaneously in five areas thousands of kilometers apart, or somehow migrated over vast expanses of water? Continental drift provides a more plausible explanation that the organism lived in a single, geographically restricted area and the rocks in which its remains are now found subsequently were separated and moved in several directions to their current location. This explanation gains strength when it can be,

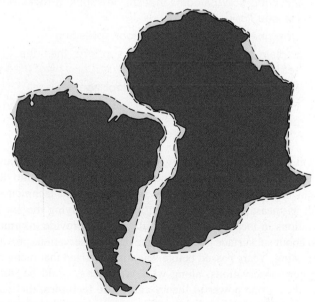

Figure 3.1

The jigsaw-puzzle fit of South America and Africa suggests that they were once joined together and subsequently drifted apart. (Blue shaded area is the continental shelf, the shallowly submerged border of a continent.)

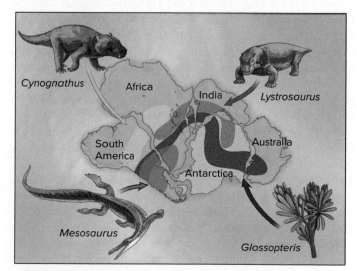

Figure 3.2

Four types of fossils that we commonly use as evidence for moving continents. Wegener cited both *Glossopteris* and *Mesosaurus* along with other creatures not shown here.

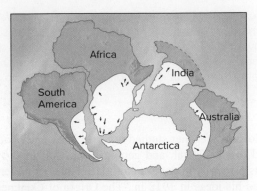

Figure 3.3
Ancient glacial deposits across the southern continents suggest they were once joined and covered by extensive ice. Arrows indicate directions of ice flow.

and has been, applied to other organisms such as the freshwater reptile *Mesosaurus,* also shown in **figure 3.2**.

Even without fossils, various types of sedimentary rocks may preserve evidence of the ancient climate that existed when the sediments were deposited. Let us recall that climate is to a great extent a function of latitude which strongly influences surface temperatures. Wegener had studied meteorology and was a polar researcher, so it is understandable that he took particular interest in the deposits of ancient glaciers. He found matching tillites of the same age in places now located in the tropics, including parts of Australia, southern Africa, and South America (**figure 3.3**). The rocks also contained scratch marks (striations) that preserve the direction of ice movement. Was Earth largely covered in ice at this time period to explain both of these phenomenon? A more comprehensive explanation is that the southern continents were once together in a high latitude location and were extensively glaciated; this eliminates an attempt to explain the impossible movement of glacial ice from the ocean toward the continent in multiple locations at the same time. Wegener also used sedimentary rocks that indicate warm, wet, plant-rich land environments (coal in Antarctica) and those that indicate deserts sands in places that are now moist and temperate to show distinct changes in latitude for each continent with time. These observations cannot be explained as the result of global climatic changes, for the continents don't all show the same warming or cooling trends at the same time.

Wegener utilized distinct sequences of rocks and rock units, as defined by their age and composition, to show that the coastlines of South America and Africa were once together. Even data from the Northern Hemisphere supported the idea of Pangaea; the Appalachian Mountains of North America continue into the counterpart the Caledonides of Greenland and the British Isles. The continents were reassembled like the pieces of a puzzle: not only did the pieces fit together (the outlines) but also the picture (the fossils and rocks) matched up. If two now-separate continents were once part of the same landmass, then the geologic features presently found at the margin of one should have counterparts at the corresponding edge of the other.

Several other prominent scientists found the idea of continental drift intriguing, and the paleobotanists were

particularly pleased. Even though he had studied geophysics, Wegener did not provide a satisfactory mechanism of continental movement. There were obvious physical objections to solid continents plowing through solid ocean basins, and there was no evidence of the expected resultant damage in crushed and shattered rock on continents or sea floor. To a great many reputable scientists, these were insurmountable obstacles to accepting the idea of continental drift. Add to that the facts that Wegener was known as a meteorologist rather than a geologist, and most of his data came from the unfamiliar Southern Hemisphere. The continental drift hypothesis was temporarily shelved.

You have already been introduced to this recurring theme in science; despite strong and varied evidence for his idea, Wegener was lacking the additional relevant evidence needed to move his hypothesis forward because it was simply undiscovered or unrecognized at the time. In the meantime, scientists began to accumulate scientific data of many different kinds that indicated that the continents have indeed moved and continue to do so. The drifting of continents turned out to be just one consequence of processes encompassed by the broader theory of plate tectonics.

3.2 More Evidence for Moving Continents

Learning Objectives

- List and locate seafloor features used to support the movement of continents
- Describe how paleomagnetism develops in rocks on the sea floor
- Describe the process of seafloor spreading
- Explain how paleomagnetism supports the idea of seafloor spreading
- Recognize the patterns of seafloor age
- Explain polar wander curves and how they support plate tectonic theory

Through the twentieth century, geologists continued to expand their knowledge of Earth, extending their observations into the ocean basins and applying new instruments and techniques, such as measuring magnetism in rocks or studying the small variations in local gravitational pull that can provide information about subsurface geology. Some of the observations proved surprising. Years passed before scientists realized that many of the new observations, along with Wegener's, could be integrated into one powerful theory call **plate tectonics,** the idea that Earth's crust is broken into pieces, or plates, that move in relationship to each other. This theory is not attributed to a single scientist but rather to a sequence of researchers whose work came together to form the underlying concept of geology. In the following, we will explore the evidence and groundbreaking (literally) work behind the theory's development.

Figure 3.4

Global relief map of the world. Seafloor topography is indicated by shade of blue—lighter means a higher elevation/shallower water. Look for examples of ocean ridges and trenches.

NOAA National Geophysical Data Center (NGDC)

The Topography of the Sea Floor

The natural philosophers and scientists who first speculated about possible drifting continents could not examine the sea floor for any relevant evidence. Many assumed that the sea floor was simply a vast, monotonous plain on which sediments derived from the continents accumulated. Once topographic maps of the sea floor became available, however, a number of striking features were revealed (**figure 3.4**). There were long ridges, some thousands of kilometers long and rising as much as 2 kilometers (1.3 miles) above the surrounding plains, thus rivaling continental mountain ranges in scale. A particularly obvious example occurs in the Atlantic Ocean, running north–south about halfway between North and South Americas to the west, and Europe and Africa to the east. Some ocean basins are dotted with hills (most of them volcanoes), some wholly submerged, others poking above the sea surface as islands, for example, the Hawaiian Islands. In other places, along ocean basin margins, there are trenches several kilometers deep. Examples are located along the western edge of South America, just south of the Aleutian Islands, and east of Japan. Studies of the ages and magnetic properties of seafloor rocks, as described below, provided the keys to the significance of the ocean ridges and trenches.

Magnetism in Rocks

The basaltic rocks of the ocean floor, and many rocks on the continents, are rich in ferromagnesian minerals. Most iron-bearing minerals are at least weakly magnetic at surface temperatures. The iron oxide mineral magnetite is strongly magnetic, and small amounts of it are commonly present in iron-rich igneous rocks. Each magnetic mineral has a **Curie temperature,** the temperature below which it remains magnetic, but above which it loses its magnetic properties. The Curie temperature varies from mineral to mineral but it is always below the mineral's melting temperature; for magnetite, the Curie temperature is 585°C (1085°F) whereas its melting temperature is 1538°C. A hot magma is therefore not magnetic, but as it cools and solidifies and iron-bearing magnetic minerals such as magnetite crystallize from it, those magnetic crystals tend to line up in the same direction. Like tiny compass needles, they align themselves parallel to the lines of force of Earth's magnetic field, and they point to the magnetic north pole (**figure 3.5**). They retain their internal magnetic orientation unless they are heated again. This is the basis for the study of **paleomagnetism.**

Magnetic north has not always coincided with its present position. In the early 1900s, scientists investigating the direction of magnetization of a sequence of volcanic rocks in France discovered some flows that appeared to be magnetized in the opposite direction from the rest: their magnetic minerals pointed south instead of north. Confirmation of this discovery in many places around the world led to the suggestion in the late 1920s that Earth's magnetic field had flipped or reversed polarity; that is, that the north and south poles had switched places. During the time those igneous rocks had crystallized, a compass needle would have pointed to the magnetic south pole.

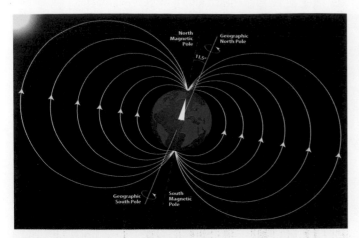

Figure 3.5
We can envision the lines of force from Earth's current magnetic field by imagining a bar magnet inside of Earth tilted at 11.5° away from the axis of rotation.

Milagli/Shutterstock

Today, the phenomenon of magnetic reversals is well documented. Rocks crystallizing at times when Earth's field was in the same orientation as it is at present are said to be *normally* magnetized; rocks crystallizing when the field was oriented the opposite way are described as *reversely* magnetized. Over the history of Earth, the magnetic field has reversed hundreds of times at variable intervals. Sometimes, the polarity remained constant for millions of years before reversing, while other reversals are separated by only a few tens of thousands of years. Through the combined use of magnetic measurements and age determinations on the magnetized rocks, geologists have been able to reconstruct the reversal history of Earth's magnetic field in some detail.

The explanation for magnetic reversals is related to the origin of the magnetic field. The outer core is a metallic fluid, consisting mainly of iron. Motions in an electrically conducting fluid can generate a magnetic field, and scientists believe this to be the origin of Earth's field. The simple presence of iron in the core is not enough to account for the magnetic field, as core temperatures—over 3500°C—are far above the Curie temperature of iron, 770°C. Perturbations or changes in the fluid motions could account for reversals of the field. What we do know is that Earth's magnetic field is not stagnant; it weakens and strengthens, the location of the poles drifts, and the poles switch places. We are still working to understand why these changes occur and how these changes affect Earth's other systems, particularly the biosphere.

Paleomagnetism and Seafloor Spreading

The ocean floor is made up largely of basalt, a volcanic rock rich in ferromagnesian minerals. During the 1950s, the first large-scale surveys of the magnetic properties of the sea floor produced an entirely unexpected result. As ships equipped with magnetometers tracked across a ridge, they recorded what seemed to be regions of alternately stronger and weaker magnetism below. At first, this seemed so incredible that it was assumed that the instruments or measurements were faulty. However, other studies consistently obtained the same results. For several years, geoscientists strove to find a convincing explanation for these startling observations.

Then, in the early 1960s, researchers realized that the magnetic "stripes" could be explained as a result of alternating bands of normally and reversely magnetized rocks on the sea floor. Adding the magnetism of normally magnetized rocks to Earth's current field produced a little stronger magnetization; reversely magnetized rocks would counter to Earth's (much stronger) field somewhat, resulting in apparently lower net magnetic strength.

These bands of normally and reversely magnetized rocks were also parallel to and symmetrically arranged on either side of the seafloor ridges. These data were combined in the concept of **seafloor spreading,** the parting of seafloor at the ocean ridges. As the oceanic crust splits and moves apart, a rift begins to open, and mantle material from below can flow upward. The resulting lowering of pressure allows extensive melting to occur in this material (as will be explored in chapter 5), and magma rises up into the rift to form new sea floor. As the magma cools and solidifies to form new basaltic rock, that rock becomes magnetized in the prevailing direction of Earth's magnetic field. If the crusts on either side of the ridge continue to move apart, the new rock will also split and part, making way for more magma to form still younger rock, and so on.

When the polarity of Earth's magnetic field reverses during the course of seafloor spreading, the rocks formed after the reversal are polarized oppositely from those formed before it. The ocean floor is a continuous sequence of basalts formed over tens or hundreds of millions of years, during which time there have been numerous polarity reversals. The basalts of the sea floor have acted as a sort of magnetic tape recorder throughout that time, preserving a record of polarity reversals in the alternating bands of normally and reversely magnetized rocks. The process is illustrated schematically in **figure 3.6**.

Normal magnetic polarity

Reversed magnetic polarity

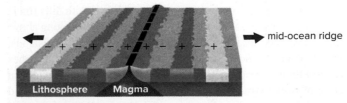

Figure 3.6
Seafloor spreading leaves a record of normally and reversely magnetized rocks on the sea floor, commonly referred to as stripes. Different colors of normally polarized rocks are used here to emphasize their different crystallization ages.

After USGS publication, This Dynamic Earth by W. J. Kious and R. I. Tilling.

Age of the Ocean Floor

The ages of seafloor basalts themselves lend further support to this model of seafloor spreading. Specially designed research ships can sample sediment from the deep-sea floor and drill into the basalt beneath.

We can determine the time at which an igneous rock crystallized from its magma by methods described in appendix A. When we do this for many samples of seafloor basalt, a pattern emerges. The rocks of the sea floor are youngest close to the ocean ridges and become progressively older the farther away they are from the ridges on either side (see **figure 3.7**). Like the magnetic stripes, the age pattern is symmetric across each ridge. As seafloor spreading progresses, previously formed rocks are continually spread apart and moved farther from the ridge, while fresh magma rises to form new sea floor at the ridge. The oldest rocks recovered from the sea floor, well away from active ridges, are rarely more than 200 million years old.

Ages of sediments from the ocean basins reinforce this age pattern. Logically, sea floor must form before sediment can accumulate on it, and unless the sediment is disturbed, younger sediments accumulate on older ones. Collected cores of oceanic sediment show us not only that sediment thickness steadily increases with distance from the ridge, but also that the deepest sediments are older at greater distances from the seafloor ridges

and that only quite young sediments have been deposited close to the ridges.

Polar-Wander Curves

Evidence for plate movements does not come only from the sea floor. Much older rocks are preserved on the continents than in the ocean so longer periods of Earth history can be investigated through continental rocks. Studies of paleomagnetic orientations of continental rocks can span many hundreds of millions of years and yield quite complex data.

The lines of force of Earth's magnetic field not only run north–south, they vary in dip angle with latitude: vertical at the magnetic poles, horizontal at the equator, at varying intermediate dips in between, as is shown in **figure 3.8**. When the orientation of magnetic minerals in a rock is determined in three dimensions, one can determine not only the direction of magnetic north, but (magnetic) latitude as well, or how far removed that region was from magnetic north at the time the rock's magnetism assumed its orientation.

Magnetized rocks of different ages on a single continent may point to very different apparent magnetic pole positions. The magnetic north and south poles may not simply be reversed, but may be rotated or tilted from the present magnetic north and south. When the directions of magnetization and latitudes of many rocks of various ages from one

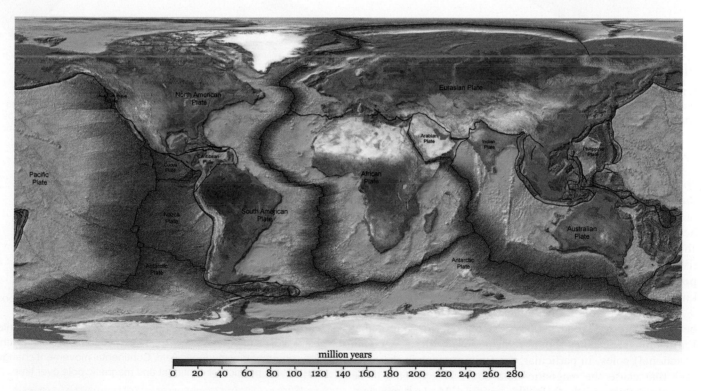

million years

0 20 40 60 80 100 120 140 160 180 200 220 240 260 280

Figure 3.7

Age distribution of the sea floor superimposed on a shaded relief map, with some of the plate boundaries that we now recognize. Wider color bands in the Pacific indicate faster spreading rates at the East Pacific Rise compared to the rates at the Mid-Atlantic Ridge.

Source: Mr. Elliot Lim, CIRES and NOAA/National Geophysical Data Center.

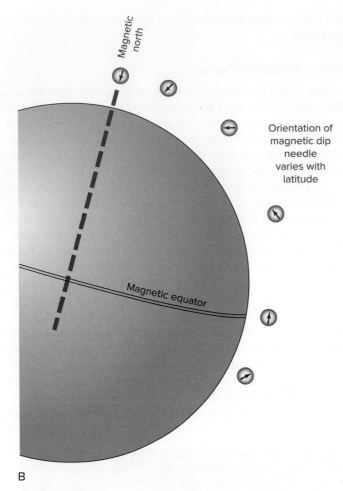

Figure 3.8

A magnetic mineral in an igneous rock will dip, or be deflected from the horizontal, more steeply near the poles, just as the lines of forces of Earth's field do. (See **figure 3.5**.)

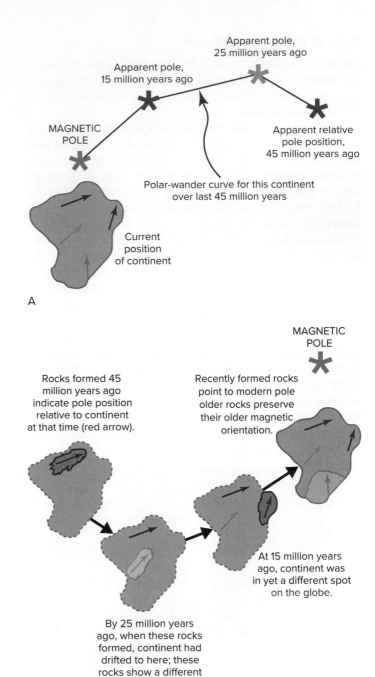

Figure 3.9

(A) Assuming a stationary continent, the shifting relative pole positions would suggest polar wandering. Actually, polar wander curves reflect wandering continents attached to moving tectonic plates. (B) As rocks crystallize, their magnetic minerals align with the contemporary magnetic field. Continental movement changes the relative position of continent and magnetic pole over time.

continent are determined and plotted on a map, it appears that the magnetic poles have meandered far over the surface—*if* the position of the continent is assumed to have been fixed on Earth throughout time. The resulting curve, showing the apparent movement of the magnetic pole relative to the continent as a function of time, is called the **polar-wander curve** for that continent (**figure 3.9**). We know now that it isn't the poles that have wandered so much.

Scientists were a bit perplexed by polar-wander curves because there are good geophysical reasons to believe that Earth's magnetic poles should remain close to the geographic (rotational) poles. In particular, the fluid motions in the outer core that cause the magnetic field could be expected to be strongly influenced by to Earth's rotation. Modern measurements do indicate some shifting in magnetic north relative to the north pole (geographic north), but not nearly as much as indicated by some polar-wander curves.

Additionally, the polar-wander curves for different continents do not match. Rocks of exactly the same age from two

different continents may seem to point to two magnetic poles in two different locations (**figure 3.10**). We can eliminate this confusion by assuming that the magnetic poles have always remained close to the geographic poles and the continents have moved and rotated. The polar-wander curves then provide a way to map the

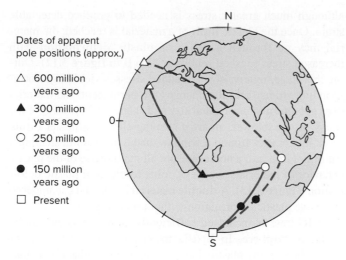

Figure 3.10

Examples of polar-wander curves. The apparent position of the magnetic pole relative to the continent is plotted for rocks of different ages, and these data points are connected to form the curve for that landmass. The present positions of the continents are shown for reference. Polar-wander curves for Africa (solid line) and Arabia (dashed line) suggest that these landmasses moved quite independently up until about 250 million years ago, when the polar-wander curves converge.

After M. W. McElhinny, Paleomagnetism and Plate Tectonics. Copyright @ 1973 by Cambridge University Press, New York, NY. Reprinted by permission.

directions in which the continents have moved through time, relative to the stationary poles and relative to each other.

Efforts to reconstruct the ancient locations and arrangements of the continents have been relatively successful, at least for the not-too-distant geologic past. For example, the data indicate that a little more than 200 million years ago, there was indeed a single, great supercontinent, the one that Wegener envisioned and named Pangaea. The present seafloor spreading ridges are the lithospheric scars of the subsequent breakup of Pangaea. **Figure 3.11** shows one of the early reconstructions of continental movements through time. There exist different and more current interpretations of these movements including projections into a future that might once again have a supercontinent.

3.3 Plate Tectonics—Underlying Concepts

Learning Objectives

- Compare different types of stress
- Distinguish elastic deformation from plastic deformation
- Contrast brittle and ductile behavior in rocks
- Contrast lithosphere and asthenosphere
- Recognize tectonic plates and their boundaries on a map

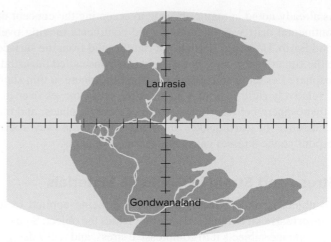

A 200 million years ago

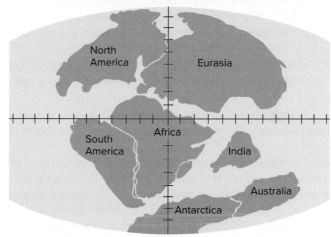

B 100 million years ago

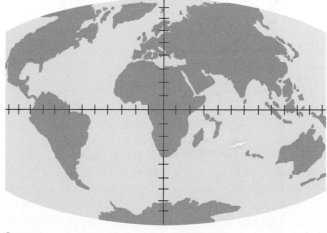

C Today

Figure 3.11

Reconstructed plate movements during the last 200 million years. *Laurasia* and *Gondwanaland* are names given to the northern and southern portions of Pangaea, respectively.

After R. S. Dietz and J. C. Holden, "Reconstruction of Pangaea," Journal of Geophysical Research, 75:4939–4956, 1970, copyright by the American Geophysical Union.

As already noted, a major obstacle to accepting the concept of continental drift was imagining solid continents moving over solid Earth. However, Earth is not rigidly solid from the surface to the center of the core. In fact, a plastic zone called the asthenosphere lies close to the surface. The lithosphere, a thin shell of relatively rigid rock that includes the crust, moves over the asthenosphere below. The existence of tectonic plates, and the occurrence of earthquakes in them, reflect the way rocks respond to the stresses caused by these movements.

Stress and Strain in Geologic Materials

An object is under **stress** when force is being applied to it. **Compressive stress** squeezes or compresses the object. Squeezing a sponge subjects it to compressive stress, and rocks deep in the crust are under compressive stress from the weight of rocks above them. **Tensile stress** stretches an object or pulls it apart. A stretched guitar or violin string being tuned is under tensile stress, as are rocks at a seafloor spreading ridge as plates move apart. A **shearing stress** is one that tends to cause different parts of the object to move in different directions across a plane or to slide past one another, as when a deck of cards is spread out on a tabletop by a sideways sweep of the hand.

Strain is deformation resulting from stress. It may be either temporary or permanent, depending on the amount and type of stress and on the physical properties of the material. If **elastic deformation** occurs, the amount of deformation is proportional to the stress applied (straight-line segments in **figure 3.12**), and the material returns to its original size and shape when the stress is removed. A gently stretched rubber band or a tennis ball that is squeezed or hit by a racket shows elastic behavior: the object returns to its original dimensions when the stress is released. Rocks, too, may behave elastically,

although much greater stress is needed to produce detectable strain. Once the **elastic limit** of a material is reached, the material may go through a phase of **plastic deformation** with increased stress (dashed section of line B in **figure 3.12**). During this stage, relatively small added stresses yield large corresponding strains, and the changes are permanent: the material does not return to its original size and shape after removal of the stress. A glassblower, an artist shaping clay, a carpenter fitting caulk into cracks around a window, and a blacksmith shaping a bar of hot iron into a horseshoe are all making use of the plastic behavior of materials. In rocks, folds result from plastic deformation (**figure 3.13**). A **ductile** material is one that can undergo extensive plastic deformation without breaking.

If stress is increased sufficiently, most solids eventually break, or **rupture.** In **brittle** materials, rupture may occur before there is any plastic deformation. Brittle behavior is characteristic of most rocks at the low temperatures and pressures near Earth's surface. It leads to faults and fractures (**figure 3.14**), which will be explored further in chapter 4. Graphically, brittle behavior is illustrated by line A in **figure 3.12**.

Different types of rocks may tend to be more or less brittle or ductile, but other factors influence their behavior as well. One is temperature. All else being equal, rocks tend to behave more plastically at higher temperatures than at lower ones. We can observe the effect of temperature in the behavior of cold and hot glass. A rod of cold glass is brittle and snaps under stress, while a warmed glass rod may be bent and twisted without breaking. Pressure is another factor influencing rock behavior. *Confining pressure* is that uniform pressure which surrounds a rock at depth. Higher confining pressure also tends to promote more-plastic, less-brittle behavior in rocks. Confining pressure increases with depth, as the weight of overlying rock increases; temperature likewise generally increases with depth. Thus, rocks such as the gneiss of **figure 2.21C**, metamorphosed deep in the crust, commonly show the folds of plastic deformation.

Rocks also respond differently to different types of stress; most are far stronger under compression than under tension. Rocks may be weakened by fracturing or weathering. The term *strength* has no single simple meaning when applied to rocks. The physical properties of rocks and how they will respond to additional stresses we may put on them fall under the consideration of the geological engineer in projects that range from tunnel building to slope stabilization to bridge construction.

Time is a further, very important factor in the physical behavior of a rock. Materials may respond differently to given stresses, depending on the rate of stress, the period of time over which the stress is applied. A ball of putty will bounce elastically when dropped on a hard surface but deform plastically if pulled or squeezed more slowly. Rocks can likewise respond differently depending on how stress is applied, showing elastic behavior if stressed suddenly, as by passing seismic waves from an earthquake, but ductile behavior in response to prolonged stress, as from the weight of overlying rocks or the slow movements of plates over the longer span of geologic time.

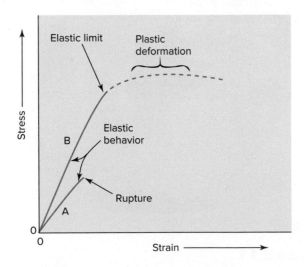

Figure 3.12

A stress-strain diagram. Elastic materials deform in proportion to the stress applied (straight-line segments). A very brittle material (curve A) may rupture before the elastic limit is reached. Other materials subjected to stresses above their elastic limits deform plastically until rupture (curve B).

A

A

B

Figure 3.13

(A) Folding such as this occurs while rocks are deeper in the crust, at elevated temperatures and pressures. (B) The tightly folded marbles and slates of this roadcut were originally flat-lying layers of limestone and shale, later metamorphosed and deformed. The fracture across the bottom of this picture could have been created during blasting of the roadcut, when the rocks were at the surface, colder and more brittle.

(A) M.R. Mudge/USGS; (B) ©Carla Montgomery

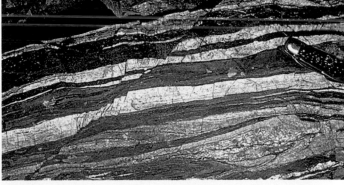

B

Figure 3.14

(A) This large fault (left) in Cook Inlet, Alaska, is easy to see due to the offset layers; it formed during a collision of continental landmasses. (B) This rock suffered brittle failure under stress, which produced a much smaller fault.

(A) Photograph by N.J. Silberling, USGS Photo Library, Denver, CO;
(B) Photograph by W.B. Hamilton, USGS Photo Library, Denver, CO

Lithosphere and Asthenosphere

The crust and uppermost mantle are somewhat brittle and elastic. Together they make up Earth's outer solid layer called the **lithosphere.** The lithosphere varies in thickness from place to place It is thinnest underneath the oceans, where it extends to a depth of about 50 kilometers (about 30 miles). The lithosphere under the continents is both thicker on average than is oceanic lithosphere, and more variable in thickness, extending in places to about 250 kilometers (over 150 miles).

The layer below the lithosphere is the **asthenosphere.** The asthenosphere extends to an average depth of about 300 kilometers (close to 200 miles) in the mantle. Its lack of strength or rigidity results from a combination of high temperatures and moderate confining pressures that allows the rock to flow plastically under stress. Below the asthenosphere, as pressures increase faster than temperatures with depth, the mantle again becomes more rigid and elastic.

The asthenosphere was discovered by studying the behavior of seismic waves from earthquakes. Its presence makes the concept of continental drift more plausible. The continents need not scrape across or plow through solid rock; instead, we can picture them as sliding over a softened, deformable layer underneath the lithospheric plates. The relationships among crust, mantle, lithosphere, and asthenosphere are illustrated in **figure 3.15**.

Recognition of the existence of the plastic asthenosphere provided a way to help move tectonic plates, but it was not proof. For most scientists, acceptance of plate tectonics required the gathering of much additional information, such as that described later in this chapter.

Locating Plate Boundaries

The location of earthquake and volcanic activity as plotted on Earth's surface indicates that these phenomena are not randomly distributed, as we will explore in chapters 4 and 5. These phenomena are concentrated in belts or linear chains. This is consistent with and strong evidence for the idea that the rigid shell of lithosphere is broken up into pieces or plates; the volcanoes and earthquakes are concentrated at the boundaries of these plates, where they jostle or scrape against each other.

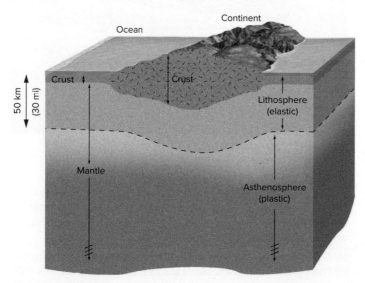

Figure 3.15

The outer zones of Earth (not to scale). The terms *crust* and *mantle* have compositional implications (see chapter 1); *lithosphere* and *asthenosphere* describe physical properties. The lithosphere includes the crust and uppermost mantle. The asthenosphere lies entirely within the upper mantle. Below it, the rest of the mantle is more rigidly solid.

Geologists have identified fewer than a dozen very large lithospheric plates (**figure 3.16**). As research continues, many smaller ones have been recognized.

3.4 Types of Plate Boundaries

Learning Objectives

- Recall geologic processes and features that mark divergent, convergent, and transform boundaries
- Describe the process of continental rifting
- Describe subduction and subduction zones
- Exemplify divergent, convergent, and transform boundaries

Different processes occur at the boundaries between two lithospheric plates, depending on their relative motions and whether continental or oceanic lithosphere is involved. Here we will summarize the features and processes that mark the different types of plate boundaries, and provide examples of each.

Divergent Plate Boundaries

At a **divergent plate boundary,** lithospheric plates are subject to tensional stress and move apart. As the asthenosphere moves upward into the newly created gap, decompression causes some of it to melt and forms basaltic (mafic) magma, which wells up to form new crust. These rift zones, whether on the ocean floor or within a continent, are marked by both volcanic activity (not necessary volcanoes themselves) and earthquakes.

Seafloor spreading ridges are the most common type of divergent boundary worldwide, and we have already noted the formation of new oceanic lithosphere at these ridges. We refer to them as *constructive* plate boundaries because new lithosphere is created here. As seawater circulates through this fresh, hot lithosphere, it is heated and reacts with the rock, becoming metal-rich. As it gushes back out of the sea floor, cooling and reacting with cold seawater, it may precipitate potentially valuable mineral deposits. People are very interested in the vast amounts of resources created in this way on the sea floor even though they are economically unprofitable to mine now.

Continents also can be rifted apart and most ocean basins likely originated through continental rifting, as shown in **figure 3.17**. The process may be initiated either by tensional forces pulling the plates apart, or by rising hot asthenosphere along the rift zone. In the early stages of continental rifting, volcanoes may erupt along the rift or great flows of basaltic lava may pour out through the fissures in the continent. If the rifting continues, a new ocean basin will eventually form between the pieces of the continent. This is happening now in northeast Africa, where three rift zones meet in a *triple junction* (**figure 3.18**). Along two branches, continental separation has proceeded to the point of formation of oceanic lithosphere to create the Red Sea and the Gulf of Aden. The third rift is less

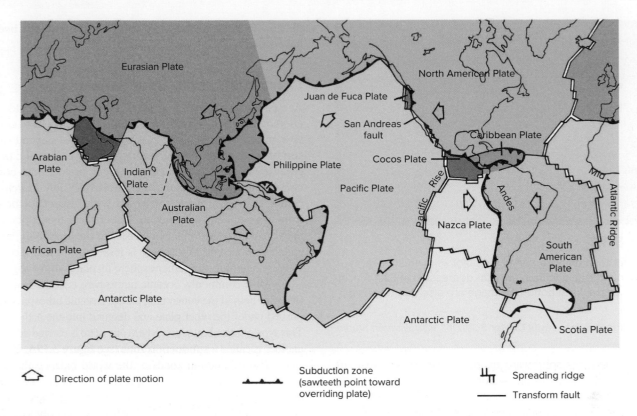

Direction of plate motion

Subduction zone
(sawteeth point toward
overriding plate)

Spreading ridge

Transform fault

Figure 3.16

The principal lithospheric plates and plate boundaries. Arrows denote approximate relative directions of plate movement. The types of plate boundaries are discussed later in the chapter. How does the information in **figures 3.4** and **3.7** help you to understand plate boundaries?

Source: After W. Hamilton, U.S. Geological Survey.

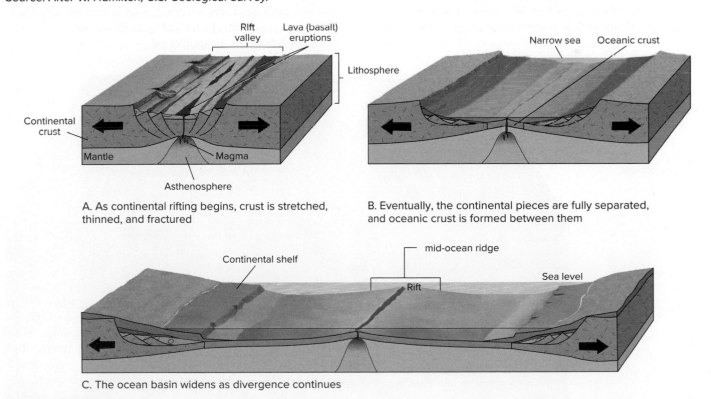

A. As continental rifting begins, crust is stretched, thinned, and fractured

B. Eventually, the continental pieces are fully separated, and oceanic crust is formed between them

C. The ocean basin widens as divergence continues

Figure 3.17

Formation of divergent plate boundaries through rifting.

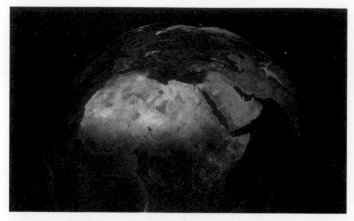

Figure 3.18

Continental crust has been rifted, and blocks of crust have dropped into the gap to form the Afar depression in East Africa. North–south cracks at bottom of image are also due to tension and rifting.

NASA/Goddard Space Flight Center Scientific Visualization Studio

mature and extends southward in the volcanically active rift valley of the Afar depression. Water has flooded into low elevations as marked by various lakes but the continent has not yet separated as in the other nearby rifts. Another example of continental rifting occurred in central North America but was more limited and halted before the process was complete, creating the New Madrid fault zone which will be discussed in chapter 4. A currently active North American rift occurs in the Southwest—the Rio Grande Rift. It is marked by the flow of the upper Rio Grande River south through New Mexico along

the newly formed depression which could eventually become a new ocean basin.

Convergent Plate Boundaries

At a **convergent plate boundary,** as the name indicates, plates are moving toward each other, subjecting the rocks to compressive stress. The details of just what happens depend on what sort of lithosphere is at the leading edge of each plate; the three types are ocean-ocean, ocean-continent, or continent-continent convergence. Continental crust is relatively low in density, so continental lithosphere is buoyant with respect to the dense, iron-rich mantle and it tends to float on the asthenosphere. Oceanic crust is more similar in density to the underlying asthenosphere, so oceanic lithosphere is less buoyant and more easily forced down into the asthenosphere as plates move together.

Most commonly, oceanic lithosphere is at the leading edge of one or both of the converging plates. Oceanic lithosphere may be pushed under the other plate and descend into the asthenosphere. This type of plate boundary, where one plate is carried down below another is called a **subduction zone** (see **figure 3.19A, B**).

The subduction zones of the world balance the seafloor equation. If new oceanic lithosphere is constantly being created at spreading ridges, an equal amount must be destroyed (perhaps recycled) somewhere or Earth would simply keep getting bigger. We refer to subduction zones as *destructive* plate boundaries because this is where sea floor is consumed. The subducted plate is heated by the hot asthenosphere. Fluids released from the ocean crust induce some of the overlying mantle to melt to form magma. As the dense, cold plate continues downward, it eventually breaks off and sinks deeper into the mantle.

Figure 3.19

(A) At ocean-continent convergence, sea floor is consumed, and volcanoes form on the overriding continent. Here the trench is partially filled by sediment.

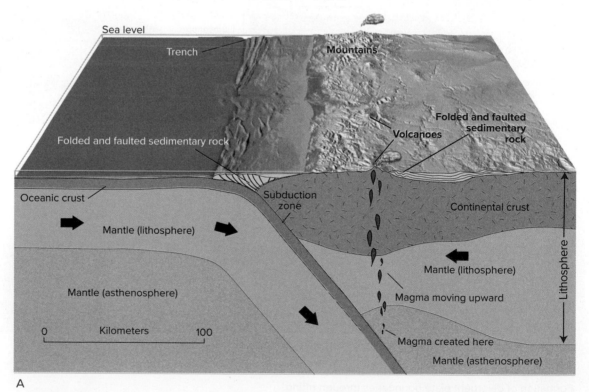

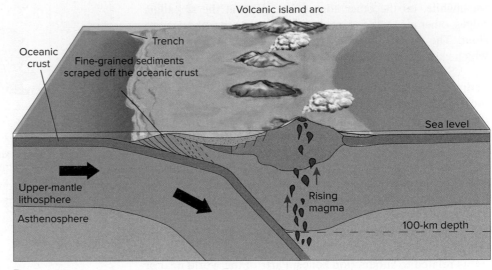

Volcanic island arc

Trench

Oceanic crust

Fine-grained sediments scraped off the oceanic crust

Sea level

Upper-mantle lithosphere

Asthenosphere

Rising magma

100-km depth

B

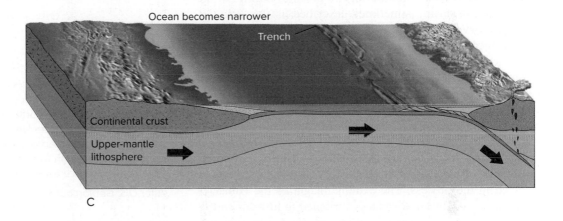

Ocean becomes narrower

Trench

Continental crust

Upper-mantle lithosphere

C

Figure 3.19 *Continued*

(B) Volcanic activity at ocean-ocean convergence creates a curved string of volcanic islands.

(C) As subduction continues, the subducting plate brings a continental mass closer and closer to the continent on the overriding plate.

(D) The resulting continental collision creates a great thickness of continental lithosphere, as in the Himalayas.

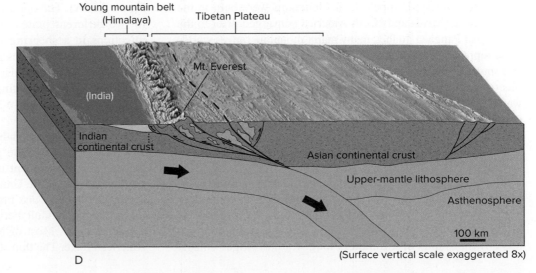

Young mountain belt (Himalaya)   Tibetan Plateau

Mt. Everest

(India)

Indian continental crust

Asian continental crust

Upper-mantle lithosphere

Asthenosphere

100 km

(Surface vertical scale exaggerated 8x)

D

Meanwhile, on the other edge of the plate at the spreading ridges, other magmas rise, cool, and crystallize to make new sea floor. The oceanic lithosphere is constantly being recycled, which explains why few very ancient seafloor rocks are known.

Subduction zones are, geologically, very active places. Sediments eroded from the continents may accumulate in the trench formed by the down-going plate, and some of these sediments may be carried down into the asthenosphere to contribute to magmas being produced there, as described further in chapter 5. Volcanoes form where molten material rises up through the overlying plate to the surface. At an ocean-ocean convergence, the result is commonly a curved line of volcanic islands, an **island arc** (see **figure 3.19B**). The great stresses involved in convergence and subduction give rise to numerous earthquakes. The bulk of mountain building, and its associated volcanic and earthquake activity, is related to subduction zones. Parts of the world near or above modern subduction zones are therefore prone to both volcanic and earthquake activity. These include the Andes region of South America, western Central America, parts of the northwest United States and Canada, the Aleutian Islands, China, Japan, Indonesia, and much of the rim of the Pacific Ocean basin, sometimes described as the "Ring of Fire" for its volcanism.

If there is also continental lithosphere on the plate being subducted at an ocean-continent convergent boundary, consumption of the subducting plate will eventually bring the continental masses together (**figure 3.19C**). The two landmasses collide, crumple, and deform. One may partially override the other, but the buoyancy of continental lithosphere ensures that neither sinks deep into the mantle and a very large thickness of continent may result. Earthquakes are frequent during continent-continent collision as a consequence of the large stresses involved in the process. The extreme height of the Himalaya Mountains and Tibetan Plateau is attributed to just this sort of collision. India was not always a part of the Asian continent. Paleomagnetic evidence indicates that it moved northward over tens of millions of years until it ran into Asia (recall **figure 3.11**), and the Himalayas were built up in the process (**figure 3.19D**). Earlier, the ancestral Appalachian Mountains were built in the same way, as Africa and North America converged prior to the breakup of Pangaea. In fact, many major mountain ranges worldwide represent sites of sustained plate convergence in the past, and much of the western portion of North America consists of bits of continental lithosphere pasted onto the continent in this way (**figure 3.20**). This process accounts for the juxtaposition of rocks that are quite different in age and geology.

Transform Boundaries

The structure of an oceanic divergent boundary is more complex than a single, straight crack. It does help to remember here that Earth is a sphere, not a flat surface. A close look at a mid-ocean spreading ridge reveals that it is not a continuous rift thousands of kilometers long. Rather, ridges consist

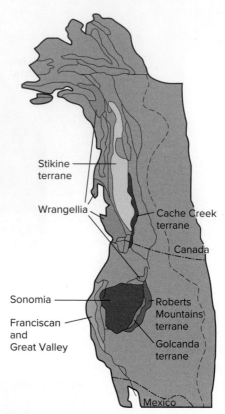

Figure 3.20

A *terrane* is a region of rocks that share a common history, distinguished from nearby, but genetically unrelated, rocks. Geologists studying western North America have identified and named many different terranes, some of which are shown here. A terrane that has been shown—often by paleomagnetic evidence—to have been transported and added onto a continent from some distance away is called an *accreted terrane*.

After U.S. Geological Survey Open-File Map 83-716.

of many short segments slightly offset from one another by a break in the lithosphere called a **transform fault** (see **figure 3.21A**). The opposite sides of a transform fault belong to two different plates, and these are moving in opposite directions in a shearing motion. As the plates scrape past each other, earthquakes occur along the transform fault. Transform faults may also occur between a trench (subduction zone) and a spreading ridge (see the Cascadian subduction zone below in **figure 3.22**),or between two trenches, but these are less common.

The famous San Andreas fault in California is an example of a transform fault. The East Pacific Rise disappears under the edge of the continent in the Gulf of California. Along the northwest coast of the United States, subduction is consuming the small Juan de Fuca Plate (**figure 3.22**). The San Andreas is the transform fault between the subduction zone and the spreading ridge. Most of North America is part of the North American Plate. The thin strip of California on the west side of

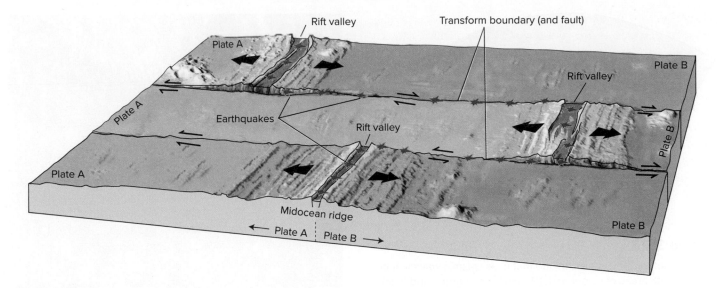

Figure 3.21

Seafloor spreading-ridge segments are offset by transform faults. Red asterisks show where earthquakes occur in the ridge system.

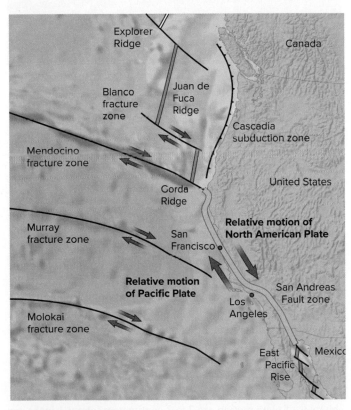

Figure 3.22

The San Andreas fault connects the East Pacific Rise to the Cascadian subduction zone with multiple fracture zones (the scars of former transform faults) in-between.

the San Andreas fault, however, is moving northwest with the Pacific Plate. The stress resulting from the shearing displacement across the fault leads to ongoing earthquake activity.

3.5 Moving the Plates

Learning Objectives

- Recall techniques used to monitor and calculate plate motion
- Explain how hot spots confirm plate motion
- List proposed mechanisms that cause plate movement

Geologic and topographic information together allow us to identify the locations and nature of the major plate boundaries shown in **figure 3.16**. We are still exploring the questions of how and why plates move, and for how much of geologic history plate-tectonic processes have operated. Recent studies suggest that features characteristic of modern plate-tectonic processes are in rocks at least 3 billion years old, by which time the volume of continental material was similar to today. Long before Pangaea, then, supercontinents were evidently forming and breaking up as a consequence of plate-tectonic activity. Below we will explore questions surrounding the how far, how fast, how long, and how come of plate motion, and how hot spots confirm our understanding of plate tectonics.

Past Motions, Present Velocities

We can calculate the rates and directions of plate movement in a variety of ways. As previously discussed, polar-wander curves from continental rocks show the general location of continents at different times. Seafloor spreading is another way of determining plate movement. The direction of seafloor spreading is usually obvious: away from the ridge. We can determine rates of seafloor spreading by very simply dating rocks at different

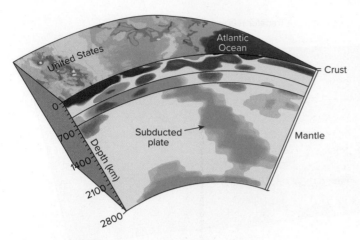

Figure 3.23

Seismic tomography is a technique that uses seismic-wave velocities to locate areas of colder rock (shown here in green and blue) and warmer rock (red and orange). The green area labeled as "Subducted plate" is interpreted as a slab of lithosphere subducted beneath the western United States some 30 million years ago, and is still somewhat cooler than the surrounding mantle.

After image by Suzan van der Lee and Steve Grand, courtesy EarthScope.

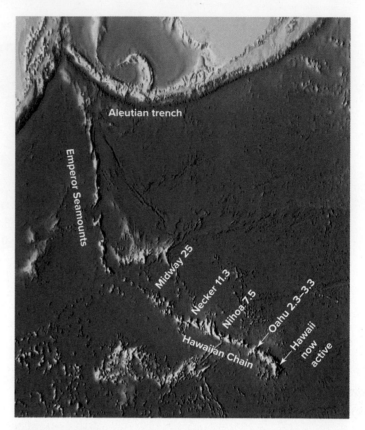

Figure 3.24

The Hawaiian Islands and other volcanoes in a chain formed over a hot spot. Numbers indicate ages of volcanic eruptions, in millions of years. Movement of the Pacific Plate has carried the older volcanoes far from the hot spot, now under the active volcanic island of Hawaii.

NOAA

distances from the spreading ridge and dividing the distance moved by the rock's age (the time it has taken to move that distance from the ridge at which it formed). For example, if a 10-million-year-old piece of sea floor is collected at a distance of 100 kilometers from the ridge, this represents an average rate of movement over that time of 100 kilometers/10 million years, or about 1 centimeter (<1/2 inch) per year.

Another way to monitor rates and directions of plate movement is by using mantle **hot spots.** These are isolated areas of volcanic activity usually not associated with plate boundaries. They are attributed to columns of warm mantle material *(plumes),* perhaps originating at the base of the mantle, that rise up through colder, denser mantle. We recognize them using techniques that reveal temperature differences within the crust and mantle (**figure 3.23**). Reduction in pressure as a plume rises can lead to partial melting, and the resultant magma can work its way up through the overlying plate to create volcanic activity. If we assume that mantle hot spots remain fixed in position while the lithospheric plates move over them, the result should be a trail of volcanoes of differing ages with the youngest closest to the hot spot.

A good example is in the north Pacific Ocean (see **figure 3.24**). The map shows a V-shaped chain of volcanic islands and submerged volcanoes. When rocks from these volcanoes are dated, they show a progression of ages, from about 75 million years at the northwestern end of the chain, to about 40 million years at the bend, through progressively younger islands to the still-active volcanoes of the island of Hawaii, located at the eastern end of the Hawaiian Island group. From the distances and age differences between pairs of points, we can again determine the rate of plate motion. For instance, Midway Island and Hawaii are

about 2700 kilometers (1700 miles) apart. The volcanoes of Midway were active about 25 million years ago. Over the last 25 million years, then, the Pacific Plate has moved over the mantle hot spot at an average rate of 2700 kilometers/25 million years, or about 11 centimeters (4.3 inches) per year. The orientation of the volcanic chain shows the direction of plate movement—west-northwest. The kink in the chain aged at about 40 million years ago indicates a change in either the direction of movement of the Pacific Plate or the location of the hot spot at that time.

Satellite-based technology now allows us to directly measure modern plate movement (**figure 3.25**). Looking at many such determinations from all over the world, geologists find that average rates of plate motion are 2 to 3 centimeters (about 1 inch) per year, though as the figure shows, they can vary quite a bit. This seemingly trivial amount of motion does add up through geologic time. Movement of 2 centimeters per year for 100 million years means a shift of 2000 kilometers, or about 1250 miles!

Why Do Plates Move?

Scientists are still debating the driving force, or forces, of plate tectonics. Several of the proposed mechanisms are shown in

figure 3.26. For many years, the most widely accepted explanation was that the plates were moved by large **convection cells** slowly churning in the plastic asthenosphere. According to this model, hot material rises at the spreading ridges; some magma escapes to form new lithosphere, but the rest of the rising asthenospheric material spreads out sideways beneath the lithosphere, slowly cooling in the process. As it flows outward, it drags the overlying lithosphere outward with it, thus continuing to open the ridges. When it has cooled somewhat, the flowing material is dense enough to sink back deeper into the asthenosphere—for example, under subduction zones. However, the existence of convection cells in the asthenosphere has not been proven definitively, even with sophisticated modern tools such as seismic tomography. There is some question, too, whether flowing asthenosphere could exercise enough drag on the lithosphere above to propel it laterally and force it into convergence.

One alternative explanation, that is backed with considerable evidence, is that the weight of the dense, down-going slab of lithosphere in the subduction zone pulls the rest of the trailing plate along with it, opening up the spreading ridges and leaving a space for hot, less dense asthenosphere to move into. Mantle convection might then result from lithospheric drag, not the reverse. This is

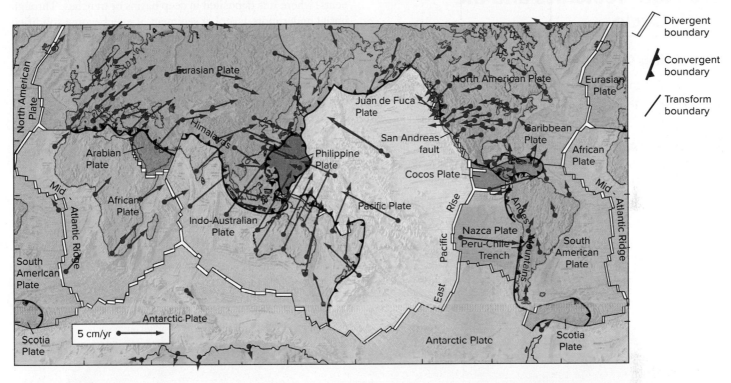

Figure 3.25

Motion of points on the surface is now measurable directly by satellite using GPS technology. Orientation of each arrow shows direction of motion; length shows rate of movement (scale at lower left).

Source: Image courtesy NASA.

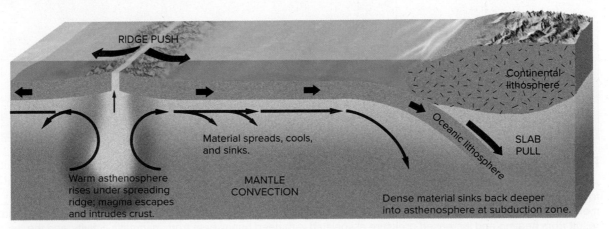

Figure 3.26

Mechanisms that may drive plate tectonics include mantle convection, ridge push, and slab pull.

described as the *slab-pull* model. Another model involves slabs of lithosphere sliding off the topographic highs formed at spreading ridges and rift zones by rising warm asthenospheric mantle, and perhaps dragging some asthenosphere along laterally in the process of *ridge-push*. Many geoscientists now believe that a combination of these mechanisms is responsible for plate motion. There may be contributions from other mechanisms not yet considered. That lithospheric motions and flow in the asthenosphere are coupled seems likely, but which one may drive the other is still unclear.

3.6 Plate Tectonics and the Rock Cycle

Learning Objective

- Discuss the rock cycle in terms of plate tectonic processes

In chapter 2, we noted that all rocks are related by the concept of the rock cycle. We can also look at the rock cycle in a plate-tectonic context, as illustrated in **figure 3.27**. New igneous rocks form from magmas rising out of the asthenosphere at spreading ridges or in subduction zones. The heat radiated by the cooling magmas can cause metamorphism, with recrystallization at an elevated temperature changing the texture and/or the mineralogy of the surrounding rocks. Some of these surrounding rocks may themselves melt to form new igneous rocks. The forces of plate collision at convergent margins also contribute to metamorphism by increasing the pressures acting on the rocks. Weathering and erosion on the continents wear down preexisting rocks of all kinds into sediment. Much of this sediment is eventually transported to the edges of the continents, where it is deposited in deep basins or trenches. Through burial under more layers of sediment, it may become solidified into sedimentary rock. Sedimentary rocks, in turn, may be metamorphosed or even melted by the stresses and the igneous activity at the plate margins. Some of these sedimentary or metamorphic materials may also be carried down with subducted oceanic lithosphere, to be melted and eventually recycled as igneous rock. Plate-tectonic processes thus play a large role in the formation of new rocks from old that proceeds continually on Earth, recycling and transforming both continents and ocean floor.

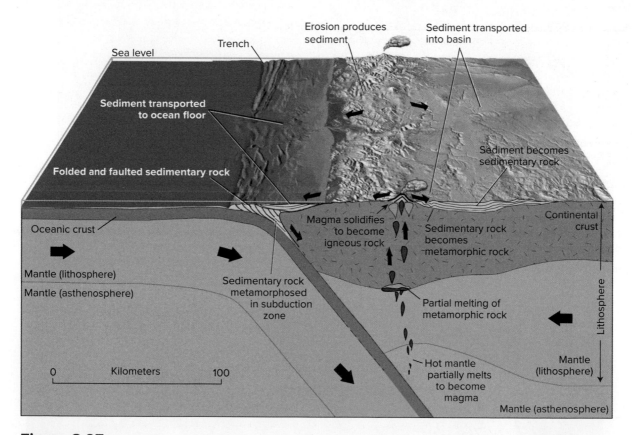

Figure 3.27

An ocean-continent convergence zone illustrates some of the many ways in which plate-tectonic activity transforms rocks. For example, continents erode, producing sediment to make sedimentary rocks; asthenosphere melts to make magma to form new igneous rock, as it also does at spreading ridges. Regional metamorphism can result from heating and increased pressure associated with deep burial, heat from magmatic activity, and the compressive stress of convergence; close to shallow magma bodies, contact metamorphism may also occur. No rocks are preserved indefinitely; all are caught up in this cycle of continuous change.

A New Theory for Old—Geosynclines and Plate Tectonics

In the mid-nineteenth century, studies of the Appalachians inspired the development of a theory of mountain building that would prevail for over a century. This was the *geosyncline theory*. A *syncline* is a trough-shaped fold in rocks (and this term is still used today). The term *geosyncline* was coined to indicate a very large syncline, on the order of the length of a mountain range. Briefly, the model was as follows: Mountain ranges began with the development of a geosyncline, generally at the margin of a continent. The trough deepened as sediments accumulated in it. The geosyncline was divided into a *miogeosyncline,* on the continental side, where sediments that would become sandstones, shales, and limestones were deposited in relatively shallow water, and a deeper *eugeosyncline,* on the oceanic side, in which a more complex package of materials, including deepwater sediments and volcanics, was deposited. As the geosyncline deepened, rocks on the bottom would be metamorphosed and, in the deepest part of the eugeosyncline, melted. The geosyncline would become unstable; folding and faulting would occur, and magmas would rise to form plutonic rocks deep in the eugeosyncline and to feed volcanoes at the surface. After a period of such upheaval, the transformed geosyncline would stabilize, and when the rocks were uplifted, erosion would begin to shape the topography of the mountain range (**figure CS 3.1.1**).

Geosyncline theory explained a number of characteristics of the Appalachians in particular and mountain ranges in general. Many mountain ranges occur at continental margins. Often, the rock types contained in them are consistent with having originated as the kinds of sediments and volcanics described by the theory. Mountains commonly show a more intensely metamorphosed and intruded core (the former eugeosyncline in this model, the depth of which would provide the necessary heat and pressure for metamorphism and melting) with less-deformed, less intensely metamorphosed rocks flanking these (the shallower former miogeosyncline). However, geosyncline theory had some practical problems. For example, many

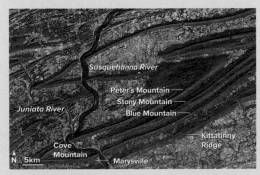

Figure CS 3.1.1

Uplift and erosion have exposed ridges of weathering-resistant rock shaped into the great folds that are characteristic of the Appalachians in Pennsylvania.

NASA image by Robert Simmon, based on Landsat data from the University of Maryland Global Land Cover Facility

mountain ranges are in continental interiors (the Urals and Himalayas, for instance), and it is much more difficult to explain how a geosyncline could form in the middle of a continent, given the relative buoyancy of continental crust. What would push the trough down into the much denser mantle? The weight of sediment would not be enough, for most sediments are even less dense—and would therefore be more buoyant—than typical continental crust. Indeed, this objection applies, to some extent, even to the formation of geosynclines at continental margins.

Plate-tectonic theory was not developed particularly to explain mountain building; it is much broader in scope, as we have seen. Yet it also succeeds far better than geosyncline theory in accounting for the formation of mountain ranges. The trench of a subduction zone provides a site for accumulation of thick sediment, while subduction provides a mechanism for creating the trench. The magma to feed volcanoes and form plutonic rocks is generated in the asthenosphere above the down-going slab. The stresses associated with convergence and the heat from rising magmas, together with burial, supply the necessary heat and pressure to account for the metamorphism in mountain ranges, and the same stresses account for the extensive folding and faulting. Intracontinental mountain ranges can be explained as products of continent-continent convergence. And as we have also seen, plate-tectonic theory is consistent with a wealth of geologic evidence (much of which was not available to those who first developed geosyncline theory). So, in the latter third of the twentieth century, geosyncline theory as a model of mountain building was abandoned, supplanted by plate tectonics.

In the early twenty-first century, the basic framework of plate tectonics is well established and accepted, but the theory is not static in all its details, and questions certainly remain. For example, much current research is being applied to questions about mantle plumes and hot-spot volcanoes. More data are needed to document definitively the locations and behavior of mantle plumes. There is now paleomagnetic evidence that at least some hot spots are *not* stationary, though whether they are deflected by churning mantle or by some other mechanism is not known. Seismic tomography suggests that many do originate at the core/mantle boundary; but if that is so and the mantle is convecting, how do the plumes retain their coherence all the way to the base of the lithosphere to produce the hot spots that do appear stationary? Or consider plate motions: If slab-pull is a significant force driving plate movement, how did it begin in the early Earth—what drove the first slab of lithosphere down into the mantle?

This is the nature of science: that even a very well-documented theory such as plate tectonics will continue to be tested by inquiring minds examining new and existing data, and refined as the data demand new interpretations. Nor, in this case, is it just a matter of intellectual curiosity among geoscientists, for, as noted, earthquakes and volcanoes are closely linked to plate-tectonic activity. And the better we can understand those, the better able we are to recognize the hazards they present and to minimize our risks from those hazards.

Summary

Rocks subjected to stress may behave either elastically or plastically. At low temperatures and confining pressures, they are more rigid, elastic, and often brittle. At higher temperatures and pressures, or in response to stress applied gradually over long periods, they tend more toward plastic behavior.

The outermost solid layer of Earth is the 50- to 100-kilometer-thick lithosphere, which is broken up into a series of rigid plates. The lithosphere is underlain by a plastic layer of the mantle, the asthenosphere, over which the plates can move. This plate motion gives rise to earthquakes and volcanic activity near the plate boundaries. At seafloor spreading ridges, which are divergent plate boundaries, new sea floor is created from rising asthenosphere that partially melts. The sea floor moves away from the ridges in a conveyor-belt fashion, ultimately to be destroyed in subduction zones, a type of convergent plate boundary, where oceanic lithosphere is carried down into the asthenosphere. It may eventually be remelted, or sink as cold slabs down through the mantle. Magma rises up through the overriding plate to form volcanoes above the subduction zone. Where continents ride the leading edges of converging plates, continent-continent collision may build high mountain ranges.

Evidence for seafloor spreading includes the age distribution of seafloor rocks and the magnetic striped patterns on the ocean floor. Continental movements are demonstrated by such means as polar-wander curves, fossil distribution among different continents, and evidence of ancient climates revealed in the rock record, which can show changes in a continent's latitude. Past supercontinents can be reconstructed by fitting together modern continental margins, matching similar geologic features from continent to continent, and correlating polar-wander curves from different landmasses.

Present rates of plate movement average a few centimeters a year. One possible driving force for plate tectonics is slow convection in the asthenosphere; another is gravity pulling cold, dense lithosphere down into the asthenosphere, dragging the asthenosphere along. Plate-tectonic processes appear to have been more or less active for much of Earth's history. They play an integral part in the rock cycle—building continents that weather into sediments, carrying rock and sediment into the warm mantle to be melted into magma that rises to create new igneous rock and metamorphoses the lithosphere through which it rises, subjecting rocks to the stress of collision—assisting in the making of new rocks from old.

Key Terms and Concepts

asthenosphere	divergent plate boundary	paleomagnetism	strain
brittle	ductile	plastic deformation	stress
compressive stress	elastic deformation	plate tectonics	subduction zone
continental drift	elastic limit	polar-wander curve	tectonics
convection cell	hot spot	rupture	tensile stress
convergent plate boundary	island arc	seafloor spreading	transform fault
Curie temperature	lithosphere	shearing stress	

Test Your Learning

1. Describe the concept of plate tectonics, and how continental drift and seafloor spreading are related to it.

2. Briefly explain two kinds of paleomagnetic evidence supporting the theory of plate tectonics.

3. Cite at least three kinds of evidence, other than paleomagnetic evidence, for the occurrence of plate tectonics.

4. Describe how elastic and plastic materials differ in their response when stress is applied to them.

5. Define the terms *lithosphere* and *asthenosphere,* and indicate where in the Earth each is located.

6. Explain how a subduction zone forms and what occurs at such a plate boundary.

7. Describe how continent-continent convergence comes about, and what happens when it occurs.

8. Briefly explain what is meant by the term *hot spot,* and how these help to determine rates and directions of plate movements.

9. Describe how convection in the asthenosphere may drive the motion of lithospheric plates. Alternatively, explain how ridge-push or slab-pull may move plates and, in turn, churn the plastic asthenosphere.

10. Describe the rock cycle in terms of plate tectonics, making specific reference to the creation of new igneous, metamorphic, and sedimentary rocks.

Exploring Further

1. The Atlantic Ocean is approximately 5000 km wide. If each plate has moved away from the Mid-Atlantic Ridge at an average speed of 1.5 cm per year, how long did the Atlantic Ocean basin take to form? What would the spreading rate be in miles per hour?

2. The Moon's structure is very different from that of Earth. For one thing, the Moon has an approximately 1000-kilometer-thick lithosphere. Would you expect plate-tectonic activity similar to that on Earth to occur on the Moon? Why or why not?

3. At one time, scientists proposed that the separation of South America and Africa was simply due to an expanding Earth, cracking and separating segments of its rocky shell. What scientific evidence do we now have that argues against this model? What other questions might you ask to challenge it?

4. Many early scientists used land bridges between continents as an alternative hypothesis to explain fossil distribution over widely separated continents. Describe strengths and weaknesses with the land bridge explanation. Can you come up with alternative hypotheses for similar fossil distribution?

CHAPTER 4

Earthquakes

Many thousands of earthquakes happen every year. Fortunately, most are small. The large ones can be deadly, and the twenty-first century has seen some particularly devastating examples. Over 200,000 people died as a result of the 2004 Indian Ocean earthquake, and more than 300,000 in Haiti in 2010. Damages were over $300 billion in Japan in 2011. These earthquakes, like most, struck without warning.

Figure 4.1 presents a sampling of significant earthquakes in history, and that will be discussed in this chapter. The human consequences of an earthquake depend on many factors beyond

---

Satellites can provide dramatic and informative views of features on Earth's surface including this coastal California image of the Point Reyes National Seashore. The northwest-southeast running San Andreas Fault separates the Pacific plate on the left from the North American Plate on the right. These plates move past each other at an average rate of 3–5 centimeters a year in a less-than-smooth process. When the plates get stuck, energy builds up that is eventually released in an earthquake.

NASA Earth Observatory images by Jesse Allen

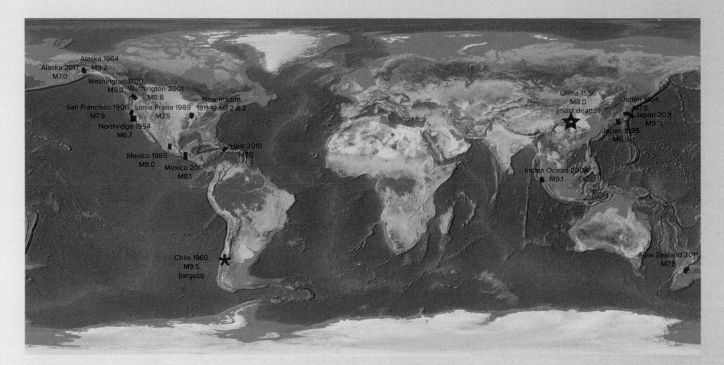

Figure 4.1

Some significant earthquakes in history. Chile 1960 was the largest magnitude ever recorded at 9.5, and China 1556 was the most deadly, causing an estimated 830,000 deaths. Every earthquake, and the hazards associated with it, is unique as it reflects a combination of natural and human factors. Visit the NOAA Natural Hazards Viewer to learn more about these and other significant earthquakes. https://www.ncei.noaa.gov/maps/hazards/

Jan Rysavy/E+/Getty Images

the energy released; they include the geology of the affected area, population density, human preparedness, and economic standing. Damages and casualties will continue to occur so long as people build, and rebuild, in areas at risk. Property damage can be reduced with improved engineering, but many existing structures predate the best modern design standards (recall **figure 1.6**, for example). As we develop ever-greater understanding of why and how earthquakes happen, the outstanding goal is to reduce casualties significantly by developing tools to make consistent, accurate earthquake prediction possible.

Chapter Outline

4.1 Earthquakes—Terms and Principles

4.2 Seismic Waves and Earthquake Severity

4.3 Earthquake Related Hazards and Their Reduction

4.4 Prediction, Forecasting, and Warnings

4.5 Earthquake Control

4.6 Future Earthquakes in North America

4.1 Earthquakes—Terms and Principles

Learning Objectives

- Define fault, creep, focus, and epicenter
- Describe the behavior of rocks before, during, and after an earthquake
- Recognize different types of faults
- Explain how strike and dip are used to record the orientation of a fault
- Relate earthquakes to plate boundaries
- Explain epicenter depth patterns at subduction zones

Earthquake science, or **seismology,** is likely a new field of study for you. It has commonly used terms that will be applied throughout this chapter. In the following, we will provide definitions and examples of how we use them to explain earthquakes. With this new knowledge, you will be better prepared to deepen your understanding of how earthquakes affect people.

Basic Terms

Major earthquakes dramatically demonstrate that Earth is a dynamic, changing system. Earthquakes represent a sudden release of built-up stress. They occur primarily in the lithosphere along **faults,** planar breaks in rock along which there is displacement of one side relative to the other. The stress can produce new faults or fractures or cause slipping along existing faults. When movement along faults occurs gradually and relatively smoothly, it is called **creep** (**figure 4.2**). Creep, sometimes termed *aseismic slip* meaning fault displacement without significant earthquake activity, can be inconvenient but rarely causes serious damage or loss of life.

When friction between rocks on either side of a fault prevents the rocks from slipping easily or when the rock under stress is not already fractured, some elastic deformation will occur before failure. When the stress at last exceeds the rupture strength of the rock or the friction along a preexisting fault, a sudden movement occurs to release the stress. This is an **earthquake** or *seismic slip*. Then the rocks snap back elastically to their previous dimensions. This behavior is called **elastic rebound** (**figure 4.3**; recall the discussion of elastic behavior from chapter 3).

Faults vary in length, from microscopically small to thousands of kilometers long. Likewise, earthquakes range from tremors so small that even sensitive instruments can barely detect them to massive shocks that can level cities. In fact, the *aseismic* movement of creep is characterized by microearthquakes so small that they are typically not felt at all. The amount of damage associated

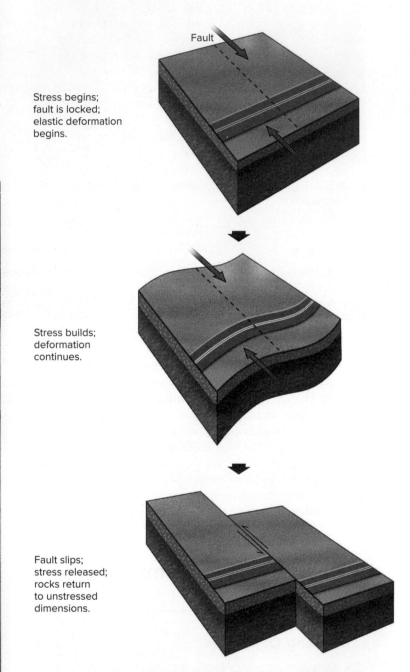

Fault

Stress begins; fault is locked; elastic deformation begins.

Stress builds; deformation continues.

Fault slips; stress released; rocks return to unstressed dimensions.

Figure 4.2

These offset sidewalk curbs and tilted sidewalk slabs are due to creep along the Calaveras Fault zone in northern California.

John A. Karachewski/The McGraw-Hill Companies

Figure 4.3

The phenomenon of elastic rebound. Stress is applied, the rocks deform, the fault slips, the rocks are displaced but they spring back elastically to an undeformed condition. A rubber band stretched to the point of breaking shows similar behavior.

with an earthquake is partly a function of the amount of accumulated energy released as the earthquake occurs. For clarification, we often discuss earthquakes as being "large" or "big" which is a reference to the amount of energy involved rather than the length of the fault or how long the earthquake lasts.

The point on a fault at which the first movement or break occurs during an earthquake is called the earthquake's **focus,** or *hypocenter* (**figure 4.4**). In the case of a large earthquake, a section of fault many kilometers long may slip, but there is always a point at which the first movement occurred, and this point is the focus. Earthquakes are sometimes characterized by their focal depth: shallow earthquakes have a focus of 0–70 km; intermediate earthquakes have a focal depth in the 70–300 km range, and deep quakes are those with a focus greater than 300 km. The point on the surface directly above the focus is the **epicenter.** When you see a map showing where an earthquake occurred, you are looking at the earthquake's epicenter. Some faults extend to the surface and leave a scar, or **fault scarp,** that is commonly evidenced by an abrupt change in elevation and a lack/change of vegetation.

Types of Faults

Faults are described in terms of the nature of the displacement—how the rocks on either side move relative to each other—which commonly reflects the nature of the stresses involved. Certain types of faults are characteristic of particular plate-tectonic settings. In describing the orientation of a fault or the planar features affected by a fault, geologists use two measures (**figure 4.5**). The *strike* is the compass orientation of the line of intersection of the plane of interest with the surface (e.g., the fault trace in **figure 4.4**). The *dip* of the fault is the angle the plane makes with the horizontal, a measure of the steepness of slope of the plane.

A **strike-slip fault,** then, is one along which the displacement is parallel to the strike (horizontal). A transform fault, such as shown in **figure 3.21**, is a type of strike-slip fault and reflects stresses acting horizontally. The San Andreas is a strike-slip fault. A **dip-slip fault** is one in which the displacement is vertical, up or down in the direction of dip. A dip-slip fault in which the block above the fault has moved down relative to the block below (as in **figure 4.4**) is called a *normal fault.* Rift valleys, whether along seafloor spreading ridges or on continents, are commonly bounded by steeply sloping normal faults, resulting from the tensional stress of rifting (recall **figures 3.17** and **3.18**). A dip-slip fault in which the block above has moved up relative to the block below the fault is a *reverse fault,* and indicates compressional stress. Convergent plate boundaries are often characterized by **thrust faults,** which are just reverse faults with relatively shallowly dipping fault planes (**figure 4.6**).

Earthquake Locations

As shown in **figure 4.7**, the locations of major earthquake epicenters are concentrated in linear belts. These belts correspond to plate boundaries; recall **figure 3.16**. Most, but not all, earthquakes occur at plate boundaries. These areas are where

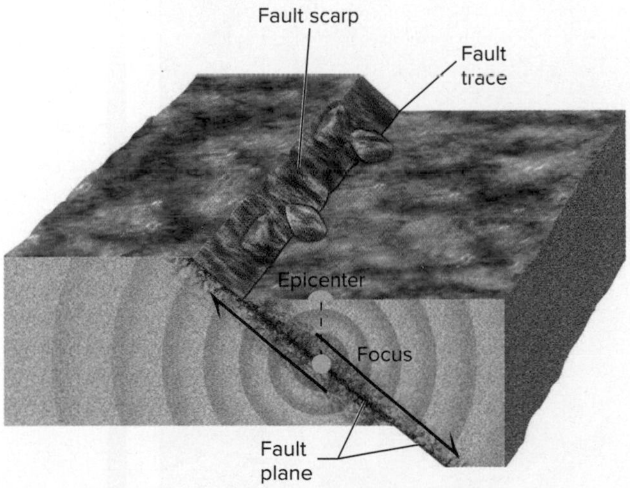

Figure 4.4

This simplified diagram of a fault shows the focus and epicenter of an earthquake. Seismic waves dissipate the energy released by the earthquake as they travel away from the fault zone in all directions. A fault scarp is a cliff formed along the fault plane at ground surface. A fault trace is the line along which the fault plane intersects the ground surface.

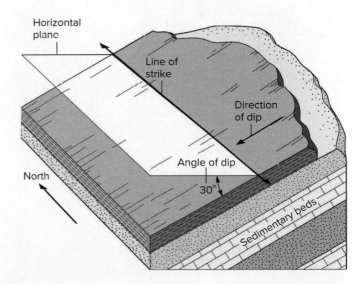

Figure 4.5

Strike and dip. To better conceive strike, imagine a straight line of seats on the angled side of a stadium or concert hall. For dip, it is the direction and steepness of the stairs as you move downward from your seats. This particular plane (bed) would be described as having a north–south strike and dipping 30° west.

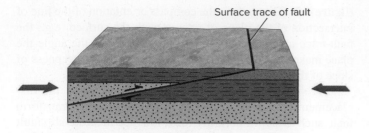

Figure 4.6

Along a thrust fault, compressional stresses (indicated by red arrows) push one block up and over the other.

plates jostle, collide with, or slide past each other; where relative plate movements may build up very large stresses; where major faults or breaks may already exist on which further movement may occur. On the other hand, intraplate earthquakes certainly occur and may be quite severe, as explored later in the chapter.

The pattern of earthquake depths is a good indication of what type of plate boundary exists in an area. Shallow earthquakes are the most common because the behavior of rocks necessary for earthquakes occurs where temperatures and pressures are relatively low, for example, in the lithosphere. Here rocks are elastic, capable of storing energy as they deform before they suddenly rupture or slip. In the plastic asthenosphere, material simply flows under stress. Divergent plate boundaries are dominated by shallow-focus earthquakes, whereas deep-focus earthquakes represent subduction zones, the places where the lithosphere is moving deep into the mantle.

It is the distribution of earthquake foci at subduction boundaries that has helped geologists recognize and understand the subduction process. For example, look at the earthquake depth patterns in western South America and near Japan in **figure 4.7**. Adjacent to each seafloor trench is a region in which earthquake foci are progressively deeper with increasing distance from the trench. This pattern is called a *Benioff zone*, named for the scientist who first extensively mapped these dipping planes of earthquake foci that we now realize reveal subducting plates. More recently, scientists have used the location of earthquake foci at subduction zones to better understand the geometry and dynamics of the subducting plate, showing that not all plates subduct in the same way.

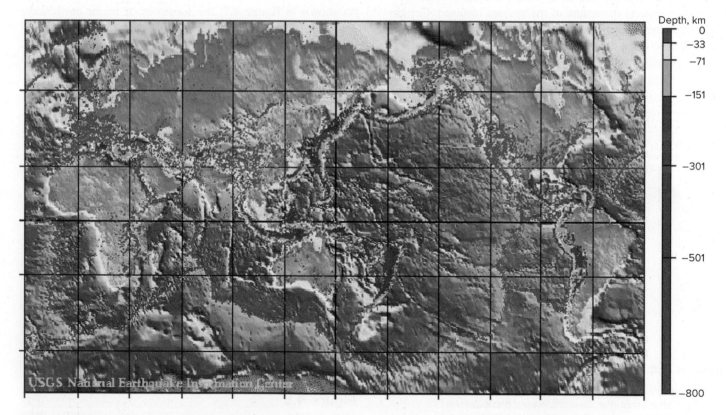

Figure 4.7

World seismicity, 1979–1995. Color of symbol indicates depth of focus. Would the addition of recent earthquake epicenters change the patterns already shown on this map?

U.S. Geological Survey Department of the Interior/USGSU.S. Geological Survey

4.2 Seismic Waves and Earthquake Severity

Learning Objectives

- Differentiate seismic waves
- Describe how to locate the epicenter of an earthquake
- Compare Richter and moment magnitudes
- Recognize the Modified Mercalli Intensity scale

In the broadest sense, a *wave* is a disturbance that travels through a medium whether it's air, water, rock, or magma. Drop a pebble into a still pond, and the water is disturbed. We see the effects at the surface as ripples traveling away from that spot, but water below the surface is affected too. When sound waves travel through air, the air is alternately compressed and expanded, though we cannot see this happening. You can create a visible analog with a Slinky toy. Stretch the coiled spring out on a smooth surface, hold one end in place, push the other end in quickly and then hold it still also. A pulse of compressed coil travels along the length of the Slinky. Waves can involve other types of motions, too. Take the Slinky and stretch it as before, but now twitch one end sideways, perpendicular to its length. The wave that travels along the Slinky this time will involve a shearing motion, loops of coil moving side-to-side relative to each other rather than moving closer together and farther apart. Rocks are affected by different types of waves such as this. Below we will review the different types of seismic waves and how we use them to measure the location, size, and severity of an earthquake.

Seismic Waves

When an earthquake occurs, it releases the stored-up energy in **seismic waves** that travel in all directions away from the focus. There are several types of seismic waves. **Body waves** (P waves and S waves) travel through the interior of Earth. **P waves** are compressional waves. As P waves travel through matter, the matter is alternately compressed and expanded. **S waves** are shear waves, involving a side-to-side motion of molecules. Because of the different type of motion, S waves are slower than P waves and they cannot travel through liquids. Geologists have taken advantage of seismic wave behavior to create images of Earth's internal layers.

Seismic **surface waves** are somewhat similar to surface waves on water, discussed in chapter 7. That is, they cause rocks and soil to be displaced in such a way that the ground surface ripples or undulates. Surface waves also come in two types. Some cause vertical ground motions, like ripples on a pond, while others cause horizontal shearing motions, the ground surface rocking side-to-side. The surface waves are larger in

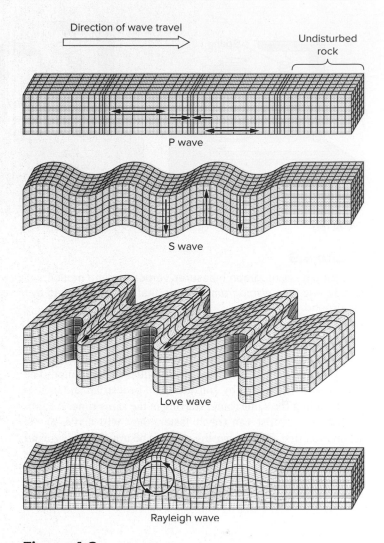

Figure 4.8

P waves and S waves are types of seismic body waves, one compressional, one shear. Love and Rayleigh waves are surface waves that differ in the type of ground motion they cause: Love waves involve a shearing motion on the surface, while Rayleigh waves resemble ripples on a pond.

amplitude than the body waves from the same earthquake. Therefore, most of the shaking and resultant structural damage from earthquakes is caused by the surface waves.

Figure 4.8 compares the actions of different types of seismic waves on rock.

Locating the Epicenter

Earthquake epicenters can be located using seismic body waves. Both types of body waves cause ground motions that are detectable using a **seismograph** (**figure 4.9**). Because P waves travel faster through rocks than S waves, at points some distance from an earthquake, the first P waves arrive before the first S waves. A helpful way to remember this is to use the alternative names for body waves—*primary* and *secondary* waves—which reflect the arrival-time differences.

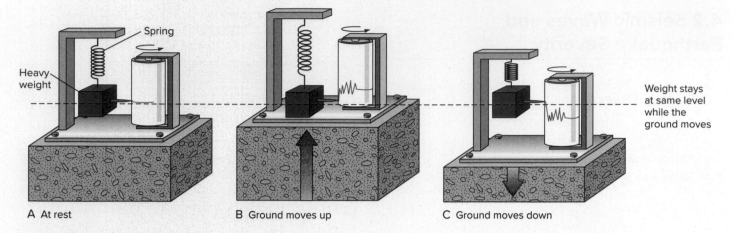

Spring

Heavy weight

A At rest   B Ground moves up   C Ground moves down

Weight stays at same level while the ground moves

Figure 4.9

This simple seismograph measures vertical ground motion. As the ground shifts up and down, the weight holds the attached pen steady, and the *seismogram* showing that motion is traced by the pen on the paper on the rotating cylinder, which is moving with the ground.

The difference in arrival times of the first P and S waves is a function of distance to the earthquake's epicenter. The effect can be illustrated by considering a pedestrian and a bicyclist traveling the same route, starting at the same time. Assuming that the cyclist can travel faster, they will arrive at the destination first. The longer the route to be traveled, the greater the difference in time between the arrival of the cyclist and the later arrival of the pedestrian. Likewise, the farther the receiving seismograph is from the earthquake epicenter, the greater

the time lag between the first arrivals of P and S waves. The general principle is illustrated graphically in **figure 4.10A**. When at least three recording stations have determined their distances from the epicenter in this way, the epicenter can be located on a map (**figure 4.10B**).

In practice, complicating factors such as inhomogeneities in the crust make epicenter location somewhat more difficult than described above. Computers assist in reconciling the data, and results are reviewed by seismologists.

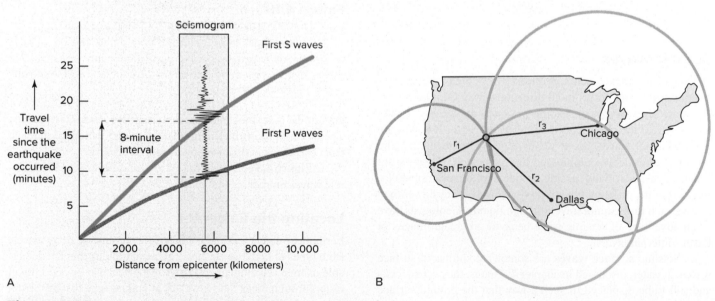

Figure 4.10

Use of seismic waves in locating earthquakes. (A) Difference in times of first arrivals of P and S waves is a function of the distance from the earthquake epicenter. This seismogram has a difference in arrival times of 8 minutes; the earthquake occurred about 5500 km away. (B) Triangulation using data from several seismograph stations allows location of an earthquake's epicenter. Note: The data from A does not coincide with that shown on B.

Magnitude and Intensity

All seismic waves represent energy release and transmission; they cause the ground shaking that people associate with earthquakes. The amount of ground motion is related to the **magnitude** of the earthquake. Historically, earthquake magnitude has been reported in North America using the *Richter magnitude scale,* named after geophysicist Charles F. Richter who developed it.

A Richter magnitude number is assigned to an earthquake on the basis of the amount of ground displacement or shaking that it produces near the epicenter. The amount of ground motion is measured by a seismograph, and is adjusted for the particular type of instrument and the distance of the station from the earthquake epicenter. The Richter scale is logarithmic, which means that an earthquake of magnitude 4 causes ten times as much ground movement as one of magnitude 3, one hundred times as much as one of magnitude 2, and so on. The amount of energy released rises even faster with increased magnitude, by a factor of about 30 for each unit of magnitude. An earthquake of magnitude 4 releases approximately thirty times as much energy as one of magnitude 3, and nine hundred times as much as one of magnitude 2. Although we only hear of the very severe, damaging earthquakes, there are hundreds of thousands of earthquakes of all sizes each year. **Table 4.1** summarizes the frequency and energy release of earthquakes in different Richter magnitude ranges.

As seismologists studied more earthquakes in more places, they realized that the Richter magnitude scale had some limitations. The scale was developed in California, where the earthquakes are all shallow-focus, and mostly the result of shear stress acting on strike-slip faults. Earthquakes in other areas, with different rock types that respond differently to seismic waves, different focal depths, and different stress regimes, may not be readily compared on the basis of Richter magnitudes. For very large earthquakes especially, a better measure of relative energy release is *moment magnitude* (denoted M_W). Moment magnitude takes into account the area of break on the fault surface, the displacement along the fault during the earthquake, and the strength of the rock. The U.S. Geological Survey now uses moment magnitude rather than Richter magnitude in reporting on modern earthquakes, for example, in press releases. Earthquake specialists have developed still other magnitude scales. The differences in assigned magnitude are generally significant only for the largest earthquakes. For example, the largest recorded earthquake (which occurred in Chile in 1960) had a Richter magnitude of 8.9, but an estimated moment magnitude of 9.5. What all the magnitude scales have in common is a logarithmic character, like the Richter scale.

An alternative way of describing the size of an earthquake is by the earthquake's **intensity.** Intensity is a measure of the earthquake's effects on humans and on surface features. The surface effects produced by an earthquake of a given magnitude vary considerably as a result of such factors as local geologic conditions, quality of construction, and distance from the epicenter. A single earthquake, then, can produce effects spanning a range of intensities in different places (**figure 4.11**), although it will have only one magnitude assigned to it. Intensity is a somewhat subjective measure typically based on observations by or impressions of individuals; different observers on the same spot may assign different intensity values to a single earthquake. Nevertheless, intensity is a more direct indication of the impact of a particular seismic event on humans in a given place than is its magnitude. The extent of damage at each intensity level is, in turn, related to the maximum ground velocity and acceleration experienced, so even in uninhabited areas, intensities can be estimated if the latter data have been measured. Several dozen intensity scales are in use worldwide. The most widely applied intensity scale in the United States is the Modified Mercalli Scale, summarized in **table 4.2.**

Table 4.1	Frequency of Earthquakes of Various Magnitudes		
Descriptor	**Magnitude**	**Number per Year**	**Approximate Energy Released per Earthquake (Joules)**
Great	8 and over	1	over 6.3×10^{16}
Major	7–7.9	15	$2–45 \times 10^{15}$
Strong	6–6.9	134	$6–140 \times 10^{13}$
Moderate	5–5.9	1320	$2–45 \times 10^{12}$
Light	4–4.9	13,000	$6–140 \times 10^{10}$
Minor	3–3.9	130,000	$2–45 \times 10^{9}$
Very minor	2–2.9	1,300,000 (estimated)	$6–140 \times 10^{7}$

Source: Frequency data and descriptors from National Earthquake Information Center, based on data since 1900.

Because energy release increases by a factor of 30 per unit increase in magnitude, most of the energy released by earthquakes each year is released not by the hundreds of thousands of small tremors, but by the handful of earthquakes of magnitude 7 or larger. For reference, exploding 1 ton of TNT releases about 4.2×10^9 joules of energy.

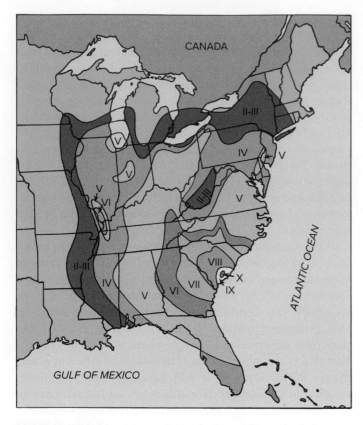

Figure 4.11

Zones of different Mercalli intensity for the Charleston, South Carolina, earthquake of 1886. You can help the USGS create such maps for future earthquakes using "Did You Feel It?"

Source: Data from U.S. Geological Survey

4.3 Earthquake-Related Hazards and Their Reduction

Learning Objectives

- List the ways the ground surface moves because of an earthquake
- Recall the factors involved in structural damage from earthquake shaking
- Describe how the ground fails because of an earthquake
- Describe the formation of a tsunami from earthquake activity
- Discuss the problems with broken utility lines following an earthquake

Earthquakes have a variety of direct and indirect harmful effects. As noted earlier, earthquakes of the same magnitude occurring in two different places can cause very different amounts of damage, depending on variables such as the nature of the local geology, whether the area affected is near the coast, and whether the terrain is steep or flat. Below we will summarize some historic and more apparent and dangerous earthquake hazards, and include some contemporary concerns.

Ground Motion

Movement along the fault is an obvious hazard. The offset between rocks on opposite sides of the fault can break power

Table 4.2	Modified Mercalli Intensity Scale (abridged)
Intensity	**Description**
I	Not felt.
II	Felt by persons at rest on upper floors.
III	Felt indoors—hanging objects swing. Vibration like passing of light trucks.
IV	Vibration like passing of heavy trucks. Standing automobiles rock. Windows, dishes, and doors rattle; wooden walls or frame may creak.
V	Felt outdoors. Sleepers wakened. Liquids disturbed, some spilled; small objects may be moved or upset; doors swing; shutters and pictures move.
VI	Felt by all; many frightened. People walk unsteadily; windows and dishes broken; objects knocked off shelves, pictures off walls. Furniture moved or overturned; weak plaster cracked. Small bells ring. Trees and bushes shaken.
VII	Difficult to stand. Furniture broken. Damage to weak materials, such as adobe; some cracking of ordinary masonry. Fall of plaster, loose bricks, and tile. Waves on ponds; water muddy; small slides along sand or gravel banks. Large bells ring.
VIII	Steering of automobiles affected. Damage to and partial collapse of ordinary masonry. Fall of chimneys, towers. Frame houses moved on foundations if not bolted down. Changes in flow of springs and wells.
IX	General panic. Frame structures shifted off foundations if not bolted down; frames cracked. Serious damage even to partially reinforced masonry. Underground pipes broken; reservoirs damaged. Conspicuous cracks in ground.
X	Most masonry and frame structures destroyed with their foundations. Serious damage to dams and dikes; large landslides. Rails bent slightly.
XI	Rails bent greatly. Underground pipelines out of service.
XII	Damage nearly total. Large rock masses shifted; objects thrown into the air.

Figure 4.12

Bends in the Trans-Alaska Pipeline allow some flex, and the fact that it sits loosely on slider beams where it crosses the Denali Fault also helped prevent rupture of the pipeline from a magnitude-7.9 earthquake in 2002.

U.S. Geological Survey/Photograph courtesy Alyeska Pipeline Service Company and U.S. Geological Survey.

lines, pipelines, buildings, roads, bridges, and other structures that cross the fault. Such offset can be large: in the 1906 San Francisco earthquake, maximum strike-slip displacement across the San Andreas fault was more than 6 meters. Planning and thoughtful design can reduce the risks. For example, power lines and pipelines can be built with extra slack where they cross a fault zone, or they can be designed with other features to allow some give as the fault slips and stretches them. Such considerations had to be taken into account when the Trans-Alaska Pipeline was built, for it crosses several known, major faults along its route. The engineering proved its value in the magnitude-7.9 2002 Denali Fault earthquake (**figure 4.12**). Fault displacement aside, the ground shaking produced as accumulated energy is released through seismic waves causes damage to and sometimes complete failure of buildings, with the surface waves responsible for most of this damage. Shifts of even a few tens of centimeters can be devastating, especially to structures made of weak materials such as adobe or inadequately reinforced concrete (**figure 4.13**). As shaking continues, damage may become progressively worse. Such effects are generally most severe on or very close to the fault, so the simplest strategy would be not to build near fault zones. However, many cities have already developed near major faults. Some have been repeatedly leveled by earthquakes and rebuilt again and again nevertheless. This raises the need for designing earthquake-resistant buildings to minimize damage.

Engineers have studied how well different types of building withstand real earthquakes. Scientists conduct laboratory experiments on scale models of skyscrapers and other buildings, subjecting them to small-scale shaking designed to simulate the kinds of ground movement to be expected during an earthquake. On the basis of their findings, special building codes for earthquake-prone regions can be developed. This approach, however, presents many challenges.

There are a limited number of records of just how the ground does move in a severe earthquake. To obtain the best such records, sensitive instruments must be in place near the fault zone beforehand, and those instruments must survive the earthquake. Increasingly, geoscientists are studying video captured by security cameras and private citizens, using the recorded motions of structures and furnishings to deduce the ground motions during the earthquake. It is vital to know the largest earthquakes to be expected on a given fault. Two reasons the 2011 Japanese earthquake was so devastating is that it was far larger than historical records led planners to expect, and shaking lasted for over four minutes. (See Case Study 4.1.) Even with good records from actual earthquakes, laboratory experiments may not precisely simulate real earthquake conditions, given the number of variables involved. Such uncertainties raise major concerns about the safety of dams and nuclear power plants near active faults. In an attempt to circumvent the limitations of scale modeling, the United States and Japan set up in 1979 a cooperative program

A

B

C

D

Figure 4.13

The nature of building materials, as well as design, influences how well buildings survive earthquakes. (A) A 100-year-old adobe structure completely collapsed in the 2010 Chilean earthquake. (B) This concrete frame office building was reduced to rubble in the 1995 Kobe earthquake, while an adjacent building stands nearly intact. (C) Building damage in Christchurch New Zealand was intensified because of previous structural damage from a nearby quake in 2010. (D) This building has pancaked with one concrete floor collapsing on another; Port-au-Prince, Haiti, 2010 earthquake.

(A) U.S. Geological Survey/Photograph by Walter D. Mooney; (B) Dr. Roger Hutchison, courtesy NOAA; (C) New Zealand Defence Force, courtesy NOAA/NGDC; (D) Eric Quintera, IFRC, courtesy NOAA/NGDC

to test earthquake-resistant building designs. The tests have included experiments on full-sized, multistory structures, in which ground shaking is simulated by using hydraulic jacks. The experiments are ongoing, and should continue to lead to improvements in design for earthquake resistance.

A further complication is that the same building codes cannot be applied everywhere. For example, not all earthquakes produce the same patterns of ground motion. The 1994 Northridge earthquake underscored the dependence of reliable design on accurately anticipating that motion. The epicenter was close to that of the 1971 San Fernando earthquake that destroyed many freeways (**figure 4.14A**). They were rebuilt to be earthquake-resistant, assuming the predominantly strike-slip displacement that is characteristic of the San Andreas. The Northridge quake, however, had a strong dip-slip component because it occurred along a previously unmapped, buried thrust fault, and many of the rebuilt freeways collapsed again (**figure 4.14B**).

It is important for us to consider not only how structures are built but also what materials lie underneath them. Buildings built directly on bedrock seem to suffer far less damage than those built on thick soil or other unconsolidated materials. In the

A

Figure 4.14

Freeway damage from the 1971 San Fernando earthquake (A) was repeated in the 1994 Northridge quake (B) because the design of the rebuilt overpasses did not account for the ground motion associated with the responsible fault.

(A) U.S. Geological Survey Department of the Interior/USGSU.S. Geological Survey; (B) U.S. Geological Survey/Photograph by M. Celebi

B

1906 San Francisco earthquake, buildings erected on filled land reclaimed from San Francisco Bay suffered up to four times more damage from ground shaking than those built on bedrock. The same general pattern was repeated in the 1989 Loma Prieta earthquake (**figure 4.15**). The extent of damage in Mexico City resulting from the 1985 and 2017 Mexican earthquakes was partly a consequence of the city being underlain by weak layers of volcanic ash and clay. Most smaller and older buildings lacked the deep foundations necessary to reach more-stable sand layers at depth. Many of these buildings collapsed completely. In the 1985 quake, Acapulco suffered far less damage even though it was much closer to the earthquake's epicenter because it stands firmly on bedrock. Ground shaking is commonly amplified in unconsolidated materials (**figure 4.16**). The fact that underlying geology influences surface damage is demonstrated by the observation that intensity is not directly correlated with proximity to the epicenter (**figures 4.11** and **4.17**)

  The characteristics of the earthquakes in a particular region also must be taken into account. For example, severe earthquakes are generally followed by many **aftershocks,** generally smaller earthquakes that follow the main shock and

Figure 4.15

Car crushed under third story of a failed building when the first two floors collapsed and sank. Some of the most severe damage in San Francisco from the 1989 Loma Prieta earthquake was here in the Marina district, built on filled land subject to liquefaction.

J.D. Nakata/U.S. Geological Survey

represent rock adjusting to the primary displacement. The main shock usually causes the most damage, but when aftershocks are many and are nearly as strong as the main shock, they may also cause serious destruction, particularly to structures already

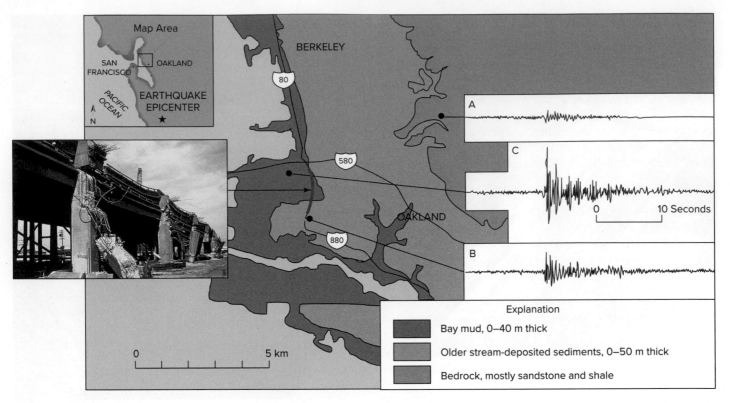

Figure 4.16

Seismograms show that ground shaking from the Loma Prieta earthquake, only moderate in bedrock (trace A), was amplified in younger river sediments (B) and especially in more recent, poorly consolidated mud (C). Inset photo is collapsed double-decker section of I-880 indicated by red line on map.

Source: R. A. Page et al., Goals, Opportunities, and Priorities for the USGS Earthquake Hazards Reduction Program, U.S. Geological Survey Circular 1079. Inset: D. Keefer/U.S. Geological Survey.

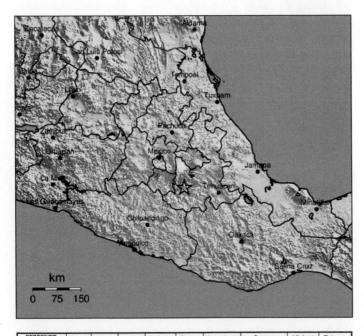

PERCEIVED SHAKING	Not felt	Weak	Light	Moderate	Strong	Very strong	Severe	Violent	Extreme
POTENTIAL DAMAGE	none	none	none	Very light	Light	Moderate	Moderate/Heavy	Heavy	Very Heavy
PEAK ACC.(%g)	<.17	.17-1.4	1.4-3.9	3.9-9.2	9.2-18	18-34	34-65	65-124	>124
INSTRUMENTAL INTENSITY	I	II-III	IV	V	VI	VII	VIII	IX	X+

Figure 4.17

Distribution of intensities for the 2017 magnitude-7.1 earthquake in Mexico (epicenter marked by star). These intensities are determined instrumentally from the maximum measured ground acceleration, reported as a percentage of g, Earth's gravitational acceleration constant.

Image courtesy U.S. Geological Survey Earthquake Hazards Program.

weakened by the main shock. And aftershocks can be very large. The magnitude-7.1 earthquake that struck Mexico City in 2017 may actually have been an aftershock of a magnitude-8.1 event that occurred offshore to the south 11 days earlier.

The duration of an earthquake also affects how well a building survives it. In reinforced concrete, ground shaking leads to the formation of hairline cracks, which then widen and develop further as long as the shaking continues. A concrete building that can withstand a one-minute main shock might collapse in an earthquake in which the main shock lasts three minutes. Many building codes are designed for a less-than-30-second main shock but experience shows us that they can last ten times that long.

Even the best building codes are typically applied only to new construction. Adobe is no longer used for construction in Chile, but older adobe buildings remain at risk; recall **figure 4.13A**. When a major city is located near a fault zone, thousands of vulnerable older buildings may already have been built in high-risk areas. Retrofitting is expensive. In the San Francisco region, since 1989 in a never-ending process, $2.5–2.8 billion is spent annually to retrofit or replace buildings. Many legislative bodies are reluctant to require such efforts, even for municipal buildings constructed in fault zones.

How do you make a building safer from earthquake shaking?

- Build it with ductile materials.
- Ensure good drainage.
- Give it is flexible foundation.
- Add dampers to decrease vibrations.
- Add structural reinforcement.

There is also the matter of public understanding of what "earthquake-resistant" construction means. It is common in areas with building codes or design guidelines that address earthquake resistance in detail to identify levels of expected structural soundness. For most residential and commercial buildings, "substantial life safety" may be the objective: the building may be extensively damaged, but casualties should be few (see **figure 4.19**). The structure, in fact, may have to be razed and rebuilt. Designing for "damage control" means that in addition to protecting those within, the structure should be repairable after the earthquake. This will add to the up-front costs, though it should save both repair costs and down time after a major quake. The highest (and most expensive) standard is design for "continued operation," meaning little or no structural damage at all; this would be important for hospitals, fire stations, and other buildings housing essential services, and also for facilities that might present hazards if the buildings failed (nuclear power plants, manufacturing facilities using or making highly toxic chemicals). This is the sort of standard to which the Trans-Alaska Pipeline was designed; it was built with a possible magnitude-8 earthquake on the Denali fault in mind, as shown in **figure 4.12**. It came through the magnitude-7.9 earthquake in 2002, with its 14 feet of horizontal fault displacement, without a break. If you live or buy property in an area where earthquakes are a significant concern, you would be well-advised to find out just what level of earthquake resistance is promised in a building's design.

Additional factors can leave some populations at particular risk. Haiti is an economically disadvantaged country where other pressing problems mean little attention is paid to developing building codes for earthquake resistance, few contractors are qualified to build structures appropriately, and most people cannot afford even minimal added building costs for enhanced earthquake safety. Further, because hurricanes are a regular threat, concrete roofs have been viewed as superior to lighter sheet-metal roofs that are more easily blown off by high winds. Unfortunately, in the 2010 earthquake, buildings collapsing under those heavier concrete roofs resulted in unusually high numbers of casualties for a quake that size; see **figure 4.13D**.

Figure 4.18

This 55-story building in Oskaka, Japan, was retrofitted after the 2011 earthquake to make it more resilient to shaking.

Mehmet Çelebi, USGS

Figure 4.19

The concept of substantial life safety, and its limitations, are illustrated by this relatively modern building damaged in the 2008 Great Sichaun, China, earthquake. Although the first story did not fare well, the majority of the structure remained intact, meeting the goal. Cases in which this goal clearly was not met are illustrated in such examples as **figure 4.13B** and **D** and the inset photo in **figure 4.16**.

USGS photo by Ying Ying Huang

A

B

Figure 4.20

(A) The 2002 Denali earthquake triggered this landslide in Alaska. (B) Pan-American Highway in El Salvador blocked by a landslide from a 2001 earthquake.

(A) USGS Earthquakes Hazards Program; (B) Photo by E.L. Harp, U.S. Geological Survey

Ground Failure

Landslides (**figure 4.20**) can be a serious secondary earthquake hazard in hilly areas. As we will cover in chapter 8, earthquakes are one of the major events that trigger failures on unstable slopes. The best solution is not to build in such areas. Even if a whole region is hilly, detailed engineering studies of rock and soil properties and slope stability may make it possible to avoid the most dangerous sites. Visible evidence of past landslides is another indication of especially dangerous areas.

Ground shaking may cause a further problem in areas where the ground is very wet such as filled-in wetlands near the coast or areas with a high water table. This problem is **liquefaction.** When wet soil is shaken by an earthquake, the soil particles may be jarred apart, allowing water to seep in between them, greatly reducing the friction between soil particles that gives the soil strength, and causing the ground to become somewhat like quicksand. When this happens, buildings can just topple over or partially sink into the liquefied soil; the soil has no strength to support them. The effects of liquefaction were dramatically illustrated in Niigata, Japan, in 1964. One multistory apartment building tipped over to settle at an angle of 30 degrees to the ground while the structure remained intact! (See **figure 4.21A**; again, "substantial life safety" was achieved, though the building was no longer habitable.) Liquefaction was likewise a major cause of damage from the Loma Prieta, Kobe, and Christchurch earthquakes (**figure 4.21B**). Telltale signs of liquefaction include sand boils, formed as liquefied soil bubbles to the surface during the quake. In some areas prone to liquefaction, improved underground drainage systems may be installed to try to keep the soil drier, but little else can be done about this hazard beyond avoiding the areas at risk. Not all areas with wet soils are subject to liquefaction; the nature of the soil or fill plays a large role in the extent of the danger.

Tsunamis and Coastal Effects

Coastal areas, especially around the Pacific Ocean basin where so many large earthquakes occur, may also be vulnerable to **tsunamis.** These are seismic sea waves; the name derives from the Japanese for "harbor wave," which is descriptive of their behavior. When an undersea or near-shore earthquake occurs, sudden vertical movement of the sea floor can create waves traveling away from that spot (**figure 4.22**). In the open sea, tsunamis have very long wavelengths and travel extremely rapidly at speeds up to 1000 km/hr (about 600 mph). Because they travel so fast, as tsunamis approach land, the water tends to pile up against the shore. The tsunami may come ashore as a very high, very fast-moving wall of water or it may develop into large breaking waves, just as ordinary ocean waves become breakers as the undulating waters touch bottom near shore. Tsunamis, however, can easily be over 15 meters high in the case of larger earthquakes. Several such waves or water surges may wash over the coast in succession. Between waves, the water may be pulled swiftly seaward, emptying a harbor or bay, and perhaps pulling unwary onlookers along. Tsunamis can also travel long distances in the open ocean. Tsunamis set off on one side of the Pacific may still cause noticeable effects on the other side of the ocean.

Given the speeds at which tsunamis travel, little can be done to warn those near the earthquake epicenter but people living some distance away can be warned in time to evacuate, saving lives, if not property. In 1948, two years after a devastating tsunami hit Hawaii, the U.S. Coast and Geodetic Survey established the Pacific Tsunami Early Warning System, based in Hawaii. Whenever a major earthquake occurs in the Pacific region, sea-level data are collected from a series of monitoring stations around the Pacific. If a tsunami is detected, data on its source, speed, and estimated time of arrival can be relayed to areas in danger, and people can be evacuated as necessary.

C

Figure 4.21

(A) Effects of soil liquefaction during an earthquake in Nilgata, Japan, 1964. The buildings, which were designed to be earthquake-resistant, simply tipped over intact. (B) Liquefaction blew holes in roads during the 2011 earthquake in Christchurch, New Zealand; this van became trapped in one. (C) Roads built on wet, sediment-rich lowlands liquified in the November 2018 Anchorage (AK) earthquake.

(A) National Oceanic and Atmospheric Administration (NOAA); (B) Steve Taylor (Ray White), courtesy NOAA/National GeophysicalData Center. (C) Rob Witter Research Geologist/U.S. Geological Survey

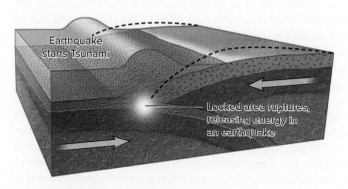

Figure 4.22

Tsunamis are created at subduction zones when the overlying plate moves suddenly in an earthquake, pushing the water above it upward. The deeper the water, the bigger the wave.

Unfortunately, individuals' responses to such warnings may be variable: some ignore the warnings, and some even go closer to shore to watch the waves, often with tragic consequences.

There was no tsunami warning system in the Indian Ocean prior to the devastation of the tsunami from the 2004 Indian Ocean earthquake (see Case Study 4.1). Even so, seismologists outside the area, noting the size of the quake, alerted local governments to the possibility that such an enormous earthquake might produce a sizeable tsunami. The word did not reach all affected coastal areas in time. Because the epicenter was so close to the coast, the first tsunami waves arrived within 20 minutes in some areas. In other cases, local officials chose not to issue public warnings, in part for fear of frightening tourists. Such behavior likely contributed to the death toll.

Five years after the Sumatran disaster, a tsunami warning system was established for the Indian Ocean. A similar warning system for the Caribbean region was established in 2010 by the

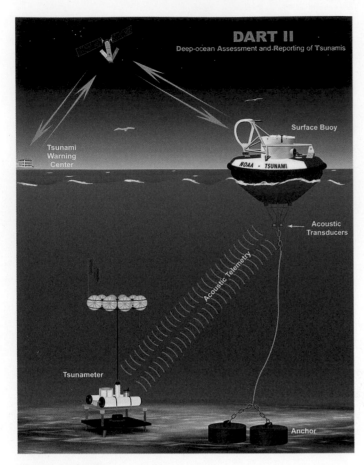

Figure 4.23

The Deep-ocean Assessment and Reporting of Tsunamis system. The tsunameter detects pressure changes on the sea floor caused by a passing tsunami. It relays the information to the buoy which sends it via satellite to the warning center.

NOAA

U.S. National Weather Service. In addition to surface buoys long used to detect the sea-surface undulations that might be due to a tsunami, warning systems now also use equipment that is sensitive to deep-water pressure changes that could indicate a tsunami passing through (**figure 4.23**). All the data are then relayed by satellite to stations where computers can quickly put the pieces together for prompt detection of the existence, location, and velocity of a tsunami so that appropriate warnings can be issued to areas at risk. The newest tsunami detectors will also use direct measurements of seafloor displacement to provide the timeliest warnings (see section 4.4).

Overall, the world is better prepared for tsunamis than we were 20 years ago based on our advances in technology. Making sure that decision makers disseminate necessary information and invest both in infrastructure and educating the general public will help to minimize future casualties. In response to the 2011 tsunami, Japan has invested billions in a network of massive seawalls that are designed to not necessarily stop tsunami waves but provide people the necessary time to move to a safe location.

Figure 4.24

Flooding in Portage, Alaska, due to tectonic subsidence during the 1964 earthquake.

U.S. Geological Survey Department of the Interior/USGSU.S. Geological Survey/Photograph by G. Plafker

Even in the absence of tsunamis, there is the possibility of coastal flooding from sudden subsidence as plates shift during an earthquake. Areas that were formerly dry land may be permanently submerged and become uninhabitable. Conversely, uplift of sea floor may make docks and other coastal structures useless. The 1964 Great Alaskan earthquake is infamous for many reasons, but one of them is the significant vertical change in land elevations (up to 10 feet) that occurred which scientists soon realized was strong evidence for plate tectonics, a fairly new concept at the time.

Broken Lines— Fire, Disease, and Power Loss

A secondary hazard of earthquakes in cities is fire, which may be more devastating than ground movement (**figure 4.25**). In the 1906 San Francisco earthquake, 70% of the damage was due to fire, not simple building failure. Fires occur because fuel lines and tanks and power lines are broken, touching off flames and fueling them. At the same time, water lines also are broken, leaving no way to fight the fires effectively, and streets fill with rubble, blocking fire-fighting equipment. In the 1995 Kobe earthquake, broken pipelines left firefighters with only the water in their trucks' tanks to battle more than 150 fires. Putting numerous valves in all water and fuel pipeline systems helps to combat these problems because breaks in pipes can then be isolated before too much pressure or liquid is lost. Broken water and sewer lines can also lead to a secondary hazard—waterborne disease. The CDC reports that over 820,000 people contracted cholera following the 2011 Haiti earthquake; nearly 10,000 died.

A

B

C

Figure 4.25

(A) Classic panorama of San Francisco in flames, five hours after the 1906 earthquake. (B) Five hours after the 1995 Kobe earthquake, the scene is similar, with many fires burning out of control. (C) In 2011, when the tsunami hit Otsuchi, Japan, it swept away the gas station, starting a fire that destroyed the entire town.

(A) NOAA/NGDC; (B) Dr. Roger Hutchison/ NOAA/NGDC; (C) Japanese Red Cross; all courtesy NOAA/National Geophysical Data Center.

One of the more recent concerns with future earthquakes is a loss of power due to broken utility lines. California's HayWired Earthquake Scenario is based on a hypothetical 7.0 magnitude along the Hayward fault; it estimates a disruption of the power grid and water supply for weeks, and that 22,000 people could be trapped in stalled elevators for days. An actual magnitude-7.0 earthquake in Mexico in September 2021 left 1.6 million people in a blackout. Depending on the time of the year, a power outage that lasts just a few days can be life-threatening. Any widespread retrofitting of power systems in large cities will likely take decades, so many municipalities encourage people to prepare for such outages themselves. Earthquake resilience is just one of the needs for a redesigned power grid. We will address some of the other in chapters 14 and 15.

4.4 Prediction, Forecasting, and Warnings

Learning Objectives

- Explain the danger of a seismic gap
- List different earthquake precursors
- Summarize earthquake forecasting status
- Describe the earthquake cycle concept
- Describe a slow slip earthquake
- Explain earthquake early warnings and their efficacy
- List ways to improve public understanding of earthquake hazards

The number of people living in earthquake hazard zones nearly doubled from 1975 to 2015, with over 2.7 billion people exposed to significant earthquake risk. The ever-expanding population is causing people to live in places that are not best-suited for habitation for many reasons. Regarding these numbers, the prediction of major earthquakes could certainly save lives. But the fact remains that we cannot predict earthquakes like we can the weather, or even flooding. Scientists understand where earthquakes will occur and have made progress in how often they occur. In the following sections, we will review the newest techniques used to understand earthquake behavior and where scientists stand in forecasting and predicting earthquakes, and issuing warnings.

Seismic Gaps

Maps of earthquake epicenters along major faults across the globe show that there are stretches with little or no seismic activity, while earthquakes continue along other sections of the same fault zone. Such quiescent sections of otherwise-active fault zones are called **seismic gaps.** They represent locked sections of faults along which friction is preventing slip. These areas may be sites of future serious earthquakes. On either side of a locked section, stresses are being released by earthquakes. In the seismically quiet section, the lack of slippage means the stresses are simply building up. The concern is that the accumulated energy will become so great that when that locked section of fault finally does slip again, a very large earthquake will result.

Recognition of these seismic gaps makes it possible to identify areas in which large earthquakes may be expected in the future. The subduction zone megathrust boundary marked by the Sunda Trench (see **figure CS 4.1.1**) had been relatively quiet seismically for many decades before the December 2004 Indian Ocean quake and tsunami; areas farther south that are still locked may be the next to let go catastrophically. The 1989 Loma Prieta, California, earthquake occurred in what had been a seismic gap along the San Andreas fault. A series of three major earthquakes in 2020 and 2021 occurred in the Shumagin Gap region along the Aleutian Trench (**figure 4.26**). Seismologists have long been concerned that this locked area could produce the next great Alaska earthquake and tsunami. It appears that much of the stress has been released through this recent activity. The study of how much stress can build up and how long that takes is a tool for anticipating major earthquakes that will fill such gaps abruptly, and is discussed more fully later in this section.

Earthquake Precursors and Prediction

In its early stages, earthquake prediction was based particularly on the study of earthquake **precursor phenomena,** recordable events that happen or rock properties that change prior to an earthquake. Many different possibilities have been examined. For example, the ground surface may be uplifted and tilted prior to an earthquake. Geophysical properties such as electrical resistivity or seismic-wave velocities in rocks near the fault may change before an earthquake. Changes have been observed in

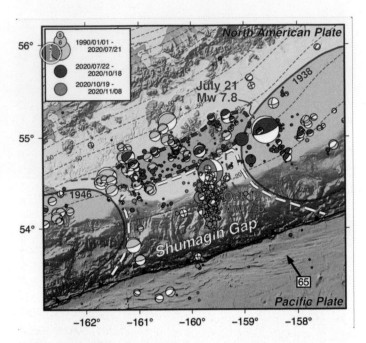

Figure 4.26

This USGS storymap shows M3.5 and larger seismic events in the Shumagin Gap region of the Aleutians Islands from 1900 to August 2021. Diameter of each circle represents the relative magnitude of the earthquake. Blue and orange circles represent the two major 2020 earthquakes. https://earthquake.usgs.gov/storymap/index-ak2020-21.html

K. Wood, L. Stern/USGS

groundwater levels and in the subsurface concentrations of radon gas. Sensitive instruments can measure elastic strain accumulating across a fault.

The hope, with the study of precursor phenomena, has been that one could identify patterns of change that could be used with confidence to issue earthquake predictions precise enough to allow precautionary evacuations or other preparations. Scientists have, up until very recently, not been very successful in interpreting the data patterns that might indicate an imminent earthquake. New approaches and techniques that involve machine learning, citizen science, global navigation satellite systems (GNSS), and the deployment of more instruments on the sea floor are offering compelling evidence for what we can learn from, and how we can use, precursors.

In late 2018, researchers from Europe and the United States used a not-so-new idea in an attempt to breakthrough the precursor stagnancy. They started a competition in which teams would create machine learning models in order to predict when simulated earthquakes would occur on a laboratory fault-making machine. Multiple teams succeeded in doing just that, and some with great accuracy. The models were even able to filter out and learn from seismic noise created by human activities. Will we be able to apply such models to real earthquake prediction? Machine learning is already being used to study creeping faults in the Cascadian subduction zone. What seismologists previously thought as a lack of earthquake precursors might rather have been our inability to detect and see patterns hidden in huge reams of highly variable data.

Other recent studies are taking advantage of the citizen science platform Zooniverse. Volunteer researchers listened to records of seismograms converted to audible frequencies and were asked to identify specific seismic events. Their accuracy at doing so was greater than that of the algorithm used by the leading scientist, and has helped to identify characteristic earthquake waveforms and signals, information which could later be used to indicate the imminent nature of a large quake.

Scientists are taking advantage of the ever-growing network of satellites to better understand the deformation of Earth's crust. In places such as the Cascadian Fault Zone and the Japan Trench, areas known to create devastating earthquakes and tsunamis, researchers are deploying GNSS-based seismic monitoring systems on the sea floor that will accurately measure crustal deformation. The scientists can then use this information to calculate slip rates along the megathrust boundaries. Knowing the mechanics of fault motion will allow us to determine what distinct movements are indicative of impending slippage and how much strain accumulates before it is released. They will also have access to detailed information as ruptures occur, immediately knowing the earthquake magnitude based on displacement.

Any information gathered immediately before or during an earthquake that might give us just a few more minutes to seek cover, get out of a building, slow down trains, or turn off gas lines is successful prediction, even if it is very short-term (see the "warning" section below). In a final example, scientists have figured out that the magnetic field generated by a tsunami can be detected one minute before the actual waves are. We are not only gathering more precursor information, but we are finding better, creative, and more efficient ways at interpreting the data to better protect people.

Current Status of Earthquake Prediction

Very few countries have government-sponsored earthquake prediction programs. Such programs typically involve intensified monitoring of active fault zones to expand the observational data base, coupled with laboratory experiments designed to increase understanding of precursor phenomena and the behavior of rocks under stress. Even with these active research programs, scientists cannot monitor every area at once. Groups such as the Collaboratory for the Study of Earthquake Predictability are working to promote research on earthquake prediction and help government agencies assess the feasibility of prediction, all through data sharing and collaborative work across borders. They currently have testing regions and centers in Japan, New Zealand, the United States, and Europe.

The director of the U.S. Geological Survey has had the authority to issue warnings of impending earthquakes and other potentially hazardous geologic events (volcanic eruptions, landslides, and so forth). A National Earthquake Prediction Evaluation Council (NEPEC) reviews scientific evidence that might indicate an earthquake threat and makes recommendations to the director regarding the issuance of appropriate public statements. These statements could range in detail and immediacy from a general notice to residents in a fault zone of the existence and nature of earthquake hazards there, to a specific warning of the anticipated time, location, and severity of an imminent earthquake. In early 1985, the panel made its first endorsement of an earthquake prediction; the results are described in Case Study 4.2.

Earthquake prediction with the help of citizen volunteers is not a new idea. In the 1970s in the People's Republic of China, tens of thousands of amateur observers and scientists worked on earthquake prediction. In February 1975, after months of smaller earthquakes, radon anomalies, and increases in ground tilt followed by a rapid increase in both tilt and microearthquake frequency, Chinese scientists predicted an imminent earthquake near Haicheng in northeastern China. The government ordered several million people out of their homes into the open. Nine and one-half hours later, a major earthquake struck, and lives were saved because people were not crushed by collapsing buildings. The next year, they concluded that a major earthquake could be expected near Tangshan, but could only say that the event was likely to occur sometime during the following two months. When the earthquake—magnitude over 8.0 with aftershocks up to magnitude 7.9—did occur, there had been no sudden change in precursor phenomena to permit a warning, and hundreds of thousands of people died. Did using macroscopic phenomena that included evidence such as odd animal behavior and fireballs in the sky actually work to predict an earthquake in the first case?

In the United States, earthquake predictions have not yet become reliable enough to prompt such actions as large-scale evacuations. We are just not able to say something like "There will be a magnitude-6.2 earthquake along this fault on Tuesday next week in the afternoon." In the near term, it seems that the more feasible approach is earthquake *forecasting*, identifying levels of earthquake probability in fault zones within relatively broad time windows. Such an example of a forecast would be "There is a 70% chance of a magnitude-7.0 earthquake in this region in the next 30 years." This at least allows for long-term preparations, such as structural improvements. Ultimately, precise predictions that save more lives will require better recognition and understanding of precursory changes discussed above, though recent studies have suggested that very short-term early *warnings* may have promise. "An earthquake has been detected. Shaking will begin in 1 minute. Move to safety immediately." We will explore early warning systems in more depth below.

The Earthquake Cycle and Forecasting

Studies of the dates of large historic and prehistoric earthquakes along major fault zones suggest that they are broadly periodic, occurring at more-or-less regular intervals (**figure 4.27**). This is interpreted in terms of an **earthquake cycle** that would occur along a non-creeping fault segment: a period of stress buildup,

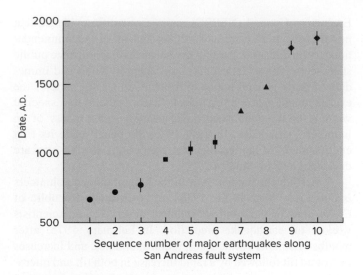

Figure 4.27

Periodicity of earthquakes assists in prediction/forecasting efforts. Long-term records of major earthquakes on the San Andreas show both broad periodicity and some tendency toward clustering of two to three earthquakes in each active period. Prehistoric dates are based on carbon-14 dating of faulted peat deposits.

Source: Data from U.S. Geological Survey Circular 1079.

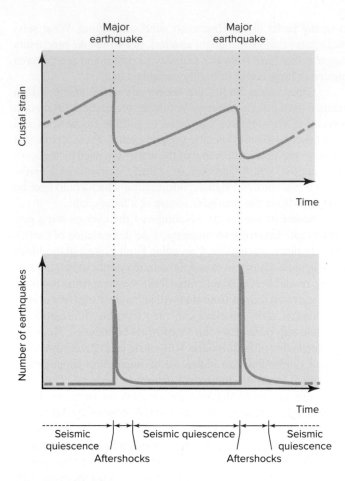

Figure 4.28

The earthquake cycle concept.

Source: After U.S. Geological Survey Circular 1079, modified after C. DeMets (pers. comm. 2006).

sudden fault rupture in a major earthquake, followed by a brief interval of aftershocks reflecting minor lithospheric adjustments, then another extended period of stress buildup (**figure 4.28**). The fact that major faults tend to break in segments is well documented; examples include the San Andreas, the thrust fault along the Sunda Trench, and the Anatolian fault zone in Turkey. The rough periodicity can be understood in terms of two considerations. First, assuming that the stress buildup is primarily associated with the slow, ponderous, inexorable movements of lithospheric plates, which at least over decades or centuries will move at fairly constant rates, one might reasonably expect an approximately constant rate of buildup of stress or accumulated elastic strain. Second, the rocks along a given fault zone will have particular physical properties, which allow them to accumulate a certain amount of strain energy before failure, or fault rupture, and that amount would be approximately constant from earthquake to earthquake. Those two factors together suggest that periodicity is a reasonable expectation. Once the pattern for a particular fault zone is established, one may use that pattern, together with measurements of strain accumulation in rocks along fault zones, to project the time window during which the next major earthquake is to be expected along that fault zone and to estimate the likelihood of the earthquake's occurrence in any particular time period. Also, if a given fault zone or segment can store only a certain amount of accumulated energy before failure, the maximum size of earthquake to be expected can be estimated.

This is, of course, a simplification of what can be a very complex reality. For example, though we speak of "the San Andreas fault," it is not a single simple, continuous slice through the lithosphere, nor are the rocks it divides either identical or homogeneous on either side. Both stress buildup and displacement are distributed across a number of faults and small blocks of lithosphere. Earthquake forecasting is correspondingly more difficult. Nevertheless, the U.S. Geological Survey's Working Group on California Earthquake Probabilities publishes forecasts for different segments of the San Andreas (**figure 4.29**). Note that in this example, the time window in question is a 30-year period. Such projections, longer-term analogues of the meteorologist's forecasting of probabilities of precipitation in coming days, are revised as large and small earthquakes occur and more data on slip in creeping sections are collected. For example, after the 1989 Loma Prieta earthquake, which substantially shifted stresses in the region, the combined probability of a quake of magnitude 7 or greater either on the peninsular segment of the San Andreas or on the nearby Hayward fault (locked since 1868) was increased from 50% to 67%. Forecasts are also being refined through recent research on fault interactions and the ways in which one earthquake can increase or decrease the likelihood of slip on another segment, or another fault entirely.

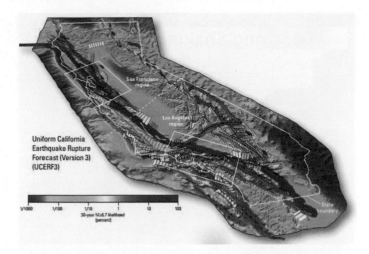

Figure 4.29

Earthquake forecast map issued in 2014 by the Working Group on California Earthquake Probabilities, indicating the 30-year probability of a magnitude-6.7 or greater earthquake.

Image after U.S. Geological Survey Fact Sheet 2015-3009.

Slow-Slip Earthquakes

A new kind of fault behavior was discovered at the end of the twentieth century, one occurring primarily in subduction zones. The original work on this subject hypothesized that rocks at depths of approximately 30–40 kilometers exist in an intermediate zone where they neither behave plastically and easily slide like they do in the high temperature and pressure conditions below, nor brittlely by either creeping or locking in the lower temperatures and pressures above. Instead, here rocks would stick for a while, then release a great deal of energy over a period of days to weeks. Slip rates were far faster than the creep of plate motion, but so much slower than a sudden earthquake that they could release energy equivalent to a magnitude 7 or greater quake without any of the usual seismicity and catastrophic damage. The term **slow-slip earthquake** is used to describe this phenomenon, which doesn't fit our usual image of an earthquake at all.

Intense interest in this field has deepened our understanding of slow-slip earthquakes. A recent project in the Hikurangi subduction zone off the coast of New Zealand's North Island demonstrated that the mechanics of slow-slip likely are affected by the nature of the rocks involved, the juxtaposition of soft sediments and hard rocks, and the presence of pressurized fluids coming from the rocks (**figure 4.30**). This research also showed that slow-slip can occur in much shallower conditions (less than 15 kilometers) than previously thought.

As with other seismic phenomena, we can determine the amount of movement with precise GNSS measurements. These events also are commonly accompanied by an unusual type of seismic tremor. We have detected them in many subduction zones around the Pacific Ocean basin. They seem to occur periodically in a given fault zone, at intervals of months to years, and they may last longer than originally thought. Further

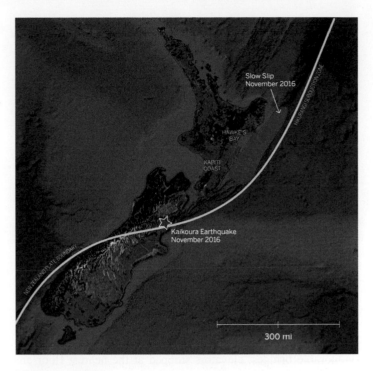

Figure 4.30

Location of the slow-slip area in the Hikurangi subduction zone, New Zealand.

The University of Texas at Austin Jackson School of Geosciences

research has shown a wide range of behaviors in different fault zones. Not all show periodic slow-slip events at regular intervals, and not all slow-slip earthquakes are accompanied by seismic tremors. We also don't know precisely what controls or triggers a slow-slip event. Perhaps the most pressing question about slow-slip earthquakes is how they are related to megathrust earthquakes in the same fault zone. Certainly the slip shifts stresses in the fault zone, which may bring locked sections closer to rupture, but while some slow-slip events may have been followed by major damaging earthquakes, some are triggered by such events. We still have a lot to learn before slow-slip events might be used to predict the next megathrust earthquake.

Earthquake Early Warnings

When hurricanes threaten, public officials agonize over issues such as when or whether to order evacuation of threatened areas; how, logistically, to make it happen; and how to secure property in evacuated areas. Similar issues will arise if and when scientific earthquake predictions become feasible. In the meantime, a new idea has surfaced for saving lives and property damage. Earthquake early warnings (EEWs) can be issued a few seconds, or perhaps a few minutes, before damaging seismic waves arrive. Tsunami warnings may provide even a longer time to move to higher and safer ground.

What good can that do? Potentially, quite a lot. Trains can be slowed and stopped, traffic lights adjusted to get people off

Figure 4.31

A warning of a few seconds is enough to protect yourself from falling objects.

Shutterstock/Pixel-Shot

Strong Shaking Expected

28

VI 7.8

On your screen: ShakeAlert

1 Real-time tracking of seismic waves from quake's epicenter.
2 Real-time tracking of the fault rupture (updates intensity).
3 Your current location tracked by GPS.
4 Seconds remaining before seismic waves reach you.
5 Expected intensity of quake at your current location.
6 Estimated magnitude of quake.
7 Intensity scale.

Figure 4.32

Example earthquake warning using the MyShake app.

Source: Earthquake Hazards/USGS

vulnerable bridges, elevators stopped at the nearest floor and their doors opened to let passengers out. Automated emergency systems can shut off valves in fuel or chemical pipelines and initiate shutdown of nuclear power plants. Surgeons can lift scalpels away from patients. People in homes or offices can quickly take cover under sturdy desks and tables. Those in coastal areas at risk of tsunamis can seek higher ground.

The idea is based on the relative travel times of seismic waves. The damaging surface waves travel more slowly than the P and S waves used to locate earthquakes. Therefore, earthquakes can be located seconds after they occur, and, at least for places at some distance from the epicenter, warnings transmitted to areas at risk before the surface waves hit. However, to decide when these warnings should be issued and automated-response systems activated, one wants to know the earthquake's magnitude as well. Scrambling to react to every one of the thousands of earthquakes that occur each year would create chaos. New seismic systems based on GNSS will provide immediate assessment of earthquake magnitude based on displacement along the fault. Crowdsourcing will also play a role as billions of smartphones have the potential to act as sensing platforms with real-time telemetry.

There are currently at least 12 countries/geographic areas that have early warning systems, and there are more in development. The most advanced program is in Japan. The Japan Meteorological Agency has issued public alerts for larger quakes since 2007 using a network of thousands of seismic instruments. This system issues warnings to public citizens and other entities such as train systems, factories, and government agencies. The EEW worked successfully in the 2011 Tohoku earthquake and is credited with saving thousands of lives. The first P waves hit the network about 23 seconds after the quake occurred, and public warnings were issued 8 seconds later. This gave Tokyo residents about a minute before the surface waves hit, and the slower-moving tsunami waves reached the coast more than 20 minutes

later. The Japanese have made improvements to their warning system in response to technical limitations of the algorithms used in the 2011 event. They have since added 2 new ground motion prediction methods and the S-net, an ocean-bottom seismic network that will result in even earlier warnings.

EEW systems are currently used in Washington, California, and Oregon. Earthquake Warning California is a publicly available statewide warning system that uses the MyShake app and wireless emergency alerts (WEAs, e.g., a text message). This system provides people with seconds to tens of seconds to "Drop, Cover, and Hold On." It also uses data collected from phones to provide more precise warnings to the people who need it (**figure 4.32**). At the same time, it gathers data to better understand earthquake dynamics and allows people to report on damage incurred which is used to estimate the earthquake's intensity.

Public Response to Earthquake Hazards

A key step toward making earthquake-hazard warning systems work is increasing public awareness of earthquakes as a hazard. In the People's Republic of China, vigorous public-education programs and several major modern earthquakes have made earthquakes a well-recognized hazard. Earthquake drills, which stress orderly response to an earthquake warning, are held in Japan on the anniversary of the 1923 Tokyo earthquake, in which nearly 150,000 people died. The Great Shake Out is a multi-organizational effort that began in California to help

Figure 4.33

The Great Shake Out encourages people to prepare for an earthquake, and provides the information necessary to do so.

Southern California Earthquake Center (SCEC)

people prepare, survive, and recover from earthquakes. In 2021, nearly 31 million people from schools, religious groups, hospitals, businesses, and governmental groups worldwide registered for the annual drill (**figure 4.33**).

Why do people choose to live in areas that are at high risk from earthquake hazards? Why would someone ignore a warning? There are many reasons including ignorance (they are unaware), a false sense of security, apathy, no past experience with the phenomenon, a distrust of the science or entity (person or group) providing the information, and increasingly, the inability to critically analyze the vast amounts of information available to us. These reasons apply to every hazard we will discuss throughout this text. Personal risk assessment is just that—personal.

In Japan in 2011, many people who were well aware of the danger of tsunamis did not evacuate when the warning was issued because their communities had built protective seawalls. In 1990, a non-geological scientist "predicted" a major earthquake on a specific day on the New Madrid fault based on the alignment of the Sun, Moon, and Earth; it resulted in panic, a run on emergency supplies, and school closings. The earthquake did not occur. In 2012, seven members of Italy's Civil

Protection Department were tried and convicted on charges of manslaughter for not accurately using foreshocks to predict and warn people of a 6.3-magnitude earthquake (in 2009) that killed 308 people. The verdict was appealed and all but one, a government official, was acquitted. Clearly, education is the key to improving public understanding of hazards in order to save lives. We all need a working knowledge of Earth processes so we can make sound decisions that affect ourselves and our communities.

Earthquake prediction aside, changes in land use, construction practices, and siting could substantially reduce the risk of property damage from earthquakes. Earthquake-prone areas need comprehensive disaster-response plans. Within the United States, California is a leader in taking necessary actions in these areas. However, even there, most laws aimed at earthquake-hazard mitigation have been passed in spurts of activity following sizable earthquakes. It has been almost 30 years since a large quake affected the west coast.

4.5 Earthquake Control

> ### Learning Objectives
>
> - List human activities that cause earthquakes
> - Describe how fluid injection would induce earthquakes
> - Recall the purpose of a seismic shield

Since earthquakes are ultimately caused by forces strong enough to move continents, human efforts to stop earthquakes from occurring would seem to be futile. However, there has been speculation about the possibility of moderating some of earthquakes' most severe effects. If locked faults, or seismic gaps, represent areas in which energy is accumulating, perhaps enough to cause a major earthquake, then releasing that energy before too much has built up might prevent a major catastrophe. Humans can and have created earthquakes as a by-product of other activities. But can we actually control them in order to make earthquake zones safer?

Induced Seismicity

In the mid-1960s, the city of Denver began to experience small earthquakes. In time, geologist David Evans suggested a connection with an Army liquid-waste-disposal well at the nearby Rocky Mountain Arsenal. The Army denied any possible link, but a comparison of the timing of the earthquakes with the quantities of liquid pumped into the well at different times showed a very strong correlation (**figure 4.34**). The earthquake foci were also concentrated near the arsenal. The increased fluid pressure in the cracks and pore spaces in the rocks resulting from the pumping-in of fluid evidently decreased the shear strength, or resistance to shearing stress, along old faults in the bedrock. This was an example of **induced seismicity,** seismic

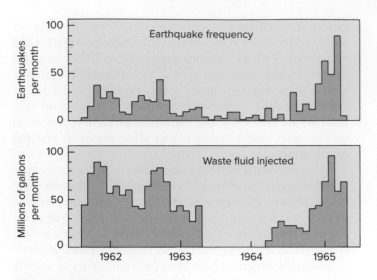

Figure 4.34

Correlation between waste disposal at the Rocky Mountain Arsenal (bottom diagram) and the frequency of earthquakes in the Denver area (top).

Source: David M. Evans, "Man-made Earthquakes in Denver," Geotimes, 10(9): 11–18 May/June 1966. Used with permission.

activity caused by human activity. Many activities can trigger earthquakes including mining, dam construction with reservoir filling, and fluid injection into the subsurface.

A sudden increase in seismicity began in 2009 in Oklahoma, a place far removed from current plate boundaries. Much like the case stated above, initially the suggestion was that human activities, this time associated with oil and gas drilling, were not involved. Eventually, geoscientists were able to demonstrate that injection of wastewater was indeed reactivating some deep faults that were hydraulically connected to the injection wells by increasing the fluid pressure in the faults. Most of the earthquakes were small, but the frequency of the earthquakes raised concerns. Recently, some wells associated with seismic events have been closed. Earthquake frequency has decreased, but Oklahoma remains one of the most likely places to experience an earthquake in the United States (see **Figure 4.35**). Induced seismicity has contributed to an increase in the risk shown for the midcontinent on recent versions of the U.S. seismic hazard map. The U.S. Geological Survey has begun to issue annual forecasts of damage from earthquakes in this country, and distinguishes between natural and induced earthquakes in so doing (**figure 4.35**).

The idea of using fluid injection along locked sections of major faults to release built-up stress has been reintroduced numerous times over the past 50 years. Unfortunately, there is no guarantee that only small earthquakes would be produced. In an area where a fault had been locked for a long time, injecting fluid along that fault could lead to the release of all the stress at once, in a major, damaging earthquake. If a great deal of energy had accumulated, it is highly unlikely that all that energy could be released entirely through small earthquakes (recall **table 4.2**). The possible casualties and damage are a tremendous concern, as are the legal and political consequences of a serious, human-induced earthquake.

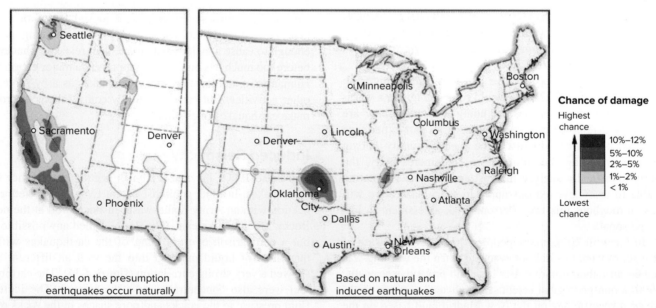

Figure 4.35

One-year forecast of earthquake damage for 2017. The high-risk area in Oklahoma reflects primarily induced earthquakes. Earthquakes of magnitude 3+ in this area increased from about 50 per year in 2010–2012 to nearly 900 in 2015.

Seismic Barriers

There are numerous ways to design a building to make it more resistant to damage from earthquake shaking. We reviewed a few of these in section 4.4. A newer idea to protect structures is to create a barrier or seismic shield. These would be independent of the building they are meant to protect. Depending on the type of surface wave involved, such barriers would trap, dampen, or deflect seismic wave energy. Seismic cloaking has yet to be deployed, remaining the subject of theoretical papers and a few field tests. While there remains no way for us to stop an earthquake from occurring, and we will likely never find a way to do so even if we determined it was a good idea, we may eventually be able to make our structures invisible to the effects of seismic waves.

4.6 Future Earthquakes in North America

Learning Objectives

- Describe Alaska, San Francisco, Los Angeles, and the Pacific Northwest as high-risk earthquake areas
- Explain the seismic risks of various intraplate areas including the central United States and Charleston, S.C.

North America has a number of distinct seismic hazard zones. Some are associated with plate boundaries, others with mountains, a few with ancient buried faults, and one is distinctly related to human activities. We have already introduced most of these areas in terms of historic seismic activity, particular risks or studies, or as examples in previous sections of this chapter. Below we will go into more depth about the areas that pose the most risk to people in the future.

Widely Recognized Risks at Plate Boundaries

Southern Alaska sits above a subduction zone. On 27 March 1964, it was the site of the second-largest earthquake of the twentieth century, estimated at M9.2. The main shock lasted three to four minutes. More than 12,000 aftershocks, many over magnitude 6, occurred during the following year. The associated uplift was up to 12 meters on land and over 15 meters in places on the sea floor, leaving some harbors high and dry and destroying habitats of marine organisms. Downwarping of 2 meters or more flooded other coastal communities. Tsunamis were responsible for about 90% of the 115 deaths. The tsunamis traveled as far as Antarctica and were responsible for sixteen deaths in Oregon and California. Landslides, both submarine and aboveground, were widespread and accounted for most of the remaining casualties (see **figure 4.36**). The death toll might have been far higher had the earthquake not occurred in the early evening on a holiday, before peak tourist or fishing

A

B

Figure 4.36

Examples of property damage from the 1964 earthquake in Anchorage, Alaska. (A) Landslide in the Turnagain Heights area, Anchorage. Notice the houses amid the jumble of down-dropped blocks in the foreground. (B) Wreckage of Government Hill School, Anchorage.

U.S. Geological Survey Photo Library, Denver, CO.

season. Alaskan earthquakes are not limited to the subduction zone, as was emphasized by the 2002 quake on the strike-slip Denali fault. Alaska experiences one magnitude-7 to -8 earthquake annually yet currently does not have an EEW system.

When the possibility of another large earthquake in San Francisco is raised, geologists debate not whether it will occur but rather when. At present, that section of the San Andreas fault has been locked for a relatively long time. The last major earthquake there was in 1906 when movement occurred along at least 300 kilometers, and perhaps 450 kilometers of the fault. If another earthquake of the size of the 1906 event were to strike the San Francisco area, an estimated 1800 deaths would occur, with tens of thousands more injured. One could expect $200 billion in property damage from ground shaking and fire alone. As we saw in section 4.4, the likelihood of a magnitude-6 or greater earthquake in this area in the next 10 years is rather high.

Events at the end of the twentieth century might suggest a greater near-term risk along the southern San Andreas, near Los Angeles, than previously thought. In the spring and summer of 1992, a set of three significant earthquakes occurred east of the southern San Andreas. Then the 1994 Northridge earthquake occurred on a previously unrecognized buried thrust fault. The southern San Andreas in this region, which last broke in 1857, has remained locked through all of this, so the accumulated strain there has not been released by this cluster of earthquakes. Some seismologists believe that their net effect has been to increase the stress and the likelihood of failure along this section of the San Andreas. Moreover, in the late 1800s, there was a significant increase in numbers of moderate earthquakes around the northern San Andreas, followed by the 1906 San Francisco earthquake. The 1992–1994 activity may represent a similar pattern of increased activity building up to a major earthquake along the southern San Andreas. The likelihood of a magnitude-7 or higher quake in southern California by 2044 is estimated by some scientists at 75%. Time will tell.

In the Pacific Northwest, the seismic risk is associated with subduction (**figure 4.37**). This region is the Cascadian subduction zone, and historical evidence shows that some past earthquakes there have been very large. In the mid-1990s, seismologists examining evidence for past major earthquakes in Japan found that a tsunami had occurred in 1700, but there was no corresponding record of a local earthquake. Seismologists in the United States, meanwhile, had identified signs of a huge

earthquake along the northwest Pacific coast at about that time. Putting all the data together led to the conclusion that the cause of both sets of observations was an earthquake in the Cascadian subduction zone, with magnitude estimated at about 9.

Subsequent studies of distinct sediment layers both in the coastal and offshore environments indicate that large, tsunami-inducing earthquakes have occurred often in the past 10,000 years, roughly every 250 years, but not since the January 1700 event. Recent measurements among the mountains of Olympic National Park show crustal shortening and uplift that indicate locking along the subduction zone to the west. Some scientists estimate a 33% probability of a major earthquake here in the next 50 years. Another earthquake the size of the event in 1700 would cause tremendous damage and loss of life among the more than 10 million people in the Seattle/Portland area and beyond. One million would be displaced and utilities would be broken for months. A bit of a wakeup call in this regard was a magnitude-6.8 earthquake that shook the region in 2001. Though no lives were lost, over $1 billion in property damage resulted.

Monitoring of the Cascadian subduction zone has revealed periodic slow-slip earthquakes there, occurring every 11 to 15 months. The largest to date lasted over six weeks, beginning near Vancouver Island and extending as far south as Seattle before it stopped. Monitoring continues as scientists strive to understand the extent to which the observed slow-slip events may signal the coming of the next major Cascadia quake.

Intraplate Earthquake Hazard Zones

People who live in the areas described above are more likely to perceive their earthquake hazard risk in a way to take some action. A more dangerous situation might exist for those people who are not conditioned to regard their area as hazardous or earthquake-prone. The central United States is an example. Though most Americans think of California when they hear the word earthquake, some of the strongest and most devastating earthquakes ever in the contiguous United States occurred in the vicinity of New Madrid, Missouri, during 1811–1812. The three strongest shocks are each estimated to have had magnitudes around 8; they were spaced over almost two months, from 16 December 1811 to 7 February 1812. Towns were leveled or drowned by flooding rivers. Tremors were felt from Quebec to New Orleans and along the east coast from New England to Georgia. Lakes were uplifted and emptied, while elsewhere the ground sank to make new lakes. Aftershocks continued for more than ten years. Total damage and casualties have never been accurately estimated. It is the severity of those earthquakes that makes the central United States a high-risk area on the U.S. seismic-risk map (**figure 4.38**).

The potential damage from even one earthquake as severe as the worst of the New Madrid shocks is enormous: over 12 million people now live in the area. Yet very few of those people are fully aware of the nature and extent of the risk. The big earthquakes there were a long time ago, beyond the memory of anyone now living. Why do earthquake occur here, so far

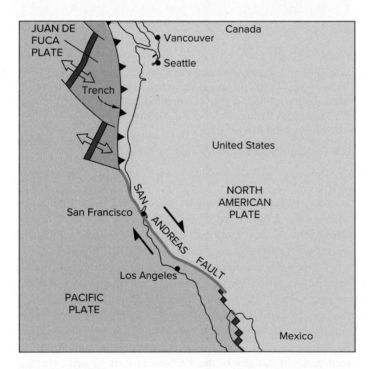

Figure 4.37

At the northern end of the San Andreas fault lies a subduction zone where the Juan de Fuca plate is moving under North American plate in the Pacific Northwest.

Modified from U.S. Geological Survey.

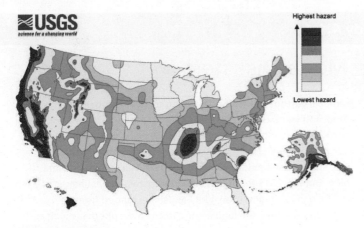

Figure 4.38

U.S. Seismic Hazard Map. The map shows peak ground accelerations having a 2 percent probability of being exceeded in 50 years, for a firm rock site. It is based on the most recent USGS models for the conterminous United States (2018), Hawaii (1998), and Alaska (2007).

Earthquake Hazards/USGS

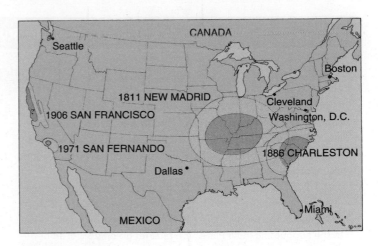

Figure 4.39

Part of the danger that midcontinent, intraplate earthquakes represent is related to the much broader areas likely to be affected, given the much more efficient transmission of seismic energy in the compact rocks that underlie these areas. Inner (dark tan) zone in each case experienced Mercalli intensity VII or above; outer (light tan), intensity VI or above.

Source: After U.S. Geological Survey Professional Paper 1240-B.

away from a plate boundary? Beneath the midcontinent is a failed rift, a place where the continent began to rift apart then stopped. These long and deep faults in the lithosphere are a zone of weakness in the continent that continues to move. This area remains a probable site for more major tremors. The effects may be far-reaching, too, when the next major earthquake comes. Seismic waves propagate much more efficiently in the solid, cold, unfractured rocks of the continental interior than in the younger, warmer, complexly faulted, jumbled rocks around the San Andreas fault (**figure 4.39**). When the next major earthquake occurs in the New Madrid fault zone, the consequences will be widespread.

There are still other hazards from buried faults in continental interiors. The 1886 Charleston, South Carolina, earthquake was a clear indication that severe earthquakes can occur in the east, as is reflected in the seismic-hazard map and in **figure 4.39**. A surprise magnitude-5.8 earthquake in Virginia in 2011 resulted in damage to the Washington Monument as well as many other structures. Much of the eastern and southeastern United States, however, is blanketed in sediment, which somewhat complicates assessment of past activity and future hazards. The mountainous western third of the United States contains numerous earthquake hazards zones associated with faults. One such is the Wasatch Fault, which runs for 390 km in Utah and Idaho. Over 2.3 million people in Utah live within 15 miles of this fault. The USGS estimates a 57% chance of a magnitude-6.0 or greater quake in the Wasatch Front region by 2066. A large quake would result in thousands of deaths, tens of thousands displaced, utilities disrupted for months, and billions in economic damages.

Figure 4.38 clearly shows that Hawaii is a very-high-risk area for earthquakes. Recall that Hawaii represents a mantle plume within a tectonic plate where magma is moving upward through shallow crustal rocks. The magma pushes on the

surrounding rocks and causes earthquakes, very few of which are large enough to cause significant damage We will discuss Hawaii in more detail in chapter 5 in terms of volcanic activity, but for now, consider that one of the best indicators of volcanic activity is an increase in the frequency and magnitude of earthquakes. The mantle plume underneath Yellowstone National Park in northwestern Wyoming not only creates the geological phenomena the park is known for, but also the seismic waves which scientist use to study the workings of the hot spot. In neither of these areas is seismic activity the primary danger to people but rather just one of the hazards that may indicate, be connected to, and/or cause a second and even more dangerous hazard.

Canada is less seismically active than the United States. Still, each year about 300 earthquakes generally less than magnitude 4 occur in eastern Canada, where a zone of crustal weakness exists in the continental interior. Most Canadian earthquakes, and the largest ones, occur in western Canadian in association with the Cascadian subduction zone and the connected Queen Charlotte transform fault that runs northward through the Haida Gwaii. Severe Canadian earthquakes are rare, excepting the January 1700 event described above. Notable earthquakes include: a magnitude-7.2 event, complete with tsunami, off the Grand Banks of Newfoundland in 1929; an earthquake swarm with largest event magnitude-5.7 in central New Brunswick in 1982; a magnitude-7.3 quake under Vancouver Island in 1946; a magnitude-7.7 earthquake in the Queen Charlotte Islands area off western Canada in 2012 that triggered a tsunami warning for Hawaii; and a total of seven quakes of magnitude 6 or greater in the area of southwestern Canada in the last century. Still, the potential for significant earthquakes clearly continues to exist, while public awareness, particularly in eastern Canada, is likely to be low.

Megathrusts Make Mega-Disasters

The tectonics of southern and southeast Asia are complex. East of the India and Australia plates lies the Sunda Trench, marking a long subduction zone. The plate-boundary fault is a *megathrust,* a very large thrust fault typical of subduction-zone boundaries.

For centuries, the fault at the contact between the Burma (micro)Plate and the India Plate had been locked (**figure CS 4.1.1**). The edge of the Burma Plate had been warped downward, stuck to the subducting India Plate. The fault let go on 26 December 2004, in an earthquake with moment magnitude of 9.1, making it the third-largest earthquake in the world since 1900. When the fault ruptured, the freed leading edge of the Burma Plate snapped upward by as much as 5 meters. The resulting abrupt shove on the water column above set off a deadly tsunami. It reached the Sumatran shore in minutes, but took hours to reach India and, still later, Africa.

How high onshore the tsunami surged varied not only with distance from the epicenter but also with coastal geometry. The highest runup was observed on gently sloping shorelines and where water was funneled into a narrowing bay. Detailed studies after the earthquake showed that in some places the tsunami waves reached heights of 31 meters (nearly 100 feet). Destruction was profound and widespread. An irony of tsunami behavior is that fishermen well out at sea were unaffected by, and even unaware of, the tsunami until they returned to port. Most of the estimated 230,000 deaths (estimates vary up to 280,000) from this earthquake were in fact due to the tsunami, not building collapse or other consequences of ground shaking. Thousands of people never found or accounted for were presumed to have been swept out to sea and drowned. Tsunami deaths occurred as far away as Africa.

Northern Japan also sits along a subduction zone. On 11 March 2011, a large section of that megathrust fault slipped by an estimated 50–60 meters, producing a magnitude-9.0 earthquake (**figure CS 4.1.2**). The quake caused severe ground shaking, some landslides, local liquefaction, fires, and a massive tsunami. The maximum run-up height of that tsunami was measured at nearly 38 meters (125 feet) in Japan, and it was eventually detected all around the Pacific basin (**figure CS 4.1.3**), still several meters high when it reached South America some 20 hours after the quake.

Predictably, damage was extensive (**figure CS 4.1.4**), but some was more serious than might have been expected. Perhaps the worst surprise occurred at Fukushima (see location in **figure CS 4.1.2**). The nuclear-power station there was planned with earthquake hazards in mind. Scientists had studied the region's earthquake history including tsunami deposits over a thousand years old, and had projected the largest likely earthquake and tsunami. The nuclear plants at Fukushima were accordingly designed to withstand an earthquake of magnitude 8.2, and the tsunami wall protecting them was built 5.7 meters high. But the 2011 earthquake was far larger, and estimates of the tsunami's height when it hit Fukushima ranged from 8 to 14 meters. And, the land along the coast dropped as much as 0.6 meters in elevation. The six reactors lost electrical power to circulate vital cooling water; several suffered core meltdown and/or hydrogen explosions; considerable radiation was released, and hundreds of thousands of people were evacuated from areas most affected by that radiation. The power station has since been permanently decommissioned. Nuclear power is explored further in chapter 15. Here we can note that the Fukushima disaster prompted a closer look at reactor design

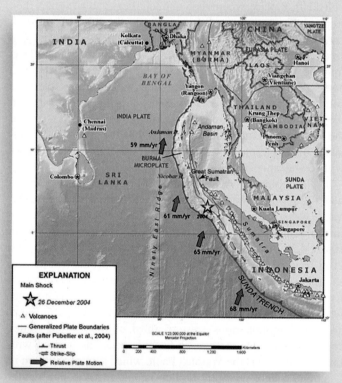

Figure CS 4.1.1

Location of the Sunda Trench.

Eric Geist, Geophysicist, USGS Pacific Coastal and Marine Science Center

and safety around the world, and caused not a few countries to move away from nuclear power.

Meanwhile, the United States has its own megathrust to worry about: the Cascadian subduction zone, discussed in this chapter. And even in subduction zones, the thrust fault is not the only danger. The earthquake that hit Mexico City in 2017 was an intraplate earthquake, occurring on a rupture in the down-going slab, not along the thrust fault proper.

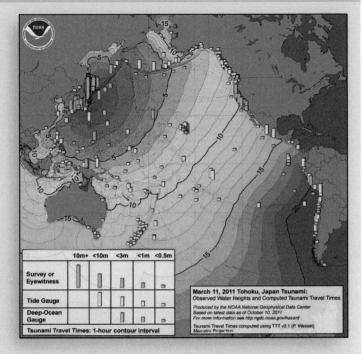

Figure CS 4.1.3

Travel times and heights for the tsunami produced by the 11 March 2011 Japan earthquake.

Map by NOAA/National Geophysical Data Center.

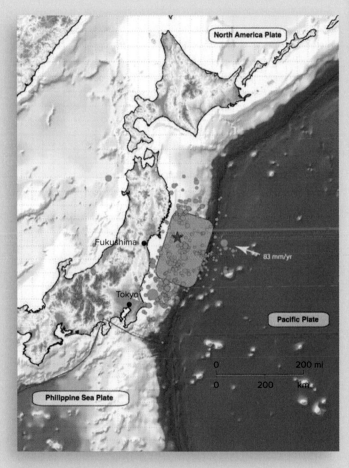

Figure CS 4.1.2

Location and size of the section of thrust fault plane that slipped in the magnitude-9.1 Japan earthquake; red star marks the epicenter.

Map after U.S. Geological Survey National Earthquake Information Center.

Figure CS 4.1.4

Wakuya Japan (March 15, 2011). A Japanese search and rescue team searches the rubble near a high-rise building.

Mass Communication Specialist 3rd Class Alexander Tidd/US Navy/DoD

Understanding Faults Better—Parkfield and SAFOD

In 1985, the USGS NEPEC made what looked like a very straightforward prediction. They agreed that an earthquake of Richter magnitude about 6 could be expected on the San Andreas fault near Parkfield, California, in January 1988 ± 5 years. The prediction was based, in large part, on the cyclic pattern of recent seismicity in the area, consistent with the earthquake-cycle model. On average, major earthquakes had occurred there every twenty-two years, and the last had been in 1966 (**figure CS 4.2.1**). Both the estimated probability and the panel's degree of confidence were very high. Yet the critical time period came and went—and no Parkfield earthquake, even fifteen years after the targeted date. Meanwhile, local residents were angered by perceived damage to property values and tourism resulting from the prediction.

The long-awaited quake finally happened in late September 2004. Despite the fact that this fault segment has been heavily instrumented and closely watched since the mid-1980s in anticipation of this event, there were no clear immediate precursors to prompt a short-term warning of it. Seismologists began combing their records in retrospect to see what subtle precursory changes they might have missed, and they launched a program of drilling into the fault to improve their understanding of its behavior.

Enter SAFOD, the San Andreas Fault Observatory at Depth. Funded by the National Science Foundation as part of a plan of geologic studies known as EarthScope, the idea is to drill through the San Andreas fault near Parkfield, collecting rock and fluid samples and installing instruments to monitor the fault zone over the long term, with a goal of better understanding the fault and how the earthquake cycle operates along it. The pilot hole was actually drilled in 2002, before the last significant Parkfield quake. The main hole makes use of technology now also used in the petroleum industry, which allows the angle of drilling to change at depth (**figure CS 4.2.2**). Thus, one starts with a vertical hole beside the fault, then angles toward it to drill across the fault zone.

In the target section, much of the fault is actually creeping, but a small section within the creeping area ruptures every few years with small (magnitude-2) earthquakes. Seismologists want to understand how that is possible. Along the fault is a zone of finely pulverized material known as fault *gouge*. Samples from the creeping zone contain small amounts of soft sheet silicates such as clay and talc that might facilitate slip. It has been suggested that the microearthquakes occur in more-coherent rock in response to increases in fluid pressure. While more drilling awaits funding, laboratory studies of samples continue. Meanwhile, monitoring instruments allow scientists to track strain, fluid pressure, and other properties along the fault zone as some parts creep and others rupture, giving real-time information about what goes on at depth in a fault zone.

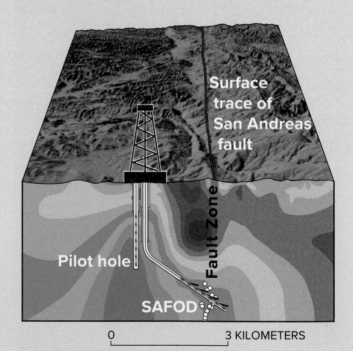

Figure CS 4.2.1

Based on the generally regular intervals separating the previous six significant earthquakes on this segment of the San Andreas, the next was predicted for 1988, plus or minus five years. But it did not occur until 2004, for reasons yet to be determined.

Figure CS 4.2.2

Cross-section of the region drilled for SAFOD. Colors reflect electrical resistivity; lower-resistivity areas (warmer colors) may contain more pore fluid to facilitate creep. Short black segments branching off low end of main hole are sites of rock cores collected; white dots are foci of minor earthquakes.

Summary

Earthquakes result from sudden slippage along fault zones in response to accumulated stress. Most earthquakes occur at plate boundaries and are related to plate-tectonic processes. Earthquake hazards include ground rupture and shaking, liquefaction, landslides, tsunamis, coastal flooding, and fires. The severity of damage is determined not only by the size of an earthquake but also by underlying geology and the design of affected structures. While earthquakes cannot be stopped, their negative effects can be limited by (1) designing structures in active fault zones to be more resistant to earthquake damage; (2) identifying and, wherever possible, avoiding development in areas at particular risk from earthquake-related hazards; (3) increasing public awareness of and preparedness for earthquakes in threatened areas; (4) refining and expanding tsunami warning systems and public understanding of appropriate response; and (5) learning enough about patterns of seismicity over time along fault zones, and about earthquake precursor phenomena, to make accurate and timely predictions of earthquakes and thereby save lives. Until precise predictions on a short timescale become feasible, longer-term earthquake forecasts may serve to alert the public to general levels of risk and permit structural improvements and planning for earthquake response, while very-short-term warnings allow automated emergency responses that save lives and property.

Key Terms and Concepts

aftershocks	epicenter	magnitude	slow-slip earthquake
body waves	fault	precursor phenomena	strike-slip fault
creep	fault scarp	P waves	surface waves
dip-slip fault	focus	seismic gap	S waves
earthquake	induced seismicity	seismic waves	thrust fault
earthquake cycle	intensity	seismograph	tsunami
elastic rebound	liquefaction	seismology	

Test Your Learning

1. Describe the process of fault creep and its relationship to the occurrence of damaging earthquakes.

2. Explain why rocks must behave elastically in order for earthquakes to occur.

3. Define strike-slip and dip-slip faults.

4. Indicate in what kind of plate-tectonic setting you might find (a) a strike-slip fault and (b) a thrust fault.

5. Define an earthquake's focus and its epicenter, and describe how they are related. Explain why deep-focus earthquakes are concentrated in subduction zones.

6. Name the two kinds of seismic body waves, and explain how they differ.

7. Explain the basis on which an earthquake's magnitude is assigned. Explain the concept of earthquake intensity, and why an earthquake may have a single magnitude but a range of intensities.

8. List at least three kinds of earthquake-related hazards, and describe what, if anything, can be done to minimize the danger that each poses.

9. Define a seismic gap, how it is recognized, and why it is a cause for concern.

10. Describe the concept of earthquake cycles and its usefulness in forecasting earthquakes.

11. Explain the distinction between earthquake prediction and earthquake forecasting, and describe the kinds of precautions that can be taken on the basis of each.

12. Describe the basis for issuing earthquake early warnings as is currently done in Japan. Cite two actions it makes possible that would reduce casualties, and one limitation on its usefulness.

13. Evaluate fluid injection as a possible means of minimizing the risks of large earthquakes.

14. Areas identified as high risk on the seismic-risk map of the United States may not have had significant earthquake activity for a century or more. Explain why they are nevertheless mapped as high-risk regions.

15. Indicate an area that has recently been mapped as one of increased risk due to induced seismicity, and explain what is causing that seismicity.

16. Explain why Los Angeles won't fall off into the Pacific Ocean during the next large earthquake there.

Exploring Further

1. Check out current/recent seismicity for the United States and Canada, or for the world, over a period of (a) one week and (b) one month. Tabulate by magnitude and compare with **table 4.2**. Are the recent data consistent with respect to relative frequencies of earthquakes of different sizes? Comment. (The USGS website **http:// earthquake.usgs.gov/ provides links to relevant data**.)

2. Look up your birthday, or other date(s) of interest, in "Today in Earthquake History" (**http://earthquake.usgs .gov/learn/today**) and see what you find! Investigate the tectonic setting of any significant earthquake listed.

3. Investigate the history of any modern earthquake activity in your own area or in any other region of interest. What geologic reasons are given for this activity? How probable is significant future activity, and how severe might it be? (The U.S. Geological Survey or state geological surveys might be good sources of such information.)

4. Consider the issue of earthquake prediction from a public-policy standpoint. How precise would you wish a prediction to be before making a public announcement? Ordering an evacuation? What kinds of logistical problems might you anticipate?

5. Watch an earthquake disaster movie or movie trailer such *San Andreas*. Was the science of seismology used correctly in the movie? What particular phenomena just didn't seem correct to you, and why?

CHAPTER 5

Volcanoes

On the morning of 18 May 1980, thirty-year-old David Johnston, a volcanologist with the U.S. Geological Survey (USGS), was watching the instruments monitoring Mount St. Helens in Washington State. He was one of many scientists keeping a wary eye on the volcano, expecting some kind of eruption, perhaps a violent one, but uncertain of its probable size. His observation post was more than 9 kilometers from the mountain's peak. Suddenly, he radioed to the control center, "Vancouver, Vancouver, this is it!"

It was indeed. Seconds later, the north side of the mountain blew out in a massive lateral blast that would ultimately cost nearly $1 billion in damages and twenty-five lives, with another thirty-seven persons missing and presumed dead. The mountain's elevation was reduced by more than 400 meters (**figure 5.1**). David Johnston was among the

casualties. His death illustrates that even the experts still have much to learn about the ways of volcanoes. And, although this eruption is the most destructive in U.S. history, it is relatively inconsequential compared to others we will discuss in this chapter.

Chapter Outline

5.1 Magma Sources and Types

5.2 Locations and Styles of Volcanic Activity

5.3 Hazards Related to Volcanoes

5.4 Issues in Predicting Volcanic Eruptions

5.5 More on Volcanic Hazards in the United States

The Fagradalsfjall volcano in Iceland created a "tourist eruption" in March 2021 when a fissure opened and lava erupted following an 800-year dormant period and then three weeks of intense seismic activity and ground deformation.

Daniel Freyr Jónsson/Alamy Stock Photo

Figure 5.1

The 1980 eruption of Mount St. Helens abruptly changed the topography of the mountain, flattening adjacent forests and blanketing them in ash.

Volcano Hazards/USGS

Like earthquakes, volcanoes are associated with a variety of local hazards, and in some cases can have global impacts. The dangers from any particular volcano depend on the kind of magma it erupts and on its geologic and geographic settings. This chapter examines different kinds of volcanic phenomena and ways to minimize the dangers.

5.1 Magma Sources and Types

Learning Objectives

- Recall the factors that affect magma composition
- Compare silicic to mafic magma
- Recall what conditions cause the asthenosphere to melt
- Recall the ways magma composition can change
- Explain the types of magmas formed at divergent boundaries, subduction zones, and hot spots

Volcanoes and **lava** play an influential role in stories ranging from classical myths, creation narratives both ancient and modern, disaster and sci fi movies, and even children's games. Yet lava is just molten rock, or magma, that flows out at Earth's surface. The majority of magmas originate in the upper mantle, at depths between 50 and 250 kilometers (30 and 150 miles), where the temperature is high enough and the pressure is low enough that the rock can melt, wholly or partially. Temperatures at the surface are generally too low to melt rock. In this section, we will review the compositional differences in magma

and how they influence volcanic eruptions; the processes involved in melting rock; how a magma can change once it is formed; and the connection between magma type and plate tectonic setting.

Magma composition and behavior

The composition of a magma is determined by the source (parent) material being melted and the extent of melting. The tectonic setting in which the magma is produced controls both what raw materials are available to melt and how the process proceeds, so each tectonic setting tends to be characterized by certain magma compositions. Knowing the composition of a magma is important because it influences its physical properties, which determine the way it erupts and therefore the particular types of hazards it represents.

The major compositional variables for magmas are the proportions of silica (SiO_2), iron, and magnesium. In general, the more silica-rich magmas are poorer in iron and magnesium, and vice versa. Recall in chapter 2 we used the term *mafic* to describe an igneous rock relatively rich in iron and magnesium (and ferromagnesian silicates). That term applies to magma, too. On the other hand, we use *silicic* or *felsic* for magmas that

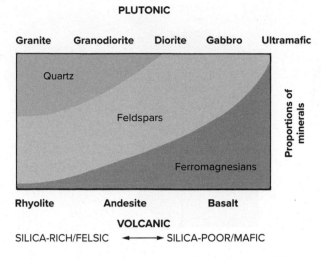

PLUTONIC

Granite   Granodiorite   Diorite   Gabbro   Ultramafic

Quartz

Feldspars

Ferromagnesians

Proportions of minerals

Rhyolite   Andesite   Basalt

VOLCANIC

SILICA-RICH/FELSIC ◄──────► SILICA-POOR/MAFIC

Figure 5.2

Common volcanic rock types (bottom labels) and their coarser-grained plutonic equivalents (top). The rock names reflect vary-ing proportions of silica, iron, and magnesium, and thus of common silicate minerals. Rhyolite is the fine-grained, volcanic compositional equivalent of granite, and so on.

have a relatively high silica content, with *felsic* being represen-tative of the igneous rocks dominated by feldspar and quartz. **Figure 5.2** relates the various common volcanic rock types and their plutonic equivalents. Note that in this context, "silica-poor" still means 45% to 50% SiO_2; all of the principal minerals in most volcanic rocks are silicates. The most silica-rich mag-mas may be up to 75% SiO_2.

The terms geologists use for magmas and the igneous rocks they create overlap. So, not only does the chart above tell us that the rock basalt does not contain quartz, but it also shows us that it formed from mafic lava. And because we are discuss-ing volcanoes in particular here, we commonly use those terms—rhyolitic, andesitic, basaltic—for the associated lavas erupted at the surface.

Magmas vary in physical properties based on their com-position and temperature. Mafic lavas are characteristically higher in temperature and low in **viscosity,** so they flow very easily. The silica-rich lavas are more viscous, thicker, and flow very sluggishly, also being lower in temperature. Magmas also contain dissolved water and gases. As magma moves up toward Earth's surface, pressure on it is reduced; water may turn suddenly to steam, and it and other gases begin to escape. Gas escapes relatively easily from the more fluid, silica-poor magmas. The viscous, silica-rich magmas tend to trap the gases, which may lead to an explosive eruption. The physical properties of magmas and lavas influence the kinds of volcanic structures they build, as well as their eruptive style, as we will explore later in the chapter.

Creating and altering magma

There are three ways in which rocks melt to form magma. The first is the most apparent, and it is what you would do if you

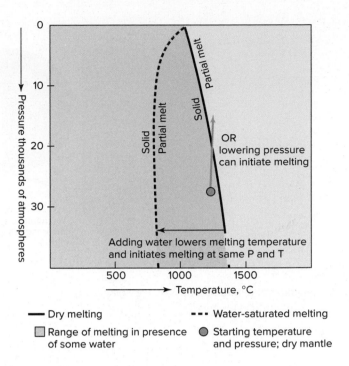

Adding water lowers melting temperature and initiates melting at same P and T

Pressure thousands of atmospheres

Temperature, °C

—— Dry melting   --- Water-saturated melting
▢ Range of melting in presence   ● Starting temperature
of some water   and pressure; dry mantle

Figure 5.3

In mantle material close to its melting point, reducing the pres-sure from overlying rock can cause melting. So can adding water; mantle rock in wet conditions melts at a lower tempera-ture than does dry mantle at the same pressure.

wanted to melt a solid—you would add heat to increase the temperature. Where does that heat come from? Often it is from a nearby magma. The heat could also come from the heat of Earth itself which increases with depth, but that would depend on the second factor involved in creating magma—pressure. **Decompression melting** occurs when rock that is very close to its melting temperature, like that in the asthenosphere, experi-ences a slight drop in pressure as it moves up toward the surface (**figure 5.3**). This is how oceanic crust is created at spreading ridges: the hot, solid, asthenosphere moves upward into the gap left by the rifting plates and partially melts due to a drop in pressure. Decompression melting also accounts for the creation of a hot spot as the upward-moving plume of hot mantle mate-rial begins to melt with a reduction in pressure. The third factor that produces melting is the addition of fluids to hot rock. A subducting oceanic plate carries water down with it, in the sea-floor rocks themselves which interacted with seawater as they formed and in the sediments subducted along with the litho-spheric plate. As the subducted rocks and sediments go down, they eventually become hot enough that water-bearing minerals break down and the fluids are driven off, rising into the asthe-nosphere above. This water weakens the bonds in the solid rock, lowering their melting point.

The creation of magma, no matter how it is done, is often discussed in terms of **partial melting** because it is rare for a parent material to completely melt. Rocks are made of different minerals, and each mineral has its own melting point. The melt

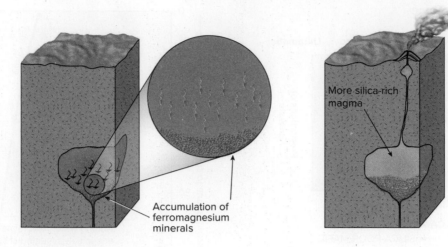

Figure 5.4

Differentiation of a magma body. As a magma body cools, minerals crystallize and settle to the bottom of the magma chamber, leaving behind a magma richer in silica.

created from a rock that has not completely melted will have a different composition, one that is higher in silica content, because silicate minerals are easier to melt (require less energy) than the ferromagnesian minerals. We will cover some examples of this process in the plate settings section below.

Once enough rock is melted, the less-dense material begins to move up out of the remaining solid asthenosphere. From there, a magma's composition can be changed in three ways. Along the way to the surface, magma will commonly accumulate for a time in a magma chamber. While there, it may cool a bit; some high-temperature minerals will crystallize and settle out in the process of *fractional crystallization* (**figure 5.4**). The remaining magma is more silicic. Rising magma interacts with the crust through which it passes. In some cases, it may *assimilate* quite a lot of crustal material, incorporating and melting bits of it to mix with the original magma. How this changes the magma's composition depends on how much material is assimilated, and the relative compositions of magma and crust. Magmas of two distinct compositions can also interact and *mix,* forming a new magma with a makeup intermediate between the two.

Magma and plate settings

Magmas are typically generated in one of three plate-tectonic settings: (1) at divergent plate boundaries, both ocean ridges and continental rift zones; (2) over subduction zones; and (3) at hot spots, isolated areas of volcanic activity that are not associated with current plate boundaries.

As discussed above, the primary factor affecting magma composition is the parent material, and that very often is the asthenosphere. The asthenosphere is *ultramafic,* meaning it is extremely rich in ferromagnesian minerals. Applying the rule of partial melting to the asthenosphere results in mafic magma. At ocean divergent boundaries, a lot of new, basaltic ocean floor is

created through the partial, decompression melting of the asthenosphere where passage of the melt to the surface is relatively unrestricted. At a continental rift zone, the asthenosphere-derived melt is also mafic, but it must work its way up through the thicker, more granitic continental crust, allowing more time for assimilation or fractional crystallization. The lava that eventually erupts may form basalt if it has an unhindered passage to the surface; **rhyolite** if the magma become silica-rich through differentiation and assimilation; or **andesite,** intermediate in composition between the mafic basalt and felsic rhyolite. A large amount of silicic magma never reaches the surface because it has such a high viscosity and instead forms the plutonic rock granite.

The magma typical of a hot-spot volcano in an ocean basin is similar to that of a seafloor spreading ridge: mafic magma derived from partial melting of the ultramafic mantle will remain basaltic even if it should interact with the basaltic sea floor on its way to the surface. The mafic magma associated with a continental hot spot may spill out at the surface in extensive basalt flows, or produce more-felsic magmas as described above for continental rift zones.

Magmatic activity in a subduction zone can be complex (**figure 5.5**). The bulk of magma generated is initially mafic, derived from partial melting of the asthenosphere as water is introduced from the subducted ocean crust below. In a continental setting, the less-dense, mafic magma moves upward and is hot enough to partially melt the continental lithosphere it encounters, adding more silica and resulting in an intermediate type of magma. A lot of heat can produce silicic magmas, but again, those are usually too viscous to reach the surface. When the subduction zone involves two oceanic plates in an island-arc setting, the overriding plate is not silica-rich, so the lavas erupted at the surface are basaltic to andesitic in composition, with andesite more common as a result of fractional crystallization.

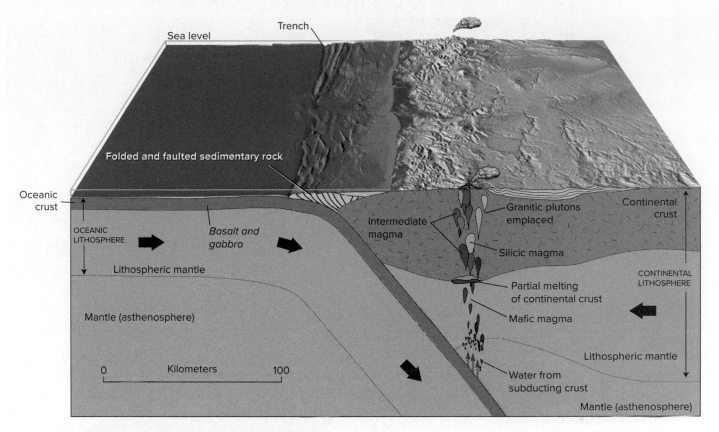

Figure 5.5

In a subduction zone, water and other fluids from the down-going slab and sediments promote melting of the overlying asthenosphere; the hot, rising magma, in turn, may assimilate or simply melt overlying lithosphere, producing a variety of melt compositions.

5.2 Locations and Styles of Volcanic Activity

Learning Objectives

- Relate volcano locations to plate-tectonic setting
- Summarize a continental fissure eruption
- Discuss hot spot volcanoes
- Describe the size, shape, and composition of a shield volcano, cinder cone, and composite volcano
- Recognize different types of pyroclastics

Most volcanic activity on Earth occurs at the seafloor-spreading ridges, where magma fills cracks in the lithosphere and crystallizes close to the surface to form basaltic ocean crust. Spreading ridges separate at the rate of only a few centimeters per year, but there are some 50,000 kilometers (about 30,000 miles) of these ridges presently active in the world. Most of this activity is under the oceans where it is largely unnoticed, involving quietly erupting mafic magma that presents no dangers to people.

In this section we will review volcanic activity associated with continental fissures, subduction zones, and hot spots, those interactions being the ones that have had the most apparent influence on the biosphere throughout geologic time and more recently, creating the most hazardous conditions for people.

Volcano Location

Many people think of volcanoes as eruptions from the central vent of some sort of mountainlike feature. **Figure 5.6** is a map of such volcanoes presently or recently active in the world. Notice the particularly close association between volcanoes and subduction zones. Proximity to an active subduction zone is a strong indicator of both volcanic and earthquake hazards. The Ring of Fire, the collection of volcanoes rimming the Pacific Ocean, is really a ring of subduction zones. A few volcanoes are associated with continental rift zones, such as Kilimanjaro and Nyiragongo of the East African Rift system. Other volcanoes are located far away from a plate boundary, and are the result of hot-spot activity

We categorize volcanoes by the kinds of structures they build. The structure, in turn, reflects the kind of volcanic material erupted and commonly the tectonic setting. These correlations provide us some indication of what to expect for future eruptions.

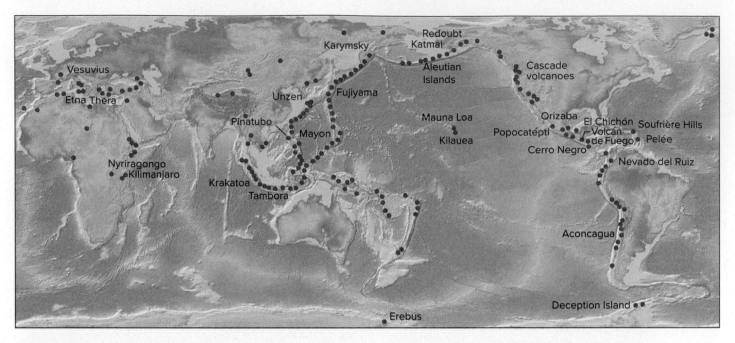

Figure 5.6

Map showing locations of recently active volcanoes. Which of these is the most active on the planet, and also not located on a plate boundary?

Continental Fissure Eruptions

The outpouring of magma at oceanic spreading ridges is an example of **fissure eruption,** the eruption of magma out of a crack in the lithosphere, rather than from a single pipe or vent. Fissure eruptions also occur on the continents (**figure 5.7**). One example in the United States is the Columbia Plateau, an area of over 150,000 square kilometers (60,000 square miles) in Washington, Oregon, and Idaho, covered by layer upon layer of basalt, piled up nearly a mile deep in places (**figure 5.8**). Scientists link such fissure eruptions to either mantle plumes and/or the initiation of continental rifting, which in this case ultimately stopped before the land ripped apart. These huge basalt outcrops remind us of how large a volume of magma can come welling up from the asthenosphere when zones of weakness in the lithosphere provide suitable openings. Even larger examples of continental flood basalts covering more than a million square kilometers are found in India, Brazil, and Siberia. The timings of some of these basalt flows coincide with mass extinction events. These long-lasting events, some continuing for a few million years, would have pumped tremendous amounts of climate-altering gases such as CO_2 into the atmosphere.

Hot-Spot Volcanoes

Some volcanoes are not associated with plate boundaries. Hawaii is likely the most familiar, a hot spot in the middle of the Pacific plate, as discussed in chapter 3. Other hot spots lie under the Galápagos Islands, Iceland, and Yellowstone National Park (**figure 5.9**). What accounts for the hot spots is unclear. The most common source cited is a mantle plume, although some

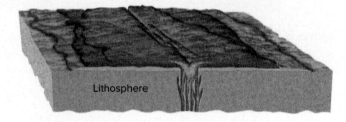

Figure 5.7

A continental fissure eruption. Magma pathways through the lithosphere may be complex and involve many passages. Notice the runny nature of the mafic lava involved.

suggest they originate from a deeper source or a subducted slab. How would a hot spot be generated in the first place? Possible mechanisms include rifting, subduction, or an impact event. Scientists are also debating the commonly held view that the molten plumes responsible for hot spots are fixed in place, even though some appear to be hundreds of millions of years old. The hot spot responsible for the Hawaiian Islands and the Emperor Seamount chain is at least 70 million years old, but we are still trying to determine whether the Pacific plate changed direction or the hot spot moved approximately 43 million years ago to create the distinct change in orientation of the linked features (see **figure 3.24**).

Shield Volcanoes

Many people unfamiliar with volcanoes think of a tall, cone-shaped feature with a single, central crater at the top. But the

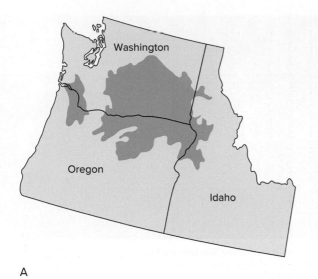

A

B

Figure 5.8

Flood basalts. (A) Areal extent of Columbia River flood basalts. (B) Multiple lava flows, one atop another, can be seen at Palouse Falls in Washington state; note additional flows stretching into the distance. For scale, the falls are 200 feet (60 meters) high.

(B) ©Carla Montgomery

Figure 5.9

Selected prominent hot spots around the world. Some coincide with plate boundaries; most do not.

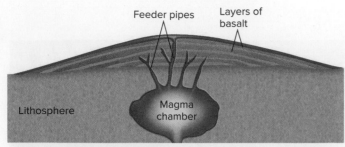

A Simplified diagram of a shield volcano in cross section.

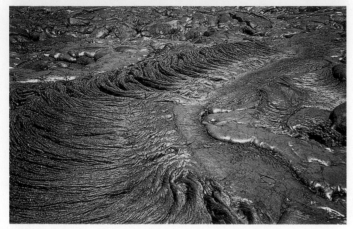

C Fluidity of Hawaiian lavas is evident even after they have solidified. This ropy-textured surface is termed *pahoehoe* (pronounced "pa-hoy-hoy").

Figure 5.10

Shield volcanoes and their characteristics.

(B) Volcano Hazards/USGS; (C) ©Carla Montgomery; (D) USGS

B Very fluid lava flows, like these on Kilauea in Hawaii, are characteristic of shield volcanoes.

D The Big Island of Hawaii has multiple shield volcanoes, the largest being Mauna Loa in the center.

largest volcanoes on the planet have a much different shape that is directly related to the chemical makeup of the magma involved. Mafic lavas have a relatively low viscosity, so they flow very freely and far when erupted. The kind of volcano they build is flat and low in relation to its diameter, and is large in areal extent. This broad, shield-like shape has led to the term **shield volcano** for such a structure (**figure 5.10**). Though the individual lava flows may be thin, a few meters or less in thickness, the buildup of hundreds or thousands of flows through time produces large structures. The Hawaiian Islands are all shield volcanoes. Mauna Loa, the largest peak on the island of Hawaii, rises 3.9 kilometers (about 2.5 miles) above sea level (**figure 5.10D**). If measured from its true base of the sea floor, it is much more impressive at about 10 kilometers (6 miles) high and 100 kilometers in diameter. It is the largest active volcano on Earth, last erupting in 1984. Lava erupts from this huge volcano out of craters (vent openings) and along fissures, both clearly evident in the aerial photo below. The largest known volcano in the solar system, Olympus Mons, is a shield volcano on Mars, which makes Earth volcanoes look puny in

comparison. Shield volcanoes commonly are associated with hot-spot activity and divergent boundaries.

Cinder Cones and Pyroclastics

As magma wells up toward the surface, the pressure on it is reduced, and dissolved gases try to bubble out of it and escape. The effect is much like popping the top off a carbonated beverage. The liquid contains carbon dioxide gas under pressure, and when the pressure is released by removing the cap, the gas comes bubbling out. Sometimes, the built-up gas pressure in a rising magma is released more suddenly and forcefully, flinging bits of lava and rock out of the volcano. The lava may freeze into solid pieces before falling to the surface. While basaltic magmas generally erupt as fluid lava flows, they sometimes produce small volumes of chunky volcanic cinders in this way. These cinders, also known as *scoria,* fall close to the vent from which they are thrown and may pile up into a very symmetric cone-shaped heap known as either a **cinder cone** or scoria cone (**figure 5.11**). Cinder cones can be found in association with shield volcanoes,

A

B

Figure 5.11

(A) *Pu'u 'Ō'ō* cinder cone was created by gassy eruptions of basaltic magma on Kilauea. (B) This aerial view of Cinder Cone at Lassen Volcanic National Park, California, shows its classic shape. Notice the lava flows at its base representing a less-gassy stage of eruption.

(A) J.D. Griggs/USGS; (B) D.R. Crandell/U.S. Geological Survey

A

B

Figure 5.12

Volcanic ash from the eruption of Mount St. Helens (A) and bombs from Mauna Kea (B). Bombs are molten, or at least hot enough to be plastic, when erupted, and may assume a stream-lined shape in the air.

*(A) U.S. Geological Survey/Photograph by David E. Wieprecht;
(B) U.S. Geological Survey/Photograph by J. P. Lockwood*

individually or in groups with hot-spot activity, and even on the sides of the more silicic-rich volcanoes discussed below. They are relatively small and short-lived.

The rocky fragments that make up a cinder cone are just one example of energetically erupted bits of volcanic material described collectively as **pyroclastics.** The most violent pyroclastic eruptions are more typical of volcanoes with thicker, more viscous silicic lavas because the thicker lavas tend to trap more gases. Gas escapes more readily and quietly from the more fluid basaltic lavas, though even shield volcanoes may sometimes put out quantities of ash and small fragments as their dissolved gases rush out of the volcano. The fragments of pyroclastic material vary considerably in size (**figure 5.12**), from very fine, flourlike, gritty volcanic ash through cinders ranging

up to golf-ball-sized pieces. Blobs of liquid lava may also be thrown from a volcano; these are volcanic bombs. A really violent explosion can fling out chunks of old volcanic rock that may be the size of a car or larger.

Composite Volcanoes

Many volcanoes erupt somewhat different materials at different times. They may emit some pyroclastics, then some lava, then more pyroclastics, and so on. Volcanoes built up in this layer-cake fashion are called **stratovolcanoes,** or, alternatively, **composite volcanoes,** because they are composed of layers of more than one kind of material (**figure 5.13**). The mix of lava and pyroclastics allows them to grow larger than cinder cones. Most of the potentially dangerous volcanoes worldwide are

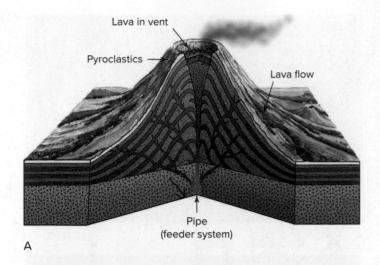

composite volcanoes, which makeup about 60% of all volcanoes on Earth. All tend to have fairly viscous, gas-charged andesitic (intermediate) lavas that sometimes flow and sometimes trap enough gases to erupt explosively with a massive blast and a rain of pyroclastic material. Composite volcanoes are associated with subduction zones.

When not erupting explosively, the slow-flowing rhyolitic and andesitic lavas tend to ooze out at the surface like thick toothpaste from a tube, piling up close to the volcanic vent, rather than spreading freely. The resulting structure is a compact and steep-sided **lava dome** (**figure 5.14**). Often, a lava dome will form in the crater of a composite volcano after an explosive eruption, as with Mount St. Helens since 1980, as the volcano rebuilds itself.

The relative sizes of these three types of volcanoes are shown in **figure 5.15**.

Figure 5.13

(A) Schematic cutaway view of a composite volcano (stratovolcano), formed of alternating layers of lava and pyroclastics. (B) Two composite volcanoes of the Cascade Range: Mount St. Helens (foreground) and Mt. Rainier (rear); photograph predates 1980 explosion of Mount St. Helens.

(B) D.R. Mullineux/U.S. Geological Survey (USGS) Library, Special Collections

Figure 5.14

Closeup of multiple lava domes that have formed in the summit crater of Mount St. Helens since the 1980 eruption. The crater is approximately 2 km by 3 km in size.

U.S. Geological Survey/Photograph by Matt Logan/ Julie Griswold

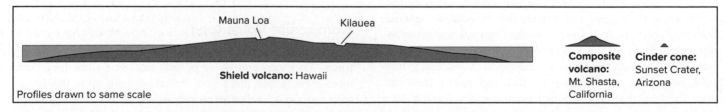

Figure 5.15

The voluminous, low-viscosity lavas of shield volcanoes can build huge feature; composite volcanoes and cinder cones are much smaller, as well as steeper.

5.3 Hazards Related to Volcanoes

Learning Objectives

- Generalize the danger of lava to people
- Describe the dangers of pyroclastics
- Explain why a pyroclastic flow is so dangerous
- Describe the creation of a lahar
- List dangerous gases associated with volcanoes
- Understand how different volcanic gases affect climate

The list of hazards associated with volcanoes is extensive, and it seems each one is more worrisome than the next, ranging from igniting, burying, and destroying everything in their path to emitting deadly gases. But these same hazards are associated with processes that are essential in the connection between Earth's lithosphere, hydrosphere, biosphere, and atmosphere. As our population continues to expand, we find ourselves living ever closer to active volcanoes. Remote sensing data collected from satellites, planes, helicopters, and drones is increasingly used to replace or support on-the-ground measurements and reach inaccessible places, deepening our understanding of how volcanoes operate so we can avoid the most dangerous events of active volcanoes.

Lava

Many people instinctively assume that lava is the principal hazard presented by a volcano. Yes, lava is very hot at temperatures typically over 850°C (over 1550°F), and basaltic lavas can be over 1100°C (2000°F). But most lava flows are not life-threatening. They advance at speeds of only a few kilometers an hour or less, so a person can usually evade the advancing lava on foot. Combustible materials such as buildings and forests do burn at these temperatures, but other structures may become engulfed in lava and then the subsequent igneous rock. (**figure 5.16**).

Lavas, like all liquids, flow downhill, so one way to protect property is simply not to build close to the slopes of the volcano. Throughout history, however, people have built on or near volcanoes for many reasons. They may not expect the volcano to erupt again and choose to ignore scientifically based advise. Or, they may make a living off of the very fertile land provided by the volcanic soils. Sometimes, land near a volcano is the only land available, such as in Hawaii and Iceland today.

Geologists have a fairly good idea of where lava flows will move, and how fast they move. By studying the extent and thickness of past lava flows, they create maps that are used in hazard prediction (see Case Study 5.1). Samples of older lava rock, and of fresh lava if possible, are analyzed to determine viscosity to better understand how fast and how far they will flow.

There is really very little people can do to stop a lava flow from reaching a certain location. People have tried to divert lava flows by digging trenches, erecting concrete barriers, and using bombs to create large craters, all futile efforts. There is a single successful story of lava diversion that occurred on the island of Heimaey, Iceland. In 1973, a volcanic fissure opened unexpectedly. The residents were evacuated within hours of the start of the event. In the ensuing months, homes, businesses, and farms were set afire by falling hot pyroclastics or buried under pyroclastic material or lava. When the harbor, the economic backbone for the fishing industry on the island, was threatened by encroaching lava, officials took the unusual step of deciding to fight back. They pumped ocean water onto the flowing lava to cool and slow it, and successfully kept the harbor from being cut off. After the eruption stopped, the islanders used the heat from the cooling lava to generate hot water and electricity, and new volcanic rock to extend the airport runway and to fill land for new home construction.

Mount Nyiragongo in the Democratic Republic of Congo erupts some of the fastest moving and hottest lava on Earth (**figure 5.17**). The city of Goma with a population of over 1.5 million people is situated approximately 6000 feet directly below this very active stratovolcano. The steepness of the volcano along with the low viscosity of the lava allows it to move very quickly toward the city. An eruption in 2002 reached the city, destroyed thousands of homes leaving over 120,000 people homeless, and displaced over 250,000 people. The most recent eruption in May 2021 resulted in the evacuation of 400,000 people and the destruction of over 3500 homes and buildings. A few hundred people did die from these two eruptions, but not directly from the lava flows. The situation that makes Mount Nyiragongo so much more dangerous is that the volcano is located in a place of political unrest, where scientific equipment is commonly stolen or destroyed, and there is often little money to either replace the monitors or support the scientists that use them.

Lava flows are hazardous to property, but they are at least predictable, flowing from high elevation to low and stopping once they reach a flat area or a body of water. Dealing with lava is a way of life in Hawaii but we will see that science and common sense does not always prevail (see Case Study 5.1).

Figure 5.16

The Church of San Juan Parangaricutiro was buried by the Paricutin lava flow, Mexico.

Natursports/Shutterstock

Figure 5.17

NASA's Landsat 7 satellite captured this false-color image of Nyiragongo in January 2003, about a year after a devastating lava flow reached Goma. The pink dot at its summit indicates unusually warm surface temperatures. The nearby shield volcano Nyamuragira has more extensive lavas that flow in the opposite direction from the city.

NASA image created by Jesse Allen, using Landsat data provided by the University of Maryland's Global Land Cover Facility

Pyroclastics

Pyroclastics are fragments of hot rock and spattering lava of all sizes. They are often more dangerous than lava flows as they erupt more suddenly and explosively, and spread faster and farther. The largest blocks and volcanic bombs present an obvious danger because of their size and weight. For the same reasons, they usually fall quite close to the volcanic vent, so they affect a relatively small area.

The sheer volume of the finer ash and dust particles can make them as severe a problem, and they can be carried over a much larger area. Ashfalls are not confined to valleys and low places. Instead they blanket the countryside like snow. The 18 May 1980 eruption of Mount St. Helens was a relatively small eruption, but the ash from it blackened the midday skies more than 150 kilometers away, and measurable ashfall was detected halfway across the United States. Even in areas where only a few millimeters of ash fell, transportation ground to a halt as drivers skidded on slippery roads and engines choked on airborne dust. Volcanic ash is also a health hazard, both uncomfortable and dangerous to breathe as it is tiny glass particles. The cleanup effort required to clear the debris strewn about by Mount St. Helens was enormous. An estimated 600,000 tons of ash landed on the city of Yakima, Washington, more than 100 kilometers (60 miles) away. In a more recent example, the 2021 Cumbre Vieja eruption on La Palma in the Canary Islands lasted more than three months and created more than 200 million cubic meters of ash and lava combined; many of the 1300+ homes destroyed are engulfed in ash.

Figure 5.18

The 1980 eruption of Mount St. Helens blew off the top 400 meters (over 1300 feet) of mountain, as shown in **figure 5.1**. Twenty years later, the landscape was still blanketed in thick ash, gullied by runoff water.

©Carla Montgomery

Volcanic ash poses a special hazard to air travel. The often-glassy fragments are typically sharp and abrasive, damaging aircraft components. Worse yet, andesitic ash melts at temperatures close to the operating temperatures of jet engines. The ash can melt and then resolidify in the engines, causing them to fail. When the Icelandic volcano Eyjafjallajökull erupted in 2010, spewing out copious ash plumes (**figure 5.19**) that wafted with the jet stream over Europe, concern for passenger safety caused officials to ground tens of thousands of flights. Air travel over northern Europe was essentially halted from 15 to 23 April, in the largest air-traffic shutdown there since World War II, and subject to intermittent disruptions over the next month. Millions of travelers' plans were affected; the airline industry lost an estimated $1.7 billion. Questions have since been raised about whether officials overreacted, and tests are now under way to try to determine just what density of volcanic ash in air a jet engine can safely tolerate.

History provides many dramatic examples of explosive eruptions of other volcanoes, often with huge volumes of pyroclastics. The 1815 explosion of Tambora in Indonesia blew out an estimated 100 cubic kilometers of debris, and the prehistoric explosion of Mount Mazama, in Oregon, which created the basin for the modern Crater Lake, was larger still. Ash is also dangerous in large amounts because it can hinder sunlight from reaching Earth's surface, causing global cooling, a subject we explore later in this section.

Pyroclastic Flows

The most dangerous volcanic hazard involves a sudden outburst of a denser-than-air mixture of hot gases and fine ash, quite

Figure 5.19

Two ash plumes drift away from Eyjafjallajökull—a lower-altitude, more diffuse plume and a denser one, shot higher into the air, that is casting a shadow on the lower plume.

NASA image by Jeff Schmaltz, MODIS Rapid Response Team

different in character from a fall of ash and small cinders. A **pyroclastic flow** is very hot with temperatures over 1000°C in the interior, and it can rush down the slopes of the volcano at more than 100 kilometers per hour (60 miles per hour), charring everything in its path. Such pyroclastic flows accompanied the major eruption of Mount St. Helens in 1980 and are an ongoing hazard at Soufrière Hills volcano on the Caribbean island of Montserrat (**figure 5.20**), among other places. They develop when rising hot gas lifts ash and magma droplets to mix with air above the volcano; if the resultant cloud becomes too dense, it collapses and flows swiftly down the volcano's slopes. Such a cloud of hot gas and ash can also form within the volcano's crater, then spill suddenly over the edge and downhill.

Perhaps the most famous such event in the twentieth century occurred during the 1902 eruption of Mont Pelée on another Caribbean island, Martinique. The volcano had begun erupting weeks before, emitting both ash and lava, but was believed by many to pose no imminent threat to surrounding towns. Then, on the morning of 8 May, with no immediate advance warning, a pyroclastic flow emerged from that volcano and swept through the nearby town of St. Pierre and its harbor. In a period of about three minutes, an estimated 25,000 to 40,000 people died or were fatally injured, burned and/or suffocated. The two reported survivors in the town were a prisoner

A

B

Figure 5.20

(A) Pyroclastic flow from Mount St. Helens. (B) Pyroclastic flows (tan) have filled valleys and stream channels radiating from the still-steaming summit of Soufrière Hills volcano.

(A) U.S. Geological Survey/Photograph by P. W. Lipman;
(B) NASA image by Robert Simmon using data provided by the NASA EO-1 team

in the town dungeon and a man in a boat in a sea cave. **Figure 5.21** gives some idea of the devastation.

Many volcanoes have a history of pyroclastic flows. Composite volcanoes, with their felsic lava and steep slopes, are most often associated with these deadly events. While the emergence of a pyroclastic flow may be sudden and unheralded by special warning signs, it is not generally the first activity shown by a volcano during an eruptive stage. Steam had issued from Mont Pelée for several weeks before the day St. Pierre was destroyed, and lava had been flowing out for over a week. This suggests the best strategy for avoiding pyroclastic flows: when a volcano known or believed to be capable of such eruptions shows signs of activity, leave.

Figure 5.21

St. Pierre, Martinique, West Indies, was destroyed by a nuée ardente from Mont Pelée, 1902.

Underwood and Underwood, courtesy Library of Congress

Guatemala's Volcán de Fuego, another composite volcano, has been very active since 2002, with more than a dozen significant eruptive events in 2017 alone, and activity continuing at a high level into 2021. The volcano sometimes casts glowing bombs and blocks into the air near the summit; sometimes shoots plumes of ash into the air, occasionally dusting nearby towns; and sends tongues of lava a half mile or more down its flanks. Casualties from all this usually are minimal. However, on 3 June 2018, a major explosion was accompanied by large pyroclastic flows that swept swiftly down the slopes, traveling as much as 8 km (5 miles), with internal temperatures over 700°C (1300°F). There was no distinctive warning that these pyroclastic flows were imminent. Within minutes, over 100 people were killed and hundreds more injured, many suffering severe burns. Activity at Volcán de Fuego continues, as do efforts to develop ways to anticipate these events so that timely evacuations can save lives.

Lahars

Volcanic ash and water can also combine to create a fast-moving volcanic mudflow called a **lahar.** Victims are engulfed and trapped in dense mud, which may also be hot from the heat of the ash. Such mudflows, like lava flows, flow downhill. They tend to follow stream channels, clogging them with mud and causing floods of stream waters. Flooding produced in this way was a major source of damage near Mount St. Helens in 1980. Following the 1991 eruption of Mount Pinatubo in the Philippines, drenching typhoon rains soaked the ash blanketing the mountain's slopes, triggering devastating lahars (**figure 5.22**). Rain is not an essential ingredient; with a volcano capped by snow or glaciers,

A

B

Figure 5.22

(A) Aerial view of Abacan River channel in Angeles City near Clark Air Force Base, Philippines. A mudflow has taken out the main bridge: makeshift bridges allow pedestrians to cross the debris. (B) Some houses were swamped by lahars. Typhoon rains dropping soaking ash on roofs also contributed to casualties when the overburdened roofs collapsed.

(A) U.S. Geological Survey/Photograph by T. J. Casadevall;
(B) U.S. Geological Survey/Photograph by R. P. Hoblitt

Figure 5.23

The town of Armero was destroyed by lahars from Nevado del Ruiz in November 1985; more than 23,000 people died. The ruins emerged as the flows moved on and water levels subsided. This tragedy led to development of the Volcano Disaster Assistance Program, an international venture supported jointly by the U.S. Geological Survey and USAID.

U.S. Geological Survey Cascade Volcano Observatory/ Photograph by R. Janda

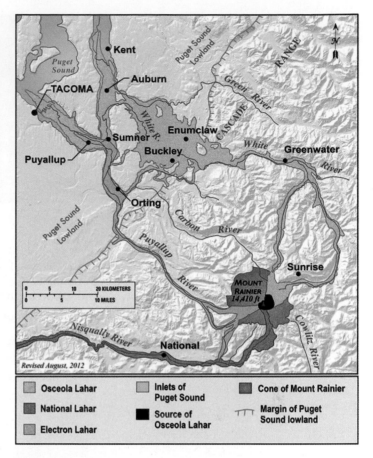

Figure 5.24

This USGS map shows Mt. Rainier lahar pathways from three major events in the past 10,000 years.

Volcano Hazards/USGS

the heat of falling ash or pyroclastic flows can melt snow and ice on the mountain, producing a lahar. The 1985 eruption of Nevado del Ruiz, in Colombia, was a deadly example (**figure 5.23**). The swift and sudden mudflows that swept down its steep slopes after its summit glaciers were melted by hot pyroclastics were the principal cause of deaths in towns below the volcano.

Mount Rainer (Tahoma or Tacoma to the Puyallup) in the Pacific Northwest has a history of producing lahars (**figure 5.24**). This tall mountain, like many composite volcanoes, is high enough for year round snow and glaciers that can be melted and mixed with ash layers to produces volcanic mud flows. Also, the heat from the volcano works together with water to alter feldspar-containing rocks into much weaker and slippery clays. Combine all this on the volcano, give it a bit of a shake during a small earthquake, and you have a very dangerous situation. Fortunately, communities such as Orting that surround Mt. Rainier have taken the risk seriously by installing a warning system based on acoustic flow monitors and having evacuation plans that they practice at least once a year. Such a warning would give people approximately 40 minutes to find safety on higher ground, off the floors of the river valleys. Mt. Rainier National Park currently is planning to expand the network of monitors.

Even after the immediate danger is past, when the lahars have flowed, long-term impacts remain. They often leave behind stream channels partially filled with mud, their water-carrying capacity permanently reduced, which aggravates ongoing flood hazards in the area. To dredge this mud to restore channel capacity can be very costly, and areas far from the volcano can be affected. Expensive bridges in remote areas are destroyed, causing time-consuming and expensive rerouting of traffic.

Toxic Gases

Volcanoes emit a variety of gases. Water vapor, the primary component, is not a direct threat but several other gases are. Carbon dioxide is nontoxic yet is dangerous at high concentrations, as described below. Other gases, including carbon monoxide, various sulfur gases, and hydrochloric acid, are actively poisonous. In the 1783 eruption of Laki in Iceland, toxic fluorine gas emissions apparently accounted for many of the 10,000 casualties. Many people have been killed by volcanic gases even before they realized the danger. During the A.D. 79 eruption of Vesuvius, fumes overcame and killed unwary observers.

A terrible disaster caused by volcanic gases occurred in Cameroon, Africa, in 1986. Lake Nyos is one of a series of lakes that lie along a rift zone that is the site of intermittent volcanic activity. On 21 August 1986, a massive cloud of carbon dioxide gas from the lake flowed out over nearby towns and claimed 1700 lives. The victims were not poisoned; they were suffocated. Carbon dioxide is about 50% denser than air,

A

Figure 5.25A

The Mammoth Mountain tree kills that began in 1990 could not be explained by drought or insect damage, prompting researchers to investigate soil gases and discover the high CO_2 levels.

U.S. Geological Survey/Photograph by Dina Venezky

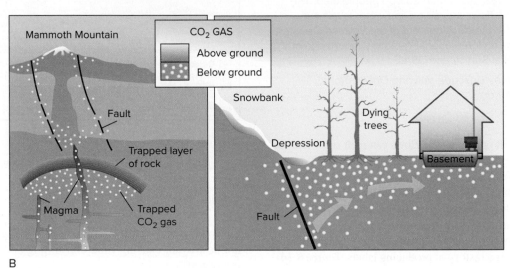

B

Figure 5.25B

Magma-derived CO_2 can seep toward the surface along faults, then accumulate in pores in soil, in basements, or even in surface depressions and under snowpack.

so such a cloud tends to settle near the ground and to disperse only slowly. Anyone engulfed in that cloud would suffocate from lack of oxygen, and without warning, for CO_2 is both odorless and colorless. This is an ongoing threat. The source of the gas in this case is believed to be near-surface magma below. The magmatic CO_2 apparently seeps up into the lake, where it accumulates in lake-bottom waters. Seasonal overturn of the lake waters as air temperatures change, or a disturbance such as an earthquake, can lead to its release. Currently, both Lake Nyos and nearby Lake Monoun (the site of a similar CO_2 event) are being studied and monitored closely, and efforts to harmlessly de-gas these two lakes is ongoing, with concerns that this is not occurring quickly enough.

Carbon dioxide can be deadly to vegetation as well as to people. In 1990, tree kills began in the vicinity of Mammoth Mountain, California, near the Long Valley Caldera discussed later in the chapter. More than 100 acres of trees died over the next decade (**figure 5.25A**). Evidently, CO_2 from magma below was rising along a fault and, being denser than air, accumulating in the soil. The tree roots need to absorb some oxygen from soil gas, but in the area of the kills, the soil gases had become up to 95% CO_2, suffocating the trees. Residents of and visitors to the area need to be aware, too, that the CO_2 might settle in low areas such as valleys and basements, where it could reach hazardous concentrations (**figure 5.25B**).

Steam Explosions

Some volcanoes are deadly not so much because of any characteristic inherent in the particular volcanoes but because of where they are located. In the case of a volcanic island, for example, large quantities of seawater may seep down into the rock, come close to the hot magma below, turn to steam, and blow up the volcano like an overheated steam boiler. This is called a **phreatic eruption** or explosion, and can occur anywhere abundant water comes too close to magma.

The classic example is Krakatoa (Krakatau) in Indonesia, which exploded in this fashion in 1883. The force of its explosion was estimated to be comparable to that of 100 million tons of dynamite, and the sound was heard 3000 kilometers away in Australia. The island literally blew apart. Some of the dust shot 80 kilometers into the air, causing red sunsets for years afterward, and ash was detected over an area of 750,000 square kilometers. Furthermore, the shock of the explosion generated a tsunami that built to over 40 meters high as it came onshore. Krakatoa itself was an uninhabited island, yet its 1883 eruption killed an estimated 36,000 people, mostly in low-lying coastal regions inundated by tsunamis.

A phreatic explosion that created a pyroclastic flow was responsible for the death of 36 hikers on Japan's Mount Ontake in 2014. Even though this volcano is heavily monitored, the sudden flash of water heated by nearby magma that caused the explosion could not be predicted.

Landslides and Collapse

The very structure of a volcano can become unstable and become a direct or indirect threat. As rocks weather and weaken, they may come tumbling down the steep slopes of a composite volcano as a landslide that may bury a valley below or dam a stream to cause flooding. Just removing that weight from the side of the volcano may allow gases trapped under pressure to blast forth through the weakened area. This is what happened at Mount St. Helens in 1980 when a relatively strong earthquake removed the newly formed bulge on the side of the volcano, which in turn allowed the volcano to erupt in a lateral blast.

With an island or coastal volcano, a large landslide crashing into the sea could trigger a tsunami. Such an event in 1792 at Mt. Mayuyama, Japan, killed 15,000 people on the opposite shore of the Ariaka Sea. At Kilauea, there is concern that a combination of weathering, the weight of new lava, and undercutting by erosion along the coast may cause breakoff of a large slab from the island of Hawaii, which could likewise cause a tsunami. Underwater imaging has confirmed huge coastal slides in the past but such a catastrophe does not appear imminent. There is a block of Kilauea's south flank, bounded by faults, that has been moving slowly seaward at up to 8–10 cm/yr (3–4 in/yr) since GPS monitoring of it began in 1993. When Hawaii has experienced one of its larger earthquakes, the block has slipped a few meters at once. But even if it slid suddenly into the ocean, it is small enough that it would cause only a local tsunami, not a Pacific-wide disaster.

Secondary Effects: Climate and Atmospheric Chemistry

A single volcanic eruption can have a global impact on climate, although the effect may be only brief. Intense explosive eruptions put large quantities of volcanic dust high into the atmosphere. The dust can take several months or years to settle, and in the interim, it partially blocks out incoming sunlight, causing measurable cooling. After Krakatoa's 1883 eruption, worldwide temperatures dropped nearly half a degree Celsius, and the cooling effects persisted for almost ten years. The larger 1815 eruption of Tambora in Indonesia caused still more dramatic cooling: 1816 became known around the world as the "year without a summer." Scientists have suggested a link between the explosive eruption of the Peruvian volcano Huaynaputina in 1600 and widespread cooling in Europe and Asia, believed responsible for acute crop failures and severe famine in Russia from 1601 through 1603.

The meteorological impacts of volcanoes are not confined to the effects of volcanic dust. The 1982 eruption of the Mexican volcano El Chichón did not produce a particularly large quantity of dust, but it did shoot volumes of unusually sulfur-rich gases into the atmosphere. These gases produced clouds of sulfuric acid droplets (*aerosols*) that spread around the planet. Not only do the acid droplets block some sunlight, like dust, but in time they fall back to Earth as acid rain, the consequences of which are explored in later chapters. Similar to El Chichón except on a larger scale, the 1783–1784 eruption of Laki in Iceland emitted so much sulfur to cause at least a 1°C drop in temperature in the Northern Hemisphere the next year. This event too has been linked to widespread crop failure, famine, and even social unrest.

The 1991 eruptions of Mount Pinatubo in the Philippines (**figure 5.26**) had long-lasting, global effects on climate that were not only measurable by scientists but noticeable to a great many other people around the world. In addition to quantities of visible dust, Pinatubo pumped an estimated 20 million tons of sulfur dioxide (SO_2) into the atmosphere. While ash falls out of the atmosphere relatively quickly, the mist of sulfuric-acid aerosols formed from stratospheric SO_2 stays airborne for several years. This acid-mist cloud slowly encircled Earth, most concentrated close to the equator near Pinatubo's latitude (**figure 5.27**). At its densest, the mist scattered up to 30% of incident sunlight and decreased by 2% to 4% the amount of solar radiation reaching Earth's surface to warm it. Average global temperatures decreased by about 0.5°C (1°F), but the decline was not uniform; the greatest cooling was observed in the intermediate latitudes of the Northern Hemisphere. Mount Pinatubo thus received the blame for the unusually cool summer of 1992 in the United States. The gradual natural removal of the ash and SO_2 from the atmosphere resulted in a rebound of temperatures in the lower atmosphere by 1994 (**figure 5.28**).

Another, more subtle impact of this SO_2 on atmospheric chemistry may present a different danger. The acid seems to

Figure 5.26

Ash and gas from Mount Pinatubo was shot into the stratosphere, and had an impact on climate and atmospheric chemistry world-wide. Eruption of 12 June 1991.

Karin Jackson/U.S. Air Force

aggravate ozone depletion (see discussion in chapter 18), which means increased risk of skin cancer to those exposed to strong sunlight.

5.4 Issues in Predicting Volcanic Eruptions

Learning Objectives

- Recall how volcanoes are classified by activity
- Generalize the current state of volcano monitoring
- Use the Volcanic Explosivity Index to rank volcanic eruptions
- List volcanic eruption precursors
- Recall evacuations that did or did not occur in response to eruption predictions

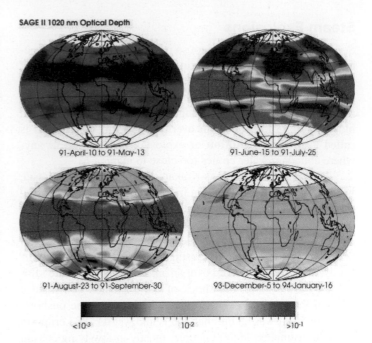

SAGE II 1020 nm Optical Depth

91-April-10 to 91-May-13   91-June-15 to 91-July-25

91-August-23 to 91-September-30   93-December-5 to 94-January-16

<10⁻³   10⁻²   >10⁻¹

Figure 5.27

These false-color images represent aerosol optical depth in the stratosphere before, during and after the June 1991 Pinatubo eruption. Red show the highest values, while dark blue shows the lowest values, which are normal. The images were created using the Stratospheric Aerosol and Gas Experiment II flying aboard NASA's Earth Radiation Budget Satellit.

NASA Langley Research Center Aerosol Research Branch

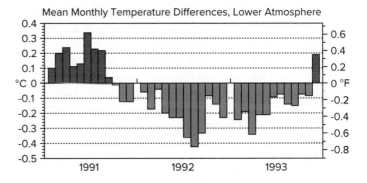

Mean Monthly Temperature Differences, Lower Atmosphere

Figure 5.28

Effect of 1991 eruption of Pinatubo on near-surface air temper-atures. Removal of ash and dust from the air was relatively rapid; sulfate aerosols persisted longer. The major explosive eruption occurred in mid-June 1991.

Source: Data from National Geophysical Data Center

Like with earthquakes, the start of volcanic activity is preceded by numerous types of physical changes or patterns that we may be able to interpret in time to provide accurate warning and evacuations. Seismic activity itself is our strongest tool in understanding the movement of materials under a volcano that indicate it is likely to erupt. Volcanology is employing different types of remote sensing to understand changes in volcano geometry, temperature, and gas content. Even though our scientific ranking of volcano activity is relatively subjective, the science of predicting volcanic eruptions and related hazards continues to succeed in saving people. We will summarize our current understanding of volcanic eruption prediction, and peoples' response to those predictions, in the section below.

Classification of Volcanoes by Activity

In terms of their activity, volcanoes are divided into three somewhat subjective categories: active, dormant, and extinct. A volcano is generally considered *active* if it has erupted within recent history. When the volcano has not erupted recently but is not too eroded or worn down, it is regarded as *dormant,* inactive for the present but with the potential to become active again. Historically, we consider a volcano to be *extinct* if it has no recent eruptive history and appears very much eroded, or is very unlikely to erupt again.

Unfortunately, precise rules for assigning a particular volcano to one category or another do not exist. As volcanologists learn more about the frequency with which volcanoes erupt, it has become clear that volcanoes differ widely in their normal patterns of activity. Long quiescence is no guarantee of extinction. Just as knowledge of past eruptive style allows anticipation of the kinds of hazards possible in future eruptions, so knowledge of the eruptive history of any given volcano is critical to knowing how likely future activity is, and on what scales of time and space.

The first step in predicting volcanic eruptions is monitoring, keeping an instrumental eye on the volcano. There are not nearly enough personnel or ground-based instruments available to monitor every volcano all the time. USGS scientists estimate there are approximately 1350 active volcanoes on Earth with about 500 erupting in historical time, and 161 of those in the United States. Remote sensing is allowing us to monitor more volcanoes more frequently, but we still need scientists to interpret the data. Commonly, intensive monitoring is undertaken only after a volcano shows some signs of near-term activity (see the discussion of volcanic precursors below). Dormant volcanoes might become active at any time. In principle, they also should be monitored as they can sometimes become very active very quickly. Can we safely ignore extinct volcanoes? That would assume we can distinguish an extinct volcano from one that has been dormant for a long time. Vesuvius had been regarded as extinct until it destroyed Pompeii and Herculaneum in A.D. 79. Mount Pinatubo had been classed as dormant prior to 1991. It had not erupted in more than 400 years when, within a matter of two months, it progressed from a few steam emissions to a major explosive eruption. The Chilean volcano Chaitén had not erupted for 9400 years prior to May 2008 when it suddenly began a pyroclastic eruptive phase complete with explosions, ash fall, lahars, and pyroclastic flows that continued for nearly two years (**figure 5.29**). Determining the detailed eruptive history of volcanoes over the long term is difficult, and we simply have not had the time or resources to do so for every possible destructive volcano.

Figure 5.29

Hazard awareness prompted many residents of Chaitén town to "self-evacuate" when earthquakes began. Had they not done so, many would have died when these lahars swept through town less than a week later.

U.S. Geological Survey Volcano Disaster Assistance Program

The Volcanic Explosivity Index

Two volcanologists developed the **Volcanic Explosivity Index** (VEI) as a way to characterize the relative sizes of explosive eruptions. The index takes into account the volume of pyroclastics produced, how high into the atmosphere they rose, and the length of the eruption. The VEI scale, like earthquake magnitude scales, is exponential, meaning that each unit increase on the VEI scale represents about a tenfold increase in size/severity of eruption. The two hazard scales also share the pattern of small events far exceeding large events in frequency.

Several eruptions discussed in this chapter are plotted on the VEI diagram in **figure 5.30**. Knowing the size of eruption of which a given volcano is capable is helpful in anticipating future hazards. At one time, scientists thought that the VEI could also be used to project the probable climatic impacts of volcanic eruptions. However, it seems that the volume of sulfur dioxide emitted is a key factor in an eruption's effect on climate, and that is neither included in the determination of VEI nor directly related to the size of the eruption as measured by VEI.

Volcanic Eruption Precursors

What information do scientists collect and study that might indicate a volcano is going to erupt in the near future? The most widely used method that indicates magma is moving is seismic activity. As magma and gas rise up through the lithosphere beneath a volcano they stress and fracture the rocks, and the process may produce months or years of small, and occasionally large, earthquakes. Increased seismicity at Mt. Pinatubo was a major factor prompting evacuations. The 19 September 2021 eruption on La Palma was preceded by four years of precursor seismic activity that changed in character preceding the

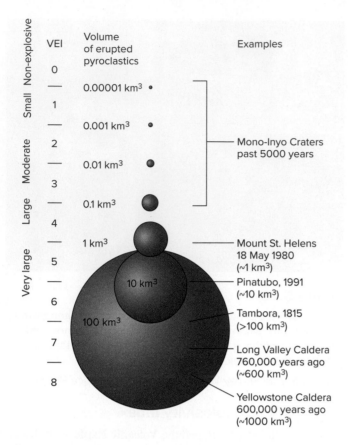

Figure 5.30

The Volcanic Explosivity Index indicates relative sizes of explosive eruptions.

Source: After USGS Volcano Hazards Program

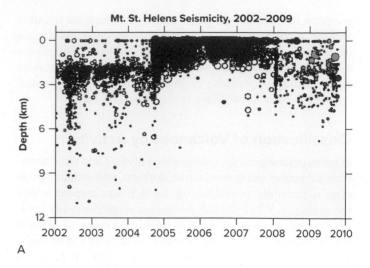

A

Figure 5.31A

Depths of earthquake foci below Mount St. Helens as a function of time, 2002–2009. Size of circle represents size of quake; virtually all were below magnitude 4. Green circles are events of the last year; red, of the last month.

Data from Pacific Northwest Seismograph Network

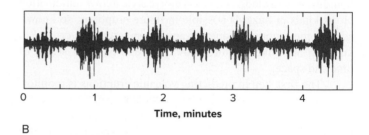

B

Figure 5.31B

Harmonic-tremor signal.

After U.S. Geological Survey

eruption; one week before the eruption earthquakes moved distinctly to the west, and immediately before the eruption they become distinctly shallower. In fall 2004, when earthquakes at Mount St. Helens became both more frequent and shallower (**figure 5.31A**), volcanologists took notice. Monitoring intensified; aircraft were diverted around the mountain in case of ash eruptions. Some small ash emissions occurred intermittently thereafter, but the eruptive phase mainly involved renewed lava-dome building in the summit crater, the results of which were shown in **figure 5.14**. Studies further indicated that this was not new magma replenishing the crustal reservoir under the volcano, but just more of the older magma rising into the crater. In late 2007, the seismicity and dome-building subsided again.

Mount St. Helens is also one of the first places where *harmonic tremors* were studied. These infrasounds are continuous, rhythmic tremors (**figure 5.31B**) that are very different in character from earthquakes' seismic waves, although likely not unrelated. Current research hypothesizes that they are created by the formation, movement, and bursting of gas bubbles in the magma as it makes its way to the surface. Scientists in Iceland, home to 30 active volcanoes (or volcanic systems), comprehensively monitor harmonic tremors along with earthquakes as they wait for their next eruption. While we

investigate their precise origin and significance, they seem to indicate magmatic activity close to the surface and very possibly an impending eruption.

Bulging, tilt, or uplift of the volcano's surface is also a warning sign. Deformation often indicates the presence of a rising magma mass, the buildup of gas pressure, or both. On Kilauea, eruptions are preceded by an inflating of the volcano as magma rises up from the asthenosphere to fill the shallow magma chamber. Unfortunately, it is not possible to specify exactly when the swollen volcano will crack to release its contents. That varies from eruption to eruption with the pressures and stresses involved and the strength of the overlying rocks, and is different for every volcano. Historically, tilt was monitored via individual tilt meters placed at a limited number of points on the volcano. Now, satellite-based interferometric synthetic aperture radar allows a richer look at the regional picture (**figure 5.32**). Interferograms are constructed using radar images taken on two different dates, showing the deformation that occurred during that time frame.

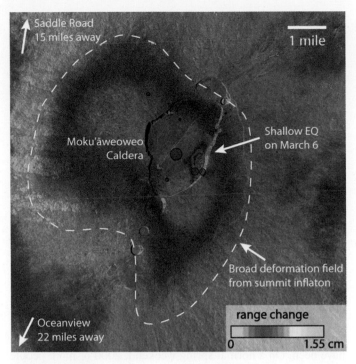

Figure 5.32

This interferogram of Mauna Loa Volcano shows deformation that occurred from November 2020 to March 2021. Each cycle of colors from blue to green represent a change in the distance between the ground and the satellite of 1.55 cm.

Hawaiian Volcano Observatory/USGS

Changes in the mix of gases coming out of a volcano may give clues to impending eruptions. The SO_2 content of the escaping gas shows promise as a precursor, perhaps because more SO_2 can escape as magma nears the surface. Gas composition can also indicate if the magma is coming from a deeper or shallower source, which aides in understanding its viscosity and how it will erupt. Surveys of ground surface temperatures may reveal especially warm areas where magma is particularly close to the surface and about to break through. Surface temperature measurements can be especially dangerous, so again, scientist rely on thermal cameras and infrared satellite sensors that provide data from a safe distance (**figure 5.33**).

More work is needed before the exact timing and nature of major volcanic eruptions can be anticipated consistently. The recognition of impending eruptions of Mount St. Helens, Pinatubo, Soufrière Hills, and Chaitèn was obviously successful in saving many lives through evacuations and restrictions on access to the danger zones. Those danger zones were effectively identified by a combination of mapping of debris from past eruptions and monitoring precursors including volcano geometry and seismic events. Mount Pinatubo displayed increasing evidence of forthcoming activity for two months before its major eruption of 15 June 1991: local earthquakes, increasing in number, and becoming concentrated below a bulge on the volcano's summit a week before that eruption; preliminary ash emissions; a sudden drop in gas output that suggested a plugging of the volcanic plumbing that would tend to increase internal gas pressure and explosive potential. Those

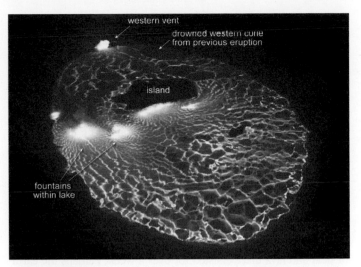

Figure 5.33

Thermal image of Halema'uma'u crater at the summit of Kīlauea on 30 September 2021. Lava fountains are the brightest (hottest) sections at temperatures around 1150°C (2100°F). A cooler island of lava is at the center. The crater is roughly 2800 m x 1800 m in size, although collapses continue to occur increasing its area.

USGS image by M. Patrick

warnings, combined with aggressive and comprehensive public warning procedures and, eventually, the evacuation of over 80,000 people, greatly reduced the loss of life. Geoscientists monitoring Soufrière St. Vincent in the Lesser Antilles issued the successful evacuation of 18,000 people on 8 April 2021 after weeks of increased activity in the composite volcano that included the growth of the dome in the crater, earthquakes, and gas and steam emissions. A series of explosions started on 9 April 2021 that resulted in extensive ashfall and pyroclastic flows (**figure 5.34**).

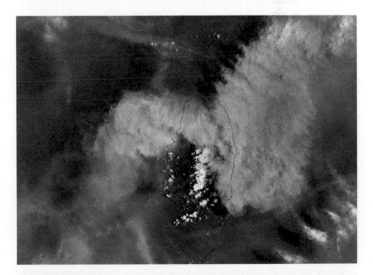

Figure 5.34

This satellite images shows the ash cloud from Soufrière St. Vincent hours after its eruption on 9 April 2021. The volcano itself is obscured.

NASA Earth Observatory/Lauren Dauphin

These successes clearly demonstrate the potential for reduction in fatalities when precursory changes occur and local citizens are educated to the dangers. On the other hand, the exact moment of an eruption may not be known until seconds beforehand, and it is not currently possible to tell whether or just when a pyroclastic flow might emerge from a volcano like Volcán de Fuego. Nor can volcanologists readily predict the volume of lava or pyroclastics to expect from an eruption or the length of the eruptive phase, although they can anticipate the likelihood of an explosive eruption or other eruption of a particular style based on tectonic setting and past behavior.

Scientists continue to discuss another possible link between seismicity and volcanoes. In May 2006, a magnitude-6.4 earthquake occurred in Java, Indonesia. Several days later, the active Javanese volcanoes Merapi and Semeru showed increased output of heat and lava that persisted for over a week. Perhaps the earthquake contributed to the rise of additional magma into these already primed volcanoes, enhancing their eruptions. A magnitude-5.2 earthquake near Chaitén may have accounted for its abrupt reawakening, by opening a crustal fissure through which magma could rise 5 km (3 miles) to the surface in a matter of hours. We might want to be paying a bit more attention to any active or dormant volcano that gets shaken by strong earthquakes.

Evacuation as Response to Eruption Predictions

When data indicate an impending eruption that might threaten populated areas, the safest course is evacuation until the activity subsides. However, a given volcano may remain more or less dangerous for a long time. An active phase, consisting of a series of intermittent eruptions, may continue for months, years, or even decades. In these instances, either property must be abandoned for prolonged periods or inhabitants must be prepared to move out, not once but many times. The Soufrière Hills volcano on Montserrat began its latest eruptive phase in July 1995. By the end of 2000, dozens of eruptive events that included ash venting, pyroclastic flows, ejection of rocks, and building or collapse of a dome in the summit had occurred. Eruptive activity continued while more than half the island remained off-limits to inhabitants. The economic and social disruption were enormous, but many lives were saved. By the time the eruptive phase subsided in 2013, the capital city of Plymouth had been completely destroyed. Many communities were abandoned and over 60% of the people left the island permanently.

Accurate prediction and assessment of the hazard is particularly difficult with volcanoes reawakening after a long dormancy because historical records for comparison with current data are sketchy or nonexistent. Such volcanoes may not even be monitored, because no significant threat is anticipated, or perhaps because the country in question has no resources to spare for volcano monitoring. And while some volcanoes, such as Mount St. Helens or Mount Pinatubo, may produce weeks or months of warning signals as they emerge from dormancy to full-blown eruption allowing time for scientific analysis, public education, and reasoned response, not all volcanoes' awakenings are so leisurely.

Given the uncertain nature of eruption prediction at present, some precautionary evacuations will continue to be shown, in retrospect, to have been unnecessary or unnecessarily early. There are other volcanoes in the Caribbean similar in character to Mont Pelée and Soufrière Hills. In 1976, La Soufrière on Guadeloupe began its most recent eruption. Some 70,000 people were evacuated and remained displaced for months, but only a few small explosions occurred. Government officials and volcanologists were blamed for disruption of people's lives.

On the other hand, it was fortunate that officials took the 1980 threat of Mount St. Helens seriously after its first few signs of reawakening. The immediate area near the volcano was cleared of all but a few essential scientists and law enforcement personnel. Access to hundreds of square kilometers of terrain around the volcano was severely restricted for both residents and workers (mainly loggers). Others with no particular business in the area were banned altogether. Many people grumbled at being forced from their homes. A few flatly refused to go. Numerous tourists and reporters were frustrated in their efforts to get a close look at the action. When the major explosive eruption came, there were casualties, but far fewer than there might have been. On a normal spring weekend, 2000 or more people would have been on the mountain, with more living or camping nearby. Considering the influx of the curious once eruptions began and the numbers of people turned away at roadblocks leading into the restricted zone, it is possible that, with free access to the area, tens of thousands could have died. Many lives were likewise saved by timely evacuations near Mount Pinatubo in 1991, made possible by careful monitoring and supported by intensive educational efforts by the government.

The better we become at recognizing and interpreting precursors, the more precisely targeted such evacuations can become. There may be other cases in which permanent evacuation is the best strategy. After the eruption of Chaitén, officials decided that the site of Chaitén town was so vulnerable to future lahars and pyroclastic flows that, rather than rebuild the town where it had previously stood, they would select a new site for it and encourage its residents to rebuild and relocate there.

5.5 More on Volcanic Hazards in the United States

Learning Objectives

- Describe the geology of and hazards associated with the Cascades and Aleutians volcanoes
- Summarize the reasons geologists are concerned about future eruptions of the Long Valley and Yellowstone calderas

We have already discussed the volcanoes of Hawaii at length in the various sections of this chapter, which makes sense, as Kilauea is the most active and most intensely studied volcano on the planet. We have also approached the eruption of Mount St. Helens from multiple angles, that being the most recent and most disruptive volcanic event in U.S. history. In this section, we will take a look at a few other areas in the United States that either currently have active volcanoes or show us that there have been significant eruptions in the not so distance past.

Cascade Range

Subduction in the U.S. Pacific Northwest results in the creation of a lot of magma, giving rise to numerous volcanoes including the previously mentioned Mount St. Helens (**figure 5.35**). The USGS ranks all the volcanoes shown on the map as either very high or high threat based on eruption potential and proximity to communities. Lassen Peak was last active between 1914 and 1921, not so very long ago geologically. Its products are very

similar to those of Mount St. Helens; violent eruptions are possible. Mount Baker (last eruption, 1870), Mount Hood (1865), and Mount Shasta (active sometime within the last 200 years) have shown seismic activity, and steam is escaping from these and from Mount Rainier, which last erupted in 1882. Overall, the majority of the Cascade peaks are presently showing thermal activity of some kind. The eruption of Mount St. Helens has in one sense been useful. Scientists are watching many of the Cascade Range volcanoes very closely now. There is particular concern about mudflow dangers from Mount Rainier (**figure 5.24**) as discussed early in this chapter.

The Aleutians

South-central Alaska and the Aleutian island chain sit above a subduction zone. This makes the region vulnerable not only to

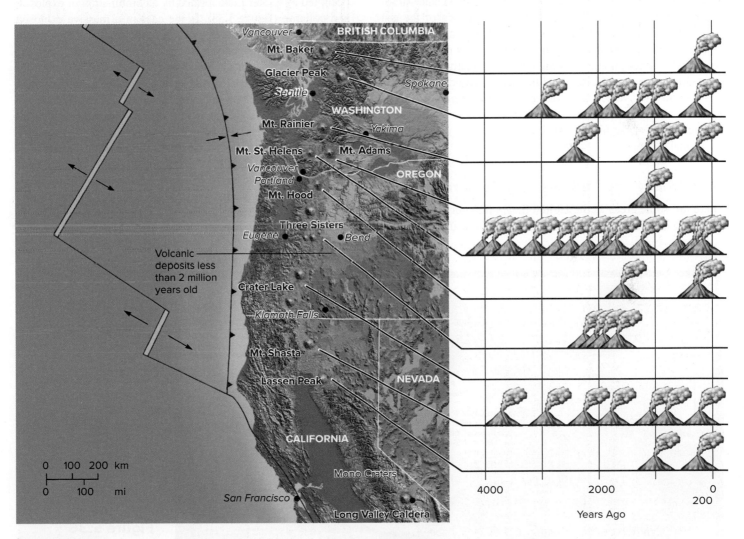

Figure 5.35

The Cascade Range volcanoes and their spatial relationship to the subduction zone and to major cities (plate-boundary symbols as in **figure 3.16**). Tan shaded areas are covered by young volcanic deposits (less than 2 million years old). The graphic on the right shows eruptive history over the past 4000 years. Visit the Cascades Volcano Observatory for the most recent activity https://www.usgs.gov/observatories/cvo.

Eruptive histories from USGS Open-File Report 94-585

earthquakes, as noted in chapter 4, but also to volcanic eruptions (**figure 5.36**). The volcanoes are typically andesitic composite cones, and pyroclastic eruptions are common. There are over 50 active volcanoes, and another 90-some that are geologically young. Thirty of the volcanoes have erupted since 1900 and collectively they average one eruption a year. In the summer of 2021, three volcanoes—Great Sitkin, Pavlof, and Semisopochni—were erupting at the same time although none created sizeable hazards.

Even though Alaska is a region of relatively low population density, these volcanoes present a significant hazard to people. As with Eyjafjallajökull, the concern is the threat that volcanic ash poses to aviation. This is especially important with respect to the Alaskan volcanoes as Anchorage is a key stopover on long-distance cargo and passenger flights through the region, and many transcontinental flights follow great-circle routes over the area. Since the hazard was first realized in the mid-1980s, a warning system has been put in place by the National Oceanic and Atmospheric Administration. However, much monitoring of the ash clouds is done by satellite, and ash and ordinary clouds cannot always be distinguished in the images (especially at night). Volcanologists now coordinate with the National Weather Service to issue forecast maps of airborne ash plumes based on volcanic activity and wind velocity at different altitudes (**figure 5.37**).

Case Study 5.2 focuses on Redoubt, cause of the most frightening of these airline ash incidents, and a great deal of additional damage.

Long Valley and Yellowstone Calderas

There are some areas in the United States that currently do not appear to be volcanically hazardous but are causing geologists

concern because of increases in seismicity or thermal activity, and because of their past histories. One of these is the Mammoth Lakes area of California. In 1980, the area suddenly began experiencing earthquakes. Within a 48-hour period in May 1980, four earthquakes of magnitude 6 rattled the region, interspersed with hundreds of lesser shocks. Mammoth Lakes lies within the Long Valley Caldera, a 13-kilometer-long oval depression formed during violent pyroclastic eruptions 700,000 years ago. Smaller eruptions occurred around the area as recently as 50,000 years ago.

A **caldera** is an enlarged volcanic crater, formed either by an explosion enlarging an existing crater or by collapse of a volcano after its magma chamber is emptied. The summit calderas of Mauna Loa and Kilauea formed predominantly by collapse as the magma chambers became depleted, and enlarged as the rocks on the rim have weathered and crumbled. The caldera now occupied by Crater Lake formed by a combination of explosion and collapse (**figure 5.38**). In the case of a massive explosive eruption such as the one that produced the Long Valley Caldera, the original volcanic cone is obliterated altogether. The concern is that future eruptions could also be on a grand scale.

Recent geophysical studies have shown that a partly molten mass of magma close to 4 kilometers across currently lies below the Long Valley caldera. Furthermore, since 1975, the center of the caldera has bulged upward more than 25 centimeters (10 inches); at least a portion of the magma appears to be rising in the crust. In 1982, seismologists realized that the patterns of earthquake activity in the caldera were disturbingly similar to those associated with the eruptions of Mount St. Helens. In May 1982, the director of the USGS issued a formal notice of potential volcanic hazard for the Mammoth Lakes/

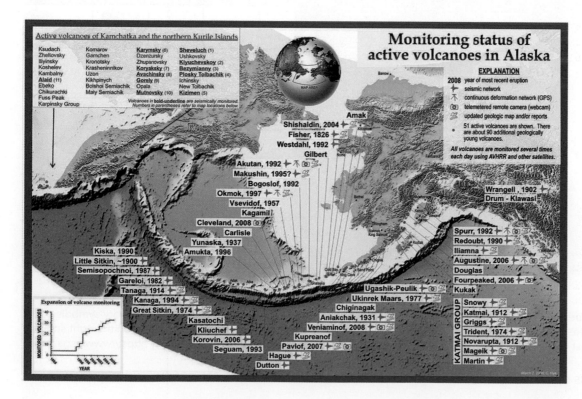

Figure 5.36

Map of the 50+ active Alaskan volcanoes with eruptions through 2008.

Image by C. J. Nye, courtesy Alaska Volcano Observatory

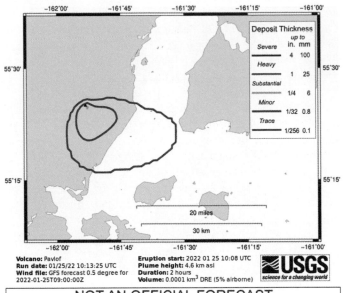

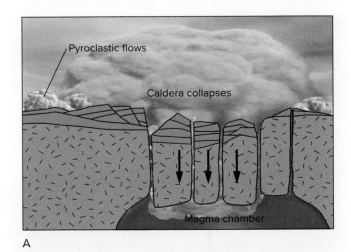

Figure 5.37

The ashfall forecast for a hypothetical eruption from Pavlof volcano. Forecasts are created for each eruptive event and include estimates for ash cloud height and load. Find actual forecasts for Pavlof at: https://www.avo.alaska.edu/activity/Pavlof.php.

USGS

Figure 5.38

(A) Formation of a caldera by collapse of rock over the partially emptied magma chamber, with ejection of pyroclastics. (B) Crater Lake, in the summit caldera of extinct Mount Mazama, is 6 miles (10 km) across. Wizard Island (foreground) is a cinder cone that formed within the caldera.

(B) Doug Sherman/Geofile

Long Valley area. A swarm of earthquakes of magnitudes up to 5.6 occurred there in January of 1983. In 1989, following an earthquake swarm under Mammoth Mountain (just southwest of the town of Mammoth Lakes), the tree-kill previously described began. So far, there has been no eruption, but scientists continue to monitor developments very closely. One of the unknown factors is the magma type in question, and therefore the probable eruptive style. Both basalts and rhyolites are found among the young volcanic rocks in the area. Meanwhile, the area has provided a lesson in the need for care when issuing hazard warnings. Media overreaction to and public misunderstanding of the 1982 hazard notice created a residue of local hard feelings, many residents believing that tourism and property values had both been hurt, quite unnecessarily. It remains to be seen what public reaction to a more-urgent warning of an imminent threat here might be.

Long Valley is not the only such feature being watched somewhat warily. As noted earlier, another area of uncertain future is Yellowstone National Park. Yellowstone, at present, is notable for geothermal features such as geysers, hot springs, and fumaroles (**figure 5.39**). Yellowstone is the site of a continental hot spot, where basaltic magma is rising up and melting the overlying crust, creating some very explosive rhyolitic magma. This magma close to the surface is responsible for the park's famous features, but also for its seismicity and massive eruptions in the relatively recent geologic past (**figure 5.40**).

Over the past 2 million years, the volcano has erupted roughly every 600,000 years spreading ash as far as Mexico and Canada (**figure 5.41**). Eruptions of this scale, with a VEI of 8, lead to the term *supervolcano*. More accurately, the park overlies a caldera created from the last eruption when the magma chamber was emptied and the volcano collapsed. This chamber is again filling with magma, but currently there are no indications of an eruption any time soon.

Scientists continue to monitor closely the surface deformation and changes in seismic and thermal activity at Yellowstone. The research goes on, there and in other volcanic areas, as we seek to understand when and why a restless volcano may become a serious threat.

Figure 5.39

Continuing thermal activity at Yellowstone National Park is extensive, with many boiling-water pools, steaming springs, and geysers.

©Gina S. Szablewski

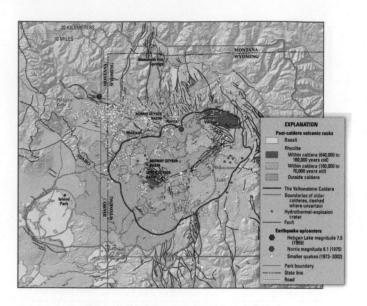

Figure 5.40

This geologic map of Yellowstone shows the extent of the magma chamber underlying the park as well as young volcanic rocks, faults, and earthquake epicenters.

Volcano Hazards/USGS

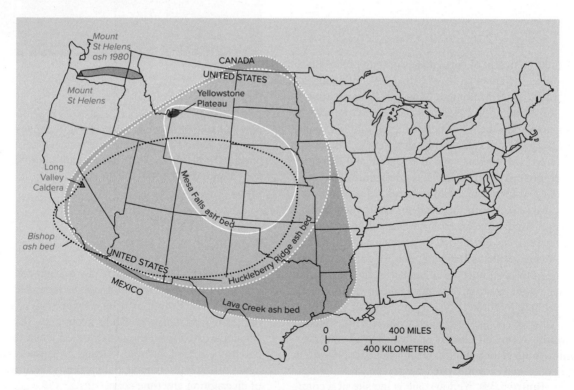

Figure 5.41

Major explosive eruptions of Yellowstone caldera produced the Mesa Falls, Huckleberry Ridge, and Lava Creek ash falls. The Lava Creek event was shown at VEI 8 in **figure 5.30**. Dashed black outline is Bishop ash fall from the Long Valley Caldera in eastern California.

Source: After USGS Fact Sheet 2005-3024

Living with Lava on Hawaii

The island of Hawaii is a complex structure built of five overlapping shield volcanoes. Even where all the terrain is volcanic, degrees of future risk can be determined from the types, distribution, and ages of past eruptions (**figure CS 5.1.1**).

The East Rift zone of Kilauea in Hawaii has been particularly active in recent years. This rift is not a spreading-ridge zone but rather one of several large fracture zones on the flanks of Mauna

Loa and Kilauea from which lava has erupted in recent history (**figure CS 5.1.2**). Less than ten years after a series of major eruptions in the 1970s, a new subdivision called Royal Gardens was begun below the East Rift. By 1983, 1500 lots had been sold primarily to mainlanders who had never visited the site, and 75 homes built. In that year, lava flows reached down the slopes of the volcano and quietly obliterated new roads and houses. Eruptions have continued intermittently for nearly four decades. The photographs in **figure CS 5.1.3** illustrate some of the results. Even those in Royal Gardens whose homes initially survived were hardly unaffected. They were isolated by several miles of lava from the nearest road to a grocery store, had no water or electricity, and were unable to insure what they had left, never mind sell the property in which they had invested. In 2012, lava finally destroyed the last house standing in Royal Gardens.

Why would anyone build in a location where the risks are well known and quite obvious? Over centuries, Hawaiians and other residents of active volcanic areas have commonly become philosophical about such activity. Anyone living in Hawaii is living on a volcano, and the nonviolent eruptive style of these shield volcanoes means that they rarely take lives. Staying out of the riskiest areas greatly reduces even the potential for loss of property.

Zone	Percent of area covered by lava since 1800.	Percent of area covered by lava in last 750 yrs.	Explanation
Zone 1	greater than 25%	greater than 65%	Includes the summits and rift zones of Kilauea and Mauna Loa, where vents have been repeatedly active in historic time.
Zone 2	15–25%	25–75%	Areas adjacent to and downslope of active rift zones.
Zone 3	1–5%	15–75%	Areas gradationally less hazardous than Zone 2 because of greater distance from recently active vents and/or because the topography makes it less likely that flows will cover these areas.
Zone 4	about 5%	less than 15%	Includes all of Hualalai, where the frequency of eruptions is lower than on Kilauea and Mauna Loa. Flows typically cover large areas.
Zone 5	none	about 50%	Areas currently protected from lava flows by the topography of the volcano.
Zone 6	none	very little	Same as Zone 5.
Zone 7	none	none	20 percent of this area covered by lava 3500–5000 years ago.
Zone 8	none	none	Only a few percent of this area covered in the past 10,000 yrs.
Zone 9	none	none	No eruption in this area for the past 60,000 yrs.

A

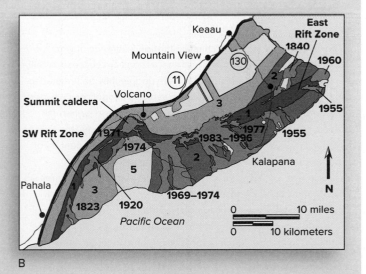

B

Figure CS 5.1.1

(A) Volcanic-hazard map of Hawaii showing the five volcanoes (labeled) and major associated rift zones (red). Degree of risk is related to topography (being downslope from known vents is worse), and extent of recent flows in the area. Zones 6–9 have experienced little or no volcanism for over 750 years. In Zone 1, 65% or more of the area has been covered by lava in that time, and more than 25% since 1800. (B) Hazard map for Kilauea's East Rift Zone.

(A) Source: C. Heliker, Volcanic and Seismic Hazards of Hawaii, U.S. Geological Survey, 1991; (B) Source: U.S. Geological Survey/Hawaii Volcanoes Observatory

(Continued)

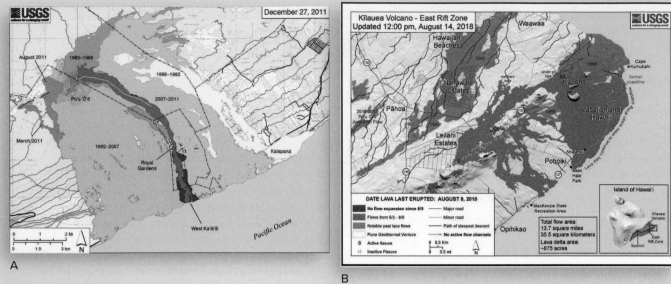

Figure CS 5.1.2

(A) Lava flows have poured down from Kilauea's East Rift for decades, with activity shifting geographically over time. Note location of Royal Gardens subdivision, which sat on a steep slope below the East Rift, and is now entirely buried under lava. Red flows are late 2011. (B) This map of the easternmost end of the East Rift Zone (see inset) shows Pahoa, threatened by flows from the west in 2014, and other developments affected by the 2018 activity. Note ages of recent previous lava flows, the line of small volcanic cones along the rift zone (shaded), and the string of fissures that first appeared in Leilani Estates. The Puna Geothermal Venture geothermal power plant suffered damage in 2018 and was shut-down for over two years.

(A) Source: *Maps courtesy of U.S. Geological Survey/Hawaii Volcanoes Observatory; (B) USGS*

The land may be relatively cheap, although insurance may be difficult to come by. Warnings and evacuations are successful. Still, as populations grow and migrate, more people may be tempted to move into areas that might better be avoided altogether. Outsiders unfamiliar with volcanoes may unknowingly settle in higher-risk areas, as with Royal Gardens. Sometimes a particularly far-reaching flow will overwhelm a town previously untouched for a century or more, where residents might have become just a little bit complacent; such was the case with Kalapana, shown in **figures CS 5.1.1B** and **CS 5.1.2A**. In all likelihood, unwise development will continue in the absence of zoning regulations.

Meanwhile, activity at Kilauea continues to ebb and flow and shift in location and character over time. In 2008, steam and small volumes of pyroclastics began to issue from the Halema`uma`u crater in the summit caldera; a lava lake rose in the crater, and it is still active. In 2014, a tongue of lava unexpectedly flowed northeast along the East Rift toward the town of Pahoa **(figure CS 5.1.2B)**, but after several nervous months for the residents, flow activity shifted back to the more usual paths for this eruptive cycle, southeastward down the cliffs below the rift toward the ocean.

Kilauea's ongoing activity continues to highlight examples of ill-advised development. In May 2018, a series of fissures suddenly opened in the Leilani Estates subdivision, built in the 1960s squarely on the lower East Rift Zone, in Hazard Zone 1 **(figure CS 5.1.2B)**. At first only steam and sulfurous gases escaped from the fissures, but then a series of lava flows and fountains began to erupt **(figure CS 5.1.4A)**. This activity was accompanied by thousands of earthquakes, mostly small, but some up to magnitude 6.9. At the same time, new things began to happen at Kilauea's summit, 30 km (20 mi) away. The lava lake in the summit crater drained rapidly, its level dropping by hundreds of meters in a matter of days. Interaction among the remaining magma, groundwater, and rocks falling into the crater from its crumbling rim produced ash plumes **(figure CS 5.1.4B)** and some small explosions; the largest plumes shot up over 30,000 feet into the atmosphere and triggered a code-red aviation warning. Hundreds of homes on the lower East Rift were overrun by lava, set afire, or isolated by lava flows; major and minor roads were cut off; and thousands of people were evacuated from the affected area due to both lava and hazardous fumes. Kilauea's current, decades-long eruptive phase shows no sign of ending anytime soon; it simply remains to be seen where its activity may next break out, and what form it may take. You can find updates at the Hawaiian Volcano Observatory https://www.usgs.gov/observatories/hvo.

A

B

C

Figure CS 5.1.3

(A) Lava flowing over a nearly new road In the Royal Gardens subdivision, Kilauea, Hawaii, 1983. Hot lava can even ignite asphalt. (B) By 1986, destruction in Royal Gardens and Kalapana Gardens subdivisions was widespread. Here, vegetation has died from heat; note engulfed trucks at rear of photograph. (C) By 1989, advancing tongues of lava had stretched to the coast and threatened more property. Here, the former Wahaula Visitor Center at Hawaii Volcanoes National Park goes up In flames; the area has been entirely covered by lava flows.

(A, B, C) U.S. Geological Survey/Photographs by J. D. Griggs

A

B

Figure CS 5.1.4

(A) Aerial view of May 2018 activity in Leilani Estates area. Some lava fountains (center) rise over 50 m (160 feet) high, dwarfing nearby trees. Note the string of steaming fissures along the East Rift (upper right), and remaining buildings isolated by lava flows (left). (B) Ash plume of 15 May 2018 at Kilauea's summit crater.

(A) Photograph courtesy U.S. Geological Survey/Volcano Helicopters; (B) Photograph courtesy U.S. Geological Survey

Redoubt Volcano, Alaska

Redoubt's location (**figure CS 5.2.1**) puts 60% of Alaska's population, and its major business centers, within reach of the volcano's eruptions. Nearby Augustine had erupted in 1986, so increasing seismicity at Redoubt in late 1989 drew prompt attention, though it was not obvious just when or how it might actually erupt. The first event, on 14 December, was an explosive one, spewing ash and rock downwind. Within a day thereafter, fresh glass shards, indicating fresh magma feeding the eruption, appeared in the ash. Redoubt would shortly wreak its costliest and perhaps most frightening havoc of this eruptive phase.

On 15 December, a KLM Boeing 747 flew into a cloud of ash from Redoubt, and all four engines failed. The plane plunged over 13,000 feet before the pilot somehow managed to restart the engines. Despite a windshield frosted and pitted by the abrasive ash, he managed to land the plane, and its 231 passengers, safely in Anchorage. However, the aircraft had suffered $80 million in damage.

Soon thereafter, a lava dome began to build in Redoubt's summit crater. Near-surface seismicity was mostly replaced by deeper events through the rest of December (**figure CS 5.2.2**). Then the regular tremors that volcanologists had come to associate with impending eruptions resumed on 29 December. Suspecting that the new lava dome had plugged the vent and that pressure was therefore building in the volcano, on 1–2 January 1990, the Alaska Volcano Observatory issued warnings of

"moderate to strong explosive activity" expected shortly. The Cook Inlet Pipeline Company evacuated the Drift River oil-tanker terminal. Strong explosions beginning on 2 January demolished

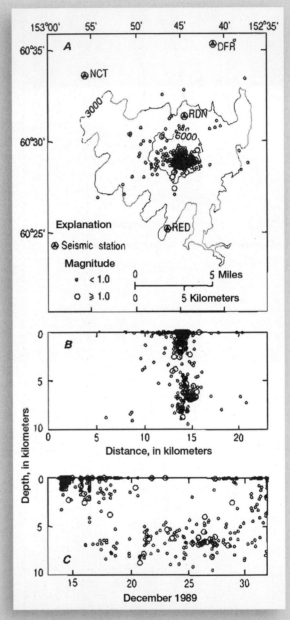

Figure CS 5.2.2

Seismicity at Redoubt, 13 December through 31 December 1989. (A) In map view, seismic activity is strongly concentrated below the summit; depth (B) is variable. (C) Depth of seismic events varies over time; mainly shallow during the first explosions, then deeper, with shallow events resuming in late December.

After USGS Circular 106

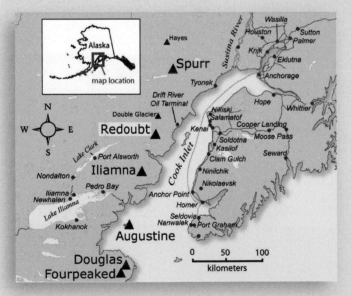

Figure CS 5.2.1

Alaska's population is concentrated around Cook Inlet; Redoubt, Augustine, and Spurr have all erupted since 1990.

J. Schaeffer and C. Cameron, courtesy Alaska Volcano Observatory/ Alaska Division of Geologic and Geophysical Survey

the young lava dome in the crater and sent pyroclastic flows across Redoubt's glaciers. Lahars swept down the Drift River valley toward the terminal (**figure CS 5.2.3**). Larger blocks of hot rock crashed nearer the summit (**figure CS 5.2.4**). After the remains of the dome were blown away on 9 January, eruptions proceeded less violently, though ash and pyroclastic flows continued intermittently, interspersed with dome-building, for several more months (**figure CS 5.2.5**). Altogether, this eruption was the second costliest volcanic eruption in U.S. history, behind only the 1980 eruption of Mount St. Helens. Between aircraft damage, disruption of oil drilling in Cook Inlet and work at the pipeline terminal, and

assorted damage due to the multiple pyroclastic eruptions, total cost was over $160 million.

In early 2009, Redoubt became active again for several months, with multiple pyroclastic eruptions. This time, new tools were available to help track the ash and keep aircraft safe (**figure CS 5.2.6**). As this is being written, the volcano is quiet, but monitoring will certainly continue, to ensure that any further eruption will not be a surprise.

Figure CS 5.2.4

USGS geologists study pyroclastic debris from Redoubt.

U.S. Geological Survey/Photograph by K. Bull, Alaska Volcano Observatory

A

B

Figure CS 5.2.3

(A) Mudflow deposits in the Drift River valley; Redoubt at rear of image. (B) Some lahars completely flooded structures in the valley.

(A, B) U.S. Geological Survey/Photographs by C. A. Gardner, Alaska Volcano Observatory

Figure CS 5.2.5

On 21 April 1990, ash rises from Redoubt. Note the classic stratovolcano shape.

R. Clucas/USGS

A

B

Figure CS 5.2.6

On 1 April 2009, ash from Redoubt drifts east across Cook Inlet and northeast toward Anchorage (see locations in **figure CS 5.2.1**). (B) Thermal infrared imaging allows tracking of ash from 23 March 2009 explosive eruption even at night. The ash shows up because it is actually *colder* than the ground, having been shot up to altitudes of up to 50,000 feet.

(A) NASA image by Jeff Schmaltz, MODIS Rapid Response Team; (B) NASA image by Jesse Allen, using data provided by the MODIS Rapid Response Team

Summary

Most volcanic activity is concentrated near plate boundaries. Addition of fluid to the asthenosphere, or reduction in pressure on it, promotes melting and the production of mafic magma. Subsequent crystallization and interaction of the magma with overlying crust can modify the magma's composition, producing more-felsic magma. Volcanoes differ in eruptive style and in the kinds of dangers they present. Those along spreading ridges and at oceanic hot spots tend to be more effusive, usually erupting a fluid, basaltic lava. Subduction-zone volcanoes are supplied by more-viscous, silica-rich, gas-charged andesitic or rhyolitic magma, so, in addition to lava, they may emit large quantities of pyroclastics and pyroclastic flows, and they may also erupt explosively. The type of volcanic structure built is directly related to the composition/physical properties of the magma; the fluid basaltic lavas build shield volcanoes, while the more-felsic magmas build smaller, steeper stratovolcanoes of lava and pyroclastics. At present, volcanologists can detect the early signs that a volcano may erupt in the near future (such as bulging, increased seismicity, or increased thermal activity), but they cannot predict the exact timing or type of eruption. Individual volcanoes, however, show characteristic eruptive styles (as a function of magma type) and patterns of activity. Therefore, knowledge of a volcano's eruptive history allows anticipation of the general nature of eruptions and of the likelihood of renewed activity in the near future.

Key Terms and Concepts

andesite	decompression melting	partial melting	shield volcano
basalt	fissure eruption	phreatic eruption	stratovolcano
caldera	lahar	pyroclastic flow	viscosity
cinder cone	lava	pyroclastics	Volcanic Explosivity Index
composite volcano	lava dome	rhyolite	

Test Your Learning

1. Define a fissure eruption and give an example.

2. Briefly explain two ways in which magma composition can be modified as the magma moves from asthenosphere to surface.

3. The Hawaiian Islands are all shield volcanoes; describe what this means, and why they are not especially hazardous to humans.

4. Describe how the same hot spot can create cinder cones and shield volcanoes.

5. Compare the eruptive style of Mount St. Helens and that of Kilauea.

6. Define *pyroclastics,* and identify a kind of volcanic structure that pyroclastics may build.

7. Describe a way in which a lahar may develop, and a way to avoid its most likely path.

8. Explain the origin and nature of a pyroclastic flow, and why a volcano known for producing these poses a special threat during periods of activity.

9. Identify the circumstances under which a phreatic eruption may occur, and give a historic example.

10. Volcanic eruptions can influence global climate; explain.

11. Volcanic ash can pose a particular hazard to aviation; explain. And, describe one strategy for reducing the risk.

12. Discuss the distinctions among active, dormant, and extinct volcanoes, and comment on the limitations of this classification scheme.

13. Describe three precursor phenomena that may precede volcanic eruptions.

14. Outline the underlying cause of present and potential future volcanic activity in the Cascade Range of the western United States.

15. Identify the origin of volcanic activity at Yellowstone, and why it is sometimes described as a "supervolcano."

Exploring Further

(For investigating volcanic activity, two excellent websites are the USGS Volcano Hazards Program site, **volcanoes.usgs.gov**, and the Global Volcanism Program site, **volcano.si.edu**. The USGS site includes links to all of the U.S. regional volcano observatories.)

1. As this is written, Kilauea continues an active phase that has lasted over four decades. Check out the current activity report of the Hawaii Volcanoes Observatory.

2. Go to the USGS Volcano Hazards Program site mentioned previously. How many volcanoes are currently featured as being watched? Which one(s) have the highest Aviation Watch color codes, and why?

3. The Global Volcanism Program site highlights active volcanoes. How many volcanoes are currently on that list? Choose a volcano from the list with which you are *not* familiar, and investigate more closely: What type of volcano is it, and what type of activity is currently in progress? What is its tectonic setting, and how does that relate to its eruptive style?

4. Before the explosion of Mount St. Helens in May 1980, scientists made predictions about the probable extent of

and kinds of damage to be expected from a violent eruption. Investigate those predictions, and compare them with the actual effects of the blast.

5. Consider a region of dry upper mantle at the pressure and temperature of the starting point in **figure 5.3**. As a warm plume of this mantle rises, the pressure on it is reduced. If the density of the upper mantle is about 3.3 g/cm^3, and 1 atmosphere of pressure approximately equals 1 kg/cm^2, how far must the plume rise before melting begins, assuming no water is added?

6. Consider the Royal Gardens or Leilani Estates developments mentioned in Case Study 5.1, and the fact that hazard maps were readily available long before it was begun. In such a case, should buyers be responsible for educating themselves about the risks? Should developers be required to inform prospective buyers of the hazards? Should the state simply prohibit development in such high-risk areas? Should taxpayers funds be used to help redevelop the area? What do you think?

CHAPTER 6

Streams and Flooding

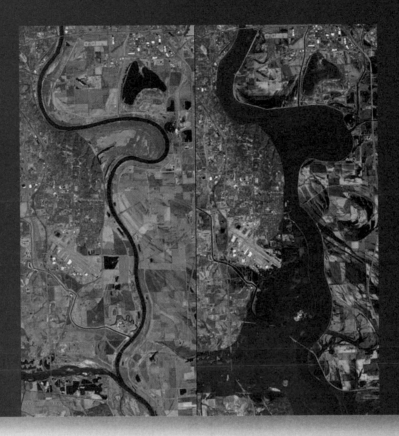

Flowing water is the single most important agent sculpting Earth's surface. Streams are found just about everywhere on the planet and take a variety of forms as they move water and sediment from high elevations to low, distinctly altering the landscape through erosion and deposition. In the hydrologic cycle, the ultimate function of a stream is to return runoff to the oceans.

Streams have long played a role in human affairs. Throughout history, streams have served as a vital source of

Chapter Outline

6.1 The Hydrologic Cycle

6.2 Streams and Their Features

6.3 Flooding

6.4 Consequences of Development in Floodplains

6.5 Strategies for Reducing Flood Hazards

Flooding on the Missouri River just southeast of Omaha, Nebraska, occurred in early spring 2019 due to a combination of factors: rapid snowmelt, ice dams, frozen ground, and intense rainfall. Extensive flooding continued throughout the Midwest through summer, creating the Great Flood of 2019 with roughly $6.2 billion in damages.

NASA Earth Observatory images by Joshua Stevens, using Landsat data from the U.S. Geological Survey

fresh water and food. Before there were motorized engines, there were boats to carry people and goods, and flowing water pushed paddlewheels that powered mills and factories. The rich soil in floodplains makes good farmland. Humans commonly settled along streams for many reasons but this has and continues to put them in the way of flooding.

Floods are one of the most widely experienced and costly natural hazards. With increasing trends in extreme precipitation events, it is perhaps not surprising that a recent study showed the cumulative impact of precipitation-related flooding from 1988 to 2017 to be $73 billion in just the United States. Flooding is often the leading cause of death from natural disasters globally, but flood deaths are decreasing and averaged about 4850 people annually from 2011 to 2020. More people are moving into floodplains, but we are also getting better at getting them out of harm's way during flood events.

Most floods are local and do not make national headlines, but they may be no less devastating to those affected by them. Some floods are the result of unusual events, such as the collapse of a dam, but the vast majority are a perfectly normal, and somewhat predictable part of the natural functioning of streams. Before discussing flood hazards and strategies for dealing with them, we will examine how water moves through the hydrologic cycle and look at the basic characteristics and behavior of streams.

6.1 The Hydrologic Cycle

Learning Objectives

- Describe the movement and storage of water in the hydrosphere

The **hydrosphere** includes all the water at and near Earth's surface. Most of it originated from outgassing of Earth's interior early in its history, when the planet's temperature was higher. Icy comets and water-rich asteroids may have contributed some, too. With only occasional minor additions from volcanoes bringing up water from the mantle, and small amounts of water returned to the mantle with subducted lithosphere, the quantity of water in the hydrosphere remains essentially constant.

All the water in the hydrosphere is contained in the **hydrologic cycle,** illustrated in **figure 6.1**. The largest single reservoir in the hydrologic cycle is the oceans, which contains over 97% of the water in the hydrosphere; lakes and streams together contain only 0.016% of the water. The main processes of the hydrologic cycle involve evaporation into and precipitation out of the atmosphere. Precipitation onto land can reevaporate directly from the ground surface or indirectly through plants by evapotranspiration, infiltrate into the ground, or run off over the ground surface. Surface runoff may occur by overland flow or become channelized into streams. Water that percolates into the ground also flows (see chapter 11). The oceans are the principal source of evaporated water because of their vast areas of exposed water surface.

The total amount of water moving through the hydrologic cycle is large, more than 100 million billion gallons per year. A portion of this water is temporarily diverted for human use, but it ultimately makes its way back into the natural global water cycle. Water in the hydrosphere may spend extended periods of time in storage in one or another of the water reservoirs shown in **figure 6.1**; a few months in a river, a few thousand years in the oceans, centuries to millennia in glaciers or as groundwater, but from the longer perspective of geologic history it is still regarded as moving continually through the hydrologic cycle. In the process, it may participate in other geologic cycles and processes. For example, streams and groundwater eroding rock and moving sediment or dissolved chemicals are also contributing to the rock cycle. We revisit the hydrologic cycle in the context of water resources in chapter 11.

6.2 Streams and Their Features

Learning Objectives

- Recall the terms stream, drainage basin, and divide
- Calculate stream discharge
- Describe how different types of sediment load move in a stream
- Define and calculate gradient
- List the influences on stream velocity and discharge
- Recall base level
- Relate stream velocity to sediment transported and deposited
- Compare a meandering stream to a braided stream
- Describe the formation of a floodplain

An individual stream can vary in character, especially if it extends for long distances. The amount of water carried by a stream and how fast it moves depends on many variables, which in turn affect sediment erosion, transportation, and deposition by the stream. In the following section we introduce many terms to help describe how a stream behaves and changes as it moves from its headwaters to it mouth.

Streams—General

A **stream** is any body of flowing water confined within a channel, although people tend to use the term *river* for a relatively large stream. Streams flow downhill, or *downstream,* from high elevation to low, carrying water over Earth's surface. The region

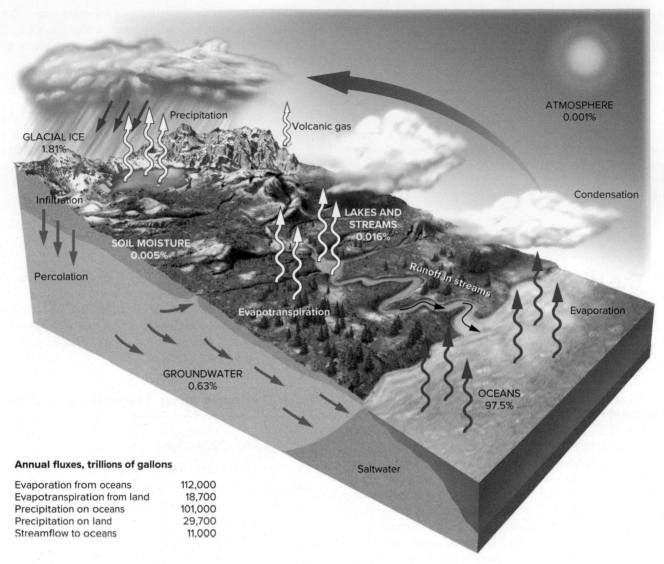

GLACIAL ICE
1.81%

Precipitation

Volcanic gas

ATMOSPHERE
0.001%

Condensation

Infiltration

LAKES AND
STREAMS
0.016%

SOIL MOISTURE
0.005%

Percolation

Evapotranspiration

Runoff in streams

Evaporation

GROUNDWATER
0.63%

OCEANS
97.5%

Saltwater

Annual fluxes, trillions of gallons

Evaporation from oceans	112,000
Evapotranspiration from land	18,700
Precipitation on oceans	101,000
Precipitation on land	29,700
Streamflow to oceans	11,000

Figure 6.1

Principal processes and reservoirs of the hydrologic cycle (reservoirs in capital letters). Straight lines show movement of liquid water; wavy lines show water vapor.

from which a stream gains water is its **drainage basin** (**figure 6.2**), or *watershed*. A *divide* separates drainage basins. Stream size is influenced by the size (area) of the drainage basin, climate (amount of precipitation and evaporation), vegetation or lack of it, and the underlying geology. A stream carves its channel, and the size of the channel is proportional to the volume of water that must typically be carried. Long-term, sustained changes in precipitation, land use, or other factors that change the volume of water customarily flowing in the stream will result in changes in channel geometry.

We describe the size of a stream by its **discharge,** the volume of water flowing past a given point in a specific length of time. Discharge is the product of channel cross-section (area) multiplied by its average stream velocity (**figure 6.3**), measured in cubic feet or meters per second. Discharge ranges from less than one cubic foot per second on a small creek to millions of cubic feet per second in a major river. For a given stream,

discharge varies by location and with season and weather, and may be influenced by human activities.

Sediment Transport

Water is a powerful agent for transporting material. Streams move material in several ways (**figure 6.4**). Heavier and larger debris is rolled, dragged, or pushed along the bottom of the stream bed as its *traction load*. Material of intermediate size is carried in short hops along the stream bed by a process called *saltation*. All this material is collectively described as the *bed load* of the stream. The *suspended load* consists of material that is light or fine-grained enough to be moved in suspension, supported by the flowing water. Suspended sediment clouds a stream and gives the water a muddy appearance, and consists mostly of silt and clay. The ions carried in solution make up the stream's *dissolved load*.

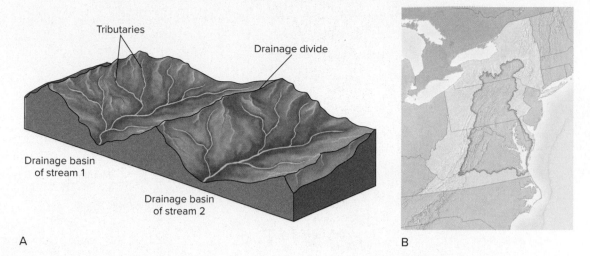

A

B

Figure 6.2

Streams and their drainage basins. (A) A topographic high creates a divide between the drainage basins of adjacent streams. These streams have dendritic drainage patterns. (B) The Chesapeake Bay watershed is 166,000 km^2 in size, with at least 150 major rivers and streams and about 100,000 smaller ones.

(B) Joshua Stevens/NASA Earth Observatory

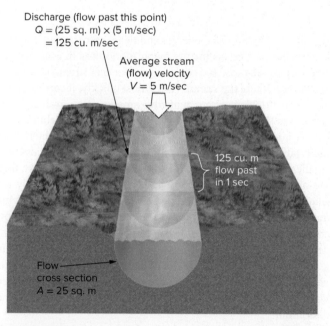

Figure 6.3

Discharge (Q) equals channel cross-section (area) times average velocity: $Q = A \times V$. In a real stream, both channel shape and flow velocity vary, so reported discharge is an average. In any cross-section, flow velocity is greatest near the center of the channel.

The total quantity of material that a stream transports by all these methods is called its **load.** Stream **capacity** is a measure of the total load of material a stream can move. Capacity is closely related to discharge: the faster the water flows, and the more water is present, the more material can be moved. How much of a load is actually transported also depends on the

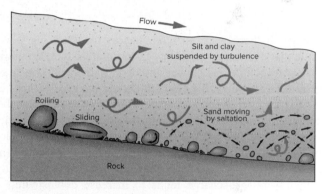

Figure 6.4

Processes of sediment transport in a stream. How material can be moved depends on its size and weight, and the flow velocity.

availability of sediments or soluble material. A stream flowing over solid bedrock will not be able to dislodge much material, while a similar stream flowing through sand or soil may move considerable material. Vegetation also influences sediment transport, by preventing the sediment from reaching the stream at all, or by blocking its movement within the stream channel.

Velocity, Gradient, and Base Level

Stream velocity is related to discharge and to the steepness of the slope down which the stream flows, but is affected by other physical parameters of the stream. The steepness of the stream channel is its **gradient.** It is calculated as the difference in elevation between two points along a stream, divided by the horizontal distance between them along the stream channel; we often describe this relationship as the rise over the run

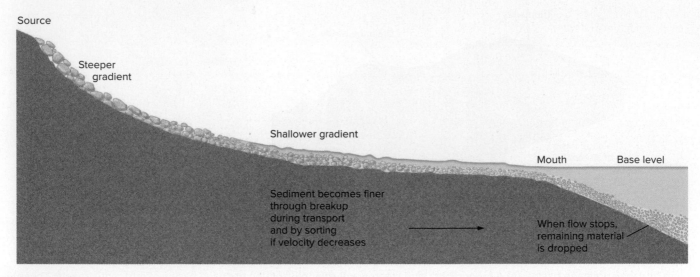

Figure 6.5

Typical longitudinal profile of a stream in a humid climate. Note that the gradient decreases from source to mouth. Particle sizes commonly decrease with breakup downstream. Total discharge, capacity, and velocity may all increase downstream despite the decrease in gradient. Conversely, in a dry climate, water loss by evaporation and infiltration into the ground may decrease discharge downstream.

(rise/run). If there were no other factors involved, we would assume that a stream would flow faster in areas of steep gradients near its source and slower with low gradients downstream (**figure 6.5**). However, the effect of decreasing gradient may be counteracted by the increased water volume as additional tributaries enter the stream, adding to the mass of water being pulled downstream by gravity. Velocity is also influenced by friction between water and stream bed, itself affected by bedload and channel geometry which both change from the source to the mouth. Overall, discharge tends to increase downstream, at least in humid climates where tributaries contribute to increased volume along the way.

As a stream approaches its end, or mouth, it no longer has the power to erode downward, like it did near its mouth, but instead is using its energy to transport an influx of sediments from the drainage basin. The stream is approaching its **base level,** which is the lowest elevation to which the stream can erode downward. For most streams, base level is the water level of the body of water into which they flow, and this is where they deposit their remaining sediment load. For streams flowing into the ocean, base level is sea level. Over time, natural streams tend toward an equilibrium between erosion and deposition of sediments; we call this a *graded* stream. The **longitudinal profile** of such a stream assumes a characteristic concave-upward shape, as shown in **figure 6.5**.

Velocity and Sediment Sorting and Deposition

Variations in a stream's velocity along its length are reflected in the sizes of sediments deposited at different points. The more

rapidly a stream flows, the larger and denser are the particles that can be moved. The sediments found motionless in a stream bed at any point are those too big or heavy for that stream to move at that point. Where the stream flows most quickly, it carries gravel and even boulders along with the smaller, fine-grained sediments. As the stream slows down, it starts leaving behind those heavy, large particles and continues to carry the lighter, finer materials. If stream velocity continues to decrease, successively smaller particles are dropped: the sand-sized particles next, then the clay-sized ones. In a very slowly flowing stream, only the finest-grained sediments and dissolved materials are still being transported. The relationship between the velocity of water flow and the size of particles moved results in **well-sorted** stream-deposited sediments, with materials deposited at a given point tending to be similar in size or weight. The size of individual particles tends to decrease downstream, becoming finer-grained with time and distance as they are subjected to collision and dissolution.

When a stream flows into a lake or ocean, the stream's flow velocity drops to zero, and all the remaining suspended sediment is dropped. If the stream is still carrying a substantial load as it reaches its mouth and flows into still waters, a **delta** may be created as the sediment is deposited (**figure 6.6**). Deltas typically are created from layers of fine-grained sand, silt, and clay. Deltas vary in shape and size depending on factors such as stream discharge, sediment load, water level in the receiving basin, wave strength, tides, various human factors, and whether or not subsidence is occurring. Similar to deltas, *alluvial fans* are formed when streams emerge from steep, mountainous canyons and deposit much of their sediment load in response to a dramatic change in gradient.

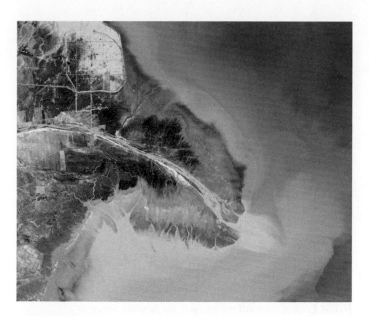
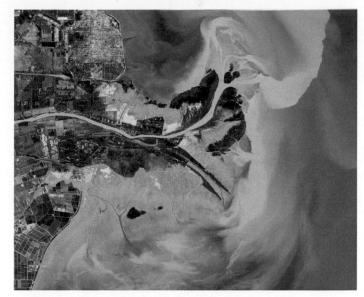

Figure 6.6

The Huang He (Yellow) River delta in China in 1989 (left) and 2020 (right). Engineers blocked the channel in 1996 forcing it to the northeast causing significant changes in both erosion and deposition of sediments. For more on this story, go to https://earthobservatory.nasa.gov/world-of-change/YellowRiver.

NASA Earth Observatory images by Lauren Dauphin

Channel and Floodplain Evolution

When a stream is flowing rapidly and its gradient is steep, down-cutting is rapid. The result is typically a steep-sided valley, V-shaped in cross section, and relatively straight (**figure 6.7**). But streams do not ordinarily flow in straight lines for very long. Small irregularities in the channel cause local fluctuations in velocity, which result in a little erosion where the water flows strongly against the side of the channel and some deposition of sediment where it slows down a bit. Bends, or **meanders,** then begin to form in the stream. Once a meander forms, it tends to enlarge and also to shift downstream. The curve is eroded on the outside and down-stream side, forming the **cut bank,** where the water flows some-what faster. Deposition of sediment, usually sand, occurs opposite the cut bank on the inside of the meander where velocity is slower to form a **point bar** (**figure 6.8**). The rates of lateral movement of meanders are geologically rapid and can range up to tens or even hundreds of meters per year, with more common rates below 10 meters/year (35 feet/year) on smaller streams (**figure 6.8B**).

Obstacles or irregularities in the channel may slow flow enough to cause localized sediment deposition there. If the sed-iment load is large, these sediment bars can build up until they reach the surface, effectively dividing the channel in a process called *braiding*. If the sediment load is very large in relation to water volume, the **braided stream** may develop a complex pat-tern of many channels that divide and rejoin, shifting across a broad expanse of sediment (**figure 6.9**). Braided streams are very commonly created at the base of glaciers, as they melt and release large amounts of sediment.

Over a period of time, the combined effects of erosion on the cut banks and deposition of point bars, downstream meander

Figure 6.7

Classic V-shaped valley carved by a rapidly downcutting stream. Grand Canyon of the Yellowstone River.

©Carla Montgomery

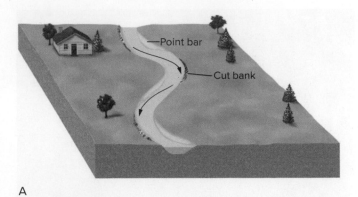

A

Figure 6.8A

Development of meanders. Channel erosion is greatest at the outside and on the downstream side of curves, at the cut bank; deposition of point bars occurs in sheltered areas on the inside of curves. In time, meanders migrate both laterally and downstream and can enlarge as well. See also **figure 6.10**.

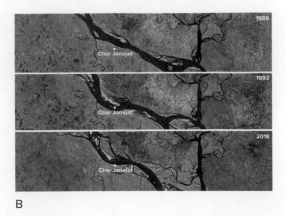

B

Figure 6.8B

The sinuosity of the Lower Padma River in Bangladesh has increased over the past 30+ years. This river is natural and flows over loose sand, making erosion relatively quick and easy.

NASA Earth Observatory images by Joshua Stevens

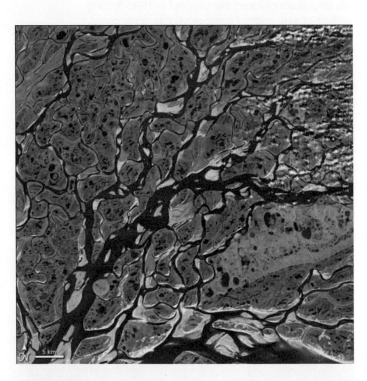

Figure 6.9

The Lena River in Russia becomes braided as it separates into numerous channels on its delta, an area of extreme ecological importance.

Earth Science and Image Analysis Laboratory, NASA Johnson Space Center

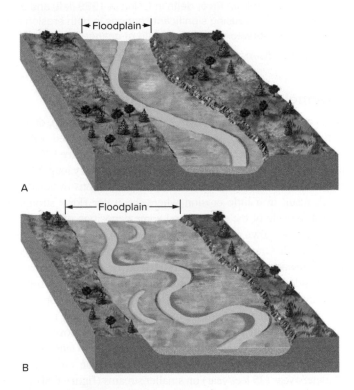

A

B

Figure 6.10

Meandering and sediment deposition during floods contribute to development of a floodplain. (A) Small bends enlarge and migrate over time. (B) A broad, flat floodplain is developed around a stream channel by meandering coupled with sediment deposition during flooding.

migration, and sediment deposition when the stream overflows its banks together produce a broad, fairly flat expanse of land covered with sediment on either side of the stream channel. This is the stream's **floodplain,** the area into which the stream spills over during floods. The floodplain is a normal product of stream evolution over time (see **figures 6.10** and **6.11A**).

Meanders do not broaden or enlarge indefinitely. Very large meanders represent major detours for the flowing stream.

A   B

Figure 6.11

(A) Floodplain of the Missouri River near Glasgow, Missouri, is bounded by steep bluffs. (B) During the 1993 flooding, the floodplain lived up to its name. The channel is barely visible, outlined by vegetation growing on levees along the banks.

(A, B) Images by Scientific Assessment and Strategy Team, courtesy U.S. Geological Survey

At times of higher discharge, especially during floods, the stream may make a shortcut, or cut off a meander, abandoning the old, twisted channel for a more direct downstream route. The cutoff meanders are called **oxbows.** These abandoned channels may be left dry, or they may be filled with standing water, making *oxbow lakes* (see **figure 6.12**).

6.3 Flooding

Learning Objectives

- List the factors that influence flood severity
- Compare upstream floods to downstream floods
- Explain how flash floods lead to fatalities
- Describe the role of hydrographs in understanding flood behavior
- Explain the creation and use of a flood-frequency curve
- Describe the use of recurrence interval in assessing flood hazards

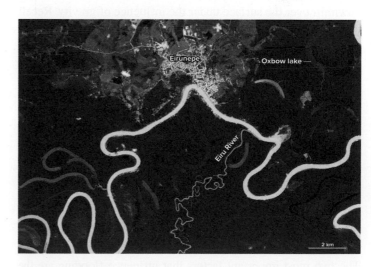

Figure 6.12

The Juruá River in the Amazon River Basin is both very sinuous and carries a large sediment load. Abandoned meanders are distinct and numerous.

NASA Earth Observatory images by Lauren Dauphin

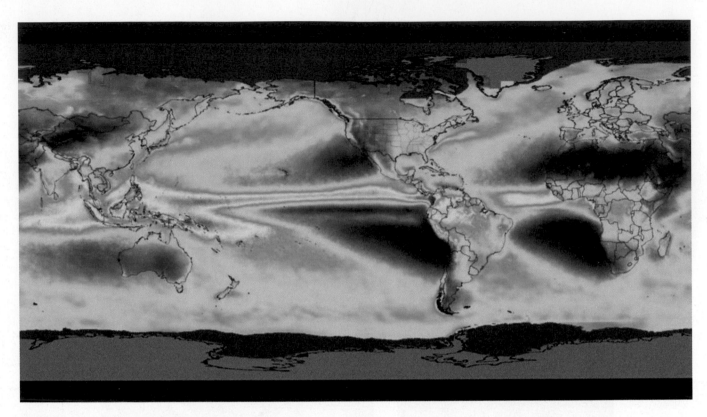

Figure 6.13

Global average precipitation from June 2000 to May 2019. Dark blue is the lowest amounts, and oranges and reds are the highest, up to 6000 mm/year. Notice the concentration of precipitation near the equator which is associated with a widespread low-pressure zone.

NASA's Scientific Visualization Studio

The vast majority of flooding events are linked to precipitation. The rainfall and snowmelt that does not **infiltrate** the ground or evaporate into the atmosphere becomes surface runoff flowing downhill over the surface under the influence of gravity. Recall **figure 6.1**. It is this surface runoff that overwhelms the capacity of a stream, which overflows its banks in a **flood.** The volume of the channel is not sufficient to accommodate the increased discharge. Most of the year, the surface of the water is well below the level of the stream bank. Flooding may occur as frequently as once every year or two. Our understanding of flood frequency is based on past stream behavior, but as climate changes, the recent past may not be a good indicator of future flood frequency or size. Below we will review what causes flooding, how we characterize floods, and how we use past behavior to forecast the location, timing, and severity of future flood events.

Factors Governing Flood Severity

The two most important factors that influence flooding are the quantity of water and the rate at which it enters the stream system. There are locations on Earth that receive relatively high amounts of precipitation, primarily controlled by latitude (**figure 6.13**). The most intense rainfall events occur in Southeast Asia, where storms have drenched the region with up to 200 centimeters

(80 inches) of rain in less than three days. To put such numbers in perspective, that amount of rain is more than double the average *annual* rainfall for the United States. In the United States, several regions are especially prone to heavy rainfall events: the southern states, vulnerable to storms from the Gulf of Mexico; the western coastal states, subject to prolonged storms from the Pacific Ocean; and the midcontinent states, where hot, moist air from the Gulf of Mexico can collide with cold air sweeping down from Canada. Streams that drain the Rocky Mountains are likely to flood during snowmelt, especially when rapid spring thawing follows a winter of unusually heavy snow. The record flooding in the Pacific Northwest in late 2021 resulted from a series of atmospheric rivers traveling from the tropics that released their precipitation when they hit land in higher latitudes (**figure 6.14**). Even worse, the same areas saw recorded breaking heat and wildfires in the previous summer, the results of which combined with the precipitation to create dangerous mudslides.

The near-surface geology of the drainage basin strongly influences how much runoff reaches a stream and how fast infiltration occurs. Soils and rocks vary in *porosity* and *permeability,* properties that are explored in chapter 11. A very porous and permeable soil allows a great deal of water to sink in relatively fast. If the soil is less permeable or is covered by artificial materials such as concrete, the proportion of water that runs off over the surface increases. Runoff will also increase if the soil

Figure 6.14

This satellite image shows estimates of rainfall totals over the 24-hour period of 14 November 2021 near Vancouver, B.C. The darkest red indicates the highest amounts, with some places receiving >10 cm (4 inches). Multiple such events lead to widespread flooding and mudslides.

NASA Earth Observatory images by Lauren Dauphin and Joshua Stevens

Figure 6.15

Even a small stream can be destructive with fast-flowing floodwaters. Northwest Colorado, June 2011.

U.S. Geological Survey/Photograph by Mark Henneberg

is already saturated with water. Even though some of the water that infiltrates and becomes groundwater likely discharges into streams, it does so much more slowly than runoff but again depends on the permeability of the soil or rock. Topography also influences the extent or rate of surface runoff. The steeper the terrain, the more readily water runs off over the surface and the less it tends to sink into the soil.

Vegetation may reduce flood hazards in several ways. The plants may simply provide a physical barrier to surface runoff, decreasing its velocity and slowing the rate at which water reaches a stream. Plant roots working into the soil loosen it, which tends to maintain or increase the soil's permeability, increasing infiltration and reducing the proportion of runoff. Plants also absorb water, using some of it to grow and releasing some slowly by evapotranspiration from foliage. Dead plant material such as leaf litter and fallen branches is effective at reducing runoff while at the same time protecting soil from erosion.

Local flood hazards can vary by season. In places where the uppermost soil freezes during the winter, a midwinter rainstorm or rapid snowmelt in the early spring may produce flooding with a quantity of runoff that could be readily absorbed by the soil in summer. The extent and vigor of vegetation varies seasonally also, as does atmospheric humidity evapotranspiration.

Flood Characteristics

During a flood, the water level of a stream is higher than usual, and its velocity and discharge increase as the greater mass of water is pulled downstream by gravity. The higher volume and velocity together produce the increased force that gives floodwaters their destructive power (**figure 6.15**). The elevation of

the water surface at any point is termed the **stage** of the stream. A stream is at *flood stage* when stream stage exceeds bank height. The magnitude of a flood can be described by either the maximum discharge or maximum stage reached. (See the Stream Hydrograph section below.) The stream is said to **crest** when the maximum stage is reached. This may occur within minutes of the influx of water, as with flooding just below a failed dam. However, in places far downstream from the water input or where surface runoff has been slowed, the stream may not crest for several days after the flood episode begins. In other words, just because the rain has stopped does not mean that the worst is over, as was made abundantly evident in the 1993 midwestern flooding, described in Case Study 6.2. The flood crested in late June in Minnesota; 9 July in the Quad Cities; 1 August in St. Louis; and 8 August in Cape Girardeau, Missouri, near the southern end of Illinois.

Flooding may afflict only a few kilometers along a small stream, or a region the size of the Mississippi River drainage basin, depending on how widely distributed the excess water is. Floods that affect only small, localized areas, or streams draining small basins, are called **upstream floods.** These are most often caused by sudden, locally intense rainstorms. Even if the total amount of water involved is moderate, the rapidity with which it enters the stream can cause it temporarily to exceed the stream channel capacity. The resultant flood is typically brief, though it can also be severe. Because of the localized nature of an upstream flood, the excess water is rapidly carried downstream, reducing the stream's stage.

Floods that affect large stream systems and large drainage basins are **downstream floods.** These floods more often result from prolonged heavy rains over a broad area or from extensive

A

B

Figure 6.16

(A) Flash flood events are common in urban environments often catching unaware drivers in quickly rising waters. (B) Antelope Canyon, Arizona, is an example of a natural setting in which dangerous flash floods are possible; steep canyon walls offer no route of quick retreat to higher ground.

(A) NOAA; (B) ©Carla Montgomery

regional snowmelt. Downstream floods usually last longer than upstream floods because the whole stream system is choked with excess water. The summer 1993 flooding in the Mississippi River basin is an excellent example of downstream flooding, brought on by many days of above-average rainfall over a broad area, with parts of the stream system staying above flood stage for weeks.

Flash Floods

Flash floods are a variety of upstream flood characterized by an especially rapid rise of stream stage, typically within a few hours of the water-input event. The water may rise feet to tens of feet in seconds to minutes, too fast for people to escape. The most common cause is the intense rainfall that accompanies thunderstorms; other causes are dam and ice jam (dam) failures. Flash floods can occur anywhere that surface runoff is rapid,

large in volume, and confined to a relatively restricted area. Even urban areas may experience flash floods, in a highway underpass, or just at a dip in the road where too much surface runoff suddenly washes over (**figure 6.16A**). Flash flooding of streams is especially common in canyons and steep-sided valleys where water from the drainage basin upstream is funneled into a narrow space (**figure 6.16B**). There is no broad floodplain into which it can spread, so the water rises quickly instead.

Flash flooding is a common cause of fatalities even in deserts and places undergoing drought. What rain does fall may come via a few intense cloudbursts. Runoff from these infrequent storms flow into *ephemeral streams,* streams that only flow occasionally when there is briefly enough water to collect in the channel. Between storms, the streambed is dry, and may not even look very much like a stream channel (**figure 6.17**). A dry streambed might seem a tempting route to hikers or drivers of off-road vehicles. However, particularly in mountainous

Figure 6.17

Ephemeral stream bed in Wyoming. The rounded shape and relatively large size of the cobbles, as well as the eroded bank, indicate that water does flow through this area.

©*Gina S. Szablewski*

A

B

Figure 6.18

(A) USGS streamgaging station 15905100 at the Atigun River, Alaska and its corresponding hydrograph from 2021 (B). Consider the trends in discharge shown, and why it is absent for part of the year. For data on your favorite stream, visit https://waterdata.usgs.gov/nwis/rt.

(A) Jeff Conaway/USGS

terrain, rain can be channeled downstream very quickly and overwhelm unwary travelers with a sudden wall of water that originated far away. The water level generally drops quickly as the rain soaks into parched ground, but the damage will already have been done.

Stream Hydrographs

A stream gaging station collects a variety of water data, some of which are used to create a **hydrograph** (**figure 6.18**). Hydrographs can be plotted with either stage or discharge on the vertical axis. Discharge is most commonly used but it may be more useful to indicate flood stage when determining what areas of a floodplain will be affected by floodwaters. Because stage and discharge are related and it is relatively easy to obtain measurements of stage, the discharge of a stream shown on a hydrograph is often one that is calculated from the stage data rather than taking a direct measurement.

Measurements of stage and discharge are also collected by field personal using traditional methods and often occur as a verification that automatic monitoring stations are operating properly. New portable instruments that use the Doppler effect are being utilized to obtain measurements of stream velocity and depth which are then used to calculate discharge. In 2021, the USGS stream monitoring system consisted of over 11,000 such stations, some of which have been operating for over 100 years. Hydrographs spanning long periods of time are especially useful in understanding the behavior of a stream, especially when considered in relation to the land-use changes that have occurred during the same time period.

Because a flood is defined as being a high discharge event, it will show up as a peak on a hydrograph. The height and width of that peak and its position in time relative to the flood-causing event depend on the relative location of the measurement within the drainage basin. A quick, sharp peak in discharge tends to occur in locations that are upstream and geographically close to the source of additional water. A broader, gentler peak occurs farther downstream, lower in the drainage basin (**figure 6.19**). The length of the event also effects the shape of the hydrograph. For example, a cloudburst would have a tall, narrow peak whereas a steady rain of several days would have a lower, spread-out peak.

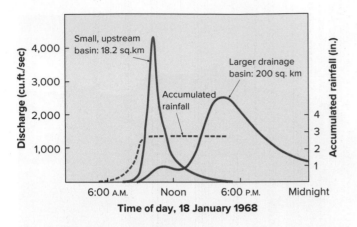

Figure 6.19

Flood hydrographs for two different points along Calaveras Creek near Elmendorf, Texas. The event was caused by heavy rainfall in the upper reaches of the drainage basin. The larger stream lower in the drainage basin responds more sluggishly to the input.

Sources: Water Resources Division, U.S. Geological Survey

Flood-Frequency Curves

Another way of looking at flooding is in terms of the frequency of flood events of differing severity. Long-term records for a particular stream or section of one make it possible to construct a curve showing discharge as a function of **recurrence interval,** which is the average of how frequently a flood of that severity occurs. The **flood-frequency curve** shown in **figure 6.20** indicates, for example, that an event producing a discharge of 750 cubic feet/second occurs once every twenty years.

Preferably, we refer to the *probability* that a flood of given size will occur in any one year; this probability is the inverse of the recurrence interval. For the stream of **figure 6.20**, a flood with a discharge of 750 cubic feet/second is called a twenty-year flood, meaning that a flood of that size occurs about once every twenty years, or has a 5% (1/20) probability of occurrence in any year.

It is important to remember that the recurrence intervals or probabilities assigned to floods are *averages*. The traditional recurrence-interval terminology is misleading in this regard, especially to the general public. Over many centuries, a fifty-year flood should occur an average of once every fifty years; or, in other words, there is a 2% chance that that discharge will be exceeded in any one year. However, that doesn't mean that two fifty-year floods couldn't occur in successive years, or even in the same year. The probability of two fifty-year flood events in one year is very low (2% × 2%, or .04%), but it is possible. It would be irresponsible to assume that just because a severe flood has recently occurred in an area that it is now safe for a while. Statistically, another such severe flood is unlikely to happen for some time, but there is no guarantee of that. Currently, both the traditional recurrence-interval terminology and more accurate probability phrasing

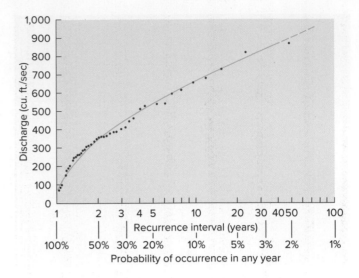

Figure 6.20

Flood-frequency curve for Eagle River at Red Cliff, Colorado. The curve was created with 46 years of data, so estimates of moderately frequent floods are likely to be fairly accurate; rare, larger floods are harder to project.

USGS Open-File Report 79-1060

(1% annual chance or 1% annual exceedance probability) are in use to describe flood hazard zones.

Flood estimation curves are useful in assessing regional flood hazards. If the severity of the 100- or 200-year (1% or 0.5%) flood can be estimated, even if such a rare and serious event has not occurred within memory, then scientists and planners can, with the aid of detailed maps and aerial photographs, project how much of the region would be flooded in such events. They can speak, for example, of the 100-year (1%) floodplain, the area that would be water-covered in a 100-year flood. Such information is useful in preparing flood-hazard maps, an exercise that is also aided by the use of aerial and satellite photography. Such maps, in turn, are helpful in siting new construction projects to minimize the risk of flood damage or in making property owners in threatened areas aware of dangers, just as can be done for earthquake and volcano hazards. Flood maps are compiled and available from the Federal Emergency Management Association in either traditional form or through GIS systems; data is also commonly available through state and local government agencies (**figure 6.21**).

Unfortunately, the best-constrained part of the flood curve is for the lower-discharge, more-frequent, higher-probability, less-serious events. Often considerable guesswork is involved in projecting the shape of the curve at the high-discharge end. Reliable records of stream stages, discharges, and the extent of past floods typically extend back only a few decades. Many areas may never have recorded a fifty- or one-hundred-year flood. Moreover, when the odd severe flood does occur, how does one know whether it was a sixty-year flood, a one-hundred-year flood, or whatever?

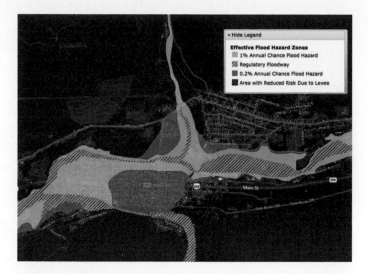

Figure 6.21

This map created using the Pennsylvania Flood Risk tool shows that a large portion of Wiconisco is at high risk for flooding in any given year. Maps such as these are commonly used for insurance purposes. Go to https://www.fema.gov/flood-maps for more information accessing flood maps.

pafloodrisk.psu.edu

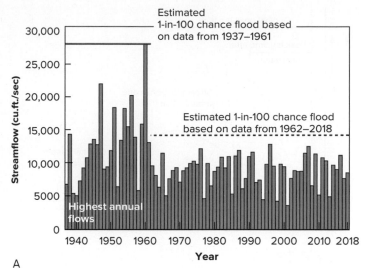

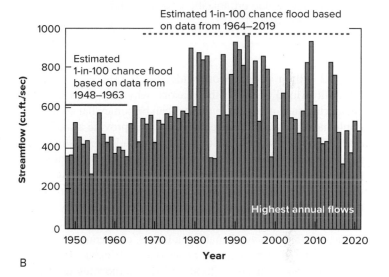

Figure 6.22

The size of the 1% annual exceedance probability (100-year flood) can change dramatically as a result of human activities. (A) For the Green River at Auburn, Washington, the projected size of this flood has been halved by the Howard Hanson Dam. (B) Rapid urban development in the Boneyard Creek area in Urbana, IL, since 1964 has nearly doubled the discharge of the projected hundred-year flood.

The high-discharge events are rare, and even when they happen, their recurrence interval can only be estimated. Considerable uncertainty can exist about just how bad a particular stream's 100- or 200-year flood might be. This problem is explored further in Case Study 6.1, "How Big Is the100-Year Flood?"

Another complication in flood frequency prediction is that streams in heavily populated areas are affected by human activities, as we shall see further in the next section. The way a stream responded to 10 centimeters of rain from a thunderstorm a hundred years ago may be quite different from the way it responds today. The flood-frequency curves are changing with time. A flood-control dam that can moderate spikes in discharge may reduce the magnitude of the hundred-year flood. Except for measures specifically designed for flood control, however, most human activities have tended to aggravate flood hazards and to decrease the recurrence intervals of high-discharge events. In other words, what may have been a 1% flood two centuries ago might be a 2% flood, or a more frequent one, today. **Figure 6.22** illustrates both types of changes.

The challenge of determining flood recurrence intervals from historical data is not unlike the forecasting of earthquakes by applying the earthquake-cycle model to historic and prehistoric seismicity along a fault zone. The results inevitably have associated uncertainties, and the more complex the particular geologic setting, the more land use and climate have changed over time, and the sparser the data, the larger the uncertainties will be.

6.4 Consequences of Development in Floodplains

Learning Objectives

- List reasons for living in a floodplain
- Relate land surface cover to lag time
- List human activities that worsen flooding

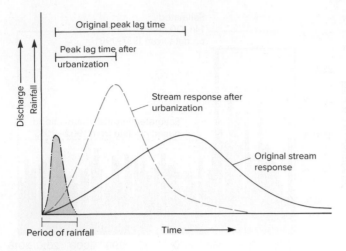

Figure 6.23

Hydrograph reflecting a change of stream response to precipitation following urbanization: peak discharge increases and lag time decreases.

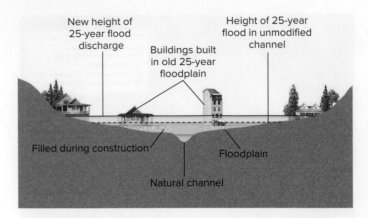

Figure 6.24

How floodplain development increases flood stage for a given discharge.

Over 2.2 billion people worldwide live within a 1% annual probability flood zone, so the reasons why they do so, when considering such a large population, can be somewhat complex. People have always settled near rivers. They provide a source of water. Floodplains contain rich soil for agriculture. A stream is often the most convenient route for transportation of people and goods. And increasingly, floodplains may be the only land available. When compared to living on a steep slope with very obvious landslide hazards, living on a floodplain may seem relatively safe. An individual may decide to live near a river because it offers a way of making a living, or because it is scenic. They may also be ignorant of the danger because they have never experienced a large flood or local authorities have not made flood education a priority. For whatever the reasons, the more people that live in floodplains, the more damage flooding will do. Unfortunately, the development of those floodplains often increases the likelihood or severity of flooding.

As mentioned earlier, two factors that affect flood severity are the amount and rate of surface runoff. The materials used extensively to cover the ground in cities, such as asphalt and concrete, are relatively impermeable and greatly reduce infiltration. When these materials cover large areas, surface runoff tends to be much more concentrated and rapid than before, increasing the risk of flooding. This is illustrated in **figure 6.23**. If *peak lag time* is defined as the time between a precipitation event and the peak flood discharge (or stage), then typically that time decreases with increased urbanization. Flash flooding is more common. The size of the peak discharge (stage) also increases, as was evident in **figure 6.22B**.

Building in a floodplain also can increase flood heights (see **figure 6.24**). The structures occupy volume that water formerly could fill when the stream flooded, and a given discharge then corresponds to a higher stage. Floods that occur are more serious. Filling in floodplain land for construction similarly decreases the volume available to stream water and further aggravates the situation.

Measures taken to drain water from low areas can likewise aggravate flooding along a stream. In cities, storm sewers are installed to keep water from flooding streets during heavy rains, and often the storm water is directed straight into a nearby stream. This works fine if the total flow is moderate enough, but by decreasing the time normally taken by the water to reach the stream channel, such measures increase the probability of stream flooding. The same is true of the use of tile drainage systems in farmland. Water that previously stood in low spots in the field and only slowly reached the stream after infiltration instead flows swiftly and directly into the stream, increasing the flood hazard for those along the banks.

Both farming and urbanization disturb the land by removing natural vegetation, thus leaving the soil exposed. The consequences are twofold. As noted earlier, vegetation can decrease flood hazards somewhat by providing a physical barrier to surface runoff, by soaking up some of the water, and through plants' root action, which keeps the soil looser and more permeable. Vegetation also can be critical to preventing soil erosion. When vegetation is removed and erosion increased, much more soil can be washed into streams. There it can fill in, or "silt up," the channel, decreasing the channel's volume and thus reducing the stream's capacity to carry water away quickly.

Both land-use and climate change are responsible for shifts in discharge, with both increasing and decreasing trends occurring concurrently. In places experiencing droughts such as in the western United States, discharges in some streams are decreasing. Other locations are wetter overall, so have increased runoff. The records for the Red River near Grand Forks, North Dakota, show an overall increase in peak discharge over the past 100 years, culminating in the catastrophic 1997 flood (**figure 6.25**). Due to its geography, underlying flat topography, and its northward flow, this river has large floods on a regular basis. Economic damages will likely never be as bad as the 1997 flood, given that over $1 billion has been invested in flood control projects along the river.

Figure 6.25

Over a century of peak-discharge records for the Red River of the North at Grand Forks, North Dakota, shows both decade-scale fluctuations (see 1914–1944) and a generally increasing long-term trend. The record 1997 discharge produced record flooding (inset photograph).

U.S. Geological Survey/Photograph by Tony Mutzenberger

6.5 Strategies for Reducing Flood Hazards

Learning Objectives

- Describe how restrictive zoning is used to reduce flood risk and damage
- Explain how retention ponds and diversion channels reduce flood risk and damage
- Explain why levees and channelization are problematic solutions to reducing flood risk
- Compare the benefits and problems with dams as flood risk reduction solutions
- Recognize the severity of U.S. flooding in the twenty-first century

The ways to reduce flood hazards can roughly be categorized into two groups: prevention and mitigation, which is keeping the floodwaters that do exist from causing damage. While we aren't going to control the amount of rain that falls, we can either avoid those places that flood in the first place by keeping all or certain types of construction or activities out of flood zones. We can also work to minimize excess runoff that quickly runs into streams. These actions are concentrated in increasing infiltration and holding onto or diverting runoff. Flood mitigation has relied heavily on engineering solutions that control flow velocity or channel storage, although not always successfully. Because many types of floods are predictable, especially when compared to other natural hazards, we have become relatively successful at providing timely warnings to people in order to save lives.

Restrictive Zoning and Floodproofing

If we choose not to avoid floodplains altogether, we can use restrictive zoning and special engineering practices to reduce the risk of flood damage. Existing structures present particular challenges, as they are typically exempt from newer regulations and may be difficult or costly to retrofit.

The first step is to identify as accurately as possible the area at risk. Careful mapping (see **Figure 6.21**) coupled with accurate

Figure 6.26

The flood-designed lower-level garage of this home along the Mississippi River is being tested out even before construction is completed.

Aaron Roeth Photography

stream discharge data allows us to recognize those areas threatened by floods of different recurrence intervals. Land that could be inundated often, perhaps by 4% annual probability (twenty-five-year) floods, might best be restricted to land uses such as parks or livestock grazing that do not involve much building.

Historically, there has been economic pressure to develop and build on floodplain land, especially in urban areas. There are various ways to design buildings to withstand flood hazards, including using stilts that raise the lowest floor above the expected 1% flood stage. This is a common practice already for homes of all sizes placed in hurricane-prone areas. Another is to create a lower level that can withstand periodic flooding, as seen

in **figure 6.26**. Floating buildings, although not a new idea, are being discussed as options not just for homes but larger buildings. While it might be better not to build there at all, such designs at least minimize interference by the structure with flood flow and also the danger of costly flood damage to the building.

Municipalities can inact floodplain zoning plans that restrict certain types of construction, prescribe design features, or ban buildings entirely, but those plans almost always apply only to new construction, leaving scores of older structures at risk. Increasingly, programs to relocate buildings or abandon and move entire towns are being used to remove people and structures from the most frequently flooded areas, and those likely to flood in the near future. Federal Emergency Management Association (FEMA), Housing and Urban Development (HUD), and the Army Corps of Engineers have all recently enacted programs using "managed retreat," with funding ranging from $700 million to $16 billion. Recent huge expenses in disaster recovery have demonstrated that some locations can no longer be protected from flooding. The risk in some areas is so high that local governments will either need to decide to force people to relocate or forfeit any federal money for all types of flood protection. These may seem like drastic solutions, but they do eliminate the long-term costs of coping with repeated flooding in the same place. Coastal areas experiencing sea level rise and hurricanes are particularly at risk to flooding hazards; we will discuss managed retreat again in chapter 7.

Retention Ponds, Diversion Channels

If open land is available, flood hazards may be greatly reduced by the use of **retention ponds** (**figure 6.27**). These ponds are basins that trap some of the surface runoff, keeping it from flowing immediately into the stream. They may be positioned

Stormwater Retention Pond

Green Stormwater Infrastructure

Figure 6.27

Comparison of retention ponds used to trap some runoff, which prevents it from reaching the stream quickly and instead allows for slow infiltration or evaporation.

Kristina Hopkins/USGS

directly along the stream itself or anywhere within a drainage basin. There are various types from elaborate artificial structures, abandoned quarries, or, in the simplest cases, fields dammed by dikes of piled-up soil. The latter option also allows the land to be used for other purposes, such as farming or parkland, except on those rare occasions of heavy runoff when it is needed as a retention pond. Retention ponds frequently are a relatively inexpensive option, provided that ample undeveloped land is available. They are commonly a requirement of newly developed areas, as a type of exchange for the land covered by impermeable substances.

A similar strategy is the use of *diversion channels or spillways* that come into play as stream stage rises. They redirect some of the water flow either into areas adjacent to the stream where flooding will cause minimal damage, or entirely around a city to reduce the discharge being carried in the natural river. Fargo, North Dakota, is currently implementing a multi-level diversion project (the FM Area Diversion) on the Red River that includes both a large retention pond upstream of the city and a 30-mile long spillway around the city. Flood gates, floodwalls, pump stations, and levees will also be part of this project to prevent all but the largest flood events.

Channelization

Channelization is a general term for various modifications of the stream channel itself that are intended to increase the velocity of water flow, the volume of the channel, or both. These modifications increase the discharge of the stream, with the goal of quickly carrying the water to another location. Channelization not only involves the widening, deepening, or straightening of the natural channel, but often covering it in an impermeable and erosion-resistant surface such as concrete (**figure 6.28**). While channelization of streams may have once seemed like a good idea, the current trend is to find other ways

to deal with flood risks and to remove existing channels and replace them with less severe alternatives.

A meandering stream tends to keep meandering or to revert to old meanders, so channelization is not a one-time effort. Constant maintenance is required to limit erosion in any straightened channel sections and to keep the river within a prescribed channel. Erosion problems also occur downstream due to channel straightening. When a meander is cutoff, the stream's gradient increases because the run over which the vertical drop occurs is shortened; the velocity respondingly increases, along with the stream's erosive power. And by causing more water to flow downstream faster, channelization increase the likelihood of flooding downstream from the alterations.

Channelization has severe ecological impacts. Wetlands may be drained as water is shifted more efficiently downstream. Streambank habitat is all but eliminated as channels are straightened and shortened, and organisms adapt poorly to new streamflow patterns and streambed configuration.

The Mississippi River system offers a grand-scale example of the difficulty of controlling a stream's choice of channel. If left to its own devices, the lower Mississippi River would wander back and forth across the floodplain, taking different paths to the Gulf of Mexico, just as it has for thousands of years. In the 1950s, the Atchafalaya River (a distributary of the Mississippi) nearly captured the flow of the Mississippi, allowing the river a steeper and shorter path to its base level. Flow increased in the Atchafalaya while water available for navigation on the low Mississippi decreased. The Old River Control structures were completed in the 1960s and continue to keep the Mississippi from changing course to the west (**figure 6.29**). The

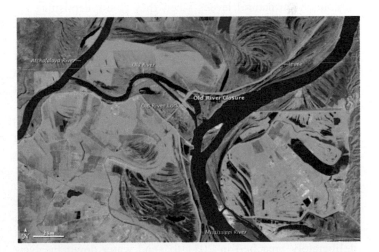

Figure 6.29

This false-color satellite image shows the location of the Old River Control structure, designed to keep 70% of the Mississippi's flow in its main channel, letting 30% into the Atchafalaya to the west. The green color indicates most of this land is vegetated/used for agriculture.

NASA Earth Observatory image by Michael Taylor and Adam Voiland

Figure 6.28

The Los Angeles River was channelized in the early 1900s after a series of devastating floods.

Studio 642/Blend Images/Alamy Stock Photo

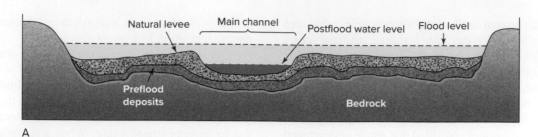

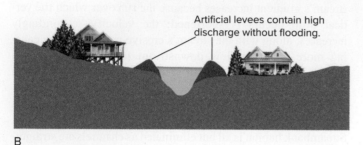

Figure 6.30

(A) When streams flood, waters slow quickly as they flow onto the floodplain, and therefore tend to deposit sediment, especially close to the channel where velocity first drops. (B) Artificial levees are designed to protect floodplain land from flooding by raising the height of the stream bank.

Mississippi is one of the most engineered rivers in the world, with expensive annual maintenance costs. There are concerns that engineered structures are not being maintained properly. For example, the Old River Control system is susceptible to failure because it does not allow sediment to flow through naturally, and the force of the sediment may overwhelm the artificial structures. A project to begin in 2023 will allow sediment through a breach in the levee system with the desired outcome of both alleviating pressure and creating new land that will help mitigate sea level rise problems.

Levees

Streams form natural levees when they flood, depositing the larger-graincd sediments next to the stream bank (**figure 6.30A**). These **levees,** raised banks along a stream channel, may be purposely enlarged or created where none exist (**figure 6.30B**). Because levees raise the height of the stream banks close to the channel, the water can rise higher without flooding the surrounding country. Levees may be used alone, but they are often just one part of more extensive systems.

However, confining the water to the channel, rather than allowing it to flow out into the floodplain, effectively shunts the water downstream faster during high-discharge events, thereby increasing flood risks downstream. It artificially raises the stage of the stream for a given discharge, which can increase the risks upstream, too, as the increase in stage is propagated above the confined section. This is why the Bird's Point levee was deliberately breached in 2011 to send water down the New Madrid floodway (see Case Study 6.6.2). Another large problem is that levees may make people feel so safe about living in the floodplain that, in the absence of restrictive zoning, development will be far more extensive than if the stream were allowed to flood naturally from time to time. If levees are overtopped by an unanticipated severe flood, or if they simply fail, far more lives and property may be lost as a result (**figure 6.31A**). Also, if levees are breached or overtopped, water is then trapped outside the stream channel, where it may stand for some time after the flood subsides (**figure 6.31B**).

Levees alter sedimentation patterns, too. During flooding, sediment is deposited in the floodplain outside the channel. If the stream and its load are confined by levees, increased sedimentation may occur in the channel. This will raise stream stage for a given discharge because the channel bottom is raised, and either the levees must be continually raised to compensate, or the channel must be dredged. Jackson Square in New Orleans stands at the elevation of the wharf in Civil War days; behind the levee beside the square, the Mississippi flows 10–15 feet *higher,* largely due to sediment accumulation in the channel. This is part of why the city was so vulnerable during Hurricane Katrina, and continues to be so.

Flood-Control Dams and Reservoirs

Another approach to moderating streamflow to prevent or minimize flooding is the construction of flood-control dams at one or more points along the stream. Excess water is held behind a dam in the reservoir formed upstream and may then be released at a controlled rate that doesn't overwhelm the capacity of the channel beyond. Additional benefits of constructing flood-control dams and their associated reservoirs may include availability of the freshwater, generation of hydroelectric power at the dam sites, and development of recreational facilities for swimming, boating, and fishing at the reservoir.

Dams used for any reason also have negative aspects. Navigation on the river, both by people and by animals, is restricted by the presence of the dams. Fish migration can be severely disrupted. Also, the creation of a reservoir necessarily floods much of the stream valley behind the dam and may destroy wildlife habitats or displace people and their settlements. China's massive Three Gorges Dam project, discussed more fully in chapter 20, displaced over a million people and flooded numerous towns. As this is being written, the issue of whether to breach a number of hydropower dams in the Pacific

A

B

Figure 6.31

The negative sides of levees as flood-control devices. (A) If they are breached, as here in the 1993 Mississippi River floods, those living behind the levees may suddenly and unexpectedly find themselves inundated. (B) Inland flooding from Hurricane Katrina in 2005 was prolonged because breached or over-topped levees dammed water behind them.

(A) U.S. Geological Survey/Photograph courtesy C. S. Melcher; (B) NOAA/Department of Commerce

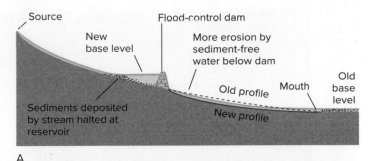

A

B

Figure 6.32

(A) Effects of dam construction on a stream: change in base level, sediment deposition in reservoir, reshaping of the stream channel. (B) Two of the four large dams and reservoirs that protect Santa Fe, California, from flash floods of the San Gabriel River as it plunges down out of the mountains. Note sediment clouding reservoir water.

(B) NASA/GSFC/METI/ERSDAC/JAROS and the U.S./Japan ASTER Science Team

Northwest in efforts to restore populations of endangered salmon continues in debate and litigation. Already, more dams are being removed than are being built each year in the United States, as their negative effects are more fully understood. Between 1912 and 2020, nearly 1800 dams were removed in the United States; 69 were removed in 2020.

If a stream normally carries a high sediment load, further complications arise. The reservoir represents a new base level for the stream above the dam (**figure 6.32**). When the stream flows into that reservoir, its velocity drops to zero, and it dumps its load of sediment. Silting-up of the reservoir, in turn, decreases its volume, so it becomes less effective as a flood-control device. Some reservoirs have filled completely in a matter of decades, becoming useless. Others have been kept clear only by repeated dredging, which can be expensive and presents the problem of where to dump the dredged sediment. At the same time, the water released below the dam is free of sediment and has more energy to erode the channel. The greatly reduced water volume downstream also alters ecosystems. After recognizing the negative long-term impacts of the Glen Canyon Dam to the Colorado River system, authorities have conducted eight controlled floods of the Grand Canyon since 1996 in order to restore streambank habitat and shift sediment in the canyon, scouring silted-up sections of channel and depositing sediment

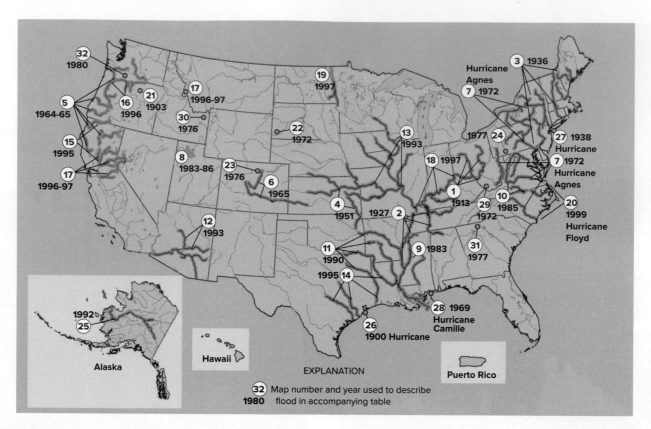

Figure 6.33

Significant U.S. floods of the twentieth century. The complete map number key is in USGS Fact Sheet 024-00. Hurricane Katrina, in 2005, affected much the same area as Camille; in 2008, Hurricane Ike descended on Galveston, like the infamous hurricane of 1900.

Modified after USGS Fact Sheet 024-00

elsewhere. The most recent planned controlled flood was canceled in 2021 due to a concern over the record low water levels in the Lake Powell reservoir caused by drought.

Some large reservoirs, such as Lake Mead behind Hoover Dam, cause earthquakes. The dam and the reservoir water represent an added load on the rocks, increasing the stresses on them, while infiltration of water into the ground under the reservoir increases pore pressure sufficiently to reactivate old faults. Since Hoover Dam was built, thousands of small earthquakes have occurred in the vicinity. Reservoir-induced earthquake activity is not uncommon. Earthquakes caused in this way are usually of low magnitude, but they have very shallow foci. This raises concerns about the possibility of catastrophic dam failure. Moreover, not all the quakes are small. In 1967, an earthquake of magnitude 6.4 in the vicinity of the Koyna Dam in India resulted in 177 deaths and considerable damage. Two even larger earthquakes have occurred near a dam on the Zipingpu River in China that was completed in 2004: a magnitude-7.9 event that left 80,000 people dead or missing, and a magnitude-6.6 quake in 2013. As more large dams are built worldwide, concern about reservoir-induced seismicity rises.

If levees increase residents' sense of security, the presence of a flood-control dam may do so to an even greater extent. People may feel that the flood hazard has been eliminated and thus neglect even basic floodplain-zoning and other practical precautions. But dams may fail; or the dam may hold, but the reservoir may fill abruptly with a sudden landslide (as in the Vaiont disaster described in chapter 8) and flooding may occur anyway from the suddenly displaced water. Floods may result from intense rainfall or runoff *below* the dam. Also, a dam/reservoir complex may not necessarily be managed so as to maximize its flood-control capabilities. For obvious reasons, dams and reservoirs often serve multiple uses. But for maximally effective flood control, one wants maximum water-storage *potential*, while for other uses, it is generally desirable to maximize actual water *storage*, leaving little extra volume for flood control. The several uses of the dam/reservoir system may thus tend to compete with, rather than to complement, each other.

Flood History, Flood Warnings

The U.S. Geological Survey (USGS) compiled an inventory of data on major U.S. floods of the twentieth century (**figure 6.33**) that confirms that severe flooding, like flooding in general, is widespread. Of the several causes noted, heavy rain or rain plus snowmelt is clearly the most common. It should be no surprise that severe flood events have continued into the twenty-first century. In 2008, the upper Midwest suffered $1.5 billion in damage from a month-long flood due to prolonged heavy rains. The Midwest again flooded in 2019 after its wettest twelve-month

Table 6.1 | Significant U.S. Floods of the Twentieth Century

Flood Type	Map No.	Date	Area or Stream with Flooding	Reported Deaths	Approximate Cost (uninflated)*	Comments
Regional flood	1	Mar.–Apr. 1913	Ohio, statewide	467	$143M	Excessive regional rain.
	2	Apr.–May 1927	Mississippi River from Missouri to Louisiana	unknown	$230M	Record discharge downstream from Cairo, Illinois.
	3	Mar. 1936	New England	150⁺	$300M	Excessive rainfall on snow.
	4	July 1951	Kansas and Neosho River basins in Kansas	15	$800M	Excessive regional rain.
	5	Dec. 1964–Jan. 1965	Pacific Northwest	47	$430M	Excessive rainfall on snow.
	6	June 1965	South Platte and Arkansas Rivers in Colorado	24	$570M	14 inches of rain in a few hours in eastern Colorado.
	7	June 1972	Northeastern United States	117	$3.2B	Extratropical remnants of Hurricane Agnes.
	8	Apr.–June 1983 June 1983–1986	Shoreline of Great Salt Lake, Utah	unknown	$621M	In June 1986, the Great Salt Lake reached its highest elevation and caused $268M more in property damage.
	9	May 1983	Central and northeast Mississippi	1	$500M	Excessive regional rain.
	10	Nov. 1985	Shenandoah, James, and Roanoke Rivers in Virginia and West Virginia	69	$1.25B	Excessive regional rain.
	11	Apr. 1990	Trinity, Arkansas, and Red Rivers in Texas, Arkansas, and Oklahoma	17	$1B	Recurring intense thunderstorms.
	12	Jan. 1993	Gila, Salt, and Santa Cruz Rivers in Arizona	unknown	$400M	Persistent winter precipitation.
	13	May–Sept. 1993	Mississippi River basin in central United States	48	$20B	Long period of excessive rainfall.
	14	May 1995	South-central United States	32	$5–6B	Rain from recurring thunderstorms.
	15	Jan.–Mar. 1995	California	27	$3B	Frequent winter storms.
	16	Feb. 1996	Pacific Northwest and western Montana	9	$1B	Torrential rains and snowmelt.
	17	Dec. 1996–Jan. 1997	Pacific Northwest and Montana	36	$2–3B	Torrential rains and snowmelt.
	18	Mar. 1997	Ohio River and tributaries	50+	$500M	Slow-moving frontal system.
	19	Apr.–May 1997	Red River of the North in North Dakota and Minnesota	8	$2B	Very rapid snowmelt.
	20	Sept. 1999	Eastern North Carolina	42	$6B	Slow-moving Hurricane Floyd.
Flash flood	21	June 14, 1903	Willow Creek in Oregon	225	unknown	City of Heppner, Oregon, destroyed.
	22	June 9–10, 1972	Rapid City, South Dakota	237	$160M	15 inches of rain in 5 hours.
	23	July 31, 1976	Big Thompson and Cache la Poudre Rivers in Colorado	144	$39M	Flash flood in canyon after excessive rainfall.
	24	July 19–20, 1977	Conemaugh River in Pennsylvania	78	$300M	12 inches of rain in 6–8 hours.
Ice-jam flood	25	May 1992	Yukon River in Alaska	0	unknown	100-year flood on Yukon River.
Storm-surge flood	26	Sept. 1900	Galveston, Texas	6000+	unknown	Hurricane.
	27	Sept. 1938	Northeast United States	494	$306M	Hurricane.
	28	Aug. 1969	Gulf Coast, Mississippi, and Louisiana	259	$1.4B	Hurricane Camille.
Dam-failure flood	29	Feb. 2, 1972	Buffalo Creek in West Virginia	125	$60M	Dam failure after excessive rainfall.
	30	June 5, 1976	Teton River in Idaho	11	$400M	Earthen dam breached.
	31	Nov. 8, 1977	Toccoa Creek in Georgia	39	$2.8M	Dam failure after excessive rainfall.
Mudflow flood	32	May 18, 1980	Toutle and lower Cowlitz Rivers in Washington	60	unknown	Result of eruption of Mount St. Helens.

After U.S. Geological Survey Fact Sheet 024-00

A

B

C

Figure 6.34

(A) Flooding along the Knife River, North Dakota, 2009. Note that the water stayed high enough long enough that ice formed on buildings and trees even though the water was moving. (B) Roads washed out in Colorado in 2013; here, water is still flowing through the gaps. (C) Houston, Texas, was drenched with over 50 inches of rain from Hurricane Harvey, causing extensive stream flooding. Nearly a week later, cars were still afloat on Houston streets.

(A) U.S. Geological Survey/Photograph by Dennis Rosenkrantz;
(B) U.S. Geological Survey/Photograph by Colorado Air Patrol;
(C) FEMA/Photograph by Dominick Del Vecchio

period on record, affecting 14 million people and causing $2.0 billion in damage just in Iowa and Nebraska. In 2009, the Red River and others in North Dakota set new flood records as the region was swamped with rain and snowmelt (**figure 6.34A**). Record-breaking rains in the Colorado Rockies in 2013 caused flooding compounded by multiple dam failures (**figure 6.34B**). In 2016, storms dropped three times as much rain on Louisiana as had Hurricane Katrina in 2005, with predictable results.

Hurricanes cause much of their damage by wind and by seawater pounding coastal regions, but they also bring rain. Long after the hurricane has weakened to tropical-storm strength or less, as it often does over land, the remaining storm system can continue to produce heavy rains and inland flooding. Much of the $108 billion in damage from Hurricane Katrina was due to freshwater flooding; Superstorm Sandy (2012) brought inland flooding to the Northeast; and in 2017, the United States was hit by several major hurricane systems that caused serious inland flooding: Harvey in the South, especially Texas; Irma in Florida and the Caribbean, including Puerto Rico and the U.S. Virgin Islands; and Maria in the Caribbean again. The damages associated with those three storms totaled more than $270 billion. The rainfall from the Hurricane Harvey

storm system set new records for the United States, with some locations in Texas receiving over 60 inches of rain (**figure 6.34C**); 39,000 people were forced to evacuate due to floodwaters.

Depending on the cause, different kinds of precautions are possible in anticipation of a flood. The areas that would be affected by flooding associated with the 1980 explosion of Mount St. Helens could be anticipated from topography and an understanding of lahars, but the actual eruption was too sudden to allow last-minute evacuation. The coastal flooding from a storm surge, discussed further in chapter 7, typically allows time for evacuation as the storm approaches, but damage to coastal property is inevitable. Dam failures are sudden; rising water from spring snowmelt, usually much slower. Rains may come as sudden cloudbursts, leading to flash floods, or be spread over days or weeks until a stream channel overflows. As long as people inhabit areas prone to flooding, property will be destroyed when the floods happen, but with better warning of impending floods, particularly flash floods, more lives can be saved.

With that in view, the USGS and National Weather Service (NWS) have established a partnership to provide such

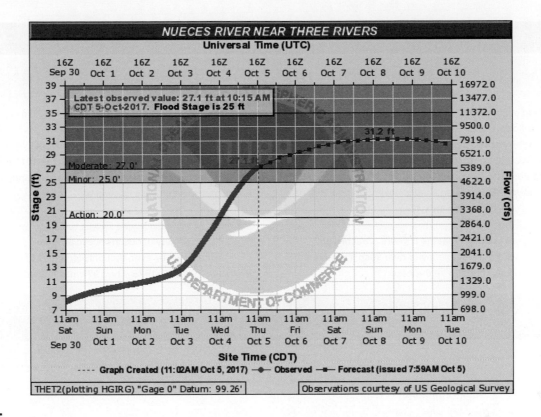

Figure 6.35

Recent past actual and near-future projected hydrograph for the Nueces River in Texas. The projection is modified over time as newer stream-stage measurements and precipitation forecasts become available. Find more data at: https://water.weather.gov/ahps/.

Source: National Weather Service

flood warnings, the Advanced Hydrologic Prediction Service (AHPS). The USGS contributes real-time stream-stage measurements and data on the relationship between stage and discharge for various streams. To this information, NWS adds historical and forecast precipitation data and applies models that relate stream response to water input. The result is a stream-stage projection (a projected hydrograph) from which AHPS can identify areas at risk of near-term flooding and issue warnings accordingly (**figure 6.35**). Later comparison of the model projections with actual water input and stream response allows refinement of the models for better, more precise future warnings. The AHPS is now experimenting with longer-term predictions, but these are clearly more difficult, given the underlying challenge of long-term weather forecasting.

Forecast stream stages are most helpful to those living very close to the relevant stream gage. To address that limitation, the USGS has developed the capability to produce regional flood-forecast maps, combining NWS flood predictions with computer models and high-precision topographic maps, and using GIS to produce maps showing both expected flood depths and time to arrival of the flood over a broad area, not just at a particular stream gage. Better yet, this can now be done quickly enough that the maps can be distributed in time for planners, emergency-services personnel, and the general public to react before the flood arrives.

Recently, attention has been given to the problem of international flood forecasting when the affected nation occupies only a small part of the drainage basin and cannot itself do all the necessary monitoring. As an extreme example, Bangladesh occupies only the lowest 7% of the Ganges–Brahmaputra–Meghna drainage basin. NASA proposed an international Global Precipitation Measurement Program (GPM), to monitor global weather, giving nations real-time access to critical precipitation data throughout the relevant drainage basin. The GPM Core Observatory satellite currently provides global measurements of precipitation every three hours, and helps to improve forecasting of extreme events.

How Big Is the 100-Year Flood?

A common way to estimate the recurrence interval of a flood of given size is as follows. Suppose the records of maximum discharge (or maximum stage) reached by a particular stream each year have been kept for N years. Each of these yearly maxima can be given a rank M, ranging from 1 to n, 1 being the largest, n the smallest. Then the recurrence interval R of a given annual maximum is defined as

$$R = \frac{(N+1)}{M}$$

This approach assumes that the N years of record are representative of typical stream behavior over such a period; the larger N, the more likely this is to be the case.

For example, table CS 6.1.1 shows the maximum one-day mean discharges of the Big Thompson River, as measured near Estes Park, Colorado, for twenty-five consecutive years, 1951–1975. If these values are ranked 1 to 25, the 1971 maximum of 1030 cubic feet/second is the seventh largest and, therefore, has an estimated recurrence interval of

$$\frac{(25+1)}{7} = 3.71 \text{ years}$$

or a 27% probability of occurrence in any year.

Suppose, however, that only ten years of records are available, for 1966–1975. The 1971 maximum discharge happens

Table CS 6.1.1	Calculated Recurrence Intervals for Discharges of Big Thompson River at Estes Park, Colorado				
		FOR TWENTY-FIVE-YEAR RECORD		FOR TEN-YEAR RECORD	
Year	Maximum Mean One-Day Discharge (cu. ft./sec)	M (rank)	R (years)	M (rank)	R (years)
1951	1220	4	6.50	3	3.67
1952	1310	3	8.67	2	5.50
1953	1150	5	5.20	4	2.75
1954	346	25	1.04	10	1.10
1955	470	23	1.13	9	1.22
1956	830	13	2.00	6	1.83
1957	1440	2	13.00	1	11.00
1958	1040	6	4.33	5	2.20
1959	816	14	1.86	7	1.57
1960	769	17	1.53	8	1.38
1961	836	12	2.17		
1962	709	19	1.37		
1963	692	21	1.23		
1964	481	22	1.18		
1965	1520	1	26.00		
1966	368	24	1.08	10	1.10
1967	698	20	1.30	9	1.22
1968	764	18	1.44	8	1.38
1969	878	10	2.60	4	2.75
1970	950	9	2.89	3	3.67
1971	1030	7	3.71	1	11.00
1972	857	11	2.36	5	2.20
1973	1020	8	3.25	2	5.50
1974	796	15	1.73	6	1.83
1975	793	16	1.62	7	1.57

Source: Data from U.S. Geological Survey Open-File Report 79-681

to be the largest in that period of record. On the basis of the shorter record, its estimated recurrence interval is (10 + 1)/1 = 11 years, corresponding to a 9% probability of occurrence in any year. Alternatively, if we look at only the first ten years of record, 1951–1960, the recurrence interval for the 1958 maximum discharge of 1040 cubic feet/second (a maximum discharge of nearly the same size as that in 1971) can be estimated at 2.2 years, meaning that it would have a 45% probability of occurrence in any one year.

Which estimate is the most accurate? Perhaps none of them, but their differences illustrate the need for long-term records to smooth out short-term anomalies in streamflow patterns, which are not uncommon, given that climatic conditions can fluctuate significantly over a period of a decade or two.

The point is further illustrated in **figure CS 6.1.1**. Not all streams have records lasting 100 years or more, so the magnitude of fifty-year, one-hundred-year, or larger floods is commonly estimated from a flood-frequency curve. Curves *A* and *B* in **figure CS 6.1.1** are based, respectively, on the first and last ten years of

data from the last two columns of **table CS 6.1.1**, the data points represented by X symbols for *A* and open circles for *B*. These two data sets give estimates for the size of the 100-year flood that differ by more than 50%. These results, in turn, both differ somewhat from an estimate based on the full 25 years of data (curve *C*, solid circles). This graphically illustrates how important long-term records can be in the projection of recurrence intervals of larger flood events. The flood-frequency curve in **figure 6.20**, based on nearly fifty years of record, inspires more confidence, assuming that recent land-use changes have been minimal.

In 1976, the stream on which this case study is based, the Big Thompson River, experienced catastrophic flash flooding (**figure CS 6.1.2**). A large thunderstorm dumped about a foot of rain on parts of the drainage basin, and the resultant flood took nearly 150 lives and caused $35 million in property damage. Research later showed that in some places, the flooding exceeded anything reached in the 10,000 years since the glaciers covered these valleys; yet at Estes Park, the gage used for the data in this case study, discharge peaked at 457 cu. ft./sec, less than a two-year flood. The Big Thompson Canyon was flooded again in 2013 due to a series of thunderstorms. This time, eight people died with an estimate $2 billion in damage.

For a further look at flood-frequency projection, see the online poster "100-Year Flood—It's All About Chance" from the USGS at pubs.usgs.gov/gip/106/.

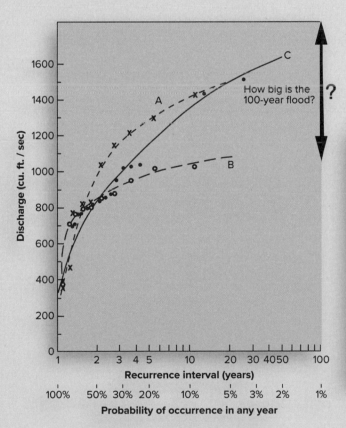

Figure CS 6.1.1

Flood-frequency curves for a given stream can look very different, depending on the length of record available and the particular period studied.

Figure CS 6.1.2

House undercut during Big Thompson Canyon flood of 1976. Behind it is a landslide also caused by undercutting of the hillside. Note how low the normal flow is in August, when this photo was taken.

U.S. Geological Survey Photographic Library/Photograph by R. R. Shroba

Life on the Mississippi: The Ups and Downs of Levees

The Mississippi River is a long-term case study in levee-building for flood control. It is the highest-discharge stream in the United States and the third largest in the world (**figure CS 6.2.1**). The first levees were built in 1717, after New Orleans was flooded. Further construction of a patchwork of local levees continued intermittently over the next century. Following a series of floods in the mid-1800s, there was great public outcry for greater flood-control measures. In 1879, the Mississippi River Commission was formed to oversee and coordinate flood-control efforts over the whole lower Mississippi area. The commission urged the building of more levees. In 1882, when over $10 million (a huge sum for the time) had already been spent on levees, severe floods caused several hundred breaks in the system and prompted renewed efforts. By the turn of the century, planners were confident that the levees were high enough and strong enough

to withstand any flood. Flooding in 1912–1913, accompanied by twenty failures of the levees, proved them wrong. By 1926, there were over 2900 kilometers (1800 miles) of levees along the Mississippi, standing an average of 6 meters high; nearly half a billion cubic meters of earth had been used to construct them. Funds totaling more than twenty times the original cost estimates had been spent.

In 1927, the worst flooding recorded to that date occurred along the Mississippi. The levees were breached in 225 places; 183 people died; more than 75,000 people were forced from their homes; and there was an estimated $500 million worth of damage. The federal government took over primary management of the flood-control measures. In the upper parts of the drainage basin, they built five major flood-control dams along the Missouri River in the 1940s in response to flooding there. But the immense volume

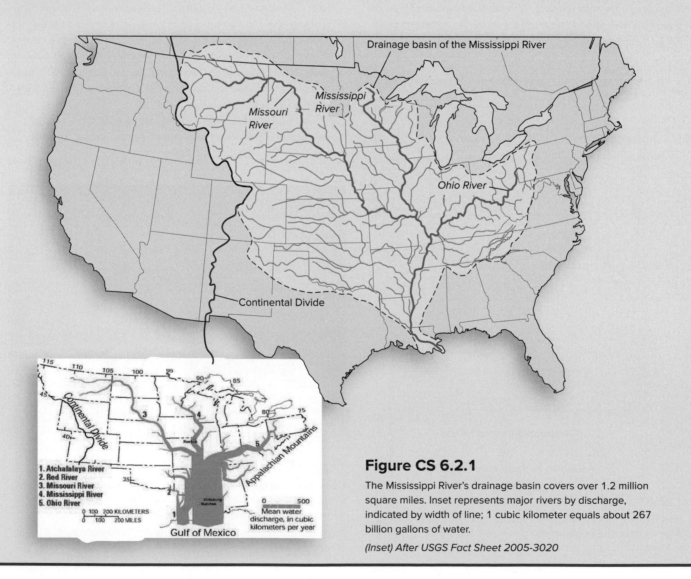

Figure CS 6.2.1

The Mississippi River's drainage basin covers over 1.2 million square miles. Inset represents major rivers by discharge, indicated by width of line; 1 cubic kilometer equals about 267 billion gallons of water.

(Inset) After USGS Fact Sheet 2005-3020

of the lower Mississippi cannot be contained in a few reservoirs so the building up and reinforcing of levees continued.

A mild and wet fall and winter followed by prolonged rain in March and April 1973 contributed to record-setting flooding in much of the Mississippi's drainage basin. The river remained above flood stage for more than three months in some places. Over 12 million acres of land were flooded, 50,000 people were evacuated, and over $400 million in damage was done, not counting secondary effects, such as loss of wildlife and the estimated 240 million tons of sediment that were washed down the Mississippi, much of it fertile soil irrecoverably eroded from farmland.

Still worse flooding was yet to come. A sustained period of heavy rains and thunderstorms in the summer of 1993 led to flooding that caused 50 deaths, over $10 billion in property and crop damage, and record-breaking flood stages and/or levee breaches at various places in the Mississippi basin. Some places remained swamped for weeks before the water finally receded (**figure CS 6.2.2**; recall also **figure 6.12**). In spring 2011, flooding

was so severe that spillways unused for decades were pressed into service and levees were deliberately breached (**figure CS 6.2.3**) to try to bring stream stages down. After nearly 300 years of ever more extensive (and expensive) flood-control efforts in the Mississippi basin, damaging floods are still happening and are facing possible increased risks from long-term climate change.

Inevitably, the question of what level of flood control is adequate or desirable arises. We could build still higher levees, bigger dams, and larger reservoirs and spillways, and (barring dam or levee failure) be safe even from a projected 500-year flood or worse, but the costs would be astronomical and could exceed the property damage expected from such events. By definition, the probability of a 500-year or larger flood is extremely low at 0.2%. Precautions against high-frequency flood events seem to make sense. But at what point, for what probability of flood, does it make more economic or practical sense simply to accept the real but small risk of the rare, disastrous, high-discharge flood events, or perhaps to leave the area at risk unoccupied altogether?

The same questions were asked about New Orleans in the wake of Hurricane Katrina, and continue to be raised whenever catastrophic flood events prompt calls for more flood protection.

A

B

Figure CS 6.2.2

(A) Flooding in Davenport, Iowa, summer 1993. (B) On a regional scale, the area around the Mississippi became a giant lake. Ironically, in some places only the vegetation growing atop ineffective levees marked the margins of the stream channel, as here.

(A) ©Doug Sherman/Geofile; (B) FEMA/Photograph by Andrea Booher

Figure CS 6.2.3

The city of Cairo, Illinois, sits at the confluence of the Mississippi (flowing in from the northwest, obviously flooding here) and the Ohio River (joining it from the northeast). Arrow indicates the Bird's Point levee, deliberately breached by the U.S. Army Corps of Engineers on 2 May 2011, allowing a flow of over 500,000 cu. ft./sec into the New Madrid floodway (broad brown area west of the river below the levee), which was created for just such an emergency after flooding in 1927 but had not been used since 1937.

NASA image created by Jesse Allen and Robert Simmon, using Landsat data provided by the U.S. Geological Survey

Summary

Streams are active agents of sediment transport. They are also powerful forces in sculpting the landscape, constantly engaged in erosion and deposition, their channels naturally shifting over Earth's surface. Stream velocity is a strong control both on erosion and on sediment transport and deposition. As velocity varies along and within the channel, streams deposit sediments that are typically well-sorted by size and density. The size of a stream is most commonly measured by its discharge. Over time, a stream will tend to carve a channel that just accommodates its usual maximum annual discharge.

Flooding is the normal response of a stream to an unusually high input of water in a short time. Regions at risk from flooding, and the degree of risk, can be identified if accurate maps and records about past floods and their severity are available. However, records may not extend over a long enough period to permit precise predictions about the rare, severe floods. Moreover, human activities may have changed regional runoff patterns or stream characteristics over time, making historical records less useful in forecasting future problems, and long-term climate change adds uncertainty as well. Strategies designed to minimize flood damage include restricting or prohibiting development in floodplains, controlling the kinds of floodplain development, channelization, and the use of retention ponds, diversion channels, levees, and flood-control dams. Unfortunately, many flood-control procedures have drawbacks, one of which may be increased flood hazards elsewhere along the stream. Flash-flood warnings can be an important tool for saving lives, if not property, by making timely evacuations possible.

Key Terms and Concepts

alluvial fan	discharge	hydrologic cycle	percolation
base level	downstream flood	hydrosphere	point bar
braided stream	drainage basin	infiltration	recurrence interval
capacity	flood	levees	retention pond
channelization	flood-frequency curve	load	stage
crest	floodplain	longitudinal profile	stream
cut bank	gradient	meanders	upstream flood
delta	hydrograph	oxbows	well sorted

Test Your Learning

1. Define stream load, and explain what factors control it.
2. Explain how a stream shows a decreased gradient and increased discharge downstream.
3. Relate sediment transport to variations in water velocity, in order to explain why stream sediments tend to be well sorted.
4. Describe how enlargement and migration of meanders contribute to floodplain development.
5. Discuss the relationship between flooding and (a) precipitation, (b) soil characteristics, (c) vegetation, and (d) season.
6. Compare upstream and downstream floods. Give an example of an event that might cause each type of flood.
7. Define the concepts of a *flood-frequency curve* and a flood *recurrence interval,* and explain two common problems with the application of these measures.
8. Describe two ways in which urbanization may increase local flood hazards. Sketch the change in stream response as it might appear on a hydrograph.
9. Channelization and levee construction may reduce local flood hazards, but they may worsen the flood hazards elsewhere along a stream. Explain.
10. Outline two potential problems with flood-control dams.
11. List several appropriate land uses for floodplains that minimize risks to lives and property.
12. Describe the kinds of information used to develop flood warnings. Discuss how the cause of a particular flood event might affect the precautions that can be taken.

Exploring Further

The first few exercises involve the USGS real-time streamflow data available on the National Water Dashboard.

1. Start at **https://dashboard.waterdata.usgs.gov** Look at the national map of current streamflow data. Note where streamflow is relatively high or low for the date, and relate this to recent weather conditions.

2. Now pick a single stream gage—click first on the dot representing the gage, then on the "site page" toggle in the upper right corner. Look at discharge over different times frames, and investigate the following

   a. How much variability do you see year to year? Are there any obvious trends over time? Make a table, and from that table try your hand at constructing a flood-frequency curve by the method described in Case Study 6.1.

b. How much variability do you see from one season to the next? Can you explain the variability? How does discharge compare to the median?

c. Use the map and site photograph (if available) to determine any human factors that might affect stream discharge.

3. If a flash-flood warning is issued by NWS in your area, look up one or more forecasts of stream stage at water.**weather.gov/ahps/forecasts.php**. Later, examine the actual stream hydrographs. How close was the forecast? What might account for any differences?

4. Consider the implications of various kinds of floodplain restrictions and channel modifications that might be considered by the government of a city through which a river flows. List criteria on which decisions about appropriate actions might be based.

5. Select a major flood event, and search the Internet for (a) data on that flooding and (b) relevant precipitation data for the same region. Relate the two, commenting on probable factors contributing to the flooding, such as local geography or recent weather conditions.

CHAPTER 7

Coastal Zones and Processes

Cooper River

Ashley River

Mount Pleasant

West Ashley

Charleston

Johns Island

James Island

5 km   N

Coastal areas vary greatly in character and in the kinds and intensities of geologic processes that occur along them. They are all dynamic, with some visibly changing through the inter-action of land and water, or they may be relatively stable. What happens in coastal zones is of direct concern to a large fraction of the people in the United States: 30 of the 50 states abut either the Pacific or Atlantic Oceans or a Great Lake, and those 30 states are home to approximately 85% of the

Chapter Outline

The population of Charleston is growing three times faster than the national average, with recent development concentrated in floodplains near the coast. With already rising sea levels, major precipitation events and hurricanes that appear to be becoming more intense will increase already high flood risks. Will their planned 12-foot seawall be enough protection?

NASA Earth Observatory images by Lauren Dauphin

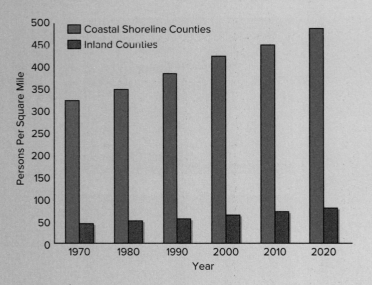

Figure 7.1

The density of population in coastal counties in the United States is far greater than in interior counties, is growing rapidly, and is projected to continue to do so.

Source: Data from National Oceanic and Atmospheric Administration

nation's population. About 40% of the U.S. population lives in coastal counties, not just in coastal states, and that population is growing rapidly (**figure 7.1**). When European settlers first colonized North America, many settled on the coast for easy access to shipping. Now, not only do those coastal cities draw people to their employment, educational, and cultural opportunities, but we also visit less-developed areas for aesthetic and recreational activities. Coastal areas also provide temporary or permanent homes for 45% of U.S. threatened or endangered species, 30% of waterfowl, and 50% of migratory songbirds. In this chapter, we will briefly review some of the processes that occur at shorelines, look at several different types of shorelines, and consider the impacts that various human activities have on them.

7.1 Nature of the Coastline

Learning Objectives

- Compare active and passive margins
- Identify features of a beach
- Describe the movement of waves in open water and at the shore
- Explain the creation of tides, including spring and neap tides
- Describe the creation of a longshore current and littoral drift
- Describe the creation and dangers of a storm surge

The nature of a coastal area is determined by a variety of factors including tectonic setting, the materials present at the shore, and the energy of the water striking the coast. Coasts are specific to marine settings and we will concentrate our discussions on those, but many of the features and processes can also be applied to inland, shoreline settings such as those around the Great Lakes.

Plate tectonics influences the underlying character and morphology of a coastline. We describe the edges of continents as either *active* or *passive* in a tectonic sense, depending upon whether or not it coincides with a plate boundary. The western margin of North America is an **active margin** because it lies

along a plate boundary marked by transform faulting and subduction (**figure 7.2**). An active margin is characterized by cliffs above the waterline, a narrow continental shelf, and a relatively steep drop to oceanic depths offshore. The eastern margin of North America is a **passive margin**, far removed from the active plate boundary at the Mid-Atlantic Ridge. It is characterized by a wide continental shelf and extensive development of broad beaches and sandy offshore islands.

A **beach** is a gently sloping surface washed over by the waves and covered by sediment (**figure 7.3**). The sand or rocks of a beach may have been produced locally by wave erosion, transported overland by wind from behind the beach, or delivered to the coast and deposited there by streams or coastal currents. The **beach face** is that portion regularly washed by the waves as tides rise and fall. The *berm* is the flatter part of the beach landward of the beach face. What lies behind the berm may be dunes, as shown in **figure 7.3A**; rocky or sandy cliffs, as in **figure 7.3B**; low-lying vegetated land, as in **figure 7.3C**; or even artificial structures such as seawalls.

Waves, Tides, and Currents

Waves and currents are the principal forces behind natural shoreline modification. Waves are induced by the flow of wind across the water surface, which sets up small undulations in that surface. The shape and apparent motion of waves reflect the changing geometry of the water surface; the actual motion of water molecules is quite different. While a wave form may seem to travel long distances across the water, in the open water

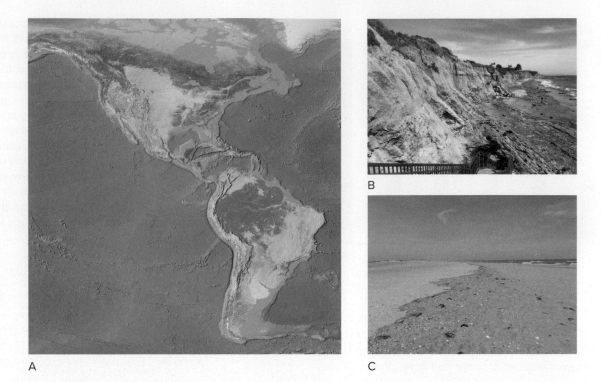

A

B

C

Figure 7.2

(A) Both North and South America have active margins on their west coasts and passive margins on their east coasts. Notice the relative size of the continental shelves. (B) Pismo Beach, California. (C) Cape Lookout National Seashore, North Carolina.

(A) Jan Rysavy/E+/Getty Images; (B, C) ©Gina S. Szablewski

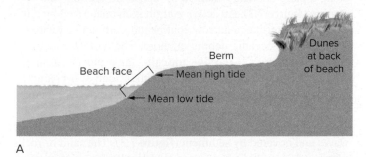

Beach face

Berm

Dunes at back of beach

Mean high tide

Mean low tide

A

B

C

Figure 7.3

(A) General beach profile. (B) The various components of the profile are clearly developed on this California beach, backed by cliffs rather than dunes. (C) The narrow Rialto Beach, Washington, is made of rounded pebbles and backed by forest.

(B) ©Carla Montgomery (C) ©Gina S. Szablewski

molecules are actually rising and falling locally in circular orbits that grow smaller with depth (**figure 7.4A**). A model we commonly use to understand this molecular motion is to think about a floating object bobbing on ocean waves; it mainly rises and falls as the wave energy moves across the surface, rather than moving horizontally forward. As waves approach the shore, the circular orbits are disrupted by the shallower bottom, and the waves develop into breakers (**figure 7.4B** and **C**).

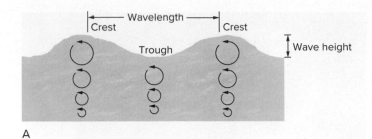

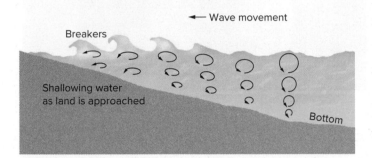

C

Figure 7.4

(A) Waves are undulations in the water surface, beneath which water moves in circular orbits. *Wave base* is the depth below which there is no orbital motion. (B) Breakers develop as waves approach shore and orbits are disrupted. (C) Evenly spaced natural breakers develop on a gently sloping shore off the California coast.

(C) ©Carla Montgomery

The most effective erosion of solid rock along the shore is through wave action, either the direct pounding by breakers or the grinding effect of sand, pebbles, and cobbles propelled by waves, which is called **milling**, or *abrasion*. Erosion is concentrated at the waterline, where breaker action is most vigorous (**figure 7.5**). The type and rate of weathering also depends on the chemistry of the water and the nature of the rock being eroded. Limestone will dissolve more readily in water that is

A

B

Figure 7.5

(A) Undercutting by wave action causes blocks of sandstone higher on this cliff to collapse into the water. Pictured Rocks National Lakeshore, Michigan. (B) Sea arch and sea stacks formed by wave action along the California coast. Erosion is most intense at the waterline, undercutting the rock above.

(A, B) ©Carla Montgomery

slightly acidic compared to seawater which is mildly alkaline. A rock that is already fractured and weakened by other geologic processes will physically weather more quickly than solid rock, no matter the chemistry of the water.

Water levels vary relative to the elevation of coastal land, over the short term, as a result of tides and storms. *Tides* are the periodic regional rise and fall of water levels caused by the gravitational pull of the Sun and Moon on the watery envelope of oceans surrounding the planet. The closer an object, the

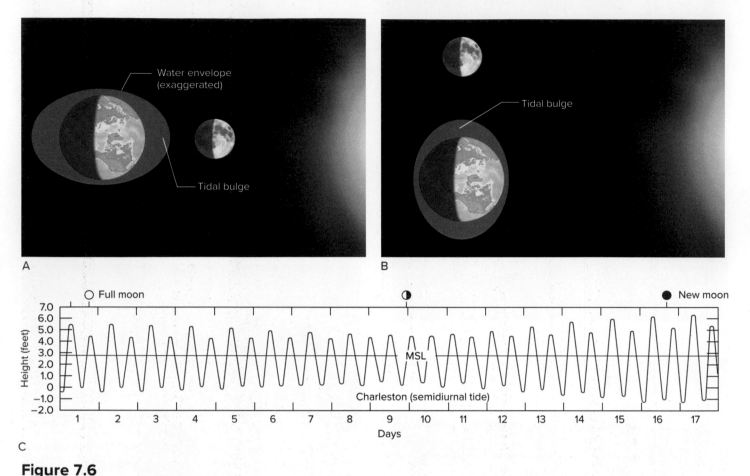

Figure 7.6

(A) Spring tides: Sun, Moon, and Earth are all aligned, near times of full and new moons. (B) Neap tides: When Moon and Sun are at right angles, tidal extremes are minimized. Note that these sketches are not to scale. (C) A typical tidal cycle, for Charleston, SC. (MSL = mean sea level.)

stronger its gravitational attraction. The Moon, though much less massive, exerts much more gravitational pull on the oceans than does the Sun. The oceans bulge on the side of Earth nearest the Moon. A combination of lesser pull on the opposite side, Earth's rotation on its axis, and overall rotation of the earth-moon system produces a complementary bulge of water on the opposite side of Earth. As the rotating Earth spins through these bulges of water each day, overall water level at a given point on the surface rises and falls twice a day. This is the phenomenon recognized as tides.

At high tide, waves reach higher up the beach face; more of the beach face is exposed at low tide. Tidal extremes are greatest when Sun, Moon, and Earth are all aligned, and the Sun and Moon are thus pulling together, at times of full and new moons. The resultant tides are *spring tides* (**figure 7.6A**). They have nothing to do with the spring season of the year. When the Sun and Moon are pulling at right angles to each other, the difference between high and low tides is minimized. These are *neap tides* (**figure 7.6B**). The shape of the shoreline and the underwater topography also affect the tides, so the frequency and magnitude of tides vary by location. Tides are

predictable; you can find them for the United States at https://tidesandcurrents.noaa.gov/tide_predictions.html.

Tides not only change water levels along the coast but they can also create *currents,* distinct bodies of water moving in a definite direction. There are different types of currents, and they can be local or stretch across ocean basins. Some are the extensive surface currents created primarily by wind and that reflect average atmospheric circulation patterns; the Gulf Stream that runs along the U.S. East Coast is an example. These surface currents redistribute heat across the globe and play a significant role in climate, which we will address later in chapters 9 and 10. Local coastal currents are created in relation to the approach of wind and waves. Ocean water also moves vertically due to differences in density. While currents may not directly affect water levels along the coast, they are part of the suite of processes that shape coastlines.

Waves and tides, along with distinct storm events, affect water levels on the short term. Over longer periods, relative water levels may shift systematically up or down as a result of tectonic processes, from changes related to shifts in climate, or from human activities such as extraction of groundwater or oil that can cause surface subsidence. These relative changes in land or sea height can produce distinct features and specific problems, as we will review in section 7.2.

Sediment Transportation and Deposition

Just as water flowing in streams moves sediment, so do waves and currents. The faster the currents and more energetic the waves, the larger and heavier the sediment particles that can be moved. As waves approach shore and the water orbits are disrupted, the water tumbles forward, up the beach face.

When waves approach a beach at an angle, as the water washes up onto and down off the beach, there is also net movement of water laterally along the shoreline, creating a **longshore current** (**figure 7.7**). Likewise, any sand caught up by and moved along with the flowing water is not carried straight up and down the beach. As the waves wash up the beach face at an angle to the shoreline, the sand is pushed along the beach as well. The net result is **littoral drift**, gradual sand movement down the beach in the same general direction as the motion of the longshore current. Currents tend to move consistently in certain preferred directions on any given beach, which means that over time they continually transport sand from one end of the beach to the other. On many beaches where this occurs, the continued existence of the beach is only assured by a fresh supply of sediment produced locally by wave erosion, delivered by streams or supplied by dunes behind the beach.

The more energetic the waves and currents, the more and farther the sediment is moved. Higher water levels during storms also help the surf reach farther inland, perhaps all the way across the berm to dunes at the back of the beach. But a ridge of sand or gravel may survive the storm's fury, and may even be built higher as storm winds sweep sediment onto it.

LONGSHORE CURRENT ⟹

Figure 7.7

When waves approach the shore at an angle, they push water and sand up the beach face at an angle, too; gravity drains water downslope again, leaving some sand behind. There is net movement of water along the coast, the longshore current; littoral drift of sand is parallel to it.

Figure 7.8

These beach ridges at Cape Krusenstern National Monument, Alaska, show a beach growing over thousands of years. The study of systems such as these helps scientists understand how beaches change.

NPS Photo/Cait Johnson

This is one way in which a *beach ridge* forms at the back of a beach (**figure 7.8**). Where sediment supply is plentiful, as near the mouth of a sediment-laden river, or where the collapse of sandy cliffs adds sand to the beach, the beach may build outward into the water, new beach ridges forming seaward of the old, producing a series of roughly parallel ridges that may have marshy areas between them. Vegetation may quickly become established on a ridge at the back of a berm, and add some stability to a beach.

Storms and Coastal Dynamics

Unconsolidated materials such as beach sand are readily moved and rapidly eroded during storms. A major reason is storm **surge,** an unusual increase in local sea level due to the presence of the storm. Storm surge is commonly discussed in association with tropical cyclones and hurricanes, but can accompany any major storm system. The primary cause of storm surge is onshore winds pushing water toward the shore, piling it up against and over the land; the low air pressure associated with the storm also contributes a bulge in the water surface. Generally, the stronger the storm, the greater the surge (**table 7.1**), but other factors, including the speed at which the storm approaches the coast and the geometry of the particular coastline, can affect the magnitude of the surge. The overall water level during a storm, the *storm tide,* is the sum of storm surge plus normal tides.

The temporarily elevated water level associated with the surge, and the unusually energetic wave action and greater wave height also caused by high winds combine to attack the coast with exceptional force. Hurricane Katrina's peak surf averaged 17 meters (55 feet) in height, with the largest waves estimated at over 27 meters (90 feet) high, and that was on top of storm surge that locally raised sea level by as much as 7.5 meters (25 feet). The gently sloping expanse of beach (berm) along the outer

Table 7.1	Saffir-Simpson Hurricane Scale		
Category	Wind Speed (in miles per hour)	Typical Storm Surge (feet above normal tides)	Evacuation
1	74–95	4–5	No.
2	96–110	6–8	Some shoreline residences and low-lying areas, evacuation required.
3	111–130	9–12	Low-lying residences within several blocks of shoreline, evacuation possibly required.
4	131–155	13–18	Massive evacuation of all residences on low ground within 2 miles of shore.
5	155	18$^+$	Massive evacuation of residential areas on low ground within 5–10 miles of shore possibly required.

Source: National Weather Service, NOAA.

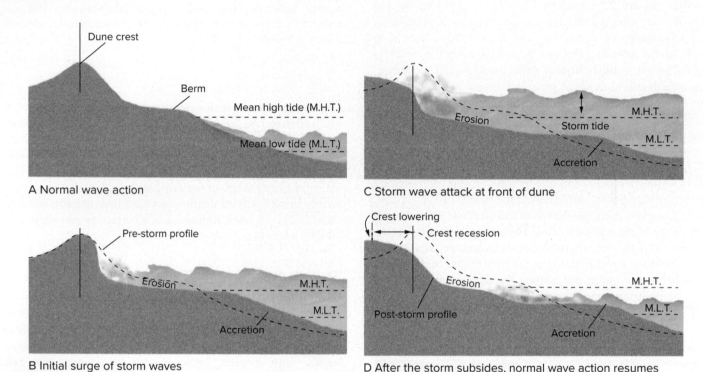

Figure 7.9

Alteration of shoreline profile due to accelerated erosion by unusually high storm tides.

Source: Shore Protection Guidelines, U.S. Army Corps of Engineers.

shore above the usual high-tide line may suffer complete over-wash, with storm waves reaching and beginning to erode dunes or cliffs beyond (**figure 7.9**). The post-storm beach profile typically shows landward recession of dune crests, as well as removal of considerable material from the zone between dunes and water (**figure 7.9D**).

To appreciate the magnitude of the effects, see table 7.1, and consider that, given the gentle slopes of many coastal regions, evacuations miles inland may be necessary in response to an anticipated surge of 10 to 20 feet. The highest storm surge ever recorded was 13 meters (42 feet) from the Bathurst Bay,

Australia, hurricane of 1899. Now hurricanes are tracked well enough that there is time to make evacuation possible. Still, the damage to immobile property can be impressive (**figure 7.10**; table 7.2). And, as was shown with Hurricanes Katrina, Sandy, and others, not everyone can or will heed a call to evacuate. Where a large city is involved, evacuation includes complex logistical problems that can be difficult to handle if an exiting plan either does not exist or is not implemented correctly and efficiently. Hurricane cleanup and recovery can be drawn out and have long-lasting physical and social effects, especially in remote areas.

A

B

Figure 7.10

(A) Hurricane Maria devastated the northeastern Caribbean; the U.S. citizens of Puerto Rico were without power for months. Notice the amount of sediment in the street. (B) One reason that storm surges are so damaging is that they can be prolonged. Here, two days after Hurricane Harvey's landfall, the Texas coast near Rockport remains submerged.

(A) Shutterstock/JEAN-FRANCOIS Manuel; (B) Army National Guard photo

Table 7.2	A Sampling of Notable U.S. Hurricanes			
Hurricane	**Date**	**Category at U.S. Landfall**	**Deaths***	**Damages (billions)†**
Galveston, TX	1900	4	8000†	$ 31
Southeast FL/MS/AL	1926	4	243	$ 84
New England	1938	3	600	$ 19
Hazel (SC/NC)	1954	4	95	$ 8
Carla (TX)	1961	4	46	$ 8
Betsy (southeast FL/LA)	1965	3	75	$ 14
Camille (MS/southeast U.S.)	1969	5	256	$ 13
Agnes (northwest FL/northeast U.S.)	1972	1	122	$ 12
Hugo (SC)	1989	4	<30	$ 11
Andrew (southeast FL/LA)	1992	4	<30	$ 38
Floyd (NC)	1999	2	57	$ 4.5
Katrina (LA, MS, AL)	2005	4	1833	$ 80
Ike (TX, LA)	2008	2	82	$ 27
Irene (eastern, northeastern U.S.)	2011	1	56	$ 10
Sandy (eastern, northeastern U.S.)	2012	1	186	$ 70
Matthew (FL, GA, NC, SC)	2016	1	43	$ 6
Harvey (TX, LA)	2017	4	76	$190
Irma (FL, GA)	2017	4	95	$100
Maria (PR, USVI)	2017	4	2975	$ 90
Laura (LA, TX)	2020	4	7	$ 19
Ida	2021	4	115	$ 75

Data for pre-1998 events from R. A. Pielke, Jr., and C. W. Landsea, "Normalized Atlantic hurricane damage 1925–1995," Weather Forecasting 13: 621–31; later data from NOAA/National Hurricane Center.

7.2 Emergent and Submergent Coastlines

Learning Objectives

- List causes of global, local, and long-term sea level change
- Compare emergent and submergent coastlines
- Describe conditions that create a wave-cut platform and a drowned valley
- Recall the current primary causes and rate of sea-level change

Many factors affect sea level, the most evident being the simple volume of water stored in the oceans compared to the other reservoirs in the hydrologic cycle. Other controlling factors include temperature, salinity, atmospheric pressure, currents, and even the uneven gravitational pull of Earth on the oceans. Over the past 30 years, NASA scientists have noticed 20–30 cm changes in sea level in the open ocean, linking them to cyclical processes such as El Niño and changes in the velocity of ocean currents. It is not these open ocean changes that we are so concerned about, but rather the relative and local increases in sea level that are linked to a warming climate and human activities. In the following section, we will review some causes of sea level change and evidence we use to understand sea level change in the recent geologic past. We will also briefly summarize current sea level trends.

Causes of Long-Term Sea-Level Change

Over the long stretches of geologic time, tectonics can play a subtle role in controlling global sea level. At times of more rapid seafloor spreading, there is more warm, young sea floor. In general, heat causes material to expand, and warm lithosphere takes up more volume than cold, so the volume of the ocean basins is reduced at such times, and water rises higher on the continents, spreading over low-lying coastal areas (**figure 7.11**). Tectonic forces can also cause gradual or sudden changes in local sea level. The mountain building that accompanies the process of subduction will cause relative increases in sea level, and subduction itself can lead to both increases and decreases in crustal elevation. Nearly instantaneous elevation changes occur when earthquakes pushup or down-drop sections of continental crust, again altering local sea level. During the 1964 Great Alaskan Earthquake discussed in chapter 4, large coastal areas experienced 2.5–25 meter elevation changes in both directions within a matter of minutes.

The lithosphere can sink or rise in response to loading or unloading, given that it buoyantly sits on top of the plastic asthenosphere. For example, in regions overlain and weighted down by massive ice sheets during the last ice age, the lithosphere was downwarped by the ice load. Tens of thousands of years later, the lithosphere is still slowly springing back to its pre-ice elevation. Where the thick ice extended to the sea, depressing the coastline, the coastline is now slowly rising relative to the sea. This is well documented, for instance, in Scandinavia, where the rebound still proceeds at rates of up to 2 centimeters (close to 1 inch) per year. Conversely, in basins being filled by sediment, like the Gulf of Mexico, the sediment loading can cause slow sinking of the lithosphere.

Glaciers also represent an immense reserve of water. As this ice melts, sea levels rise worldwide (**figure 7.12**). Global warming not only melts ice but aggravates the sea-level rise in another way; warmed water expands, and this effect, multiplied over the vast volume of the oceans, can cause more sea-level rise than the melting of the ice itself. In recent years, glacial melting and thermal expansion each account for about half of the observed rise in sea level. Its consequences will be explored further later in this chapter.

Locally, pumping large volumes of liquid from underground has caused measurable and problematic subsidence of the land. In Venice, Italy, decades of groundwater extraction caused up to 15 cm of subsidence, and that is in addition to subsidence caused by the weight of buildings, sediment compaction, and tectonic setting. In response to a current sinking rate of 8 cm/20 years and rising sea levels, Venice implemented a $6 billion barrier system in 2020 to protect itself from tidal flood damage. It is working as designed so far, but it is only used when waters are predicted to reach a certain level so not everyone and every building is protected from all floods. In some parts of Houston, Texas, which has one of the fastest

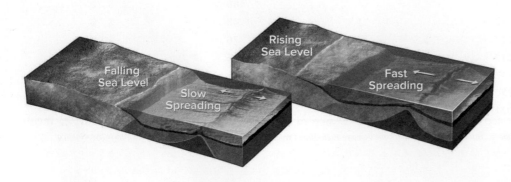

Figure 7.11

Eustastic sea level falls when seafloor spreading rates are slow, and rises with rates are fast.

McGraw Hill

Figure 7.12

Eustatic sea level falls when there is more continental ice, and rises when continental glaciers melt.

McGraw Hill

growing populations in the United States, subsidence caused by groundwater extraction is occurring at over 2 cm/year, putting this coastal city at high risk of flooding, as evidenced by the damage caused by Hurricane Harvey in 2017.

Signs of Changing Relative Sea Level

The elevation at which water surface meets land depends on the relative heights of land and water, either of which can change. From a practical standpoint, it tends to matter less to people whether *global* sea level is rising or falling than whether the water is getting higher relative to the local coast (*submergent* coastline), or the reverse (*emergent* coastline). Often, geologic evidence will reveal the local trend.

A distinctive coastal feature that develops where the land is rising and/or the water level is falling is a set of **wave-cut platforms** (**figure 7.13**). Given sufficient time, wave action tends to erode the land down to the level of the water surface. If the land rises in a series of tectonic shifts and stays at each new elevation for some time before the next movement, each rise results in the erosion of a portion of the coastal land down to the new water level. The eventual product is a series of steplike terraces. These develop most visibly on rocky coasts, rather than coasts consisting of soft, unconsolidated material. The surface of each such step, or platform, represents an old water-level marker on the continent's edge.

When global sea level rises or the land drops, the base level of a stream that flows into the ocean is raised. A portion of the floodplain near the mouth may be filled by encroaching seawater, and the fresh water backs up above the new, higher base level, forming a **drowned valley** (**figure 7.14**).

What happens when a local land surface and global sea level are rising at the same time? Features such as the beach ridges in (**figure 7.15**) will emerge if the uplift, in this case glacial rebound, is occurring faster than sea level is increasing. Not only do the beach ridges indicate former sea levels, but their emergence indicates a relative change in sea level.

Glaciers flowing off land and into water do not just keep eroding deeper underwater until they melt. Ice floats, and freshwater glacier ice floats especially readily on denser salt water. The carving of glacial valleys by ice does not extend long distances offshore. During the last ice age, when sea level

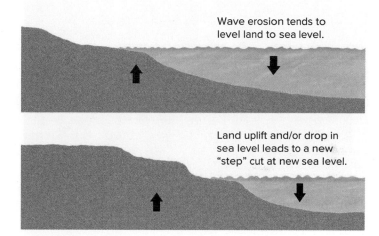

Wave erosion tends to level land to sea level.

Land uplift and/or drop in sea level leads to a new "step" cut at new sea level.

A

B

Figure 7.13

(A) Wave-cut platforms form when land is elevated or sea level falls. (B) Wave-cut platforms at Mikhail Point, Alaska.

(B) U.S. Geological Survey Photo Library, Denver, CO/Photograph by J. P. Schafer

worldwide was as much as 100 meters lower than it now is, alpine glaciers at the edges of continental landmasses cut valleys into what are now submerged offshore areas. With the recession of the glaciers and the concurrent rise in sea level, these old glacial valleys were emptied of ice and partially filled with water. That is the origin of the steep-walled fjords common in Scandinavian countries and Iceland.

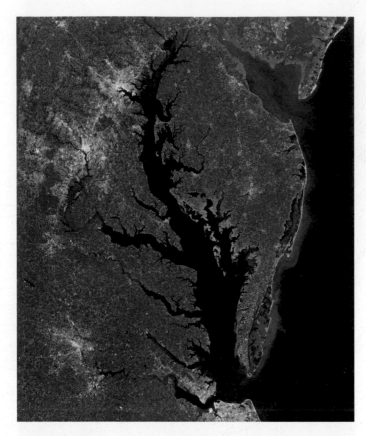

Figure 7.14

Chesapeake Bay is a drowned valley, as seen in this satellite image.

NASA/Image by Mike Taylor

Figure 7.15

The beach ridges on the south end of Akimiski Island in James Bay, Ontario, are emerging as the island rebounds. Each ridge represents wave action on the shore at former sea levels since the ice sheets melted.

NASA image created by Jesse Allen

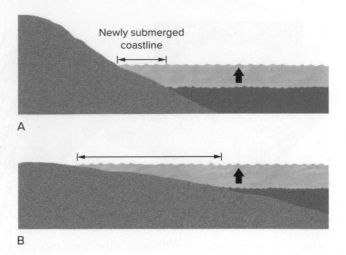

Newly submerged coastline

A

B

Figure 7.16

Effects of small rises in sea level on steeply sloping shoreline (A) and on gently sloping shore (B). The same rise inundates a much larger expanse of land in case (B).

Present and Future Sea-Level Trends

Much of the coastal erosion presently plaguing the United States is the result of gradual but sustained global sea-level rise, from the combination of melting of glacial ice and expansion of the water itself as global temperatures rise. Scientists currently estimate sea-level rise at about ⅓ meter (1 foot) per century. While this does not sound particularly threatening, we need to consider two additional factors. First, many coastal areas slope very gently or are nearly flat, so that a small rise in sea level results in a far larger inland retreat of the shoreline (**figure 7.16**). Rates of shoreline retreat from rising sea level have been measured at several meters per year in some low-lying coastal areas. A sea-level rise of ⅓ meter would erode beaches by 15 to 30 meters in the northeast, 65 to 160 meters in parts of California, and up to 300 meters in Florida. The average commercial beach is only about 30 meters wide now. Second, the documented rise of atmospheric carbon-dioxide levels, discussed in chapter 10, suggests that increased greenhouse-effect heating will melt the ice caps and glaciers more rapidly, as well as warming the oceans more, accelerating the sea-level rise; some worst-case scenario models put the anticipated rise in sea level at over 1 meter by the year 2100. Aside from the havoc this would create on beaches, it would flood 30% to 70% of U.S. coastal wetlands (**figure 7.17**).

Consistently rising sea level is a major reason why shoreline-stabilization efforts repeatedly fail. It is not just that the high-energy coastal environment presents difficult engineering problems. The problems themselves are intensifying as a rising sea level brings waves farther and farther inland, pressing ever closer to and more forcefully against more and more structures developed along the coast. And, a warmer climate makes more energy available to fuel storms, creating even more energetic waves.

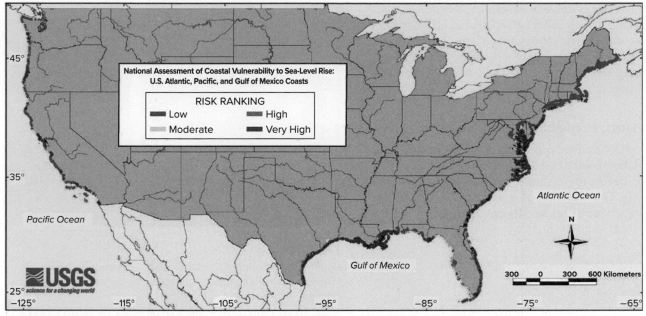

A

B

Figure 7.17

(A) Relative vulnerability of coastal regions to change resulting from rising sea level depends on factors such as topography and materials present. (B) Rising sea level causes flooding of coastal wetlands, increases salinity of estuaries; cypress in this flooded swamp died as a result.

Source: (A) After USGS National Assessment of Coastal Vulnerability to Sealevel Rise; (B) Spring Images/Alamy Stock Photo

7.3 Coastal Erosion and Attempts at Stabilization

Learning Objectives

- Identify the categories of coastline management strategies
- Describe the use of groins, breakwaters, and beach nourishment to reduce erosion
- Explain the erosion of headland by wave refraction
- Describe the effectiveness of seawalls and riprap in preventing erosion

Coastlines are inherently dynamic, and beaches especially so. In fair weather, waves washing up onto the beach carry sand from offshore to build up the beach; storms attack the beach and dunes and carry sediment back offshore. Where longshore currents are present, the beach reflects the balance between sediment added from the up-current end and sediment carried away down-current. Where sandy cliffs are eroding, they supply fresh sediment to the beach to replace what is carried away by waves and currents. Our attempts to stabilize coasts in order to develop them and protect what we build is an expensive, never-ending process that disrupts these natural, give-and-take systems.

The approaches we use in managing coastlines fall into three broad categories. We will look at examples of each. *Hard structural stabilization* involves the construction of solid structures. *Soft structural stabilization* methods include sand replenishment, dune rebuilding or stabilization, and planting and encouraging native vegetation. These methods may be preferred for their more naturalistic look, but their effects can be fleeting. Nonstructural strategies, the third type, involve land-use restrictions, prohibiting development or mandating minimum setback from the most unstable or dynamic shorelines; this is analogous to restrictive zoning in fault zones or floodplains, and may meet with similar resistance. The move of the Cape Hatteras lighthouse (**figure 7.18**) is an example of a nonstructural approach, the type of managed retreat discussed in chapter 6 that we will bring up again later in this chapter.

Beach Erosion, Protection, and Restoration

Beachfront property owners, concerned that their beaches will wash away and their buildings will flood, erect structures to try to stabilize the beach; the common result is a change in the beach's geometry. One widely used method is the construction of one or more groins or jetties (**figure 7.19**), long, narrow

Figure 7.19

When groins or jetties perpendicular to the shoreline disrupt longshore currents, deposition occurs up-current, erosion below.

Photograph courtesy U.S. Geological Survey Coastal and Marine Geology Program.

A

B

Figure 7.18

The Cape Hatteras Lighthouse is a historic landmark. (A) It had been threatened for many years; shoreline stabilization attempts included a groin and the sandbagging shown here. (B) Finally the lighthouse was moved, at a cost of $11.8 million.

(A) U.S. Geological Survey/Photograph by R. Dolan; (B) U.S. Geological Survey Center for Coastal Geology

obstacles set more or less perpendicular to the shoreline. By disrupting the longshore currents, these structures change the shoreline. Just as a stream deposits sediments when its velocity decreases, longshore currents deposit their load of sand when they encounter a barrier and are slowed.

Down-current from the barrier, the water picks up more sediment to replace the lost load, and the beach is eroded. The common result is that a formerly straight shoreline develops an unnatural scalloped shape. The beach is built out in the area up-current of the groins and eroded landward below. Beachfront properties in the eroding zone may be more severely threatened after construction of the structures than they were before. This is especially problematic if there are different property owners on either side of the groin.

Any interference with sediment-laden waters can cause redistribution of sand along beachfronts. A marina built out into such waters may cause some deposition of sand around it and possibly in the protected harbor. Farther along the beach, the now unburdened waters can more readily take on a new sediment load of beach sand. Breakwaters also cause sediment redistribution, even though they are constructed primarily to moderate wave action (**figure 7.20**).

Even modifications inland, far from the coast, can affect the beach. One notable example is the practice of damming large rivers for flood control or power generation. As indicated in chapter 6, one consequence of artificial reservoirs is that the sediment load carried by the stream is trapped behind the dam. Farther downstream, the cutoff of sediment supply to coastal beaches can lead to erosion. It can be difficult to find a readily available alternate supply of sand, or expensive to dredge and transport the sand stuck behind the dam. Flood-control dams,

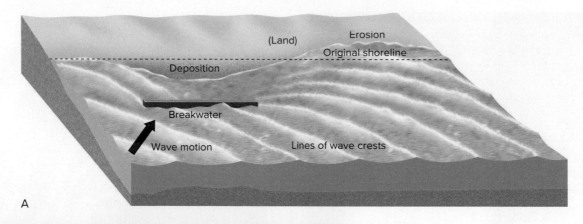

A

B

Figure 7.20

(A) Sediment is deposited behind a breakwater; erosion occurs down-current through the action of water that has lost its original sediment load. (B) Breakwaters and groins together combat beach erosion at Presque Isle State Park, Pennsylvania.

(B) U.S. Army Corps of Engineers/Photograph by Ken Winters

especially those constructed on the Missouri River, have reduced the sediment load delivered to the Gulf of Mexico by the Mississippi River by more than half, initiating rapid coastal erosion in parts of the Mississippi delta.

Where beach erosion is rapid, and development and tourism are widespread, people often choose to invest in imported replacement sand to maintain sizeable beaches (**figure 7.21**). The initial cost of such an effort can be very high. Even if steps are taken to reduce the erosion rates, the sand will have to be replenished over and over at ever-rising cost. Typically, the reason for beach nourishment or replenishment projects is that natural erosion has removed much of the sand. No one should be surprised that continued erosion may make short work of the restored beach. Sea Bright/Ocean Township, New Jersey, tops the list of beach nourishment in the United States with 15 projects spread out between 1962 and 2019 spending over $366 million for 33.8 million cubic yards of sand. This area experienced severe flooding and erosion from Hurricane Sandy in 2012.

In many cases, it has not been possible or thought necessary to duplicate the mineralogy or grain size of the sand originally lost. The result has sometimes been further environmental deterioration. When coarse-grained sands are replaced by finer ones, softer and muddier than the original, the finer materials stay suspended in the water, clouding it. This is not only unsightly but can also be deadly to organisms. Off Waikiki and Miami Beaches, delicate coral reef communities have been damaged or killed by this extra water turbidity caused by replenishment efforts. Replacing finer-grained sand with coarser may steepen the beach face making the beach less safe for beach-goers. And, as the world experiences its first shortage of sand due primarily to booms in construction and urbanization, sand is becoming more difficult to obtain, much more expensive, and associated with the unfair work practices typically tied to other mineral resources, which we will explore in chapter 13.

A soft structure stabilization solution that is gaining popularity is a living shoreline, a way to support the natural coastal systems that already exist and that are effective at absorbing and deflecting wave and storm energy. These living solutions can range from small groups of people piling oyster shells offshore and planting native marsh grasses to reef restoration along hundreds of kilometers of coastline. A report published by the USGS in 2021 demonstrated that reef restoration in southeastern Florida would prevent flooding-associated damages worth more than $232 million annually.

Figure 7.21

Peck Beach in Ocean City, NJ, has been periodically replenished with sand with the stated goal of reducing flood and storm damage. The Army Corps of Engineers estimates future nourishment at 1.1 million cubic yards every 3 years.

U.S. Army Corps of Engineers

Cliff Erosion

Both waves and currents can vigorously erode and transport loose sediments, such as beach sand, much more readily than solid rock. Erosion of sandy cliffs may be especially rapid. Removal of material at or below the waterline undercuts the cliff, leading to slumping and sliding of sandy sediments and the swift landward retreat of the shoreline.

Not all places along a shoreline are equally vulnerable. Jutting points of land, or headlands, are more actively under attack than recessed bays because wave energy is concentrated on these headlands by **wave refraction** (**figure 7.22**). Wave refraction occurs because waves touch bottom first as they approach the projecting points, which slows the waves down; wave motion continues more rapidly elsewhere. The net result is deflection of the waves around and toward the headlands, where the wave energy is focused. The long-term tendency is toward a rounding-out of angular coastline features, as headlands are eroded and sediment is deposited in the lower-energy environments of the bays (**figure 7.23**).

Various measures, often as unsuccessful as they are expensive, may be tried to combat cliff erosion. A fairly common practice is to place some kind of barrier at the base of the cliff to break the force of wave impact. The protection may be a *seawall*, or *riprap* which is usually a pile of repurposed construction debris or large boulders (**figure 7.24**). If the obstruction is placed only along a short length of cliff directly below a threatened structure, the water is likely to wash in beneath, around, and behind it, rendering it largely ineffective. Erosion continues on either side of the barrier. Wave energy may also bounce off, or be reflected from a length of smooth seawall, to

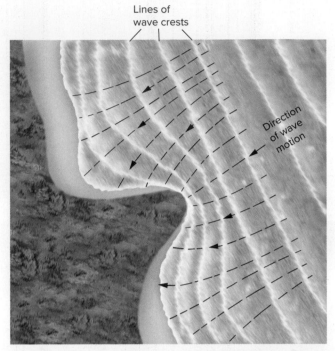

A

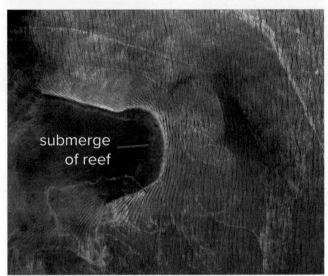

B

Figure 7.22

(A) This sketch illustrates how the energy of waves approaching a jutting point of land is focused on it by refraction. (B) Wave refraction around Bajo Nuevo reef in the western Caribbean Sea; waves are moving from right to left in image. Behind the reef is calm water, appearing darker blue. See refraction around breakwaters in **figure 7.20B**.

(A) McGraw Hill; (B) Earth Sciences and Image Analysis Laboratory, NASA Johnson Space Center.

erode the beach more quickly or attack a nearby unprotected cliff with greater force.

Seawalls and riprap can be effective at absorbing energy, but at some point, the structure will not be strong or tall enough

Figure 7.23

As waves approach this rugged section of the California coast, projecting rocks are subject to highest-energy wave action; sand accumulates in recessed bays.

©Carla Montgomery

to protect the land behind it. The Galveston seawall was built to protect the city after the deadly 1900 hurricane. It proved its value in 2008 when Hurricane Ike hit, and in 2017 with Hurricane Harvey. But a 120-year-old structure is not sufficient to handle current sea level rise and future predicted storm hazards. The "Ike Dike" coastal storm barrier system, a $29 billion dollar project that includes beach nourishment, flood gates, flood walls, levees, and pumping stations, has finished its final study phase but could take up to 20 years to be implemented; the Galveston and Houston areas remain very vulnerable in the meantime.

7.4 Especially Challenging Coastal Environments

Learning Objectives

- Explain the vulnerability of barrier islands to wave erosion
- Describe why barrier island stabilization doesn't work
- List the environmental pressures on estuaries

Many coastal environments are unstable, but some are particularly vulnerable either to natural forces or human interference. In the following, we will address the environmental pressures on barrier islands and estuaries, providing examples of successful and not so successful efforts at stabilizing these natural settings.

A

B

Figure 7.24

(A) This seawall protects an estate in Newport, Rhode Island. (B) Pieces of concrete riprap are used to protect this shoreline along Lake Michigan.

(A) ©Carla Montgomery; (B) Doug Sherman/Geofile

Barrier Islands

Barrier islands are long, low, sandy features located offshore and parallel to a coastline (**figure 7.25**). Our current understanding is that they form from the submergence of either coastal dunes or sandy depositional features associated with longshore currents. They provide important protection for the water and shore inland from them because they constitute the first line of defense against the high energy waves during a storm. The barrier islands themselves are extremely vulnerable, partly as a result of their low relief. Water several meters deep may wash right over these low-lying islands during unusually high storm tides, such as occur during hurricanes. Strong storms may even slice right through narrow barrier islands, separating people from bridges to the mainland, if the bridges survive at all (**figure 7.26**).

Because they are usually subject to higher-energy waters on their seaward sides than on their landward sides, most barrier islands are receding landward with time. Typical velocities on the Atlantic coast of the United States are 2 meters (6 feet)

A

to Houston

Texas City Dike

Bolivar
Peninsula

Intracoastal
Waterway

Galveston

Subsided
Wetlands

N

B

Figure 7.25

(A) North Carolina's Outer Banks are barrier islands. (B) One reason that Galveston is so vulnerable is that Galveston Island is a barrier island. The seawall runs along the southern edge of the island.

(A) U.S. Geological Survey/Photograph by R. Dolan; (B) Earth Sciences and Image Analysis Laboratory, NASA Johnson Space Center

Figure 7.26

Photographs of Mantoloking, New Jersey, before and shortly after Hurricane Sandy. (Top picture taken 18 March 2007; bottom picture, 31 October 2012.) Note extreme erosion, landward shift of sand, loss of houses, and destruction of bridge.

Photographs by NOAA Remote Sensing Division, courtesy NASA

per year, but rates in excess of 20 meters per year have been noted. Such settings represent particularly unstable locations in which to build, yet the aesthetic appeal of long beaches has led to extensive development on privately owned stretches of barrier islands. About 1.4 million acres of barrier islands exist in the United States, and approximately 20% of this total area has been developed. Thousands of structures, including homes, businesses, roads, and bridges, are at risk.

On barrier islands, shoreline-stabilization efforts such as groins, breakwaters, and beach replenishment tend to be especially expensive and are frequently futile. At best, the benefits are temporary. Construction of artificial stabilization structures may easily cost tens of millions of dollars and at the same time destroy the natural character of the shoreline and even the beach, which was the whole attraction for developers in the first place. The option of moving buildings and roads farther landward is just as costly, if it is an option at all. Other human activities exacerbate the plight of barrier islands. Dams on coastal rivers trap sediment, starving barrier islands of vital sand supply. Considering the inexorable rise of global sea level and that more barrier-island land is being submerged more frequently, the problems will only get worse. See also Case Study 7.1.

Estuaries

An **estuary** is a body of water along a coastline, open to the sea, in which the tide rises and falls and in which fresh and salt water meet and mix to create brackish water (**figure 7.27**). San Francisco Bay, Chesapeake Bay, Puget Sound, and Long Island Sound are examples. Some estuaries form at the lower ends of stream valleys, on deltas or in drowned valleys. Others are located in tidal basins in which the water is more salty than not.

The salinity in an estuary reflects the balance between freshwater input and salt water. It may vary with seasonal fluctuations in streamflow, and within the basin with proximity to the freshwater source(s). Over time, the complex communities of organisms in estuaries have adjusted to the salinity of the particular water in which they live. Any modifications that permanently alter this balance can have a catastrophic impact on the organisms. In many places, demand for the fresh water that would otherwise flow into estuaries is diminishing that flow. The salinity of San Francisco Bay has been rising markedly as fresh water from rivers flowing into the bay is consumed en route. Also, water circulation in estuaries is often very limited. This makes them especially vulnerable to pollution, which can easily accumulate due to a lack of flushing. Many of the world's largest coastal cities are located on estuaries that receive polluted effluent water: London, Calcutta, Hong Kong, Buenos Aires, San Francisco, and New York are some examples.

Unfortunately, many vital wetland areas are estuaries under pressure from environmental changes and human activities. For example, where land is at a premium, estuaries are frequently called on to supply more. They may be isolated from the sea by dikes and pumped dry or, where the land is lower, wholly or partially filled in. Naturally, this further restricts water flow and also generally changes the water's chemistry. In addition, pollution usually accompanies development. All of this greatly stresses the organisms of the remaining estuary. Depending on the local geology, additional problems may arise. It was already noted in chapter 4 that buildings erected on filled land rather than on bedrock suffer much greater damage from ground shaking during earthquakes. This has been demonstrated many times in San Francisco, which is also the location of the Millennium Tower, a 58-story building finished in 2009 that is currently tipping at a rate of 3 inches/year as the soils underneath it compact. Yet the pressure to create dry land for development where no land existed before continues.

One of the most ambitious projects involving land reclamation from the sea in an estuary is that of the Zuiderzee in the Netherlands (**figure 7.28**). People started reclaiming small sections of land in this area over 1000 years ago by constructing dikes and building mounds to move onto during times of high water. After extensive flooding in 1916, a 30-km long dam, or

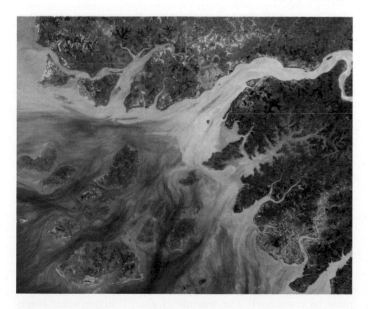

Figure 7.27

This natural color image shows the estuaries of Guinea-Bissau as they reach the Atlantic Ocean. These coastal valleys flood often and play an important role in agriculture in the region. Mangroves have been destroyed to extend rice cultivation, aggravating coastal erosion in this low-lying area.

NASA Earth Observatory images by Joshua Stevens

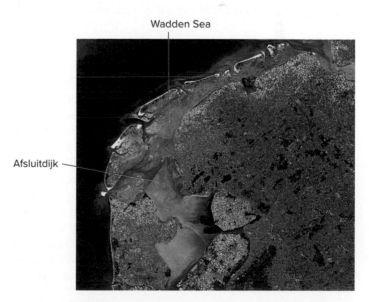

Figure 7.28

The Afsluitdijk separates two shallow artificial lakes from the intertidal Wadden Sea. The sections of land called *polders* were progressively reclaimed; most is used for agriculture.

European Space Agency

dike, called the Afsluitdijk was built across the Zuiderzee. By 1980, four *polders* totaling more than half a million acres areas of reclaimed land were created by a combination of isolation from the water, pumping, and infilling.

In the Netherlands, where land is at a premium, the benefits of filling in the Zuiderzee may have outweighed the costs, environmental and economic. The effort has been costly and the original estuary ecosystem is gone. While the Dutch are maintaining their existing dikes and dams, they are also aggressively preparing for higher sea levels through a decades-long, multi-billion-dollar plan of coastal fortification that includes urban design projects, floodwater storage solutions, resilient infrastructure, and educational programs.

Across the globe, natural estuaries need to be protected. They provide numerous essential ecosystem services that include not only critical natural habitats that act as spawning grounds and migration stops, but also support economically valuable services such as tourism, recreation, fisheries, coastal protection, and pollution filtration.

7.5 Coastal Hazards— Recognition and Costs

Learning Objectives

- Exemplify beach nourishment projects
- Discuss the use of federal disaster relief funds for coastal areas
- List indications that a coastal area is safe/unsafe from hazards
- Explain why managed retreat is problematic

If we concede the attraction of beaches and coastlines and the likelihood that people will want to continue living along them, and we decide that public funds should be used to protect both beaches and structures that are at risk, we can at least identify the most unstable or threatened areas so that the problems can be minimized. That is often possible especially given new remote sensing and mapping capabilities, but like all the other natural hazards we have discussed, it depends both on observations of present conditions and on some knowledge of the area's history (**figure 7.29**). We also need to consider that rising sea levels will make living in coastal areas more risky and more expensive. What parameters do we consider in deciding how limited funds will be used to protect coastlines, given that some building and beaches will most likely need to be abandoned?

Recognition of Coastal Hazards

The best setting for building near a beach or on an island is at a relatively high elevation; 5 meters or more above normal high tide is above the reach of most storm tides. Added protected would be in a spot with many high dunes between the proposed

A

B

Figure 7.29

(A) In northern Monterey Bay, California, the houses at the cliff base provide unintentional, and vulnerable, protection for the cliff-edge structures above. (B) This cliff at Moss Beach in San Mateo County, California, retreated more than 50 meters in a century. Past cliff positions are known from old maps and photographs; arrows indicate position of cliff at corresponding dates.

(A) U.S. Geological Survey/Photograph by Cheryl Hapke; (B) U.S. Geological Survey Photo Library, Denver, CO/Photograph by K. R. LaJoie

building site and the water. Thick vegetation, if present, will help to stabilize the beach sand. Information about what has happened in major storms in the past is very useful. Was the site flooded? Did the overwash cover the whole island? Were protective dunes destroyed? A key factor in determining a relatively safe elevation is overall water level. On a lake, not only the short-term range of storm tide heights but also the long-term range in lake levels must be considered. For example, Lake Michigan water levels change on a roughly 10-year cycle and varied between high and low elevations by 2 meters (6.5 feet) from 2012 to 2021.

On a seacoast, it is important to know if that particular stretch of coastline is emergent or submergent over the long term. The plate tectonic setting is important. Is subsidence or uplift occurring, and does the ground get shaken on a regular basis due to earthquakes? Around the Pacific Ocean basin, possible dangers

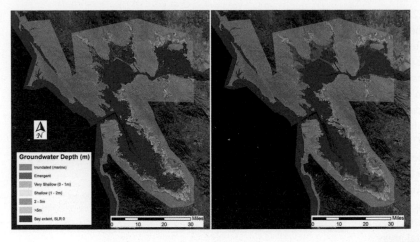

Figure 7.30

These groundwater maps of the San Francisco Bay regions are an example of the type of products created from CoSMoS. Sea level rise will cause the water table to rise, causing slow flooding and numerous infrastructure problems, and increasing the likelihood of coastal erosion.

Juliette Finzi-Hart, Patrick Barnard, Kevin Befus/USGS

from tsunamis need to be considered. Are there other factors leading to coastal slope instability, such as intense rain events or wildfires? Living along the California coastline, one would need to consider just about every one of the factors just listed.

On either beach or cliff sites, one very influential factor is the rate of coastline erosion. Historically, information regarding rates of erosion were determined using dated aerial photographs, maps, and personal experiences that gave details about past coastline configurations that could be compared to present. Scientists are now also using remote sensing data and on-the-ground measurements to create computer models that help public and private entities understand and visualize how different sea-level rise and storm scenarios will affect different areas of a coast. Flooding levels, rates of erosion, and rates of coastal retreat are all estimated, among others. Results of models such as the USGS's Coastal Storm Modeling System (CoSMoS) are being incorporated into vulnerability assessment plans in order to make scientifically based local and regional decisions on how to prepare for sea-level rise (**figure 7.30**).

One should also find out what shoreline modifications are in place or are planned, not only close to the site of interest but elsewhere along the coast. These can have impacts far away from where they are actually built. Aerial and satellite photographs make it possible to examine the patterns of sediment distribution and movement along the coast, which help in the assessment of the likely impact of any shoreline modifications; sediment transport computer models can also provide estimates. The presence of existing groins, breakwaters, or seawalls is a big warning sign that erosion is already a problem. A history of repairs to or rebuilding of structures not only suggests a very dynamic coastline, but also the possibility that protection efforts might have to be abandoned in the future for economic reasons, or they might simply fail altogether (**figure 7.31**).

Costs of Coastal Hazards—Economic and Human

Historically, the most concentrated damage in coastal areas has resulted from major storms; the number of people and value of property at risk only increase with the shift in population toward

A

B

Figure 7.31

This cliff at Bodega Bay, California, has been protected by a patchwork of different structures, which have largely proved ineffective; cliff retreat rates here have been of the order of a meter a year. (A) In this picture, taken in 2009, some buildings have already tumbled downslope (left side of photograph), while others hang on the cliff edge. (B) By 2017, several more structures had collapsed, and those that remained were condemned as unsafe to occupy.

©Carla Montgomery

the coast, illustrated in **figure 7.1**. (Recall also table 7.2.) U.S. federal disaster relief funds provided to barrier-island and other coastal areas are now running into billions of dollars after each such storm. Increasingly, we are asking if it makes sense to continue subsidizing and maintaining development in such very risky areas using public funds. From 1978 to 1982, $43 million in federal flood insurance (see chapter 19) was paid in damage claims to barrier-island residents, which far exceeded the premiums they had paid. So, in 1982 under the Coastal Barrier Resources Act (CBRA), Congress decided to remove federal development subsidies from about 650 miles of beachfront property and to eliminate in the following year the availability of federal flood insurance for these areas. It simply did not make good economic sense to encourage unwise development through such subsidies and protection. CBRA has been reauthorized numerous times and additional coastal areas have been added to the system. However, there is much land not included in the program, and structures existing before CBRA went into effect may still be eligible for federal coverage.

A continuing and growing concern is one of equity in the distribution of costs of coastal protection. While some stabilization and beach-replenishment projects are locally funded, many are undertaken by the U.S. Army Corps of Engineers. Hundreds of millions of federal taxpayer dollars have been be spent to protect a handful of private properties of individuals choosing to locate in high-risk areas. A national study done in March 2000 reported that beach renourishment must typically be redone every 3 to 7 years. For example, from 1962 to 2019, the beach at Cape May, New Jersey, was renourished 33 times at a total cost of $127 million; from 1961 to 2018, Sand Key in Florida (**figure 7.32**) was supplied with fresh sand 21 times, costing over $142 million, with roughly 60% federal support. Reported costs of beach maintenance and renourishment in developed areas along the Florida and Carolina coasts range up to $17.5 million per mile. Various beach-replenishment projects have been undertaken at Hawaii's Waikiki Beach, since at least the 1930s. The most recent project from 2002 to 2021 pumped 48,000 cubic yards of offshore sand to a 1700-foot section of Waikiki which was eroding at 1–2 feet per year; the estimated cost was $2.5 million, and the actual cost was $3.5 million. Clearly, the economic importance of tourism there means that such investments will continue in the future.

Most barrier-island areas are owned by federal, state, or local governments. Even where they are undeveloped, much money has been spent maintaining access roads and various other structures, often including beach-protection structures. It is becoming increasingly obvious that the most sensible thing to do with many such structures is to abandon the costly and ultimately doomed efforts to protect or maintain them and simply to leave these areas in their dynamic, natural, rapidly changing state, most often as undeveloped recreation areas. However, the economic success of these coveted beach areas will likely result in politicians pushing for continuing and even expanding shoreline-protection efforts.

What about hazardous areas that are developed with costly homes, such as one would encounter along the California coastline? In places such as Malibu, short-term cliff recession is forecast at 62–135 feet, and there is a lot of pressure to replenish

A

B

Figure 7.32

Beach nourishment is a continuing process at popular tourist locations such as (A) Cape May, New Jersey and (B) Sand Key, Florida.

(A) USACE; (B) Photograph by Tony Santana, courtesy U.S. Army Corps of Engineers

eroding beaches at costs that approach $50 million for 10 years worth of sand. Recall the idea of managed retreat, introduced in chapter 6 during the discussion on options in dealing with flood hazards. This idea of abandoning the most hazardous areas and paying people to leave their homes already built there is part of many local strategies in dealing with sea level rise. Governments have been using this practice retroactively for decades. However, the idea of proactively leaving a location rather than installing costly if not unsightly structures that will need maintenance is not popular with many homeowners and realtors. Having your structure labeled as being in the retreat zone would likely have costly insurance and sales-potential repercussions, and many are reluctant to be involved in something that sounds defeatist. The practice has been rebranded into "phased adaption" or "corrective shoreline planning" to avoid the negative connotations of its original name. No matter what it is called, it will be far cheaper to leave these areas than to try and protect them in perpetuity, and the overall risk to people will be decreased.

Other groups of people, such as those living on Isle de Jean Charles, Louisiana, are already involved in relocation plans. The residents here historically are of Native American ancestry, and their island has lost 98% of its landmass in 70 years due to subsidence and storm damage. In fall 2021, Hurricane Ida destroyed or damaged the remaining homes on the island. The new planned community, part of the relocation process begun in 2016, wasn't yet finished, leaving some homeless in the meantime.

Coastal residents are more at risk than ever given the increasing sea levels and storm intensities, yet some are assessing this risk just like any other and basing decisions more on their group identity rather than the science, scientists, and big picture. It is very possible that people choosing not to leave the riskiest areas before unrepairable damage is done will find themselves without insurance, and eventually will be forced to leave their expensive homes or native lands. There is no single correct approach in dealing with rising sea levels, and the use of public monies to handle inevitable coastal erosion will continue to be difficult and fraught with inequities.

Case Study 7.1

Hurricanes and Coastal Vulnerability

Even in 2005, a year notable for the number of major hurricanes that occurred, Katrina stood out. Though it was only Category 4 at landfall and weakened quickly thereafter, it was relatively large, with hurricane-force winds extending 200 km (125 miles) outward from its center. Its storm surge was locally reported at up to 10 meters (over 30 feet). Barrier islands were hard-hit. On the mainland, things weren't necessarily any better.

And then there was New Orleans. Much of the city lies below sea level, by as much as 5 meters, the result of decades of slow sinking. It has been protected by levees, not just along the Mississippi, but along the shore of Lake Pontchartrain, which is a large estuary. The levees around Lake Pontchartrain and its canals had been built with a Category 3 hurricane in mind. Unfortunately, Katrina was worse. Once some poorly constructed levees were overtopped, erosion behind them caused them to fail altogether. The result was inundation of a huge area of the city and standing water for weeks thereafter, as levees were patched and the water then pumped out (recall **figure 6.31B**).

Much of the city has been rebuilt, but thousands of buildings remain abandoned and the population in 2019 was still 19% lower than it was in April 2000. There is still disagreement as to how to proceed going forward, and discussions include issues of race and equity. Some have advocated abandoning the lowest, most flood-prone sections of New Orleans, those typically occupied by impoverished and disadvantaged citizens. Others advocate higher, stronger levees, and/or systems of gates to stem the influx of storm surge into Lake Pontchartrain. The costs of such enhanced flood-control measures are in the tens of billions of dollars. Some floodwalls have been rebuilt higher than before, but the city is as vulnerable as the weakest of the protection structures. Debate continues over how much protection is enough, and when stronger measures are no longer cost-effective. Furthermore, New Orleans continues to sink, at rates recently measured at up to 12 mm per year. Likely causes include subsidence related to groundwater extraction, draining of wetlands, loss of sediments from the Mississippi, and settling of sediments; the latter could be exacerbated by the weight of additional fill or levees.

Three years later, Hurricane Ike, slammed into Galveston. While the seawall protected the city from waves and surge, anything in front of it was lost. On the western end of Galveston Island, where there is no seawall and the land is sinking, water damage was far worse (**figure CS 7.1.1**). Here too, the risks will only increase with a rising sea level.

Hurricane Irene struck in 2011. After hitting Puerto Rico, it moved north to landfall as a Category 1 storm at North Carolina's Outer Banks, wreaking havoc on these barrier islands before moving farther north (**figure CS 7.1.2**). Though it had weakened to a tropical storm when it reached New England, Irene still brought strong winds and heavy rain which produced severe flooding and sediment runoff, much of it into the Long Island Sound estuary.

Many of the same areas of the northeastern U.S. were hit by Hurricane Sandy in 2012. It too was only a Category 1 storm at U.S. landfall, near Atlantic City, NJ. However, the hurricane merged with two other storm systems to become "Superstorm Sandy," the largest Atlantic storm on record, over 1000 miles across, and its storm surges and heavy rain produced massive damage (**figure CS 7.1.3**; recall also **figure 7.26**). Beyond repairs to damaged structures, some of the rebuilding after Sandy involved new or enhanced protection structures (**figure CS 7.1.4A**); much involved beach restoration (**figure CS 7.1.4B**), sometimes of beaches that had been wholly or partially constructed in the first place. For example, Rockaway Beach was created in the 1970s with the importation of 6.3 million cubic yards of sand. It was replenished several times. After Sandy, yet another 3.5 million cubic yards of sand had to be brought in to restore the beach. Altogether, between clearing out sand that

(Continued)

September 9, 2008

September 15, 2008

May 6, 2008

August 31, 2011

Figure CS 7.1.1

Without a seawall, Crystal Beach, west of Galveston, took the full brunt of the storm surge in addition to the wind. Houses and vegetation were swept away, and erosion and overwash pushed the shoreline significantly landward. Yellow arrows point to same structures in both photographs.

U.S. Geological Survey Coastal and Marine Geology Program.

Figure CS 7.1.2

Outer Banks near Rodan the, NC, before and after Hurricane Irene. Yellow arrows point to same structure. Afterwards, the large building on the beach is gone, along with most of the beach and Highway 12. Water is still murky with suspended sediment churned up by the storm.

U.S. Geological Survey Coastal and Marine Geology Program.

had washed inland and dredging more to clear harbors and collect more sand for beach restoration, the U.S. Army Corps of Engineers moved an estimated 27 million cubic yards of sand in the New York City area alone after Hurricane Sandy. It seems, too, that global climate change exacerbated the damage. Sea level has risen 20 cm (about 8 inches) since 1900. That may not

sound like much, but it means more inland area flooded, and a greater depth of floodwater everywhere, during a coastal storm. Researchers modeling the difference in flood damages with and without that extra 20 cm of water height have estimated that the current higher sea level meant about $2 billion in additional damage just to New York City.

Meanwhile, global sea level continues to rise, and warmer water levels suggest more energy for future hurricanes. In order to protect the ever-increasing amounts of people moving to coastal areas, we will need even more ambitious protection efforts from severe storms.

A

A

B

B

Figure CS 7.1.3

(A) Hurricane Sandy damage to a barrier island on the New Jersey coast. (B) This beach house was built on stilts to protect it from high seas, but it did not survive Sandy's surge and waves.

(A) U.S. Air Force photo by Master Sgt. Mark C. Olsen; (B) Photograph by Brandon Beach, courtesy U.S. Army Corps of Engineers

Figure CS 7.1.4

(A) New post-Sandy breakwater made of interlocking cement units. (B) Moving sand to restore Rockaway Beach after the hurricane.

U.S. Army Corps of Engineers

Summary

Coastal areas are dynamic settings and many are rapidly changing. Accelerated by rising sea levels worldwide, erosion is causing shorelines to retreat landward in most areas, often at rates of more than a meter a year. Sandy cliffs, barrier islands, and estuaries are especially vulnerable to erosion. Storms, with their associated surges and higher waves, cause especially rapid coastal change. Efforts to stabilize beaches are generally expensive, often ineffective over the long (or even short) term, and frequently change the character of the shoreline. They also cause unforeseen negative consequences to the coastal zone and it's organisms. Demand for development has led not only to construction on unstable coastal lands, but also to the reclamation of estuaries to create more land, to the detriment of ecosystem health. Increasingly, we are deciding that some locations cannot be protected from coastal hazards, and that boosting natural systems could be a very effective way at fortifying coastlines.

Key Terms and Concepts

active margin
barrier islands
beach
beach face

drowned valley
estuary
littoral drift
longshore current

milling
passive margin
surge
wave-cut platform

wave refraction

Test Your Learning

1. High storm tides may cause landward recession of dunes. Explain this phenomenon using a sketch.

2. Evaluate the use of riprap and seawalls as cliff-protection structures.

3. Explain longshore currents and how they cause littoral drift.

4. Sketch a shoreline on which a jetty has been placed to restrict littoral drift; indicate where sand erosion and deposition will subsequently occur and how this will reshape the shoreline.

5. Describe the phenomenon of storm surge; explain how it develops and how it exacerbates coastal erosion.

6. Discuss the pros and cons of sand replenishment as a strategy for stabilizing an eroding beach.

7. Describe three ways in which the relative elevation of land and sea may be altered; note the present trend in global sea level, and explain the implications for coastal-protection efforts.

8. Briefly explain the formation of (a) wave-cut platforms and (b) drowned valleys.

9. Describe what barrier islands are, why they are so named, and why they have proven to be particularly unstable environments for construction.

10. Explain the nature of an estuary, how one might form, and why estuaries constitute such distinctive coastal environments.

11. Briefly describe at least two ways in which the dynamics of a coastline over a period of years can be investigated.

Exploring Further

1. Choose any major coastal city, find out how far above sea level it lies, and determine what proportion of it would be inundated by a relative rise in sea level of (a) 1 meter and (b) 5 meters. Assess any vulnerability plan this city might have and what defensive strategies are proposed to protect threatened areas.

2. The fictional film *The Day After Tomorrow* depicts a supposed storm surge that immerses the Statue of Liberty up to her neck. In round numbers, that indicates a surge of about 250 feet. Use the data from **table 7.1** to make a graph of surge height versus wind velocity, and extrapolate to estimate the wind speed required to create a surge that large. Compare your estimate with the highest measured natural wind speed, 231 mph, and consider the plausibility of the movie's depiction.

3. Consider the vulnerability of barrier-island real estate, and think about what kinds of protections and controls you might advocate: How much should a prospective purchaser of the property be told about beach stability? Should local property owners pay for shoreline-protection structures, or should state or federal taxes be used? Should property owners be prevented from developing the land, for their own protection? What would you consider the most important factors in your decisions? What are the main unknowns?

4. Using the Our Coast, Our Future tool, choose a location (Malibu is a good place to start) and use the hazard map application to see what happens to properties along the coast at different scenarios of sea level rise and storm frequency. Propose a reasonable plan to deal with each scenario, from the viewpoint of a homeowner or a local town planner. **https://ourcoastourfuture.org/**

CHAPTER 8

Mass Movements

Gravity is the great leveler of Earth materials. The mountains and landforms built by tectonic processes are worn away by the force of gravity acting through weathering and erosion. The downhill movement of geologic materials is known by many names including **mass wasting,** or **mass movements,** or *slope failure,* and there are many different types. These movements can be slow and subtle, almost undetectable on a day-to-day basis, or sudden and swift, as in a rockslide or avalanche. Mass movements involve more than slippery slopes. They include subsidence, as from extraction of oil or groundwater, or collapse into sinkholes as described further in chapter 11.

Landslide is a general term we use for rapid mass movement. Landslide damage is widespread and expensive, and is

---

Chapter Outline

8.1 Factors Influencing Slope Stability

8.2 Types and Examples of Mass Wasting

8.3 Impact of Human Activities

8.4 Recognition, Prevention, and Monitoring

In May 2017, this slide on the California coast—the largest ever recorded in that state—buried U.S. Highway 1 under 5 million cubic yards of debris, cutting off access to the town of Big Sur from the south. Unfortunately, sliding following winter storms had previously damaged a bridge north of Big Sur on Highway 1, and that bridge had been torn out in March. The town was isolated until October, when a rush $24-million repair restored the bridge. Rebuilding Highway 1 across this slide took 14 months and $54 million. Some debate the long-term wisdom of attempting to maintain Highway 1 along this notoriously unstable stretch of coast.

U.S. Geological Survey/Photograph by Bob Van Wagenen

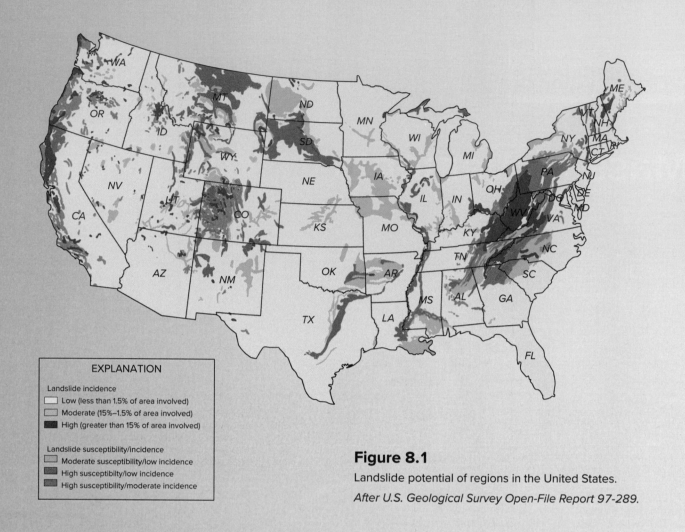

EXPLANATION

Landslide incidence
- Low (less than 1.5% of area involved)
- Moderate (15%–1.5% of area involved)
- High (greater than 15% of area involved)

Landslide susceptibility/incidence
- Moderate susceptibility/low incidence
- High susceptibility/low incidence
- High susceptibility/moderate incidence

Figure 8.1

Landslide potential of regions in the United States.

After U.S. Geological Survey Open-File Report 97-289.

very likely under-reported. In the United States alone, landslides and other mass movements cause over $3.5 billion in property damage every year, and 25 to 50 deaths. Natural landslides occur in any location where a slope becomes over-steepened. In some areas, we have taken active steps to control downslope movement or to limit its damage. On the other hand, many human activities aggravate local landslide dangers.

Large areas of this country are at risk from landslides (**figure 8.1**). It is important to realize, that given the scale of the figure, *regional* landslide hazard levels may be missed. New online, real-time, hazard assessment tools can give you a more accurate idea of your local landslide risk.

8.1 Factors Influencing Slope Stability

Learning Objectives

- Recall shearing stress and shear strength
- Describe how slope steepness and materials affect the likelihood of slope failure
- Describe how fluids and vegetation affect slope stability
- Describe how earthquake trigger slope failure
- Recognize how quick clay contribute to slope failure

Mass movements occur whenever the downward pull caused by gravity overcomes the forces resisting it. As first described in chapter 3, a **shearing stress** is one tending to cause parts of an object to slide past each other across a plane, as the plates on opposite sides of a transform fault do. The downslope pull leading to mass movements is also a shearing stress, its size related to the mass of material involved and the slope angle. With a block of rock sitting on a slope, *friction* between the block and the underlying slope counteracts the shearing stress. A single body of rock can also be subjected to shearing stress; its **shear strength** is its ability to resist being torn apart along a plane by the shearing stress. When shearing stress exceeds frictional resistance or the shear strength of the material, as applicable,

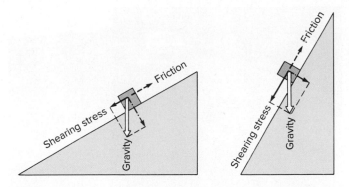

Figure 8.2

The mass of the block and thus the total downward pull of gravity are the same in both cases, but the steeper the slope, the greater the shearing stress component.

sliding occurs. Factors that increase shearing stress, decrease friction, or decrease shear strength tend to increase the likelihood of sliding, and vice versa. Application of these simple principles is key to reducing landslide hazards. In this section, we will review how the nature of a sloped surface and processes affecting that surface affect its stability.

Effects of Slope and Materials

The mass of the material involved is one key factor in slide potential. The gravitational force pulling it downward, and thus the shearing stress, is directly proportional to that mass. Anything that increases the mass increases the risk of a slide.

All else being equal, the steeper the slope, the greater the shearing stress (**figure 8.2**) and the greater the likelihood of slope failure. For dry, unconsolidated material, the **angle of repose** is the maximum slope angle at which the material is stable (**figure 8.3**). This angle varies with the material. Smooth, rounded particles tend to support only very low-angle slopes, while rough, sticky, or irregular particles can be piled more steeply without becoming unstable. Coarse-grained fragments can usually maintain a steeper slope angle than fine-grained ones. The tendency of a given material to assume a constant characteristic slope can be seen in such diverse geologic forms as cinder cones (**figure 8.4**), sand dunes (**figure 8.4**), and beach faces (recall **figure 7.3A and B**).

Solid rock can be perfectly stable even at a vertical slope but will lose its strength when it is broken up by weathering or fracturing. In layered sedimentary rocks, there may be weakness along bedding planes where different rock units are imperfectly held together; some units, such as clay-rich layers, may themselves be weak or even slippery. If these planes of weakness are parallel to the slope, the potential for failure and sliding is particularly hazardous.

Slopes may be steepened to unstable angles by natural erosion by water or ice. Erosion can undercut rock or soil, removing the support beneath a mass of material and leaving it susceptible to falling or sliding. This is a common contributing factor to landslides in coastal areas (**figure 8.5**) and along stream valleys.

Over long periods of time, slow tectonic deformation also can alter the angles of slopes and bedding planes, making them steeper or shallower. This is most often a significant factor in

Angle of repose

Finer, rounder particles can sustain only low slope angle.

Coarse, irregular particles can be more steeply piled.

A

B

Figure 8.3

(A) Angle of repose indicates an unconsolidated material's resistance to sliding. Coarser-grained and more angular particles can maintain a steeper stable slope angle. (Actual angles shown are inaccurate.) (B) The upper rock cliff in Iceland has a near-vertical slope whereas the talus slope created by broken rock fragments assumes its natural angle of repose.

(B) www.sandatlas.org/Shutterstock

Figure 8.4

The uniform slope on the face of this sand dune is maintained by slumping to the sand's angle of repose. Oregon Dunes National Recreation Area.

© Carla Montgomery

Figure 8.5

Houses on top of eroded cliffs on the Pacific Ocean coastline, Moss Beach, Fitzgerald Marine Reserve, California.

Sundry Photography/Getty Images

young, active mountain ranges, such as the Alps or the coast ranges of California. One of the underlying factors that make mass movements so common in the Himalayas is that this is an actively uplifting mountain range with layered, folded, and fractured rocks.

Effects of Fluid

The role of fluids in mass movements is variable. Adding a small amount of liquid to dry sediment may increase adhesion,

Figure 8.6

Instability in expansive clay seriously damaged this road near Boulder, Colorado.

U.S. Geological Survey Photo Library, Denver, CO.

helping the particles stick together. However, completely saturating that same material reduces the friction between the particles, decreasing the cohesion and strength, which can destabilize a slope. As was described in chapter 4, elevated pore-fluid pressure can facilitate slip in faulted rocks under tectonic stress; it is also effective in promoting sliding in rocks under stress due to gravity, reducing the force holding the rock to the plane below (vector perpendicular to slope in **figure 8.2**). And, the very mass of water in saturated soil adds extra weight, resulting in extra downward pull.

Water can greatly increase the likelihood of mass movements in other ways. It can seep along bedding planes in layered rock, reducing friction and making sliding more likely. The expansion and contraction of water freezing and thawing in cracks in rocks or in soil can act as a wedge to drive chunks of material apart in a process called *frost wedging. Frost heaving,* the expansion of wet soil as it freezes and the ice expands, loosens and displaces the soil, which may then be more susceptible to sliding when it next thaws.

Some soils rich in clays absorb water readily. One type of clay, montmorillonite, may take up twenty times its weight in water and form a very weak gel. Such material fails easily under stress. Other clays expand when wet, contract when dry, and can destabilize a slope in the process (**figure 8.6**; see also chapter 20).

Sudden, rapid slope failures commonly involve a triggering mechanism. Heavy rainfall or rapid melting of abundant snow can be a very effective trigger, quickly adding mass, decreasing friction, and increasing pore pressure (**figure 8.7**). As with flooding, the key is not simply how much water is involved, but how rapidly it is added and how close to saturated the system was already. Severe flooding and landslides in January 2009 in the Pacific Northwest were caused by heavy rainfall that rapidly melted the heavy snow that had fallen earlier in the winter. In March 2014, a rain-soaked slope partially undercut by a stream gave way and flowed over the

Figure 8.7

Laguna Beach, California, landslide of 1 June 2005.

U.S. Geological Survey/Photograph by Gerald Bawden

community of Oso, Washington, burying 45 houses and killing 43 (**figure 8.8**). Such slides are relatively common in this area, but the magnitude of the Oso event was much larger than had occurred previously and thought possible. Tropical regions subject to intense rainfall events also have frequent slides in the thick soils which tend to be clay-rich, and in severely weathered rock (**figure 8.9**).

Effects of Vegetation

Vegetation is a bit like water in affecting slope stability in that a moderate amount of it usually results in the least hazardous situation. Plant roots, especially those of trees and shrubs, can provide a strong interlocking network to hold unconsolidated materials together and prevent flow. In addition, vegetation takes up moisture from the upper layers of soil, reducing the overall moisture content of the mass and increasing its shear strength. Moisture loss through the vegetation by transpiration helps to dry out sodden soil more quickly. In general, vegetation tends to increase slope stability. However, plants can destabilize a slope if their added weight is large and the root network is limited, or if the plants take up so much water that the soil loses it adhesion.

When previously forested slopes are bared by fires or by logging, they become more prone to sliding. Illegal logging on slopes above the town of Guinsaugon in the Philippines was blamed for the February 2006 mudslide, triggered by heavy rains, that engulfed the town and killed an estimated 1800 of its 1857 residents. Wildfires in California frequently strip slopes of stabilizing vegetation, and deadly landslides commonly follow. In late 2017, a massive wildfire in southern California destroyed vegetation on the hillsides above the coastal town of Montecito. Intense January 2018 rains then mobilized soil and boulders into fast-moving debris flows that poured down hillside gullies and swept through the town, destroying or burying hundreds of homes and killing more than 20 people (**figure 8.10A**). Debris basins built over 50 years earlier in order to contain these types of flows had fallen into disrepair and were not designed to hold the volume of material associated with large storms. Although authorities attempted to empty them after the fires and before the rains came, they were unable to, leaving them well below their potential holding volume (**figure 8.10B**). The flows were also very fluid and traveled farther than the mandatory evacuation zone.

Wildfires that remove vegetation, and the slope failures that follow them, are nothing new for Californians. However, as the climate warms, places with steep slopes that haven't usually experienced wildfires are now doing just that. Recall the massive flooding that occurred in southern British Columbia in late 2021 that we discussed in chapter 7. Those floods were

Figure 8.8

The Oso, Washington, landslide. Although people were aware of the potential for landslides here, they did not think one could flow out so far from the slope face; tragically, they were proved wrong. Note cars and houses in foreground for scale.

U.S. Geological Survey/Photograph by Mark Reid

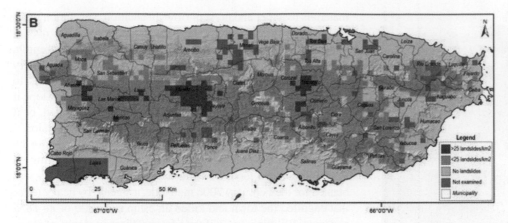

Figure 8.9

Hurricane Maria in 2017 caused at least 40,000 separate landslides in Puerto Rico, located in the tropics. This landslide density map was created from satellite and aerial images and on-the-ground information.

Supplemental Appropriations for Disaster Recovery Activities/USGS

preceded by record high temperatures that triggered massive forest fires in this usually humid environment. The resulting mud and debris flows that followed the rains cutoff major transportation routes in the region. The models that scientists use to predict such mass wasting events incorporate data about rainfall patterns, vegetation, and soil texture, and the current models that work well in California and Colorado, for example, cannot be applied to places that have rarely burnt in the past. Climate models predict a significant increase in extreme wildfires, even using moderate change scenarios. And so, the risk of slope failures will also increase in number and become more widespread. Our rapidly changing climate continues to challenge our understanding of natural processes and our ability to adapt.

Earthquakes

Landslides are a common consequence of earthquakes in steep terrain, as noted in chapter 4. Seismic waves passing through rock stress and fracture it. The added stress may be as much as half that already present due to gravity. Ground shaking also jars apart soil particles and rock masses, reducing the friction that holds them in place. The Himalayas experience frequent landslides caused directly and indirectly by earthquakes: the mountains created at this continental collision boundary are young, steep, layered and folded, making them very susceptible to weakening and failure when shaken. The 2015 Gorkha earthquake in Nepal had a IX Mercalli intensity and created more than 10,000 landslides that killed thousands, destroyed towns, blocked roads, and damaged infrastructure. The shaking also left the rocks weaker and more likely to slide. Add to that deforestation, road building, melting glaciers, monsoons, and a growing population, mass wasting hazards are becoming even more serious in this region of the world.

One of the most lethal earthquake-induced landslides occurred in Peru in 1970. An earlier landslide had already occurred below the steep, snowy slopes of Nevados Huascarán,

A

B

Figure 8.10

Santa Barbara County. (A) This home was damaged during a post-wildfire debris flow in January 2018. (B) Post debris-flow cleanup of existing basins (catchments) occurred to prepare for the next mass wasting events.

(A) Jason Kean/USGS; (B) U.S. Army Corps of Engineers

the highest peak in the Peruvian Andes, in 1962, without the help of an earthquake. It had killed approximately 3500 people. In 1970, a magnitude-7.7 earthquake centered 130 kilometers to the west shook loose a much larger debris avalanche that buried most of the towns of Yungay and Ranrachira and more than 18,000 people (**figure 8.11**). Some of the debris was estimated to have moved at 1000 kilometers per hour (about 600 miles per hour). The steep mountains of Central and South America, sitting above subduction zones, continue to suffer earthquake-induced slides (e.g., recall **figure 4.20B**).

Earthquakes in and near ocean basins may also trigger submarine landslides. Sediment from these may be churned up into a denser-than-water suspension that flows downslope to the sea floor, much as pyroclastic flows flow down from a volcano. These *turbidity currents* not only carry sediment to the deep-sea floor but are forceful enough to break the hundreds of undersea data cables that connect the global internet. In turn, the study of turbidity current deposits have been very useful in understanding the severity and frequency of prehistoric earthquakes.

Quick Clays

A geologic factor that contributed to the infamous Turnagain Heights landslides in Anchorage, Alaska, during the 1964 earthquake and that continues to add to the landslide hazards in Alaska, California, parts of northern Europe, and elsewhere is a material known as "quick" or "sensitive" clay. True **quick clays** (**figure 8.12**) are most common in northern polar latitudes. The grinding and pulverizing action of massive glaciers produce a

A

B

Figure 8.12

The landslide (perhaps a slump) triggered by the 1964 Great Alaskan Earthquake destroyed 75 homes in the Turnagain Heights neighborhood in Anchorage, built upon unconsolidated outwash underlain by quick clay.

(A) U.S. Geological Survey; (B) Alaska Earthquakes slide Collection/USGS

Figure 8.11

The Nevados Huascarán debris avalanche, Peru, 1970.

Courtesy USGS Photo Library, Denver, CO

rock flour of clay-sized particles, less than 0.02 millimeter (0.0008 inch) in diameter. When this extremely fine material is deposited in a marine environment, and the sediment is later uplifted above sea level by tectonic movements, it contains salty pore water. The sodium chloride in the pore water acts as a glue, holding the clay particles together. Fresh water subsequently infiltrating the clay washes out the salts, leaving a delicate, honeycomb-like structure of particles. Seismic-wave vibrations break the structure apart, reducing the strength of the quick clay by as much as twenty to thirty times, creating a finer-grained equivalent of quicksand that is highly prone to sliding. Failure of the Bootlegger Clay, a quick clay underlying Anchorage, Alaska, was responsible for the extent of damage from the 1964 earthquake. Nor is a large earthquake necessary to trigger failure; vibrations from passing vehicles can also do it. So-called **sensitive clays** are somewhat similar in behavior to quick clays but may be formed from different materials. Weathering of volcanic ash, for example, can produce a sensitive-clay sediment. Such deposits are not uncommon in the western United States. Underlying quick clay may have been a contributing factor to the Oso, Washington, landslide.

8.2 Types and Examples of Mass Wasting

Learning Objectives

- Recall how mass wasting is classified
- Describe and exemplify falls, slumps, slides, flows, and avalanches

Mass wasting takes on many forms. In the broadest sense, even subsidence of the ground surface is a form of mass wasting because it is gravity-driven. Here we focus on downslope movements and displacement of distinct masses of rock and soil.

When downslope movement is quite slow, we describe this process as **creep**. Soil creep, which is often triggered by frost heaving, occurs more commonly than rock creep. Though gradual, creep leaves telltale signs of its occurrence so that areas of particular risk can be avoided (**figure 8.13**). For example, building foundations may gradually weaken and fail as soil shifts; roads and railway lines may be disrupted; fences and stone walls tip downhill. Soil creep often causes serious property damage, though lives are rarely threatened. A related phenomenon, *solifluction,* describes slow movement of wet soil over impermeable material. We will consider it further in connection with permafrost, discussed in chapter 10.

There is no single standard for classifying mass movements but we commonly subdivide them on the basis of the type of material that is moving and the character of the movement. The type can range from unconsolidated, fairly

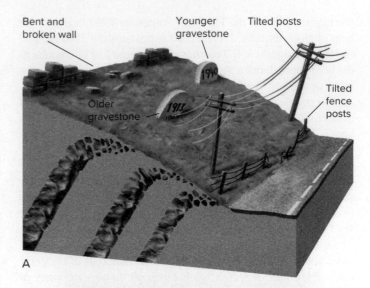

A

B

Figure 8.13

(A) Signs of creep are tilting and leaning structures; common examples are fences, stone walls, headstones, trees, utility poles. (B) Underlying bedrock is affected by creep, as evidence by the tilted layers.

(B) John A. Karachewski

fine-grained material such as snow or soil to large, solid masses of rock. A description of material motion involves both the nature and rate of motion. In general, a fast mass movement reflects a higher level of moisture, and results in more casualties. *Landslide* is a generic term people use to describe a rapid mass movement, but more precisely, *landslide* would mean a mass of unconsolidated material moving rapidly downhill while it stays in contact with the underlying surface. Some different types of mass movements are summarized in **figure 8.14**.

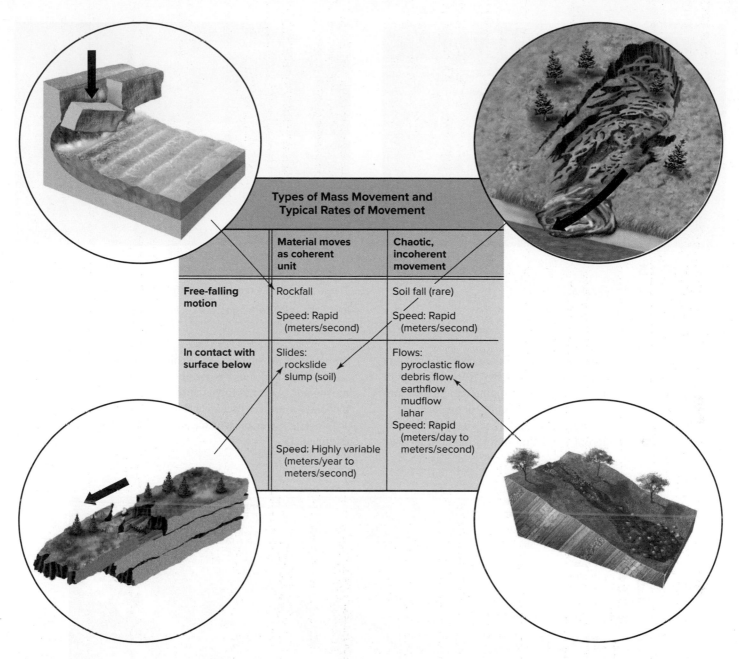

Figure 8.14

Comparison of nature of various types of mass movements.

Types of Mass Movement and Typical Rates of Movement	Material moves as coherent unit	Chaotic, incoherent movement
Free-falling motion	Rockfall Speed: Rapid (meters/second)	Soil fall (rare) Speed: Rapid (meters/second)

In contact with surface below	Slides:
	  rockslide
	  slump (soil)
	
	Speed: Highly variable (meters/year to meters/second)
		Flows:
		  pyroclastic flow
		  debris flow
		  earthflow
		  mudflow
		  lahar
		Speed: Rapid (meters/day to meters/second)

Falls

A **fall** is a free-falling action in which the moving material is not always in contact with the surface below. Falls are most often **rockfalls** (**figure 8.15**). They frequently occur on very steep slopes when rocks high on the slope, weakened and broken up by weathering, lose support as materials under them erode away. Falls are also common along rocky coastlines where cliffs are undercut by wave action and at roadcuts through solid rock (**figure 8.16**). The coarse rubble that accumulates at the foot of a slope prone to rockfalls is **talus**.

Rockfalls are very common in Yosemite National Park, where exposed granite batholiths have been carved into cliffs by glaciers. These fractured rocks are subject to many rockfall triggers including water, freeze-thaw cycles, earthquakes, and vegetation (**figure 8.17**).

Slides and Slumps

In a **slide**, a fairly cohesive unit of rock or soil slips downward along a clearly defined surface or plane. Rockslides involve movement along a plane of weakness that might result from a

Figure 8.15

Rockfall in Colorado National Monument.

©Carla Montgomery

Figure 8.16

These large rocks have recently fallen onto this mountain road.

Igor Batenev/Shutterstock

fracture, a bedding surface between sedimentary rock layers, or a zone of chemical weathering (**figure 8.18**). Rockslides can be complex, with multiple rock blocks slipping along the same plane, and they often transform into rockfalls or avalanches.

Slumps occur in both soil and rock, and like slides, stay in a roughly coherent mass that maintains contact with the underlying surface. In a soil slump, a rotational movement of the soil mass typically accompanies the downslope movement. A *scarp* forms at the top of the slump where the loosened material breaks away from the surrounding landscape. The lower part of the slump block may end in a flow (**figure 8.19**).

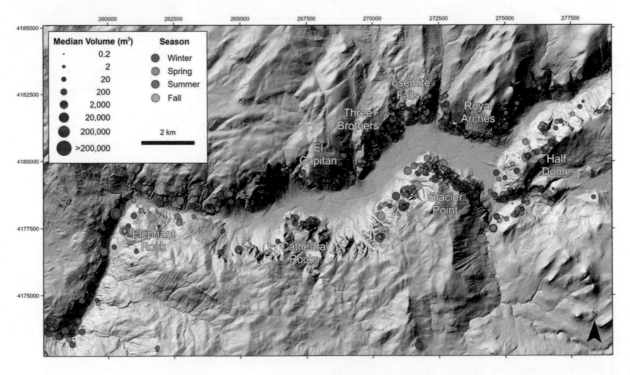

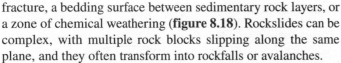

Figure 8.17

This map of Yosemite Valley shows the location and volume of rockfalls from 1857 to 2020. Visit https://www.nps.gov/yose/learn/nature/rockfall.htm for more information.

National Park Service

Figure 8.18

Rockslides occur along steeply sloping fractures in this granite in the Colorado Rocky Mountains.

©Carla Montgomery

Figure 8.19

The hillside above La Conchita, California, slumped in spring 1995, creating a debris flow that destroyed houses but no people. It again failed in 2005, remobilizing after heavy rains, killing ten people. Numerous human and natural factors ensure that slope failures will occur here again.

U.S. Geological Survey Department of the Interior//Photography by Jonathan Godt

Figure 8.20

Rock avalanches in the high peaks of Rocky Mountain National Park create large talus piles.

©Gina S. Szablewski

Flows and Avalanches

In a **flow**, the material moves in a chaotic, disorganized fashion, with particles mixing within the flowing mass. Flows of unconsolidated material are extremely common, and can involve substances other than soil. Snow avalanches and rock avalanches are two kinds of flow (**figure 8.20**); pyroclastic flows (see chapter 5) are another. *Earthflows* involve relatively dry soil (**figure 8.21**) as compared to *mudflows* which are saturated with water. *Lahars* are a type of mudflow specific to volcanoes. A flow involving a wide variety of materials such as soil, rocks, and trees is a **debris flow** or **debris avalanche**, the former term more often used for water-saturated debris (recall **figures 8.10** and **8.11**). Regardless of the nature of the materials moved, all flows have in common the chaotic, incoherent movement of the particles or objects involved.

Mass Wasting Examples

The Thistle, Utah, landslide is a good example of how a mass movement can be both hard to classify and can change over time. Heavy rains and snows in Utah in 1982–1983 caused an old 3-million-cubic-meter complex of stacked earthflows near

Thistle to reactivate. The 1983 failure was really a landslide, involving fine-grained, loose materials that mainly stayed in contact with the underlying surface as it moved at rates of up to 6 feet/hour, much faster than it had previously. The new landslide not only included previously failed materials but also grew in width by 100 to 450 feet. This larger, faster failure blocked Spanish Fork Canyon and damned the river (**figure 8.22**). The

town was evacuated as a large lake filled behind the slide, and transportation routes were cutoff for months. The rails and roads were rebuilt at a huge cost, the small town abandoned, and the lake is gone, but the slide doesn't appear to be done moving.

The January 2010 landslide in Attabad, Pakistan is similar to the Thistle slide. The landslide buried the town of Attabad, Pakistan, killing 20 and blocking both the Hunza River and the Karakoram Highway, the only road linking Pakistan and China, thereby disrupting trade between the two countries (**figure 8.23**). It created a lake that eventually grew to 23 km (nearly 15 miles) long, drowning several villages and

Figure 8.21

Recent earthflows in grassland, Gilroy, California.

Doug Sherman/Geofile

A

Figure 8.22

The Spanish Fork Canyon slide. Repairing the damage cost over $200 million (in 1984 dollars), making this at the time the single most expensive landslide in U.S. history.

U.S. Geological Survey Photo Library, Denver, CO.

B

Figure 8.23

(A) The Attabad slide (lower left of figure) created what has now become known as Attabad Lake. (B) Dust rises from the slide debris just after the event. Note the spectator on the road.

(A) NASA Earth Observatory image created by Jesse Allen using data provided by NASA EO-1 team; (B) U.S. Geological Survey/Photograph by Inayat Ali

displacing 6000 people. The accumulated water began to flow over the slide a few months after the event, flooding towns downstream as well. Plans were made to cut a spillway through the slide to drain the lake, but this has not occurred because the new Attabad Lake has become a tourist attraction. Concerns about the ability of the natural material to hold back the water have now lead to an artificial dam plan to add reinforcement. Transportation is still an issue for the 25,000 people in the valley isolated by the slide and lake, and there are valid fears that someday the slide may abruptly wash out.

Caraballeda, Venezuela, is just one city located on the north coast of Venezuela (state of Vargas) that is built on top of an alluvial fan at the base of steep mountain slope drained by narrow stream channels (**figure 8.24**). In December 1999, this area received nearly 1.2 meters of rain (almost 4 feet) within less than 3 weeks, causing flash floods and debris avalanches that mixed to form debris flows (**figure 8.25**). Many people were successfully evacuated but another 10,000-30,000 died in the combined floods and debris flows; the estimates are so wide because reports indicate many of the people who died lived in temporary housing and were undocumented. While the hazardous nature of the area is clear, and the risk remains high, there is very little other land for people to occupy in this area.

View Downstream   View Upstream

Figure 8.24

Downstream view: Quebrada San Julián (foreground) and Caraballeda (rear) were built, respectively, in a floodplain and on an alluvial plain; but the mountainous terrain (upstream view) leaves little choice.

U.S. Geological Survey/Photographs by M. C. Larsen

8.3 Impact of Human Activities

Learning Objectives

- Recall ways in which human activities contribute to mass wasting

Given the factors that influence slope stability, it is fairly easy for us to recognize the ways in which human activities both directly and indirectly increase the risk of mass movements. We have already discussed how the absence of vegetation leads to slope destabilization. Clear-cutting and fires expose sloping soils, creating more frequent and more severe

A

B

Figure 8.25

Damage in Caraballeda: (A) Overview of Los Corales sector. (B) A close look at this apartment building shows that at their peak, debris flows flowed through the lowest three floors. The largest boulders moved were estimated to weigh up to 400 tons.

(A) U.S. Geological Survey/Photographs by M. C. Larsen; (B) Matthew C. Larsen/USGS

mud- and debris flows than would occur naturally without our intervention.

Many types of construction lead to oversteepening of slopes. Highway roadcuts, quarrying or open-pit mining operations, and construction of stepped home-building sites on hillsides are among the activities that can cause problems (**figure 8.26**). Where dipping layers of rock are present, removal of material at the bottom ends of the layers may leave large masses of rock unsupported, held in place only by friction between layers. Slopes cut in unconsolidated materials at angles higher than the angle of repose of those materials are by nature unstable, especially if there is no attempt to plant stabilizing vegetation (**figure 8.27**). In addition, putting a house above a naturally unstable or artificially steepened slope adds mass to the slope, increasing the shear stress acting on the slope. Other activities connected with the presence of housing developments on hillsides can increase the risk of landslides in more subtle ways. Watering the lawn, using a septic tank for sewage disposal, and installing an in-ground swimming pool from which water can seep slowly out are all activities that increase the moisture content of the soil and render the slope more susceptible to slides. On the other hand, the planting of suitable vegetation can reduce the risk of slides.

Building-code restrictions have limited development in a few unstable hilly areas, and some measures have been taken to correct past practices that contributed to landslides. Chapter 19 includes an illustration of the effectiveness of legislation to restrict building practices in Los Angeles County. However, landslides do not necessarily cease just because ill-advised practices that caused or accelerated them have been stopped. Once activated or reactivated, slides may continue to slip for decades. The area of Portuguese Bend in the Palos Verdes Peninsula in Los Angeles County is a case in point.

In the 1950s, housing developments were begun in Portuguese Bend. In 1956, a 1-kilometer-square area began to slip, though the slope in the vicinity was less than 7 degrees, and within months, there had been 20 meters of movement. What activated this particular slide is unclear. Most of the homes used cesspools for sewage disposal, which added fluid to the ground, potentially increasing fluid pressure. The county highway department had added a lot of weight by building a road across what became the top of this slide. No matter the cause, the resultant damage was extensive: cracks developed within the sliding mass; houses on and near the slide were damaged or destroyed; a road built across the base of the slide had to be rebuilt repeatedly. Over $10 million in property damage resulted from the slippage. Worse, slow movement has continued for decades at a rate of about 3 meters per year. Some portions of the slide have moved 70 meters and recent (2019) surveys of the area show extensive fracturing in the surface materials. Current plans to manage the slide are concentrated on drainage of water.

Human activities can increase the hazard of landslides in still other ways. Irrigation and the use of septic tanks increase the flushing of water through soils and sediments. In areas underlain by quick or sensitive clays, these practices may hasten the washing-out of salty pore waters and the destabilization of the clays, which may fail after the soils are drained. Even cleanup after one slide may reduce stability and contribute to the next slide. Often, the cleanup involves removing the toe of a slump or flow—for example, where it crosses a road. This removes support

Figure 8.27

Failure of steep slope along a roadcut now threatens the cabin above.

©Carla Montgomery

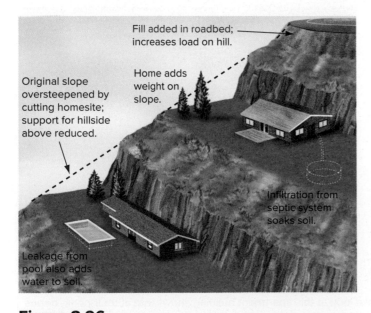

Original slope oversteepened by cutting homesite; support for hillside above reduced.

Fill added in roadbed; increases load on hill.

Home adds weight on slope.

Infiltration from septic system soaks soil.

Leakage from pool also adds water to soil.

Figure 8.26

Effects of construction and human habitation on slope stability.

for the land above and may oversteepen the local slope, too. Artificial reservoirs may cause not only earthquakes, but landslides, too. As the reservoirs fill, pore pressures in rocks along the sides of the reservoir increase, and the strength of the rocks to resist shearing stress can be correspondingly decreased. The case of the Vaiont reservoir disaster (Case Study 8.1) is infamous in this respect. Chapter 20 describes other examples of ways in which engineering activities may influence land stability.

8.4 Recognition, Prevention, and Monitoring

Learning Objectives

- Recognize indicators of high potential for slope failure
- List ways to reduce slide potential
- Exemplify techniques to reduce slide potential
- Describe ways in which the USGS and NASA are monitoring landslides
- Explain how landslide warnings are issued

Recognizing and then avoiding the most landslide-prone areas altogether would greatly limit damages, but as is true with fault zones, floodplains, and other hazardous settings, developments may already exist in areas at risk, and economic pressure for more development can be strong. In areas such as the Venezuelan coast, there may be little buildable land *not* at risk. Population density, too, coupled with lack of safe locations, pushes development into unsafe areas. In this section, we will review how we recognize landslide hazards, how we attempt to reduce the potential for slope failure, and the current status of landslide warning systems.

Recognizing the Hazards

Mass movements occur as a consequence of the basic characteristics of a region's climate, topography, and geology, independent of human activities. Where mass movements are naturally common, they tend to recur in the same places. Recognition of past mass movements in a region indicates a need for caution and serious consideration of the hazard in any future development. Such recognition can also save lives.

Even when recent activity is limited, we can often clearly see the potential for failure. The granite domes of Yosemite have been shaped by the exfoliation of massive curving slabs of rock (**figure 8.28A**). Rockfalls occur regularly; about 70 are recorded per year, and most are small and don't involve people. The rockfalls that involve larger pieces are particularly dangerous. A 90,000 ton piece of granite fell in July 1996, killing one park visitor and injuring 14. A more recent 30,000 ton piece fell in September 2018, killing one climber. The rockfalls in Yosemite usually have a clear cause, such as an earthquake or the intense freeze/thaw cycles in March and November. Recent

studies using crack meters that show slabs of rock moving in and out by 1 cm or more in a single day suggest that the warmer summer temperatures may be thermally inducing slides. The over 4 million people who visit this park every year have more than bears to worry about (**figure 8.28B**).

Recognizing past rockfalls can be quite simple in vegetated areas. The large chunks of rock in talus are inhospitable to most vegetation, so rockfalls tend to remain barren of trees and plants. The few trees that do take hold may only contribute to further breakup by root action. Lack of vegetation may also mark the paths of past debris avalanches or other soil flows or slides (**figure 8.29**). These scars on the landscape point plainly to slope instability.

A

B

Figure 8.28

(A) A close look at the granite domes of Yosemite shows curved slabs of rock at the surface; some above valleys have nearly vertical fractures (B) In 2017, a 1300-ton block of granite fell from El Capitan, killing one visitor, and the next day a still larger block fell; that event is shown here.

(A) ©Carla Montgomery; (B) National Park Service

A

B

Figure 8.29

Areas prone to landslides may be recognized by the failure of vegetation to establish itself on unstable slopes. (A) Rock Creek Valley, Montana; note road snaking across slide-scarred surfaces. (B) This road cut in California likely contributed to the sliding now revealed by the fresh, raw slope.

(A, B) ©Carla Montgomery

Figure 8.30

Tilted and curved tree trunks are a consequence of creep, as seen here on a weathered slope in Bryce Canyon National Park.

©Carla Montgomery

Landslides are not the only kinds of mass movements that recur in the same places. Historical records of past snow avalanches can pinpoint particularly risky areas; or their tracks may be marked by swaths of downed trees. Records of the character of past volcanic activity and an examination of a volcano's typical products can similarly be used to recognize a particular volcano's tendency to produce pyroclastic flows or lahars.

Very large slumps and slides may be less obvious, especially when viewed from the ground. The coherent nature of rock and soil movement in many slides means that vegetation growing atop the slide may not be greatly disturbed by the movement. Remote sensing and arial photography can be helpful here. In a regional overview, the mass movement often shows up very clearly, revealed by a scarp at the head of a slump or an area of hummocky, disrupted topography relative to surrounding, more-stable areas (recall **figure 8.19**).

With creep or gradual soil flow, individual movements are short-distance and the whole process is slow, so vegetation may continue to grow in spite of the slippage. Curved tree trunks indicate slow creep has occurred over a considerable amount of time (**figure 8.30**). Cracks that occur across the slope of a surface, as well as the tilted structures shown in **figure 8.13**, are also easily-noticed indicators of gradual downslope movements.

A prospective home buyer can look for additional signs that might indicate unstable land underneath (**figure 8.31**). Ground slippage may have caused cracks in driveways, garage floors, freestanding brick or concrete walls, or buildings; cracks in walls or ceilings are especially suspicious in newer buildings that would not yet normally show the settling cracks common in old structures. Doors and windows that jam or do not close properly may reflect a warped frame due to differential movement in

A

B

Figure 8.31

Evidence of slope instability in the Pacific Palisades of California includes broken walls (A) and, in extreme cases, (B) scarps.

(A, B) U.S. Geological Survey Photo Library, Denver, CO/Photographs by J. T. McGill

the soil and foundation. Sliding may have caused leaky swimming or decorative pools, or broken utility lines or pipes. If movement has already been sufficient to cause obvious structural damage, it is probable that the slope cannot be stabilized adequately, except perhaps at very great expense. Possible sliding should be investigated particularly when a site has more than 15% slope, it has much steeper slopes above or below it, or in any area where landslides are a recognized problem. Given appropriate conditions, slides may occur on quite shallow slopes.

How would you find out if landslides have occurred in an area? NASA maintains a Global Landslide Catalog based on mainly English language news reports that cover disastrous landslides in populated areas. This database is being updated through a citizen-science project called the Cooperative Open Online Landslide Repository with the goal of recording first-hand accounts of smaller events or those that don't affect many people or structures, that is, the slope failures that don't usually make the news. Not only will such a database provide global information for anyone who wants to access it, but also necessary information for scientists to use in landslide susceptibility models, as described later in this section. The USGS and states such as Kentucky and California also maintain online, accessible landslide inventories.

Slope Stabilization

Once a slope is recognized as both unstable and hazardous, those involved decide on the appropriate actions to take given money available and distinct characteristics of the situation. Sometimes, we resign ourselves to the existence of a mass-movement problem and take steps to limit the resulting damage that will eventually occur. In places where the structures to be protected are few

or small, and the slide zone is narrow, it may be economically feasible to bridge structures and simply let the slides flow over them. This may be done to protect a railway line or road running along a valley from avalanches (**figure 8.32**). This solution would be far too expensive on a large scale, and no use at all if the ground on which the structure was built were also sliding.

If a slope is too steep to be stable under the load it carries, any of the following steps will reduce slide potential: (1) reduce the slope angle, (2) place additional supporting material at the foot of the slope to prevent a slide or flow at the base of the slope, or (3) reduce the load (weight, shearing stress) on the slope by removing some of the rocks soil, or artificial structures high on the slope (**figure 8.33**). These measures may be used in combination. Depending on just how unstable the slope, they may need to be executed cautiously. If earthmoving equipment is being used to remove soil at the top of a slope, for example, the added weight of the equipment and vibrations from it could possibly trigger a landslide.

To stabilize exposed near-surface soil, ground covers or other vegetation may be planted, preferably ones that are fast-growing and have sturdy root systems. But sometimes plants are insufficient, there may be lots of large rocks involved, and the other preventive measures already described are impractical. Then, retaining walls or other stabilization structures can be used built against the slope to try to hold it in place (**figure 8.34**) while still allowing drainage. Given the distribution of stresses acting on retaining walls, the greatest successes of this kind have generally been low, thick walls placed at the toe of a fairly coherent slide to stop its movement. High, thin walls have been less successful (**figure 8.35**).

Since water can play such a major role in mass movements, the other principal strategy for reducing landslide

A

B

Figure 8.32

Avalanche-protection structures reduce the consequences of slope instability. (A) Shelter built over railroad track or road in snow avalanche area diverts snow flow. (B) Example of avalanche-protection structure in the Canadian Rockies.

(B) ©Carla Montgomery

hazards is to decrease the water content or pore pressure of the rock or soil. This might be done by covering the surface completely with an impermeable material and diverting surface runoff above the slope. Alternatively, subsurface drainage might be undertaken. Systems of underground boreholes can be drilled to increase drainage, and pipelines installed to carry the water out of the slide area (**figure 8.36**). All such moisture-reducing techniques naturally have the greatest impact where rocks or soils are relatively permeable. Where the rock or soil is fine-grained and drains only slowly, hot air may be blown through boreholes to help dry out the ground. Such moisture reduction reduces pore pressure and increases frictional resistance to sliding.

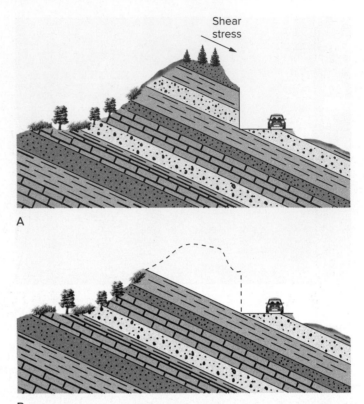

A

B

C

Figure 8.33

Slope stabilization by slope reduction and removal of unstable material along roadcut. (A) Before: Roadcut leaves steep, unsupported slope. If the shale layers (gray) are relatively impermeable, fluid may accumulate along bedding planes, promoting sliding. (B) After: Material removed to reduce slope angle and load. (C) This roadcut near Denver, Colorado, has been "terraced," cut into steps to break up the long slope while reducing the overall slope angle.

(A) Doug Sherman/Geofile; (B) ©Carla Montgomery

Figure 8.34

(A) Erosion control mats are used to increase slope stabilization in many settings (B) Gabions, rock rubble encased in wire mesh, are used for support and to encourage drainage at this site in New York (C). Here in the Colorado Rockies, chain-link fencing draped over a roadcut won't prevent rockfalls, but it will keep the falling rocks at the edge of the road, minimizing danger to motorists. (D) Not far from (C), more-serious slope-stabilization efforts involve concrete blocks bolted into the hillside.

(A) ThamKC/Shutterstock; (B) ©Doug Sherman/Geofile; (C, D) ©Carla Montgomery

Other slope-stabilization techniques that have been tried include the driving of vertical piles into the foot of a shallow slide to hold the sliding block in place. The procedure works only where the slide is comparatively solid, with thin slides so that piles can be driven deep into stable material below the sliding mass, and on low-angle slopes.

The use of rock bolts to stabilize rocky slopes has had greater success (**figure 8.37**). Rock bolts have long been used in tunneling and mining to stabilize rock walls. They can be driven through rocks at risk of sliding, anchoring them into more stable materials below the slip plane. Again, this works best on thin slide blocks of very coherent rocks on low-angle slopes.

Procedures occasionally used on unconsolidated materials include hardening unstable soil by drying and baking it with heat (this procedure works well with clay-rich soils) or by treating with portland cement. By far the most common strategies are modification of slope geometry and load, dewatering, or a combination of these techniques. The more ambitious engineering efforts are correspondingly expensive and usually reserved for large construction projects, not individual homesites.

Landslide Monitoring

In early 1982, severe rain-triggered landslides in the San Francisco Bay area killed 25 people and caused over $66 million in

B

Figure 8.35

(A) This crib structure both supports the toe of the slope and catches loose falling rocks. (B) Slapping concrete sheets on a steep, rubbly hillside in Rock Creek Valley, Montana, is not altogether effective as it shows slipping, buckling, and telltale road repair after a few years.

(A) Doug Sherman/Geofile; (B) ©Carla Montgomery

damages. In response to this event, the U.S. Geological Survey (USGS) began to develop a landslide warning system. The basis of the warning system was the study of quantitative relationships among rainfall intensity, storm duration, and a variety of slope and soil characteristics relating to slope stability. These relationships were formulated using statistical analyses of observational data on past landslides. For a given slope, scientists were able to approximate threshold values of storm

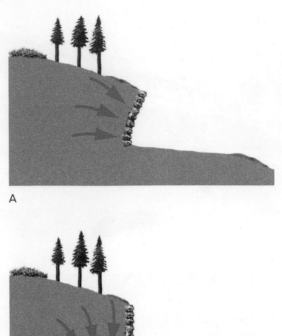

A

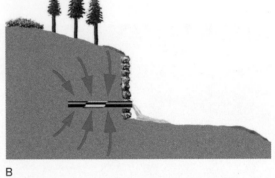

B

Figure 8.36

Improved drainage enhances slope stability by reducing pore pressure. (A) Before: Water trapped in wet soil causes movement, pushing down retaining wall. (B) After: Water drains through pipe, allowing wall to keep slope stable.

intensity and duration above which landslides were likely, given how saturated the ground was prior to the particular storm of concern. A later study in Seattle confirmed these relationships in understanding slide probability (**figure 8.38**).

Today scientists observing particularly unstable slopes continue to obtain soil and rainfall data by collecting soil samples and installing monitoring stations to track the site-specific relationship of pore pressure to rainfall (**figure 8.39**). Currently, 22 sites in the United States are subject to such detailed monitoring. Landslide hazards are assessed globally by using models such as NASA's Landslide Hazard Assessment for Situational Awareness that couples existing landslide susceptibility maps based on natural and human factors with real-time rainfall data collected by satellites. These global hazard maps are updated every 30 minutes, and can be used in conjunction with site-specific data to create regional landslide warnings (**figure 8.40**). Other new landslide models are also using satellite data, taking advantage of machine learning to sift through huge radar datasets in order to better understand patterns of displacement.

The USGS currently conducts post-fire debris-flow hazard assessments for select burned areas in the western United

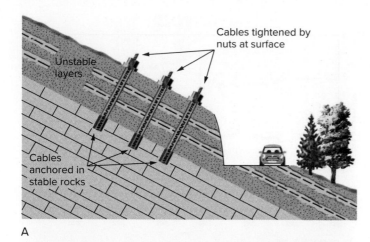

A

B

Figure 8.37

(A) Installation of rock bolts to stabilize a slope. The bolts are steel cables anchored in cement. Tightening nuts at the surface pulls unstable layers together and anchors them to the bedrock below. (B) Rock bolts used to stabilize a rock face above a tunnel in Alaska. See also **figure 8.34.**

(B) ©Carla Montgomery

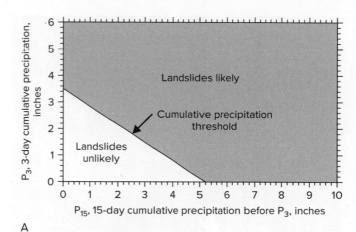

A

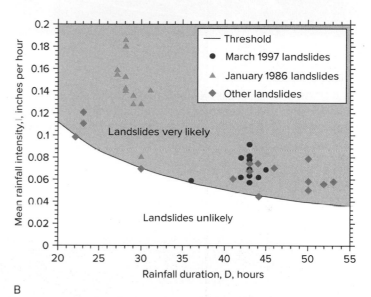

B

Figure 8.38

Studies in Seattle yielded a general relationship between rainfall and probability of landslides. (A) Probability of landslides depends both on how wet the soil is from prior precipitation (represented by P_{15}) and on recent rainfall (P_3). (B) The more intense the rainfall, the shorter the duration required to trigger landslides.

After USGS Fact Sheet 2007-3005.

States, creating maps that show the likelihood and estimated volume of debris flows based on a peak 15-minute rainfall intensity of 24 millimeters per hour (mm/h) (**figure 8.41**). The assessments are based on predictive models that use historical debris flow data, rainfall measurements, soil and slope characteristics, and burn severity data. In 2021, approximately 65 burn sites were assessed.

Slope monitoring isn't always about rainfall or residences. The Bingham Canyon mine in Utah, the world's largest human excavation, is 1.2 km (0.75 miles) deep. In early 2013, just months after installing a interferometric radar system to monitor the slopes for instability, mine operators detected signs of increasing strain. A warning was issued to successfully evacuate the mine, and seven hours later, the largest nonvolcanic landslide in North America crashed into the pit in two stages (**figure 8.42**). Seismic networks detected the earthquakes created by the rock avalanches, providing scientists with data to better understand landslide physics.

The ultimate goal in collecting data and creating landslide susceptibility models is to provide timely warnings for imminent slope failures. How does this happen and who provides the warnings? First and foremost, if you do live in an area susceptible to landsliding, prepare yourself for such an

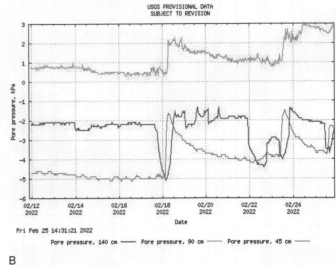

A

B

Figure 8.39

(A) A USGS scientist installs a monitoring system in North Carolina. (B) The pore pressure readings from this same station over a 2-week period; the three colors represent different measurement depths. Data from other monitoring stations can be obtained at https://www.usgs.gov/programs/landslide-hazards/science/monitoring-stations.

(A) York Lewis/USGS

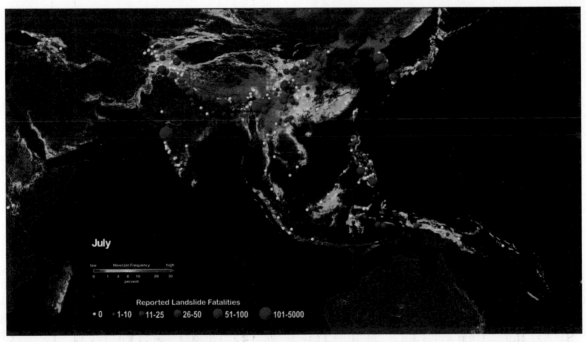

Figure 8.40

The NASA Landslide Hazard Assessment for Situational Awareness model considers (and shows) historic fatalities from 2007 to 2018 in determining risk for any given July in southeast Asia. Red represent high landslide probability at 30%.

NASA's Scientific Visualization Studio

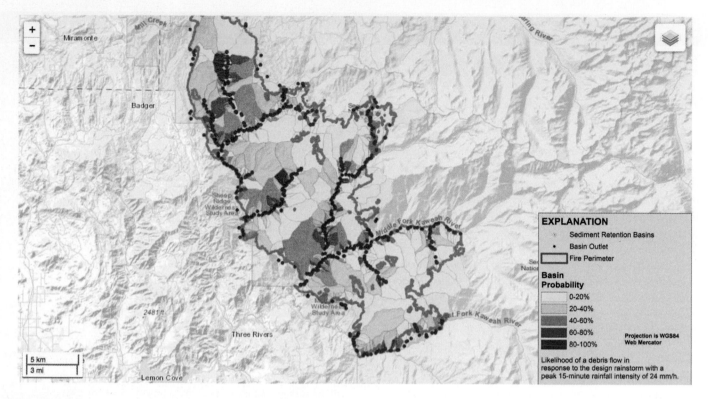

Figure 8.41

The KNP Complex Fire started in September 2021, and burned about 350 km^2 in the area around Sequoia and Kings Canyon National Parks. This debris flow hazard map shows the probability of debris flows by individual stream. Visit the USGS Emergency Assessment of Post-Fire Debris-Flow Hazards site for more information: https://landslides.usgs.gov/hazards/postfire_debrisflow/

U.S. Geological Survey

Figure 8.42

The Bingham Canyon mine slide of 2013. To begin to appreciate the scale of the event, note scale bar at lower right, and realize that 1600 meters is about 1 mile.

NASA Earth Observatory image by Jesse Allen and Robert Simmon, using EO-1 ALI data from the NASA EO-1 team

occurrence. Many state and regional agencies provide materials on how to do so. Because the trigger for many slope failures is large rainfall events, warnings come through NOAA's National Weather Service, just as you might get for a flood or a tornado. The local Weather Forecast Office will have the information from the USGS regarding the amount and intensity of rainfall that will cause slope failure (see **figure 8.38**) for susceptible areas in their region. There is no national-scale warning system for all landslides, although there have been prototypes. The most current comprehensive system is for the post-fire debris-flow hazards as described above. With the passing of the National Landslide Preparedness Act in January 2021, which calls for more interagency coordination, programs regarding landslide hazard reduction, loss reduction, preparedness, and warnings will be developed or expanded.

The Vaiont Dam—Reservoirs and Landslides

The Vaiont River flows through an old glacial valley in the Italian Alps. The valley is underlain by a thick sequence of sedimentary rocks, principally limestones with some clay-rich layers, that were folded and fractured during the building of the Alps. The sedimentary units of the Vaiont Valley are in the form of a syncline, or trough-shaped fold, so the beds on either side of the valley dip down toward the valley (**figure CS 8.1.1**). The rocks are relatively weak and particularly prone to sliding along the clay-rich layers. Evidence of old rockslides can be seen in the valley. Extensive solution of the carbonates by groundwater further weakens the rocks.

The Vaiont Dam was built for power generation. It is made of concrete and stands over 265 meters (875 feet) high. Modern engineering methods were used to stabilize the rocks in which it is based. The capacity of the reservoir behind the dam was initially 150 million cubic meters.

The obvious history of landslides in the area originally led some to object to the dam site. Later, the building of the dam and subsequent filling of the reservoir aggravated an already-precarious situation. As the water level in the reservoir rose following completion of the dam, pore pressures of groundwater in the rocks of the reservoir walls rose also. This tended to buoy up the rocks and to swell the clays of the clay-rich layers, further decreasing their strength, making sliding easier.

In 1960, a block of 700,000 cubic meters of rock slid from the slopes of Monte Toc on the south wall into the reservoir. Creep was noted over a still greater area. A set of monitoring stations was established on the slopes of Monte Toc to track any further movement. Measurable creep continued, occasionally reaching 25 to 30 centimeters (10 to 12 inches) per week, and the total volume of creeping rock was estimated at about 200 million cubic meters.

Late summer and early fall of 1963 were times of heavy rainfall in the Vaiont Valley. The rocks and soils were saturated. The reservoir level rose by over 20 meters, and pore pressures in the rocks increased. By mid-September 1963, measured creep rates were consistently about 1 centimeter per day. No one recognized that the rocks were slipping not in a lot of small blocks, but as a single, coherent mass. On 1 October, animals that had been grazing on those slopes of Monte Toc moved off the hillside and reportedly refused to return.

The rains continued, and so did the creep. Creep rates grew to 20 to 30 centimeters per *day*. Finally, on 8 October, the engineers realized that all the creep-monitoring stations were moving together. Worse yet, they also discovered that a far larger area of hillside was moving than they had thought. They tried to reduce the water level in the reservoir by opening the gates of two outlet tunnels. But it was too late, and the water level continued to rise. The silently creeping mass had begun to encroach on the reservoir, decreasing its capacity significantly. Engineers watched as the imminent disaster struck on the night of 9 October, in the midst of another downpour. No one downstream was evacuated.

A chunk of hillside, over 240 million cubic meters in volume, had slid into the reservoir (**figure CS 8.1.2**). That immense movement, occurring in less than a minute, set off a shock wave of wind that rattled the town of Casso and drew the water 240 meters *upslope* out of the reservoir after it. Displaced water crashed over the dam in a wall 100 meters (over 325 feet) above the dam crest and rushed down the valley below. The water wave was still over 70 meters high some 1.5 kilometers downstream at the mouth of the Vaiont Valley, where it flowed into the Piave River. The energy of the rushing water was such that some of it flowed *upstream* in the Piave for more than 2 kilometers. Within about five minutes, nearly 3000 people were drowned and whole towns wiped out.

It is a great tribute to the designer and builder of the Vaiont Dam that the dam itself held through all of this, resisting forces far beyond the design specifications; no longer in use, it still stands today (**figure CS 8.1.3**). It is also very much to the discredit of those who chose the site that the dam and reservoir were ever put in such a spot. With more thorough study of the rocks' properties and structure, the abundant evidence of historic slope instability, and the possible effects of rising water levels, the Vaiont tragedy could have been avoided.

History often seems destined to repeat itself. In the late 1970s, the Tablacha Dam on Rio Mantaro, Peru, was threatened by a similar failure (**figure CS 8.1.4**). Foliated metamorphic rocks dip toward the dam and reservoir. A portion of an ancient slide was reactivated, most likely by the increasing pore pressure due to reservoir filling. The situation is much like that of the Vaiont Dam. In this case, however, the Peruvian government spent $40 million on stabilization efforts—building a 460,000-cubic-meter earth buttress at the toe of the slide block, and adding drainage, rock bolts, and other measures. So far, they seem to be succeeding.

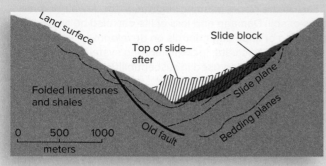

Figure CS 8.1.1

Geologic cross section of the Vaiont Valley.

G. A. Kiersch, "The Vaiont Reservoir Disaster," Mineral Information Service 18(7): 129–138, 1965. State of California, Division of Mines and Geology

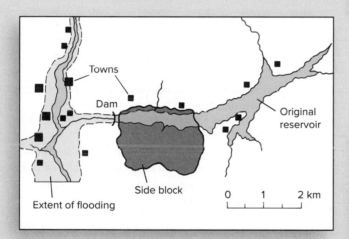

Figure CS 8.1.2

The landslide at the Vaiont reservoir and resultant flood damage.

G. A. Kiersch, "The Vaiont Reservoir Disaster," Mineral Information Service 18(7): 129–138, 1965. State of California, Division of Mines and Geology

Figure CS 8.1.3

After the Vaiont disaster: The dam (foreground) still stands despite the massive landslide that occurred just behind it.

U.S. Geological Survey Department of the Interior/USGSU.S. Geological Survey

Figure CS 8.1.4

Reactivated slide block (identified by red arrows) above Tablacha Dam and reservoir, Peru.

U.S. Geological Survey Department of the Interior/USGSU.S. Geological Survey/Photograph by J. T. McGill

Summary

Mass movements occur when the shearing stress acting on rock or soil exceeds the shear strength of the material to resist it or the friction holding it in place. Gravity provides the force behind the shearing stress. The basic susceptibility of a slope to failure is determined by a combination of factors, including geology (nature and strength of materials), geometry (slope steepness), the mass of material involved, and moisture content. Sudden failure usually involves a triggering mechanism, such as vibration from an earthquake, addition of moisture, steepening of slopes, or removal of stabilizing vegetation. Landslide hazards may be reduced by such strategies as modifying slope geometry, reducing the weight acting on the slope, planting vegetation, or dewatering the rocks or soil. Damage and loss of life can also be limited by restricting or imposing controls on development in areas at risk. Site-specific data, local monitoring systems, and global databases are being used in susceptibility models to help deliver timely landslide warnings.

Key Terms and Concepts

angle of repose	flow	rock flour	slide
creep (rock or soil)	landslide	rockfall	slump
debris avalanche	mass movement	sensitive clay	talus
debris flow	mass wasting	shear strength	
fall	quick clay	shearing stress	

Test Your Learning

1. Explain in general terms why landslides occur, and what two factors particularly influence slope stability.

2. Earthquakes are one landslide-triggering mechanism; water can be another. Evaluate the role of water in mass movements.

3. Explain the nature of quick clays, and how they can promote slope failure.

4. Describe at least three ways in which development on hillsides may aggravate landslide hazards.

5. List two ways to recognize soil creep and one way to identify sites of past landslides.

6. Briefly explain the distinctions among falls, slides, and flows.

7. Common slope-stabilization measures include physical modifications to the slope itself or the addition of stabilizing features. Choose any three such strategies and briefly explain how they work.

8. Describe, in general terms, how the duration and intensity of precipitation are related to the likelihood of a landslide.

9. List the parameters used in monitoring and modeling landslide susceptibility.

Exploring Further

1. Experiment with the stability of a variety of natural unconsolidated materials (soil, sand, gravel, and so forth). Pour the dry materials into separate piles; compare angles of repose. Add water, or put an object (load) on top, and observe the results.

2. Demonstrate the role of pore pressure: Spread a thin layer of water on a smooth, slightly tilted surface. Empty a well-chilled 12-oz beverage can and place it upside down in the pool of water. As the can and the air inside it warm, expansion of that air raises the fluid pressure under the can, and it should begin to slide.

3. Consider how you might respond to a known local landslide hazard or a warning. How definite a warning would you look for before evacuating? Would you take any precautions prior to the issuance of an official warning if conditions seemed to be increasing the risk?

4. Go to the USGS landslide-monitoring page, **https://www .usgs.gov/natural-hazards/landslide-hazards/science/ monitoring-stations?qt-science_center_objects=0#qt-science_center_objects**. Select one of the present or past monitoring sites; examine the kinds of data being collected, and, if the site is currently active, check its current landslide status.

5. Go to Google Earth and view the area of La Conchita, CA, from space. Make a list of the natural and human factors that are contributing to mass wasting there. Can you find other indications of slope failure along the nearby coast?

6. Go to the NASA Cooperative Open Online Landslide Repository to search for landslides near you or a place you may have visited. **https://gpm.nasa.gov/applications/ disasters/help-create-largest-landslide-database**

CHAPTER 9

Ice and Glaciers, Wind and Deserts

The topics we will cover in this chapter and the next are fundamentally related to Earth's climate system which is driven by solar heat. Solar heat powers the hydrologic cycle, including the evaporation and precipitation needed to make glaciers. A relative lack of solar heat allows existing glaciers to grow. Differential solar heating of land and water, and thus of the air above them, drives the winds. Seasons, and their variable temperature and precipitation patterns, are caused the changing intensity of incident sunlight as Earth, tilted on its axis, revolves around the Sun over the course of a year (**figure 9.1**). Over longer periods, a variety of geologic and biologic factors lead to warming or cooling trends worldwide, and to movement of climate zones.

Glaciers have a dramatic affect on the landscapes they move through and also provide a variety of resources for

humans. Ice represents the largest supply of fresh water on Earth. Past glacial activity has left behind sediments that supply gravel for construction, store subsurface water for future consumption, and contribute to farmland fertility.

As an agent of change of Earth's surface, wind is considerably less efficient than water or ice. It can play a

Chapter Outline

9.1 Glaciers and Ice Ages

9.2 Wind and Its Geologic Impacts

9.3 Deserts and Desertification

The Lucy Glacier is on the edge of the West Antarctica ice sheet, as it flows outward from its center and through the Transantarctic Mountains. Antarctica is a desert, with its very arid climate created by the continent's location over the South Pole.

©Gina S. Szablewski

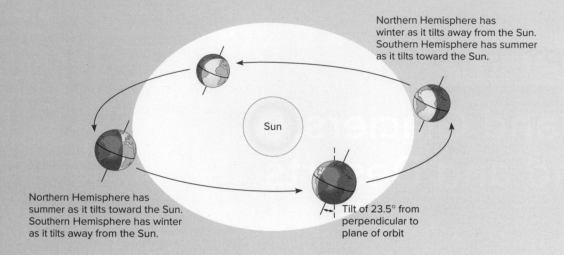

Northern Hemisphere has winter as it tilts away from the Sun. Southern Hemisphere has summer as it tilts toward the Sun.

Sun

Northern Hemisphere has summer as it tilts toward the Sun. Southern Hemisphere has winter as it tilts away from the Sun.

Tilt of 23.5° from perpendicular to plane of orbit

Figure 9.1

Seasonal warming and cooling depend on whether your hemisphere is tilted toward or away from the Sun at that time; the effect is most pronounced near the poles. If Earth's axis were not tilted relative to the ecliptic plane, we would not have seasons.

significant local role, particularly with regard to erosion. The effects of wind are most often visible in deserts, where little vegetation or development is present to mask or counter its effects.

Over a human lifetime, the extent of glaciation, the expanse of deserts, and the overall global climate generally vary little although there may be local exceptions. Over the longer term, significant changes do occur. We have evidence in the geologic record for a number of episodes during which vast ice masses covered major portions of the continents. Today, human activities are demonstrably changing the

chemistry of our atmosphere, which is causing detrimental changes in climate. In chapter 7, we have noted two consequences: melting of glaciers worldwide and thermal expansion of oceans, with resultant rise in net sea level, inundating low-lying coastal regions. There is also concern that on a warmer globe, arid lands will be more extensive. Desertification caused by human activities is already a serious concern of those who wonder how Earth's still-growing population will be fed. Local or regional drought aggravated by global warming will compound the problem. Chapter 10 will explore the subject of climate more broadly.

9.1 Glaciers and Ice Ages

Learning Objectives

- Summarize the formation of a glacier
- Compare alpine and continental glaciers
- Describe how glaciers move
- Describe how glaciers erode landscapes
- Characterize different types of glacial deposits
- List possible causes of ice ages

A **glacier** is a mass of ice that moves over land under its own weight, through the action of gravity. The amount of snow needed to form a glacier doesn't fall in a single winter, so for glaciers to develop, the climate must be cold enough that some snow and ice persist year-round. This requires an appropriate combination of altitude and latitude. Glaciers tend to be associated with the extreme cold of polar regions. However, temperatures also generally decrease at high altitudes (**figure 9.2**),

Figure 9.2

The snow line on a mountain range reflects decreasing temperature at higher elevation. Olympic Range, Washington.

©Carla Montgomery

so glaciers can exist in mountainous areas even in tropical or subtropical regions. Three mountains in Mexico have glaciers, but these glaciers are all at altitudes above 5000 meters (15,000 feet). Similarly, there are glaciers on Mount Kenya and Mount Kilimanjaro in east Africa, within a few degrees north or south of the equator, but only at high altitudes. Here we will discuss the formation and movements of glaciers, their affect on the landscape, and the possible causes of ice ages.

Glacier Formation

For glaciers to form, there must be sufficient moisture in the air to provide the necessary precipitation. Glaciers are, after all, large reservoirs of frozen water. In addition, the amount of winter snowfall must exceed summer melting so that the snow accumulates year by year. Most glaciers start in the mountains as snowfields that survive the summer. Slopes that face the poles (north-facing in the Northern Hemisphere, south-facing in the Southern Hemisphere), protected from the strongest sunlight, favor this survival. So do gentle slopes on which snow can pile up thickly instead of plummeting down periodically in avalanches.

As the snow accumulates, it is gradually transformed into ice. The weight of overlying snow packs it down, drives out much of the air, and causes it to recrystallize into a coarser, denser form called *firn,* an intermediate texture between snow and ice, and finally to a compact solid mass of interlocking ice crystals (**figure 9.3**). The conversion of snow into solid glacial ice may take only a few seasons or several thousand years, depending on such factors as climate and the rate of snow accumulation at the top of the pile, which hastens compaction. Eventually, the mass of ice becomes large enough to either flow downhill or outward from the thickest accumulations, depending on its setting. The moving ice is then a true glacier.

Types of Glaciers

Glaciers are divided into two types on the basis of size and occurrence. The most numerous today are the **alpine glaciers,** also known as mountain or valley glaciers (**figure 9.4**). As these names suggest, they are glaciers that occupy valleys in mountainous terrain, most often at relatively high altitudes. Most of the estimated 70,000 to 200,000 glaciers in the world today are alpine glaciers.

The larger and rarer **continental glaciers** are also known as *ice caps* (generally less than 50,000 square kilometers in area) or *ice sheets* (larger). They can cover whole continents and reach thicknesses of a kilometer or more. Though they are far fewer in number than the alpine glaciers, the continental glaciers comprise far more ice. At present, the two principal continental glaciers are the Greenland and the Antarctic ice sheets. The West Antarctic ice sheet is so large that it could easily cover the forty-eight contiguous United States. The geologic record indicates that, at several times in the past, even more extensive ice sheets existed on Earth.

Movement and Change of Glaciers

A glacier does not behave as a rigid, unchanging lump of ice. Its flow is plastic and different parts move at different rates. At the base of a glacier, or where it scrapes against and scours valley walls, the ice moves more slowly. In the central portions and closer to the surface, it flows more freely and rapidly (**figure 9.5**). Where it is cold enough for snow to fall, fresh material accumulates, adding to the weight of ice and snow that pushes the glacier downhill. Parts of a glacier may locally flow up an incline for a short distance. However, the net flow of the glacier must be downslope, under the influence of gravity. The downslope movement may be either down the slope of the underlying land surface, or in the downslope direction of the *glacier's* surface, from

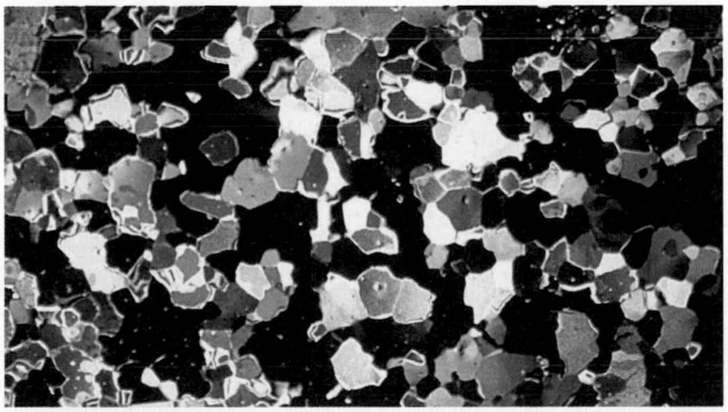

Figure 9.3

Thin slice of glacier ice viewed under a microscope in polarized light. The interlocking ice crystals resemble the compact texture of an igneous or metamorphic rock.

A. Gow, U.S. Army Corps of Engineers, NOAA/National Geophysical Data Center

Figure 9.4

The Aletsch Glacier in Switzerland is the largest glacier in the Alps.

Daniel Haller/iStock/Getty Images

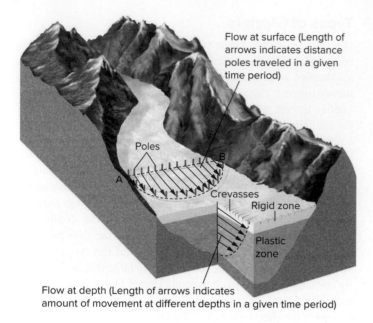

Flow at surface (Length of arrows indicates distance poles traveled in a given time period)

Poles

Crevasses

Rigid zone

Plastic zone

Flow at depth (Length of arrows indicates amount of movement at different depths in a given time period)

Figure 9.5

Glacier flow is mainly plastic; flow is fastest at the center surface. Crevasses form as the brittle surface ice cracks. Poles set into the ice have been used historically to measure movement.

areas of thicker ice to thinner, regardless of the underlying topography, as is typical for ice sheets (**figure 9.6**).

Flow is typically slow, a few tens of meters per year. However, alpine glaciers may *surge* at rates of several tens of meters per day for a few days or weeks. In addition to flowing plastically under its own weight, a glacier slides on any meltwater accumulated at the base of the ice. In some cases, dammed subglacial meltwater that is suddenly released in large quantities may cause surges.

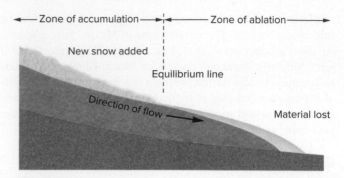

Zone of accumulation — Zone of ablation

New snow added

Equilibrium line

Direction of flow

Material lost

Figure 9.7

Schematic of longitudinal cross section of glacier. In the zone of accumulation, the addition of new material exceeds loss by melting or evaporation; the reverse is true in the zone of ablation, where there is a net loss of ice.

At some point, the advancing edge of the glacier terminates. It may flow out over water and break up, creating icebergs by a process known as **calving**, or more commonly, it flows to a place that is warm enough that ice loss by melting or evaporation is at least as rapid as the rate at which new ice flows in (**figure 9.7**). The set of processes by which ice is lost from a glacier is collectively called **ablation**. The **equilibrium line** is a changeable boundary between the zones of accumulation and ablation.

Over the course of a year, the size of a glacier varies somewhat (**figure 9.8**). In winter, the rate of accumulation increases, and melting and evaporation decrease. The glacier becomes thicker and longer; it is then said to be advancing. The rate at which it does so can be measured physically (as with the survey poles shown in **figure 9.5**) or remotely, as by satellite. In summer, snowfall is reduced or halted and melting accelerates; ablation exceeds accumulation. We then often say the glacier is

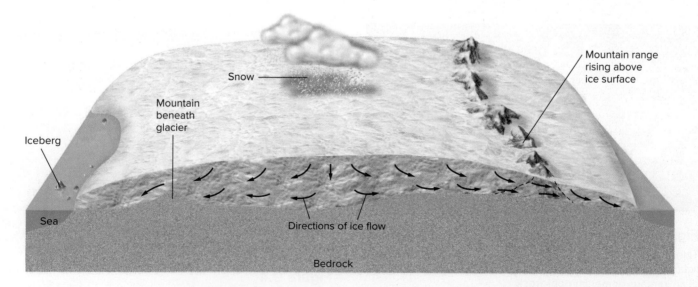

Snow

Mountain range rising above ice surface

Mountain beneath glacier

Iceberg

Sea

Directions of ice flow

Bedrock

Figure 9.6

An ice sheet flows outward from its thickest point.

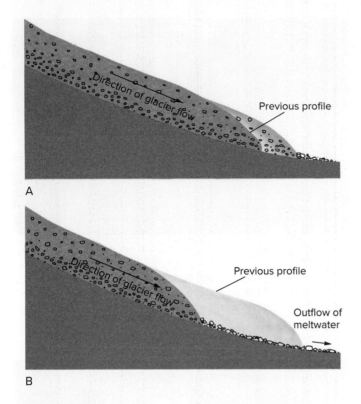

Figure 9.8

Glacial advance and recession. (A) Advance of the glacier moves both ice and sediment downslope. (B) Ablation, especially melting, causes apparent recession of glacier, accompanied by deposition of sediment from the lost ice volume.

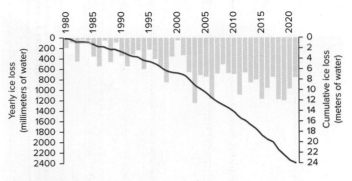

Figure 9.9

Several decades of measurements on a set of 41 reference alpine glaciers worldwide shows a generally accelerating trend of ice loss. Some of the possible consequences are examined in Case Study 9.1.

Source: https://wgms.ch/global-glacier-state/

retreating, although because it not flowing backward, *receding* is perhaps a better descriptor. The leading edge simply melts back faster than the glacier moves forward. Over many years, if the climate remains stable, the glacier achieves a dynamic equilibrium, in which the winter advance just balances the summer recession, and the average extent of the glacier remains constant from year to year.

More significant are the sustained increases or decreases in glacier size that result from consistent cooling or warming trends. An unusually cold or snowy period spanning several years would be reflected in a net advance of the glacier over that period, while a warm period would produce a net retreat. Such phenomena have been observed in recent history. From the mid-1800s to the early 1900s, a worldwide recession of alpine glaciers far up into their valleys occurred. Beginning about 1940, the trend then seemed to reverse, and glaciers advanced, which may have been due to cooling associated with air pollution, as will be explored in later chapters. More recently, a clear global warming trend has resumed, with corresponding shrinkage of many glaciers. It is important to bear in mind in all discussions of global climate changes that considerable local climate variations can exist as a result of local topography, geography, precipitation, winds and water currents, and many other factors. So, for example, though the great majority of alpine glaciers worldwide are currently receding, and over

several decades that trend has strengthened so that the net mass of ice in these glaciers is decreasing (**figure 9.9**), some glaciers may still be stable or advancing. Similarly, while the edges of the Greenland ice sheet are melting back rapidly and the overall mass of ice is decreasing, the center is thickening as a result of increased precipitation.

Glacial Erosion and Deposition

A glacier's great mass and solidity make it especially effective at altering any landscape it may move through. Glaciers pick up, transport, and deposit huge amounts of rocks and sediments. They leave distinctive features we can use to track their past extent and movements, long after the ice is gone. While some of these features are distinct to alpine settings, there are others that can be found in association with both alpine and continental glaciers because they are created by the same processes that don't vary by setting. Once you know what to look for, these features are hard to miss.

Alpine glaciers often flow through existing V-shaped valleys carved by streams, eroding them into more of a U-shape (**figure 9.10A**). On a grand scale, thick continental glaciers can gouge out a series of large troughs in the rock over which they move. The Great Lakes, and the Finger Lakes of upstate New York (**figure 9.10B**), occupy valleys deepened by the advance and retreat of a continental glacier up to two miles thick. On a very small scale, bits of rock that have become frozen into the ice at the base of a glacier act like sandpaper, polishing and making fine, parallel scratches, called **striations**, into underlying bedrock (**figure 9.11**). Such signs not only indicate the past presence of glaciers but also the direction in which they flowed. Erosion by the scraping of ice or included sediment on the surface underneath is **abrasion**. Water can also seep into cracks in rocks at the base of the glacier and refreeze, attaching rock fragments to the glacier. As the ice moves on, small bits of the rock are torn away in a

A

Figure 9.11

Striations on an Alaskan rock surface show direction of glacial flow. Notice that the scratches, parallel to the pen, extend continuously across changes in underlying rock type.

©Carla Montgomery

B

Figure 9.10

The erosional work of glaciers is apparent. (A) This U-shaped valley is just one of the clear indicators that alpine glaciers recently occupied the Rocky Mountains of Colorado. (B) A continental ice sheet gouged out depressions later filled by water to form the Great Lakes and the Finger Lakes (in the southeast corner of the photo).

(A) ©Gina S. Szablewski; (B) National Aeronautics and Space Administration

A

B

Figure 9.12

Erosional features of alpine glaciers include (A) this cirque in the Canadian Rockies and (B) an arête in Mt. Rainier National Park, Washington.

(A) Dianne Leeth/Alamy Stock Photo; (B) ©Gina S. Szablewski

process known as *plucking*. Plucking at the head (upper end) of an alpine glacier contributes to the formation of a *cirque,* a bowl-shaped depression (**figure 9.12A**). Where alpine glaciers flow side by side, the wall of rock between them may be thinned to a sharp ridge, an *arête* (**figure 9.12B**); erosion by several glaciers around a single peak produces a *horn* such as the Matterhorn in the Swiss Alps.

Glaciers also transport and deposit sediment very efficiently. Debris that has been eroded from the sides of a glacier's valley, that has fallen onto the glacier from peaks above, or that has been plucked from below tends to be carried along on, in, or under the ice, regardless of the size, shape, or density of the sediment. It is not likely to be deposited until the ice

Figure 9.13

Till deposited by an Alaskan glacier shows characteristically poor sorting; fragments are also angular and blocky, not rounded. This pile of till is about two meters high.

©Carla Montgomery

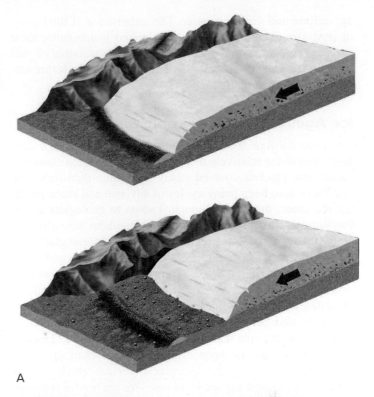

A

B

Figure 9.14

(A) Formation of end moraine occurs primarily through repeated annual cycles of deposition of till by melting ice. An advancing glacier deposits till in the zone of ablation and pushes material ahead of the ice front; then the glacier retreats, leaving a ridge of till behind. (B) The terminal moraine of the Penny Ice Cap, Baffin Island, Nunavit, Canada.

(B) Michael Studinger/NASA

carrying it melts or evaporates, and then all the debris, from large boulders to fine dust, is dropped together. Sediment deposited directly from the ice is called **till**. Till is characteristically angular and poorly sorted material (**figure 9.13**), as contrasted with stream-deposited sediment that is rounded and well sorted. Some of the till may be transported and redeposited by the meltwater, creating **outwash**. Till and outwash together are two varieties of glacial **drift**.

A landform made of till is a **moraine**. There are many different types of moraine, depending on where they were formed relative to the glacier. One of the most distinctive is a curving ridge of till piled up at the toe (lower end) of a glacier. Some of it may have been pushed along ahead of the glacier as it advanced. Much of it, however, is accumulated at the glacier's end by repeated annual advances and retreats; if the net extent of the glacier remains fairly constant for some years, the effect is like that of a conveyor belt carrying sediment to the end of the glacier. This is an *end moraine*. When the glacier begins a long-term retreat, a *terminal moraine* is left to mark the farthest advance of the ice (**figure 9.14**). Moraines are useful in mapping the extent of past glaciation on both a local and a continental scale.

Glaciers are responsible for much of the topography and near-surface geology of northern North America. The Rocky Mountains and other western mountain ranges show abundant evidence of carving by alpine glaciers. Continental glaciers deposited moraines and other features full of useful sand and gravel. Nearly all of the millions of natural lakes in Canada and the northern United States are glacial in origin, ranging in size from a small pond to Lake Michigan. Canada's Hudson Bay occupies a depression formed by the weight of the ice sheets that flooded with seawater after the ice melted away. Much of the basic drainage pattern of the Mississippi River

was established during the surface runoff of the enormous volumes of meltwater from retreating ice sheets; the sediments carried along with that same meltwater are partially responsible for the fertility of midwestern farmland (see "Loess" later in the chapter). Infiltration of a portion of the meltwater resulted in groundwater supplies tapped today for agriculture and drinking water. The importance of the legacy of past glaciers should not be overlooked just because their extent at present is more limited. Better understanding of past widespread glaciation may also help in modeling future climate changes and consequences.

Ice Ages and Their Possible Causes

The term *Ice Age* is used by most people to refer to a time beginning in the relatively recent geologic past when extensive continental glaciers covered vastly more area than they now do. There have been many cycles of advance and recession of the ice sheets during this epoch, known to geologists as the Pleistocene, which spanned the time from about 2.5 million to 10,000 years ago. At its greatest extent, Pleistocene glaciation in North America covered essentially all of Canada and much of the northern United States (**figure 9.15**). The ice was well over a kilometer thick over most of that area. Climatic effects extended well beyond the limits of the ice, with cooler temperatures and more precipitation common in surrounding lands, and worldwide sea levels lowered as much as 122 meters (400 feet) as vast volumes of water were locked up in ice sheets. During the warmer interglacial periods, the ice sheets shrank and receded and sea level rose. We are now in just such an interglacial time, though answering the question of how

Figure 9.15

Extent of glaciation in North America during the Pleistocene epoch. Arrows show direction of ice movement. Note that not every bit of land within the area colored in blue was necessarily ice-covered, and the maximum ice extent shown is for the whole Ice Age, not each glacial advance.

After C. S. Denny, National Atlas of the United States, U.S. Geological Survey

soon another ice-sheet advance might occur is complicated by the addition of human influences to natural climate-change cycles.

We can study Pleistocene glaciation in some detail because it was so recent, geologically speaking, and the evidence left behind is fairly well preserved. It is not, however, the only large-scale continental glaciation in Earth's past. There have been at least five major ice ages over Earth's history, going back over a billion years, each involving multiple advances and recessions at the margins of the ice sheets. As previously noted, the reassembly of the Gondwana continents during the development of the continental drift hypothesis was aided by matching up features produced by past ice sheets on what are now separate continents.

What factors are involved in creating the conditions necessary to form such immense ice masses? We will discuss a variety of processes that cause climate to change and loosely group them into those that affect *insolation,* the amount of solar energy reaching Earth's surface, and those that occur separate from insolation. As we introduce these below, keep in mind that ice ages likely don't start or end because of just a single cause but rather from an overlapping of processes that push climate in the same direction.

The external driver of Earth processes is the energy received from the Sun, and there are a few different ways that varies. The most direct is that the Sun's energy output changes; this is indicated by the 11-year cycle of sunspot activity that cause variations in the Sun's brightness. Historical reconstructions of these cycles show that they can be strong or weak, and that there are times when sunspots all but disappear for decades. One of these solar minimums called the Maunder Minimum, when intensity was particularly weak, overlaps the Little Ice Age which occurred from 1450 to 1850. Prehistorical reconstructions of solar activity based on cosmogenic isotopic data preserved in tree rings and ice cores reach back 9400 years and provide even more evidence for scientists to understand solar cycles. Scientists estimate that solar intensity may be responsible for about 1% of the warming that has occurred in the last 170 years, so it is unlikely that minimal solar output is solely responsible for the creation of any ice age.

Insolation is also affected by the tilt of Earth's axis, its orientation in space, and the shape of Earth's orbit around the Sun, all of which vary over tens to hundreds of thousands of years (**figure 9.16**). Such changes do not significantly change the total amount of sunlight reaching Earth, but they do affect its distribution, the amount of sunlight falling at different latitudes (and therefore the relative surface heating expected), and the time of year at which a region receives its maximum sunlight. Could these *Milankovitch cycles,* named after the scientist who recognized them in the early twentieth century, account for ice advances and recessions by changing the relative heating of the poles, equator, and midlatitudes?

Climate records (see chapter 10) indicate temperature variations do correlate with Milankovitch cycles over the past

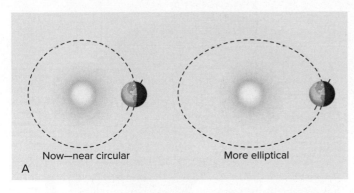

Figure 9.16

(A) Earth's orbit varies in shape from almost circular to more elliptical and back, in *eccentricity* cycles of about 100,000 years. (B) The angle of tilt of Earth's axis relative to its orbital plane (the ecliptic), now 23.5°, varies between 22.1° and 24.5° in *obliquity* cycles of about 41,000 years. In addition, the axis wobbles as a spinning top does when slowing down, pointing toward different stars (North Star at present); this *precession* cycle is about 26,000 years long.

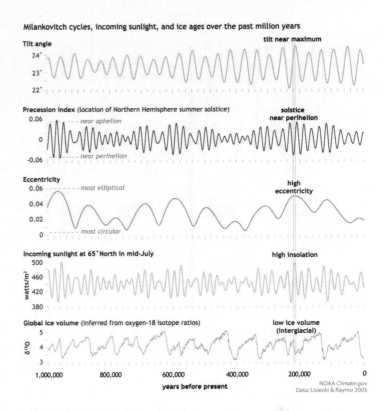

Figure 9.17

Milankovitch cycles over the past 1 million years paired with total solar irradiance and global ice volume (taken from oxygen isotope ratios in seafloor sediments). The gray column indicates the overlapping conditions that coincide with a warming period about 220,000 years ago (opposite of those discussed above for a glacial period).

National Oceanic and Atmospheric Administration

1 million years. In particular, the conditions that create cold summers in the high latitudes of the Northern Hemisphere seem to be key to ice ages; these allow the snow and ice created in the winter to not completely melt. Solar insolation is minimized when tilt is small, eccentricity is extreme, and the summer solstice occurs when Earth is farthest from the Sun (aphelion) (**figure 9.17**). These correlations are not perfect. And while Milankovitch cycles may account for ice fluctuations during the Ice Age, it is less clear that they can account for initiation of an ice age.

Insolation can also change if the radiation coming from the Sun is blocked by substances in Earth's atmosphere. That material can come from space or from Earth itself. Researchers have linked at least one ice age, in the Ordovician, to the breakup of extraterrestrial material from an asteroid belt object (**figure 9.18**). More often, scientists point to volcanic eruptions that emit ash and sulfuric-acid droplets. The eruption of Mount Pinatubo in 1991 caused measurable, widespread cooling that lasted for two to three years. Larger prehistoric eruptions, such

as those that occurred in association with the Yellowstone caldera or the explosion that produced the crater in Mount Mazama that is now Crater Lake, would have cause widespread cooling, but not enough to initiate an ice age. There have been periods in Earth's past with more intense or frequent sustained volcanism. The cumulative effects of multiple eruptions during such an episode might have included a long cooling trend and, ultimately, ice-sheet formation. A recent study into the Late Permian Ice Age (360–260 mya) found that explosive volcanism was three to eight times more frequent at the peak of that ice age, and that the massive amount of aerosols from these eruptions would have sustained the ice age not only by decreasing insolation but also by stimulating phytoplankton and thereby increasing their uptake of carbon dioxide.

The relationship of carbon dioxide (and other greenhouse gases) and climate is likely the one you are most familiar with, one we will cover in more depth in chapter 10 especially the role of humans in disrupting the natural carbon cycle. Greenhouse gases in the atmosphere vary naturally, and multiple lines of evidence over long periods of geologic time show a direct correlation between carbon dioxide and temperature: high CO_2, high

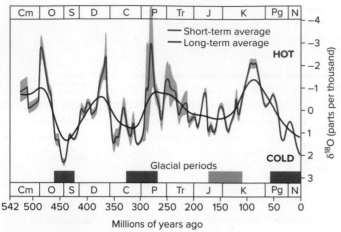

Phanerozoic Climate Change

Figure 9.18

The ice that started in the Ordovician Period and lasted from approximately 460–430 million years ago is just one of the three ice ages identified in the Phanerozoic Eon, and is shown by the darker blue boxes on the lower axis.

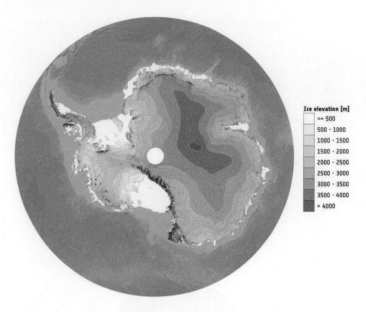

Figure 9.19

This map of Antarctica ice elevation was created by the European Space Agency's CryoSat mission; over 250 million measurements were taken over a span of 6 years.

European Space Agency

temperature. The two primary ways that levels of atmospheric CO_2 change are through biology and geology, processes that are difficult both to compartmentalize and to separate from the hydrosphere and atmosphere. An example from biology is the evolution of land plants, which became widespread beginning about 440 million years ago. This would have reduced atmospheric CO_2 in several ways. For example, the plants would consume CO_2 directly through photosynthesis. Also, their presence would accelerate weathering of continental rocks, putting calcium and magnesium into the oceans to combine with dissolved carbonate, making carbonate rocks and allowing more atmospheric CO_2 to dissolve. Less atmospheric CO_2 leads to cooling.

The weathering of continental rocks is accelerated during times of mountain building, which is just one way in which plate tectonics is linked to ice ages. In particular, scientists have shown strong evidence in which suture zones (continental collisions) in the tropics preceded the three most recent ice ages. Weathering would not only have occurred faster because of the steep slopes in the uprising mountains, but also because of the warm and wet conditions, sequestering CO_2 as described just above. Mountain ranges also affect precipitation patterns through the *orographic effect,* which creates distinct wet and dry zones on either side of a mountain range; the initiation of an ice age would involve a lot of snow.

There are other ways that plate tectonics affects climate that have to do with the size and location of landmasses. Ice ages tend to occur when there is a large landmass centered on or near a pole, just as Antarctica is now located over the South Pole. The ocean currents that surround Antarctica isolate it from the warming effects of water from lower latitudes, keeping it very cold. Antarctica is also very dry and is classified as a desert, but the snow it does get doesn't melt, resulting in the large ice sheet

(actually 2 of them) that covers the continent and has persisted in its present state for about 15 million years (**figure 9.19**). And, once a large ice sheet forms, it reflects solar energy rather than absorbing it, positively feeding back into the cooling climate.

Climate science is complex because there are so many interrelated processes that must be factored in, making it difficult to pinpoint one specifically as the cause of an ice age. Perhaps there is no single cause, and there could be a few different combinations that would not only create a cooling climate but push it into a sustained ice age. We will need to do more research, not only to discover how Earth behaved in the past but perhaps more importantly to understand how, and how quickly, climate will change in the future. We will explore Earth's ever changing climate, and our role in that, in chapter 10.

9.2 Wind and Its Geologic Impacts

Learning Objectives

- Recognize Earth's principal atmospheric circulation patterns
- Exemplify wind erosion
- Characterize a dune
- Summarize dune migration
- Describe loess

The combination of glaciers and winds in the same chapter may seem a bit odd, but they both are the result of processes that control Earth's climate. Winds redistribute moisture and heat in the atmosphere in large-scale circulation cells. More locally, they erode landscapes and distribute sediments. We will review these large-scale patterns and smaller-scale features associated with wind in the section below.

Atmospheric Circulation and Wind

Air moves from place to place over Earth's surface mainly in response to differences in pressure, which are often related to differences in surface temperature. Solar radiation falls most intensely near the equator, so solar heating of the atmosphere and surface is more intense there. The Sun's rays are more dispersed near the poles. Imagine a non-tilted, not-rotating Earth covered in a uniform material. That surface would be heated more near the equator and less near the poles; correspondingly, the air over the equatorial regions would be warmer than the air over the poles (**figure 9.20**). Because warm, less-dense (lower-pressure) air rises, warm near-surface equatorial air would rise, while cooler, denser polar air would move into the low-pressure region. The rising warm air would spread out, cool, and sink. Large circulating air cells would develop, cycling air from equator to poles and back. These would be atmospheric convection cells, analogous to the mantle convection cells associated with plate motions.

The actual picture is considerably more complicated. Earth rotates on its tilted axis, which produces curving patterns of circulating winds. Land and water are heated differentially by sunlight, with surface temperatures on the continents generally fluctuating much more than temperatures of adjacent oceans. The pattern of distribution of land and water influences the distribution of high- and low-pressure regions and further modifies air flow. In addition, Earth's surface is not flat; terrain irregularities introduce further complexities in air circulation. Friction between moving air masses and land surfaces can also alter wind direction and speed; tall, dense vegetation may reduce near-surface wind speeds by 30% to 40% over vegetated areas. A generalized view of actual global air circulation patterns is shown in **figure 9.21**. Different regions of Earth are characterized by different prevailing wind directions. Most of the United States is in a zone of westerlies, in which winds from the west dominate. Local weather conditions and the details of local geography produce regional deviations from this pattern on a day-to-day basis.

Air and water have much in common as agents shaping the land. Both can erode and deposit material, move material more effectively the faster they flow, and move particles by rolling, saltation, or in suspension (see chapter 6). Because water is far denser than air, it is much more efficient at eroding rocks and moving sediments, and has the added abilities to dissolve geologic materials and to attack them physically by freezing and thawing. On average worldwide, wind erosion moves only a small percentage of the amount of material moved by stream erosion. And like glaciers, winds can be very important in individual locations particularly subject to their effects.

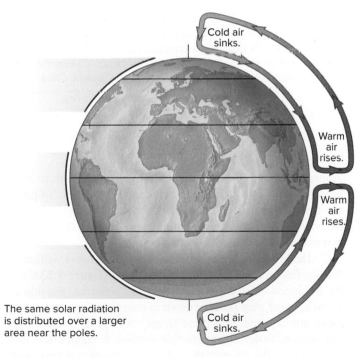

Figure 9.20

The non-rotating Earth model of atmospheric circulation. Heating at the equator and cooling at the poles produce a single, large convection cell in each hemisphere. Air rises at the equator and sinks at the poles. Note that these cells encircle the globe.

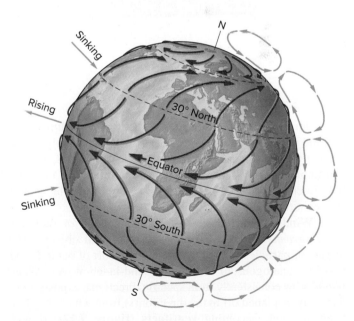

Figure 9.21

Principal present atmospheric circulation patterns. Red arrows represent surface wind flow.

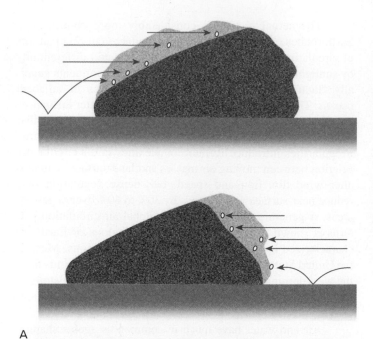

A

C

B

Figure 9.22

Wind abrasion on low-lying rocks results in the planing of rock surfaces. (A) If the wind is predominantly from one direction, rocks are planed or flattened on the upwind side. With a persistent shift in wind direction, additional facets are cut in the rock. (B) Example of a ventifact: rock polished and faceted by windblown sand at San Gorgonio Pass near Palm Springs, California. (C) Granite boulder undercut by wind abrasion, Llano de Caldera, Chile.

(B) Doug Sherman/Geofile; (C) U.S. Geological Survey/ Photograph by K. Segerstrom

Wind Erosion

Like water, wind erosion acts more effectively on sediment than on solid rock. Wind-related processes are especially significant where the sediment is exposed in areas such as deserts, beaches, and unplanted (or incorrectly planted) farmland. In dry areas like deserts, wind may be the major or even the sole agent of sediment transport.

Wind erosion consists of either abrasion or deflation. *Wind abrasion* is the wearing-away of a solid object by the impact of particles carried by wind. It is a sort of natural sandblasting, analogous to milling by sand-laden waves. Where winds blow consistently from specific directions, exposed boulders may be planed off in the direction(s) from which they have been abraded, becoming **ventifacts** (**figure 9.22**). If wind velocity is too low to lift the largest transported particles very high above the ground, tall rocks may show undercutting close to ground level (**figure 9.22C**). Abrasion can also cause serious

property damage. Desert travelers caught in windstorms have been left with cars stripped of paint and windshields so pitted and frosted that they could no longer be seen through. Abrasion likewise scrapes paint from buildings and can erode construction materials such as wood or soft stone.

Deflation is the wholesale removal of loose sediment, usually fine-grained sediment, by the wind (**figure 9.23A and B**). In barren areas, a combination of deflation and erosion by surface runoff may proceed down to some level at which larger rocks are exposed, and these larger rocks protect underlying fine-grained material from further erosion. Such an effect may contribute to the maintenance of a surface called *desert pavement* (**figure 9.23C**). A desert pavement surface, once established, can be quite stable. Finer sediment added to the surface may be swept between coarser chunks, later settling below them, while the pavement continues to protect the fine material against loss. The disturbance of the protective coarse-rock layer

A

B

C

Figure 9.23

(A) Deflation has selectively removed surface sand, leaving coarser rocks perched on the surface. Indiana Dunes National Lakeshore. (B) In this part of Death Valley, surface sand remains only where clumps of vegetation block wind and slow it down. (C) More fully developed desert pavement elsewhere in Death Valley. Coarser rocks at surface protect finer sediment below from wind erosion.

(A, B) ©Carla Montgomery; (C) U.S. Geological Survey/ Photograph by J. R. Stacy

by construction or resource exploration will expose the finer sediment below to rapid wind erosion.

The importance of vegetation in retarding erosion was demonstrated especially dramatically in the United States during the early twentieth century. After the Civil War, there was a major westward migration of farmers to the flat or gently rolling land of the Great Plains. Native prairie grasses and wildflowers were removed and the land plowed and planted to crops. While adequate rainfall continued, all was well. Then, in the 1930s, several years of drought killed the crops. The native prairie vegetation had been suitably adapted to the climate, while many of the crops were not. Once the crops died, there was nothing left to hold down the soil and protect it from the west winds sweeping across the plains. This was the Dust Bowl period. We look further at the Dust Bowl and the broader problem of minimizing soil erosion on cropland in chapter 12.

Wind Deposition

Where sediment is transported and deposited by wind, the principal depositional feature is a **dune**, a low mound or ridge, usually made of sand. Dunes start to form when sediment-laden winds slow down. The lower the velocity, the less the wind can carry. The coarser and heavier particles are dropped first. The deposition, in turn, creates an obstacle that constitutes more of a windbreak, causing more deposition. Once started, a dune can grow very large. A typical dune is 3 to 100 meters high, but dunes over 200 meters exist. What ultimately limits a dune's size is not known, but it is probably some aspect of the nature of the local winds.

When the word *dune* is mentioned, most people's reaction is *sand dune*. However, the particles can be any size, ranging from sand down to fine dust, and they can even be snow or ice crystals, although sand dunes are by far the most common. The particles in a given set of dunes tend to be similar in size. Eolian, or wind-deposited, sediments are well-sorted like many stream deposits, and for the same reason: the velocity of flow controls the size and weight of particles moved. The coarser-grained the particles, the stronger was the wind forming the dunes. The orientation of the dunes reflects the prevailing wind direction (if winds show a preferred orientation), with the more shallowly sloping side facing upwind (**figure 9.24A**). Sand dunes also differ in shape depending on sediment supply, wind, and topography (**figure 9.24B**).

Dune Migration

Dunes move if the wind blows predominantly from a single direction. As noted previously, a dune assumes a characteristic profile in cross-section, gently sloping on the windward side, steeper on the downwind side. With continued wind action, particles are rolled or moved by saltation up the shallower slope. They pile up a bit at the peak, then tumble down the steeper face, or **slip face**, which tends to assume a slope at the angle of repose for whatever size of particle is involved; recall **figure 8.4**. The net effect of these individual particle shifts is that the dune

A

B

Figure 9.24

Dune formation. (A) Huge sand dunes at Great Sand Dunes National Monument result as windswept sand piles up against the Sangre de Cristo mountains. (B) The dunes of Nili Patera indicate active geologic processes and winds are ongoing on Mars. Scientists have been challenged in interpreting the various shapes of dunes created with materials and by conditions distinctly different from those on Earth.

(A) ©Carla Montgomery; (B) NASA/JPL-Caltech/Univ. of Arizona

A

B

Figure 9.25

Dune migration and its consequences. (A) A schematic of dune migration. (B) The huge dune named Mt. Baldy marches through full-grown trees at Indiana Dunes National Lakeshore.

(B) ©Carla Montgomery

moves slowly downwind (**figure 9.25A** and **B**). As layer upon layer of sediment slides down the slip face, slanted *crossbeds* develop in the dune (**figure 9.26**).

Migrating dunes, especially large ones, can be a real menace as they march across roads and even through forests and over buildings. During the Dust Bowl era, farmland and buildings were buried under shifting, windblown soil. The costs to clear and maintain roads along sandy beaches and through deserts can be high because dunes can move several meters or more in a year. The usual approach to dune stabilization is to plant vegetation. However, since many dunes exist because the terrain is too dry to support vegetation, such efforts are often

futile. Aside from any water limitations, young plants may be difficult to establish in shifting dune sands because their tiny roots may not be able to secure a hold.

Loess

Rarely is the wind strong enough to move sand-sized or larger particles very far or very rapidly. Fine silt, on the other hand,

Figure 9.26

Crossbeds in a sand dune. Here, several sets of beds meeting at oblique angles show the effects of shifting wind directions, changing sediment deposition patterns. Note also wind-produced ripples in foreground. Oregon Dunes National Recreation Area.

©Carla Montgomery

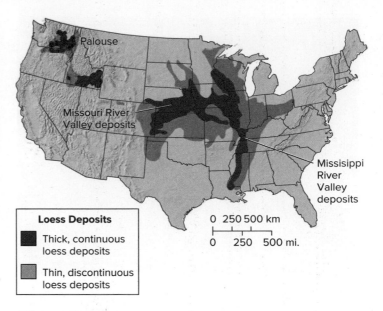

Loess Deposits

■ Thick, continuous loess deposits

■ Thin, discontinuous loess deposits

Figure 9.28

Much of the loess distribution in the central United States is close to principal stream valleys supplied by glacial meltwater. Loess to the southwest may have been derived from western deserts.

is more easily suspended in the wind and can be carried many kilometers before it is dropped. A deposit of windblown silt is known as **loess** (**figure 9.27**). The fine mineral fragments in loess are in the range of 0.01 to 0.06 millimeter (0.0004 to 0.0024 inch) in diameter.

The principal loess deposits in the United States are in the central part of the country, and their spatial distribution provides a clue to their source (**figure 9.28**). They are concentrated

Figure 9.27

Example of loess. Norton County, Kansas.

Photograph by L.B. Buck, U.S. Geological Survey

around the Mississippi River drainage basin, particularly on the east sides of major rivers of that basin. Those same rivers drained away much of the meltwater from retreating ice sheets in the last ice age. Glaciers grinding across the continent were the original producers of this sediment. The sediment was subsequently washed down the river valleys, and the lightest material was further blown eastward by the prevailing west winds.

As noted previously, the great mass of ice sheets enables them to produce quantities of finely pulverized *rock flour*. Because dry glacial erosion does not involve as much chemical weathering as stream erosion, many soluble minerals are preserved in glacial rock flour. These minerals provide some plant nutrients to the farmland soils now developed on the loess. Because newly deposited loess is also quite porous and open in structure, it has good moisture-holding capacity. These two characteristics together contribute to making the farmlands developed on midwestern loess particularly productive.

Not all loess deposits are of glacial origin. Loess deposits form wherever there is an abundant supply of very fine sediment. For example, loess derived from the Gobi Desert covers large areas of China (**figure 9.29**).

Loess does have drawbacks with respect to applications other than farming. While its light, open structure is reasonably strong when dry and not heavily loaded, it may not make suitable foundation material. Loess is subject to hydrocompaction, a process by which it settles, cracks, and becomes denser and more consolidated when wetted, to the detriment of structures built on top of it. The very weight of a large structure can also cause settling and collapse.

Figure 9.29

The Loess Plateau in Shanxi, China.

Pixtal/AGE Fotostock

Figure 9.30

The Mojave Desert in Joshua Tree National Monument is arid and supports little plant life besides cacti, yuccas, and the Joshua trees for which the monument is named (foreground).

©Carla Montgomery

9.3 Deserts and Desertification

Learning Objectives

- Define desert
- Explain how air currents, mountains, and proximity to the oceans create deserts
- Describe the role of plants in combating desertification
- Exemplify regions experiencing desertification and their response

Many of the features of wind erosion and deposition are readily observed in **deserts**. The term doesn't have a precise definition: it is commonly used to describe a region that receives less than 25 cm precipitation yearly, is hot, and has distinct yet limited vegetation. Deserts may be a sea of sand, or rugged and rocky (**figure 9.30**). Some deserts are cold. Antarctica is a desert; with an average yearly precipitation of about 50 mm, it is the driest continent. In lower latitudes, deserts are characterized by very little precipitation, and may be consistently hot, cold, or variable in temperature depending on the season or time of day. Deserts are often described as *arid* regions, which refers to a climate zone that loses more water through evapotranspiration than receives through precipitation. The distribution of the arid regions of the world (exclusive of polar deserts) is shown in **figure 9.31**. Below we will review the conditions that create deserts, and then discuss the process of desertification.

Causes of Natural Deserts

There exist a variety of ways to produce the arid conditions necessary for a desert. Many deserts are located in lower latitudes that naturally have higher temperatures, but deserts can also exist in higher latitudes and high altitudes due to reasons other than having a warm climate.

Looking closely at **figure 9.31**, you can see that many deserts are located at or near the 30° North and South latitudes. What processes are occurring there to create arid conditions? The explanation goes back to the atmospheric circulation patterns shown in **figure 9.21**, which govern the availability of precipitation. To understand these, we need to recall some general facts about the temperature, pressure, and moisture content of air masses: warm air holds more moisture than cold air, an increase in pressure allows air to hold more moisture, and warm air rises. Let us consider conditions at the equator. Here warm, saturated (full of moisture) air rises, and as it does, both pressure and temperature decrease with altitude. The cooler, decompressed air can't hold as much moisture, resulting in condensation and high precipitation rates in the tropics. Now this relatively cold, dry air moves laterally and eventually sinks at about 30° latitude. As it moves toward Earth's surface, it warms and experiences higher pressures; this allows it to hold more water, which it takes from the surface through evaporation, creating arid conditions. Many deserts such as the Sahara and those in Australia exist in this subtropical zone of high atmospheric pressure.

Topography also plays a role in controlling the distribution of precipitation. High mountains located along the path of principle air currents can create an **orographic effect**, controlling the distribution of rain on either side of the mountain range. As moisture-laden air approaches the mountains, it is forced upward, where it encounters less pressure and colder temperatures. This air can't hold as much moisture, so precipitation occurs on the upwind side of the mountains. The air is much drier as it moves to lower elevations on the downwind side of the range, creating the arid conditions of a **rain shadow** (**figure 9.32**). Rain shadows cast by the Sierra Nevada of California and to a lesser extent,

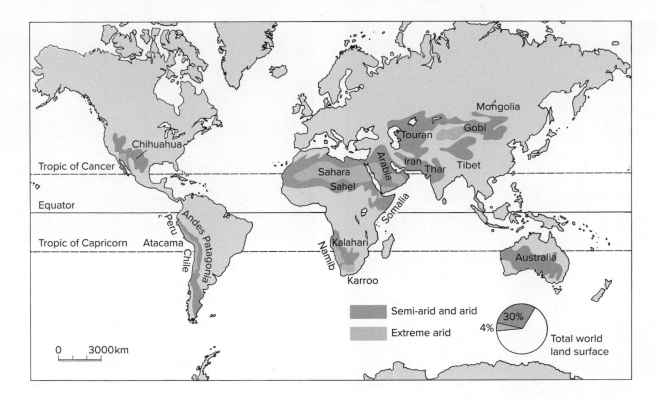

Figure 9.31

Distribution of the world's arid lands.

Data from A. Goudie and J. Wilkinson, The Warm Desert Environment. Cambridge University Press

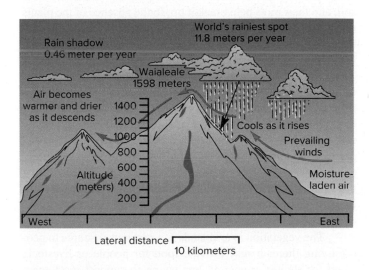

Figure 9.32

Rain shadows can occur even on an island in the ocean. Mt. Waialeale, on the Hawaiian island of Kauai, is described as the world's rainiest spot, receiving over 450 inches of rain a year. But downwind from it, in its rain shadow, annual rainfall is just 18 inches.

by the southern Rockies, contribute to the dryness and desert regions of the western United States.

Because the oceans are the major source of the moisture in the air, distance from the ocean can itself be a factor contributing to the formation of a desert. The longer an air mass is in transit over dry land, the greater chance it has of losing some of its moisture through precipitation. This contributes to the development of deserts in continental interiors. On the other hand, even coastal areas can have deserts under special circumstances. If the land is hot and the adjacent ocean cooled by cold currents, the moist air coming off the ocean will be cool and carry less moisture than air over a warmer ocean. As that cooler air warms over the land and becomes capable of holding still more moisture, it causes rapid evaporation from the land rather than precipitation. This phenomenon is observed along portions of the western coasts of Africa and South America in the Namib and Atacama Deserts, respectively (**figure 9.33**). Polar deserts can also be attributed to the differences in moisture-holding capacity between warm and cold air. Air traveling from warmer latitudes to colder near-polar ones will tend to lose moisture by precipitation, so less remains to fall as snow near the poles, and the limited evaporation from cold high-latitude oceans contributes little additional moisture to enhance local precipitation. Thick polar ice caps reflect effective preservation of what snow does fall, rather than heavy precipitation.

Desertification

Deserts, or arid regions, are just one of the many types of climate zones on Earth. Climate zones are impermanent, and they change in response to not only global temperature fluctuations,

Figure 9.34

Dead trees and erosion after severe drought in northern Sahel near Banh, Burkina Faso, Africa.

Robert_Ford/iStock/Getty Images Plus/Getty Images

Figure 9.33

The Namib Desert lies along the west coast of Namibia (see **figure 9.31**). It receives an average of 63 mm (about 2½ in.) of rain per year, plus some additional moisture as fog off the adjacent Atlantic Ocean.

NASA/GSFC/METI/ERSDAC/JAROS and the U.S./Japan ASTER Science Team

but also changes in topography and locations of landmasses. Amid these modifications, new deserts develop in areas that previously had more extensive vegetative cover. The term **desertification** is generally applied to the relatively rapid development or expansion of deserts caused or accelerated by the impact of human activities.

The exact definition of the lands at risk is difficult. *Arid* lands are commonly defined as those with annual rainfall of less than 250 millimeters (about 10 inches); *semiarid* lands, 250 to 500 millimeters (10–20 inches). The extent to which vegetation will thrive in low-precipitation areas depends on such additional factors as temperature and local evaporation and transpiration rates. Many of the arid lands border true desert regions. Desertification does not necessarily involve the advance of

desert into non-desert regions as a result of forces originating within the desert. Rather, desertification is often a patchy conversion of dry-but-habitable land to uninhabitable desert as a consequence of land-use practices, and perhaps accelerated by such natural factors as drought (**figure 9.34**).

The limited vegetation of arid and semiarid lands is critical in preventing their conversion to deserts. Plants shade the soil, and their roots help break it up. In the absence of the plants under the baking sun, the soil may crust over and harden, becoming less permeable. This decreases infiltration of what little rain does fall and increases water loss by surface runoff. That, in turn, decreases reserves of soil moisture on which future plant growth depends. Vegetation also shields the soil from erosion by wind. This is key to preserving soil fertility, for it is typically the topmost layer of soil that is richest in organic matter. The loss of vegetation leads to soil degradation that permanently diminishes the ability of the land to support future plant growth. Natural drought cycles may be short enough that the land will recover when the rains return. As human activities increase the pressure on these fragile arid regions, the degradation may be irreversible.

The vegetation in arid lands is a precious resource to people living there. It may provide food for people or livestock, wood for shelter or energy. Just trying to support too large a population in such a region may result in stripping the vegetative cover to the point of initiating desertification. As it progresses, the land can support ever-fewer people, and the situation worsens.

On land used for farming, native vegetation is routinely cleared to make way for crops. The native vegetation would have adapted to the dry conditions; the crops often require more moisture or lack the ability to survive a natural local drought cycle. While the crops thrive, all may be well. If the crops fail, perhaps during a drought, the land is left bare of vegetation and vulnerable to the types of degradation described earlier. It becomes harder to grow future crops, and desertification progresses.

Similar results follow from the raising of numerous livestock. In drier periods, vegetation may be reduced or stunted. Yet it is precisely during those periods that livestock, needing the vegetation not only for food but also for the moisture it contains, put the greatest grazing pressure on the land. The soil may again be stripped bare, and the deterioration of desertification follows.

Desertification is cause for concern because it effectively reduces the amount of arable land on which the world depends for food. The United Nations estimates that 500 million people live in areas that have experienced desertification since the 1980s. Because land degradation is occurring at a rate 30 to 35 times the historical rate, the UN also estimates that 50 million people could be displaced by desertification by 2030 (**figure 9.35**).

The Gobi Desert region in China and the Sahel in Africa (see **figure 9.31**) are regions currently undergoing large-scale desertification. The Great Green Wall initiative is underway in both these areas attempting to restore degraded land, sequester carbon, stop the expansion of arid lands, and create jobs primarily through the planting of trees and other vegetation (**figure 9.36**). These wide-reaching projects have had somewhat mixed success initially; 18 million hectares of land have been restored in the Sahel and 350,000 jobs were created. Information from China indicates that the amount of forested land has increased significantly to over 20%, the frequency of sandstorms has dropped, and advancing dunes have been stabilized. The massive efforts in China will continue to 2050, but that might not be enough to hold back desertification as indicated by multiple major sandstorms to hit Beijing in spring 2021. A recent funding pledge of $14 billion through the Great Green Wall Accelerator in Africa has the goal of better coordinating efforts in restoring 100 million hectares of degraded land.

Though desertification makes news most often in economically disadvantaged countries with large deserts, the process is ongoing elsewhere. In the United States, a large portion of the country is potentially vulnerable: much of the western half of the country can be classified as semiarid on the basis of its low precipitation. An estimated one-million-plus square miles of land, more than a third of this low-rainfall area, has undergone severe desertification, characterized by loss of desirable native vegetation, seriously increased erosion, and reduced crop yields. The problems are caused or aggravated by intensive use of surface water and overgrazing, and the population of these areas is growing.

Why does it appear there is so little concern in this country? The main reason is that the impact on humans has been modest. Other parts of the country can supply needed food. The erosion and associated sediment-redistribution problems are not yet highly visible to most people. Groundwater supplies are being tapped both for water supplies and for irrigation, creating

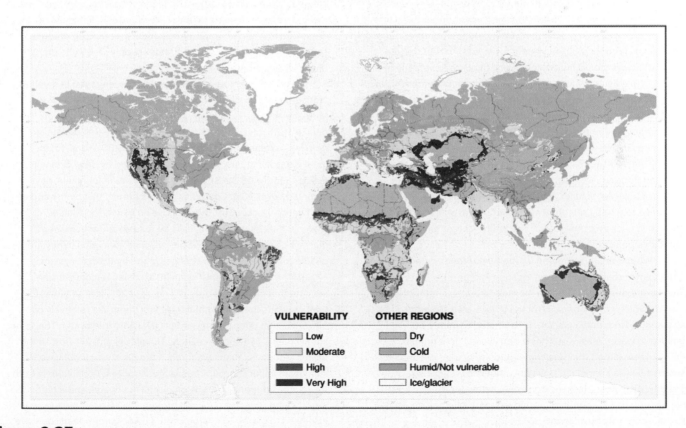

Figure 9.35

Lands vulnerable to desertification, and the degree of vulnerability. (Gray areas are already deserts.)

Map from U.S. Department of Agriculture Natural Resources Conservation Service

Figure 9.36

Women and children in Khartoum, Sundan are planting trees as part of Africa's Great Green Wall initiative to combat desertification.

Xinhua/Alamy Stock Photo

the illusion of adequate water; but as will be seen in chapter 11, groundwater users may be living on borrowed time, or at least borrowed water that is destined to run out. As that happens, it will become painfully obvious that too much pressure from human use was brought to bear on dry lands.

Finally, global climate change threatens to increase the extent of vulnerable arid lands. Higher temperatures and changing precipitation patterns may shift some now-temperate regions in that direction. Chapter 10 explores more fully the evidence for, and possible consequences of, such climate change.

Case Study 9.1

Vanishing Glaciers, Vanishing Water Supply

Glaciers are not only large masses of ice that drastically shape the landscapes they move through, but also important freshwater reserves. Approximately 75% of the fresh water on Earth is stored as ice in glaciers. And glacier meltwaters are the primary source of fresh water for over a billion people for at least part of the year.

Glaciers create meltwater even if they are not shrinking or receding, particularly during the spring and summer seasons. In regions where glaciers are large or numerous, glacial meltwater is the principal source of summer streamflow. But most alpine glaciers have been receding for at least 50 years, and the rate of retreat has increased (**Figure CS 9.1.1**). A rapidly melting glacier will, for the short-term, provide a lot of freshwater; that may result in spring flooding and decreased meltwater flow later in the season. However, smaller alpine glaciers that have already shrunk may have reached their peak meltwater stage, meaning they will provide less or even no water in the future. Larger glaciers may not reach this peak until later in the century.

What does this all mean for meltwater fresh water supplies? That answer really depends on location and population density. The most extensive glaciers in the United States are those in Alaska, which cover about 3% of the state's area. Summer streamflow from these glaciers is estimated at nearly 50 trillion gallons of water. Alaska still has extensive glaciers remaining, and many of these feed their meltwater into the ocean, not into local streams so their shrinking does not influence water availability directly. And most Alaskans don't use meltwater as their fresh water source; the abundant groundwater supply they do use is recharged partly by meltwaters.

The water-supply impact is potentially more serious elsewhere. In less-glaciated states like Montana, Wyoming,

California, and Washington, streamflow from summer meltwater amounts to tens or hundreds of billions of gallons. Here, glaciers, icefields, and seasonal snowpack areas are shrinking in many areas due to warmer temperatures and long-standing drought conditions. Storage in California water supply reservoirs are consistently below average.

The U.S. Geological Survey has been engaged in detailed monitoring of the state of the glaciers in Glacier National Park, Montana, and the results are striking. Of 150 named glaciers in the park in 1850, only 26 still exist at all, and those that remain are both less extensive and thinner. Between 1966 and 2015, the areal extent of the park's glaciers decreased by 34%. Grinnell Glacier (**Figure CS 9.1.2**), a popular destination for hikers, has shrunk by over 90%. Computer models predict that by the end of the twenty-first century, there will be no glaciers in Glacier National Park. Not only will there be less water for humans as glaciers disappear, some species will face extinction if they are unable to adapt quickly to changes in the hydrologic system.

New studies (published in 2022) using remote sensing indicate that melting glaciers will result in 20% less sea level rise than previous thought. That may at first seem like relatively good news, but that does also mean less available freshwater. The same study reveals 37% more ice in the Himalayas and 27% less in the Andes Mountains than measured in the past, which will have significant implications for freshwater supplies. Agriculture in both these areas depends heavily on glacial meltwater, and a reduction would mean reductions in crop productivity. Glacier meltwater is also used to generate hydroelectric power, and Andean countries have already experienced significant reductions in electricity generation due to lower discharges from mountain streams.

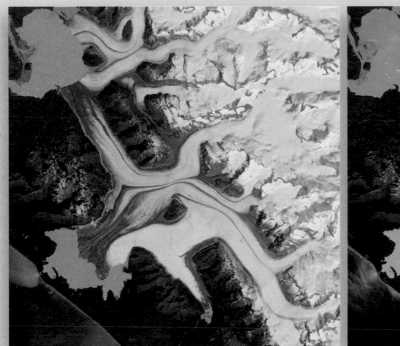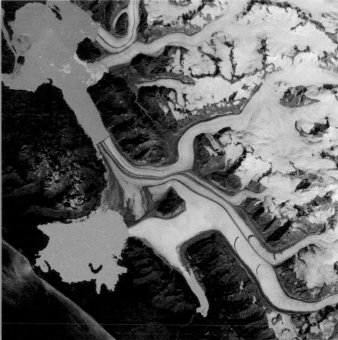

Figure CS 9.1.1

Grand Plateau Glacier in Glacier Bay National Park, Alaska in 1984 and 2019. Not only has the glacier receded but it has also thinned, and the entire flow of the glacial system has changed.

NASA Earth Observatory images by Joshua Stevens

A   B   C   D

Figure CS 9.1.2

Grinnell Glacier from the summit of Mount Gould, Glacier National Park. Photographs taken in 1938 (A), 1981 (B), 1998 (C), and 2009 (D). By 2009, most of what remained of the glacier had disintegrated into icebergs in the meltwater lake.

(A) Glacier National Park Archives/Photograph by T. J. Hileman; (B) U.S. Geological Survey/Photograph by C. Key;
(C) U.S. Geological Survey/Photograph by D. Fagre; (D) U.S. Geological Survey/Photograph by L. Bengtson

Figure CS 9.1.3 exemplifies changes occurring the Himalayas, where the equilibrium (snow) line is moving upward and glaciers are melting even in the winter due to warmer conditions. Less snow is falling, zones of ablation are increasing in size, and the amount of dark surface cover, capable of absorbing heat rather than reflecting it, is increasing. Scientists estimate that ice loss rates in the Himalayas are ten times higher than average rates since the Little Ice Age, and the glaciers have lost 40% of their area.

Consider the amount of people living in the drainage basins on Himalayan Rivers, and the estimates of population growth in

(Continued)

these areas for the near future. Currently over 800 million people depend on seasonal runoff for drinking water, agriculture, and hydroelectric power. Water shortages are certain to occur in the coming decades as these glaciers continue to shrink and the population continues to rise.

A

B

Figure CS 9.1.3

These false-color images compare conditions near the Rolwaling Glacier in October 2020 (left) to January 2021 (right). Brightest blue is snow, darker blue is ice and, darkest blue is meltwater. Rocks and vegetation are left in their natural colors.

(A) Source: NASA Earth Observatory images; (B) Source: NASA Earth Observatory

Summary

Glaciers past and present have sculptured the landscape not only in mountainous regions but also over wide areas of the continents. They leave behind U-shaped valleys, striated rocks, piles of poorly sorted sediment (till) in a variety of landforms (moraines), and outwash. Most glaciers currently are alpine glaciers, and at present, the majority of these are receding. The two major areas with remaining ice sheets are Greenland and Antarctica. Meltwater from glaciers supplies surface water and groundwater. The loss of ice mass and associated water-storage capacity can locally have serious water-supply implications.

There are many natural processes that result in a cool climate necessary for glacier formation, by either controlling the amount of solar energy reaching Earth or through the geologically and biologically controlled process of weathering and its affect on atmospheric CO_2. Scientists have yet to determine that a single of one of these is responsible for the initiation of an ice age, and it is likely that more than one is needed to sustain an ice age.

Wind moves material much as flowing water does, but less forcefully. As an agent of erosion, wind is less effective than water but may have significant effects in dry, exposed areas, such as beaches, deserts, or farmland. Wind action creates well-sorted sediment deposits, as dunes or in blankets of fine loess. The latter can improve the quality of soil for agriculture. Features created by wind are most obvious in deserts, which are dry and sparsely vegetated. The extent of unproductive desert and arid lands is increasing through desertification brought on by intensive human use of these fragile lands.

Key Terms and Concepts

ablation	desert	loess	striations
abrasion	desertification	moraine	till
alpine glacier	drift	orographic effect	ventifact
calving	dune	outwash	
continental glacier	equilibrium line	rain shadow	
deflation	glacier	slip face	

Test Your Learning

1. Briefly describe the formation of an alpine glacier, and the controls on its movement.

2. Describe what a *moraine* is and explain how moraines can be used to reconstruct past glacial extent and movements.

3. Choose any two proposed causes of past ice ages and evaluate the plausibility of each. (Is the effect on global climate likely to have been large or long enough? Is there any geologic evidence to support the proposal? Is the mechanism specific to one particular ice age, or applicable to any?)

4. Describe two ways in which glaciers store water, and explain how the recession of alpine glaciers can cause both flooding and drought.

5. Explain how sunlight falling on Earth's surface affects wind circulation.

6. Name and describe the two principal processes by which wind erosion occurs.

7. Briefly describe the way in which dunes form and migrate.

8. Define *loess,* explain how most loess deposits in the United States originated, and indicate another way that loess can be derived.

9. Assess the significance of loess to farming.

10. Define *desertification,* and describe two ways in which human activities contribute to the process.

Exploring Further

1. During the Pleistocene glaciation, a large fraction of Earth's land surface may have been covered by ice. Assume that 10% of the 149 million square kilometers was covered by ice averaging 1 kilometer thick. How much would sea level have been depressed over the 361 million square kilometers of oceans? This site is a good resource: **https://www.antarcticglaciers.org/glaciers-and-climate/**

2. Global mean sea level was about 9 cm higher in 2020 than in 1993, with about 6 cm of that change coming from meltwater. Using ocean-area data from problem 1, estimate the volume of ice melted.

3. Investigate photos and stories at **https://earthobservatory.nasa.gov/** related to dust and sand. Can you find one that has affected you either directly or indirectly?

4. Information on the U.S. Geological Survey's monitoring of the glaciers in Glacier National Park, and on their repeat-photography project, is at **www.usgs.gov/centers/norock/science/repeat-photography-project** Investigate the techniques the researchers are using; examine the changes shown by the repeat photography.

5. Contrast the Great Green Wall projects in Africa and China in terms of progress, jobs created, future outlook, and challenges. What successful strategies might be deployed from one place to another?

CHAPTER 10

Climate—Past, Present, and Future

We all experience and respond to *weather* daily, the atmospheric conditions that can vary considerably from day to day or even hour to hour. You start the day in a rain coat and sweater and end it in shorts, t-shirt, and sun hat. *Climate* also changes, but is less variable on a human timescale; it is the weather averaged over longer periods of time. The geologic rock record clearly shows that climate has changed many times over the long span of Earth's history. The ice ages we reviewed in the last chapter are just one example of natural climate fluctuation.

Both weather and regional climate affect the operation or intensity of the surface processes examined in the previous four chapters, directly and indirectly impacting human activities. While we can't do much about a day's weather, it has become increasingly evident that humans can influence climate on a regional or global scale, intentionally or otherwise. Remote sensing is again providing us with information

Chapter Outline

10.1 The Greenhouse Effect and Global Temperature

10.2 Climate and Ice Revisited

10.3 Oceans and Climate

10.4 Other Aspects of Climate Change

10.5 Evidence of Climates Past

10.6 Feedbacks, Uncertainly, and Geoengineering

Wildfires in Australia burned over 8.5 million hectares of land from October 2019 to January 2020, killing or displacing an estimated 3 billion animals and causing an estimated $100 billion in damages. Researchers tracked the iron-rich plumes of smoke emitted from the fires and linked them to phytoplankton blooms in the South Pacific and Southern oceans. These same blooms took up much of the carbon dioxide released by the fires. Understanding and systemizing links such of these are important in building accurate climate models.

NASA Earth Observatory images by Joshua Stevens

10.1 The Greenhouse Effect and Global Temperature

Learning Objectives

- Explain the greenhouse effect
- Identify the 150-year trend of atmospheric carbon dioxide gas and temperatures

Climate is the result of interplaying factors. The main source of energy input to Earth is sunlight, which warms the land surface and in turn radiates heat back into the atmosphere. Globally, how much surface heating occurs is related to the Sun's energy output and how much of the sunlight actually reaches Earth's surface. Incoming sunlight may be absorbed or blocked in the atmosphere by gases, cloud cover, or dust and sulfuric acid droplets. Heat (infrared rays) radiating outward from Earth's surface can be trapped by certain atmospheric gases, a phenomenon known as the **greenhouse effect**.

On a sunny day, it is much warmer inside a greenhouse than outside it. Light enters through the glass and is absorbed by the ground, plants, and pots inside. They, in turn, radiate heat: infrared radiation, not visible light. Infrared rays, with longer wavelengths than visible light, cannot readily escape through the glass panes; the rays are trapped, and the air inside the greenhouse warms up. You might experience the same effect in a car on a bright, cold day.

In the atmosphere, molecules of various gases, especially water vapor and carbon dioxide, act similarly to the greenhouse's glass. Light reaches Earth's surface, warming it, and the surface radiates infrared rays back. But the longer-wavelength infrared rays are trapped by these gas molecules, and a portion of the radiated heat is trapped in the atmosphere, creating the greenhouse effect (see **figure 10.1**). As a result of the greenhouse effect, the atmosphere stays warmer than it would if that heat radiated freely back out into space. In moderation, the greenhouse effect makes life as we know it possible. Without it, average global temperature would be closer to −17°C (about 1°F) than the roughly 15°C (59°F) it now is. However, one can have too much of a good thing.

The evolution of a technological society has meant rapidly increasing energy consumption. Historically, we have relied most heavily on carbon-rich fuels such as wood, coal, oil, and natural gas to supply that energy. These will continue to be important energy sources for several decades at least. One combustion by-product that all these fuels have in common is carbon dioxide gas (CO_2).

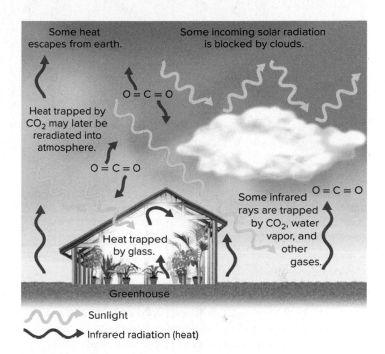

Sunlight

Infrared radiation (heat)

Figure 10.1

The greenhouse effect. Both glass and air are transparent to visible light. Like greenhouse glass, CO_2 and other greenhouse gases in the atmosphere trap infrared rays radiating back from the sun-warmed surface of Earth.

A greenhouse gas is one that traps infrared rays and thus promotes atmospheric warming. Water vapor is the most abundant greenhouse gas in Earth's atmosphere, but human activities don't substantially affect its abundance, and it is in equilibrium with surface water and oceans. Excess water in the atmosphere readily falls out as rain or snow. Some of the excess carbon dioxide is removed by geologic processes (see chapter 18), but since the start of the Industrial Age in the mid-nineteenth century, the amount of carbon dioxide in the air has increased by 50%, and its concentration continues to climb (**figure 10.2**).

So far, the temperature rise has been much more moderate than expected if carbon dioxide were the only substance or process involved; Earth's climate system is complex. Measuring a global temperature increase directly, and separating it from the background noise of local weather variations, has historically been difficult. Locally, at any one time, a wide variety of climates can exist simultaneously—icy glaciers, hot deserts, steamy rain forests, temperate regions—and a given region may be subject to wide seasonal variations in temperature and rainfall. One of the challenges in determining global temperature trends is deciding

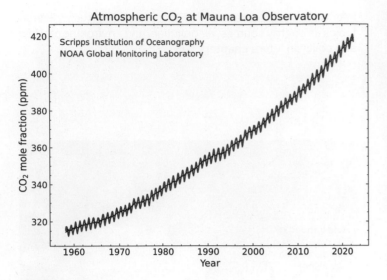

Figure 10.2

Rise in atmospheric CO_2 over the past sixty years is clear. Zigzag pattern reflects seasonal variations in local uptake by plants. Preindustrial levels in ice cores were about 280 parts per million.

Source: Data from https://gml.noaa.gov/ccgg/trends/

just how to measure global temperature at any given time. Commonly, scientists use several different kinds of data including air temperatures at various altitudes over land and sea, and sea surface temperatures (SSTs). Satellites help greatly by making it possible to survey large areas quickly; see Case Study 10.1.

Altogether, since the start of the Industrial Age, global surface temperature has risen about 1.1°C (2.0°F) (**figure 10.3**). It is already having obvious and profound impacts in many parts of the world. The warming is not uniform everywhere. The Arctic, in particular, is experiencing rapid changes. And the probability that warming will continue is quite high, so the primary

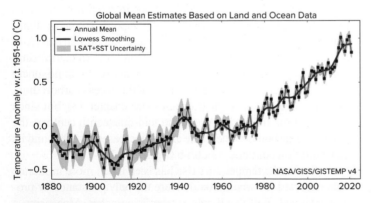

Figure 10.3

Global temperature rise emerges from background noise; gray shading indicates uncertainties, which have decreased with time as completeness of data has improved. Visit https://data.giss.nasa.gov/gistemp/graphs_v4/ for the most up-to-date data.

Image by NASA Goddard Institute for Space Studies.

concerns include the rate and degree of warming, and the influence and response of humans.

10.2 Climate and Ice Revisited

Learning Objectives

- Describe the effect of melting ice sheets on sea level
- Recall ways in which melting Arctic sea ice affects marine ecosystems
- List the effects of melting permafrost

High latitudes are experiencing rising temperatures, and their consequences, at much higher rates than elsewhere on Earth. The melting of continental-based ice sheets and glaciers is resulting in global sea level rise. More regionally, the melting of sea ice in the Arctic is disrupting ecosystems and allowing more insolation to warm Earth's surface. Warm temperatures in the Arctic are also melting permafrost, releasing a very effective greenhouse gas. In the following, we will summarize the current state of melting ice and permafrost, and explore some possible future scenarios for sea level rise.

Melting Ice, Rising Seas

Early discussions of climate change related to increasing greenhouse-gas concentrations in the atmosphere tended to focus heavily on the prospect of global warming and the resultant melting of Earth's reserves of ice, especially the remaining ice sheets. If *all* the ice melted, sea level could rise by 60–70 meters (190–230 feet) from the added water alone, leaving aside thermal expansion of that water. About 20% of the world's land area would be submerged. Many millions, perhaps billions, of people now living in coastal or low-lying areas would be displaced.

Such large-scale melting of ice sheets would take time, perhaps several thousand years. On a shorter timescale, the problem would still be significant. Both the Greenland and Antarctic ice sheets are losing mass at a collective rate of over 420 gigatons a year, resulting in a 1.2-mm annual rise in sea level since 2002 (**figure 10.4**). Complete melting of the West Antarctic ice sheet, which is more vulnerable because it is grounded below sea level, could occur within a few hundred years. The resulting 5- to 6-meter rise in sea level would be enough to flood many coastal cities and ports, along with most of the world's beaches; see, for example, **figure 10.5**. Even though displaced inhabitants would have decades to adjust to the changes, this situation would also create large numbers of environmental refugees. The World Bank estimates that there could be 200 million people displaced in the next thirty years due to a variety of environmental issues including sea level rise.

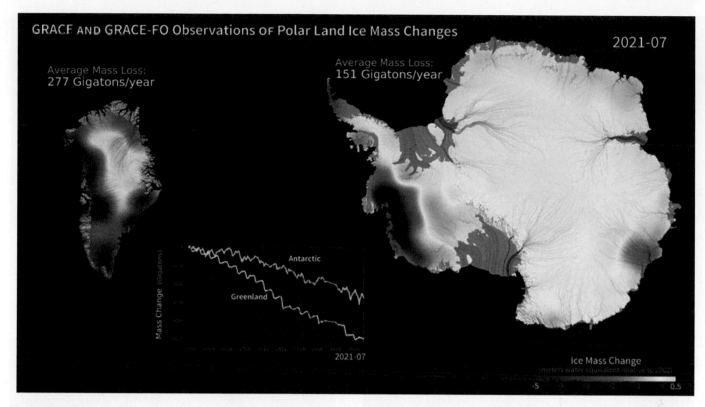

GRACE AND GRACE-FO Observations of Polar Land Ice Mass Changes

2021-07

Average Mass Loss:
277 Gigatons/year

Average Mass Loss:
151 Gigatons/year

Antarctic

Greenland

Mass Change (Gigatons)

2021-07

Ice Mass Change
(meters water equivalent relative to 2002)

-5   0.5

Figure 10.4

The Greenland ice sheet is losing mass at a higher rate than the Antarctic ice sheets, which include the smaller West and much larger East parts. The West Antarctic ice sheet is behaving much differently than the East ice sheet.

NASA and JPL/Caltech

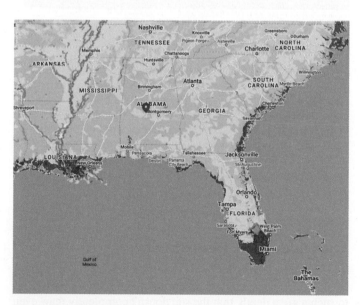

Figure 10.5

Impact on the southeastern United States if melting of the West Antarctic ice sheet resulted in 5 meters of sea level rise. Every location in red is below the tide line. Go to https://coastal.climatecentral.org/ to use the interactive map.

Climate Central

Meanwhile, the smaller alpine glaciers are clearly dwindling, as we discussed in the chapter. As sea level rises, too, the base levels of streams draining into the ocean are also raised, which alters the stream channels and could cause significant flooding along major rivers; and this would not require nearly such a large sea-level rise.

Arctic sea ice is also demonstrably shrinking, in both area (**figure 10.6**) and thickness. It is vulnerable because it floats on the circulating, warming ocean rather than long-chilled rock. The good news is that melting sea ice will not cause a rise of sea level, just as melting ice in a glass does not cause your beverage to overflow. Arctic ecosystems are being affected by less seasonal ice. For species such as the polar bear that rely on seasonal ice for hunting, less ice is making it more difficult for them to access their traditional food sources. For other species, like the orca, more open water means more opportunities for hunting success. And if less sea ice means more light for photosynthetic microorganisms, the base of some food chains may increase in productivity, but that also depends on the temperature of the water and the supply of nutrients. The Arctic is warming at a rate twice as fast as the rest of the planet; why? The lose of ice means less reflectivity and more absorption of

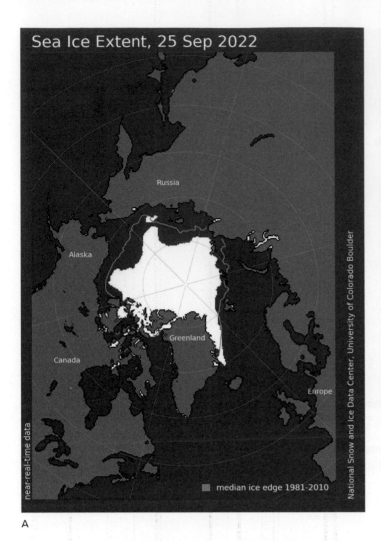

Sea Ice Extent, 25 Sep 2022

Russia

Alaska

Greenland

Canada

Europe

near-real-time data

National Snow and Ice Data Center, University of Colorado Boulder

■ median ice edge 1981-2010

A

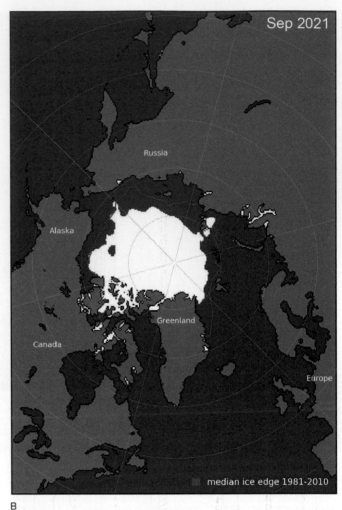

Sep 2021

Russia

Alaska

Greenland

Canada

Europe

■ median ice edge 1981-2010

B

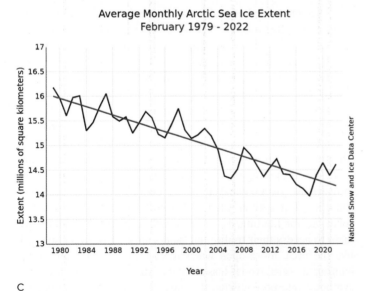

Average Monthly Arctic Sea Ice Extent
February 1979 - 2022

Extent (millions of square kilometers)

Year

National Snow and Ice Data Center

C

Figure 10.6

Average maximum Arctic Sea ice extent (A) has declined only moderately (here, September 2022, as compared to the median for 1981–2010, indicated by the yellow line), but the annual minimum (B) has declined sharply (here, September 2021, compared to the 1981–2010 median, indicated by the pink line). Thus, on average, annual ice extent (area) has been declining (C). So has the ice thickness, meaning a net loss of large volumes of ice in the Arctic.

(A, B, C) National Snow and Ice Data Center

The Hidden Ice: Permafrost

Warming is affecting more than the visible ice of cold regions. In alpine climates, winters are so cold, and the ground freezes to such a great depth, that the soil doesn't completely thaw even in the summer. The permanently frozen zone is **permafrost** (**figure 10.7**). The meltwater from the thawed layer cannot infiltrate the frozen ground below, so the terrain is often marshy (**figure 10.8**). Structures built on or in permafrost may be very stable while it stays frozen but sink into the muck when it thaws; vehicles become mired in the sodden ground, so travel may be impossible when the upper soil layers thaw.

energy (from darker colored water), resulting in even warmer temperatures, and more ice melting. This is the positive climate feedback we introduced in chapter 9. We will review more feedback examples at the end of this chapter.

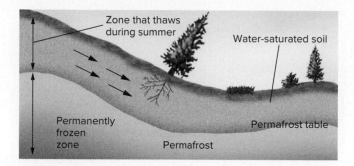

Figure 10.7

In cold climates, permafrost can range from 1 meter to over 1000 meters thick.

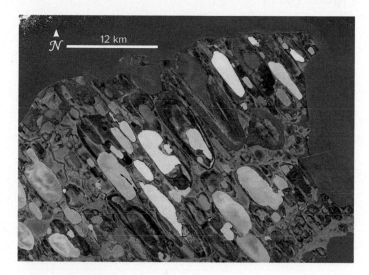

Figure 10.8

A portion of the north coast of the National Petroleum Reserve in Alaska. Abundant lakes reflect soggy soil over ice. Note elongated topographic features, a result of ice-sheet flow during the last ice age.

NASA/GSFC/METI/ERSDAC/JAROS, and the U.S./JapanASTER Science Team

Permafrost is being lost as the world warms. Where some permafrost remains, the thaw penetrates deeper, and the frozen-ground season is shorter. Some areas are losing their permafrost altogether, and lakes drain abruptly when the last of the ice below melts away. Thawing permafrost exposes a feast for microorganisms that release methane and carbon dioxide. Researchers have found sinkholes in the Arctic landscape that they hypothesize formed when methane reserves built up and exploded. Along the north coast of Alaska, a combination of melting permafrost and loss of protection from waves as sea ice dwindles has led to the collapse of coastal cliffs, and shoreline retreat of tens of meters per year (**figure 10.9**). All these changes are having effects on ecosystems, animal migrations, and the traditional lifestyles of some native peoples.

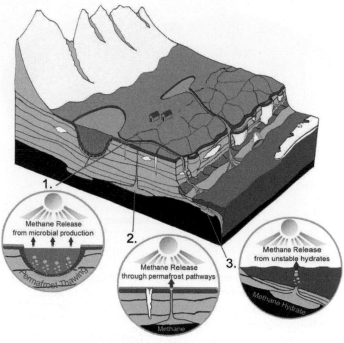

A

B

Figure 10.9

(A) There are multiple ways that methane can escape into the atmosphere from rapidly warming Arctic environments.
(B) Coastal erosion rates along Alaska's north coast average 1.4 meters/year, with some areas as high as 45 feet/year.

(A) Source: USGS; (B) U.S. Geological Survey Department of the Interior/USGSU.S. Geological Survey/Photograph by Benjamin Jones

10.3 Oceans and Climate

Learning Objectives

- Recall that oceans store and transport heat and moderate climate
- Explain thermohaline circulation
- Describe the conditions that result in an El Niño event
- Recall that there are many ocean oscillations

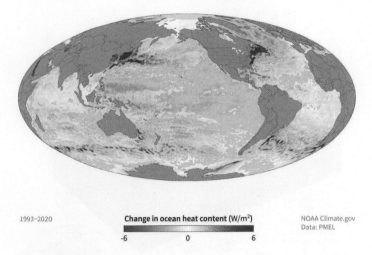

1993–2020

Change in ocean heat content (W/m²)

-6   0   6

NOAA Climate.gov
Data: PMEL

Figure 10.10

Modeled increase in the heat content of the upper 700 meters of the oceans, based on observations from 1993 to 2020. Thegray shading indicate heat changes that were not statistically significant.

National Oceanic and Atmospheric Administration

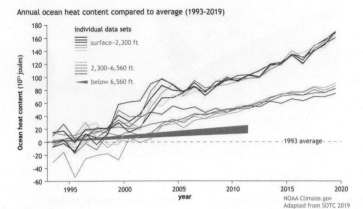

Annual ocean heat content compared to average (1993-2019)

NOAA Climate.gov
Adapted from SOTC 2019

Figure 10.11

Ocean heat content is rising faster in shallow ocean water, from 0 to 700 meters (2,300 feet). It is also rising in deeper water from 700–2,000 meters (6,650 feet) and below 6,650 feet (2,000 meters), but not as quickly.

National Oceanic and Atmospheric Administration

One reason that there is no simple correlation between atmospheric greenhouse-gas concentrations and land surface temperatures is the oceans. That huge volume of water represents a much larger thermal reservoir than the atmosphere. Coastal regions may have relatively mild climates for this reason. As air temperatures drop in winter, the ocean releases some heat to the atmosphere; as air warms in summer, the ocean water absorbs some of the heat. The oceans store and transport a tremendous amount of heat around the globe. A thorough discussion of the role of the oceans in global and local climate is beyond the scope of this chapter, but it is important to appreciate the magnitude of that role. Below we will discuss the essential climate role of the ocean in storing and transporting heat around the globe, and use the El Niño phenomenon as an example of ocean-atmospheric oscillations that have an affect on weather and climate.

Ocean Heat

The most recent studies of the balance between the energy received from the Sun and the energy radiated back into space have confirmed an excess of energy absorbed, most of which is passed into the oceans at an average rate of 0.58–0.78 watts per square meter per year (**figure 10.10**). This is a tremendous amount of heat given that the ocean covers more than 360 million square kilometers. Much of this absorbed energy is being stored in the shallow ocean. Water deeper than 700 meters is also warming, although not as quickly as shallower water (**figure 10.11**). The ability of the ocean to store large amounts of heat may account for the apparent hiatus in the rise of global surface temperatures early in the twentieth century, which is now evidently past (recall **figure 10.3**).

We have already discussed that the oceans supply most of the water vapor to make rain and snow. Water absorbs heat during vaporization, so warmer ocean water will provide not only more energy but also increase the amount of water evaporated. Heat from sea surface water supplies the energy to create tropical storms and hurricanes. But more energy doesn't necessarily mean more storms; more energy might result in more intense storms or larger storms that bring more rain. Modeling extreme weather events is difficult given that small-scale processes are involved in their formation and life, so determining the precise effect of warmer water is also difficult. Scientists have only fairly recently collected the types of data necessary in order to see trends in tropical storm intensity. For now, it appears that tropical storms may be either becoming more intense or more numerous in response to warming temperatures. Researchers will continue to sort through the large amount of data collected through remote sensing in the hopes of better understanding the future risks of hurricanes.

Thermohaline Circulation

Oceanic circulation is driven by a combination of winds, which push surface currents, and differences in density, related to temperature and salinity, that drive deeper currents. Cold water is denser than warm, and at a given temperature, density increases with increasing salinity. The roles of temperature and salinity are reflected in the name **thermohaline circulation** given to the large-scale circulation of the oceans (**figure 10.12**). One of the most familiar currents involved is the Gulf Stream, which carries warmed equatorial water to the North Atlantic and moderates climate in places like the British Isles. Thermohaline circulation is initiated in polar regions by the salty, dense water that forms at the surface when water gets cold enough to form ice. The salty water sinks, and more water is pulled in to replace it, setting up the *ocean conveyor* circulation system that transports heat around the globe.

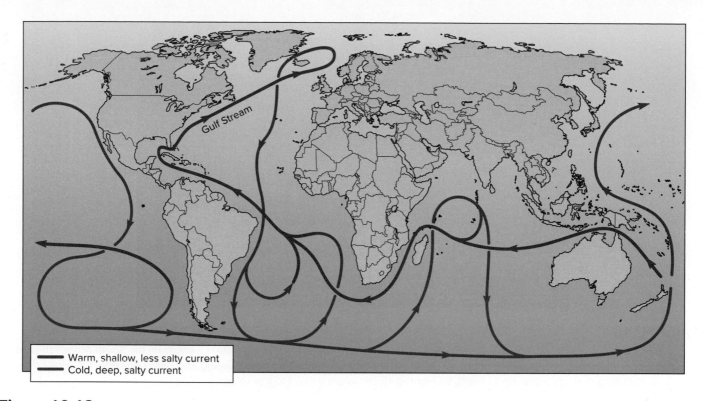

Figure 10.12

The thermohaline circulation, moving water and heat around the globe.

Legend:
— Warm, shallow, less salty current
— Cold, deep, salty current

Map label: Gulf Stream

Scientists have found evidence of abrupt climate change occurring in Earth's past. One proposed mechanism for sudden cooling of the North Atlantic is disruption of the thermohaline circulation in this region, reducing the influx of warm water from lower latitudes. More specifically, extensive melting of the Greenland ice sheet could create a pool of cool, fresh surface water that would not be dense enough to sink, so no warmer surface water would be drawn in. Even extensive melting of alpine glaciers could create a large influx of freshwater into the ocean with the same results. Researchers think that the Younger Dryas, a short period of abrupt cooling in the Northern Hemisphere that started about 13,000 years ago, was caused by such processes, and that signs are strong that circulation in the Northern Atlantic is destabilizing again. Such a change in the thermohaline circulation would have wide-ranging effects especially in locations that have tropical monsoon seasons. Another possible cause of abrupt climate change is increased water evaporation around a warmer equator; this could increase the salinity of the surface water so much that it would sink rather than flowing northward. Much more research is needed to understand the consequences of a destabilized ocean conveyor belt that might change significantly as quickly as in a few decades or perhaps within the next century.

El Niño

Most of the vigorous circulation of the oceans is confined to the near-surface waters. Only the shallowest waters, within 100 to 200 meters of the surface, are well mixed by waves, currents, and winds, and warmed and lighted by the Sun. The average temperature of this layer is about 15°C (60°F).

Below the surface layer, temperatures decrease rapidly to about 5°C (40°F) at 500 to 1000 meters below the surface. Below this is the so-called *deep layer* of cold, slow-moving, rather isolated water. The temperature of this bottommost water is close to freezing and may even be slightly below freezing due to high salinity and pressure. This cold, deep layer originates largely in the polar regions and flows very slowly toward the equator.

When winds blow offshore, they push the warm surface waters away from the coastline. This movement creates a region of low pressure that results in *upwelling* of deep waters to replace the displaced surface waters (**figure 10.13**). The deeper waters are enriched in dissolved nutrients, in part because few organisms live in the cold, dark depths to consume those nutrients. When the nutrient-laden waters rise into the warm, sunlit zone near the surface, they support abundant plant life and, in turn, animal life that feeds on the plants. Many rich fishing grounds are located in zones of coastal upwelling. The west coasts of North and South Americas and of Africa are subject to especially frequent upwelling events.

From time to time, for reasons not precisely known, the winds that usually blow from east to west in the tropical Pacific weaken, allowing warm surface waters from the western South Pacific to extend eastward toward South America (**figure 10.14A**). The warm water suppresses upwelling of fertile cold

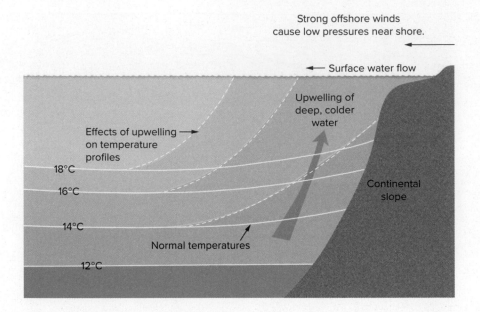

Figure 10.13

Warm surface water normally overlies colder seawater. When offshore winds blow warm waters away from the South American shore and create local low pressure, upwelling of colder deep waters may occur (dashed lines). During an El Niño episode, the winds die down, and upwelling is suppressed (solid lines), so warm surface waters extend to the coast.

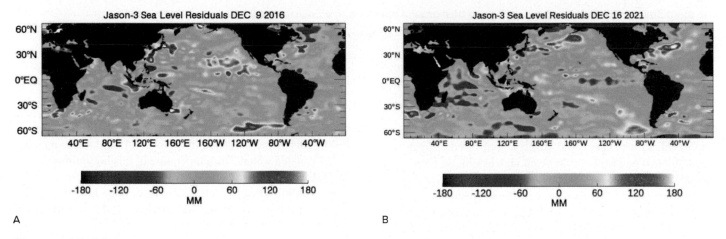

A   B

Figure 10.14

(A) Relatively warm water appears as higher sea level elevations in the eastern and central tropical Pacific during El Niño, the warm phase of El Niño–Southern Oscillation (ENSO). (B) Relatively cooler water appears as lower sea level elevations in the eastern and central tropical Pacific during La Niña, the cool phase of ENSO.

(A, B) National Aeronautics and Space Administration

water, having a catastrophic effect on the Peruvian anchovy industry. We call this event *El Niño* because it commonly occurs in winter, near the Christmas season.

El Niño events are cyclic in nature, occurring every two to seven years as part of the El Niño–Southern Oscillation (ENSO). The opposite of an El Niño, a situation with unusually cold surface waters off western South America, is *La Niña* (**figure 10.14B**). The extensive shifts in seawater surface temperatures that occur during ENSO events cause large-scale changes in evaporation,

precipitation, and wind-circulation patterns (**figure 10.15**). Climatic shifts generally include changes in frequency, intensity, and paths of Pacific storms, short-term droughts and floods in various regions of the world, and changes in the timing and intensity of the monsoon season in India. Heavy rains associated with El Niño events have been blamed for major landslides along the Pacific coast of the United States; La Niña produces particularly heavy monsoon rains that contribute to disastrous flooding in south Asia. There is already some evidence that El Niño and La Niña events

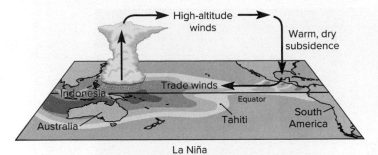

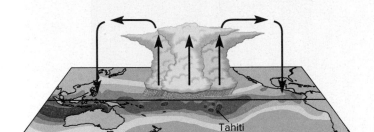

Figure 10.15

Under La Niña conditions, warm, wet air over the western Pacific dumps monsoon rains on southeast Asia as it rises and cools; when the air sinks and warms again, over the eastern Pacific, it is drying. During El Niño, surface waters in the eastern Pacific are warmer (red colors), which shifts the patterns of evaporation and precipitation; the monsoon season is much drier, while more rain falls on the western United States.

are becoming more frequent, intense, and/or prolonged, and this can have especially devastating consequences to the nations that depend on rainy seasons. Based upon data collected by buoys in the equatorial Pacific, ENSO events are predictable up to a year in advance, helping people prepare for the likely associated conditions that might cause hardships.

ENSO is not the only naturally occurring oceanic-atmospheric oscillation. At least 10 others are currently recognized including the Atlantic Multidecadal Oscillation, the Indian Ocean Dipole, the Antarctic Oscillation, and the Pacific Decadal Oscillation (PDO). The PDO has warm and cold phases that last twenty to thirty years, and that do affect the frequency of ENSO events (**figure 10.16**). Our current understanding of ocean oscillations is incomplete. Continued research is necessary to determine the role and frequency of natural changes in the atmosphere and oceans so that we can better model how our climate will change in the future.

10.4 Other Aspects of Climate Change

Learning Objectives

- Identify the benefits and downsides of precipitation redistribution due to climate change
- Exemplify extreme weather event related to a warming atmosphere
- Describe the relationship of plants and phytoplankton to changing atmospheric CO_2 levels
- Identify the effects of ocean acidification
- Describe the link between climate change and the spread of disease

As we have already explored in this chapter, *global warming* is an inaccurate phrase to describe the variety of changes in weather and climate associated with a warming climate. Some places may cool, and there are repercussions for oceanic circulation, patterns of storms, the volume and distribution of glacial

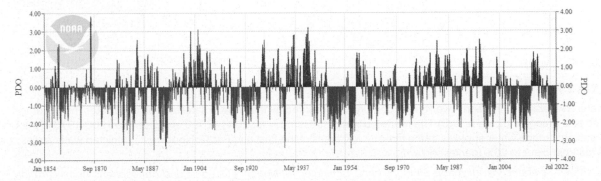

Figure 10.16

The Pacific Decadal Oscillation (PDO) has a positive value when SSTs are warm along the Pacific coast and cool in the interior North Pacific. The PDO has a negative value when the pattern is reversed: cool along the Pacific coast, warm in the interior. To access these data directly, go to https://www.ncdc.noaa.gov/teleconnections/pdo/.

National Oceanic and Atmospheric Administration

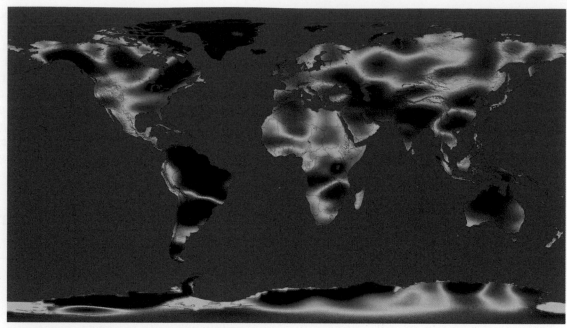

A

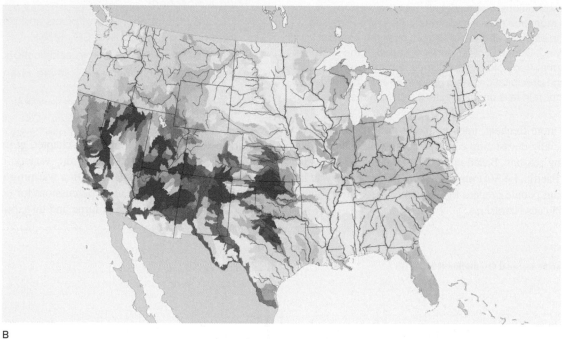

B

Figure 10.17

(A) Freshwater availability in the early twenty-first century as measured by the GRACE satellites; blue indicates more storage than average, red and orange indicate less (B) The modeled water stress index for 2041–2060 compared to 1900-1970 in the United States shows increasing surface water supplies in the already wet areas of the East and Pacific Northwest. Predicted water stress change ranges from -20% (darkest blue) to 20% (darkest brown).

(A) NASA/ NASA's Scientific Visualization Studio/Trent L. Schindler; (B) National Oceanic and Atmospheric Administration

ice, and more, as we will review below. And climate change does not affect all parts of the world equally or in the same ways.

Changes in wind-flow patterns and amounts and distribution of precipitation will cause differential impacts in areas not equally resilient as indicated by recent trends in water availability shown in **figure 10.17A**. A region of temperate climate and ample rainfall may not be seriously harmed by a temperature change of a few degrees, one way or the other, or several inches

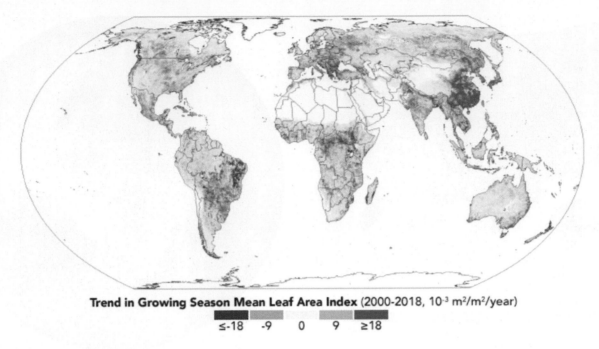

Trend in Growing Season Mean Leaf Area Index (2000-2018, 10^{-3} m^2/m^2/year)

≤-18   -9   0   9   ≥18

Figure 10.18

The greenness of places such as China and India has increased with more agriculture and efforts of reforestation, while other locations are greening because of higher temperatures and CO_2 concentrations.

NASA Earth Observatory images by Joshua Stevens

more or less rainfall. A modest temperature rise in colder areas may actually increase plant vigor and the length of the growing season, and some dry areas may enjoy increased productivity with increased rainfall. For example, the current greening trend in Arctic and boreal regions is attributed to both higher temperatures and CO_2 levels, showing as much as a 30% increase in greenness associated with a 1.8°C increase in summer temperatures from 1986 to 2015 (**figure 10.18**). These high latitudes areas are also expected to experience the greatest change in greenness through 2100.

Warming and moisture redistribution also have potential downsides. In chapter 9, we noted the impact of the loss of alpine glaciers on water supplies. **Figure 10.17** suggests potential water shortages in areas not relying now on glacial meltwater. In many parts of the world, agriculture is already only marginally possible because of hot climate and little rain. A temperature rise of only a few degrees, or loss of a few inches of rain a year, could easily make living and farming in these areas impossible. Some projections suggest that summer soil-moisture levels in the Northern Hemisphere could drop by up to 40% with a doubling in atmospheric CO_2. This is a sobering prospect for farmers in areas where rain is barely adequate now, for in many places, irrigation is not an option. In the United States, both temperature and precipitation have been increasing, but the effects are unevenly distributed, with potentially problematic results (**figure 10.19**).

Recent evidence and computer modeling indicate that global warming and associated climate changes is already producing more extreme events: catastrophic flooding from torrential storms, devastating droughts, killer heat waves, and stronger hurricanes, tornadoes, and thunderstorms. We don't have to go back farther than 2021 for examples of all these hazards in just the United States. Hurricane Ida was the most expensive of these events at $75 billion. The storm intensified quickly to a category 4 when it hit Louisiana on August 29 bringing along dangerous winds, flooding, tornadoes, and a storm surge. The cold wave and winter storm that hit the northwestern, central, and eastern parts of the United States in early February was the costliest on record (**figure 10.20**). The extension of abnormally low temperatures into southern states was the result of a destabilized jet stream that allowed polar continental air to migrate south, itself caused by the rapid warming of the Arctic. In Texas, the interconnected energy system had not been prepared for the understandably large demand on heating systems and it failed, leaving at one point over a million people stranded in their homes in frigid conditions; some didn't have power for days. A drought occurred throughout 2021 in western states, with a heat wave that created temperatures as high as 116°F in Portland, Oregon. This drought led to wildfires in June through December causing damages over $10 billion; the Dixie Fire was California's second largest ever, burning 960,000 acres and destroying over 1,000 structures. December also saw an outbreak of severe weather in the Midwest which created the first December *derecho* on record and at least fifty tornadoes including Minnesota's first ever in that month. Again, power outages lasted

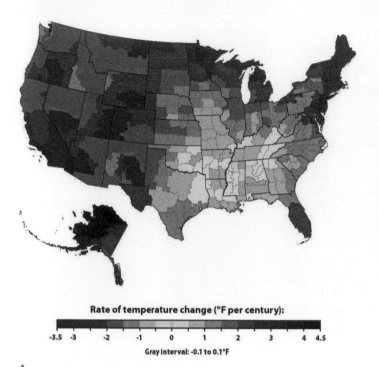

Rate of temperature change (°F per century):

-3.5 -3 -2 -1 0 1 2 3 4 4.5

Gray interval: -0.1 to 0.1°F

A

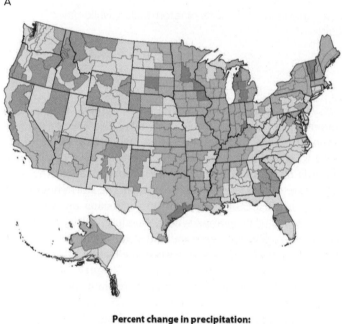

Percent change in precipitation:

-30 -20 -10 -2 2 10 20 30

B

Figure 10.19

Rate of change from 1901 to 2020 in temperature (A) and precipitation (B).

(A, B) National Oceanic and Atmospheric Administration

for days with temperatures below freezing. These events are just five of the twenty listed by NOAA as weather and climate disasters occurring in the United States in 2021 that cost over a billion dollars a piece. The trends seem fairly clear: a warmer

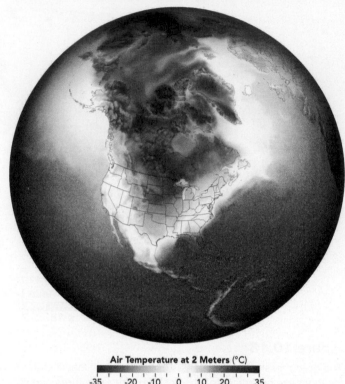

Air Temperature at 2 Meters (°C)

-35 -20 -10 0 10 20 35

Figure 10.20

Arctic air spills into the South on February 15, 2021. Notice the shape of the air mass, and that locations at much higher latitudes such as Maine and Alaska are 10°C warmer than Texas.

National Aeronautics and Space Administration

climate is leading to more intense weather events that are sometimes occurring outside of their expected schedule. This shouldn't be surprising given what we know about relationship among heat, storms, and precipitation.

As discussed above, it appears that rising atmospheric CO_2 levels are beneficial in promoting plant growth. After all, photosynthesis involves using CO_2 and water and solar energy to manufacture more complex compounds and build the plant's structure. However, controlled experiments have shown that not all types of plants grow significantly more vigorously given higher CO_2 concentrations in their air, and even those that do may show enhanced growth only for a limited period of time. Even if a land plant does benefit from more CO_2, if the area it is growing in experiences a distinct change in temperature and/or provided water, the deficit or excess of either could surpass the boost from extra CO_2.

Determining the effect of increasing temperatures and carbon dioxide in the oceans on phytoplankton productivity is a bit more complicated. Data collected in the twenty-first century from buoys, ships, and satellites consistently shows the SSTs are warming just about everywhere (**figure 10.21**). Many types of phytoplankton increase productivity in warmer waters. But recent studies indicate that not all types of

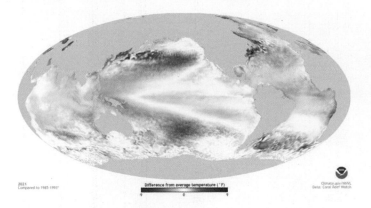

Figure 10.21

Global, yearly sea surface temperature difference in 2020 compared to average from 1981 to 2010.

National Oceanic and Atmospheric Administration

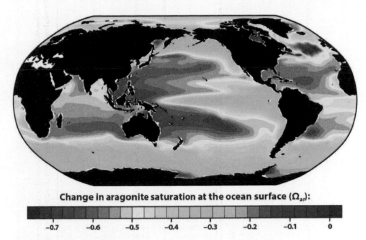

Figure 10.22

Ocean acidity can be estimated using aragonite saturation levels. Aragonite is a type of calcium carbonate that many marine organisms use to build their shells. Shells are difficult to build when levels are lower. A negative change represents a decrease in saturation. This map shows the change in aragonite saturation from the 1880s to levels between 2006 and 2015.

U.S. Environmental Protection Agency

phytoplankton will respond the same way to predicted higher SSTs, and that temperatures near the equator will likely get too high, resulting in less phytoplankton biomass. Phytoplankton productivity is also dependent on the correct levels of light and nutrients. Phytoplankton depend on the upwelling of nutrient-rich cold waters, which may be suppressed by warm-water layers. The effect is like that of El Niño, but much more widely distributed. Currently, levels of productivity appear to be increasing in the Arctic Ocean as water temperatures and CO_2 levels increase and sea ice coverage decreases. Because phytoplankton form the base of marine ecosystems, it is reasonable to assume those ecosystems will change as the amount of biomass either increases or decreases, having a cascading effect on larger marine animals and on humans who depend upon ocean fisheries.

The ocean acts as a sink for atmospheric CO_2. With more CO_2 in the air, more dissolves in the oceans, making more carbonic acid (**figure 10.22**). This acidification makes it more difficult for corals and other marine organisms to build their calcium-carbonate skeletons. The combination of acidification and warming waters puts severe stress on coral reefs, for the warming causes another problem. Corals owe their colors to symbiotic algae; when water warms just a few degrees, the algae are expelled, and bare white coral skeletons result, a phenomenon known as *coral bleaching*. Without the algae, the corals also suffer from the loss of nutrients the algae provided. If the water cools soon enough, the algae return and the reef can recover; if the elevated water temperatures persist too long, the coral dies. The combination of rising ocean temperatures overall and more-intense El Niño events is causing more coral bleaching and reef death worldwide. And coral reefs are not merely beautiful; they support thousands of species in marine ecosystems.

Public-health professionals have identified yet another threat related to global change. Climate influences the occurrence and distribution of a number of diseases.

Mosquito-borne diseases, such as dengue, are influenced by conditions that broaden or reduce the range of their particular carrier species; the incidence has increased in recent years and is expected to expand further (**figure 10.23**). Similar concerns have been raised with respect to hanta virus and plague, both associated with rodents. The expansion may be vertical, too. Researchers have found malaria occurring at higher altitudes in Ethiopia and Colombia as warming comes to the population of the highlands. Waterborne parasites and bacteria can thrive in areas of increased rainfall. Efforts to anticipate and plan responses to the climate-enhanced spread of disease are growing. Still other insect pests that die in winter in temperate climates—termites, for instance—may also expand their ranges in a warmer world.

U.N. studies further suggest that less economically developed nations, with fewer resources, will be much less able to cope with the stresses of climate change. Yet many of these nations are suffering the most immediate impacts. The very existence of some low-lying Pacific island nations is threatened by sea-level rise. The Maldives are both exploring options to purchase land in other countries while at the same time enhancing their resilience to sea-level rise that could inundate the country as early as 2050. Reduced soil moisture and higher temperature quickly affect the world's drylands where many poorer countries are located. The tropical diseases are spreading most obviously in economically disadvantaged countries where medical care is limited. These differential impacts may have implications for global political stability.

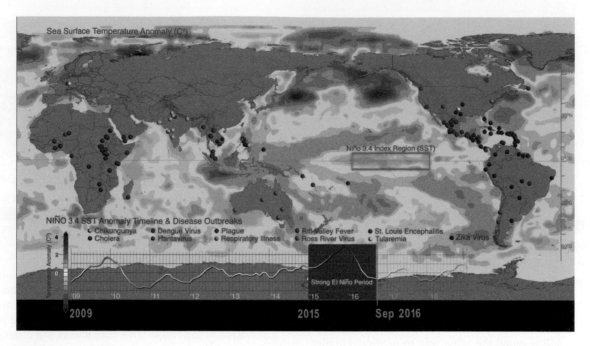

Figure 10.23

The 2015–2016 La Nina event demonstrated a strong relationship with various disease outbreaks across the world. For more information and a short animation demonstrating trends from 2009 to 2018, visit https://svs.gsfc.nasa.gov/4785.

NASA's Scientific Visualization Studio

10.5 Evidence of Climates Past

Learning Objectives

- Summarize direct evidence of climate change
- List climate proxies
- Explain how we use isotopes to study past climate change
- Explain how we use fossils to study past climate change

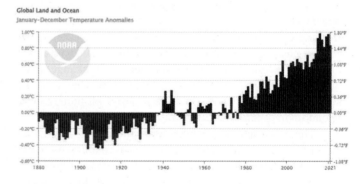

Figure 10.24

2021 was the sixth warmest year on record at 1.51°F (0.83°C) above the century's average. The Northern Hemisphere's land temperatures were the third highest recorded since 1880 at 2.77°F (1.54°C) above average.

National Oceanic and Atmospheric Administration

Our direct evidence of Earth's climate only goes back about 140 years when people starting using modern equipment to measure the physical parameters of the atmosphere and oceans (**figure 10.24**). Yet scientists discuss Earth's climate much earlier than the late 1800s. Where do they obtain this information regarding ancient climates? Data that we use to give us indirect measurements of parameters such as sea level and temperature are contained in *proxies* allowing researchers to reconstruct past climates. Proxies generally fall into three groups that sometimes overlap: evidence of past life, chemical evidence, or cyclical natural processes. Proxies vary in how far back they can take us, and we will review some of them in this section.

Humans have recorded details about local past climates in written records, stories, art, and oral histories. A painting from the 1500s may contain information about what people were wearing, how extensively a river froze in winter, and types of flora and fauna in a particular location. Ship logs from the 1800s contain information about weather, ice extent, and surface currents. *Phenology* documents that record natural seasonal variations go back hundreds or thousands of years and commonly include the date of the spring flowering of certainly plant species. Stories passed through generations may not contain as specific as information, but they can be useful in corroborating other evidence of climate change.

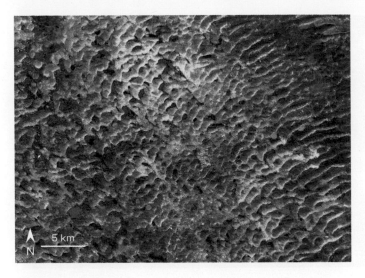

Figure 10.25

The Nebraska Sand Hills formed during the Ice Age: Sediment ground off the Rocky Mountains by the ice sheets became outwash in the plains. The dunes developed later in dry conditions. Now many are vegetated, and small lakes have accumulated in low spots.

NASA/GSFC/METI/ERSDAC/JAROS, and U.S./JapanASTER Science Team

Figure 10.26

Paleobotanists use the size, shape, and smoothness of plant leaves such as this one from the Laramie Formation to deduce ancient climate. Additional evidence comes from comparison with modern relatives.

©Gina S. Szablewski

We can deduce aspects of local climate change from the geologic record, especially from young, surficial deposits and much further back using the sedimentary rock record. The now-vegetated dunes of the Nebraska Sand Hills (**figure 10.25**) are remnants of an arid sand sea in the region 18,000 years ago, when conditions must have been much drier than they are now. We have already noted that much older glacially deposited rocks are now located in tropical regions, and this was part of the evidence for continental drift/plate tectonics. When such deposits were widespread globally, they indicate an ice age at the corresponding time. Distinctively warm- or cold-climate animals and plants identified in the fossil record provide evidence of the local environmental conditions at the time they lived (**Figure 10.26**). Any sign of life in the rock record provides climate information, and these proxies include evidence from pollen, tree rings, insect feeding patterns, and pack rat middens constructed from foraged vegetation. Buried and lithified ancient soils called *paleosols* are rich in climate information through their thickness, color, fossils, and organic content, and relationship with other sedimentary layers above and below. Even without fossils, sedimentary deposits and rocks are in and of themselves a record of the distinct environment in which they formed.

Marine sediments and the fossils of shelled sea creatures they contain provide evidence of ocean chemistry and temperatures. The proportion of calcium carbonate in some sediments can reflect water temperature because its solubility is strongly temperature-related. Isotopes of carbon and oxygen are used to reconstruct ocean temperatures as far back at 66 million years, although most ocean sediment samples reach back a few 100,000 years. Distinct chemicals such as iron can reflect times when chemical weathering on land was prevalent in warm, humid conditions. (Recall here that many ocean sediments have a land source.) The types, distribution, diversity, and abundance of microfossils are strong indicators of ocean conditions, and are a very common source of climate data gathered from marine sediment cores.

The oxygen isotopes just mentioned are a powerful tool in climate reconstruction. Scientists analyze the chemistry of not only ocean sediments and shells, but other proxies including annual lake deposits, seawater, corals, cave deposits, or snow and ice. This process involves variations in the proportions of oxygen isotopes. By far the most abundant oxygen isotope in nature is oxygen-16 (^{16}O); the heaviest is oxygen-18 (^{18}O). Because they differ in mass, so do molecules containing them, and the effect is especially pronounced when oxygen makes up most of the mass, as in H_2O. Certain natural processes, including evaporation and precipitation, produce fractionation between ^{16}O and ^{18}O, meaning that the relative abundances of the two isotopes will differ between two substances, such as ice and water, or water and water vapor. (See **figure 10.27**.) As water evaporates, the lighter $H_2^{16}O$ evaporates preferentially, and the water vapor will then be isotopically lighter (richer in ^{16}O, poorer in ^{18}O) than the residual water. Conversely, as rain or snow condenses and falls, the precipitation will be relatively enriched in the heavier $H_2^{18}O$. As water vapor evaporated from equatorial oceans drifts toward the poles, depositing $H_2^{18}O$-enriched precipitation, the remaining water vapor becomes progressively lighter isotopically, and so does subsequent precipitation; snow falling near the poles has a much lower $^{18}O/^{16}O$ ratio than tropical rain, or seawater (**figure 10.27B**).

A

Water vapor is relatively depleted in ^{18}O, enriched in ^{16}O.

Isotopically "lighter" $H_2{}^{16}O$ evaporates preferentially.

Condensing rain or snow is richer in ^{18}O than coexisting water vapor; extent of fractionation is temperature-dependent.

Remaining vapor is isotopically "lighter."

Next precipitation will have lower $^{18}O/^{16}O$ ratio because vapor was depleted in ^{18}O.

Near the poles, atmospheric vapor is increasingly depleted in ^{18}O.

Snow in the interior of Antarctica has 5 percent less ^{18}O than ocean water.

Heavy, ^{18}O-rich water condenses over mid-latitudes.

Meltwater from glacial ice is depleted in ^{18}O

Water, slightly depleted in ^{18}O, evaporates from warm subtropical waters.

B

Figure 10.27

(A) Mass differences between ^{18}O and ^{16}O lead to different $^{18}O/^{16}O$ ratios in coexisting water and water vapor; the size of the difference is a function of temperature. (B) As the water vapor evaporated near the equator moves poleward and rain and snow precipitate, the remaining water vapor becomes isotopically lighter.

(B) Image by Robert Simmon, courtesy NASA.

The fractionation between ^{18}O and ^{16}O in coexisting water vapor and rain or snow is temperature-dependent, more pronounced at lower temperatures. At a given latitude, variations in the $^{18}O/^{16}O$ ratio of precipitation reflect variations in temperature. Similarly, oxygen isotopes fractionate between water and minerals precipitated from it. This fractionation, too, is temperature-dependent. Variations in the $^{18}O/^{16}O$ ratio in the oxygen-rich carbonate ($CaCO_3$) or silica (SiO_2) skeletons of marine microorganisms give evidence of variations in the temperature of the near-surface seawater from which these skeletons precipitated, and they have been used to track past El Niño cycles.

The longevity of the massive continental glaciers makes them useful in preserving evidence of both air temperature and atmospheric composition extending hundreds of thousands of years. The oxygen-isotope composition of the ice reflects seawater O-isotope composition and the air temperature at the time the parent snow fell. Tiny bubbles of air are trapped as the snow is converted to dense glacial ice, preserving samples of the atmosphere. If those recreated temperatures can be correlated with the greenhouse gas content of the air bubbles or the presence of volcanic ash layers in the ice from prehistoric explosive

eruptions, we gain strong evidence about possible causes of past climatic variability (**figure 10.28**) that can be used to refine climate-prediction models.

The longest continuous ice-core records span over 800,000 years and come from Antarctica (**figure 10.29**). Researchers have more recently collected much older Antarctic ice samples (from blue ice) that they radiometrically dated at 2.7 million years old, and are hopeful that they will be able to obtain climate signatures between 1 and 2.7 million years and from even older samples. Records from the Greenland ice sheet go back about 160,000 years, and over that period show a pattern very similar to the Antarctic results, indicating that the cores do capture global climate trends. The data show substantial temperature variations, spanning a range of about 10°C (20°F), some very rapid warming trends, and obvious correlations between temperature and the concentrations of carbon dioxide (CO_2) and methane (CH_4). Correlation does not show a cause-and-effect relationship, however. The warming might be a result of sharp increases in greenhouse gases in the atmosphere, or the cause (see discussion of climate feedbacks in the next section). One fact is clear: the *highest* greenhouse-gas concentrations in the ice

Figure 10.28

This ice core from West Antarctica contains annual layers of ice (the dark and light bands) and a deposit of volcanic ash from an eruption about 21,000 years ago.

Heidi Roop, NSF

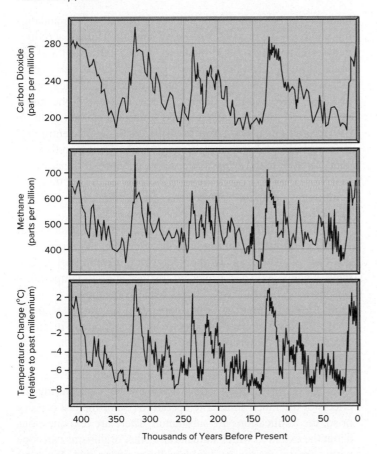

Figure 10.29

Ice-core data from Greenland and Antarctica reveal strong correlations between concentrations of the greenhouse gases carbon dioxide and methane, and temperature changes as indicated by oxygen isotopes.

Data from NOAA.

cores are about equal to the levels in our atmosphere just prior to the start of the Industrial Age. Since then, we have pushed CO_2 concentrations more than 140 parts per million (ppm) higher still, and the numbers continue to climb.

Proxies are best used to reconstruct climate when they come from many sources and many places. These correlations allow us to see that climate has varied quite a lot through Earth's history, at different times being somewhat cooler or substantially warmer than at present, and that the entire planet doesn't respond in the same way at the same time. The geologic record does indicate far higher atmospheric CO_2 levels during the much warmer periods of Earth's history, long before the time represented by the ice cores. But those occurred tens of millions of years or more before humans occupied the planet, with distinctive ecosystems that presumably adapted to those warmer conditions as global temperatures rose. Modern CO_2 levels are unprecedented in the history of modern humans.

10.6 Feedbacks, Uncertainty, and Geoengineering

Learning Objectives

- Exemplify positive and negative climate feedbacks
- Give examples that explain the uncertainties in climate modeling
- Explain why most geoengineering plans to address a warming climate should be addressed with caution

We have already introduced some of the many short-term responses of Earth systems to global warming associated with increasing greenhouse gases. Trying to project the longer-term trends is complicated because so many interrelated processes are involved in climate and that changes in part of the climate system can either reinforce a climate trend or weaken it, taking it in the other direction. We will provide a few examples here to illustrate the challenges faced by those trying to predict future climate change.

We already know that loss of ice and glacial cover is a consequence of rising temperatures. This decreases the average *albedo* of Earth's surface, which should mean more heating of the land by sunlight; and more infrared radiation back out into the atmosphere to be captured by greenhouse gases, and thus more heating, more ice melting, and so on. This is an example of positive feedback, a warming trend reinforced.

But wait; if air temperatures rise, evaporation of water from the oceans would tend to increase. Water vapor is a greenhouse gas, so that could further increase heating. However, more water in the air also means more clouds. More clouds, in turn, would reflect more sunlight back into space before it could reach and heat the surface, and that effect would be likely to be more significant (figure 10.30). This is an example of negative

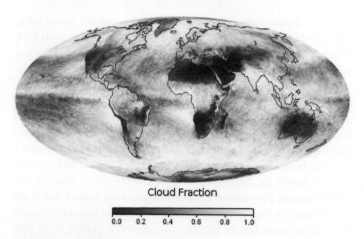

Figure 10.30

Cloud fraction in September 2021, showing what areas were cloudy on average for the month. Blue is no clouds and white is totally cloudy.

National Aeronautics and Space Administration

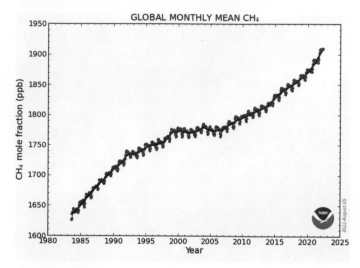

Figure 10.31

Globally averaged, monthly mean atmospheric methane abundance determined from marine surface sites

National Oceanic and Atmospheric Administration

feedback, in which warming induces changes that tend to cause cooling. But clouds aren't that simple. Many modelers identify cloud cover as the single largest source of uncertainty in projections of greenhouse-effect heating because their effects depend on many variables: day or night, thin or thick, high or low, icy or not-so-icy, lots of air particulates or cleaner air. The most recent models trend toward clouds created in a warmer atmosphere as enhancing warming, a positive feedback.

Now consider the oceans. Atmospheric carbon dioxide can dissolve in seawater, and more dissolves in colder water. So if global warming warms sea-surface water enough, it may release some dissolved CO_2 back into the atmosphere, where it can trap more heat to further warm the surface waters (positive feedback). Or will melting of polar and glacier ice put so much cool, fresh water on the sea surface that there will be more solution of atmospheric CO_2, removing it from the atmosphere and reducing the heating (negative feedback)? The recent observation that phytoplankton productivity generally declines in warmer surface waters indicates another positive-feedback scenario. Phytoplankton are plants, and photosynthesis consumes CO_2. Sharply reduced phytoplankton abundance, then, means less CO_2 removed from the atmosphere, and more left there to absorb heat.

Still more complexities exist. For instance, as permafrost regions thaw, large quantities of now-frozen organic matter in the soil begin to decompose, adding CO_2 and CH_4 to the atmosphere, another positive-feedback scenario. But the thawing of soil will allow other organisms to expand their range, so more plants would mean less CO_2 and a negative feedback to the system.

Beyond water vapor, CO_2, and CH_4, there are still more greenhouse gases produced by human activity, including nitrous oxide (N_2O), a product of internal-combustion engines, and the synthetic chlorofluorocarbons. Many of these are even more effective at greenhouse-effect heating of the atmosphere than is CO_2. So is methane (**figure 10.31**). Methane is produced naturally during some kinds of decay of organic matter and during the digestive processes of many animals. Human activities that increase atmospheric methane include extraction and use of fossil fuels, raising of domestic livestock (cattle, dairy cows, goats, camels, pigs, and others), and the growing of rice in rice paddies, where bacterial decay produces methane in the standing water. Methane is also released from the stagnant water in dam impoundments and from landfills. Methane concentrations have also risen but currently are far less abundant than CO_2; its concentration in the atmosphere is below 2 ppm. If the amounts of minor but potent greenhouse gases continue to rise, they will complicate climate predictions tied primarily to CO_2 projections.

In the near term, the dominant feedbacks are positive, for we see continued warming as CO_2 levels continue to rise. A 2014 assessment report by the Intergovernmental Panel on Climate Change (IPCC) projected further global temperature increases of 0.3 to 4.8°C (0.5–8.6°F) from already-elevated (2005) temperatures by 2100, depending on model assumptions; the lower projections require sharp reductions in global greenhouse-gas emissions. More recent IPCC reports have concentrated on the actions necessary to remain below a 1.5°C rise in temperature by 2100 compared to pre-industrial levels, and the consequences of exceeding that increase. Considering the effects of the much smaller temperature rise of the last century or so, the implications are substantial. There are real questions about the extent to which humans and other organisms can cope with climate change as geologically rapid as this. We have already noted some of the observed impacts.

Geoengineering

The strategy most widely discussed for minimizing human-induced global warming and climate change is reduction in greenhouse-gas emissions. This is proving to be a challenge, both practically and politically. *Geoengineering* is the idea of reducing warming by tinkering with Earth systems, either by removing carbon dioxide from the atmosphere or reducing the amount of sunlight striking Earth's surface. While some of these may seem in the realm of science fiction, we are in a situation in which all possible remediations need to be explored.

Carbon capture and storage (CCS) is a suite of technologies used to capture carbon dioxide from industrial plants or directly from the air. Captured CO_2 is either used to create products or services or is pumped into deep geologic formations both onland and offshore. CCS plants have been around since the 1970s and there are currently twenty-seven in operation worldwide. The interest in CCS in dealing with a warming climate expanded tremendously in 2021, with sixty-six more plants in advanced development and another ninty-seven plants either announced or in planning stages. Why so much interest now when the technologies have been around for decades? The financial and business aspects of CCS do play a part, but the largest factor is that CCS is necessary in order to reach the goal of net zero carbon emissions set for the near future.

Other geoengineering approaches are mostly still in the research phase. Methods to reduce incoming sunlight include installing giant reflectors into the space above Earth or injecting clouds of sulfate aerosols into the stratosphere. The latter would mimic the cooling effects of explosive volcanic eruptions, as discussed in chapter 5. However, both modeling and observations of the after-effects of natural eruptions show that the aerosols can cause redistribution of precipitation, producing floods in some areas and droughts in others, and might contribute to acid rain as well. Another suggestion is to spray seawater or other aerosols into the air over the oceans, promoting the formation of more bright clouds to reflect more sunlight, but again, there are questions with respect to effects on precipitation. Yet another proposal is to fertilize the oceans to stimulate phytoplankton growth, the resulting photosynthesis to consume more atmospheric CO_2.

Other than CCS, these geoengineering proposals share certain issues. One country cannot geoengineer its own climate in isolation; these strategies will have widespread, typically global, impacts, which may not be positive for all those affected. There are political questions of who will choose and carry out what is to be done, and who will fund these activities. And while we have ever-better models to examine the probable effects of various strategies, Earth's climate system is sufficiently complex that we may not correctly anticipate all the consequences of a given action. For the near future, then, we will likely focus our attention on other approaches: promoting the capture of carbon dioxide either by natural systems or technologies, energy sources that do not produce CO_2 (explored further in chapter 15), research aimed at better understanding of the global carbon cycle and ways to moderate atmospheric CO_2 (chapter 18), and regulations and international agreements to limit greenhouse-gas emissions (chapter 19).

Taking Earth's Temperature

Suppose that you were given the assignment of determining Earth's temperature so as to look for evidence of climate change. How would you do it?

Our main concern as humans is clearly temperature at or near the surface, where we and other organisms reside. Thus we can disregard Earth's interior; fortunately, as we can't probe very far into it and even deep-crustal drill holes are costly. It is also helpful that we focus on trends in temperature over time rather than instantaneous temperatures. We know from experience that, especially in the midlatitudes, temperatures can fluctuate wildly over short time periods; they can be very different over short distances at the same moment; and they can be quite different from year to year on any given date. If we are looking at global or regional trends, we will be more concerned with averages and how they may be shifting.

Land surface temperatures are the simplest to address. Many weather stations and other fixed observation points have been in place for over a century. Moreover, research has shown that temperature anomalies and trends tend to be consistent across broad regions. For instance, if your city is experiencing a heat wave or acute cold spell, so are other cities and less-populated areas in the region. So, data from a modest number of land-based sites should suffice to characterize land surface temperatures.

The oceans obviously present more of a challenge. There have been shipboard temperature measurements for centuries, but relatively far fewer, and typically not repeated at the same locations year after year. There are now stationary buoys that monitor ocean temperatures, but relatively few. Satellites have come into play here. For some decades, SSTs have been estimated using, as a proxy, radar measurements of sea surface elevation, as in **figure 10.14**. This is actually a reflection of the temperature through the uppermost layer of seawater, as it is expansion and contraction of that layer as it warms or cools that produces the surface-elevation anomalies. More recently, a more sophisticated approach has been implemented (**figure CS 10.1.1**). NASA's Moderate-Resolution Imaging Spectroradiometer (MODIS) detector not only measures directly the infrared radiation (heat)

Figure CS 10.1.1

Sea surface temperatures, December 2021, as measured by the Moderate-Resolution Imaging Spectroradiometer aboard NASA's Aqua satellite. Temperatures range from −2°C (dark) to 35°C (light).

NASA image by Jesse Allen based on data provided by the MODIS Ocean Team and the University of Miami Rosenstiel School of Marine and Atmospheric Science Remote Sensing Group

coming from the sea surface. It also measures the concentration of atmospheric water vapor through which the radiated heat is passing—important because water vapor, as a greenhouse gas, will absorb some of that heat on its way to the satellite, a particular problem in the humid tropics. MODIS can correct for this, and also for interference from such sources as clouds and dust. The result is exceptionally accurate measurement of SSTs around the globe.

The average temperature for a given point over the course of a year can then be determined, and those data compared over longer periods of time, or averaged over the earth or a region of it. Each year, NASA publishes the resultant trend of global average temperatures (as in **figure 10.3**) and a map of temperature anomalies around the world (**figure CS 10.1.2**). The baseline used for reference is the average for the period 1951–1980, a period recent enough that plentiful data are available, long enough to smooth out a good deal of short-term variability, and long enough ago to allow us to look at trends in the last several decades. Clearly, global temperature is rising. In 2021, it was 0.85°C (1.5°F) above that baseline. Though the magnitude of the average increase may not yet look impressive, the anomalies are far from uniform around the globe; the much greater warming of the Arctic is particularly concerning.

Is this really due to the greenhouse effect? Still other measurements suggest that it is. Satellites can also measure atmospheric temperatures at various altitudes. Over the past few decades, temperatures in the troposphere (the lowest 10 km of the atmosphere) have been rising, while those in the stratosphere above it have been falling (**figure CS 10.1.3**). And the thin transition layer between the troposphere and the stratosphere, the tropopause, is increasing in height. Greenhouse gases concentrate in the troposphere so increasing concentrations mean more heat trapped in the troposphere, an increased height of the tropopause, and less heat escaping to warm the stratosphere (or radiate into space). The data are thus consistent with surface temperatures rising as a consequence of the observed increase in greenhouse gases.

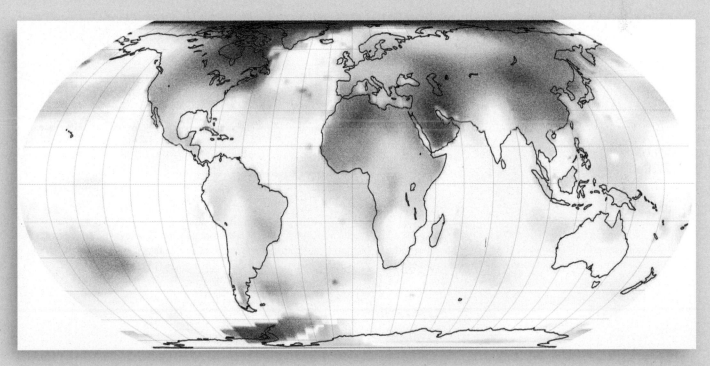

Figure CS 10.1.2

2021 global surface temperature anomalies, relative to 1951–1980 averages. Anomalies range from about −2°C (light blue) to 1°C (yellow) to ≥4°C (red).

NASA Earth Observatory images by Joshua Stevens

(Continued)

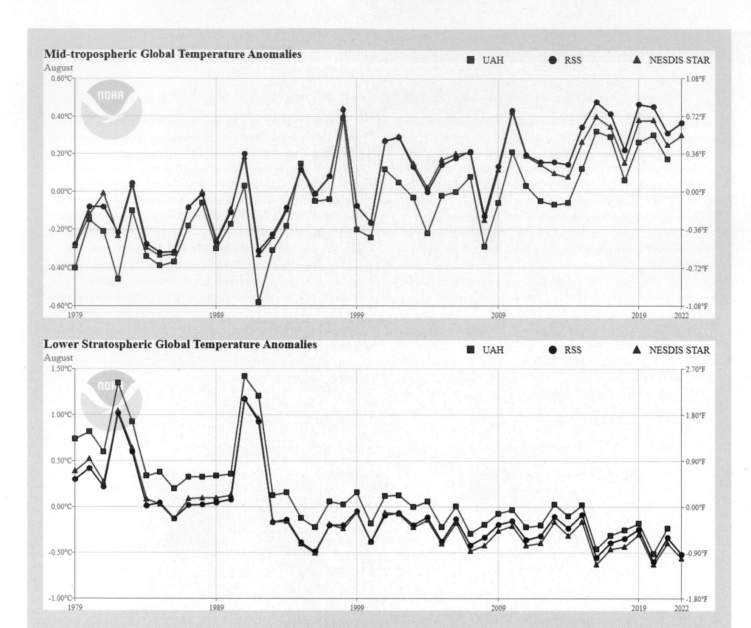

Figure CS 10.1.3

Temperature changes from January 1979 to December 2022 in the middle troposphere, about 5 kilometers (3 miles) above the surface, and in the lower stratosphere, 18 kilometers (11 miles) above the surface. The apparent stratospheric temperature anomalies near the poles may be related to special meteorological conditions of polar regions.

NASA image created by Jesse Allen, using data provided courtesy of Remote Sensing Systems.

Summary

Modern burning of fossil fuels has been increasing the amount of carbon dioxide in the atmosphere. The resultant greenhouse-effect heating is melting alpine glaciers, parts of the Greenland and Antarctic ice sheets, Arctic sea ice, and permafrost, and contributing to the thermal expansion of seawater, causing a rise in global sea level. Coastal areas are beginning to experience the resulting flooding. Global warming may influence the oceans' thermohaline circulation and the occurrence of ocean-atmosphere oscillation events such as El Niño. Other probable consequences involve changes in global weather patterns, including the amount and distribution of precipitation, which may have serious implications for agriculture and water resources; more episodes of extreme temperature; and more-intense storms. Depressed productivity of phytoplankton in warmer seawater affects both the marine food web and atmospheric CO_2; ocean acidification and warming put stress on coral reefs. Warming appears to be linked to the expansion of certain diseases, thus affecting human health.

Cores taken from ice sheets provide data on past temperature and greenhouse-gas fluctuations over 800,000 years. They document a correlation between the two, and the fact that current CO_2 levels in the atmosphere are far higher than at any other time during this period.

Climatic projections for the future are complicated by the number and variety of factors to be considered, including other greenhouse gases, a variety of positive and negative climate feedback mechanisms, and complex interactions between oceans and atmosphere. The current trend is clearly warming. The rock record shows that Earth has been subject to many climatic changes. Global temperatures have been more extreme than at present, and other instances of geologically rapid climate change occurred long before human influences. However, such changes have inevitably affected biological communities and, more recently, human societies. We already observe significant impacts from the warming and associated climatic changes of recent decades, and note that these impacts are much more pronounced in some parts of the globe; some of the results are likely to be irreversible. While CCS projects are expanding, various other proposed geoengineering solutions have significant scientific, practical, and political concerns to be addressed before they could be implemented.

Key Terms and Concepts

greenhouse effect

permafrost

thermohaline circulation

Test Your Learning

1. Describe the greenhouse effect and its relationship to modern industrialized society.

2. Explain why Arctic sea ice is more vulnerable to melting than many land-based glaciers.

3. Define *permafrost,* and indicate why traveling in permafrost regions is more difficult during the summer thaw.

4. Identify what drives the thermohaline circulation in the oceans, and how this conveyor affects climate around the North Atlantic.

5. Describe what happens during an El Niño event, and how it influences both precipitation patterns in the Pacific and the productivity of coastal fishing grounds.

6. Explain how global warming is related to the spread of certain diseases.

7. Indicate how phytoplankton productivity would be affected by changes in SSTs, ice cover, and thermohaline circulation.

8. Describe any three kinds of evidence in the sedimentary-rock record for past global climate.

9. Explain how ice cores are used to investigate relationships between atmospheric greenhouse gases and temperature, and what is measured as the proxy for temperature.

10. Briefly describe one positive and one negative feedback process relating to global warming.

11. Water vapor and methane are both greenhouse gases. Suggest why the focus is so often on carbon dioxide instead.

12. Choose any proposed geoengineering scheme; note both positive and negative aspects related to economics, politics, and technology.

Exploring Further

1. NASA's Goddard Institute for Space Studies identified 2021 as the sixth warmest year since 1880. (See **figure CS 10.1.2** in Case Study 10.1) Examine the map, noting where the most marked warming occurred. Consider what the cause(s) and consequences of such pronounced warming in that region might be. Also, see if you can determine from the map if 2021 was an El Niño or La Niña year.

2. NASA's Goddard Institute for Space Studies maintains a comprehensive global-temperature website at **https://data.giss.nasa.gov/gistemp/**. Use it to try the following:

   a. Via Global maps, make your own temperature-anomaly maps for some period(s) of interest. Compare a map for the years of the twenty-first century with the 1980s or 1990s; make a map for a single month and compare it with the corresponding full year, and so on.

   b. Use the Animations link to watch the temperature anomalies shift around the world over time. Where is the temperature variability greatest? least? Watch for the rise and fall of El Niño/La Niña events or other events that may reflect shifting ocean currents.

   c. Look at the Zonal Means data, showing the anomalies by latitude through time. At what latitudes does warming appear earliest? What latitudes first show consistently warm anomalies, and starting when? When does the earth first show warming from pole to pole?

3. Global temperature does not necessarily increase proportionately with atmospheric CO_2 levels, as we have noted, but both have tended to increase linearly over the last thirty to forty years (see **figures 10.2** and **10.3**). Use those figures to estimate the average annual rate of increase of CO_2 and of temperature over that period, and calculate the resultant apparent rate of temperature increase for every 10-ppm increase in CO_2.

CHAPTER 11

Groundwater and Water Resources

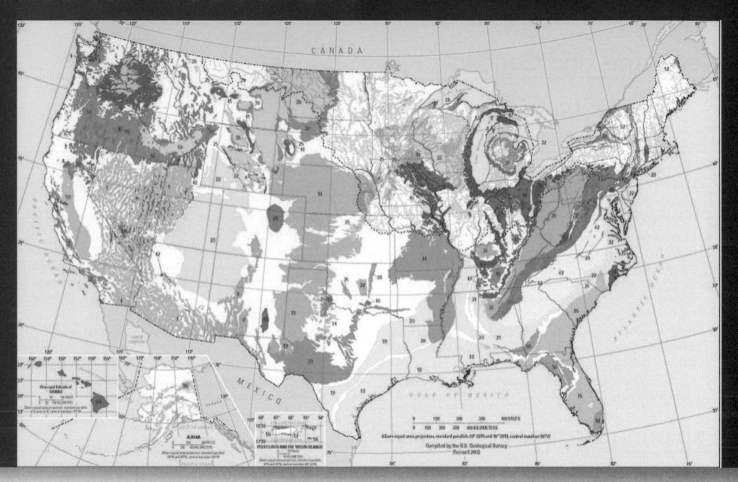

We were introduced to the hydrologic cycle in chapter 6 with our discussion of surface-water flow in streams. We now take a wider view of water on Earth, to consider water as a resource. The importance of water availability for domestic use, agriculture, and industry is apparent. What may be less obvious is that water, or the lack of it, may control the extent to which we can develop certain other resources, such as fossil fuels. To begin, we will look at Earth's water supply and its distribution in nature. This involves considering not only surface water but also groundwater and ice.

Figure 11.1 shows how the water in the hydrosphere is distributed. Several points emerge immediately from the data. One is that there is relatively little fresh liquid water on Earth. More than half of the fresh water is locked up as ice, mainly in the large polar ice caps. And most of the unfrozen fresh water is underground, making it unfamiliar to most of us. Groundwater is an important source of fresh water especially in areas without surface waters reserves. The geologic setting in which it occurs influences not only how much is locally available, but also how accessible it is. From the long-term

---

This USGS map shows the Principal Aquifers of the United States, highlighting those geologic units that have the potential to serve as a source of drinking water.

Source: U.S. Geological Survey

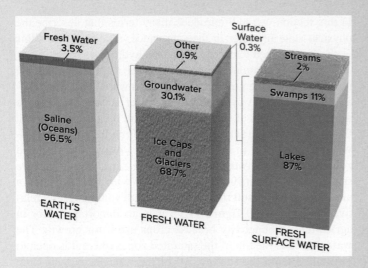

Figure 11.1

The distribution of water on Earth according to the USGS. This does not include water bound in minerals or in the deep interior of Earth.

geologic perspective, water is a renewable resource, but local supplies may be both inadequate and nonrenewable in the short term. These facts underscore the need for restraint in our use of fresh water, especially considering our growing population.

11.1 Porosity and Permeability

Learning Objectives

- Compare porosity and permeability
- Exemplify rocks and sediments with different porosities and permeabilities

We will start this chapter by introducing two physical properties of geologic materials that are important for both their ability to retain fluids and the ease with which the fluids move through them. **Porosity** is the proportion of void space in the material and is a measure of how much fluid the material can store. Porosity is usually expressed as a percentage. These empty spaces exist between individual mineral grains or can be cracks in the rock as a whole; they may be occupied by gas, liquid, or a combination of the two. **Permeability** is a measure of how readily fluids pass through the material. It is related to how well the pores or cracks are interconnected, and to their size; larger pores have a lower surface-to-volume ratio so there is less frictional drag to slow the fluids down. Porosity and permeability of geologic materials are influenced by *texture,* that is the shape and size range of the mineral grains or rock fragments and how well they fit together (**figure 11.2**).

Igneous and metamorphic rocks consist of tightly interlocking crystals, and they usually have low porosity and permeability unless they have been broken up by fracturing or weathering. The same is true of many of the chemical sedimentary rocks, unless cavities have been produced in them by dissolution. Clastic sediments, however, accumulate as layers of

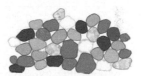

A. Low porosity in an igneous rock with crystalline texture.

B. A well-sorted sandstone with rounded grains has ample pore space.

C. The smaller grains fill up pore space in a poorly-sorted sediment.

D. Packing of plates of clay in a shale may result in high porosity but low permeability.

Figure 11.2

Porosity and permeability vary with texture.

loosely piled particles. Even after lithification, they may contain open pore space. Well-rounded equidimensional grains of similar size can produce sediments of relatively high porosity and permeability. Many sandstones have these characteristics; consider, for example, how rapidly water drains into beach sands. In materials containing a wide range of grain sizes (poorly sorted), finer materials can fill the gaps between coarser grains. Porosity can thereby be reduced, though permeability may remain high. Clastic sediments consisting predominantly of flat, platelike grains such as clay minerals or micas may be

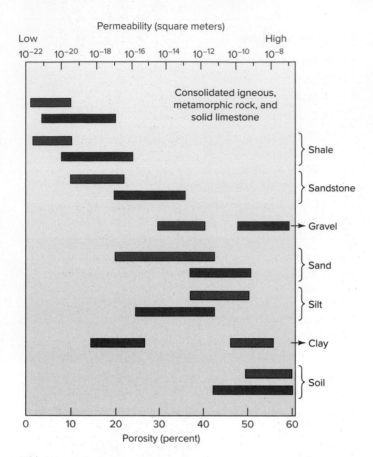

Figure 11.3

Typical ranges of porosities (blue bars) and permeabilities (brown) of various geologic materials.

porous, but because the flat grains can be packed closely together parallel to the plates, these sediments may not be very permeable, especially in the direction perpendicular to the plates. Shale is a rock commonly made from such sediments. The low permeability of clays can readily be demonstrated by pouring water onto a slab of artists' clay.

Figure 11.3 gives some representative values of porosity and permeability. These properties are relevant not only to discussions of groundwater availability and use, but also to issues such as stream flooding, petroleum resources, water pollution, and waste disposal.

11.2 Subsurface Waters

Learning Objectives

- Identify unsaturated zone, water table, and saturated zone
- Describe variabilities in the water table
- Compare groundwater recharge and discharge
- Contrast an aquifer with an aquitard; exemplify both

In this section, we will introduce many terms that describe the physical state of water in the subsurface, describing how this water is connected to other parts of the hydrologic cycle.

Water accumulates in the subsurface as precipitation and runoff infiltrate the ground, moving downward through permeable materials due to gravity. The water accumulates above impermeable rocks or sediments, filling all the pore spaces and fractures within the **saturated zone**, or *phreatic zone*. **Groundwater** is the water in the saturated zone. Above the saturated zone is the **unsaturated zone**, or *vadose zone,* where the pore spaces are filled partly with water, partly with air. The water in unsaturated soil is **soil moisture**. Although volumetrically relatively minor (recall **figure 11.1**), it is an important factor in agricultural productivity because crops use it for growth. The **water table** is the top of the saturated zone, where it is open to infiltration from the surface. All of the water occupying pore spaces below the ground surface is *subsurface water.* These relationships are illustrated in **figure 11.4**. Usually, subsurface water is located within a few kilometers of the surface. In the deep crust and below, pressures on rocks are so great that compression closes up any pores that water might fill.

The water table is not a flat surface. It often is a subdued expression of the ground surface above it, and will change as it moves laterally from one geologic unit to another. A water table can be high, meaning close to the surface. It is highest when the ratio of input water to water removed is greatest, typically in the spring, when rain is heavy or snow and ice accumulations melt. In dry seasons, or when local human use of groundwater is intensive, the water table drops and the amount of available groundwater remaining decreases. Where the topography dips down, the water table will intersect the land surface to form lakes, streams, springs, or wetlands.

Groundwater flows laterally through permeable soil and rock, from higher elevations to lower, from areas of higher pressure or potential to lower, from areas of abundant infiltration to drier ones, or from areas of little groundwater use toward areas of heavy use. Depending on the topography and climate, groundwater may contribute to streamflow, or water from a

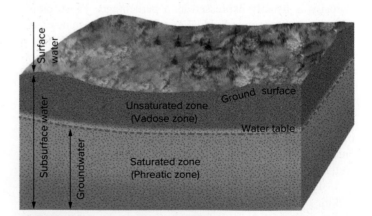

Figure 11.4

Nomenclature of surface and subsurface waters. Groundwater is water in the saturated zone, below the water table.

stream may replenish groundwater (**figure 11.5**). The replenishment of groundwater from infiltration or lateral flow is called **recharge**. Groundwater **discharge** occurs where groundwater flows into a stream, escapes at the surface in a spring, moves laterally away, or is pumped out by humans. The term *discharge* (recall from our study of streams) also applies to the measurement of water moving through an aquifer; see the discussion of Darcy's Law later in the chapter.

What do we look for in a geologic material if we are interested in obtaining groundwater? A rock or sediment that holds enough water and transmits it rapidly enough to be useful as a source of water is an **aquifer**. That is, we would look for a geologic material that has both high porosity and high permeability. But this would also depend on much water you were interested in pumping. An owner of a single home could use a

A

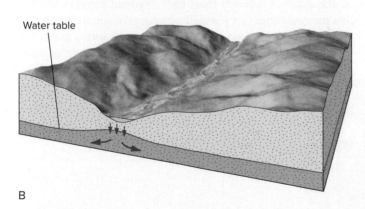

B

Figure 11.5

(A) In this natural system where the groundwater table and stream surface are intimately connected, groundwater is moving from high elevation to low in both directions and discharging into the stream. (B) In this arid setting, the stream and groundwater are disconnected, and the stream is recharging the groundwater.

rock with lower porosity to install a well in compared to, say, a large chocolate candy-making factory. Independent of the amount of water, the aquifer would need a high permeability so that the water is easy to obtain. Many of the most productive aquifers are sandstones or other coarse clastic sedimentary rocks, but any other type of rock may serve if it is sufficiently porous and permeable—a porous or fractured limestone, fractured basalt, or weathered granite. Conversely, an **aquitard** is a rock or sediment layer that may store a considerable quantity of water, but in which water flow is slowed, or severely hindered; that is, its permeability is very low, regardless of its porosity. Clays and shales are common aquitards.

11.3 Aquifer Geometry and Groundwater Flow

Learning Objectives

- List the properties of unconfined and confined aquifers
- Explain an artesian system
- Identify the parameters used in Darcy's Law
- Recognize a perched water table

If we are interested in obtaining water from aquifers for use, we need to understand the physical parameters of both the geologic material and the fluid within them to calculate the flow. A thorough assessment of the subsurface geology is essential in determining the direction and rate at which groundwater, and any contaminant it may be carrying, is flowing. Below, we will review the two primary types of aquifers and how we go about assessing the movement and amount of groundwater within them.

Confined and Unconfined Aquifers

The behavior of groundwater is controlled to some extent by the geology and geometry of the particular aquifer in which it is found. When the aquifer is directly open to infiltration from the surface, it is an **unconfined aquifer** (**figure 11.6**). An unconfined aquifer is largely recharged from infiltration. When a well is drilled into an unconfined aquifer, the water will rise in the well to the same height as the water table in the aquifer. The water must be actively carried up or pumped to the ground surface.

A **confined aquifer** is bounded above and below by aquitards. Water in a confined aquifer may be under considerable pressure as a consequence of lateral differences in elevation within the aquifer: water at a higher elevation exerts pressure on water deeper (lower) in the same aquifer. The confining layers prevent the free flow of water to relieve this pressure. At some places within the aquifer, the water level in the saturated zone may be held above or below the level it would assume if the aquifer were unconfined and the water allowed to spread freely (**figure 11.7**).

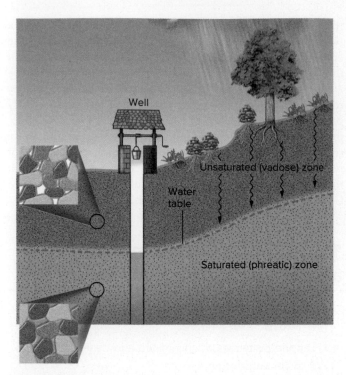

Figure 11.6

An unconfined aquifer. Water rises in an unpumped well just to the height of the water table.

When a well is drilled into a confined aquifer, the water can rise above its level in the aquifer because of this extra hydrostatic (fluid) pressure. This is called an **artesian system**. The water in an artesian system may or may not rise all the way to the ground surface; some pumping may still be necessary to bring it to the surface for use. In such a system, rather than describing the height of the water table, geologists refer to the height of the **potentiometric surface**, which represents the height to which the water's pressure would raise the water if the water were unconfined. This level will be somewhat higher than the top of the confined aquifer where its rocks are saturated, and it may be above the ground surface, as shown in **figure 11.7**.

Advertising claims may try to convince you otherwise, but this is all that *artesian water* means; it is no different chemically from, and no purer, better-tasting, or more wholesome than any other groundwater. Artesian water is just under natural pressure. Water towers create the same effect artificially. After water has been pumped into a high tower, gravity increases the fluid pressure in the water-delivery system so that water rises up in the pipes of individual users' homes and businesses without the need for many pumping stations.

The overlying aquitard also means that a confined aquifer cannot be recharged by infiltration from above. Water extracted can be replenished only by lateral flow from elsewhere in the aquifer, and modern recharge to the aquifer overall may be limited or nonexistent. This has significant implications for water use from confined-aquifer systems.

Darcy's Law and Groundwater Flow

How readily groundwater can move through rocks and soil is governed by permeability, but where and how rapidly it actually does flow is also influenced by differences in **hydraulic head** (potential energy) from place to place. Groundwater flows from areas of higher hydraulic head to lower. The height of the water table or of the potentiometric surface reflects the hydraulic head at each point in the aquifer. All other factors being equal, the greater the difference in hydraulic head between two points, the faster groundwater will flow between them. This is a key result of **Darcy's law**, which can be expressed this way:

$$Q = K \cdot A \cdot \frac{\Delta h}{\Delta l}$$

where Q = discharge; A = cross-sectional area; K is a parameter known as *hydraulic conductivity* that takes into account both the permeability of the rock or soil and the viscosity and density of the flowing fluid; and ($\Delta h/\Delta l$) is the *hydraulic gradient,* the difference in hydraulic head between two points (Δh) divided by the distance between them (Δl). Hydraulic conductivities, like permeabilities, vary widely among geologic materials; for example, K may exceed 1000 ft/day in a well-sorted gravel but be less than 0.00001 ft/day in a fresh limestone or unfractured

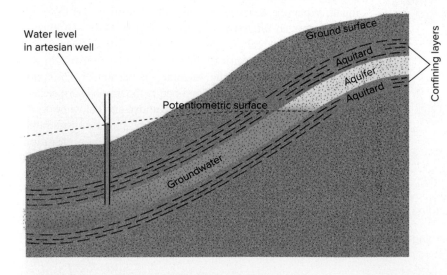

Figure 11.7

Natural internal pressure in a confined aquifer system creates artesian conditions, in which water may rise above the apparent (confined) local water table. The aquifer here is a sandstone; the confining layers, shale.

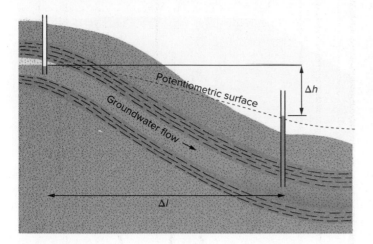

Figure 11.8

Darcy's law relates groundwater flow rate, and thus discharge, to hydraulic gradient. Height of water in wells reflects relative hydraulic head.

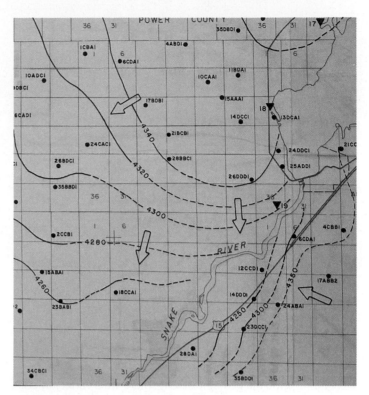

Figure 11.9

This traditional ground water contour map from 1984 shows shallow groundwater discharging into the Snake River in Idaho. Wells are represented by dots.

From the American Geographical Society, The University of Wisconsin-Milwaukee Libraries

basalt. The hydraulic gradient of groundwater is analogous to a stream's gradient. (See **figure 11.8**.)

Note that this relationship works not only for water, but for other fluids, such as oil; the value of K will change with changes in viscosity, but the underlying behavior is the same. Also, Darcy's law somewhat resembles the expression for stream discharge, with flow velocity represented by $(K \frac{\Delta h}{\Delta l})$.

There are many applications where one would want to know the direction and speed of groundwater flow in an area. Typically, the data needed to determine these parameters is collected from multiple wells drilled into the same aquifer. Water levels within the wells are gathered either manually or remotely. Samples of aquifer materials are collected when the wells are installed and sent to laboratories to determine site-specific hydraulic conductivity. Researchers can then analyze groundwater flow by creating a groundwater contour map (**Figure 11.9**) or inputting values into numerical models that create 3-dimensional representations of the subsurface. Increasingly, groundwater elevation data such as these are available in public databases. The USGS National Ground-Water Monitoring Network is a good place to start.

Other Factors in Water Availability

More-complex local geologic conditions can make it difficult to determine the availability of groundwater without thorough study. For example, locally occurring lenses or patches of relatively impermeable units within otherwise permeable ones may result in a *perched water table* (**figure 11.10**). Immediately above the aquitard is a local saturated zone, far above the true regional water table. Someone drilling a well above this area could be deceived about the apparent depth to the water table and might also find that the perched saturated zone contains very little total water. The quantity of water available would be especially sensitive to local precipitation levels and fluctuations.

Using groundwater over the longer term might well require drilling down to the regional water table, at a much greater cost.

In order to more completely assess water availability in an aquifer, we also should know the location of recharge zones, groundwater basin divides, and the amount of water the aquifer

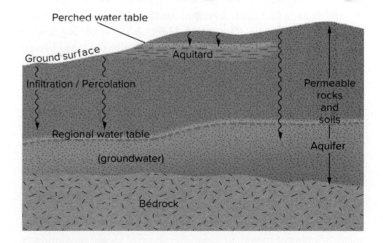

Figure 11.10

A perched water table may create the illusion of a shallow regional water table.

can store. For example, if water is being consumed very close to the recharge area and consumption rate exceeds recharge rate, the stored water may be exhausted rather quickly. If the point of extraction is far down the flow path from the recharge zone, a larger reserve of stored water may be available to draw upon. The *specific yield* of an aquifer is the measurement of how much water can be pumped, and it's always less than the pore volume, as some of the water will be retained due to molecular and surface tension forces.

11.4 Consequences of Groundwater Withdrawal

Learning Objectives

- Describe the creation of a cone of depression
- Identify consequences of lowering the water table
- Describe saltwater intrusion

Depending upon the geologic setting, there are a variety of consequences to pumping groundwater out of an aquifer at faster rate than what can be naturally recharged. All of these are problematic, often difficult or impossible to reverse, and ultimately costly to the people involved.

Lowering the Water Table

When groundwater is pumped from an aquifer, the rate at which water flows in from surrounding rock to replace that which is extracted is generally slower than the rate at which water is taken out. In an unconfined aquifer, the result is a circular lowering of the water table immediately around the well, creating a **cone of depression** (**figure 11.11A**). When there are many closely spaced wells, the cones of depression of adjacent wells may overlap, further lowering the water table between wells (**figure 11.11B**). A cone of depression also alters the direction of water flow. Wells may need to be drilled deeper or abandoned, especially if the new flow regime is now allowing non-potable water to flow into the area. Cones of depression can develop in potentiometric surfaces. When artesian groundwater is withdrawn at a rate exceeding the recharge rate, the potentiometric surface can be lowered.

These should be warning signs, for the process of deepening wells to reach water cannot continue indefinitely, nor can a potentiometric surface be lowered without limit. Both unconfined and confined aquifers can have small saturated thicknesses, meaning they are not vertically extensive. Groundwater flow rates are highly variable, but in many aquifers they are of the order of only meters or tens of meters per year. Recharge of significant amounts of groundwater, especially to confined aquifers with limited recharge areas, can require decades or centuries.

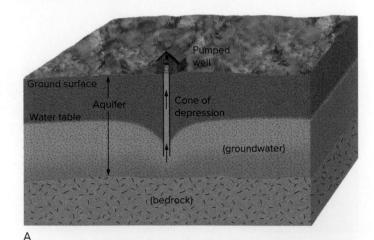

A

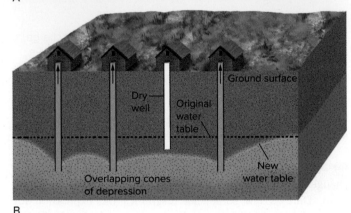

B

Figure 11.11

The formation and effect of cones of depression. (A) Lowering of the water table in a cone of depression around a pumped well in an unconfined aquifer. (B) Overlapping cones of depression lead to net lowering of the water table; shallower wells may run dry.

Where groundwater is being depleted by too much withdrawal too fast, one can speak of mining groundwater (**figure 11.12**). The idea is not necessarily that the water will never be recharged but that the rate is so slow on the human timescale as to be insignificant. From the human point of view, we may indeed exhaust groundwater in some heavy-use areas. Also, as we will see in the next section, human activities may themselves reduce natural recharge, so groundwater consumed may not be replaced, even slowly.

Compaction and Surface Subsidence

Lowering of the water table has secondary consequences. The water in saturated rocks and soils provides buoyancy. Aquifer rocks no longer saturated with water can become compacted from the weight of overlying rocks. This decreases their porosity, permanently reducing their water-holding capacity, and

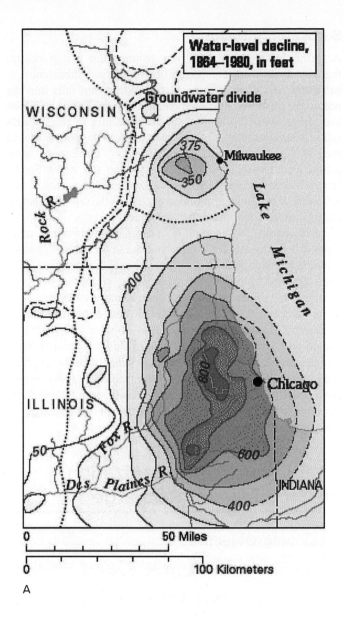

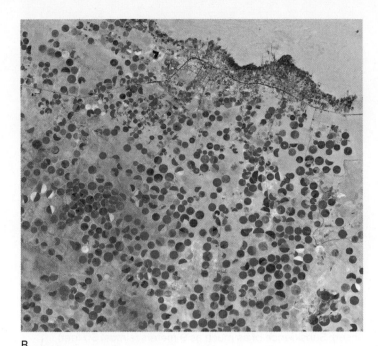

B

Figure 11.12

(A) Major cities' groundwater consumption can create huge cones of depression. (B) Agriculture in Wadi-ad-Dawasir is supported by groundwater that is 30,000 years old. Water levels are dropping so fast that the aquifer could be depleted in a few decades.

Sources: (A) From USGS Fact Sheet 103-03; (B) NASA Earth Observatory images by Joshua Stevens

also their permeability. At the same time, as the rocks below compact and settle, the ground surface itself subsides. Where water depletion is extreme, particularly if the water has been extracted from saturated soils, which would typically be more compressible than rocks, the surface subsidence can be several meters. Lowering of the water table/potentiometric surface also may contribute to *sinkhole* formation, as described in a later section.

At high elevations or in inland areas, this subsidence mostly causes structural problems as building foundations are disrupted. In low-elevation coastal regions, the subsidence may lead to extensive flooding, as well as to increased rates of coastal erosion. As noted in chapter 7, the city of Venice, Italy, is one such slowly drowning coastal area. Many of its historical, architectural, and artistic treasures are threatened by the combined effects of the gradual rise in worldwide sea levels, the tectonic sinking of the Adriatic coast, and surface subsidence from extensive groundwater withdrawal. The groundwater

withdrawal began about 1950. Since 1969, no new wells have been drilled, groundwater use has declined, and subsidence from this cause has apparently stopped, but the city remains awash. Closer to home, the Houston/Galveston Bay area has suffered subsidence of up to several meters from a combination of groundwater withdrawal and, locally, petroleum extraction. Some 80 sq. km (over 30 sq. mi.) are permanently flooded, and important coastal wetlands have been lost. In the San Joaquin Valley in California, half a century of groundwater extraction, much of it for irrigation, produced as much as 9 meters of surface subsidence between 1925 and 1977; continued groundwater use has led to even more-rapid recent subsidence (**figure 11.13**). The city of Corcoran is expected to sink up to 11 more feet by 2040.

In such areas, simple solutions, such as pumping water back underground, are unlikely to work. The rocks likely have been permanently compacted. This is the case in Venice, where subsidence appears to have been halted but rebound is not expected because clayey rocks in the aquifer system are irreversibly compacted. Compaction has occurred in the San Joaquin Valley, too. Also, where would the water come from to pump back underground? If groundwater use is heavy, it may be

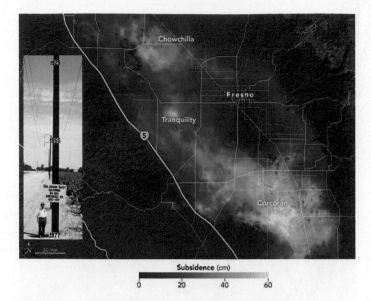

Subsidence (cm)
0   20   40   60

Figure 11.13

(Inset): Subsidence of as much as 9 meters (almost 30 feet) occurred between 1925 and 1977 in the San Joaquin Valley, California, as a consequence of groundwater withdrawal. Signs on pole indicate former ground surface elevation. (Main figure): Continued groundwater use during recent drought has led to serious subsidence: Studies over the period 2006–2016 have revealed some areas subsiding at 0.6 meter (about 2 feet) *per year.*

Sources: (Main figure) NASA Earth Observatory map by Joshua Stevens; (Inset) U.S. Geological Survey Department of the Interior/USGSU.S. Geological Survey/Photograph courtesy Richard O. Ireland

because there is no great supply of fresh surface water. Pumping in salt water, in a coastal area, will in time make the rest of the water in the aquifer salty, too. And presumably, a supply of fresh water is still needed for local use, so groundwater withdrawal often cannot be easily or conveniently curtailed.

Saltwater Intrusion

A further problem arising from groundwater use in coastal regions, aside from the possibility of surface subsidence, is **saltwater intrusion (figure 11.14)**. When rain falls into the ocean, the fresh water promptly mixes with the salt water. However, fresh water falling on land does not mix so readily with saline groundwater at depth because water in the pore spaces in rock or soil is not vigorously churned by currents or wave action. Fresh water is also less dense than salt water. So the fresh water accumulates in a lens, which floats above the denser salt water. If water use approximately equals the rate of recharge, the freshwater lens stays about the same thickness.

However, if consumption of fresh groundwater is more rapid, the freshwater lens thins, and the denser saline groundwater, laden with dissolved sodium chloride, moves up to fill in pores emptied by removal of fresh water. Wells that had been tapping the freshwater lens may begin pumping unwanted salt water instead, as the limited freshwater supply gradually decreases. Moreover, once a section of an aquifer becomes tainted with salt, it cannot readily be made fresh again. Saltwater intrusion destroyed useful aquifers beneath Brooklyn, New York, in the 1930s and is a significant problem in many coastal areas nationwide: the southeastern and Gulf coastal states, some densely populated parts of California, and Cape Cod in Massachusetts.

11.5 Impacts of Urbanization on Groundwater Recharge

Learning Objectives

- Identify ways in which urbanization affects groundwater recharge
- Recognize the role of artificial recharge basins

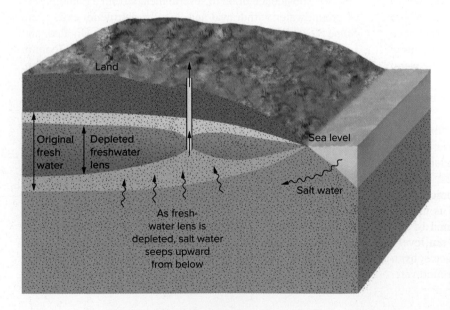

Figure 11.14

Saltwater intrusion in a coastal zone. If groundwater withdrawal exceeds recharge, the lens of fresh water thins, and salt water flows into more of the aquifer system from below. Upconing of saline water also occurs below a cone of depression. Similar effects may occur in an inland setting where water is drawn from a fresh groundwater zone underlain by more-saline waters.

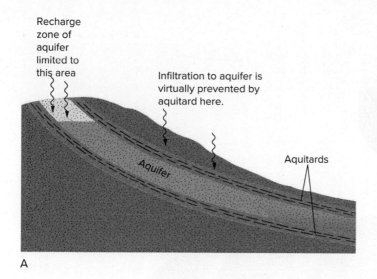

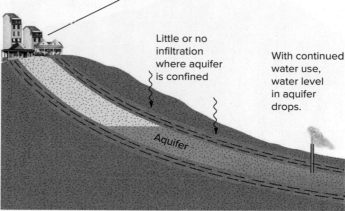

A

B

Figure 11.15

Recharge to a confined aquifer. (A) The recharge area of this confined aquifer is limited to the area where permeable rocks intersect the surface. (B) Recharge to the confined aquifer may be reduced by placement of impermeable cover over the limited recharge area. (Vertical scale exaggerated.)

Obviously, an increasing concentration of people means an increased local demand for water. We saw in chapter 6 that urbanization often involves extensive modification of surface-water runoff patterns and stream channels. By influencing the amount of infiltration, urbanization also has an affect on groundwater systems.

Impermeable cover—buildings, asphalt and concrete roads, sidewalks, parking lots, airport runways—over one part of a broad area underlain by an unconfined aquifer has relatively little impact on that aquifer's recharge. Infiltration will continue over most of the region. In the case of a confined aquifer, however, the available recharge area may be very limited, since the overlying confining layer prevents direct downward infiltration in most places (**figure 11.15**). If impermeable cover is then built over the recharge area of a confined aquifer, recharge is considerably reduced, aggravating the water-supply situation. Permeable concrete and other paving surfaces have been designed to help address this problem, but a great deal of impermeable cover is typically already in place in urban areas.

Filling in wetlands is a common way to provide more land for construction. This practice interferes with recharge, especially if surface runoff is rapid elsewhere in the area. A marsh or swamp in a recharge area, holding water for long periods, can be a major source of infiltration and recharge. Filling it in so water no longer accumulates there, and, worse yet, topping the fill with impermeable cover, again may greatly reduce local groundwater recharge and increase the risk of flooding. Recharging aquifers is just one of the ecosystem services provided by wetlands, which globally are estimated to be worth over $45 billion annually.

In steeply sloping areas or those with low-permeability soils, well-planned construction that includes artificial recharge

basins can aid in increasing groundwater recharge (**figure 11.16A**). The basin acts similarly to a flood-control retention pond in that it is designed to catch some of the surface runoff during high-runoff events. Trapping the water allows more time for infiltration and thus more recharge in an area from which the fresh water might otherwise be quickly lost to streams and carried away.

Recharge basins are a partial solution to the problem of areas where groundwater use exceeds natural recharge rate, but, of course, they are effective only where there is surface runoff to catch, and they rely on precipitation, an intermittent water source. Increasingly, artificial recharge involves diverting streams, as shown in **figure 11.16B**.

11.6 Karst and Sinkholes

Learning Objectives

- Define karst
- Describe groundwater in a karst setting
- Describe the creation of a sinkhole

Broad areas of the contiguous United States are underlain by rocks that are extremely soluble in fresh water. These include the carbonate rocks limestone and dolomite, and the evaporites rock salt and gypsum, all formed in shallow marine settings (**figure 11.17**). As groundwater flows through these rocks, they are dissolved, causing subsidence and the collapse of the ground surface above it into the resultant cavities, creating a distinctive terrain called **karst** (**figure 11.18**).

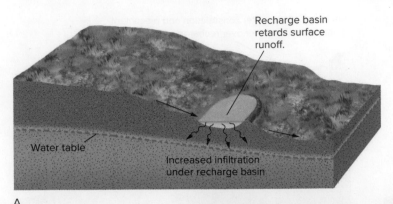

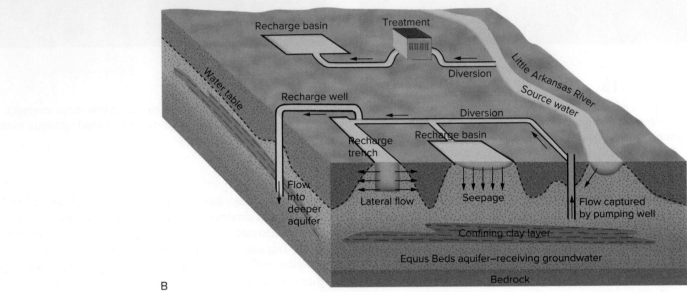

Figure 11.16

(A) Artificial recharge basins can aid recharge by slowing surface runoff. (B) The USGS artificial-recharge project near Wichita, Kansas, was designed to stockpile water to meet future needs.

(B) After U.S. Geological Survey.

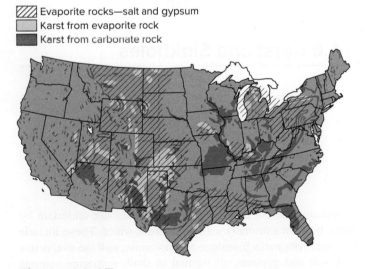

Figure 11.17

Highly soluble rocks underlie more than half of the contiguous United States.

From USGS Fact Sheet 165-00.

Underground, the solution process forms extensive channels, voids, and even large caverns (**figure 11.19**). This creates large volumes of space for water storage and, where the voids are well interconnected, makes rapid groundwater movement possible. Karst aquifers can be sources of plentiful water; in fact, 40% of the groundwater used in the United States for drinking water comes from karst aquifers. However, the news in this regard is not all good. First, the irregular distribution of large voids and channels makes it more difficult to estimate available water volume or to determine groundwater flow rates and paths. Second, because the water flow is typically much more rapid than in non-karst aquifers, recharge and discharge occur on shorter timescales, and water supply may be less predictable. While flow in non-karst aquifers may be of the order of centimeters per day, water in a karst aquifer with well-connected voids can be of the order of tens of meters per day or more. Contaminants can spread rapidly to pollute the water in karst aquifer systems, and the natural filtering that occurs in many non-karst aquifers with finer pores and passages and slower water flow may not occur in karst systems.

A

B

Figure 11.18

(A) Uplift of the carbonate rocks that now make up Croatia's Biokovo mountain range has exposed them to solution and erosion, producing the strikingly pitted karst surface shown here. (B) The Shilin "stone forest" in China is a different type of karst produced by a solution of limestone along vertical fractures.

Sources: (A) Earth Sciences and Image Analysis Laboratory, NASA Johnson Space Center; (B) Eric PHAN-KIM/Moment/Getty Images

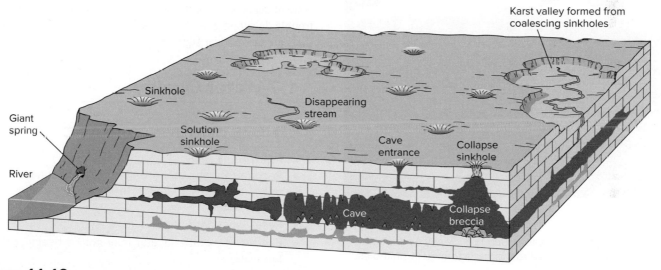

Figure 11.19

Dissolution of soluble rocks underground creates large voids below the surface and, often, leads to surface subsidence above.

Given enough time, groundwater can dissolve large volumes of soluble rocks, slowly enlarging underground caverns that remove support for the land above if they extend close to the surface. There may be little obvious evidence at the surface of what is taking place until the ground collapses abruptly into the void, producing a **sinkhole** (**figure 11.20A**). The collapse of a sinkhole may be triggered by a drop in the water table as a result of drought or water use that leaves rocks and soil that were previously buoyed up by water pressure unsupported. Failure may also be caused by the rapid input of large quantities of water from heavy rains that wash overlying soil down into the

cavern, or by an increase in subsurface water flow rates. Other sinkholes can be created by human activities, such as the injection of wastewater into soluble rocks (**figure 11.20B**).

Sinkholes come in many sizes. The larger ones are quite capable of swallowing many houses at a time: They may be over 50 meters deep and cover several tens of acres. If one occurs in a developed area, a single sinkhole can cause millions of dollars in property damage. Sudden sinkhole collapses beneath bridges, roads, and railways have occasionally caused accidents and even deaths. The most often cited sinkhole incident in the United States occurred in 1981 in Winter Park, Florida, where a

A

B

Figure 11.20

(A) Over 100 sinkholes developed in the Dover, Florida, area during a cold spell in January 2010 as farmers pumped groundwater to use in protecting their plants. (B) This sinkhole in Daisetta, Texas, formed in 2008 when wastewater was injected into a salt dome.

(A) Ann Tihansky/USGS; (B) Randall Orndorff/USGS

Figure 11.21

The karst topography in eastern Florida is clearly evidenced from above, showing aligned round lakes formed in sinkholes.

National Aeronautics and Space Administration

100-meter diameter collapse occurred in early May, caused in part by drought, resulting in an estimated $2 million in damage at that time.

Sinkholes are rarely isolated phenomena. As noted, limestone, gypsum, and salt beds deposited in shallow seas can extend over broad areas. Where there is one sinkhole in a region underlain by such soluble rocks, there are likely to be others. An abundance of sinkholes in an area is a strong hint that more can be expected. The regional geology and topography are the major factors and the first to be considered when determining the hazard risk (**figure 11.21**). In such an area, the subsurface situation should be investigated before buying or building a home or business. Circular patterns of cracks on the ground or conical depressions in the ground surface may be early signs of trouble developing below. Cracks in building walls and foundations are also a clear sign.

11.7 Water Quality

Learning Objectives

- List ways to express water quality
- Recognize areas of the United States that have naturally occurring radioactive elements in the groundwater
- Recall what hard water is
- Explain why bottled water may not be healthier than tap water

As noted earlier, most of the water in the hydrosphere is in the very salty oceans, and almost all of the remainder is tied up in ice. That leaves relatively little surface or subsurface water for potential freshwater sources. Moreover, much of the water on and in the continents is not strictly fresh. Even rainwater, the standard for clean water for long, contains dissolved chemicals of various kinds, especially in industrialized areas with substantial air pollution. Once precipitation reaches the ground, it reacts with soil, rock, and organic debris, dissolving still more chemicals naturally, and any pollution generated by human activities. Water quality must be a consideration when evaluating water supplies.

Pollution of both surface water and groundwater, and some additional measures of water quality used particularly in describing pollution, will be discussed in chapter 17. Here, we focus on aspects of natural water quality.

Water quality is described in a variety of ways. A common approach is to express the amount of a dissolved chemical substance present as a concentration in parts per million (ppm) or, for very dilute substances, parts per billion (ppb). These units are analogous to weight percentages (which are really parts per hundred) but are used for lower (more dilute) concentrations. For example, if water contains 1 weight percent salt, it contains one gram of salt per hundred grams of water, or one ton of salt per hundred tons of water, or whatever unit one wants to use. Likewise, if the water contains only 1 ppm salt, it contains one gram of salt per million grams of water, and so on. For comparison, the most abundant dissolved constituents in seawater can be measured in parts per thousand (magnesium, sulfate) or even percent (sodium, chloride).

Another way to express overall water quality is in terms of *total dissolved solids* (TDS), the sum of the concentrations of all dissolved solid chemicals in the water. How low a level of TDS is required or acceptable varies with the application. Standards might specify a maximum of 500 or 1000 ppm TDS for drinking water; 2000 ppm TDS might be acceptable for watering livestock; industrial applications where water chemistry is important (in pharmaceuticals or textiles, for instance) might need water even purer than normal drinking water.

Yet describing water in terms of total content of dissolved solids does not present the whole picture. At least as important as the quantities of impurities present is what those impurities are. If the main dissolved component is calcium carbonate from a limestone aquifer, the water may taste fine and be perfectly wholesome with well over 1000 ppm TDS in it. If iron or sulfur is the dissolved substance, even a few parts per million may be enough to make the water taste bad, though it may not be actually unhealthful. Many synthetic chemicals that have leaked into water through improper waste disposal are toxic even at concentrations of 1 ppb or less.

Other parameters also may be relevant in describing water quality. One is pH, which is a measure of the acidity or alkalinity of the water. The pH of water is inversely related to acidity: the lower the pH, the more acid the water. Water that is neither acid nor alkaline has a pH of 7. For health reasons, concentrations of certain bacteria may also be monitored in drinking-water supplies.

One current water-quality concern is the presence of naturally occurring radioactive elements in groundwater that may present a radiation hazard to the water consumer. Uranium, which can be found in most rocks including those serving commonly as aquifers, decays through a series of steps. Several of the intermediate decay products pose special hazards. One of those, radium, behaves chemically much like calcium and tends to be concentrated in the body in bones and teeth; recall the periodic table from chapter 2. Another,

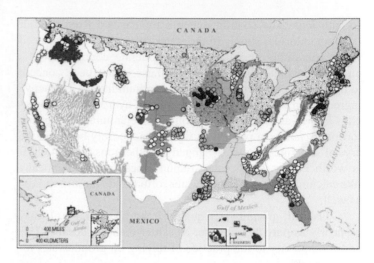

Figure 11.22

Background colors represent different aquifer systems. Dots are wells sampled, with dot colors indicating radium levels: white, <1 pCi/liter; yellow, 1–4.99 pCi/liter; red, ≥5 pCi/liter. A pCi is a *picocurie,* one-trillionth of a curie.

After USGS Fact Sheet 2010–3013.

radon, is a chemically inert gas but is radioactive itself and decays to other radioactive elements in turn. Radon leaking into indoor air from water supplies contributes to indoor air pollution (see chapter 18). High concentrations of radium and/or radon in groundwater may result from decay of uranium in the aquifer itself or, in the case of radon, from seepage out of adjacent uranium-rich aquitards, especially shales. Uranium content of rocks varies considerably, so local geology strongly influences the radium content of groundwater (**figure 11.20**). We will discuss other elemental water-quality concerns in chapter 17.

Aside from the issue of health, water quality may be of concern because of the particular ways certain dissolved substances alter water properties. In areas where water supplies have passed through soluble carbonate rocks such as limestone, the water may be described as hard. **Hard water** simply contains substantial amounts of dissolved calcium and magnesium. When calcium and magnesium concentrations reach or exceed the range of 80 to 100 ppm, the hardness may become objectionable.

Perhaps the most irritating routine problem with hard water is the way it reacts with soap, preventing the soap from lathering properly, causing bathtubs to develop rings and laundered clothes to retain a gray soap scum. Hard water or water otherwise high in dissolved minerals may also leave mineral deposits in plumbing and in appliances, such as coffeepots and steam irons. Primarily for these reasons, many people in hard-water areas use water softeners, which remove calcium, magnesium, and certain other ions from water in exchange for added sodium ions. The sodium ions are replenished from the salt (sodium chloride) supply in the water softener. While softened water containing sodium ions in moderate concentration is unobjectionable in taste or household use, it may be of concern to those on diets involving restricted sodium intake. The active ingredient in water softeners is a group of hydrous silicate minerals known as *zeolites.* The zeolites have an unusual capacity for *ion exchange,* a process in which ions loosely bound in the crystal structure can be exchanged for other ions in solution. It is worth noting that because radium does behave chemically like calcium and magnesium, water softeners do remove radium as well.

Overall, groundwater quality is highly variable. It may be nearly as pure as rainwater or saltier than the oceans. This adds complexity to the practice of relying on groundwater for water supply.

One additional practical note: many people drink bottled water on the assumption that it is safer or purer than their tap water. This may or may not be the case. Bottled water is not subject to EPA drinking-water regulations or to the Safe Drinking Water Act. It falls instead under Food and Drug Administration jurisdiction, and the FDA does not even require bottlers to publish water-quality analyses of their product as municipalities must do for their water supplies. Often, bottled water is just local tap water, bottled.

11.8 Water Use, Water Supply

Learning Objectives

- Summarize the U.S. water budget
- Compare water withdrawal to water consumption
- List the reasons why groundwater supplies are more reliable than surface water
- List the principal categories of water use in the United States and their relative withdrawals
- Recognize the general patterns of water use for irrigation in the United States

Where do we get our water in the United States, and how do we use it? Maybe surprisingly, our use of water has continued to decrease as we find smarter ways to use the water available to us. We will review our water use patterns here, and then discuss options for extending our water supply later in section 11.10.

General U.S. Water Use

Inspection of the U.S. water budget overall would suggest that ample water is available for use (**figure 11.23**). Some 4200

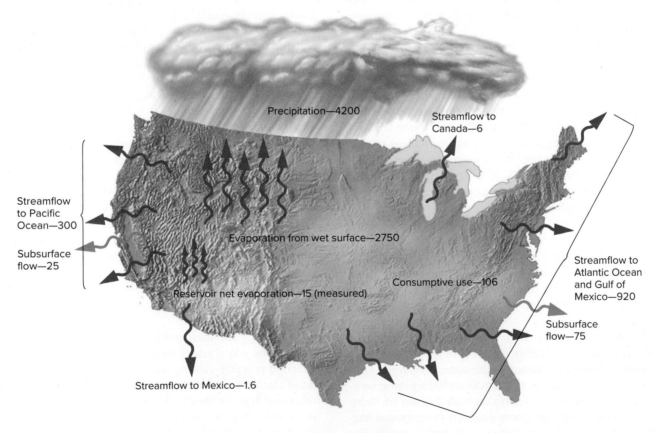

Precipitation—4200

Streamflow to Canada—6

Streamflow to Pacific Ocean—300

Subsurface flow—25

Evaporation from wet surface—2750

Reservoir net evaporation—15 (measured)

Consumptive use—106

Streamflow to Atlantic Ocean and Gulf of Mexico—920

Subsurface flow—75

Streamflow to Mexico—1.6

Figure 11.23

In terms of gross water flow, U.S. water budget seems ample. Figures are in billions of gallons per day.

Source: Data from The Nation's Water Resources 1975–2000, U.S. Water Resources Council.

billion gallons of precipitation fall on this country each day; subtracting 2750 billion gallons per day lost to evapotranspiration still leaves a net of 1450 billion gallons per day for streamflow and groundwater recharge. Water-supply problems arise, in part, because the areas of greatest water availability do not always coincide with the areas of concentrated population or greatest demand, and also because a portion of the added fresh water quickly becomes polluted by mixing with impure or contaminated water.

People in the United States use a large amount of water and for many different purposes in our modern society. Biologically, each of us requires about a gallon of water per day which would be about 325 million gallons per day (given the population during the last complete water census in 2015). In 2015, we withdrew 322 billion gallons of water each day to be used in households for cooking and cleaning, in industrial processes and power generation, for irrigation and livestock support, along with a few other activities. Some of the water withdrawn was *consumed,* meaning that is was not returned as wastewater to the original source or that it was no longer available for reuse. Of all the water used for irrigation, which was 118 billion gallons per day, 62% was consumed mainly through evaporation and incorporation into plant material.

It might seem easier to use surface waters rather than subsurface waters for water supplies. Why, then, worry about using groundwater at all? One basic reason is that there is little or no surface water available in some places, while there may be a substantial supply of water deep underground. Tapping the groundwater supply allows us to live and farm in otherwise uninhabitable areas.

Another reason is that streamflow varies seasonally. During dry seasons, the surface water supply may be inadequate. Dams and reservoirs allow a surplus of water to accumulate during wet seasons for use in dry times, but we have already seen some of the negative consequences of reservoir construction and some droughts are lasting a relatively long time (chapter 6). Furthermore, if a region is so dry at some times that dams and reservoirs are necessary, then the rate of water evaporation from the broad, still surface of the reservoir may itself represent a considerable water loss and aggravate the water-supply problem. Lake Mead serves as an infamous example of a large surface water source shrinking due to drought conditions. The reservoir was created by the damming of the Colorado River and is currently at its lowest level since it was filled in the 1930s. It is expected to drop even further. This dammed water serves as both a source of water for multiple states and also generates electricity for the area as it flows through the dam (**figure 11.24**).

Precipitation (the prime source of abundant surface runoff) varies widely geographically (**figure 11.25**), as does population density. In many areas, the concentration of people far exceeds what can be supported by available local surface waters, even during the wettest season.

Also, streams and large lakes have historically been used as disposal sites for untreated wastewater and sewage, which makes the surface waters decidedly less appealing as drinking

Figure 11.24

Light-colored rocks along the shore of Lake Mead mark the high water level and show just how low the lake level has dropped.

superjoseph/Shutterstock

waters. A lake, in particular, may remain polluted for decades after the input of pollutants has stopped if there is no place for those pollutants to go or if there is a limited input of fresh water to flush them out. Groundwater from the saturated zone, on the other hand, has passed through the rock of an aquifer and has been naturally filtered to remove some impurities such as soil or sediment particles and even the larger bacteria, although it can still contain many dissolved chemicals that may be long-lasting.

Finally, groundwater is by far the largest reservoir of unfrozen fresh water. For a variety of reasons, then, we often prefer groundwater as a freshwater source.

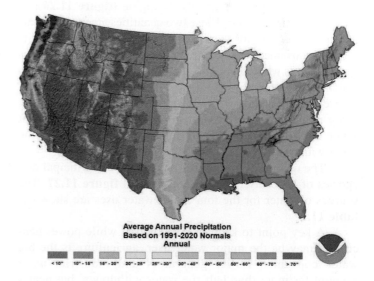

Average Annual Precipitation
Based on 1991-2020 Normals
Annual

| < 10" | 10" - 15" | 15" - 20" | 20" - 25" | 25" - 30" | 30" - 40" | 40" - 50" | 50" - 60" | 60" - 70" | > 70" |

Figure 11.25

Average annual precipitation in the contiguous United States. (One inch equals 2.5 cm.)

National Oceanic and Atmospheric Administration

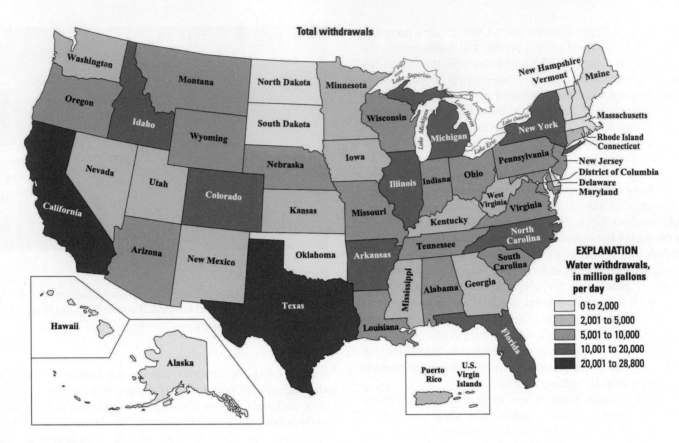

Total withdrawals

EXPLANATION
Water withdrawals, in million gallons per day

- 0 to 2,000
- 2,001 to 5,000
- 5,001 to 10,000
- 10,001 to 20,000
- 20,001 to 28,800

Figure 11.26

U.S. variations in water withdrawals by state in 2015. Compare regional patterns with **figure 11.25**.

Source: USGS

Regional Variations in Water Use

Water withdrawal varies greatly by region (**figure 11.26**), as does water consumption. These two quantities are not necessarily directly related, for the fraction of water withdrawn that is actually consumed depends, in part, on the principal purposes for which it is withdrawn. This, in turn, influences the extent to which local water use may deplete groundwater supplies. Of the fresh water withdrawn, only about 20% is groundwater; but many of the areas of heaviest groundwater use rely on that water for irrigation, which consumes more than half of the water withdrawn.

The quantities withdrawn by each of the principal categories of water users are summarized in **figure 11.27**. The sources of water for the four major water uses are shown in **table 11.1**.

A key point to keep in mind is that while power generation may be the major water user, agriculture is the big water consumer. Power generators and industrial users account for more than half the water withdrawn, but nearly all their wastewater is returned as liquid water at or near the point in the hydrologic cycle from which it was taken. Most of these users are diverting surface waters and dumping used water back into the same lake or stream. Together, industrial

users and power generation *consume* only about 10% of the water withdrawn.

A much higher fraction of irrigation water is consumed: lost to evaporation, lost through transpiration from plants, or lost because of leakage from ditches and pipes (**figure 11.28**). Most of the water lost to evapotranspiration drifts out of the area to come down as rain or snow somewhere far removed from the irrigation site. It then doesn't contribute to the recharge of aquifers or to runoff to the streams from which the water was drawn. And about 48% of water used in irrigation nationwide is groundwater, coming from aquifers that may experience no modern recharge at all.

The regional implications of all this can be examined through **figures 11.26** and **11.29**. The former indicates the relative volumes of water withdrawn, by state. The latter figure shows that many states with high overall water withdrawal use their water extensively for irrigation. Where irrigation use of water is heavy, water tables have dropped by tens or hundreds of meters and streams have been drained nearly dry, while we have become increasingly dependent on the crops.

Water supply is an issue that is not confined to the western United States. It is an international problem as well. Large

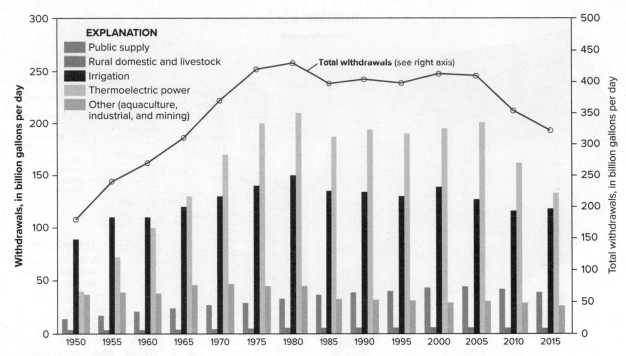

Figure 11.27

Water withdrawal by sector, 1950–2015. Note the drops in 2010 and 2015, despite the fact that U.S. population continues to grow. Water-supply pressures have encouraged increased application of conservation/recycling techniques such as those discussed later in the chapter; in some cases, less water is being used because drought has made water less available.

Source: U.S. Geological Survey, Estimated use of water in the United States in 2015 (USGS Circular 1441).

Table 11.1	US Freshwater Withdrawals by Use, 2015			
			Source, %	
Use	Water Withdrawn, Million Gallons per Day	Change from 2010	Surface Water	Groundwater
Thermoelectric power generation*	133,000	−18%	100	0
Irrigation	118,000	+2%	52	48
Public supply	39,000	−7%	61	39
Self-supplied industrial	15,000	−9%	81	19
Total**	322,000	−9%	61	39

Source: U.S. Geological Survey Circular 1441.

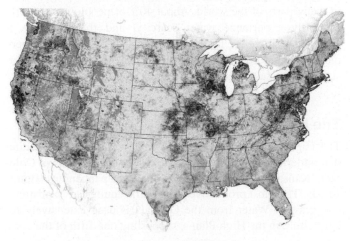

Figure 11.28

The evaporative stress index (ESI) for March-May 2021 shows many areas were in drought conditions. The ESI is a measurement of evapotranspiration, determined by land surface temperatures and leaf area index as measured by two satellites. ESI data is now openly available on the individual field scale to better manage water budgets in western states through OpenET.

National Oceanic and Atmospheric Administration

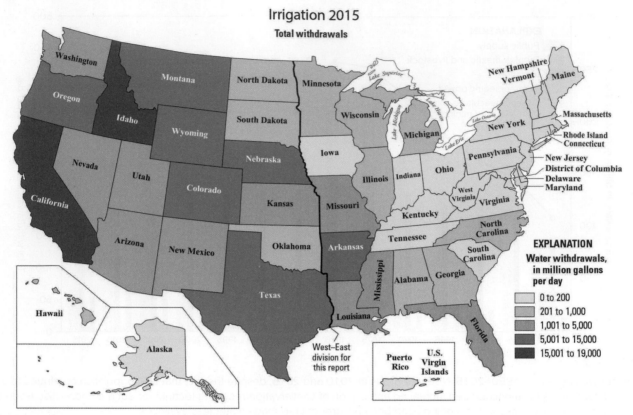

Irrigation 2015
Total withdrawals

EXPLANATION

Water withdrawals, in million gallons per day

- 0 to 200
- 201 to 1,000
- 1,001 to 5,000
- 5,001 to 15,000
- 15,001 to 19,000

West–East division for this report

Figure 11.29

Water use for irrigation, by state.

Source: U.S. Geological Survey

areas of China and India, the world's two most populous countries, face severe, imminent shortages of fresh water. So do many countries in the Middle East and northern Africa, where populations are growing quickly. Water becomes a political issue between nations when it crosses national boundaries. In many cases, the central issues relate either to water availability or to the quality of the available water. One country's water-use preferences and practices can severely disadvantage another, especially if they share a common water source. The Water, Peace, and Security partnership has created a Global Early Warning Tool based on water availability in order to identify risks so that water-related conflicts can be prevented.

11.9 Case Studies in Water Consumption

Learning Objectives

- Describe the physical aspect of the High Plains aquifer
- Explain the drop in water level in the High Plains aquifer
- Relate the near disappearance of the Aral Sea to human activity
- Explain the drop in water levels in Lake Chad

There is no lack of examples of water-supply problems; they may involve lakes, streams, or groundwater. Problems of surface-water supply may be either aggravated or alleviated as precipitation patterns shift in reflection of normal climate cycles and/or global change induced by human activities. Where groundwater is at issue, however, it is often true that relief through recharge is not in sight, as modern recharge may be insignificant. Many areas are drawing on fossil groundwater recharged long ago. For example, the upper Midwest uses groundwater recharged over 10,000 years ago as continental ice sheets melted. Water availability and population distribution and growth together paint a sobering picture of coming decades in many parts of the world. About 40% of people in the world currently experience water scarcity, and that estimate is expected to increase, likely displacing tens of millions of people in near future (**figure 11.30**). We will review a few water-supply issue cases below and in Case Study 11.1.

The High Plains (Ogallala) Aquifer System

The High Plains or Ogallala is a shallow, unconfined aquifer that underlies most of Nebraska and sizeable portions of Colorado, Kansas, and the Texas and Oklahoma panhandles (**figure 11.31**). The most productive units of the aquifer are sandstones and gravels. Water from the Ogallala is used extensively for agriculture in the High Plains, supporting one-fifth of the corn,

Freshwater withdrawals as a share of internal resources, 2017

Annual freshwater withdrawals refer to total water withdrawals from agriculture, industry and municipal/domestic uses. Withdrawals can exceed 100% of total renewable resources where extraction from nonrenewable aquifers or desalination plants is considerable.

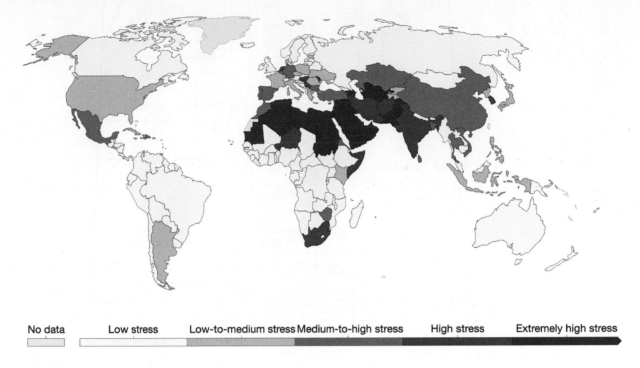

| No data | Low stress | Low-to-medium stress | Medium-to-high stress | High stress | Extremely high stress |

Source: UN Food and Agriculture Organization (FAO)   OurWorldInData.org/water-access-resources-sanitation/ • CC BY

Figure 11.30

Water stress is the ratio of water use to water availability; water scarcity is when there is not enough fresh water to meet demand. Note the coincidence of high water stress and high population.

Our World In Data

cotton, wheat, and cattle produced in the United States. More than 15 million acres of land are irrigated with water pumped from the Ogallala. The majority of people in this region also get their drinking water from the aquifer.

The groundwater was, for the most part, stored during the melting of the Pleistocene continental ice sheets. Present recharge is negligible over most of the region. Researchers estimate the original recoverable volume of groundwater to have been 3.2 billion **acre-feet**. Estimated depletion from predevelopment times through 2015 is over 273 million acre-feet. Recent studies put groundwater consumption at around 10.7 million acre-feet from 2013 to 2015, which is about a 25% increase in the annual rate of consumption compared to the previous 65 years. Soon after development in the plains began, around 1950, the water table began to drop noticeably. In areas of heavy use it had dropped over 100 feet by 1980, and over 200 feet by 2015. The remaining saturated thickness is less than 200 feet over two-thirds of the aquifer's area, and less than 100 feet over much of this southern region. Recent drought in the area has exacerbated the problem (**figure 11.32**). Because the aquifer covers such a large area and both rates of use and precipitation vary, depletion of the aquifer is not occurring uniformly.

Historically there has been little incentive to restrain water use given the large volume of water in the aquifer. And the fact that federal policies with price supports encourage the growing of water-thirsty crops such as cotton. The government cost-shares in soil-conservation programs and provides for crop-disaster payments, thus giving compensation for the natural consequences of water depletion. Federal tax policy provides for groundwater depletion allowances (tax breaks) for High Plains area farmers using pumped groundwater, with larger breaks for heavier groundwater use. The results of all these policies aren't surprising.

Still, farmers and ranchers are employing conservation techniques to use less water from the aquifer while improving the overall health of their land, and being financially successful while doing so. Improving soil health, which we will discuss later in chapter 12, is a key to retaining water and lessening the need for irrigation. No-till and intermixing crops with pasture grasses are two ways to do this. Farmers are also using more technical data, such as that that comes from remote sensing, or direct measurements of soil moisture so that they can use water more precisely and just when their crops need it, which is also a money-saving technique. Recent

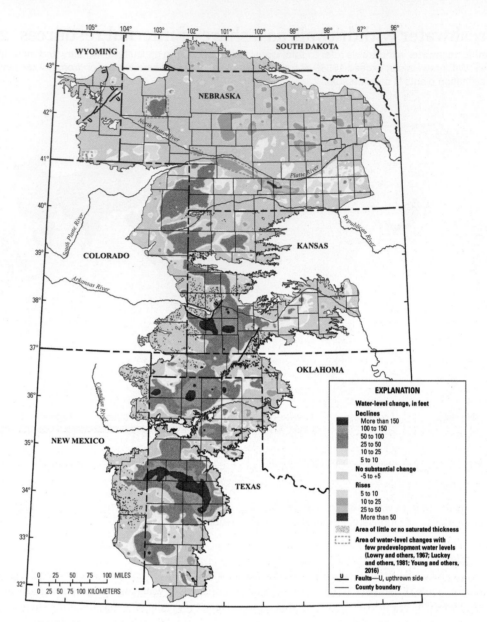

Scientific Investigations Report 2017–5040

Figure 11.31

Changes in water levels in the High Plains aquifer, predevelopment to 2015. Only in a few places, such as along the Platte River, has the water table risen significantly, in response to short-term precipitation increases; the overall long-term trend is clearly down.

U.S. Department of the Interior/U.S. Geological Survey

initiatives run through the USDA were very successful in assisting agriculturists in becoming more productive and sustainable at the same time through nutrient management, installation of irrigation management systems, conversion of cropland to dryland or retirement, and improving irrigation efficiency. All that wasn't cheap, with a total investment of over $26 million. The efforts are working for now at slowing down the depletion of the aquifer. However, the amount of irrigated land in Nebraska grew by 1.6 million acres between

1997 and 2017 and may continue to grow as agriculture shifts eastward from the dry western states.

The Aral Sea

The Aral Sea lies on the border of Kazakhstan and Uzbekistan. For decades, water from rivers draining into the Aral Sea has been diverted to irrigate land for growing rice and cotton. From 1973 to 1987, the Aral Sea dropped from fourth- to sixth-largest

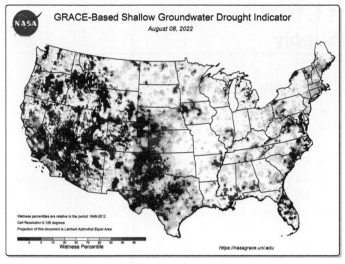

A

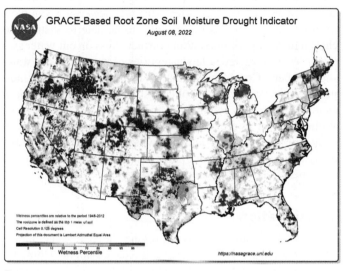

B

Figure 11.32

(A) Groundwater storage and (B) root-zone soil moisture (water in the top meter of soil) in August 2022 compared with the average for 1948–2012. Maps are based on data from satellites that detect changes in water storage by resultant small changes in Earth's gravitational field, combined with ground-based measurements. Warm colors indicate conditions dryer than normal, blues are wetter than normal.

(A, B) Source: NASA

lake in the world. Its area shrank by more than half, its water volume by over 60%. Its salinity increased from 10% to over 23%, and a local fishing industry failed; the fish could not adapt to the rapid rise in salt content. By now, the South Aral Sea is all but gone (**figure 11.33**). As the lake has shrunk, former lake-bed sediments—some containing residues of pesticides washed off of the farmland—and salt have been exposed. Meanwhile, the climate has become drier, and dust storms now regularly scatter the salt and sediment over the region. Ironically, the salt

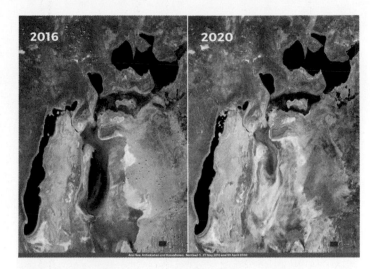

Figure 11.33

The shrinking of the Aral Sea. The original shoreline can be approximated by following the outline of the lighter-colored sediments.

European Union, Copernicus Sentinel-2 imagery

deposited on the fields by the winds is reducing crop yields. Respiratory ailments are also widespread as the residents breathe the pesticide-laden dust. With the loss of the large lake to provide some thermal mass to stabilize the local climate, the winters have become colder, the summers hotter. Here, then, there are serious issues even beyond those of water supply, which are acute enough.

Some recent efforts at water management have led to limited local improvement. A dam completed in 2005 now traps more water in the North Aral Sea, keeping it from draining down to the south. This has allowed recovery of some fisheries in the North Aral Sea. Another dike is in the planning to further improve the lake, but it is unlikely that it will ever be restored to its former extent.

Lake Chad

Lake Chad, on the edge of the Sahara Desert in West Africa, is another example of a disappearing lake. It was once the sixth largest lake in the world, close in area to Lake Erie. But like Lake Erie, it is relatively shallow; Lake Chad was only 5 to 8 meters deep before the depletion of recent decades so the volume of water was modest from the outset. A combination of skyrocketing demand for irrigation water from the four countries bordering the lake (Chad, Niger, Nigeria, and Cameroon), plus several decades of declining rainfall, shrank Lake Chad to about one-twentieth its former size (**figure 11.34**). While it still supports vegetation in its former lake bed, the volume of usable water remaining in Lake Chad is severely limited; yet 30 million people still depend on it, and the population is growing rapidly.

A

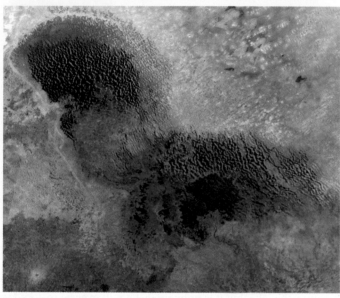

B

Figure 11.34

Lake Chad (A) close to highest measured level in 1963 compared to (B) 2021; blue is open water, green is wetlands, and desert is tan-colored.

(A, B) Source: USGS

The lake level has recovered a little since its lowest level in the late 1980s, likely due to increased precipitation from a shift in the Atlantic Multidecadal Oscillation mentioned in the last chapter. However, the lake is still well below its levels of the 1960s. Researchers think there is a minimum lake level supported by local groundwater, but they cannot presently examine the system directly because of political upheaval and violence in the region. Satellites remain the only safe way to monitor the lake.

11.10 Extending the Water Supply

Learning Objectives

- List ways in which to conserve water
- Summarize recent trends in irrigation water conservation
- Describe interbasin water transfer
- Summarize the process of desalination through distillation and filtration
- Recognize the disadvantages of desalination

Water supplies change along with climate, so as we find ourselves with large, growing population centers located in arid areas that are getting even drier, what practices can we employ, from personal to industrial, in order to be able to sustain our chosen lifestyles without putting further stress on our water systems? We will review a few categories of extending our water supplies here. Consider, too, that no matter what we do as individuals, using our water supplies sustainably requires the application of necessary regulations on the state and federal level. While many of us are choosing to support companies that voluntarily conserve water and following water-saving practices, many other entities will not make that change until it is monetarily beneficial or they are forced to do so.

Conservation

The most basic approach to improving the U.S. water-supply situation is conservation. Conservation means not only thoughtfully using less water but also technically reducing the amount of water lost. In response to soaring water demand in the 1980s, officials in Boston focused on fixing leaks in the water-supply system, together with promoting low-flow plumbing fixtures and improving efficiency of industrial processes; municipal water use declined over 30% in the next two decades. We often describe water as being *wasted,* but a better description is that we use it without thinking of the consequences of that use, especially if the water is readily available and cheap. A person living in the humid Northeast using a private groundwater source to keep their tomatoes healthy likely thinks a lot differently about how they use water compared to someone in the dry, drought-prone Southwest using municipal water for the same activity. We likely all use water injudiciously in our homes at some points, whether it is taking extra long hot showers on a cold day, using inefficient appliances, or watering an extensive lawn just to keep it green. There are other lifestyle choices that involve water that may not be as apparent, like choosing what we eat. Raising large livestock for meat requires far more water per pound of protein than growing vegetables for protein. Still, municipal and rural water uses (excluding irrigation) together account for only about 10% of total U.S. water consumption.

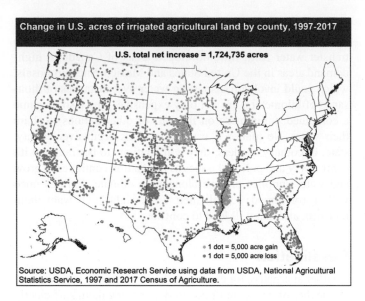

Change in U.S. acres of irrigated agricultural land by county, 1997-2017

U.S. total net increase = 1,724,735 acres

• 1 dot = 5,000 acre gain
• 1 dot = 5,000 acre loss

Source: USDA, Economic Research Service using data from USDA, National Agricultural Statistics Service, 1997 and 2017 Census of Agriculture.

Figure 11.35

Texas and California each lost more than 1 million acres of irrigated land from 1997 to 2017, while Arkansas and Nebraska increased by that same amount during that time period.

Source: USDA

Industrial use is only 5%, and has decreased by over 40% in the last 30 years due to less manufacturing overall, regulations, and conservation practices.

The big water drain is clearly irrigation, which already has decreased withdrawals in the last 40 years by over 20% while at the same time the amount of land irrigated increased by 20%. This decrease in the intensity of irrigation is a reflection of technological improvements and north- and eastward shifts to areas with more readily available water (**figure 11.35**). The raising of crops that require a great deal of water is already shifting to areas where natural rainfall is adequate to support them. Use of pressurized irrigation systems (sprinkler and micro/drip) that generally are more efficient at using water increased from 37% to 72% of all irrigated land from 1984 to 2018; less water is lost through evaporation, deep percolation, and runoff when water is delivered directly to plants compared to gravity systems. In an orchard, pipes can carry water to individual trees or groups of trees and release it slowly into the ground close to the plant roots, with much less waste. The most efficient methods are typically more expensive. They have become more attractive as water prices have been driven up by shortages, but they may not be practical for low-priced field-grown crops for which land is plowed up each year, such as corn, wheat, or soybeans. The newest technologies based on remote sensing, coupled with government policies that provide incentives to use less, are the next steps in irrigation conservation.

If you are interested in decreasing your domestic use of water, there are a variety of ways to do that. For example, lawns can be watered morning or evening when evaporation is less rapid than at midday; or you can remove your lawn entirely and

Figure 11.36

Using plants that are native to your area lessens the need for irrigation and the costs that come along with it while improving the environment for other creatures.

Debra Wiseberg/E+/Getty Images

plant it with native vegetation that benefits bees, birds, and butterflies (**figure 11.36**). Purchase the most efficient appliances you can when your old ones are no longer working. To further decrease your water use, make a conscious effort to find out how much water goes into making the products you purchase and the food you eat. In most cases, if you are thinking about "reduce, reuse, recycle" when going about your activities, you are also actively conserving water. On a larger scale, municipalities and industries are practicing water conservation because it saves money and prepares for a future where water resources may be more precious than they currently are. We will be discussing the recycling of wastewater later in chapter 16).

Interbasin Water Transfer

In the short term, conservation alone will not resolve the imbalance between demand and supply. New sources of supply are needed. Part of the supply problem, of course, is purely local. People settle in areas that may have enough water initially, but then population, industry, and/or agriculture expands; or water sources are contaminated; or an area experiences a drought. And then more water is needed than is naturally available, so distant sources are sought. This is the idea behind interbasin transfers—moving surface waters from one stream system's drainage basin to another's where demand is higher.

California pioneered the idea with the controversial Los Angeles Aqueduct project. The aqueduct was completed in 1913 and carried nearly 150 million gallons of water per day from the Owens Valley in the eastern slopes of the Sierra Nevada to Los Angeles. In 1958, the system was expanded to bring water from northern California to the southern part of the state. Bringing water to the city was a popular idea with city goers but not so much with residents of the Owens

Figure 11.37

Diversion of surface water that formerly flowed into Mono Lake, California, to supply water to Los Angeles caused lake level to drop, exposing sedimentary formations originally built underwater. Lake levels are still 36' lower that they were before the diversion, and 11' below goal level.

Steve Heap/Shutterstock.com

Valley that lost their water, resulting in the Los Angeles Water Wars (**Figure 11.37**) More and larger projects have been undertaken since. Bringing water from the Colorado River to southern coastal California required the construction of over 300 kilometers (200 miles) of tunnels and canals. Other water projects have transported water over whole mountain ranges. Such projects are not confined to the drier west. For example, New York City draws on several reservoirs in upstate New York.

Interbasin transfers of surface water continue to be proposed and implemented in the United States and worldwide. Political problems are common even when the transfer involves diverting water from one part of a single state to another. A proposal to expand the California aqueduct system was introduced in 1982, but voters in the northern part of the state overwhelmingly opposed it and the proposition lost. The Great Lakes, with over 20% of the world's surface fresh water, can be very tempting to more arid states with ideas of water diversions. In the 1990s, officials in states around the Great Lakes objected to a suggestion to divert some lake water to states in the South and Southwest. A large-scale diversion project from Lake Michigan to Waukesha, Wisconsin, is currently in its construction stage. In the late 1980s, the city's public groundwater supply had too much radium and the diversion is the decades-long answer to the search for another source which involved the creation of a legal compact before the water could even be requested. In the agreement, all of the water used, which will be between 5 and 8 million gallons per day, has to be returned to Lake Michigan. Not surprisingly, the process of requesting the water and approving that request took longer than expected and involved long-standing animosities between local politicians.

The problems may be far greater when transfers between nations are involved. Various proposals have been made to transfer water from little-developed areas of Canada to high-demand areas in the United States and Mexico. Such proposals, which could involve transporting water over distances of thousands of kilometers, are not only expensive, they also presume a continued willingness on the part of other nations to share their water. No matter if water diversions occur over county, state, or country borders, they will continue to be fraught with controversy and political issues as population continues to grow and water supplies change along with climate. Entities once willing (or forced) to sell their water rights will want them returned, and that will lead to conflict.

Desalination

Another alternative for extending the water supply is to improve the quality of waters not now used, purifying them sufficiently to make them usable. Desalination allows people in coastal regions to tap the vast ocean reservoirs, and those farther inland to use groundwaters that contain excessive concentrations of dissolved materials. There are a variety of ways to desalinate water that fall into two categories: either passing the water through a membrane or filter, or using a heat source in distillation (**figure 11.38**).

Reverse osmosis is the most widely used method of purifying water with the use of a filter. Pressure is employed to force water through a semi-permeable membrane which captures the undesirable compounds. An advantage of this method is that it can rapidly filter great quantities of water. For example, the Carlsbad plant in San Diego, California, produces up to 50 million gallons a day and the Victorian plant in Melbourne, Australia, can purify 460 million liters (121 million gallons) a day. Reverse osmosis can also operate on a very small scale providing drinking water for small, remote communities using the Sun as a source of energy. A major disadvantage of reverse osmosis is that the membranes need to be flushed out on a regular basis, and this may include using chemicals that get washed back into the original water source.

Distillation involves heating or boiling water full of dissolved minerals. The water vapor driven off is pure water, while the minerals stay behind in what remains of the liquid. Distillation requires a heat source which traditionally would be fossil fuels. The Sun is an alternative possible heat source, and some solar desalination facilities already exist; in fact, using the Sun to purify water is a process that humans have employed for a long time. But because the Sun provides low-intensity heat, large desalination plants employing this technique are impractical considering that a large city would need a facility covering thousands of square kilometers.

Desalination is a relatively expensive process that can increase the cost of water by more than 75%. Such an increase in a water bill would hit farmers a lot harder than homeowners given the amount used, adding costs that might be hard to pass on to consumers while still remaining competitive domestically and abroad. In areas such as the arid Middle East that are rich in

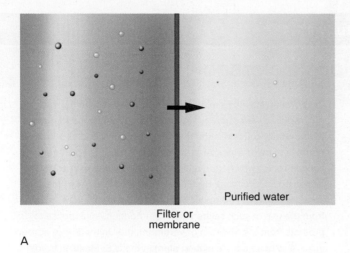

A

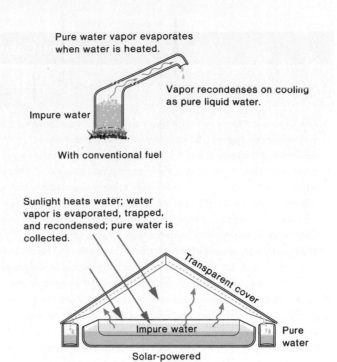

B

Figure 11.38

Methods of desalination. (A) Filtration (simplified schematic): Dissolved and suspended material (represented by dots at left) is screened out by very fine filters. (B) Distillation: As water is heated, pure water is evaporated, then recondensed for use. Dissolved and suspended materials stay behind.

energy resources, the more-acute need for irrigation water has made desalinated seawater a viable agricultural option economically. In most other areas, desalinated water is simply too expensive for large-scale and/or long-term irrigation use. Continued improvements in technologies that make the desalination process more efficient, meaning it uses less energy, are needed so agriculture doesn't continue to drain limited and dwindling surface- and groundwater supplies. Government subsidies to support desalination also likely will be needed to keep costs within reach.

There are even more issues related to desalinated water. Care must be taken to adjust the chemistry of the resultant water, not only for human consumption, but for plants as well. Large amounts of briny water are created that need to be disposed of, raising concerns about ecosystem health related to the chemistry of the water and how the injection of dense water might affect thermohaline circulation. Some people are viewing the briny wastewater as a economic opportunity, as it may be used as a source of desirable compounds such as uranium or rare metals.

The Contentious Colorado

The Colorado River's drainage basin drains portions of seven western states (**figure CS 11.1.1**). Many of these states have extremely dry climates, and it was recognized over a century ago that some agreement would have to be reached about which region was entitled to how much of that water. Intense negotiations during the early 1900s led in 1922 to the adoption of the Colorado River Compact, which apportioned 7.5 million **acre-feet** of water per year each to the Upper Basin (Colorado, New Mexico, Utah, and Wyoming) and the Lower Basin (Arizona, California, and Nevada). No provision was made for Mexico, into which the river ultimately flows.

Rapid development occurred throughout the region. Huge dams impounded enormous reservoirs of water for irrigation or for transport out of the basin (**figure CS 11.1.2**). In 1944, Mexico was awarded by treaty some 1.5 million acre-feet of water per year, but with no stipulation concerning water quality. Mexico's share was to come from the surplus above the allocations within the United States or, if that was inadequate, equally from the Upper and Lower Basins.

Heavy water use has led to a reduction in both water flow and water quality in the Colorado River. The reduced flow results not only from diversion of the water for use but also from large evaporation losses from the numerous reservoirs in the system. The reduced water quality is partly a consequence of that same evaporation, concentrating dissolved minerals, and of selective removal of fresh water. Also, many of the streams flow through soluble rocks, which then are partially dissolved, increasing the dissolved mineral load. By 1961, the water delivered to Mexico contained up to 2700 ppm TDS, and partial crop failures resulted from the use of such saline water for irrigation. In response to protests from the Mexican government, the United States agreed in 1974 to build a desalination plant at the U.S.-Mexican border to reduce the salinity of Mexico's small share of the Colorado.

Repeated analysis shows that sufficient water flow simply does not exist in the Colorado River basin even to supply the full U.S. allocations. The streamflow measurements made in the early 1900s occurred during an unusually wet period and not are representative of current conditions. The average flow for the twentieth century was 15.2 million acre-feet, and the current flow is down about 20% at 12.4 million acre-feet. The most recent research by the USGS has annual evaporative loss from Lake Mead at just under 1 million acre-feet, and that is just one of the many reservoirs along the river.

Given the long drought in the west and the clear trend of water available from the Colorado, states in the Upper and Lower Basins worked on drought contingency plans in order to avoid disastrous water supply issues in the future. Actions were based on water levels in Lake Power in the Upper Basin and Lake Mead

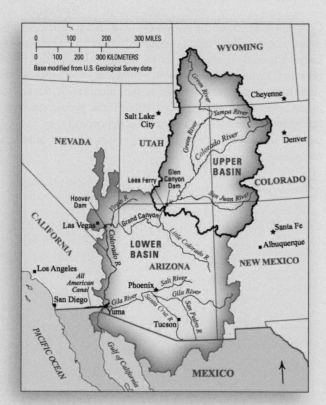

Figure CS 11.1.1

The Colorado River Basin.

Source: U.S. Geological Survey Water Supply Circular 1001.

Figure CS 11.1.2

When this portion of the fertile Colorado River floodplain on the Fort Mojave reservation was leased to agribusiness, population in the area grew, and so did the demand for irrigation water in this obviously dry region where Nevada, Arizona, and California meet.

Source: NASA

A

B

Figure CS 11.1.3

Lake Mead levels fell nearly 130 feet between September 2000 (left) and August 2021 (right). It was last close to its full capacity of 1220' (measured near Hoover Dam) in 1999. In June 2022 the level was 1045', the lowest level ever recorded since it first filled, and is continuing to drop. Check where it is now at USBR's Lake Mead at Hoover Dam, end of month elevation

(A, B) Source: U.S. Geological Survey

in the Lower Basin, and were approved by Congress in 2019. For the given water elevation reached in the lakes, 8 tiers of water usage curtailment were outlined. When levels reached between 1067 and 1068' in August 2021 in Lake Mead, a Tier 1 water shortage was declared. According to agreed-upon drought contingency, that calls for Arizona to reduce their allocation by 512,000 acre-feet and Nevada by 21,000. California has water rights seniority, and will not need to cut back until a Tier 3 event, when Lake Mead falls below 1,045*. Because most water from the Colorado is used for agriculture and municipal use is preferred, cuts will likely impact agriculture the hardest.

California has already been compelled to reduce it use of Colorado River water due to a lawsuit from Arizona over how much water each state is allowed. California is very dependent on this water, as the Colorado River Aqueduct, completed in 1941, alone carries over 1.2 million acre-feet of water per year from Lake Havasu to supply Los Angeles and San Diego, and California's 4.4 million acre-feet is already the lion's share of the Lower Basin's allocation.

Back to the level of the reservoirs. The combination of increased demand and increasing drought is draining the reservoirs. The largest of these is Lake Mead behind Hoover Dam, which provides water supply and hydropower for millions of people. But the lake level has been dropping; it is now at its lowest level since the reservoir first began to fill in the 1930s. The lake has

* This level was reach in June 2022.

shrunk considerably (**figure CS 11.1.3**), reducing the water reserve and threatening hydropower generation, and leaving recreational docks high and dry far from the new shore. Lake Powell levels are also historically low, being over 40' lower in March 2022 than at the same time in 2021, and about 90 feet lower that in 2018.

There are a few other ventures to consider when determining future use of the water. One of these is any expansion of mining, which can be a water-intensive process. With the current trend in coal used for electricity generation falling sharply, it is unlikely coal related use will increase. Mining oil shale is also water intensive, and there is a lot of it in the Colorado basin, but the expansion of oil shale will continue to be tenuous in the near future, as we will discuss in more depth in chapter 14. The most likely and nearest future expansion of water use is related to expanding populations in the southwest in places such as Washington County, Utah which has grown 80% in the last 20 years. The proposed $2 billion Lake Powell pipeline would bring 86,000 acre-feet annually from Lake Powell 140-miles to a location that already has a relatively high per capita usage of water (Do a quick search of golf courses in St. George, Utah). Utah argues that they have a right to the water as they are not using their full 23% allotment of the Colorado River. Local tribes and conservationists, as well as officials from other states within the basin have already voiced concerns, objections, and lawsuits.

Summary

Most of the water in the hydrosphere at any given time is in the oceans; most of the remaining fresh water is stored in ice sheets. Relatively little water is in lakes and streams. Humans use both surface water and groundwater as freshwater sources. The availability of groundwater is influenced by such factors as the presence of suitable aquifers, water quality, and rate of recharge relative to rate of water use. Where groundwater is plentiful and subsurface rocks are soluble, dissolution may create caves below ground and sinkholes at the surface, producing a distinctive karst landscape. Karst aquifers may be productive, with rapid water flow, but the supply may be erratic and easily polluted.

   In the United States, more than half of water withdrawals are used for public supplies and thermoelectric power generation. These applications generally consume very little water. Most of the water actually consumed in the United States is associated with irrigation practices. Groundwater accounts for only about 25% of all withdrawals, but half of that is consumed and often much faster than it can be replaced by recharge. Adverse consequences of such rapid groundwater consumption include lowering water tables, surface subsidence, and saltwater intrusion. Conservation, interbasin transfers of surface water, and desalination are possible ways to extend the water supplies of high-demand regions. Desalinated water is presently too expensive to use for irrigation on a large scale except in countries that are energy-rich and water-poor, which means that it is unlikely to have a significant impact on the largest consumptive U.S. water use for some time to come. The chemistry of water is critical to the health of crops as well as people; with desalinated water, key nutrients may need to be added back to the water before use.

Key Terms and Concepts

aquifer	discharge	porosity	soil moisture
aquitard	groundwater	potentiometric surface	unconfined aquifer
artesian system	hard water	recharge	unsaturated zone
cone of depression	hydraulic head	saltwater intrusion	water table
confined aquifer	karst	saturated zone	
Darcy's Law	permeability	sinkhole	

Test Your Learning

1. Explain the importance of porosity and of permeability to groundwater availability for use.

2. Define the following terms: *groundwater, water table,* and *potentiometric surface.*

3. Describe how an artesian aquifer system is formed.

4. State Darcy's law, and explain how the various parameters in the equation affect groundwater flow.

5. Explain how sinkholes develop in a karst landscape.

6. Describe how karst aquifers differ from non-karst aquifers, in the context of water supply.

7. List three characteristics used to describe water quality.

8. Define *hard water;* indicate why it is often considered undesirable, and how the problem can be addressed.

9. Describe two possible consequences of groundwater withdrawal exceeding recharge.

10. Explain the process of saltwater intrusion.

11. Indicate two ways in which urbanization can affect groundwater recharge.

12. Industry and power generations are the big water *users,* but agriculture is the big water *consumer.* Explain.

13. Compare filtration and distillation as desalination methods, noting advantages and drawbacks of each.

Exploring Further

1. Where does your water come from? What is its quality? How is it treated, if at all, before it is used? Is a long-term shortage likely in your area? If so, what plans are being made to avert it?

2. Compare **figures 11.26** and **11.29**. Speculate on the reasons for some of the extremes of water withdrawal and consider the likelihood that this withdrawal is predominantly *consumption* in the highest-use states.

3. Some seemingly minor activities are commonly cited as great water-wasters. One is letting the tap run while brushing one's teeth. Try plugging the drain while doing this; next, measure the volume of water accumulated during that single toothbrushing (a measuring cup and bucket may be helpful). Consider the implications for water use, bearing in mind the roughly 330 million persons in the United States. Read your home water meter before and after watering the lawn or washing the car, and make similar projections.

4. The USGS now maintains a groundwater site: **groundwaterwatch.usgs.gov/** Examine the site for areas of unusually high or low groundwater levels, and compare with streamflow data for the same areas. Consider what factors might cause groundwater and streamflow levels to be correlated (both high or both low), or not. You may find areas in which some wells show abnormally high levels, and some show abnormally low levels; how can this be?

5. Consider a confined sandstone aquifer with a hydraulic gradient of 0.2 and hydraulic conductivity of 10^{-5} cm/sec, and porosity of 12%. Use Darcy's law to determine the discharge, in liters per day, through a cross section of 1 square meter. (Watch units as you work this problem!)

6. Model artesian water flow by holding up a water-filled length of rubber tubing or hose at a steep angle. The water is confined by the hose, and the water in the low end is under pressure from the weight of the water above it. Poke a small hole in the top side of the lower end of the hose, and a stream of water will shoot up to the level of the water in the hose.

CHAPTER 12

Weathering, Erosion, and Soil Resources

Although it may look like it, soil is not just dirt. It is a complex substance created by the interplay of Earth's lithosphere, biosphere, hydrosphere, and atmosphere. Each type of soil is a reflection of the parent materials it formed from, the climate it formed in, and the weathering processes that broke it down. Soil is an essential resource on which we depend for the production of the major portion of our food. It may not strike many people as a resource requiring special care for its preservation. It most places, a substantial quantity seems to be underfoot. And problems associated with soil such as erosion, loss of fertility, and sediment pollution of surface waters may simply go unnoticed to the untrained eye. Unfortunately, we are losing soil at a rate higher than it is being naturally replenished. As we disturb more

places, and fail to take care of the soil we do have, soil erosion is becoming a significant and very expensive problem. Soils vary in character and so in their suitability for agriculture, construction, and other purposes. In this chapter, we will examine the nature of soil, its formation and properties, aspects of the soil erosion problem, and some strategies for reducing erosion.

Chapter Outline

12.1 Soil Formation

12.2 Chemical and Physical Properties of Soils

12.3 Soils and Human Activities

The thick, dark, organic-rich surface horizon is clearly visible in this mollisol, one of the most productive agricultural soils in the world. It is the soil of much of the Great Plains in the United States, and combined with the water from the Great Plains aquifer, is the reason why so many crops are grown in this region.

USDA Natural Resources Conservation Service

12.1 Soil Formation

Learning Objectives

- Recall the different definitions of soil
- Compare mechanical and chemical weathering
- Exemplify different types of weathering
- Identify the soil horizons in a generalized profile
- Describe the variability in soil horizons and boundaries

Soil is defined in different ways for different purposes. Engineering geologists define soil very broadly to include all unconsolidated material overlying bedrock. Soil scientists restrict the term *soil* to those materials capable of supporting plant growth and distinguish it from *regolith,* which includes all unconsolidated material at the surface, fertile or not. For example, the loose material on Mars is regolith. In addition to rock and mineral fragments, soils generally must contain organic matter. Conventionally, the term *soil* implies little transportation away from the site where it formed, while the term *sediment* indicates matter that has been transported and redeposited by wind, water, or ice.

Soil is produced by *weathering,* a term that includes a variety of chemical, physical, and biological processes acting to break down rocks and minerals. It may be formed directly from bedrock or from further breakdown of transported sediment such as glacial till. The relative importance of the different kinds of weathering processes is largely determined by climate. Climate, topography, the composition of the material from which the soil is formed, the activity of organisms, and time govern a soil's final composition. In this section, we will review the weathering processes involved in the creation of soil, and characterize the different soil horizons that make-up a soil profile.

Soil-Forming Processes: Weathering

Mechanical weathering, also called *physical weathering,* is the physical breakup of rocks without changes in the rocks' composition. Rocks physically weather in response to changes in temperature and pressure; the abrasive action of wind, water, and ice; biological action; and wedging caused by ice and minerals. In a cold climate, with temperatures that fluctuate above and below freezing, water in cracks repeatedly freezes and expands, forcing rocks apart. Crystallizing salts in cracks may have the same wedging effect. Whatever the cause, the principal effect of mechanical weathering is the breakup of large chunks of rock into smaller ones. In the process, the total exposed surface area of the particles is increased (**figure 12.1**).

Chemical weathering involves the breakdown of minerals by chemical reaction with water, with other chemicals dissolved in water, or with gases in the air. Minerals differ in the kinds of chemical reactions they undergo. Calcite tends to dissolve completely, leaving no other minerals behind in its place. Calcite dissolves rather slowly in pure water but more rapidly in acidic water. Many natural waters are slightly acidic; acid rainfall or acid runoff from coal strip mines (see chapter 14) is more

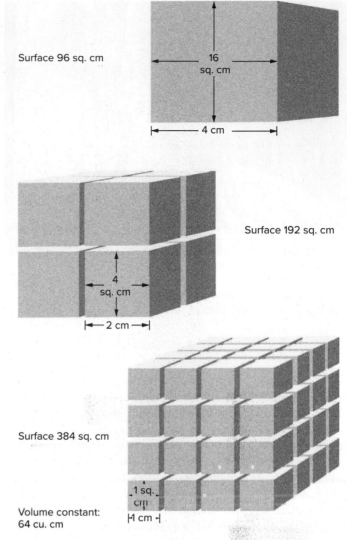

Figure 12.1

Mechanical breakup into smaller pieces increases the surface area and the surface-to-volume ratio.

so and causes more rapid dissolution. This is a serious problem where limestone and its metamorphic equivalent marble, are widely used for outdoor sculptures and building stone (**figure 12.2**). Calcite dissolution is gradually destroying delicate sculptural features and eating away at the very fabric of many buildings in urban areas and where acid rain is common.

Silicates tend to be less susceptible to chemical weathering and leave other minerals behind when they are attacked. Feldspars principally weather into clay minerals, an important component of many soils. Ferromagnesian silicates leave behind insoluble iron oxides and hydroxides and some clays, with other chemical components being dissolved away. Those residual iron compounds are responsible for the reddish or yellowish colors of many soils (**figure 12.3**). In most climates, quartz is extremely resistant to chemical weathering, dissolving only slightly. Representative weathering reactions are shown in table 12.1.

Figure 12.2

This 170-year-old outdoor sculpture in Segovia, Spain, has weathered due to dissolution.

DanielCz/Shutterstock

Figure 12.3

This vivid red Hawaiian soil is derived from weathering of iron-rich basalt flows, exposed here in Waimea Canyon.

©Carla Montgomery

The degree of susceptibility of many silicates to chemical weathering can be inferred from the conditions under which the silicates formed. Given several silicates that have crystallized from the same magma, those that formed at the highest temperatures tend to be the least stable, or most easily weathered, and vice versa. A rock's tendency to weather chemically is determined by its mineral composition. For example, a gabbro (the coarsely crystalline equivalent of basalt), formed at high temperatures and rich in ferromagnesian minerals, generally weathers more readily than a granite rich in quartz and low-temperature feldspars.

Climate plays a major role in the intensity of chemical weathering. Most of the relevant chemical reactions involve water. All else being equal, the more water, the more chemical weathering. Also, most chemical reactions proceed more rapidly at high temperatures than at low ones. Therefore, warm climates are more conducive to chemical weathering than cold ones.

The rates of chemical and mechanical weathering are interrelated. Chemical weathering may speed up the mechanical breakup of rocks if the minerals being dissolved are holding the rock together by cementing the mineral grains, as in some sedimentary rocks. Increased mechanical weathering may, in turn, accelerate chemical weathering through the increase in exposed surface area, because it is only at grain surfaces that minerals, air, and water interact. The higher the ratio of surface area to volume, the more rapid the chemical weathering. The vulnerability of surfaces to attack is also shown by the tendency of angular fragments to become rounded by weathering (**figure 12.4**).

The organic component of soil is particularly important to its fertility and influences other physical properties as well.

Table 12.1	Some Chemical Weathering Reactions

Solution of calcite (no solid residue)

$CaCO_3 + 2\,H^+ = Ca^{2+} + H_2O + CO_2$ (gas)

Breakdown of ferromagnesians (possible mineral residues include iron compounds and clays)

$FeMgSiO_4$ (olivine) $+ 2\,H^+ = Mg^{2+} + Fe(OH)_2 + SiO_2{}^*$

$2\,KMg_2FeAlSi_3O_{10}(OH)_2$ (biotite) $+ 10\,H^+ + \frac{1}{2}\,O_2$ (gas) $= 2\,Fe(OH)_3 + Al_2Si_2O_5(OH)_4$ (kaolinite, a clay) $+ 4\,SiO_2{}^* + 2\,K^+ + 4\,Mg^{2+} + 2\,H_2O$

Breakdown of feldspar (clay is the common residue)

$2\,NaAlSi_3O_8$ (sodium feldspar) $+ 2\,H^+ + H_2O = Al_2Si_2O_5(OH)_4 + 4\,SiO_2{}^* + 2\,Na^+$

Solution of pyrite (making dissolved sulfuric acid, H_2SO_4)

$2\,FeS_2 + 5\,H_2O + \frac{15}{2}\,O_2$ (gas) $= 4\,H_2SO_4 + Fe_2O_3 \cdot H_2O$

Notes: Hundreds of possible reactions could be written; the above are only examples of the common kinds of processes involved.

All ions (charged species) are dissolved in solution; all other substances, except water, are solid unless specified otherwise.

Commonly, the source of the H+ ions for solution of calcite and weathering of silicates is carbonic acid, H_2CO_3, formed by solution of atmospheric CO_2.

*Silica is commonly removed in solution.

Figure 12.4

(A) Angular fragments are rounded by weathering. Corners are attacked on three surfaces, edges on two. In time, fragments become rounded as a result. (B) Granite at Joshua Tree National Park, California, first breaks up along fractures into rectangular blocks (see outcrop at rear); the blocks gradually weather into rounded boulders (foreground).

(B) ©Carla Montgomery

Figure 12.5

Tree roots working into cracks in this granite help break up the rock mass in Rocky Mountain National Park.

©Gina S. Szablewski

Biological processes are also important to the formation of soil. Biological weathering effects can be either mechanical or chemical. Among the mechanical effects is the action of tree roots in working into cracks to split rocks apart (**figure 12.5**). Chemically, many organisms produce compounds that may react with and dissolve or break down minerals. Plants, animals, and microorganisms develop more abundantly and in greater variety in warm, wet climates. Mechanical weathering is generally the dominant process in areas where climatic conditions have limited the impact of chemical weathering and biological effects; cold regions at high altitudes and in high latitudes, and arid regions.

Soil Profiles, Soil Horizons

The result of mechanical, chemical, and biological weathering, together with the accumulation of decaying remains from organisms living on the land and any input from the atmosphere, is the formation of a blanket of soil between bedrock and atmosphere. A cross section of this soil blanket usually reveals a series of zones of different colors, compositions, and physical properties. The number of recognizable zones and their thicknesses vary. A basic, generalized soil profile as developed directly over bedrock is shown in **figure 12.6A**.

At the very top is the **O horizon,** consisting wholly of organic matter, living or decomposed. Below that is the **A horizon.** It consists of the most intensively weathered rock material, being the zone most exposed to surface processes, mixed with organic debris from above. Unless the local water table is exceptionally high, precipitation infiltrates down through the A horizon and below. In so doing, the water may dissolve soluble minerals and carry them away with it. This process is known as **leaching,** and it may be especially intense just below the A horizon, as acids produced by the decay of organic matter seep downward with percolating water. The **E horizon,** below the A horizon, is therefore also known as the **zone of leaching.** Fine-grained minerals, such as clays, may also be washed downward through this zone.

Many of the minerals leached or extracted from the E horizon accumulate in the layer below, the **B horizon,** also known as the **zone of accumulation.** Soil in the B horizon has been somewhat protected from surface processes. Organic matter from the surface is largely absent from the B horizon. It may contain relatively high concentrations of iron and aluminum oxides, clay minerals, and, in drier climates, even soluble minerals such as calcite. Below the B horizon is a zone consisting principally of very coarsely broken-up bedrock and little else. This is the **C horizon,** which does not resemble our usual idea of soil at all.

	Organic matter
O	Organic matter mixed with rock and mineral fragments
A E	**Zone of leaching** dissolved or suspended materials are carried downward by water
B	**Zone of accumulation** accumulation of iron, aluminum, and clay leached down from the E horizon; contains soluble minerals like calcite in drier climates
C	
Bedrock	Weathered parent material, partially broken down

A

B

C

Figure 12.6

(A) A generalized soil profile. Individual horizons can vary in thickness. Some may be locally absent, or additional horizons or subhorizons may be identifiable. (B) This Colorado roadcut illustrates well-developed soil horizons; note the dark color of the organic-rich layer at the top—and how thin it is. (C) Soil developed over this sandstone along the California coast shows a thicker A horizon.

(B, C) ©Carla Montgomery

12.2 Chemical and Physical Properties of Soils

Learning Objectives

- Identify the different influences on soil properties
- Recall soil color reflects composition
- Explain how soil texture affects drainage
- Define soil structure
- Compare pedalfer and pedocal soils
- Use the U.S. comprehensive soil classification

The boundaries between adjacent soil horizons may be sharp or indistinct. In some instances, one horizon may be divided into several recognizable subhorizons. Subhorizons also may exist that are gradational between A and B or B and C. It is also possible for one or more horizons locally to be absent from the soil profile. Some of the diversity of soil types is illustrated in the next section. All variations in the soil profile arise from the different mix of soil-forming processes and starting materials found from place to place. The overall total thickness of soil is partly a function of the local rate of soil formation and partly a function of the rate of soil erosion. The latter reflects the work of wind and water, the topography, and often the extent and kinds of human activities. Where slopes are too steep and erosion too rapid, soil formation may not occur at all.

The chemical and physical properties of a soil are in large part a reflection of the material from which it formed and the type and extent of weathering it experiences. If the bedrock or parent sediment is low in certain critical plant nutrients, the soil produced from it will also be low in those nutrients, and fertilizers may be needed to grow particular crops in that soil even if it has never been farmed before. Chemical weathering tends to remove elements from rock or soil, so extensive weathering would result in further depletion of certain elements in a soil. The weathering processes also influence the mineralogy of the soil, which not only affects its chemical nature but also its physical properties. Organic content also plays an important role in the nature of a soil. Below we will review the physical and chemical parameters used to characterize and classify soils.

Color, Texture, and Structure of Soils

Soil *color* tends to reflect compositional characteristics. Soils rich in organic matter tend to be black or brown, while those poor in organic matter are paler in color, often white or gray. When iron is present and has been oxidized by reaction with oxygen in air or water, it adds a yellow or red color (recall **figure 12.3**).

Soil *texture* is related to the sizes of fragments in the soil. The U.S. Department of Agriculture recognizes three size components: sand (grain diameters 2–0.05 mm), silt (0.05–0.002 mm), and clay (less than 0.002 mm). Soils are named on the basis of the dominant grain size(s) present (**figure 12.7**). An additional term, *loam,* describes a soil that is a mixture of all three particle sizes in similar proportions (10 to 30% clay, plus nearly equal amounts of sand and silt). Soils consisting of a mix of two particle sizes are named accordingly: for example, a silty clay would be a soil consisting of about half silt-sized particles, half clay-sized particles. The significance of soil texture is primarily the way it influences drainage. Sandy soils, with high permeability, drain quickly. In an agricultural setting, this may mean a need for frequent watering or irrigation. In a flood-control context, it means relatively rapid infiltration. Clay-rich soils, by contrast, may hold a great deal of water but be relatively slow to drain because of their much lower permeability. Satellite and ground-based surveys of soil over the contiguous United States show the variability of soils across the country (**figure 12.8**).

Soil *structure* relates to the soil's tendency to form lumps or clods of soil particles. These clumps are technically called *peds*. A soil that clumps readily may be more resistant to erosion; some of the fine-grained soils developed on loess (silt-rich sediments) in the United States and in China may erode very readily. On the other hand, soil consisting of very large peds with large cracks between them may be a poor growing medium for small plants with fine roots. Abundant organic matter may promote the aggregation of soil particles into crumblike peds especially conducive to good plant growth. Mechanical weathering can break up larger clumps into smaller, just as it breaks up rock fragments.

Soil Classification

Soil classification provides us with a convenient way to discuss soils of similar composition or structure. It also may indicate something about a soil's origins, which in turn may have implications for its suitability for agriculture or construction, or its vulnerability to degradation.

Early soil classification schemes emphasized compositional differences among soils and principally reflected the effects of chemical weathering. The resultant classification into two broad categories of soil was basically a climatic one. The **pedalfer** soils were characteristic of more humid regions. Where the climate is wetter, there is naturally more extensive leaching of the soil, and what remains is enriched in the

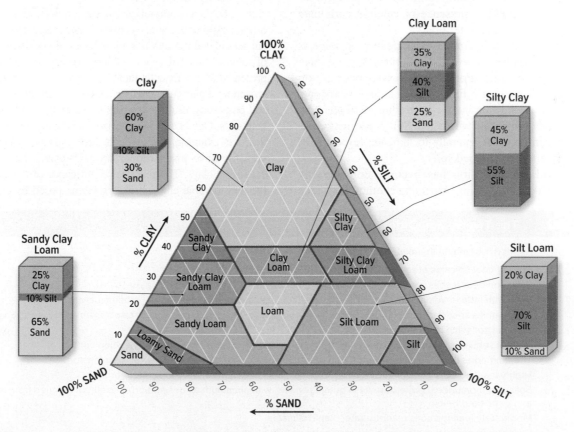

Figure 12.7

Soil-texture terminology reflects particle size, not mineralogy.

Figure 12.8

The proportions of soil components vary widely across the United States. Even at this scale, some distinctive areas stand out. For example, the Nebraska Sand Hills (dark sand-rich patch near center of map) and the silt-rich areas of the Mississippi River drainage basin.

NASA Earth Observatory image by Joshua Stevens

less-soluble oxides and hydroxides of aluminum and iron, with clays accumulated in the B horizon. In North America, pedalfer-type soils are found in higher-rainfall areas such as the eastern and northwestern United States and most of Canada. Pedalfer soils are typically acidic. Where the climate is drier, such as in the western and especially the southwestern United States, leaching is much less extensive. Even quite soluble compounds like calcium carbonate remain in the soil, especially in the B horizon. From this observation came the term **pedocal** for the soil of a dry climate. The presence of calcium carbonate makes pedocal soils more alkaline.

One problem with this simple classification scheme is that, to be strictly applied, the soils it describes must have formed over suitable bedrock. For example, a rock poor in iron and aluminum, such as a pure limestone or quartz sandstone, does not leave an iron- and aluminum-rich residue, no matter how extensively leached or intensely weathered. Still, the terms *pedalfer* and *pedocal* can be used generally to indicate, respectively, more and less extensively leached soils.

Modern soil classification has become considerably more sophisticated and complex. Various schemes may take into account characteristics of present soil composition and texture, the type of bedrock on which the soil was formed, the present climate of the region, the degree of maturity of the soil, and the extent of development of the different soil horizons. Different countries have adopted different schemes. The United Nations Education, Scientific, and Cultural Organization world map uses 110 different soil map units. This is, in fact, a small number compared to the number of distinctions made under other classification schemes.

The U.S. comprehensive soil classification, known as the Seventh Approximation, has twelve major categories (orders), which are subdivided through five more levels of classification into a total of some 12,000 soil series. A short summary of the twelve orders is presented for reference in table 12.2. Some of the orders are characterized by a particular environment of formation and the distinctive soil properties resulting from it; for example, the histosols are wetland soils. Others are characterized principally by physical properties; for example, the entisols lack horizon zonation, and the vertisols have mixed upper layers because these soils contain expansive clays. The practical significance of the vertisols lies in the considerable engineering problems posed by expansive clays

Table 12.2	The Twelve Soil Orders
Alfisols	Formed in semiarid to moist areas; deeper layers enriched in clay
Andisols	Soils of cool volcanic areas with moderate to heavy precipitation
Aridisols	Common desert soils, dry and rich in soluble minerals such as calcite, halite, or gypsum
Entisols	Soils with little development of horizons, formed on recently deposited or rapidly eroding parent materials
Gelisols	Soils with near-surface permafrost or that show evidence of mixing by frost action; also known as cryosols
Histosols	Wetland soils rich in organic matter
Inceptisols	Soils of semiarid to humid climates with limited evidence of weathering or horizon development
Mollisols	The common soils of relatively dry grasslands, with a high content of organic matter near the surface
Oxisols	Highly leached soils of tropical and subtropical climates, which often have indistinct horizons
Spodosols	Soils common under conifer forests in humid regions, with aluminum oxides and organic matter leached from the surface and deposited below
Ultisols	Moderately leached soils of temperate, humid climates
Vertisols	Soils rich in expansive clays (clays that expand when wet, shrink and cause cracking when dry)

Source: After U.S. Department of Agriculture

(see chapter 20). Mollisols are particularly well-suited for agriculture, and are the predominant soils of the productive farmlands of the central and western United States. Most of the oxisols and some ultisols, on the other hand, are soils of a type that has serious negative implications for agriculture; these are lateritic soils which we discuss in the next section. **Figure 12.9** illustrates some of the variety possible among soils. **Figure 12.10** shows, broadly, the distribution of soil types worldwide.

Figure 12.9

A spectrum of soils. Though we may not notice it when looking down from above, soils in cross-section vary widely in color, texture, structure, and mineralogy. Soil (A) is an entisol, a low-fertility, sandy soil developed on glacial till in northern Michigan; (B) an inceptisol, forming on layered volcanic ash in Japan; (C) an aridisol, topped with desert pavement, Arizona; (D) a mollisol with visible white carbonate nodules in the B and C horizons; (E) a spodosol with pale, leached A horizon, in upstate New York; and (F) a histosol of decomposed mud over peat, southern Michigan.

Photographs from Marlbut Memorial Slides, courtesy Soil Science Society of America

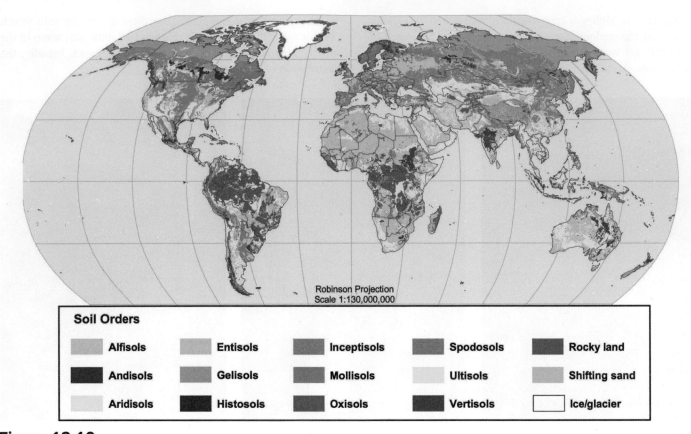

Figure 12.10

Distribution of major soil types worldwide.

Map from USDA Natural Resources Conservation Service

12.3 Soils and Human Activities

Learning Objectives

- Explain why laterite isn't a good farmland soil
- Recall the nature of wetland soils
- Recall the consequences of soil erosion
- List activities that exacerbate soil erosion
- Recall techniques used to reduce soil erosion
- Discuss the cost obstacle in soil erosion control
- Recognize areas of greatest concern for soil degradation
- List causes of soil degradation

Soils and their characteristics are not only critical to agricultural activities; their properties influence the stability of construction projects. In the area of soil erosion, human activities may aggravate or moderate the problems, and sometimes may be influenced by the erosion in turn. This section reviews several relationships between soil and human activities.

Lateritic Soil

Lateritic soil is fairly common and poses special agricultural challenges. A **laterite** may be regarded as an extreme kind of pedalfer. Lateritic soils develop in tropical climates with high temperatures and heavy rainfall, so they are severely leached. Even quartz may have been dissolved out of the soil under these conditions. Lateritic soil may contain very little besides the insoluble aluminum and iron compounds. Indeed, lateritic weathering has produced useful mineral deposits, as described in chapter 13. Soils of the lush tropical rain forests are commonly lateritic, which seems to suggest that lateritic soils have great farmland potential. Surprisingly, however, the opposite is true, for two reasons.

The highly leached character of lateritic soils is one reason. Even where the vegetation is dense, the soil itself has few soluble nutrients left in it. The forest holds a huge reserve of nutrients, but there is no corresponding reserve in the soil. Further growth is supported by the decay of earlier vegetation. As one plant dies and decomposes, the nutrients it contained are quickly taken up by other plants or leached away. If the forest is cleared to plant crops, most of the nutrients are cleared away with it, leaving little in the soil to nourish the crops. Many people living in tropical climates practice a slash-and-burn agriculture, cutting and burning the jungle to clear the land. Some of the nutrients in the burned vegetation settle into the topsoil temporarily, but relentless leaching by the warm rains makes the soil nutrient-poor and infertile within a few growing seasons. Nutrients could, in principle, be added through synthetic chemical fertilizers.

Figure 12.11

These bright red soils in Honduras are a good example of a laterite.

Steven P. Lynch

Figure 12.12

Northern Madagascar. Originally, most of Madagascar was covered by jungle and lush vegetation. Deforestation from slash-and-burn agriculture has bared most of the island, exposing its predominantly lateritic soil. Now, where slopes are steep, heavy rains wash the soil away even before it can bake hard. The Betsiboka Estuary (upper left) is so choked with the eroded soil that large oceangoing ships cannot sail into it.

Image by Earth Observations Laboratory, Johnson Space Center, courtesy NASA.

However, many of the nations in regions of lateritic soil are among the poorer developing countries, and vast expenditures for fertilizer are simply impractical.

Even with fertilizers available, a second problem with lateritic soils remains. A clue to the problem is found in the term *laterite* itself, which is derived from the Latin for "brick." A lush rain forest shields lateritic soil from the drying and baking effects of the Sun, while vigorous root action helps to keep the soil well broken up. When its vegetative cover is cleared and it is exposed to the baking tropical sun, lateritic soil can quickly harden to a solid, brick-like consistency that resists infiltration by water or penetration by crops' roots. What crops can be grown provide little protection for the soil and do not slow the hardening process very much. Within five years or less, a freshly cleared field may become completely uncultivatable as well as infertile. Often, the only recourse is to abandon each field after a few years and clear a replacement. This results, over time, in the destruction of vast tracts of rain forest where only a moderate expanse of farmland is needed (**figure 12.12**). Also, once the soil has hardened and efforts to farm it have been abandoned, it may revegetate only very slowly, if at all.

In Indochina, in what is now Cambodian jungle, are the remains of the Khmer civilization that flourished from the ninth to the sixteenth centuries. There is no clear evidence of why that civilization vanished. A major reason may have been the difficulty of agriculture in the region's lateritic soil. (It is clear that the phenomenon of laterite solidification was occurring at that time. Chunks of hardened laterite were used like bricks to construct temples at Angkor Wat and elsewhere.) Perhaps the Mayas, too, moved north into Mexico to escape the problems of lateritic soils. In Sierra Leone in west Africa, increased clearing of forests for firewood, followed by deterioration of the exposed soil, has reduced the estimated carrying capacity of the land to twenty-five persons per square kilometer; the actual population is about 110 persons per square kilometer. Part of the concern over deforestation associated with harvesting of hardwoods in

South America relates to loss of habitat and possible extinction of unique species (loss of biodiversity), but in part it relates to the fact that what is left behind is barren lateritic soil (**figure 12.13**). Today, some countries with lateritic soil achieve agricultural success only because frequent floods deposit fresh, nutrient-rich soil over the depleted laterite. The floods, however, cause enough problems that flood-control efforts are in progress or under serious consideration in many of these areas of Africa and Asia. Unfortunately, successful flood control could create an agricultural disaster for such regions.

Wetland Soils

The soils of wetlands are nearly opposite of lateritic soils as they tend to be soft and rich in accumulated organic matter. (Recall **figure 12.9F**.) Wetlands provide vital habitat for waterbirds and distinctive ecological niches for other organisms and can act as natural retention ponds for floodwaters. Some act as settling ponds, reducing sediment pollution and water pollution from contaminants carried on the sediment in water passing through them. Unfortunately, wetlands have not always been properly appreciated; many of these swampy areas have been drained for farmland or for development, or simply to provide water wanted elsewhere for irrigation. In drier climates, reduced precipitation can exacerbate wetland loss, as in the case of Lake Chad, described in chapter 11. As the value of wetlands has become more widely recognized, concerted regulatory efforts to preserve, protect, or restore wetlands have expanded in the United States (**figure 12.14**) and elsewhere. Direct human

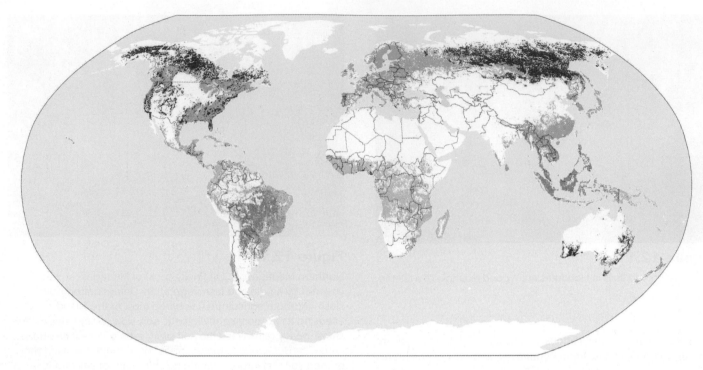

Figure 12.13

This map shows the dominant forms of tree cover loss or change from 2001–2020: yellow (agriculture), green (forestry), red (wildfire), purple (urbanization), and brown (commodity-driven deforestation). Compare this map to that in **figure 12.10**.

NASA

Figure 12.14

Even where undeveloped land is plentiful, wetlands can serve special roles. This wetland, near Anchorage, Alaska, is protected as a wildlife refuge.

©Carla Montgomery

Figure 12.15

Peat cutting in the Shetland Islands, Scotland.

David Chapman/Alamy Stock Photo

impacts aside, however, there is an added threat to many coastal wetlands: rising sea level, as noted in chapter 8.

Wetlands play an important role in the carbon cycle, and as such, there is much interest in protecting and restoring them to encourage their natural capacity to sequester carbon. In these wet conditions, the roots of dying plants aren't exposed to readily available oxygen and instead of decomposing, they accumulate. Peat is a particularly organic-rich material created in wetlands that historically have been used as an energy source and more recently

as a garden additive. Because peat no longer works to sequester carbon once it is removed from its natural environment and dries out, there are calls to end its harvest (**figure 12.15**).

There are many types of wetlands and they don't all function in the same way, so our management of them in the future will need to consider their unique conditions. They could also play an important role in ensuring clean water supplies and providing food. By identifying the locations and types of wetlands, the National Wetlands Inventory which includes a GIS wetlands mapper is an integral step in the process of protecting them. The next report on the status of U.S. wetlands covering years 2009–2019 is due out in 2022.

Soil Erosion

Weathering is the breakdown of rock or mineral materials in place; *erosion* involves physical removal of material from one place to another. Soil erosion is caused by the action of water and wind. Rain striking the ground helps to break soil particles loose (**figure 12.16**), and the harder it falls, the more soil can be dislodged. Surface runoff and wind together carry away loosened soil. The faster the wind and water travel, the larger the particles and the greater the load they move. High winds cause more erosion than calmer ones, and fast-flowing surface runoff moves more soil than slow runoff. Steep and unobstructed slopes are particularly susceptible to erosion by water, for surface runoff flows more rapidly over them. Flat, exposed land is correspondingly more vulnerable to wind erosion, as surface runoff is slower and obstacles to deflect the wind are lacking. The physical properties of the soil also influence its vulnerability to erosion.

We can estimate the rates of soil erosion in a variety of ways. Over a large area, erosion due to surface runoff may be judged by estimating the sediment load of streams draining the area. On small plots, runoff can be collected and its sediment load measured. Controlled laboratory experiments can be used to simulate wind and water erosion and measure their impact. Because wind is harder to monitor comprehensively, especially over a range of altitudes, the extent of natural wind erosion is more difficult to estimate. A common way to measure the amount of soil lost to horizontal wind erosion is through the use of traps set at different heights. Generally, wind erosion loss is less significant than through surface-water runoff, except under drought conditions. The Dust Bowl of the 1930s is an infamous example of excessive loss of soil due to wind erosion and exacerbated by human activity.

The Dust Bowl area proper, although never exactly defined, comprised close to 100 million acres of southeastern Colorado, northeastern New Mexico, western Kansas, and the Texas and Oklahoma panhandles. The farming crisis there during the 1930s resulted from an unfortunate combination of factors: clearing or close-grazing of natural vegetation, which had been better adapted to the local climate than were many of the crops; drought that then destroyed those crops; sustained winds; and poor farming practices.

Once the crops had died, there was nothing to hold down the soil. The action of the wind was most dramatically illustrated during the fierce dust storms that began in 1932. The storms were described as black blizzards that blotted out the Sun (**figure 12.17**). Black rain fell in New York, black snow in Vermont, as wind-blown dust moved eastward. People choked on the dust, some dying of suffocation or of a dust pneumonia similar to the silicosis miners can develop from breathing rock dust.

As the dust from Kansas and Oklahoma settled on their desks, politicians in Washington, DC, realized that something had to be done. In April 1935, a permanent Soil Conservation Service was established to advise farmers on land use, drainage, erosion control, and other matters. By the late 1930s, concerted efforts by individuals and state and federal government agencies to improve farming practices to reduce wind erosion, together with a return to more-normal rainfall, had considerably reduced the problems. Renewed episodes of drought and/or high winds in the 1950s, the winter of 1965, and the mid-1970s led to similar, if less extreme, dust storms and erosion. In recent years, the extent of wind damage has been limited somewhat by the widespread use of irrigation to maintain crops in dry areas and dry times. But, as sources of that irrigation water are depleted, future spells of combined drought and wind may yet produce more scenes like that in **figure 12.17**. Erosion by both wind and water, while declining, remains a significant problem over much of the United States (**figure 12.18**).

Figure 12.16
The impact of a single raindrop loosens many soil particles.
USDA Soil Conservation Service

Figure 12.17
A 1930s dust storm in eastern Colorado.
USDA Natural Resources Conservation Service

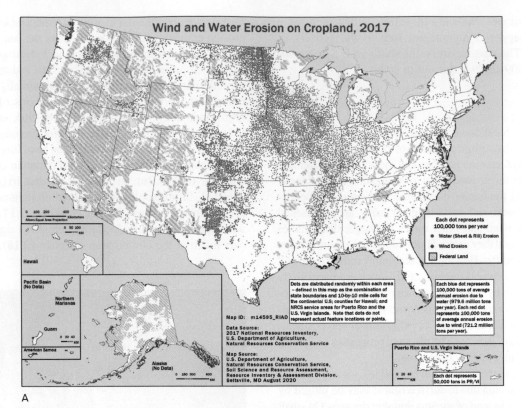

A

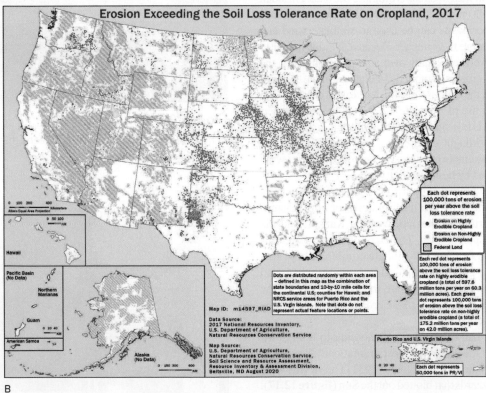

B

Figure 12.18

(A) Not surprisingly, wind erosion is more severe in the drier plains of the West. (B) The USDA Natural Resources Conservation Service defines the maximum soil loss tolerance rate as the "maximum rate of annual soil loss that will permit crop productivity to be sustained economically and indefinitely on a given soil." Red dots represent areas where soil quality is being degraded and productivity is threatened. Green dots represent areas where the soil loss is spread over a greater area.

USDA NRCS National Resources Inventory 2017

If all the soil in an area is lost, farming clearly becomes impossible. Long before that point, however, erosion becomes a cause for concern. The topsoil of the A horizon, with its higher content of organic matter and nutrients, is especially fertile and suitable for agriculture, and it is the topsoil that is lost first as the soil erodes. Fertilizer is added, at substantial expense, to compensate for loss of fertile topsoil. In addition, the organic-matter-rich topsoil usually has the best structure for agriculture—it is more permeable, more readily infiltrated by water, and retains moisture better than deeper soil layers. The cycle of using synthetic fertilizers, many of them created from petroleum products, to replace nutrients lost to erosion is tough to break because it doesn't allow for natural replenishment of the soil.

Soil erosion from cropland leads to reduced crop quality and reduced agricultural income. The loss of topsoil is measured annually in the billions of dollars for just the Midwest agricultural industry. Even when the nutrients required for adequate crop growth are added through fertilizers, other chemicals that contribute to the nutritional quality of the food grown may be lacking; in other words, the food itself may be less healthful. Also, the soil eroded from one place is deposited, sooner or later, somewhere else. If a large quantity is moved to other farmland (and it is just as likely to not be deposited on farmland), the crops on which it is deposited may be stunted or destroyed, although small additions of fresh topsoil may enrich cropland and make it more productive.

A series consequence of soil erosion in some places has been increased persistence of toxic residues of herbicides and pesticides in the soil. The loss of nutrients and organic matter through topsoil erosion may decrease the activity of soil microorganisms that normally speed the breakdown of some toxic agricultural chemicals. Many of these chemicals, which contribute significantly to water pollution, also pollute the soils. They likely also destroy the microorganisms, the importance and roles in ecosystem services such as carbon sequestration, we are just beginning to measure and understand.

Another major problem related to soil erosion is sediment pollution (**figure 12.19**). In the United States, about 750 million tons per year of eroded sediment end up in lakes and streams. This decreases the water quality and may harm wildlife. The problem is still more acute when those sediments contain toxic chemical residues, as from agricultural herbicides and pesticides. The sediment is then both a physical and a potential chemical pollutant. A secondary consequence of this sediment load is the infilling of stream channels and reservoirs, restricting navigation, and decreasing the volume of reservoirs and thus their usefulness for their intended purposes, whether for water supply, hydropower, or flood control. In 2022, the Army Corps of Engineers is currently planning on hundreds of dredging projects supported by newly appropriated Congressional funding, many of them associated with harbors, costing approximately $14 billion.

In recent years, there has been an increase in several kinds of activities that may increase soil erosion in places other than farms and cities. One is strip mining, which leaves behind readily erodable piles of soil and broken rock, at least until the land is reclaimed. Another is the use of off-road recreational vehicles (ORVs). This last is a special problem in dry areas where vegetation is not very vigorous or easily reestablished. Fragile plants can easily be destroyed by a passing ORV, leaving the land bare and vulnerable to intensified erosion (**figure 12.20**).

Figure 12.21 shows that erosion during active urbanization is considerably more severe than erosion on any sort of undeveloped land. Erosion during highway construction in the Washington, DC, area, for example, was measured at rates up to 75 tons/acre/year. The amount land being newly developed in the United States is only about 0.5 million acres each year, as compared to nearly 380 million devoted to cropland. Also, construction disturbance is of relatively short duration. Once the land is

Figure 12.19

Post-storm runoff carries a heavy sediment load from this Tennessee farm field to a drainage ditch leading to a nearby stream.

USDA Natural Resources Conservation Service/Photograph by Tim McCabe.

Figure 12.20

Fragile vegetation on this Oregon dune does not survive disturbance by off-road vehicles, leaving exposed sand more vulnerable to erosion.

©Carla Montgomery

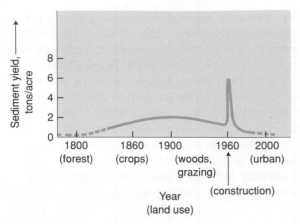

Figure 12.21

The impact of urbanization and other human activities on soil erosion rates. Construction may cause even more intensive erosion than farming, but far more land is affected by farming, and for a longer time. In this study of the Washington, DC, area, erosion rates rose sharply during a highway construction boom around 1960; once construction was completed, erosion became negligible.

From "A Cycle of Sedimentation and Erosion in Urban River Channels," M. G. Wolman from Geografiska Annaler, Vol. 49, series A, figure 1, by permission of Scandinavian University Press.

covered by buildings, pavement, or established landscaping, erosion rates typically drop to below even natural, predevelopment levels. (Whether this represents an optimum land use is another issue.) The problems of erosion during construction may be very intensive but are typically localized and short-lived.

Soil Erosion Versus Soil Formation

Estimates of the total amount of soil erosion in the United States vary widely, but the U.S. Department of Agriculture Natural Resources Conservation Service (NRCS) put the figure at close to 1.7 billion tons per year from cropland alone. To a great extent, human activities, such as construction and farming, account for the magnitude of soil loss (**figure 12.22**). On cropland, average erosion losses from wind and water together are most recently estimated at 4.6 tons per acre per year. The rate of erosion from land under construction might be triple that. Erosion rates from other intensely disturbed lands, such as unreclaimed strip mines, might be at least as high as those for construction sites.

How this compares with the rate of soil *formation* is difficult to generalize because the rate of soil formation is so sensitive to climate, the nature of the parent rock material, and other factors. We can determine the upper limits on soil formation rates by looking at soils in areas of the northern United States last scraped clean by the glaciers tens of thousands of years ago. Where the parent material was glacial till, which should weather more easily than solid rock, perhaps 1 meter (about 3 feet) of soil has formed in 15,000 years in the temperate, fairly humid climate of the upper Midwest. The corresponding average rate

A

B

C

Figure 12.22

Examples of erosion of soil. (A) Gully erosion of soil bank exposed during construction. (B) Gullying caused by surface runoff downhill along furrows on Iowa farmland. (C) This severely wind-eroded field dramatically illustrates the role plant roots can play in holding soil in place.

(A, B) USDA Natural Resources Conservation Service/ Photographs by Lynn Betts; (C) USDA Natural Resources Conservation Service

of 0.007 centimeter (0.002 inch) of soil formed per year is an order of magnitude less than the average cropland erosion rate. Soil formation in colder and drier areas, and over more-resistant bedrocks, is slower still. In some places in the Midwest and Canada, virtually no soil has formed on glaciated bedrocks, and in the drier Southwest, soil formation is also typically very much slower. It seems clear that erosion is stripping away farmland far faster than the soil is being replaced. In other words, soil in general—and high-quality farmland—is a nonrenewable resource in many populated areas.

Strategies for Reducing Erosion

The wide variety of approaches for reducing erosion on farmland basically involve either reducing the velocity of an eroding agent or protecting the soil from its effects. Under the latter heading come such practices as leaving stubble in the fields after a crop has been harvested rather than clearing and tilling the field, and planting cover crops in the off-season between cash crops. In all such cases, the plants' roots help to hold the soil in place, and the plants themselves, to some extent, shield the soil from wind and rain (**figure 12.23**).

The lower the wind or water runoff velocity, the less material carried. Surface runoff may be slowed on moderate slopes by contour plowing (**figure 12.24A**). Plowing rows parallel to the contours of the hill, perpendicular to the direction of water flow, creates a ridged land surface so that water does not rush downhill as readily (contrast with **figure 12.22B**). Other slopes may require terracing (**figure 12.24B**), whereby a single slope is terraced into a series of shallower slopes, or even steps that slant backward into the hill. Again, surface runoff makes its way down the shallower slope more slowly, if at all, and carries far less sediment with it.

A

B

C

Figure 12.23

On this Iowa cornfield, stubble has been left in the field over winter; soil is then cultivated and planted in one step in the spring, so it is never left exposed between harvest and planting. This is described as no-till agriculture.

USDA Natural Resources Conservation Service/Photograph by Lynn Betts.

Figure 12.24

Decreasing slope of land, or breaking up the slope, decreases erosion by surface runoff. (A) Contour-plowed field in northern Iowa. (B) Terracing creates steps of shallower slope from one long, steeper slope. (C) Terraces become essential on very steep slopes, as with these rice fields in Indonesia.

(A, B) USDA Natural Resources Conservation Service/ Photographs by Tim McCabe; (C) Jan Hruby/123RF

A

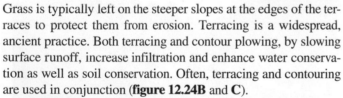

C

B

Figure 12.25

Creating near-ground irregularities decreases wind velocity.
(A) Windbreaks create wind barriers on this otherwise-flat North Dakota farmland. (B) Strip cropping also breaks up the surface topography, decreasing wind velocity. Allamakee County, Iowa. (C) Contour strip cropping of corn and alfalfa on the Iowa/Minnesota border.

(A) USDA Natural Resources Conservation Service/Photograph by Erwin Cole; (B, C) USDA Natural Resources Conservation Service/Photographs by Tim McCabe

Grass is typically left on the steeper slopes at the edges of the terraces to protect them from erosion. Terracing is a widespread, ancient practice. Both terracing and contour plowing, by slowing surface runoff, increase infiltration and enhance water conservation as well as soil conservation. Often, terracing and contouring are used in conjunction (**figure 12.24B** and **C**).

The power of the wind can be reduced by planting hedges or rows of trees as windbreaks along field borders or in rows perpendicular to the dominant wind direction (**figure 12.25A**) or by erecting low fences, like snow fences, similarly arrayed. This doesn't altogether stop soil movement, as shown by the ridges of soil that sometimes pile up along the windbreaks. However, it does reduce the distance over which soil is transported, and some of the soil caught along the windbreaks can be redistributed over the fields. Strip cropping (**figure 12.25B**), alternating crops of different heights, slows near-ground wind by making the land surface more irregular. Combining strip cropping and contouring (**figure 12.25C**) may help reduce both wind and water erosion.

A major obstacle to erosion control on farmland is cost. Many of the recommended measures such as terracing and planting cover crops can be expensive up-front, especially on a large scale. Even though the long-term benefits of reduced erosion are obvious and may involve long-term savings or increased income, the effort may not be made if short-term costs seem too high. The benefits of a particular soil conservation measure may be counterbalanced by substantial drawbacks, such as reduced crop yields. Also, if strict erosion-control standards are imposed selectively on farmers of one state (as by a state environmental agency, for instance), those farmers are at a competitive disadvantage with respect to farmers elsewhere who are not faced with equivalent expenses. Even meeting the same standards does not cost all farmers everywhere equally. Increasingly, the government is becoming involved in cost-sharing for erosion-reduction programs so that the financial burden does not fall too heavily on individual farmers. Since the Dust Bowl era, over $110 billion (currently worth over $300 billion) in federal funds have been spent on soil conservation efforts.

One simple approach to erosion reduction on farmland is to take the most susceptible land out of cultivation altogether. The USDA's Conservation Reserve Program (**CRP**) pays farmers to plant grass or trees on some seriously erodible land,

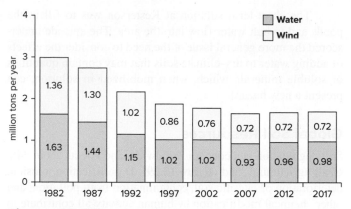

Figure 12.26

Erosion on cropland, in billions of tons, includes land removed from cultivation under the Conservation Reserve Program.

From USDA Natural Resources Conservation Service, National Resources Inventory 2017

requiring that they agree to leave the land uncultivated for ten to fifteen years. In the beginning of 2022, about 22.9 million acres were enrolled in the program, a bit under the cap set by the 2018 Farm Bill of 25 million. The amount of land included was higher just fifteen years ago, but the surge in interest in ethanol fuels produced from corn created economic pressures to opt out of the program when their contracts expired.

The USDA NRCS reports that a total of 1.7 billion tons of soil was lost from 367 million acres of cropland in 2017. Clear progress has been made in erosion reduction over the past few decades (**figures 12.26** and **12.27**), though the rate of progress slowed after 1997 and did not change from 2012 to 2017. All agricultural areas of the United States show reduced erosion rates since 1982, but still remain high in some areas, notably the Southern Plains region.

Other strategies can minimize erosion in non-farming areas. For example, in areas where vegetation is sparse, rainfall limited, and erosion a potential problem, ORVs should be restricted to

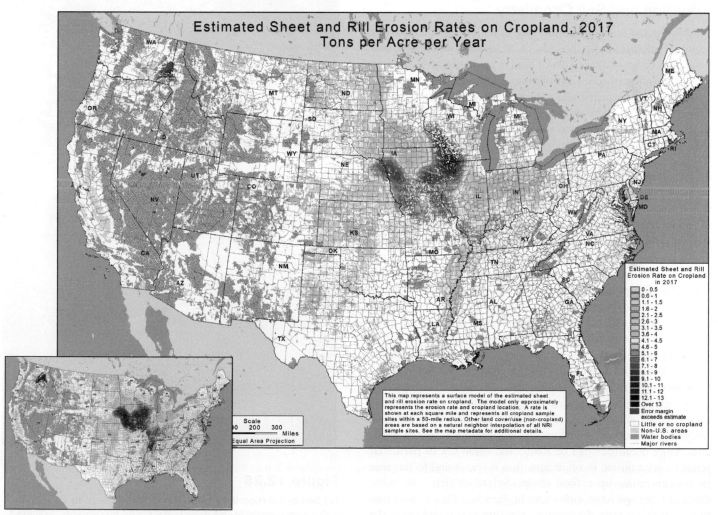

Figure 12.27

Erosion by water on cropland in 2017, in tons per acre per year. Considerable progress has been made in reducing erosion since 1982 (inset), but there are clearly broad areas where significant erosion still occurs.

After USDA NRCS National Resources Inventory 2017

prescribed trails. In the case of urban construction projects, one reason for the severity of the resulting erosion is that it is common practice to clear a whole area, such as an entire housing-project site, at the beginning of the work, even though only a small portion of the site is actively worked at any given time. Clearing the land in stages, as needed, minimizes the length of time the soil is exposed and reduces urban erosion. Mulch or temporary ground covers can protect soil that must be left exposed for longer periods. Stricter mining regulations requiring reclamation of strip-mined land (see chapter 13) are already significantly reducing soil erosion and related problems in these areas.

Where erosion cannot be eliminated, sediment pollution can still be reduced. Soil fences, and hay-lined soil traps through which runoff water must flow, may reduce soil runoff from construction sites. On either construction sites or farmland, surface runoff water may be trapped and held in ponds to allow suspended sediment to settle out in the still ponds before the clarified water is released; such settling ponds are illustrated in chapter 17.

Irrigation and Soil Chemistry

We have noted leaching as one way in which soil chemistry is naturally modified. Application of fertilizers, herbicides, and pesticides on any type or area of land is another way we change the composition of soil. Less-obvious additions can occur when irrigation water redistributes soluble minerals. In dry climates, for example, irrigation water may dissolve salts in pedocal soils; as the water evaporates near the surface, it can redeposit those salts in near-surface soil from which they had previously been leached away. As the process continues, enough salt may be deposited that plant growth suffers (**figure 12.28A**).

A different kind of chemical-toxicity problem traceable to irrigation was recognized in California's San Joaquin Valley in the mid-1980s. Wildlife managers in the Kesterson Wildlife Refuge noticed sharply higher incidence of deaths, deformities, and reproductive failures in waterfowl nesting in and near the artificial wetlands created by holding ponds for irrigation runoff water (**figure 12.28B**). In the 1960s and 1970s, that drainage system had been built to carry excess runoff from the extensively irrigated cropland in the valley, precisely to avoid the kind of salinity buildup just described. The runoff was collected in ponds at Kesterson and seemed to create beneficial habitat. What had not initially been realized is that the irrigated soils were also rich in selenium.

Selenium is an element present in trace amounts in many soils. Small quantities are essential to the health of humans, livestock, and other organisms. As with many essential nutrients, large quantities can be toxic; and selenium in particular tends to accumulate in organisms that ingest it and to increase in concentration up a food chain. Selenium-rich soils were derived from selenium-rich rocks in the area. This was no particular problem until the soluble selenium was dissolved in the irrigation water, drained into the ponds at Kesterson, and concentrated there by evaporation in the ponds, where it built up through the algae→insects→fish→birds food chain. In time, it reached levels obviously toxic to the birds.

The short-term solution at Kesterson was to fill in the ponds and to halt water flow into the area. The episode underscored the more general issue of the need to consider the effects of adding water to dry-climate soils that may contain quantities of soluble minerals which, when mobilized in solution, can present a new hazard.

Global Soil Resources

Soil degradation is a concern worldwide, and affects about 34% of all agricultural land (**figure 12.29**). Desertification, erosion, deterioration of lateritic soil, contamination from pollution, and other chemical modification by human activity all contribute to

A

B

Figure 12.28

(A) Salt buildup on land surface in Colorado, due to irrigation: Salty water wicks to the drier surface and evaporates, leaving a salty crust. (B) Kesterson Wildlife Refuge, with selenium-contaminated ponds.

(A) USDA Natural Resources Conservation Service/Photograph by Tim McCabe; (B) U.S. Geological Survey Photo Library, Denver, CO.

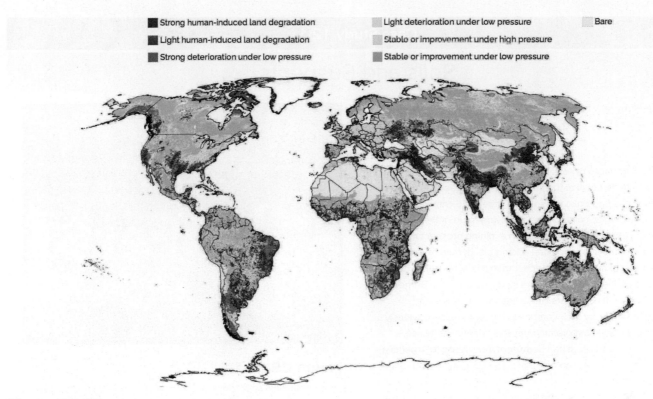

Strong human-induced land degradation
Light human-induced land degradation
Strong deterioration under low pressure
Light deterioration under low pressure
Stable or improvement under high pressure
Stable or improvement under low pressure
Bare

Figure 12.29

Trends of land degradation and deterioration, 2015. Degradation of land is caused by human activity, whereas deterioration is through natural activity. Find out more info about global trends at the Food and Agriculture Organization of the United Nations: State of Land, Soil, and Water

Food and Agriculture Organization of the United Nations

reduced soil quality, fertility, and productivity. The problems are widespread and vary in degree, and the contributing causes vary regionally. South Asia is the most heavily affected area with 41% of its land affected by human degradation, and Sub-Saharan Africa has the most degraded land at 330 million hectares.

Land area is finite. Differences in population density as well as soil quality create great disparities in per capita arable land by region of the globe, and the numbers decline as population grows and as arable land is covered up by development. The 2018 world average of arable land was 0.63 hectares, varying from 1.36 ha in South America to 0.21 in Southeast Asia. Soil degradation simply further diminishes the availability of the farmland needed to feed the world.

Soils and Suspects

Forensic geology may be said to have its roots in the Sherlock Holmes stories of Sir Arthur Conan Doyle. Holmes was a new sort of investigator, one who used careful observation of myriad small details in identifying criminals and unraveling mysteries. The concept of using traces of soils to track where a person or object had been was quite novel at the time. Since then, geologic evidence of many kinds has figured in the solution of real-life crimes.

Some are white-collar crimes, involving fraud or theft. Scam artists selling gullible investors shares in a mining prospect may "salt" samples of the supposed ore, adding bits of gold or other valuable mineral to make it seem richer than it is. (The stakes can be very high: A massive fraud of this kind involving a mine in Borneo cost investors billions of dollars.) Investigation can reveal the additions, which may, by their composition or texture, be obviously incompatible with the host ore material. In a famous theft case, a shipment of electronics that had been sent from Texas to Argentina via Miami disappeared en route, replaced by concrete blocks. Study of the sand in the concrete, and comparison with local sands at each of the shipment's stops, indicated that the substitution was made in Florida, which narrowed the search for the culprits and led to their eventual arrest. Protected cacti stolen from federal land have been linked to their source by the soil clinging to their roots.

Some of the most spectacular cases involve murder or kidnaping. A coverup originally complicated the investigation of the death in Mexico of an agent of the U.S. Drug Enforcement Agency. The body was moved from where it was originally buried to a farm unconnected with the crime. But the source location of a small sample of soil attached to the body was subsequently identified, the coverup revealed, and the murderers prosecuted.

Not all cases are straightforward. Soils and sediments differ in many ways—color, texture, mineralogy, sorting, and more (**figure CS 12.1.1**). However, not every sample is unique to a small, specific area. In the abovementioned murder case, for example, the soil contained rhyolitic ash with a distinctive mineralogy, found only in one small volcanic area in Mexico. Also, a very small sample may not be representative of the average composition of the soil or sediment from which it comes. Identifying the exact location of a very specific soil sample is not always the goal. It is often sufficient to indicate, for instance, that a suspect was in an area where that person claims never to have been, and thereby break an alibi, or to

Figure CS 12.1.1

Three distinctive soils (clockwise from lower left): a quartz-rich sand, a commercial potting soil rich in vermiculite and organic matter, and a poorly sorted soil with very fine clay and rocky gravel together. With a spoonful-sized sample you could characterize each fairly well; with only a few grains, it would be much more difficult to compare your sample accurately with others.

©Carla Montgomery

associate the suspect with a person, area, or object of interest to an investigation. Or to rule out the number of locations or lessen the land area police need to search. Analytical techniques are becoming more sophisticated over time and may now include spectroscopy, magnetics, X-ray diffraction and fluorescence, and even genomics, making the identification of geologic materials an even more powerful investigative tool.

For an introduction to forensic geology and some of the cases it has helped to solve, the interested reader is referred to *Evidence from the Earth,* a book written for the nonspecialist by experienced forensic geologist Raymond C. Murray. A summary of the most recent aspects of forensic geology are included in *An introduction to forensic soil science and forensic geology: a synthesis.*

Summary

Soil forms from the breakdown of rock materials through mechanical and chemical weathering, often accelerated by biological activity. Weathering rates are closely related to climate, with chemical weathering tending to dominate in warmer, wetter regions, and mechanical weathering in cooler, drier areas. The character of the soil reflects the nature of the parent material and the kinds and intensities of weathering processes that formed it. The wetter the climate, the more leached the soil. The lateritic soil of tropical climates is particularly unsuitable for agriculture. Not only is it highly leached of nutrients, but when exposed, it may harden to a brick-like solidity. The texture of soil is a major determinant of its drainage characteristics; soil structure is related to its suitability for agriculture. Agricultural practices may also alter soil chemistry, whether through direct application of additional chemicals or through irrigation-induced changes in natural cycles.

Soil erosion by wind and water is a natural part of the rock cycle. Where accelerated by human activity, however, it can also be a serious problem, especially on farmland, in part because of the large area involved. On a more local scale, erosion can also be rapid in areas subject to construction or strip mining. Erosion rates far exceed inferred rates of soil formation in many places. A secondary problem is the resultant

sediment pollution of lakes, streams, and nearshore ocean waters. Strategies to reduce soil erosion on farmland include terracing, contour plowing, planting or erecting windbreaks, the use of cover crops, strip cropping, and no-till farming. Leaving the most erodible land uncultivated also reduces erosion overall; but short-term economic gains over long-term benefits can encourage increased farming on marginal lands. Elsewhere, restriction of ORVs, more selective clearing of land during construction, the use of sediment traps and settling ponds, and careful reclamation of strip-mined areas could all help to minimize soil erosion. Globally, arable land per capita is declining, as potential farmland is lost to development and degradation and population grows.

Key Terms and Concepts

A horizon	E horizon	O horizon	zone of accumulation
B horizon	laterite	pedalfer	zone of leaching
C horizon	leaching	pedocal	
chemical weathering	mechanical weathering	soil	

Test Your Learning

1. Briefly explain how the rate of chemical weathering is related to (a) the amount of precipitation, (b) the temperature, and (c) the amount of mechanical weathering.

2. Sketch a generalized soil profile and indicate the A horizon, B horizon, C horizon, E horizon, O horizon, zone of leaching, and zone of accumulation.

3. Compare pedalfer and pedocal types of soil, and indicate in what kind of climate each is more common.

4. The lateritic soil of the tropical jungle is poor soil for cultivation. Explain.

5. Soil erosion during active urbanization is far more rapid than it is on cultivated farmland, and yet the majority of soil-conservation efforts are concentrated on farmland. Note why this is so.

6. Cite and briefly describe three strategies for reducing cropland erosion. Explain how interest in biofuels such as ethanol can be related to soil erosion.

7. Irrigation can, over time, cause harmful changes in soil chemistry that reduce crop yields or create toxic conditions in runoff water. Explain briefly.

8. Describe three factors that are reducing the amount of available farmland per capita around the world.

Exploring Further

1. Choose any major agricultural region of the United States and investigate the severity of any soil erosion problems there. What efforts, if any, are being made to control that erosion, and what is the cost of those efforts?

2. Visit the site of an active construction project; examine it for evidence of erosion and erosion-control activities.

3. How does your state fare with respect to soil erosion? What factors can you identify that contribute to the extent—or the lack—of soil erosion? (**The National Resources Inventory 2017** may be a place to start; your local geological survey may be additional resources.)

4. Assume that solid rock can be eroded at a rate comparable to that at which it can be converted to soil, and use the rate of 0.006 centimeter per year as representative. How long would it take to erode away a mountain 1500 meters (close to 5000 feet) high?

5. Take a sample of local soil; examine it closely with a magnifying lens and describe it as fully as you can. What features of it (if any) might be relatively unusual?

CHAPTER 13

Mineral and Rock Resources

Mineral resources occur in many settings and are mined in many ways. At this Chilean mine, lithium-rich groundwater exists in an aquifer underneath the Salar de Atacama. The briny water is pumped to the surface to shallow evaporation ponds where the naturally dry, sunny, windy weather leaves deposits of lithium behind. Lithium is a key component of rechargeable batteries used in electronics from phones to cars.

Source: NASA Earth Observatory images by Lauren Dauphin

Though we may not often think about it, virtually everything we build or create or use in modern life involves rock, mineral, or fuel resources. In some cases, this is obvious. We can easily see the steel and aluminum used in vehicles and appliances; the marble or granite of a countertop or floor; the gold, silver, and gemstones in jewelry; the copper and nickel in coins. But our dependence on these resources is much more pervasive. Asphalt highways, which themselves use petroleum resources, are built on crushed-rock roadbeds. A wooden house is built on a slab or foundation of concrete which requires sand and limestone to make; held together with metal nails; finished inside with wallboard made with gypsum; wired with copper or aluminum; given a view by glass windows made from quartz-rich sand. The coffee cup you use each morning is made with clay. Your toothpaste cleans your teeth using mineral abrasives such as quartz, calcite, corundum, or phosphate minerals; reduces cavities with fluoride from the mineral fluorite; is probably colored white with mineral-derived titanium dioxide, the same pigment as in white paint; and, if it sparkles, may contain mica flakes. The computers and mobile devices that pervade our lives would not exist without metals for circuits and semiconductors.

The bulk of Earth's crust is composed of fewer than a dozen elements, and as noted in table 1.2, eight chemical elements make up more than 98% of the crust. Many of the elements not found in abundance in Earth's crust, including industrial and precious metals, essential components of chemical fertilizers, and elements like nickel that are used to produce batteries, are vitally important to society. Some of these are found in very minute amounts in the average rock of the continental crust: copper, 100 ppm; molybdenum, 1 ppm; platinum, 4 ppb. Clearly, many useful elements must be mined from very atypical rocks. This chapter looks at the occurrences of a variety of rock and mineral resources and examines the U.S. and world supply-and-demand picture.

Chapter Outline

13.1 Resources, Reserves, and Ore Deposits

Learning Objectives

- Differentiate categories of reserves and resources
- Define ore and concentration factor
- List the factors that influence the economics of mining a particular deposit
- Recall that economic mineral deposits are unevenly distributed

In a broad sense, *resources* are all those materials that are necessary or important to human life and civilization, and that have some value to individuals and/or to society. What constitutes a resource may change with time or social context. Many of the fuels, building materials, metals, and other substances important to modern technological civilization were of no importance to prehistoric humans. As we develop into a highly technical, global society, our use and need for minerals and fuels is merging, and we find ourselves competing for a finite supply of resources that are becoming a bit more difficult to access.

In discussing supply/demand questions at any scale, we make distinctions between quantities of reserves and various other categories of resources, as summarized in **figure 13.1**.

The **reserves** are the quantity of a given geologic material that has been found and that can be recovered economically with existing technology. Usually, the term is used only for material not already consumed, as distinguished from **cumulative reserves**, which include the quantity of the material already used up in addition to the remaining (unused) reserves. The reserves represent the most conservative estimate of how much of a given metal, mineral, or fuel remains unused. Beyond that, several additional categories of resources can be identified.

Those deposits that have already been found but that cannot presently be profitably exploited are the **subeconomic resources** (also known as **conditional resources**). Some may be exploitable with existing technology but contain ore that is too low-grade (see below) or fuel that is too dispersed to produce a profit at current prices. Others may require further advances in technology before they can be exploited. Still, these deposits have at least been located, and we can estimate the quantity of material they represent fairly well.

Then there are the undiscovered resources. These are sometimes subdivided into **hypothetical resources**—additional deposits expected to be found in areas in which some deposits of the material of interest have already been found—and the **speculative resources**—those deposits that might be found in explored or unexplored regions where deposits of the material are not already known to occur. Estimates of undiscovered resources are extremely rough by nature, and it would obviously be unwise to count too heavily on deposits that have not even been found yet, especially for the near future.

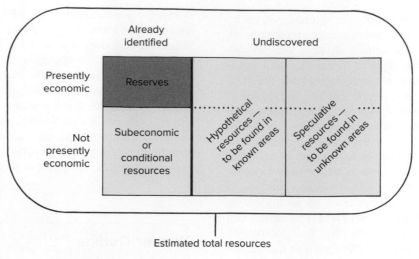

Figure 13.1

Different categories of reserves and resources.

Most resources we are interested in don't exist as pure deposits and are bound together with other undesirable elements, minerals, or rocks. An **ore** is a rock in which a valuable or useful mineral occurs at a concentration sufficiently high, relative to average rocks, that it may be economically worth mining. A given ore deposit may be described in terms of the enrichment or **concentration factor** of a metal of interest:

$$\text{concentration factor} = \frac{\text{concentration of the metal in the ore}}{\text{concentration of the metal in the continental crust}}$$

The higher the concentration factor, the richer the ore, and the less of it needs to be mined to extract a given amount of the metal.

In general, the minimum concentration factor required for profitable mining is inversely proportional to the average crustal concentration. If ordinary rocks are already fairly rich in metal, they need not be concentrated much further to make mining it economic. Metals like iron or aluminum, which make up about 6% and 8% of average continental crust, respectively, need only be concentrated by a factor of four to five times for mining to be profitable. Copper must be enriched about 100 times relative to average rock, while mercury (average concentration 80 ppb) must be enriched to about 25,000 times its average concentration before the ore is rich enough to mine profitably. The exceptions to this rule are those few extremely valuable elements, such as gold, that are so high-priced that a small quantity justifies a considerable amount of mining. At a price of several hundred dollars an ounce, gold need to be found in concentrations only a few thousand times its average 4 ppb to be worth mining. With gold consistently between $1700 and $2000/ounce (2020–2022), even the most diffuse deposits are economic to mine. This inverse relationship between average concentration and the concentration factor required for profitable mining further suggests that economic ore deposits of relatively abundant metals (such as iron and aluminum) might be more plentiful than economic deposits of rarer metals, and this is indeed commonly the case.

The value of the mineral or metal extracted and its concentration in a particular deposit are major factors determining the profitability of mining it. The economics are naturally sensitive to world demand. If demand climbs and prices rise in response, additional, not-so-rich ore deposits may be opened up; a fall in price causes economically marginal mines to close. The discovery of a new, rich deposit of a given ore may make other mines with poorer ores noncompetitive and uneconomic in the changing market. The practicality of mining a specific ore body may also depend on the mineral(s) in which a metal of interest is found, because this affects the cost to extract the pure metal. Three iron deposits containing equal concentrations of iron are not equally economic if the iron is mainly in oxides in one, in silicates in another, and in sulfides in the third. The relative densities of the iron-bearing minerals, compared to other minerals in the rock, will affect how readily those iron-rich minerals can be separated from the rest; the nature and strengths of the chemical bonds in the iron-rich phases will influence the energy costs to break them down to extract the iron, and the complexity of chemical processes required; the concentration of iron in the iron-rich minerals (rather than in the rock as a whole) will determine the yield of iron from those minerals, after they are separated. Different types of mines, too, involve different costs. These factors and others influence the economics of mining. Given the definition of reserves, fluctuating economic factors will affect whether some ore deposits are counted as reserves or put in the subeconomic resources category—and for a specific deposit, that can change over time.

By definition, ores are somewhat unusual rocks. Known economic mineral deposits are very unevenly distributed around the world because their location is controlled by geology rather than human-based factors. Just by taking a quick look at **figure 13.2**, we can see that the United States must depend on many other countries for many types of non-fuel mineral resources. The United States controls about 17% of the world's known

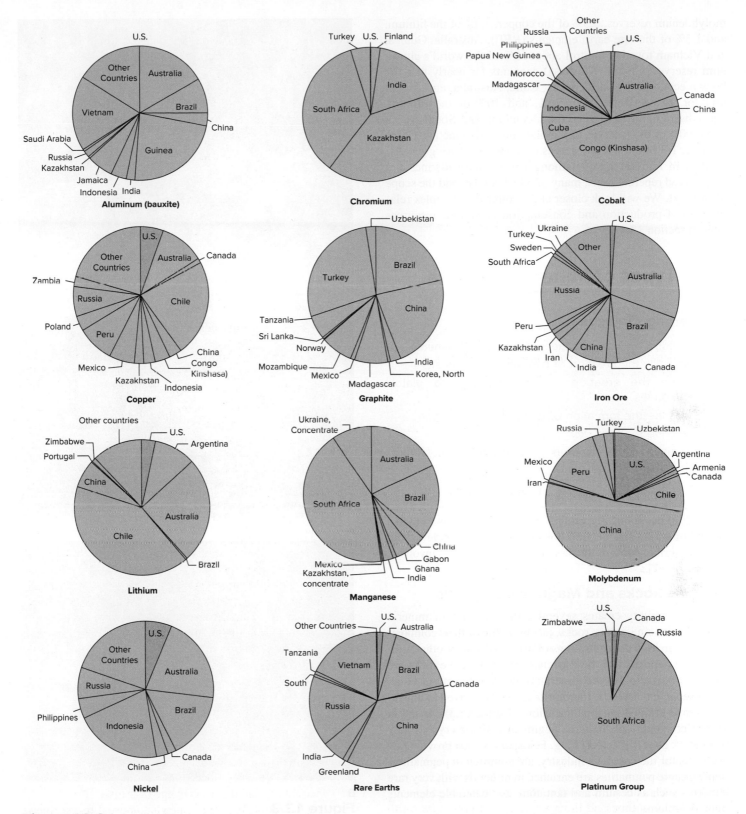

Figure 13.2

Proportions of world reserves of some nonfuel minerals controlled by various nations. Although the United States is a major consumer of most metals, it is a major producer of very few.

Source: Data from Mineral Commodity Summaries 2022, U.S. Geological Survey.

molybdenum reserves, 5.5% of the copper, 3.4% of the lithium, and 1.5% of the rare earth elements (REE). Australia, Guinea, and Vietnam together control about 58% of the world's aluminum reserves. Congo (Kinshasa) accounts for nearly half the known economic cobalt deposits; Chile, Australia, and Argentina for, respectively, 42%, 26%, and 10% of the lithium. Kazakhstan controls 40% of the chromium, and South Africa about 90% of the reserves of the platinum-group metals. China has over 1/3 of the world reserves of the REE. These great disparities in mineral wealth among nations have distinct and widespread repercussions, many of which are beyond the scope of this text. We will look closer at the somewhat complex relationship of production and consumption of various resources later in section 13.4.

13.2 Types of Mineral Deposits

Learning Objectives

- Describe how crystallization concentrates minerals
- Exemplify the creation of hydrothermal deposits
- Relate the creation of ore deposits to plate boundaries
- Describe the formation of banded iron formation
- Describe the formation of placer deposits
- Exemplify useful minerals created through metamorphic processes

Deposits of economically valuable rocks and minerals form in a variety of ways. This section does not attempt to review them all but describes some of the more important processes involved.

Igneous Rocks and Magmatic Deposits

Magmatic activity creates several different kinds of mineral deposits. Certain igneous rocks, just by virtue of their composition, contain high concentrations of useful silicate or other minerals. The deposits may be especially valuable if the rocks are coarse-grained, so that the mineral(s) of interest occur as large, easily separated crystals. **Pegmatite** is the term given to unusually coarse-grained igneous intrusions (**figure 13.3**) created in water-rich conditions. In some pegmatites, single crystals may be over 10 meters (30 feet) long. Feldspars, which provide raw materials for the ceramics industry, are common in pegmatites. Some granite pegmatites are enriched in minerals with very rare elements such as cesium and tantalum, and desirable elements such as lithium.

Other useful minerals may be concentrated within a cooling magma chamber by gravity. If they are more or less dense than the magma, they may sink or float as they crystallize, instead of remaining suspended in the solidifying silicate mush, and accumulate in thick layers that are easily mined (**figure 13.4**). Chromite, an oxide of the metal chromium, and

A

B

Figure 13.3

Pegmatite, a very coarse-grained plutonic rock, may yield large, valuable crystals or rare, desirable elements. (A) This sample has very large crystal of quartz and potassium feldspar, approximately 0.4 meters in length. (B) The large biotite (mica) crystals in this pegmatite could be used for production of electrical equipment.

Source: (A) Doug Sherman/Geofile; (B) ©Carla Montgomery

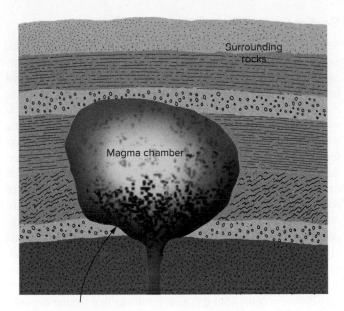

Figure 13.4

A magmatic ore deposit is formed when a dense mineral such as chromite or magnetite gravitationally settles out of a crystallizing magma.

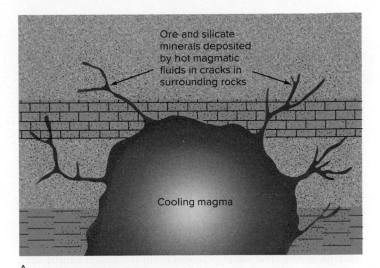

Figure 13.5

(A) Hydrothermal ore deposits form in veins around a magma chamber. (B) The quartz veins in this rock are hydrothermal deposits.

(B) John A. Karachewski

magnetite, one of the iron oxides, are both quite dense. In a magma of suitable bulk composition, rich concentrations of these minerals may form in the lower part of the magma chamber during crystallization of the melt. The dense precious metals, such as gold and platinum, may also be concentrated during magmatic crystallization. Even where disseminated throughout an igneous rock body, a very valuable mineral commodity may be worth mining. This is true of diamonds. One reason for the rarity of diamonds is that they are formed at extremely high pressures, such as those found within the mantle, and then brought rapidly up into the crust. They are mined primarily from igneous rocks called **kimberlites**, which occur as pipelike intrusive bodies. Even where only a few gem-quality diamonds are scattered within many tons of kimberlite, their high unit value makes the mining profitable.

Hydrothermal Ores

Moving, hot fluids are pervasive in Earth's crust and may be related to magma, the proximity to hot crust, or to tectonic activity. These *hydrothermal fluids* typically contain large amounts of dissolved elements that mineralize when the water cools. Magma contains a lot of fluids including water which escape from cooling magma during the later stages of crystallization. These fluids seep through cracks and poor in the surrounding rocks, carrying with them dissolved salts, gases, and metals. These warm fluids can leach additional metals from the rocks through which they pass. In time, the fluids cool and deposit their dissolved minerals, creating a **hydrothermal** ore deposit (**figure 13.5**).

The particular minerals deposited vary with the composition of the hydrothermal fluids, but worldwide, a great variety of metals occur in hydrothermal deposits: copper, lead, zinc, gold, silver, platinum, uranium, and others. Because sulfur is a common constituent of magmatic gases and fluids, the ore minerals are frequently sulfides. For example, the lead in lead ore deposits is found in galena (PbS); zinc, in sphalerite (ZnS); and copper, in a variety of copper and copper-iron sulfides ($CuFeS_2$, CuS, Cu_2S, and others).

Hydrothermal fluids don't all originate within the magma. Sometimes, circulating subsurface waters are heated sufficiently by a nearby cooling magma to dissolve, concentrate, and redeposit valuable metals in a hydrothermal ore deposit. Or the fluid involved may be a mix of magmatic and non-magmatic fluids. This may well be true of the ore-rich fluids gushing out at black smoker vents (**figure 13.6**) associated with mid-ocean ridge systems, and the fluids that deposit metal-rich sulfides in the muds at the bottom of the Red Sea rift zone. Some hydrothermal deposits are found far from any obvious magma source, and those fluids might have been of metamorphic origin.

The relationship between magmatic activity and the formation of many ore deposits suggests that hydrothermal and igneous-rock deposits should be especially common in regions of extensive magmatic activity—that is, plate boundaries. This is generally true. A striking example of this link is in **figure 13.7A**, which shows the locations of a certain type of copper and molybdenum deposit of igneous origin in North and South America. Comparison with **figure 3.16** shows that these locations correspond closely to present or recent subduction zones. The same plate-tectonic influence can be seen in **figure 13.7B**. The precious-metal deposits in the western United States are predominantly hydrothermal deposits related to magmatic activity associated with plate collision and subduction along the western margin of the continent. As was illustrated in **figures 3.27** and **5.5**, magma generated in subduction zones can rise up into the overlying plate, forming plutonic rock bodies in the crust. Fluids and heat associated with this plutonic activity contribute to the development of hydrothermal ore deposits.

A

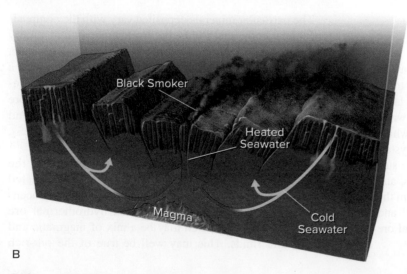

B

Sedimentary Deposits

Sedimentary processes also produce economic mineral deposits. Some such ores have been deposited directly as chemical sedimentary rocks. Layered sedimentary iron ores, called **banded iron formation**, are an example (**figure 13.8A**). In these ores, iron-rich layers (predominantly hematite or magnetite) alternate with silicate- or carbonate-rich layers. These large deposits, which may extend for tens of kilometers, are very ancient. Their formation is related to the development of Earth's atmosphere. As pointed out in chapter 1, the early atmosphere lacked free oxygen. Under those conditions, iron from the weathering of continental rocks would have been very soluble in the oceans. As photosynthetic organisms began producing oxygen, that oxygen reacted with the dissolved iron and caused it to precipitate layers of iron-rich minerals accumulating along with other products of continental weathering. If the majority of large iron ore deposits formed in this way, when Earth's surface chemistry was very different from what it is now, it follows that similar deposits are unlikely to form now or in the future.

Other sedimentary mineral deposits form from seawater, which contains a variety of dissolved salts and other chemicals. When a body of seawater trapped in a shallow sea dries up, it deposits these minerals in **evaporite** deposits (**figure 13.8B**). Some evaporites may be hundreds of meters thick. Ordinary table salt, known mineralogically as halite, is one mineral commonly mined from evaporite deposits. Others include gypsum and salts of the metals potassium and magnesium. The lithium deposits being mined in the chapter-opening photograph are evaporites.

Other Low-Temperature Ore-Forming Processes

Streams play a role in the formation of many mineral deposits. Streams don't create the deposits themselves, but rather concentrate them through erosion, transportation, and depositional processes. As noted in chapter 6, streams are very good at sorting sediments by size and density. The sorting action can effectively concentrate certain weathering-resistant, dense minerals in distinct locations along the stream channel. The currents of a coastal environment can also cause sediment sorting and selective concentration of minerals. Such deposits, mechanically concentrated by water, are called **placers**. The minerals of interest are typically weathered out of local

Figure 13.6

(A) This black smoker is created when seawater and deeper water mix and are heated from the magma below. (B) As they rise to the surface, the hot waters leach additional chemicals from the rocks they pass through. Precipitation occurs when the hot water cools suddenly at the surface and creates chimney structures rich in sulfur-bearing minerals.

(A) Image courtesy of NOAA Ocean Exploration, 2016 Deepwater Exploration of the Marianas; (B) McGraw Hill

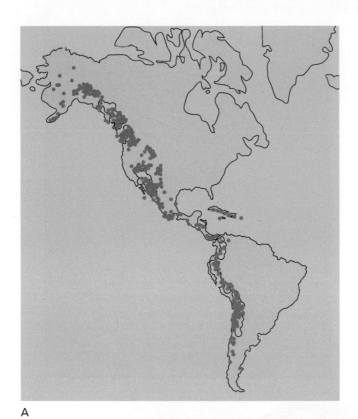

A

rocks, then transported, sorted, and concentrated while other minerals are dissolved or swept away (**figure 13.9**). Gold, diamonds, and minerals containing REE are examples of resources that have been mined from the sands and gravels of placer deposits.

Weathering alone can produce useful ores by dissolving away unwanted minerals, leaving a residue enriched in some valuable metal. As described in chapter 12, the extreme weathering of tropical climates gives rise to lateritic soils from which nearly everything has been leached except for aluminum and iron compounds. Many of the aluminum deposits (bauxite) presently being mined were enriched in aluminum to the point of being economic by lateritic weathering. (Note that the countries identified in **figure 13.2** as having substantial aluminum-ore reserves are located wholly or partially in tropical or subtropical climate zones.) The iron content of the laterites is generally not as high as that of the richest sedimentary iron ores described above. However, as the richest of those ore deposits are mined out, it has become profitable to mine some laterites for iron as well as for aluminum.

Metamorphic Deposits

The mineralogical changes caused by the heat or pressure of metamorphism can produce economic mineral deposits.

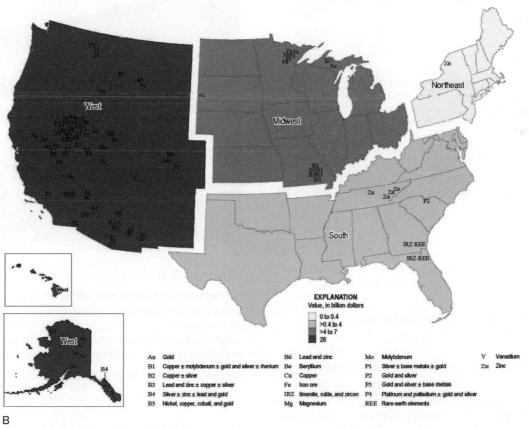

B

Figure 13.7

(A) Distribution of copper and molybdenum deposits. (B) Major metal-producing areas in the United States. Base metals are the more common metals, such as copper and lead, as distinguished from precious metals like gold and silver.

Source: (A) Data from U.S. Geological Survey Scientific Investigations Report 2010–5090. (B) Mineral Commodity Summaries 2022, U.S. Geological Survey.

A

B

Figure 13.8

Sedimentary ores are often layered. (A) This banded iron formation in Ishpeming, Michigan, consists of bands of chert and hematite (Fe_2O_3). The photo is about 0.4 meters across. (B) Rock salt in an evaporite deposit.

(A) ©Gina S. Szablewski; (B) ©Carla Montgomery

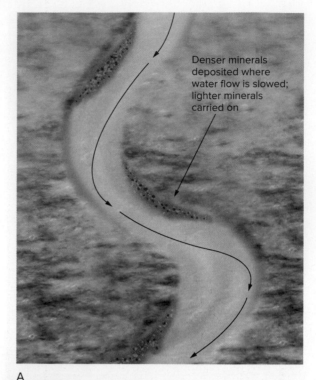

Denser minerals deposited where water flow is slowed; lighter minerals carried on

A

B

Figure 13.9

(A) Formation of a placer deposit. (B) Since placer deposits typically occur as loose gravels, mining them can be relatively easy; here, near Nevada City, California, water blasting from giant nozzles washes gold-bearing gravels to sluices, where flowing water will separate the denser gold from lighter minerals and rock fragments.

(B) U.S. Geological Survey/Photograph by Carlton Watkins

Graphite, used in batteries, as a lubricant, and for many applications where its very high melting point is vital, is usually mined from metamorphic deposits. Graphite is pure carbon, and one way in which it forms is by the metamorphism of coal. Asbestos, as previously noted, is not a single mineral but a general term applied to a group of fibrous silicates that are formed by the metamorphism of igneous rocks rich in ferromagnesian minerals, with the addition of water. The asbestos minerals are used less in insulation now but are still valuable for their heat- and fire-resistant properties, especially because their unusual fibrous character allows them to be woven into cloth. Garnet is a common mineral in many metamorphic rocks. It may be used as a semiprecious gemstone; lower-quality crystals are effective industrial abrasives.

13.3 Mineral and Rock Resources—Examples

Learning Objectives

- Recall the occurrence and principal application of various metal and nonmetal resources
- Recognize types and uses of rock resources

Dozens of minerals and rocks have some economic value. In this section, we discuss a sampling of these resources, noting their occurrences and principal applications.

Metals

The term *mineral resources* usually brings metals to mind first. Overwhelmingly, the most heavily used metal is iron. It is also one of the most common metals. Nearly all iron ore is used for the manufacture of iron and especially steel products. It is mined principally from the ancient sedimentary deposits described in the previous section but also from some laterites and from concentrations of magnetite in some igneous bodies.

Aluminum is another relatively common metal, and it is the second most widely used. Its light weight, coupled with its strength, makes it particularly useful in the transportation and construction industries (**figure 13.10A**); it is also widely used in packaging, especially for beverage cans. Aluminum is the third most common element in the crust, but there it is most often found in silicates, from which it is extremely difficult to extract. Most of what is mined commercially is bauxite, an aluminum-rich laterite in which the aluminum exists as a hydroxide. Even in this form, the extraction of aluminum is somewhat difficult and energy-intensive.

Many less common but also important metals, including copper, lead, zinc, nickel, cobalt, and others, are found in sulfide ore deposits. Sulfides occur frequently in hydrothermal deposits and may also be concentrated in igneous rocks. Copper, lead, and zinc may also be found in sedimentary ores; some laterites are moderately rich in nickel and cobalt. Clearly, these metals may be concentrated into economically valuable deposits in a variety of ways. Copper is primarily used for electrical applications because it is an excellent conductor of electricity (**figure 13.10B**); it is also used in the construction and transportation industries. An important use of lead is in batteries; among its many other applications, it is a component of many solders and is used in paints and ceramics. The zinc coating on steel cans (misnamed "tin cans") keeps the cans from rusting, and zinc is also used in the manufacture of brass and other alloys. Cobalt is used in lithium-ion batteries.

The precious metals gold, silver, and platinum have some unique practical uses. Gold is used not only for jewelry, in the arts, and in commerce, but also in the electronics industry and in dentistry. It is particularly valued for its resistance to tarnishing. Historically, silver's principal single use has been for photographic materials (e.g., film). With the expansion of digital imaging, photographic use of silver has declined sharply, but industrial and electronics applications are expanding. Platinum is an excellent *catalyst,* a substance that promotes chemical reactions. Currently, close to half the platinum used in the United States goes into automobile emissions-control systems, with the rest finding important applications in the petroleum and chemical industries, in electronics, and in medicine, among other areas. All of the precious metals can be found as native metals, most frequently in igneous or hydrothermal ore deposits. Silver also commonly forms sulfide minerals. Most gold and silver production in this country is a by-product of mining ores of more abundant metals like copper, lead, and zinc, with which small amounts of precious metals may be associated.

A

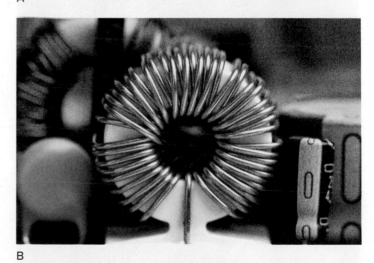

B

Figure 13.10

(A) Lightweight and strong, aluminum is a very useful metal. These alloy wheels are one of many applications of aluminum in the transportation industry. (B) Copper is commonly used in wiring, as seen in this inductor coil on a circuit board.

(A) Anna Volobueva/Alamy Stock Photo; (B) Flegere/ Shutterstock

Lithium is a soft metal that is rare in Earth's crust but has much higher concentrations in ocean water. Primary sources of lithium are fairly widely dispersed and include minerals found in granite pegmatites, and briny solutions associated with dried-up sea/lake beds though many of these are too small or the lithium in too low a concentration to be economically valuable. Lithium has experienced a five-fold increase in production in the past fifteen years due to its use in batteries, and its supply has become a major security issue with tech companies.

Nonmetallic Minerals

Another by-product of mining sulfides is the nonmetal sulfur. Sulfur may also be recovered from petroleum during refining, from volcanic deposits (sulfur is sometimes precipitated as pure native sulfur from fumes escaping from volcanic vents, as seen in **figure 13.11**), and from evaporites. The primary use of sulfur is for the manufacture of sulfuric acid for industrial purposes.

Several important minerals are recovered from evaporite deposits. The most abundant is halite, or rock salt, used principally as a source of the sodium and chlorine of which it is composed, and secondarily for road salt, either directly or through the production of other salts from it. Halite has many lower-volume applications, including, of course, seasoning food as table salt. Gypsum, essential to the manufacture of plaster, portland cement, and wallboard for construction, is another evaporite mineral. Others include phosphate rock and potassium-rich potash, key ingredients of the synthetic fertilizers on which much of U.S. agriculture depends.

As noted in chapter 2, clay is not a single mineral, but a group of layered hydrous silicates that are formed at low temperature, commonly by weathering, and that are abundant in sedimentary deposits in the United States. The diversity of clay minerals leads to a variety of applications, from fine ceramics to the making of clay piping and other construction materials, the processing of iron ore, and drilling for oil. In the last case, the clay is mixed with water and sometimes other minerals to make drilling mud, which lubricates the drill bit; the low permeability of clays also enables them to seal off porous rock layers encountered during drilling.

Rock Resources

When it comes to overall mass, the quantities used of the various metals and minerals mentioned previously shrink to insignificance besides the amount of rock, sand, and gravel used. In the United States in 2021, over 1 billion tons of sand and gravel were used in construction, especially in making cement and concrete. In addition, over 72 million tons of sand and gravel were used in industry, over half of that for fracking and drilling operations. Also, over 2.3 million tons of dimension and facing stone were sold or used. These included limestone and slate for flagstones, and various attractive and/or durable rocks such as marble, granite, sandstone, and limestone for monuments, building facings, and interior surfaces (**figure 13.12**).

Figure 13.11

Sulfur deposition around a fumarole, Kilauea, Hawaii.

©Carla Montgomery

Figure 13.12

This decorative stone would make a memorable surface in someone's kitchen. Many people would call it a "granite" countertop, but this rock certainly is not granite.

©Gina S. Szablewski

13.4 Mineral Supply and Demand

Learning Objectives

- Recall how industrial growth and standard of living improvements affect demand for minerals
- Recognize the production/consumption/project lifetime of various minerals in the United States
- List reasons supply and demand for mineral resources changes
- Define conflict mineral
- Identify ways in which subeconomic resources may become available

Our assessments of the adequacy of known supplies of various rock and mineral resources require information both on those supplies and on the rates at which they are, and will be, consumed. We have already examined some of the variables on the supply side of the calculation, including geologic and economic factors. Large uncertainties also exist on the demand side.

Attempting to summarize the historic demand for mineral resources in a single paragraph is impractical, but we can say that demand fluctuates with the economy and the development of technology and can be very specific to the mineral itself. The Great Recession in 2007–2009 paralleled significant reductions in the production and consumption of sand and gravel, gypsum, and crushed stone because of declines in the U.S. housing market; the same with copper, iron, and lead due to declines in auto manufacturing. The estimated value of U.S. metal mine production fell by 22% between 2008 and 2009. More recently, the COVID-19 pandemic caused decreases in the consumption of many nonfuel resources. The rebound in U.S. production and consumption was hindered by numerous factors including supply chain issues, lack of workers, a ship that blocked the Suez Canal for six days, and various pervasive restrictions. The estimated value of U.S. metal mine production increased by 23% between 2020 and 2021 as the economy started to recover.

Underlying economic variables, improvements in standards of living and industrial growth for billions of people continue to increase demand pressures. China's historic production and consumption of copper is often used as an example (**figure 13.13**), as this metal is essential to electronics of all kinds. A few features of this figure are noteworthy: the sharp rise in consumption at the turn of the century, the widening gap between consumption and production from 2000 to 2010, and then recent, relative stability of that gap around 20%. As China developed, their demand for copper quickly outpaced their production of it, increasing competition for the global copper supply among copper-importing nations. China's growth may have slowed down, but they still account for over 50% of global copper consumption, and just a small increase in a sector such as construction would cause a not insignificant increase in demand.

Higher levels of development are generally correlated with higher per-capita consumption of resources, suggesting considerable potential for demand growth as other nations such as India become more industrialized.

U.S. Mineral Production and Consumption

The United States consumes huge quantities of mineral and rock materials, relative to both its population and its production of most minerals. The information in **figures 13.14** and **13.15** shows us the U.S. per-capita and total consumption of a variety of resources. When considering these numbers to the ratio of the U.S. to global population, it is clear to see that Americans consume many resources at much higher rates compared to people in other countries.

The relationship of domestic consumption to domestic production can be complex. *Primary production* is production from new ore, excluding production from recycling or from partially processed ore imported from elsewhere; total production includes the latter kinds of production as well. Sometimes, the United States even imports materials of which it has ample domestic supplies, perhaps to stay on good trade terms with countries having other, more critical commodities that the United States also wishes to import or because the cost of complying with U.S. environmental regulations makes it cheaper to import the commodity from another nation. For example, in the case of molybdenum, in 2021, U.S. primary production was 48,000 metric tons, against reported consumption of 16,000 tons—yet we imported 29,000 tons; in the same year, we produced 270 metric tons of gold (including recycling) and consumed 250, but imported another 190 tons.

Of course, where production falls short of consumption, importing is necessary to meet domestic demand. Moreover, low domestic production may not necessarily mean that the United States is conserving its own resources, but that it has none to produce or, at least, no economic deposits. This is true, for example, of graphite, an essential ingredient in lithium-ion electric vehicle batteries, and manganese, which is currently

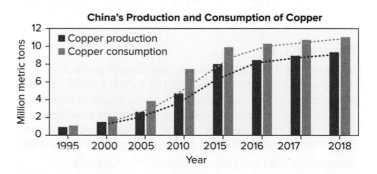

Figure 13.13

As China pushed ahead industrially, its demand for mineral resources such as copper increased sharply.

Source: Data from table comes from USGS Minerals Yearbook 0 The Mineral Industry of China (multiple years)

Crushed stone
4806 kg

Sand and gravel
3003 kg

Salt
141 kg

Gypsum
129 kg

Phosphate
75.1 kg

Iron ore
108 kg

Potash
22.2 kg

Aluminum
10.8 kg

Zinc
2.8 kg

Lead
4.8 kg

Copper
5.4 kg

A

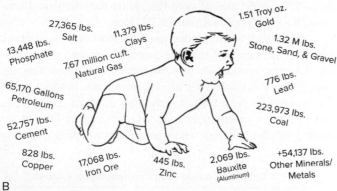

27,365 lbs.
Salt

11,379 lbs.
Clays

1.51 Troy oz.
Gold

13,448 lbs.
Phosphate

7.67 million cu.ft.
Natural Gas

1.32 M lbs.
Stone, Sand, & Gravel

65,170 Gallons
Petroleum

776 lbs.
Lead

223,973 lbs.
Coal

52,757 lbs.
Cement

828 lbs.
Copper

17,068 lbs.
Iron Ore

445 lbs.
Zinc

2,069 lbs.
Bauxite
(Aluminum)

+54,137 lbs.
Other Minerals/
Metals

B

Figure 13.14

(A) Per-capita consumption of selected mineral and rock resources in the United States, 2021 (1 kg = 2.2 lb). (B) The Minerals Education Coalition takes similar data, together with average U.S. life expectancy, to calculate the average total lifetime consumption of various rock, mineral, and fuel resources by an American born now. Note that the total is over 3 million pounds!

Sources: (A) Data from Mineral Commodity Summaries 2022, U.S. Geological Survey; (B) from Minerals Education Coalition (www.mineralseducationcoalition.org)

irreplaceable in the manufacture of steel. Currently, for materials such as these, the United States is wholly or heavily dependent on imports (**figure 13.16**), and those imports come from around the globe (**figure 13.17**).

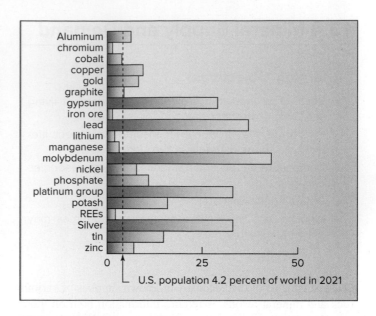

Figure 13.15

U.S. share of global consumption of selected materials. Assumes that world consumption of each commodity approximately equals world production.

Source: Data from Mineral Commodity Summaries 2022, U.S. Geological Survey

Leaving aside the unanswerable political question of how long any country can count on an uninterrupted flow of necessary mineral imports, how well supplied is the world, overall, with minerals? How imminent are the shortages? We will address these questions below.

World Mineral Supply and Demand

Certain assumptions must underlie any projections of lifetimes of mineral reserves. One is future demand. For reasons already outlined, demand can fluctuate both up and down. Furthermore, not all commodities follow the same trends. Making the assumption that world production annually approximates demand, at least for new mine production, we find that from 2011 to 2019, annual demand for iron ore and copper decreased slightly, demand for zinc increased slightly, and demand for REE and lithium increased dramatically. These same trends extended through the end of 2021 with the exception of copper, the production of which fell significantly; this is a good example of how the COVID-19 pandemic disrupted production and consumption of many materials, not all in the same way, and how the 2021 numbers may not be representative of longer-term trends. Rather than attempt to make precise demand estimates for the future, the following discussion assumes demand equal to primary production at constant 2021 levels.

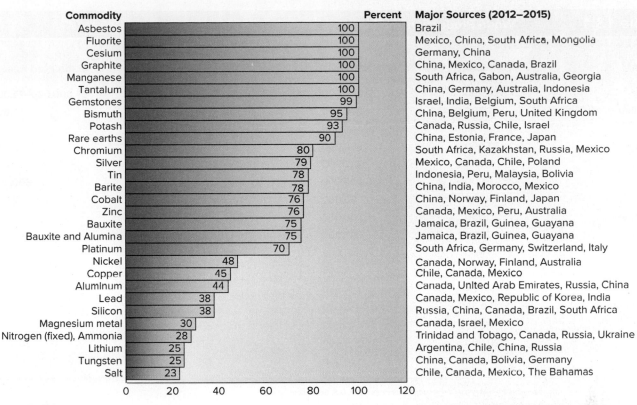

Commodity	Percent	Major Sources (2012–2015)
Asbestos	100	Brazil
Fluorite	100	Mexico, China, South Africa, Mongolia
Cesium	100	Germany, China
Graphite	100	China, Mexico, Canada, Brazil
Manganese	100	South Africa, Gabon, Australia, Georgia
Tantalum	100	China, Germany, Australia, Indonesia
Gemstones	99	Israel, India, Belgium, South Africa
Bismuth	95	China, Belgium, Peru, United Kingdom
Potash	93	Canada, Russia, Chile, Israel
Rare earths	90	China, Estonia, France, Japan
Chromium	80	South Africa, Kazakhstan, Russia, Mexico
Silver	79	Mexico, Canada, Chile, Poland
Tin	78	Indonesia, Peru, Malaysia, Bolivia
Barite	78	China, India, Morocco, Mexico
Cobalt	76	China, Norway, Finland, Japan
Zinc	76	Canada, Mexico, Peru, Australia
Bauxite	75	Jamaica, Brazil, Guinea, Guayana
Bauxite and Alumina	75	Jamaica, Brazil, Guinea, Guayana
Platinum	70	South Africa, Germany, Switzerland, Italy
Nickel	48	Canada, Norway, Finland, Australia
Copper	45	Chile, Canada, Mexico
Aluminum	44	Canada, United Arab Emirates, Russia, China
Lead	38	Canada, Mexico, Republic of Korea, India
Silicon	38	Russia, China, Canada, Brazil, South Africa
Magnesium metal	30	Canada, Israel, Mexico
Nitrogen (fixed), Ammonia	28	Trinidad and Tobago, Canada, Russia, Ukraine
Lithium	25	Argentina, Chile, China, Russia
Tungsten	25	China, Canada, Bolivia, Germany
Salt	23	Chile, Canada, Mexico, The Bahamas

Figure 13.16

Proportions of U.S. needs for selected minerals supplied by imports as a percentage of apparent material consumption, 2021. Principal sources of imports indicated at right in order of relative contribution; note the variety of nations represented.

Source: Data from Mineral Commodity Summaries 2022, U.S. Geological Survey

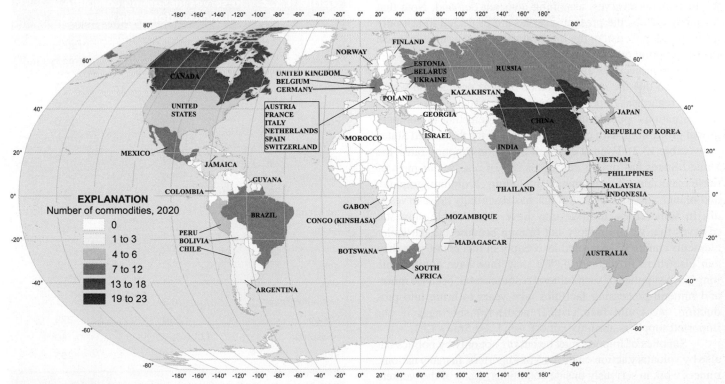

Figure 13.17

Major Sources of Nonfuel Mineral Commodities 2021.

Note that there are dozens of minerals for which the United States relies on imports for more than 50% of its consumption, far more than the examples illustrated in figure 13.16.

Source: Mineral Commodity Summaries 2022, U.S. Geological Survey

Table 13.1

Table 13.1 — World Production and Reserves Statistics, 2021*

Material	Production	Reserves	Projected Lifetime of Reserves (years)	Estimated Resources
bauxite	390,000	32,000,000	82	55,000,000–75,000,000
chromium	41,000	570,000	14	12,000,000
cobalt	170	7600	45	25,000
copper	21,000	880,000	42	5,600,000˙
graphite	1000	320000	320	800,000
iron ore	2,600,000	180,000,000	69	800,000,000
lead	4300	90,000	21	2,000,000
manganese	20,000	1,500,000	75	**
nickel	2700	95,000	35	300,000
zinc	13,000	250,000	19	1,900,000
gold	3.0	54	18	53 (US only)
silver	24	530	22	n.a.
platinum group	0.38	70	184	100,000
lithium	100	22000	220	89,000
phosphate	220,000	71,000,000	322	300,000,000
REE	280	120,000	429	17,400

Source: Production, reserves, and resources from Mineral Commodity Summaries 2022, U.S. Geological Survey.

*All production, reserve, and resource figures in thousands of metric tons.

˙ Includes 0.7 billion tons copper estimated to occur in manganese nodules; "extensive" resources of nickel are also projected to occur in these nodules.

Table 13.1 shows projections of the lifetimes of selected world mineral reserves, assuming constant demand. Note that for many metals, the present reserves are projected to last only a few decades. Consider the implications of just a small increase in consumption rates.

Such projections also presume unrestricted distribution of minerals so that the minerals can be used as needed and where needed, regardless of political boundaries or such economic factors as nations' purchasing power. It is interesting to compare global projections with corresponding data for the United States only (**table 13.2**). Even at constant 2021 consumption rates, the United States has only a decade or two of reserves of most of these materials. Of some metals, including chromium, graphite, and manganese, the United States has essentially no reserves at all. In some cases, too, we may have considerable domestic reserves that we are simply not mining because the particular elements cannot be extracted as cheaply in this country as they can elsewhere in the world. We may thus have comfortable long-term supplies of such mineral resources; but because mines and mineral-processing facilities take years to bring into production, we could face critical near-term shortages if our imported supply is abruptly cut off. See Case Study 13.1.

Supplies of imports may be cut off not only by politics but also by voluntary action on the part of importers. In many nations, miners work in seriously unsafe conditions, many of the workers are children, and workers are forced to labor in the mines. In other cases, the profits from sales of mineral resources support violent warring factions or repressive governments. The

Table 13.2 — Projected Lifetimes of U.S. Mineral Reserves (assuming complete reliance on domestic reserves)

Material	Reserves*	Projected Lifetime (years)
bauxite˙	0	0
chromium	620	1.0
cobalt	69	1.0
copper	48,000	27
iron ore	2,600,000	72
graphite	0	0
lead	5000	3.1
manganese	0	0
nickel	340	4.1
zinc	9000	9.8
gold	3	12
silver	26	4.9
platinum group	0.9	7.1
lithium	750	375
phosphate	1,000,000	40
REEs	1800	295

Source: Reserves data from Mineral Commodity Summaries 2022, U.S. Geological Survey.

*Reserves in thousands of metric tons.

˙ Note that bauxite consumption is only a partial measure of total aluminum consumption; additional aluminum is consumed as refined aluminum metal, of which there are no reserves.

commodities in question are so-called *conflict minerals*. Increasingly, governments and companies in wealthier countries are making commitments not to trade in conflict minerals. This presents challenges in two respects. The more obvious is that it reduces the number of possible sources for mineral imports. It also requires that an importer be able to trace the supply chain for imported materials all the way back to the source, and the necessary records may not always be accurate or complete. The practical supply problem can be daunting enough. For example, in 2017, a major U.S. maker of electric cars vowed to use cobalt only from North American sources (more than half the world's cobalt comes from the Congo, a major focus of the conflict-minerals issue). But analysts quickly concluded that the availability of cobalt from North America is simply insufficient to meet that demand—without even taking into account any increased demand from other companies/nations wishing to make a similar commitment. The problem of conflict minerals in the world resource picture seems unlikely to be resolved easily or soon.

Some relief in the overall supply picture can be anticipated from the economic component of the definition of reserves. For example, as currently identified reserves are depleted, the law of supply and demand will drive up minerals' prices. This, in turn, will make some of what are presently subeconomic resources profitable to mine; some of those deposits will effectively be reclassified as reserves. Improvements in mineral-processing technology can accomplish the same result. In addition, continued exploration can be expected to locate additional economic deposits, particularly as exploration methods become more sophisticated, as described later in the chapter.

The technological advances necessary for the development of some of the subeconomic resources, however, may not be achieved rapidly enough to help substantially. Also, if the move is toward developing lower- and lower-grade ore, then by definition, more and more rock will have to be processed and more and more land disturbed to extract a given quantity of mineral or metal. Some of the consequences of mining activities, which will have increasing impact as more mines are developed, are also discussed later in this chapter.

What other prospects exist for averting future mineral-resource shortages?

13.5 Future Mineral Supplies

Learning Objectives

- List ways to extend supplies of nonrenewable resources
- Identify reasons that demand for mineral resources will change in the future
- Describe different modern methods used in mineral exploration
- List different marine mineral resources
- List reasons for and challenges with recycling metallic and nonmetallic mineral resources

There are a variety of ways to extend our supply of nonrenewable resources, and it is likely that we will need to employ them all at some level in the future to support our modern way of life. If we want something to last longer, the simple answer is to reduce our consumption of it. In practice, this seems unlikely to occur, assuming that many people want to sustain their relatively wealthy live styles and many others, especially those in populous countries, want to improve theirs in a way that increases consumption. Looking at the trends in particular consumer products can give us an idea of how difficult cutting consumption might be.

The United States has one of the world's highest rates of car ownership: In 2019, 59% of U.S. households owned two or more cars, and 21%, three or more, for total household vehicle ownership of about 842 vehicles per 1000 people. Factors including necessity, more leisure for travel, lack of alternative transportation, and desires for convenience or status put upward pressure on car-ownership numbers. Other commodities that clearly reflect consumerism are personal computer and cellular phones (see Case Study 13.2), many of which are replaced after just two to three years of use. They represent consumption of both materials and energy, and like cars, ownership is based on numerous factors. While it may be natural for people to want to acquire goods that increase comfort, convenience, efficiency, or perceived quality of life, the result is an increase in already high resource-consumption rates. And consumption usually outpaces population; in the United States, from 1970 to 2020, the population grew by about 63%, while consumption of rock and mineral resources just about tripled.

Even if the technologically advanced nations were to restrain their appetite for minerals, they represent a minority of the world's people. If the developing nations are to achieve a standard of living for their citizens that is even remotely comparable to that in the most industrialized nations, vast quantities of minerals and energy will be needed. For example, recent data put total motor vehicles at only 203 per 1000 people in China and 50 per 1000 in India. Both countries have rapidly growing economies. If China's population of over 1.4 billion persons were to increase its vehicle-ownership rate to the U.S. average, that would mean more about 900 million additional vehicles. Consider the resource implications if such populous nations as China and India were to increase not only their vehicle-ownership rates but also their consumption of other consumer goods to the levels common in the most-developed countries. However, car sales are predicted to decrease in the United States as a reflection of our altered work patterns, trends that accelerated quickly during the pandemic. Other nations may also follow this trend, and (multiple) car ownership soon may not be a solid indicator of material consumption. And the switch to electrically powered cars will alter the list of desired resources to build them.

If demand for resources cannot realistically be cut, supplies must be increased or extended. We must develop new sources of minerals, in either traditional or nontraditional areas, or we must conserve our minerals better so that they will last longer. Recycling and finding substitute materials are also ways to sustain our need for materials.

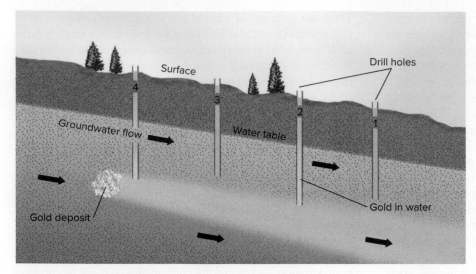

Figure 13.18

Groundwater sampling can detect trace metals in concentrations as low as a few parts per trillion; even low-solubility metals like gold can thus be detected down the flow path from an ore deposit, and mapping anomalies helps locate it.

After U.S. Geological Survey.

Modern Methods in Mineral Exploration

Most of the near-surface mineral deposits in readily accessible places have probably already been discovered, at least in developed and/or well-explored regions. Fortunately, a variety of new methods are being applied to the search. Geophysics provides some assistance. Rocks and minerals vary in density, and in magnetic and electrical properties, so changes in rock types or distribution below Earth's surface can cause small variations in the gravitational or magnetic field measured at the surface, as well as in the electrical conductivity of the rocks. Some ore deposits may be detected in this way. As a simple example, because many iron deposits are strongly magnetic, large magnetic anomalies may indicate the presence and the extent of a subsurface body of iron ore. Radioactivity is another readily detected property; uranium deposits may be located with the aid of a Geiger counter.

Geochemical prospecting is another approach with increasing applications. It may take a variety of forms. Some studies are based on the recognition that soils reflect, to a degree, the chemistry of the materials from which they formed. The soil over a copper-rich ore body, for instance, may itself be enriched in copper relative to surrounding soils. Surveys of soil chemistry over a wide area can help to pinpoint likely spots to dig in search of ores. Occasionally, plants can be sampled instead of soil. Certain plants tend to concentrate particular metals, making the plants very sensitive indicators of locally high concentrations of those metals. Sometimes just the presence of a particular plant is a clue; a variety of pandanus plant thrives especially well in the soil that develops over kimberlite pipes. Groundwater may contain traces of metals leached from ore deposits along its flow path (**figure 13.18**). Even soil gases can supply clues to ore deposits. Mercury is a very volatile metal, and high concentrations of mercury vapor have been found in gases filling soil pore spaces over mercury ore bodies. Geochemical prospecting has also been used in petroleum exploration.

Remote sensing methods are becoming increasingly sophisticated and valuable in mineral exploration. These methods rely on detection, recording, and analysis of wave-transmitted energy, such as visible light and infrared radiation, rather than on direct physical contact and sampling. Remote sensing using satellites is a quick and efficient way to scan broad areas, to examine regions having such rugged topography or hostile climate that they cannot easily be explored on foot or with surface-based vehicles, and to view areas to which ground access is limited for political reasons. Probably the best known and most long-lived Earth satellite imaging system is the Landsat system, initiated in 1972.

The sensors in the Landsat satellites do not detect all wavelengths of energy reflected from the surface. They do not take photographs in the conventional sense. They are particularly sensitive to selected green and red wavelengths in the visible light spectrum and to a portion of the infrared (invisible heat radiation, with wavelengths somewhat longer than those of red light). These wavelengths were chosen because plants reflect light most strongly in the green and the infrared. Different plants, rocks, and soils reflect different proportions of radiation of different wavelengths. Even the same feature may produce a somewhat different image under different conditions. Wet soil differs from dry; sediment-laden water looks different from clear; a given plant variety may reflect a different radiation spectrum depending on what trace elements it has concentrated from the underlying soil or how vigorously it is growing. Landsat images can be powerful mapping tools. In recent decades, additional satellites and other spacecraft have offered still more opportunities for imaging Earth's surface, and resolution has been increasing as well.

Basic geologic mapping, identification of geologic structures, and resource exploration are only some of the applications of such imagery (**figure 13.19**); we've discussed a lot of others in previous chapters, especially regarding the assessment of geologic hazards such as earthquakes, volcanic eruptions,

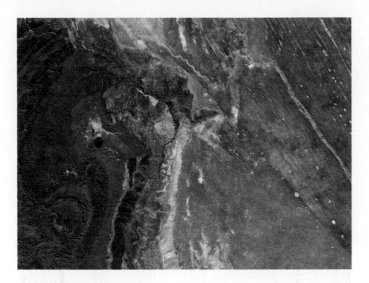

Figure 13.19

This image of Namibia was taken by the Operational Land Imager on Landsat 8 in May 2020. The Kalahari Desert is to the right, and the rugged mountains of the Central Plateau to the right. Landsat satellite images may reveal details of geology that will aid in mineral exploration. Vegetation, also sensitive to geology, may enhance the image.

NASA Earth Observatory images by Joshua Stevens

and flooding. Remote sensing is especially useful when we can support it with information gathered by direct surface examination, or *ground truthing*. If this is logistically impractical or unsafe, images of inaccessible regions can be compared with images from other regions that have been mapped and sampled directly, using the similarities in imagery characteristics to infer the actual geology or vegetation. Our sensors and our understanding continue to improve, giving us even more and more detailed information about rocks and mineral resources (**figure 13.20**). With so much data, we are turning to artificial intelligence and machine learning applications to predict areas rich in mineral resources.

Finally, advances in geologic understanding also play a role in mineral exploration. Development of plate-tectonic theory helped geologists to recognize the association between particular types of plate boundaries and the occurrence of certain kinds of mineral deposits. This recognition has directed the search for new ore deposits. For example, because geologists know that molybdenum deposits are often found over existing subduction zones, they logically explore for more molybdenum deposits in other present or past subduction zones. Also, the realization that many of the continents were once united suggests likely areas to look for more ores (**figure 13.21**). If mineral deposits of a particular kind are known to occur near the margin of one continent, similar deposits might be found at the corresponding edge of another continent once connected to it. If a mountain belt rich in some kind of ore seems to have a counterpart range on another continent once linked to it, the same kind of ore may occur in the matching mountain belt. Such

reasoning is partly responsible for the supposition that economically valuable ore deposits may exist on Antarctica. (Some of the legal/political consequences of such speculations are explored in chapter 19; fundamental concerns about the wisdom of mineral exploration in the fragile Antarctic environment have motivated international agreements to restrict resource exploration there until 2048.)

Marine Mineral Resources

There continues to be interest to seeking mineral wealth in unconventional places. In particular, the sea may provide partial solutions to some mineral shortages. Seawater contains virtually every chemical element in dissolved form, although most of it is sodium chloride. Most metals we are interested in occur in far lower concentrations, and are so diffuse that vast volumes of seawater would have to be processed to extract even small quantities. The costs would be very high, especially since existing technology is not adequate to efficiently extract a few specific metals of interest. Several types of concentrated underwater mineral deposits have greater potential.

During the last Ice Age, when a great deal of water was locked in ice sheets, worldwide sea levels were lower than at present. Much of the area of the now-submerged continental shelves was dry land, and streams running off the continents flowed out across the exposed shelves to reach the sea. Where the streams' drainage basins included appropriate source rocks, these streams might have produced valuable placer deposits. With the melting of the ice and rising sea levels, any placers on the continental shelves were submerged. Finding and mining these deposits may be costly and difficult, but as land deposits are exhausted, placer deposits on the continental shelves may well be worth seeking. Already, more than 10% of world tin production is from such marine placers; potentially significant amounts of titanium, zirconium, and chromium could also be mined from them if the economics were right.

The hydrothermal ore deposits forming along some seafloor spreading ridges are another possible source of needed metals (the purple dots in **figure 13.22A**). In many places, the quantity of material being deposited is too small, and the depth of water above would make recovery prohibitively expensive. However, the metal-rich muds of the Red Sea contain sufficient concentrations of such metals as copper, lead, and zinc that some exploratory dredging is underway, and several companies are interested in the possibility of mining those sediments. Along a section of the Juan de Fuca ridge, off the coast of Oregon and Washington, hundreds of thousands of tons of zinc- and silver-rich sulfides may already have been deposited, and the hydrothermal activity continues. In 2017, a Japanese firm began testing a seafloor-mining method on hydrothermal sulfide off Okinawa. A Canadian and a U.K.-based mining company have each secured exploration licenses for substantial areas of the territorial waters of New Zealand, Fiji, Tonga, and other parts of the southwest Pacific basin, and hope to be developing hydrothermal sulfides associated with plate boundaries in this region within a decade.

A

B

C

Figure 13.20

Three ASTER images of Saline Valley, California, taken from NASA's Terra satellite, provide different information via different wavelengths of radiation. In the visible/near-infrared range (A), vegetation shows as red, water and snow as white, and rocks in various shades of gray, brown, yellow, and blue. Different rock types are distinguished better in the short-wavelength infrared image (B); for example, limestones appear yellow-green, and purple areas are rich in kaolinite clay. In the thermal-infrared image (C), carbonate rocks appear green, mafic rocks show as purple, and varying intensity of red color reflects differences in the amount of quartz present.

(A-C) NASA, GSFC, MITI, ERSDAC, JAROS, and the U.S./Japan-ASTER Science Team

lesser but potentially valuable amounts of copper, nickel, cobalt, platinum, and other metals, the value of which may be a greater motive for mining the nodules than the manganese itself. The nodules are found over much of the deep-sea floor, in regions where sedimentation rates are slow enough not to bury them (**figure 13.22B**). At present, the costs of recovering these nodules would be high compared to the costs of mining the same metals on land, and the technical problems associated with efficiently recovering the nodules from beneath several kilometers of seawater remain to be worked out. Still, manganese nodules represent a sufficiently large metal resource that they have become a subject of International Law of the Sea conferences (see chapter 19).

With any ocean resources outside territorial waters, there are questions of ownership or of who has the right to exploit

Perhaps the most widespread undersea mineral resource is **manganese nodules** (polymetallic nodules in **figure 13.22A**). These are concretions up to about 10 centimeters in diameter, composed mostly of manganese minerals. They also contain

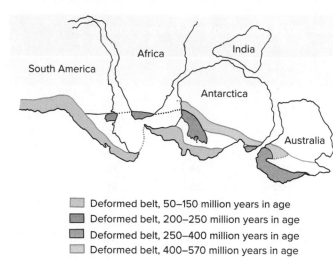

Deformed belt, 50–150 million years in age
Deformed belt, 200–250 million years in age
Deformed belt, 250–400 million years in age
Deformed belt, 400–570 million years in age

Figure 13.21

Pre-continental-drift reassembly of landmasses suggests locations of possible ore deposits in unexplored regions by extrapolation from known deposits on other continents.

Source: Data from C. Craddock et al., Geological Maps of Antarctica, Antarctic Map Folio Series, folio 12, the American Geographical Society

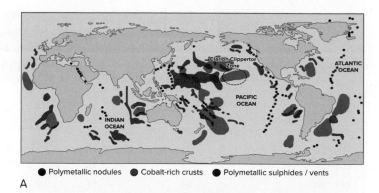

● Polymetallic nodules   ● Cobalt-rich crusts   ● Polymetallic sulphides / vents

A

B

Figure 13.22

(A) Marine mineral resources are extensive but not well understood, especially how mining them will affect marine ecosystems. (B) A crab crawls over a field of ferromanganese nodules in the North Atlantic off the U.S. east coast.

(A) IUCN World Conservation Union, https://www.iucn.org/ resources/issues-briefs/deep-sea-mining; (R) Image courtesy of NOAA Ocean Exploration

these resources. China is very interested in deep-sea mining and has been recently vigorously exploring and surveying the deep ocean. Other private companies are working on mining and processing techniques. There are serious concerns about consequences any type of seafloor mining will have on ecosystems. The International Seabed Authoring is working on regulations that would allow commercial-scale mining to begin as early as 2023, even though many scientists suggest that we don't yet know enough about seafloor creatures and deep-sea processes to determine the best course of action, if any. California has already proposed a bill to ban mining in the state's ocean territories.

Conservation of Mineral Resources

Overall need for resources is likely to increase, but for selected individual materials that are particularly scarce, perhaps demand can be moderated. One way is by making substitutions. For certain applications, we might be able to replace a very rare metal with a more abundant one. The extent to which this is likely to succeed is limited, since reserves of most metals are quite limited, as was clear from tables 13.1 and 13.2. Substituting one metal for another reduces demand for the second while increasing demand for the first; the focus of the shortage problem shifts, but the problem persists. Nonmetals are replacing metals in other applications. Unfortunately, many of these nonmetals are plastics or other materials derived from petroleum, which has simply contributed to petroleum consumption. As we will see in chapter 14, supplies of that resource are limited, too. For some applications, ceramics or fiber products can be substituted; however, sometimes the physical, electrical, or other properties required for the particular application demand the

use of metals. There are alternatives to the REE for many applications, but they generally don't work as well.

The most effective way to extend mineral reserves, for some metals at least, is through recycling. Overall use of the metals may increase, but when we find ways to reuse a significant fraction of these metals repeatedly, proportionately less new metal will need to be extracted from mines, reducing primary production. Some metals are already extensively recycled, at least in the United States (table 13.3 lists a few examples). Worldwide, recycling is less widely practiced, in part because the less-industrialized countries have accumulated fewer manufactured products from which materials can be retrieved. Still, given the limited U.S. reserves of many metals, recycling can be a very important means of extending those reserves. Additional benefits of recycling include a reduction in the volume of the waste-disposal problem, a decrease in the extent to which more land must be disturbed by new mining activities, and a decrease in the energy demand to mine and process raw materials.

Unfortunately, not all materials lend themselves equally well to recycling. Among those that work out best are metals

Table 13.3	Metal Recycling in the United States, 1986–2021 (recycled scrap as percentage of consumption)						
Metal	**1986**	**1992**	**1999**	**2005**	**2011**	**2016**	**2021**
aluminum*							
old scrap	15	30	20	16	36	31	30
total	34	47	51	44	77	73	74
chromium	21	26	20	29	40	42	20
cobalt	15	25	34	27	24	30	24
copper	22	42	33	30	35	31	32
lead	50	62	68	72	83	69	62
manganese	0	0	0	0	0	0	0
nickel	25	30	38	39	43	43	52

Sources: Mineral Commodity Summaries 1990, 1993, 2000, 2006, 2012, 2017, and 2022 U.S. Geological Survey.

*For most metals, recycling is of "old scrap," post-consumer material. Aluminum is unusual in that much of what is recycled is "new scrap" produced during the manufacture of aluminum products. The totals for aluminum include both old and new scrap.

that are used in pure form in sizeable pieces—copper in pipes and wiring, lead in batteries, and aluminum in beverage cans. The individual metals are relatively easy to recover and require minimal effort to purify for reuse. Recycling can be more energy-efficient than producing new metal by mining and refining new ore, which results in the dual benefits of saving energy and cutting costs: The U.S. Environmental Protection Agency estimates that recycling just 1 ton of aluminum cans saves energy equivalent of 1665 gallons of gasoline.

Where different materials are intermingled in complex manufactured objects, it is more difficult and costly to extract individual metals. Consider trying to separate the various metals from a refrigerator, a lawn mower, or a computer. The effort required to do that, even if it were technically possible, would make most of the materials recovered far too costly and thus noncompetitive with new production. Only in a few rare cases are metals valuable enough that the recycling effort may be worthwhile; for example, the platinum from catalytic converters in exhaust systems.

Alloys present special problems. The United States uses tens of millions of tons of steel each year, and some of it is indeed recycled scrap. However, steel is not a single chemical substance. It is iron alloyed with one or more other elements, and, often, the composition is very specific to the application. Chromium is added to make stainless steel; titanium, molybdenum, or tungsten steels are high-strength steels; other alloys are used when other properties are important. If the composition of the alloy is critical to a particular application, clearly one cannot just toss any old steel scrap into the recycling pot. Each type of steel would have to be recycled individually. Facilities for separate collection and recycling of many different compositions of steel are rare. Progress is nevertheless being made, at least with respect to vehicles, which account for about 6% of all the steel (and 14% of the aluminum) in use in the United States at any given time. When a vehicle reaches the end of its useful life, it can first go to a dismantler for removal of still-useable replacement parts along with components such as the tires, battery, and catalytic converter (for recovery of the platinum and other valuable metals). The remaining hulk can then go to a shredder that will reduce it to bits from which scrap metals can be separated and salvaged. As of 2019, about 80% of an individual vehicle is recyclable, and about 27 million cars were recycled annually. Vehicle recycling saves resources, landfill space, and energy, and also creates an industry that employs about 100,000 people in the United States and adds $25 billion to the GDP. As with aluminum, it takes much less energy to recycle old steel scrap than to manufacture new steel.

Some resources cannot be recovered or reused because of the way they are employed. The potash and phosphorus in fertilizers are strewn across the land, taken up by plants and dissolved in water, and cannot be recovered. Road salt washes off streets and into the soil and storm sewers. For these and other reasons, it is unrealistic to expect that all minerals can ever be wholly recycled. However, more than half of the U.S. consumption of certain metals already is being recovered from old scrap. Where this is possible, the benefits are substantial.

13.6 Impacts of Mining-Related Activities

Learning Objectives

- Compare the environmental effects of surface to underground mining
- Explain how acid mine drainage is created
- Describe mine reclamation
- List environmental problems associated with mineral processing

Mining and mineral-processing activities can modify the environment in a variety of ways, and few areas in the United States are untouched by mining activities (**figure 13.23**). Most obvious is the presence of the mine itself. Both underground mines and surface mines have their own sets of associated impacts. Though safety is improving, the hazards of mining activities to the miners should also not be overlooked. Mining is a dangerous occupation. Both underground and surface mines can also be sources of water pollution, as explored further in chapters 14 and 17.

Underground mines are generally much less apparent than surface mines. They disturb a relatively small area of the land's surface close to the principal shaft(s). Waste rock dug out of the mine may be piled close to the mine's entrance, but in most underground mines, the tunnels follow the ore body as closely as possible to minimize the amount of non-ore rock to be removed and, thus, mining costs. When mining activities are complete, the shafts can be sealed, and the area often returns very nearly to its pre-mining condition. Near-surface underground mines occasionally have collapsed years after abandonment, when supporting timbers have rotted away or groundwater

has enlarged the underground cavities through solution (**figure 13.24A**). In some cases, the collapse occurred so long after the mining had ended that the existence of the old mines had been entirely forgotten. On the other hand, misjudgment occasionally results in collapse of an active mine (**figure 13.24B**).

Surface-mining activities generally consist of either open-pit mining (including quarrying) or strip mining. Quarrying extracts rock to be used either intact (for building or facing stone) or crushed (as for cement-making, roadbed, etc.), and is also commonly used for unconsolidated sands and gravels (see **figure 13.25**). Open-pit mining is practical when a large, three-dimensional ore body is located near the surface (**figure 13.26**; recall also **figure 8.42**, the Bingham Canyon mine). Most of the material in the pit contains the valuable commodity and is extracted for processing. Both procedures permanently change the topography, leaving a large hole. The exposed rock may begin to weather and, depending on the nature of the ore body, may release pollutants into surface runoff water or groundwater.

As noted earlier, many metal ores are hydrothermal deposits rich in sulfides. Pyrite (iron sulfide, FeS_2) is not only

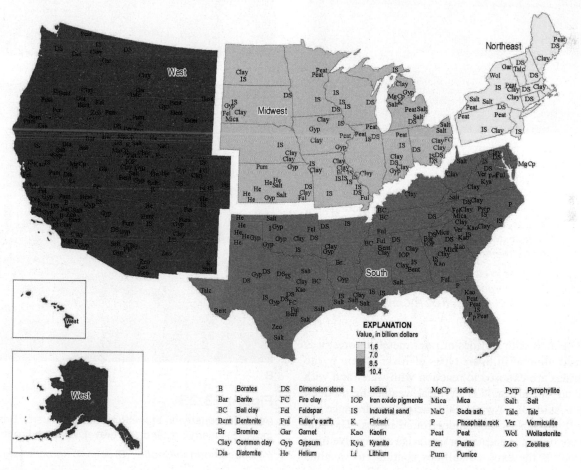

Figure 13.23

This map shows only sites of mining for industrial rock and mineral resources, but clearly shows that these activities are widespread in the United States. Still more places are affected by mining for metal ores. Recall also figure 13.7B.

Source: Mineral Commodity Summaries 2022, U.S. Geological Survey

A

B

Figure 13.24

Subsurface mining activities sometimes affect the surface. (A) Subsidence pits and troughs over abandoned underground coal mines in Sheridan County, Wyoming. (B) Remains of Lake Emma, Sunnyside Mine, Colorado. Mining too shallowly under the lake caused it to collapse into the mine workings in 1978.

*U.S. Geological Survey Photo Library, Denver, CO/
(A) Photograph by C. R. Dunrud; (B) U.S. Geological Survey
Photo Library, Denver, CO/Photograph by P. Carrara*

frequently associated with sulfides of rarer metals in ores; it is a common trace mineral in many types of rock. Where pyrite and other sulfides are exposed to reaction with water and with oxygen in the air, sulfuric acid is a typical weathering product, and the result is acid runoff water. Such acid mine drainage is a problem with both surface and underground mines. The acid drainage can continue for decades or centuries after active mining has stopped, as the exposed sulfides continue to weather. This problem will be explored further in chapter 17.

Strip mining, more often used to extract coal than mineral resources, is practiced most commonly when the material of interest occurs in a layer near and approximately parallel to the surface (**figure 13.27**). Overlying vegetation, soil, and rock are

A

B

Figure 13.25

(A) Granite quarrying at the Rock of Ages quarry, Barre, Vermont. (B) Quarrying sand and gravel, Wasatch Front, Utah.

(A) Doug Sherman/Geofile; (B) John A. Karachewski

stripped off; the coal or other material is removed; and the waste rock and soil are dumped back as a series of **spoil banks** (**figure 13.28A**). In the past, that was all that was done. The broken-up material of the spoil banks, with its high surface

A

Figure 13.27

Strip mining for manganese in South Africa.

U.S. Geological Survey Photo Library, Denver, CO

B

Figure 13.26

(A) The Hull-Rust-Mahoning open-pit iron mine in Minnesota was established in 1895 and is still in operation. (B) Mining in the Betze-Post open-pit gold mine in Nevada requires pumping thousands of gallons of groundwater per minute, some of it used in ore processing and for irrigation nearby. Vegetation appears red in this false-color image.

(A) Jim West/Alamy Stock Photo; (B) NASA Earth Observatory image by Jesse Allen

area, is very susceptible to both erosion and chemical weathering. Chemical and sediment pollution of runoff from spoil banks was common. Vegetation reappeared on the steep, unstable slopes only gradually, if at all (**figure 13.28B**). Now, much stricter laws govern strip-mining activities, and reclamation of new strip mines is required. Reclamation usually involves regrading the area to level the spoil banks and to provide a more gently sloping land surface (**figure 13.28C**); restoring the soil; replanting grass, shrubs, or other vegetation; and, where necessary, fertilizing and/or watering the area to help establish vegetation. The result, when successful, can be land on which evidence of mining activities has essentially been obliterated (**figure 13.28D**).

Naturally, reclamation is much more costly to the mining company than just leaving spoil banks behind. Costs can reach into the tens of thousand dollars per acre. Even when conscientious reclamation efforts are undertaken, the consequences may not be as anticipated. For example, the inevitable changes in topography after mining and reclamation may alter the drainage patterns of regional streams and change the proportion of runoff flowing to each. In dry parts of the western United States, where every drop of water in some streams is already spoken for, the consequences may be serious for landowners downstream. In such areas, the water needed to sustain the replanted vegetation may also be in very short supply. An additional problem is presented by old abandoned, unreclaimed mines that were operated by now-defunct companies: Who is responsible for reclamation, and who pays? (Superfund, discussed in chapter 16, was created in part to address a similar problem with toxic-waste dumps.)

Mineral processing to extract a specific metal from an ore can also cause serious environmental challenges. The waste rock removed to get at the ore can itself constitute a large volume of debris, with no obvious place to put it while mining is still in progress (**figure 13.29**). Processing generally involves finely crushing or grinding the ore. The fine waste materials, or **tailings**, that are left over may end up heaped around the processing plant to weather and wash away, much like spoil banks. Generally, traces of the ore are also left behind in the tailings. Depending on the nature of the ore, rapid weathering of the tailings may produce acid drainage and leach out harmful elements, such as mercury, arsenic, cadmium, and uranium, to contaminate surface water and groundwater. Thoughtless use of

Figure 13.28

Strip mining and land reclamation. (A) Spoil banks, Rainbow coal strip mine, Sweetwater County, Wyoming. (B) Revegetation of ungraded spoils at Indian Head mine is slow. (C) Grading of spoils at Indian Head coal strip mine, Mercer County, North Dakota. (D) Reclaimed portion of Indian Head mine one year after seeding.

(A-C) Photograph by H.E. Malde, U.S. Geological Survey; (D) Photograph by H.E. Malde, USGS Photo Library, Denver, CO

Figure 13.29

(A) As the Bingham Canyon mine (figure 8.40) is approached, huge piles of crushed waste rock stand out against the mountains. (B) Closer inspection shows obvious erosion by surface runoff.

(A-B) ©Carla Montgomery

uranium-bearing tailings as fill and in concrete in Grand Junction, Colorado, resulted in radioactive buildings that had to be extensively torn up and rebuilt before they were safe for occupation. Mine operators must now trap runoff to control the loss of eroded sediment, and process water leaving the site to contain toxic chemicals that may have weathered from the ore.

The chemicals used in processing are often hazardous also. In the last century, mercury was used to help extract gold in placer-mining operations such as that shown in **figure 13.9B**, and millions of pounds of mercury were lost into the environment in California alone. Mercury is still used in gold processing in countries such as Ghana and Peru where artisanal, unregulated mining is occurring. In large-scale operations, cyanide is commonly used to extract gold from its ore. Water pollution from poorly controlled runoff of cyanide solutions can be a significant problem. Additional concerns arise as attention turns to trying to recover additional gold from old tailings by heap-leaching, using cyanide solutions percolating through the tailings piles to dissolve out the gold. Gold is valuable, and heap-leaching old tailings is relatively inexpensive, especially in comparison to mining new ore. But unless the process is very carefully controlled, both metals and cyanide can contaminate local surface water or groundwater. Smelting to extract metals from ores may, depending on the ores involved and on emission controls, release arsenic, lead, mercury, and other potentially toxic elements along with exhaust gases and ash. Sulfide-ore processing also releases the sulfur oxide gases that are implicated in the production of acid rain (chapter 18). These same gases, at high concentrations, can destroy nearby vegetation. During the twentieth century, one large smelter in British Columbia destroyed all conifers within 19 kilometers (about 12 miles) of the operation. More-stringent pollution controls are curbing such problems today. Still, mineral processing can be a huge potential source of pollutants if not carefully executed. Worldwide, too, not all nations are equally sensitive to the hazards; residents of areas with weak pollution controls or using hazardous metal-extraction methods may face significant health risks.

Case Study 13.1

The Not-So-Rare Rare Earths

As we noted in chapter 2, certain groups of elements within the periodic table have collective names—for example, the alkali metals, those elements in the first column. The group known as the **rare-earth elements** (REE) is the set of elements with atomic numbers from 57 (lanthanum) to 71 (lutetium) (see **figure 2.2**). Their names and chemical symbols are listed in **table CS 13.1.1**. They are similar both in ionic size and in ionic charge (nearly always +3), so they behave very similarly geochemically and tend to be found together in the same minerals.

Geologically speaking, the REE are actually not all that rare. Their abundances in average continental crust are about a thousand times that of gold, and comparable to the abundances of common metals like lead or tin. They were long considered rare mainly because they are seldom found as the major cations in minerals; instead, they tend to substitute for common cations in silicates and other minerals, so their overall abundances were not realized until geologists had made detailed chemical analyses of many rocks and minerals. Even then, their usefulness was not immediately recognized.

Modern technology has changed that. REE are used in a tremendous variety of applications: as catalysts; in making and polishing glass; as components in alloys, lasers, optical fibers, permanent magnets, fluorescent lights, television screens, superconductors, rechargeable batteries (both small ones and those used in hybrid and electric vehicles), and much more. The REE have become essential to a great many industries.

Table CS 13.1.1	Chemical Symbols and Names of the Rare-Earth Elements		
La	lanthanum	Tb	terbium
Ce	cerium	Dy	dysprosium
Pr	praseodymium	Ho	holmium
Nd	neodymium	Er	erbium
Pm	prometheum	Tm	thulium
Sm	samarium	Yb	ytterbium
Eu	europium	Lu	lutetium
Gd	gadolinium		

The United States began mining REE in the mid-1960s, primarily at Mountain Pass, California. In the 1980s, China began REE mining and its share of production rose, while that of the United States and the rest of the world declined (**figure CS 13.1.1**). In 2002, owing to economic conditions and environmental issues, the Mountain Pass mine ceased operation. By 2011, China accounted for over 95% of the world's REE production.

This had been a concern for some time, but suddenly became a major global issue in the fall of 2010. First, in a political dispute with Japan, China suspended REE exports to that country;

(Continued)

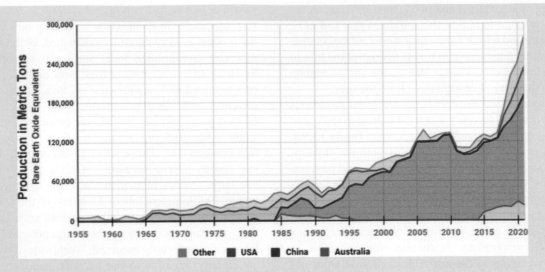

Figure CS 13.1.1

The United States is currently self-sufficient with respect to REE as it was prior to 2000, but for ten years was completely reliant on imports from China. (1 metric ton = 1000 kg = 2200 lb)

U.S. Geological Survey

within weeks, the suspension was extended to the United States and Europe. REE prices began to rise sharply. Exports to the United States and Europe were resumed within weeks, and those to Japan within months, but concerns about supply stability and cost outside China remained acute. During all this, prices of REE on world markets had soared, some rising by a factor of 30 or more, and although prices did drop once exports resumed, the whole episode motivated the United States and other technological nations to rethink their dependence on REE imports.

Because the REE substitute readily in the crystal structures of many minerals, they seldom become highly concentrated in ores. The United States has only about 1.5% of the world's known REE reserves (China controls over 35%, Brazil, Russia, and Vietnam about 18% each). In 2012, the Mountain Pass mine was reopened. It operated until 2016 but then closed again in the face of the precipitous decline in REE prices. (For example, at the end of 2011, prices ranged from about $62/kg for lanthanum to $3850/kg for europium oxide—europium is uniquely useful as the red phosphor in liquid-crystal displays, among other applications—but by the end of 2016, those prices had plunged to $2/kg and $425/kg, respectively.) It opened again in 2018, and U.S. production increased to 43,000 metric tons in 2020, more than seven times the amount consumed domestically. The relative abundances of the various REE differ between deposits (**figure CS 13.1.2**), so one nation's REE reserves may not match its demands precisely, but in gross numbers, U.S. reserves amount to about six times 2021

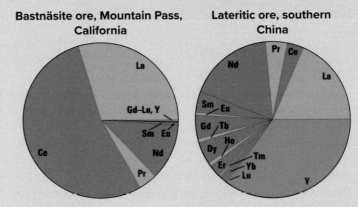

Bastnäsite ore, Mountain Pass, California

Lateritic ore, southern China

Figure CS 13.1.2

Different REE ores contain different proportions of individual REE. Yttrium (Y) is just above La in the periodic table and behaves chemically so much like it that Y is sometimes considered part of the REE. Bastnäsite is a rare mineral, an REE carbonate-fluoride.

From U.S. Geological Survey Fact Sheet 087-02.

world REE production, so we should be able to develop more domestic REE sources and continue to supply ourselves. as long as REE prices make it feasible. Other nations are not so favorably placed; Japan, for instance, has no REE reserves at all to supply its many high-technology industries.

Mining Your Cell Phone?

Consumer electronics make up a fast-growing segment of the U.S. waste stream. Unlike many consumer products that are discarded only when they cease to function, many types of electronics, such as cell phones and computers, may be replaced for other reasons—because they have become obsolete or because consumers want new features that are available only on new models. Many of the replaced items end up in landfills. This not only increases the challenge of finding adequate space for solid-waste disposal; it also adds a variety of toxic metals, from antimony and arsenic to mercury and selenium, to those landfills. From there, the toxins may escape to contaminate air and water, as will be explored in later chapters. Of course, the manufacture of new devices consumes resources, many of them imported (**figure CS 13.2.1**).

However, viewed and handled differently, these products can provide us resources. The individual cell phone may not seem a rich treasure, but collectively, cell phones represent a significant source of recyclable metal (**figure CS 13.2.1**). There are over 290 million mobile phones actively in use in the United States at present. The Environmental Protection Agency (EPA) estimates that

each year, over 150 million cell phones are taken out of service, whether discarded, recycled, or set aside. Hundreds of millions more are believed already to be in storage—stuck into closets or desk drawers or otherwise tucked away, no longer being used but not actually having been discarded. Are all those phones worth anything? Consider just five of the metals they typically contain:

This table shows that just the cell phones in active use represent over $1 billion worth of these metals alone; those in storage would constitute a substantial resource ready to be tapped. Granted, the figures do not take into account the costs of recycling, but in any event it is clear that there is value here, as well as potential for mineral-resource conservation (and reduced waste disposal and pollution into the bargain). Estimates for phone recycling rates are as low as 15%; the figures for personal computers and their components and peripherals are not much better: Though they contain not only recyclable metals but recyclable glass and plastic as well, only about 17% of electronic waste is properly recycled.

A major reason given for the low recycling rate on such products is that many people lack access to, or are simply unaware

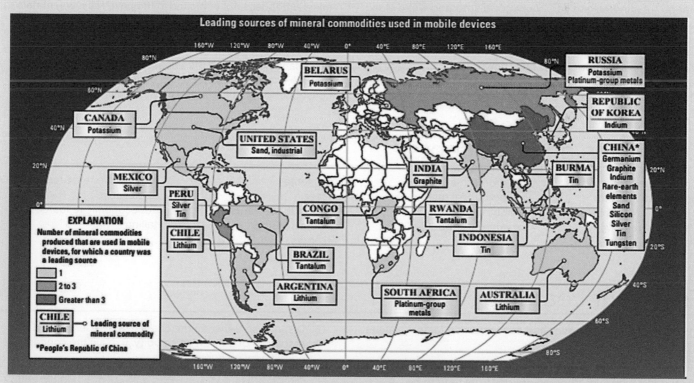

Figure CS 13.2.1

To manufacture even a small item like a cell phone can require resources from around the world.

Source: USGS General Interest Product 167.

(Continued)

Metal	Ounces per Cell Phone*	Commodity Price**	Weight in 290 Million Cell Phones	Value in 290 Million Cell Phones
Copper	0.563	$4.30/lb	5100 tons	$43.9 million
Silver	0.0113	$25.00/oz	3.3 million oz	$81.9 million
Gold	0.0011	$1800/oz	319,000 oz	$ 574 million
Palladium	0.000484	$2600/oz	140,000 oz	$364 million
Platinum	0.0000110	$1200/oz	3190 oz	$3.8 million

Except for copper, all weights and prices reported in ounces are troy ounces, as is customary for precious metals.

From USGS Fact Sheet 2006-3097.

**From Mineral Commodity Summaries 2022, U.S. Geological Survey.*

of, programs to recycle their electronic items. To help, the EPA now not only provides online information about what is sometimes called eCycling, but also provides links to other organizations that can put consumers in touch with local electronics-recycling programs. If you want to investigate the options for recycling your own used electronics, see the EPA's Electronics Donation and Recycling page. Many places that sell electronics also accept them for recycling, often with a financial incentive.

Summary

Economically valuable mineral deposits occur in a variety of specialized geologic settings. Both the occurrence of and the demand for minerals are very unevenly distributed worldwide. Projections for mineral use, even with conservative estimates of consumption levels, suggest that present reserves of most metals and other minerals could be exhausted within decades. As shortages arise, some currently subeconomic resources will be added to the reserves, but the quantities are unlikely to be sufficient to extend supplies greatly. Other strategies for averting further mineral shortages include applying newer exploration methods to find more ores, looking to undersea mineral deposits not exploited in the past, recycling metals to reduce the demand for newly mined material, and using substitutes. Carefully controlling mining activities would minimize the negative impacts of mining, such as disturbance of the land surface and the release of harmful chemicals through accelerated weathering of pulverized rock or as a consequence of mineral-processing activities.

Key Terms and Concepts

banded iron formation
concentration factor
conditional resources
cumulative reserves
evaporite

hydrothermal
hypothetical resources
kimberlites
manganese nodules
ore

pegmatite
placer
rare-earth elements
remote sensing
reserves

speculative resources
spoil banks
subeconomic resources
tailings

Test Your Learning

1. Explain how economics and concentration factor relate to the definition of an ore.

2. Describe two examples of magmatic ore deposits.

3. Explain why hydrothermal ore deposits tend to be associated with plate boundaries.

4. Give an example of a common evaporite mineral, describing how such deposits form.

5. Explain how stream action may lead to the formation of placer deposits, and why there is interest in exploring for placers on the continental shelves.

6. Distinguish between the quantities of *reserves* and of *resources* of a given commodity. Explain the concept of *conditional resources.*

7. As mineral reserves are exhausted, some resources may be reclassified as reserves. Explain.

8. Indicate how plate-tectonic theory can contribute to the search for new ore deposits.

9. Describe how satellites are contributing to ore prospecting, and note two advantages of such remote-sensing methods.

10. Identify the metallic mineral resource found over much of the deep-sea floor, and what political problem arises in connection with it.

11. Explain why aluminum and lead are comparatively easy to recycle, while steel is less so.

12. Describe one hazard associated with underground mining.

13. Identify the water-pollution problem particularly associated with mining of sulfide ore deposits.

14. Summarize the steps involved in strip-mine reclamation, and explain why land cannot always be fully restored to its pre-mining condition, even with sincere effort.

15. Give two ways in which waste rock and tailings from mineral processing pose potential environmental concerns.

Exploring Further

1. Choose an area that has been subjected to extensive surface mining (examples include the coal country of eastern Montana or Appalachia and the iron ranges of Upper Michigan). Investigate the history of mining activities in this area and of legislation relating to mine reclamation. Assess the impact of the legislation.

2. Select one metallic mineral resource and investigate its occurrence, distribution, consumption, and reserves. Evaluate the impact of its customary mining and extraction methods. (How and where is it mined? How much energy is used in processing it? What special pollution problems, if any, are associated with the processing? Can this metal be recycled, and if so, from what sorts of material? Are there practical substitutes for it?) You might start at the **USGS Mineral Resources Program**.

3. Choose an everyday object, and investigate what mineral materials go into it and where they come from. (A good place to start is with the **Minerals Education Coalition**.)

4. Select one or more recyclable metals used in a computer, and perform an analysis like that of Case Study 13.2.

CHAPTER 14

Energy Resources—Fossil Fuels

Our current technological society depends upon a ready supply of relatively cheap energy. Fossil fuels have provided the majority of that energy throughout the last and into the twenty-first century, and without them, our industrial development would not have occurred as it did. Even as our energy consumption has flattened (**figure 14.1**) and we are necessarily moving toward a much more diverse mix of energy sources, we persist in using petroleum to power our machines, vehicles, and planes. We have found ways to access fuels that we once determined were too difficult and/or too expensive to produce, and in doing so, have increased our reserves. Our continued use of fossil fuels comes with environmental costs that include carbon dioxide and methane emissions, air and water pollution, toxic spills, and landscape destruction.

Not all countries have an energy consumption profile similar to that of the United States. In very populous places, many people are using wood to heat their homes and cook their food. As countries develop, they will use more fossil fuels, which will result in more greenhouse gas emissions (**figure 14.2**). Even if we have enough fossil fuels to power the world for another 50 to 100 years, it is likely not in our best interest to use them if the result is a climate that changes so drastically that we cannot adapt to it. In this chapter and the next, we will discuss the energy resources we use now and what we will likely use in the near future, keeping in mind that any expansion of fossil fuel use (without capture of carbon dioxide) will come with serious climate repercussions.

The term *fossil* refers to any remains or evidence of ancient life. Energy is stored in the chemical bonds of the

Chapter Outline

14.1 Formation of Oil and Natural Gas Deposits

14.2 Supply and Demand for Oil and Natural Gas

14.3 Oil Seeps and Spills

14.4 Coal

14.5 Environmental Impacts of Coal

14.6 Oil Shale

The Eagle Ford Shale play is just one example of a fossil fuel that, until quite recently, was considered uneconomical, so it was left undeveloped. With rising petroleum prices and improvements in hydraulic fracturing and horizontal drilling techniques, oil and natural gas have been produced from it for over ten years, helping to make the United States independent of foreign reserves. But not without taking its toll on the environment as exemplified by the change in the landscape from 2000 to 2015.

NASA Earth Observatory images by Joshua Stevens

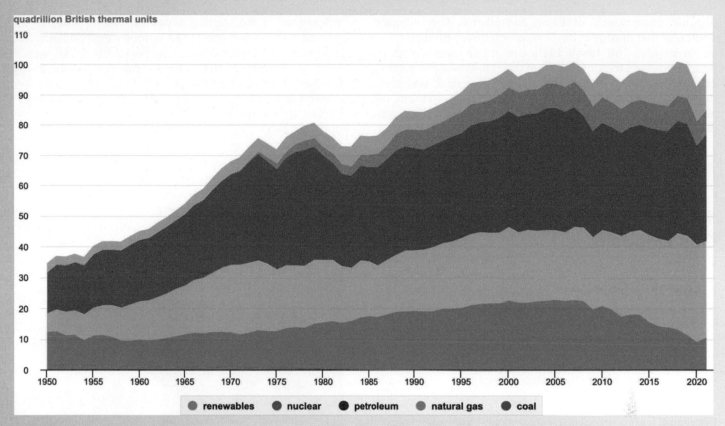

Figure 14.1

U.S. energy consumption, 1950–2020. The trend is generally upward until the Great Recession in 2008, with dips throughout due to economic downturns. Maximum consumption was reached in 2018 at over 100 quadrillion BTU, but that upward turn was disrupted by the COVID-19 pandemic.

U.S Energy Information Administration, Monthly Energy Review, Table 1.3, April 2021. preliminary data for 2020.

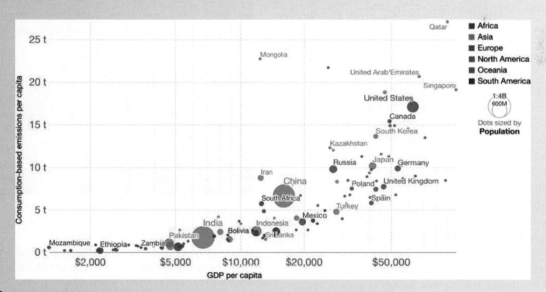

Figure 14.2

GDP has commonly been deployed as a measure of energy consumption—countries with high GDPs consume more energy per capita. Energy consumption reflects CO_2 emitted because most energy is created with fossil fuels. The global average emission of CO_2 per person is 4.8 tons per year; countries with GDPs around \$15,000–\$20,000 fall in that range. Data is from 2019.

Our World in Data based on the Global Carbon Project, Data compiled from multiple sources by World Bank OurWorldInData.org/co2-and-other-greenhouse-gas-emissions • CC BY

organic compounds of living organisms. The ultimate source of that energy is the Sun, which drives photosynthesis in plants. The **fossil fuels** are those now compact, energy sources that formed from the remains of once-living organisms. These include oil, natural gas, coal, and fuels derived from oil shale and oil sand. When we burn them, we are using that stored energy. The differences in the physical properties among the various fossil fuels arise from differences in the starting materials from which the fuels formed and in what happened to those materials after the organisms died and were buried. What the fossil fuels all have in common is that they are **nonrenewable** energy sources, meaning that the processes by which they form are so slow that they are not replaceable on a human timescale. Whatever we consume is, from our perspective, gone for good.

14.1 Formation of Oil and Natural Gas Deposits

Learning Objectives

- Describe the conditions in which oil and natural gas are created
- Compare the source and reservoir rocks for petroleum
- Recognize different petroleum traps

Petroleum is not a single chemical compound, and can technically exist as a solid, liquid, or gas, although we often use the term to describe only the liquid state. **Oil** is liquid petroleum (**figure 14.3A**), and **natural gas** is gaseous petroleum. We also refer to petroleum products as *hydrocarbons* because they are compounds made up of different proportions of carbon and hydrogen. As such, there is a variety of petroleum products. *Crude oil* is not a special type of liquid hydrocarbon, but rather another name for oil, and implies that it has not been *refined* or separated into distinct types. Methane (CH_4) is the most common type of natural gas (**figure 14.3B**). In the following, we will review how petroleum is created from organic matter in a *source* rock, and how it often migrates to a *reservoir* rock from which we are more easily able to access it.

The production of a petroleum deposit requires a large initial accumulation of organic matter, which is rich in carbon and hydrogen. Another requirement is that the organic debris be buried quickly to protect it from the air so that decay by biological means or reaction with oxygen doesn't result in complete decomposition.

The conditions to create petroleum occur in marine and lacustrine environments, which are rich in organic materials. Scientists think that most petroleum forms when microorganisms such as algae and zooplankton die and their remains settle to the bottom of the sea or lake, where they rapidly accumulate and then are buried by sediment in oxygen-poor conditions.

As burial continues, the organic matter begins to change. Pressure increases with the weight of the overlying sediment or rock; temperature increases with depth under the surface; and slowly, over long periods of time, chemical reactions take place. These reactions break down the large, complex organic molecules into simpler, smaller hydrocarbon molecules. The nature of the hydrocarbons changes with time and continued

A

B

Figure 14.3

Different types of petroleum or hydrocarbons. (A) Oil is liquid petroleum. (B) Methane is one type of natural gas. This model shows the unconnected hydrocarbon molecules as they would be in a gas; the larger atom in the middle is carbon.

(A) nevodka/Shutterstock; (B) Kid A/Shutterstock

heat and pressure. In the early stages of petroleum formation in a marine deposit, the deposit may consist mainly of *kerogen,* a complex, waxy solid mixture of longer hydrocarbon molecules.

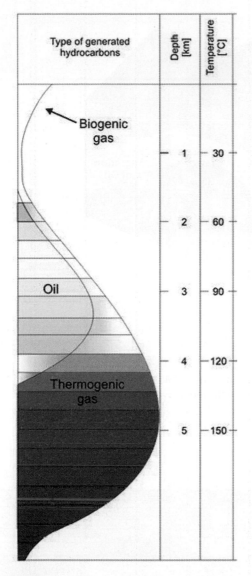

Type of generated hydrocarbons	Depth [km]	Temperature [°C]

Biogenic gas

Oil

Thermogenic gas

Figure 14.4

This simplified diagram shows how petroleum matures in a marine deposit with increased temperature and depth. As opposed to thermogenic (natural) gas, biogenic gas is immature and formed without the influence of burial.

https://infolupki.pgi.gov.pl/en/gas/thermal-maturity-organic-matter-and-gas-exploration

As the petroleum matures, and as the breakdown of large molecules continues, successively lighter hydrocarbons are produced. Thick liquids give way to thinner ones, from which are derived lubricating oils, heating oils, and gasoline. In the final stages, most or all of the petroleum is further broken down into very simple, light, gaseous molecules—natural gas. Most of the oil maturation process occurs in the temperature range of 60° to 120°C (approximately 140° to 250°F). The creation of natural gas, almost wholly methane, occurs in the *gas window* from 120° to 150°C; beyond that, the hydrocarbons are further broken down and destroyed (see **figure 14.4**).

Each given oil field yields crude oil containing a distinctive mix of hydrocarbon compounds, depending on the history of the material. The refining process separates the different types of hydrocarbons for different uses (see **table 14.1**). Some of the heavier hydrocarbons may also be broken up during refining into smaller, lighter molecules through a process called *cracking,* which allows some of the lighter compounds such as gasoline to be produced as needed from heavier components of crude oil.

Once the solid organic matter is converted to liquids and/or gases, the hydrocarbons may migrate out of the fine-grained, low-permeability, clastic sedimentary source rocks (shale) and into more permeable reservoir rocks (sandstone), where they accumulate over long spans of geologic time (**figure 14.5**). The pores, holes, and cracks in these rocks are commonly full of water. Most oils and all natural gases are less dense than water, so they tend to rise and migrate laterally through the water-filled pores of permeable rocks. Unless stopped by impermeable rocks, oil and gas may keep rising right up to the surface. At many known oil and gas seeps, these substances escape into the air or the oceans or flow out onto the ground.

Commercially, the most valuable deposits are those in which a large quantity of oil and/or gas has been concentrated and trapped by impermeable rocks in geologic structures (see **figure 14.6**). The reservoir rocks should be relatively porous and permeable to accumulate large volumes of readily accessible oil and gas. If the reservoir rocks are not naturally very permeable, it may be possible to fracture them artificially with explosives or with water or gas under high pressure to increase the rate at which oil or gas flows through them.

Table 14.1	Fuels Derived from Liquid Petroleum and Gas	
	Material	**Principal Uses**
Heavier hydrocarbons	waxes (for example, paraffin)	candles
	heavy (residual) oils	heavy fuel oils for ships, power plants, and industrial boilers
	medium oils	kerosene, diesel fuels, aviation (jet) fuels, power plants, and domestic and industrial boilers
	light oils	gasoline, benzene, and aviation fuels for propeller-driven aircraft
	bottled gas (mainly butane, C_4H_{10})	primarily domestic use
Lighter hydrocarbons	natural gas (mostly methane, CH_4)	domestic/industrial use and power plants

Source: Data from J. Watson, Geology and Man. Allen and Unwin, Inc.

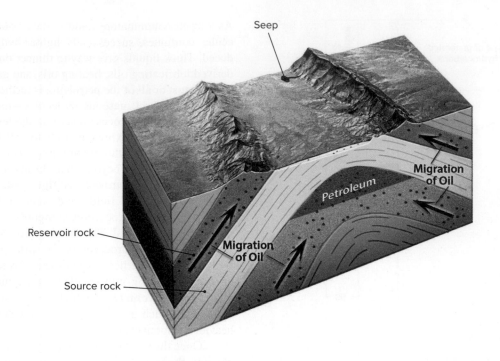

Seep

Migration
of Oil

Petroleum

Reservoir rock

**Migration
of Oil**

Source rock

Figure 14.5

Petroleum migrates out of source rocks and into more permeable reservoir rocks because of its relatively low density. It can become trapped into economically rich concentrations by geologic structures such as the anticline shown here.

McGraw Hill

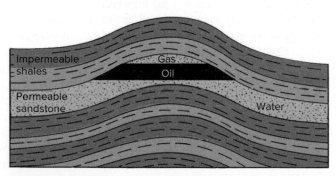

Impermeable
shales

Gas

Oil

Permeable
sandstone

Water

(A) A simple fold trap.

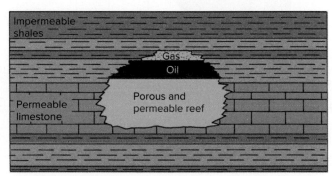

Impermeable
shales

Gas

Oil

Permeable
limestone

Porous and
permeable reef

(B) Petroleum accumulated in a fossilized ancient coral reef.

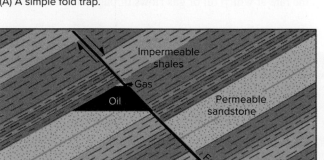

Impermeable
shales

Gas

Oil

Permeable
sandstone

Fault

(C) A fault trap.

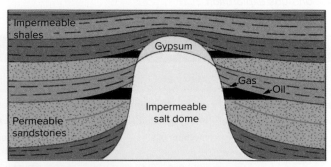

Impermeable
shales

Gypsum

Gas
Oil

Permeable
sandstones

Impermeable
salt dome

(D) Petroleum trapped against an impermeable salt dome, which has risen up from a buried evaporite deposit.

Figure 14.6

Types of petroleum traps.

We don't know the precise amount of time required for oil and gas to form. Since virtually no petroleum is found in rocks younger than 1 to 2 million years old, geologists infer that the process is comparatively slow. And since we are using petroleum at a rate much faster than it forms, we call it a *nonrenewable resource*.

14.2 Supply and Demand for Oil and Natural Gas

Learning Objectives

- Recognize the location of proven world reserves of oil, natural gas, and oil sands
- Compare historic U.S. oil and gas production and consumption
- Describe the characteristics of oil sand, shale oil, and shale gas
- Explain how fracking has increased U.S. reserves of oil and natural gas
- List methods for enhanced oil recovery
- Describe methane hydrates and the challenges of accessing and using them
- List reasons and challenges to conservation of petroleum

As with minerals, the most conservative estimate of the supply of an energy source is the amount of known **reserves,** proven accumulations that can be produced economically with existing technology. A more optimistic estimate is total **resources,** which include reserves plus known accumulations that are technologically impractical or too expensive to tap at present, plus some quantity of the substance that is expected to be found and extractable. Estimates of energy reserves are sensitive both to price fluctuations and to technological advances. In this section, we will review our current reserve estimates of various fossil fuels while addressing where they are located, how we access them, how fast we are using them, and how long they might last.

Oil

Oil is commonly discussed in units of barrels, with 1 barrel = 42 gallons. Estimated total world reserves are about 1.73 trillion barrels (**figure 14.7**). That may seem like a lot of oil until we compare that to the approximately 35 billion barrels consumed every year. Global demand will likely increase, at least in the short term, as population increases and as more countries advance technologically.

Oil supply and demand are very unevenly distributed around the world. Some low-tech countries like Angola may be producing ten to twenty times as much oil as they themselves consume. At the other extreme are countries like Japan, highly

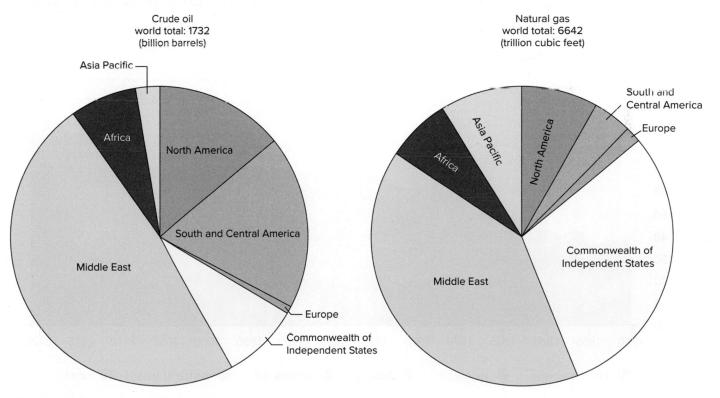

Figure 14.7

Estimated proven world reserves of crude oil and natural gas, 2020.

Source: BP Statistical Review of World Energy 2021.

Table 14.2	Proven U.S. Reserves Crude Oil and Natural Gas, 1977–2020	
Year	Crude Oil (billions of barrels)	Natural Gas (trillions of cu. feet)
1980	29.8	206
1990	26.3	178
2000	22.0	187
2005	21.8	213
2010	23.3	318
2015	32.3	324
2016	32.8	341
2017	39.2	464
2018	43.8	505
2019	44.2	495
2020	35.8	473

U.S. Energy Information Administration, Form EIA-23L, Annual Report of Domestic Oil and Gas Reserves.

industrialized and energy-hungry, with no petroleum reserves at all. The United States alone consumes 20% of the oil used worldwide, more than all of Europe together, and about 47% more than China, the next-highest individual-country consumer, although that gap has been getting smaller. As with mineral resources, the size of petroleum resources of a given nation is not correlated with geographic size. Middle East countries take up a little over 3% of the land area yet have 48% of world oil reserves.

According to the U.S. Energy Information Administration (USEIA), the United States currently has 35.8 billion barrels of oil reserves. For more than three decades prior to 2008, the United States had been consuming an amount of oil each year equal to or more than the amount of newly discovered reserves, so that the net U.S. oil reserves had been decreasing. With rising oil prices and improvements in recovery technology (see Shale Gas and Shale Oil section below), reserve estimates increased from 2008 to 2019, and then fell again in 2020 in response to the pandemic (**table 14.2**). Domestic crude oil production followed a similar path (**figure 14.8**).

Until 2019, the United States had long relied heavily on imported oil to meet part of its energy demand. At this point, even though oil production still lagged slightly behind oil consumption, the United States became energy independent (**figure 14.9**) when total energy production exceeded energy consumption for the first time since the late 1950s, in large part due to the increased production of petroleum. The United States currently consumes about 6.3 billion barrels of oil per year, to supply about 35% of all the energy used. As with minerals, the United States both imports and exports petroleum and petroleum products, with those numbers currently being roughly equal at 8.5 million barrels per day. Principal sources of U.S. imports were, until very recently, Canada, Mexico, Russia, Saudi Arabia, and Colombia, with sixty-eight other nations also contributing. As of the writing of this chapter, a recent ban on imported Russian oil has been inacted.

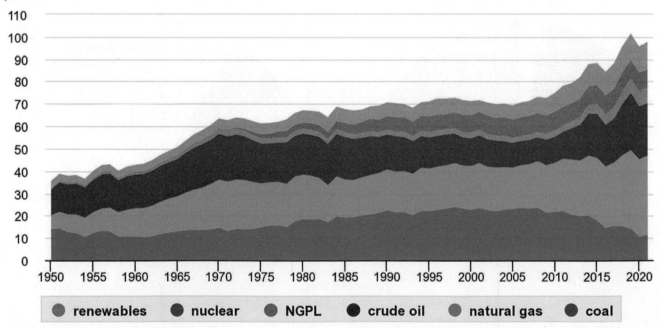

quadrillion British thermal units

● renewables ● nuclear ● NGPL ● crude oil ● natural gas ● coal

Figure 14.8

U.S. energy production by source since the mid-twentieth century.

U.S Energy Information Administration, Monthly Energy Review, Table 1.2, April 2021. preliminary data for 2020.

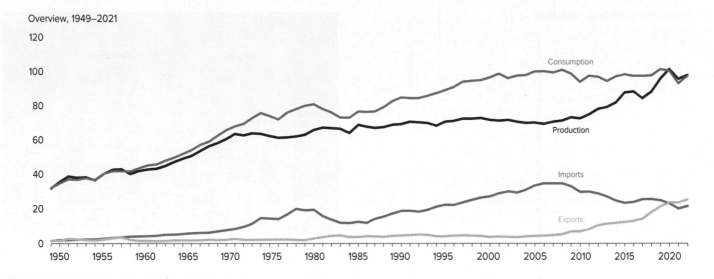

Overview, 1949–2021

Figure 14.9

U.S. energy overview from 1949 to 2021.

U. S. Energy Information Administration/Monthly Energy Review April 2022

Why does the United States import oil if we are now producing as much as or even more than we need? Recall that petroleum is refined into a variety of hydrocarbon products that have different properties and uses. Not every oil pumped out of the ground is the same, and not every country has the refining capacity to turn the crude oil into the desired products. It may be cheaper to import oil for use on the East Coast compared to having it refined somewhere else across the country and then having it shipped. And, we cannot disregard the political relationships that are, at least in part, tied to the fossil fuel industry.

Our past estimates on how long U.S. oil would last were based in the assumption that our reserves were dwindling, so we need to discuss our oil future from a slightly different viewpoint without forgetting that oil is a nonrenewable source of energy and that we will ultimately run out of it. The 19% decrease in estimated reserves between 2019 and 2020 makes the future energy outlook even a bit more complicated. Given the 2020 USEIA data of 35.8 billion barrels in reserves being consumed at 18.1 million barrels every day, U.S. reserves would last about six years. However, 2020 levels of production and consumption are not likely representative of the future; 2021 consumption is already higher, other entities estimate our reserves are higher at close to 68 billion barrels, and other liquid petroleum products such as natural gas liquids can be included in estimated reserves. The best-case scenario in all the numbers available indicates about eleven years left in oil reserves. So we either use less or find more if we decide to continue using oil.

If we do continue to rely on oil to generate a large portion of energy, we will need to find other supplies or they could be depleted within decades, especially considering the probable acceleration of world energy demands. On occasion,

exploration geologists do find the rare, very large concentrations of petroleum—the deposits on Alaska's North Slope and beneath Europe's North Sea are examples. Yet even these make only a modest difference in the long-term picture. The Prudhoe Bay oil field, for instance, represented reserves of only about 13 billion barrels, a great deal for a single region, but less than two years' worth of U.S. consumption. U.S. oil production peaked at nearly 13 million barrels per day in November 2019, which was more than double the rate ten years previous, and is currently over 11 million barrels per day. Current global demand is around 100 million barrels per day. It is likely that production and consumption of oil will rebound back to its pre-COVID-19 levels and perhaps stay there until about 2030. Globally, we use about 65% of our oil for transportation, so how soon and how quickly we see oil demand decrease will be a reflection of how quickly vehicles become electrified and how quickly we switch to other energy sources to create that electricity.

The United States has long been concerned about oil independence, especially following the Arab Oil Embargo that resulted in oil prices quadrupling from 1973 to 1974. This led to the establishment of the Strategic Petroleum Reserve (SPR) in 1977, which at its maximum held over 700,000 thousand barrels of oil. The oil is stored in four underground salt caverns near refineries around the Gulf Coast. The SPR has historically been reported in "days of net imports" but currently that metric seems to not be pertinent. The most recent release from the SPR began in April 2022 and will occur at a prescribed rate of 1 million barrels per day for six months with the goal of supplementing the cut in production that occurred due to the pandemic and to the banning of imported oil from Russia in response to the war in Ukraine. Other large drawdowns of the SPR occurred in 2011 in response to oil production disruptions in Libya, in 2005 after Hurricane Katrina, and in 1991 during

U.S. Ending Stocks of Crude Oil in SPR

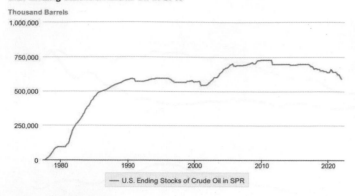

Figure 14.10

The Strategic Petroleum Reserve (SPR) was established to provide a cushion in the event of disruption of imports of oil. It is currently being used in response to decreased production and a ban on imported Russian oil.

U.S. Energy Information Administration.

Operation Desert Storm. At about 568 million barrels in March 2022, the six-month draw will lower the SPR to its lowest levels since the mid-1980s (**figure 14.10**).

Oil Sand

Oil sands are loose or semi-consolidated mixtures of sand, clay, water, and a semisolid, tar-like petroleum called **bitumen** (**figure 14.11A**). The heavy petroleum in oil sands forms in the same way described earlier for petroleum except that the light oil component is biodegraded by bacteria, leaving the heavier and more complex hydrocarbon behind. Bitumen is too thick to flow out of its source rock. Oil sand must be either mined, crushed, and heated to extract the petroleum, or mined *in situ,* where large volumes of water or steam are pumped underground to reduce the viscosity of the bitumen so it can be pumped out (**figure 14.11B**). However it is recovered, the bitumen is then refined into various fuels.

The largest reserves of oil sands are in the Orinoco Basin in Venezuela and the Athabasca Basin in Alberta, Canada. There are an estimated 260 billion barrels of oil in the Orinoco belt, making Venezuela the country with the largest oil reserves totaling over 303 billion barrels. In Canada, the oil sands contain about 165 billion barrels of recoverable oil with total resources ten times that amount. Canadian supplies of bitumen in the oil sands far exceed that nation's conventional oil deposits, and are the reason that Canada ranks third in oil reserves behind Venezuela and Saudi Arabia. Currently, about 4.7 million barrels of oil a day are being produced from Canada's oil sands, and production is high even as many investors are divesting themselves of shares in the oil sand industry. The United States has virtually no oil-sand reserves, and Canada is the single largest supplier of U.S. crude oil imports at 3.9 million barrels a day (**figure 14.12**). The Venezuelan oil industry has all

A

B

Figure 14.11

(A) Oil sand. (B) The Athabasca oil sands are located in eastern Alberta and are one of the richest sources of oil in the world.

(A) NOAA; (B) Athatbasca Oil Sands/Alamy Stock Photo

but collapsed, with production falling 79% between 2015 and 2020, paralleling the economic and infrastructure collapse of the entire country. Limited sanctions in 2019 banned exports of Venezuelan oil to U.S. refineries but as the price of oil has increased, these bans are being reconsidered.

The bitumen in oil sand is disseminated and makes up 3 to 18% of the rock volume. Large volumes of rock are mined and processed to extract the petroleum; on average, 2 tons of oil sand must be processed to yield 1 barrel of oil. Open-pit mining is employed for the near-surface deposits, and the deeper oil sands are extracted *in situ.* One method, nicknamed the "huff-and-puff" procedure, involves injection of hot steam for several weeks to warm the bitumen, after which the less-viscous fluid can be pumped out for a month or so before it chills and stiffens and reheating is necessary. Processing is energy-intensive as well as water-intensive, and the amount of waste rock after processing may be larger than the original volume of oil sand.

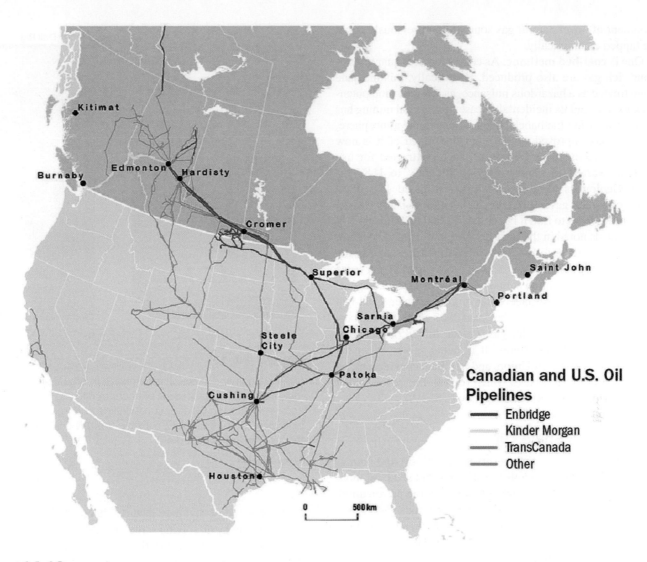

Figure 14.12

Much of the oil produced from Canadian oil sands is transported through pipelines to the United States for refining and use. A small but growing amount of it travels through the country to the Gulf Coast for exportation.

Canada Centre for Mapping and Earth Observation, Natural Resources Canada 2014.

The scale of operations is large and the negative environmental impacts of oil-sand production naturally increases along with production. These include land disturbance through strip mining; large amount of tailings; and accumulation of oil tailings that leave oil slicks on the tailings ponds which can endanger migrating birds. Processing oil sand releases two to three times as much CO_2 as extracting conventional oil via wells, along with sulfur gases, hydrocarbons, and fine particulate pollutants. Those environmental impacts cause serious concern, but the resource is too large and too valuable to ignore.

Natural Gas

The supply/demand picture for conventional natural gas is similar to that for oil (see **figures 14.1** and **14.8**). Natural gas presently supplies about 34% of the energy used in the United States. The United States has proven natural gas reserves of over 473 trillion cubic feet, and currently consumes roughly 29 trillion cubic feet of these reserves each year. Prior to the mid-1980s, U.S. natural gas production roughly equaled consumption. Then they began importing significant amounts of natural gas which accounted for about 10% of consumption. Throughout the 1980s and 1990s, less has been found in new domestic reserves than the quantity consumed with the net result of generally declining reserves (**table 14.2**). The rate of production picked up around 2005, and in 2017, the United States produced more than it consumed, with that trend widening through 2021. While the U.S. supply of conventional natural gas, like oil, could be used up in a matter of decades, technology has given a boost to our gas reserves and production, by allowing development of novel natural gas sources that previously could not be tapped economically.

One is **coal-bed methane.** As coal is formed, quantities of methane-rich gas are also produced. Historically, this methane has been treated as a hazardous nuisance, as it is toxic and potentially explosive, and its incidental release during coal mining has contributed to rising methane concentrations in the atmosphere. Technology has improved to the point that much of it is now economically recoverable. Where it can be extracted for use rather than wasted, it contributes significantly to U.S. gas reserves. Geologists have identified an estimated 12 trillion cubic feet of natural gas reserves occurring as coal-bed methane. And we know where to find it because U.S. coal beds are already well mapped, as noted in the later discussion of coal. The major technological obstacle to recovery of coal-bed methane is the need to extract and dispose of the water that typically pervades coal beds, water that is often saline and requires careful disposal to avoid polluting local surface water or groundwater. Coal-bed methane production fell by half between 2008 and 2017 and currently provides 3% of total dry gas (greater than 85% methane) production in the United States. It is being produced in the Appalachian, Eastern, Fort Union, and San Juan coal regions, an interactive map and database of which you can find at the United States Environmental Protection Agency (USEPA).

Shale Gas and Shale Oil

The primary reason why U.S. reserves of both oil and natural gas have increased dramatically is shale gas and shale oil, both considered to be in tight formations in which the petroleum is not accessible through traditional drilling methods. The petroleum stays in the organic-rich, shale source rock, accumulating in isolated, small openings (**figure 14.13A**). The low permeability of the shale also doesn't allow the oil and gas to migrate out into a reservoir rock.

The technology to increase permeability by **hydraulic fracturing,** or **fracking,** and to drill horizontally along shale beds to facilitate fracking and extraction, was introduced in the 1940s and has improved to the point that these hydrocarbons are now considered reserves (**figure 14.13B**). Fracking has dramatically changed both the natural gas and oil outlook for the United States. For example, in 2019, shale-gas reserves were estimated at over 353 trillion cubic feet, nearly double the amount in 2010. Prior to 2007, production was negligible. In 2015, it accounted 56% of U.S. natural gas production, and by 2017, gas from wells employing hydraulic fracturing accounted for more than three-fourths of U.S. gas production. U.S. production of gas now exceeds consumption, and net exports in 2021 exceeded 3800 billion cubic feet. See **figure 14.14** for locations of shale gas and shale oil resources.

In hydraulic fracturing, fluid under pressure is pumped into shales or other tight, low-permeability rocks to open fractures and increase their connectivity, improving permeability so that gas can be extracted more readily. In addition to water, the fracking fluid typically contains sand to prop open the fractures formed, and other chemicals. The latter vary with the particular

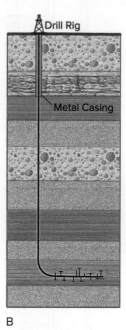
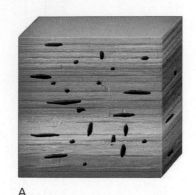

Figure 14.13

(A) Shale gas and shale oil accumulate in organic-rich, low permeability sedimentary rocks but cannot migrate out. They are accessed through (B) directional drilling and fracking, which increases the permeability, allowing the gas and oil to flow into the well and be pumped out.

McGraw Hill

well and geology, and may include chemicals intended to help start cracking by dissolving some minerals in the rock, maintain appropriate viscosity in the fluid, or prevent corrosion in the well casing. Careful handling of the copious amounts of fracking fluid and wastewater generated is necessary to prevent pollution of groundwater and surface water, which is one of the primary concerns associated with the fracking process. Fluids can leak into water supplies through mishandling, accidents, and escape into the formations they are injected into. Just as with other types of drilling, contaminated waters are often injected into very deep wells. Another concern is the induced seismicity associated with the disposal of fracking fluids, discussed in chapter 4.

Future Prospects for Oil and Gas

Many people tend to assume that, as oil and natural gas supplies dwindle and prices rise, there is increased exploration and discovery of new reserves so that we will not really run out of these fuels. The recent fluctuations in stated reserves (**table 14.2**) illustrate the influence of economics as well as of technology. But the supplies are ultimately finite, after all.

Most regions not yet explored for oil have been neglected precisely because they are unlikely to yield appreciable amounts of petroleum. The large portions of the continents underlain predominantly by igneous and metamorphic rocks

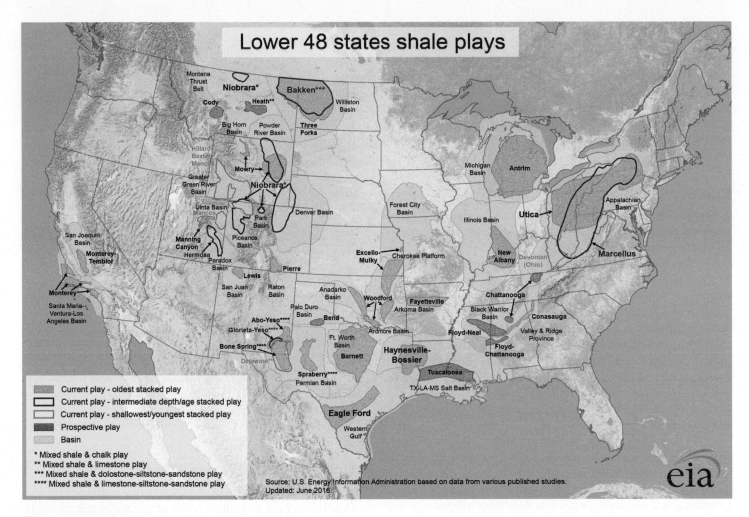

Figure 14.14

Shale plays in North America, as of 2016. The names on the map are geological formations that identify the rock units in which the shale is found.

Source: U.S. Energy Information Administration.

are, for the most part, very unpromising places to look for oil. The high temperatures involved in the formation of igneous and most metamorphic rocks would destroy organic matter, so oil would not have formed or been preserved in these rocks. Nor do these rock types tend to be very porous or permeable, so they generally make poor reservoir rocks as well, unless fractured. Other, more promising areas may be protected or environmentally sensitive.

The costs of exploration and drilling have gone up, and not simply due to inflation. Horizontal and directional wells are more expensive than conventional wells, and made up 81% of all new wells drilled in the United States in 2021. The average footage of a well has doubled in the last decade from 7300 to 15,200 feet. That means not only more piping, but also more water, more sand, more chemicals, and more labor. These shale oil and shale gas wells are economical when oil prices are above about $50/barrel, as compared to as low of $20/barrel

for their conventional well counterparts. Costs for drilling off-shore are substantially higher than for land-based drilling, too, and many remaining sites where hydrocarbons might be sought are offshore.

Despite a quadrupling in oil prices between 1970 and 1980 (*after* adjustment for inflation), U.S. proven reserves declined rather than increased. In April 2022, crude oil stands at around $100/barrel yet the United States is not ramping up production that fell during the pandemic because of supply chain issues, a lack of workers, and investors who are wary of the "boom and bust" cycles in the petroleum industry. These examples show that higher prices do not automatically lead to proportionate, or even appreciable, increases in fuel supplies and production because there are many disparate factors involved.

Many major oil companies have been branching out into other energy sources beyond oil and natural gas. At the same time, they are shifting their interest away from petroleum for fuels and instead looking at them in terms of petrochemicals. Petrochemicals are used to make fertilizers, textiles, plastics, carpet, pharmaceuticals, and more, and are pervasive in our modern lives. The demand for plastics has nearly doubled since 2000. And, as we have already discussed with other resources, advanced economies use a lot more plastic and fertilizer per person than countries with developing economies, so the potential for growth is huge.

Enhanced Oil Recovery

A few techniques have been developed to increase petroleum production from known deposits. An oil well initially yields its oil with minimal pumping, or even gushes on its own because the oil and any associated gas are under pressure from overlying rocks. Recovery using no techniques beyond pumping is *primary recovery*. When flow falls off, water may be pumped into the reservoir, filling empty pores and buoying up more oil to the well (*secondary recovery*). Primary and secondary recovery together extract an average of one-third of the oil in a given trap, though the figure varies greatly with the physical properties of the oil and host rocks in a given oil field. On average, then, two-thirds of the oil in each deposit has historically been left in the ground. Thus, additional *enhanced recovery* methods have attracted much interest.

Enhanced recovery comprises a variety of methods beyond conventional secondary recovery. Permeability of rocks can be increased by fracking, as already discussed with shale oil and shale gas. Hot water or steam may be pumped underground to warm thick, viscous oils so that they flow more easily and can be extracted more completely. Detergents or other substances can be used to break up the oil. One technique that has been expanding more rapidly than others is the use of carbon dioxide gas under pressure to force out more oil (**figure 14.15**). A recent study by the United States Geological Survey (USGS) of 3500 U.S. oil reservoirs estimated 29 billion barrels of oil that was technically recoverable using CO_2-enhanced oil recovery methods. The added benefit of this method is that the source of the CO_2 can come from capturing anthropogenic CO_2 rather than using natural gas reserves, so the enhanced recovery also results in carbon sequestration. This method has been discussed in terms of producing carbon-neutral or -negative oil, and could make carbon capture and storage more economical.

All these methods add to the cost of oil extraction, but they are becoming increasingly important as new deposits become harder to find. If the economics are right, they can be applied to old oil fields that have already been discovered, developed by conventional methods, and abandoned. Researchers in the petroleum industry believe that, from a technological standpoint, up to an additional 40% of the oil initially in a reservoir might be extractable by enhanced-recovery methods. This would substantially increase oil reserves. We do need to keep in mind that enhanced recovery will likely involve more problems such as ground subsidence and water pollution that arise with conventional recovery methods.

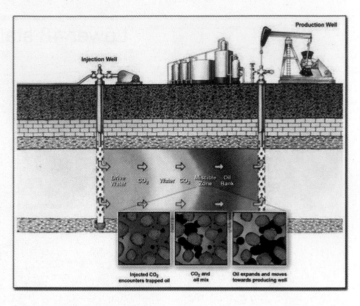

Figure 14.15

Carbon dioxide enhanced oil recovery. In 2017, globally there were approximately 166 applications, many in the United States. Between 300 and 600 kg of CO_2 is used to produce 1 barrel of oil.

United States Department of Energy, https://www.energy.gov/ fecm/science-innovation/oil-gas-research/enhanced-oil-recovery

Unconventional Natural Gas Sources

Major new potential sources of natural gas are **methane hydrates.** Gas hydrates are crystalline solids of gas and water molecules (**figure 14.16**). These hydrates have been found to be abundant in sediments within and beneath permafrost in arctic regions and in marine sediments. The USDOE's Methane Hydrate R&D Program estimated in a 2017 report that the amount of carbon found in gas hydrates is equal to that in all known fossil fuels on Earth. USGS estimates from 2008 include 85 trillion cubic feet of undiscovered, technically recoverable gas in northern Alaska, and The Bureau of Ocean Energy Management estimate more than 6000 trillion cubic feet on the outer continental shelf in the Gulf of Mexico. Methane hydrates clearly represent a huge potential natural gas resource but just how to tap methane hydrates safely for their energy is not yet clear.

In arctic regions, the hydrates are stabilized by the low temperature. If these cold sediments are warmed enough by the continued warming of the Arctic to break down some of the hydrates, or if they are disturbed in the course of extracting methane, methane gas would be released into the atmosphere, which would further increase greenhouse-effect heating. Considering that methane is a far more efficient greenhouse gas than is CO_2 and that the amount of methane locked in hydrates now is estimated to be at least 3000 times the methane currently in the atmosphere, the potential impact is large and represents yet another source of uncertainty in global climate modeling.

Despite the technical challenges, some countries are starting to explore developing methane hydrates, for different reasons.

A

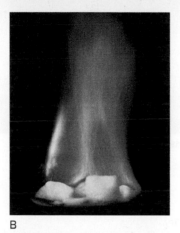

B

Figure 14.16

(A) Samples of methane hydrate (white material) are obtained through drilling in the Mackenzie Delta, Canada, conditions similar to northern Alaska. (B) Methane released from a piece of hydrate burning.

(A) USGS; (B) U.S. Geological Survey/Photograph by J. Pinkston and L. Stern

Japan looks to reduce its virtually complete dependence on imports for the fossil fuels it consumes. In 2013, they engaged in the first experimental extraction of methane from hydrates in sandy seafloor sediments off its coast; prototype methane hydrate extraction units are set to be deployed there in 2022, with the aim of using the methane as source of hydrogen to help reach their zero emissions goal by 2050. China successfully extracted methane from hydrates in the South China Sea in 2017. China needs a cleaner alternative to its abundant coal as its energy use soars. The U.S. continues with test projects in both the Alaskan arctic and the Gulf of Mexico. Analysts believe that commercial-scale development of methane from hydrates is unlikely before 2030. There are still serious technical obstacles associated with developing and efficiently using this resource. Besides the emissions issue, other concerns of mining them include underwater landslides and ecosystem destruction.

Some evidence suggests that a different type of natural gas resource exists at very great depths in the crust. Thousands of meters below the surface, conditions are sufficiently hot that any oil would have been broken down to natural gas (recall **figure 14.4**). This natural gas, under the tremendous pressures exerted by the overlying rock, may be dissolved in the water filling the pore spaces in the rock. Enormous quantities of natural gas might exist in such **geopressurized zones;** estimates of the amount of potentially recoverable gas in this form range from 150 to 2000 trillion cubic feet. Special and expensive technological considerations would be involved in developing these deposits. Geopressurized natural gas may well become an important supplementary fossil fuel in the future, though its total potential and the economics of its development are poorly known.

Conservation

Our contemporary conversations about oil and gas conservation may need to be reframed, given that our use of them as energy sources is now just as closely, if not more so, associated with our concerns about climate and the environment rather than how long we can stretch our remaining supplies. Historically, individuals, businesses, and governments practice energy conservation for both these reasons, and to save money when costs are high. The combined effects of conservation and economic recession can be seen in the flattening of the U.S. energy-consumption curve in **figure 14.17** during the mid-1970s to mid-1980s, and again in the early 2000s. The largest annual decrease in energy consumption in the last seventy years occurred in 2020, down 7% from the 2019 level in response to the economics of the pandemic. The transportation sector had the largest drop.

Energy consumption rose steadily from about 1990 until the Great Recession in 2007, and then remained below its 2007 high until 2018. Factors involved in rising energy consumption include uncontrollable events like weather and others such as air travel that are discretionary. Per capita energy consumption in the Unites States peaked in the late 1970s; it has generally decreased every year since 2000 even while population increased. Reasons for this trend include increasing efficiency in appliances, building insulation, electrical equipment, and vehicles; reduction in energy-intensive manufacturing; and financial incentives for energy-efficient investments.

Conservation can buy some much-needed time to develop alternative energy sources and an infrastructure to support them. However, world energy consumption will certainly not decrease in the near future. The USEIA predicts global energy consumption to increase 47% by 2050, driven by economic and population growth especially in Asia. Even as industrialized countries continue to become more energy efficient and switch away from fossil fuels, many non-industrialized countries will rely upon them for development. Ultimately, we would conserve petroleum for those applications where electrical power is impractical; and that electricity would be generated by means other than burning fossil fuels and letting the greenhouse gases escape into the atmosphere.

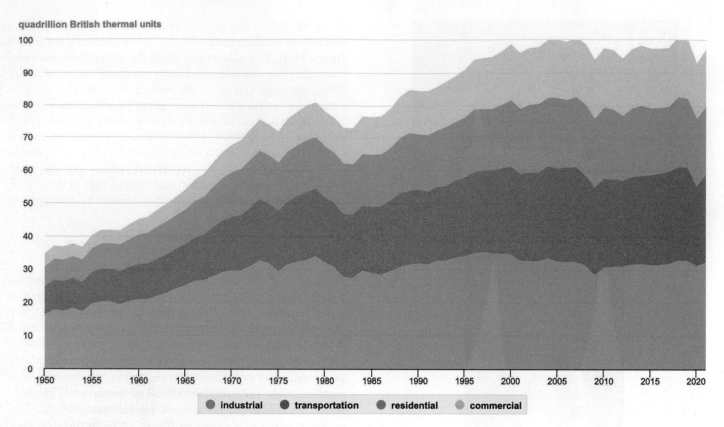

quadrillion British thermal units

● industrial ● transportation ● residential ● commercial

Figure 14.17

U.S. energy consumption by end-use sector, 1950–2020. Data for 2020 are preliminary.

U.S. Energy Information Administration, https://www.eia.gov/energyexplained/use-of-energy/

14.3 Oil Seeps and Spills

Learning Objectives

- List ways in which oil seeps or spills into the environment
- Describe how oil changes after being spilled
- Identify ways in which oils spills are detected and managed
- Summarize the Exxon Valdez and Deepwater Horizon oil spills

Petroleum has a density that is lower than its source or reservoir rocks, so when there are fractures and other pathways for it to follow, it will make its way to the surface. The LaBrea Tar Pits in Los Angeles are an infamous example of crude oil seeping out at the surface and then the lighter fractions of the oil evaporating leaving behind a viscous, tarry-type of petroleum often referred to as asphalt (**figure 14.18**). Most natural oil seeps occur on the ocean floor in predictable places marked by surface oil slicks that now can be detected using remote sensing technology. Lighter oil from these seeps rises to the ocean surface and then begins to

weather with exposure to the Sun, while heavier oil stays on the seafloor. Bacteria helps to naturally break down the hydrocarbons, and exposure to the Sun weathers it, reducing the volume by as much as 85%. Resulting tar balls, formed with the help of wind and waves, are not uncommon on beaches. Estimates of the volume of oil from natural seeps in the ocean ranges from 60 million to 600 million gallons per year, with annual amounts varying widely, and best estimates around 180 million gallons.

Oil spills represent large negative environmental impacts associated with the extraction and transportation of petroleum, although as a source of water pollution they represent only about 5% of all petroleum released to the environment. Tankers that flush out their holds at sea continually add to the oil pollution of the oceans and, collectively, are a significant source of such pollution. We tend to hear only about the occasional massive, disastrous spill created by a tanker or drilling platform accident, but there are many more that go unreported in the media that are associated with events such as leaking pipelines. The U.S. Coast Guard estimates that about 10,000 leaks and spills, large and small, occur in and around U.S. waters each year, totaling 15 to 25 million gallons of oil annually. Natural disasters add to the total. In 2005, damage to oil-drilling and processing facilities in the Gulf of Mexico from Hurricane Katrina caused about 100 spills of a total of 8 million gallons of oil.

A

B

Figure 14.18

Oil seeps and spills. (A) Many types of fossils as old as 50,000 years have been discovered in the natural oil seeps of the LaBrea Tar Pits. (B) This tar ball found on a California beach is about 10 cm across and likely associated with a natural seep on the sea floor, although could also be the result of an oil spill.

(A) Tara Wilsworth; (B) NOAA

When an oil spill occurs, the ensuing processes are the same as for a natural seep, with the exception of the volume involved. While the natural environment may be able to dissipate oil slowly seeping out at the surface, huge catastrophic spills can be disastrous and persist for months. Oil is toxic to marine life, causes water birds to drown when it soaks their feathers, may decimate fish and shellfish populations, and can severely damage the economies of beach resort areas. There are a variety of responses to cleaning up oil spills at sea as shown in **figure 14.19**. National Oceanic and Atmospheric Administration (NOAA) also provides a variety of databases and modeling, mapping, and planning tools in order to track and monitor oil spills, determining how where they may move and how they might degrade, so cleanup is based upon the most up-to-date, specific scientific information available.

The largest supertankers are over 380 meters long and can carry nearly 2 million barrels of oil. One of the largest single

Figure 14.19

The first step in responding to an oil spill at sea is often containment. From there, the oil can be skimmed, burned, or treated with dispersants. For more information, visit NOAA's Office of Response and Restoration.

NOAA

Figure 14.20

When the *Amoco Cadiz* ran aground, its entire cargo was lost, eventually making an oil slick 18 miles (30 km) wide and 80 miles (130 km) long.

NOAA Office of Response and Restoration.

marine tanker spills resulted from the wreck of the *Amoco Cadiz* near Portsall, France, in 1978 (**figure 14.20**). The bill for cleaning that up was more than $50 million, and only about 140,000 barrels of the 1.6 million spilled were recovered. The negative environmental impacts were still detectable years later.

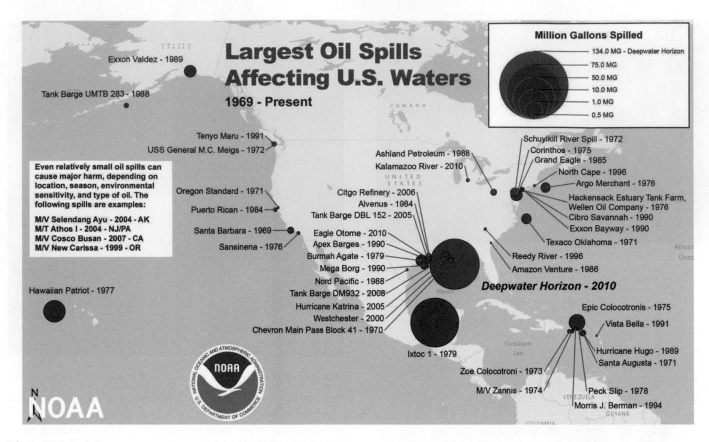

Figure 14.21

Largest oil spills affecting U.S. waters since 1969–2017.

NOAA/Office of Response and Restoration

The largest tanker accident in U.S. waters was that of the *Exxon Valdez,* in 1989. The port of Valdez lies at the southern end of the Trans-Alaska Pipeline (**figure 14.21**). In 1989, more than 1.8 million barrels of oil a day flowed through it. From Valdez, the oil is transported by tanker to refinery facilities elsewhere. Prince William Sound, on which Valdez is located, is home to a great variety of wildlife including whales, dolphins, sea otters, seals, sea lions, birds, shellfish, and economically important fish such as salmon and herring.

Early in the morning of 23 March 1989, the tanker *Exxon Valdez,* loaded with 1.2 million barrels of crude oil from Alaska's North Slope, ran aground on Bligh Island in Prince William Sound. Response to the spill was delayed for ten to twelve hours after the accident. More than 10 million gallons of oil escaped, eventually spreading out over more than 900 square miles of water. The various cleanup efforts cost Exxon an estimated $2.5 billion.

Skimmers recovered relatively little oil. Three weeks after the accident, only about 5% of the oil had been recovered. Chemical dispersants were not very successful, nor were attempts to burn the spill. Four days after the accident, the oil had emulsified into a thick, mousse-like substance, difficult to handle, too thick to skim, impossible to break up or burn effectively. Oil that had washed up on shore had coated many miles of beaches and rocky coast, not to mention marine mammals

and birds (**figure 14.22**). Thousands of birds and marine mammals died or were sickened by the oil. The annual herring season in Valdez was canceled, and salmon hatcheries were threatened.

Given the size of the spill and the cold Alaskan temperatures, leaving the oil to slowly degrade over time would certainly result in its long persistence. So, as an experiment, Exxon scientists sprayed miles of beaches in the Sound with a fertilizer solution designed to stimulate the growth of naturally occurring microorganisms that are known to consume oil. Within two weeks, treated beaches were markedly cleaner than untreated ones. Five months later, the treated beaches still showed higher levels of those microorganisms than did untreated beaches. This type of treatment isn't universally successful. The fertilizer solution runs off rocky shores, and loses its effectiveness once the oil has seeped deep into beach sands. But microorganisms may be the future treatment of choice, when feasible. Examination of the long-term effects of another strategy, hot-water washing of beaches, has suggested that it may actually do more harm than good. The heat can kill organisms that had survived the oil, the pressure of the hot-water blast can drive oil deeper into the sediments where degradation is slower, and the water treatment may wash oil onto beaches and into tidepools that were unaffected by the original spill. Over thirty years later, some of the oil remains in the sediments or beaches.

A   B

Figure 14.22

(A) Cleanup workers try pressure washing with hot water to clean a rocky beach after the *Exxon Valdez* oil spill. (B) Oiled wildlife, like this cormorant, needed more-delicate cleaning, if they survived at all.

(A) Accent Alaska.com/Alamy Stock Photo; (B) Exxon Valdez Oil Spill Trustee Council, courtesy NOAA

Precautionary measures can be increased—for instance, double-hulled tankers and tugboat escorts are now mandatory for oil tankers in Prince William Sound, and practice drills in manipulation of oil-containment booms are now regularly held—but as long as demand for petroleum remains high, some accidents are, perhaps, inevitable.

Drilling accidents have been a growing concern as more areas of the continental shelves are opened to drilling. Normally, the drill hole is lined with a steel casing to prevent lateral leakage of the oil, but on occasion, the oil finds an escape route before the casing is complete. Alternatively, drillers may unexpectedly hit a high-pressure pocket that causes a blowout.

This is what happened on 20 April 2010 to the offshore drilling rig *Deepwater Horizon* as it drilled an exploratory well about 65 km (40 miles) off the Louisiana coast (**figure 14.21**). The blowout-prevention system failed. The resulting explosion and fire killed eleven workers and injured several more. The half-billion-dollar rig burned for two days (**figure 14.23A**) before the remains of the structure collapsed and sank to the bottom of the Gulf of Mexico. Meanwhile, oil gushed from the blown-out well, creating both a huge and spreading slick on the surface (**figure 14.23B**) and underwater plumes of oil drifting away from the drill site. Oil continued to leak for twelve weeks before efforts to plug the well finally succeeded. Government estimates put the amount of oil that escaped from the well at over 4.9 million barrels.

Many of the strategies described earlier were applied to this spill. Some oil was burned in the Gulf to keep it offshore (**figure 14.24A**), since oil that reached the beaches would be much harder to remove (**figure 14.24B**) and would negatively affect more people, organisms, and ecosystems. Skimmer ships from around the world were mustered to collect as much of the oil as possible. Some 1.8 million gallons of dispersants were applied to both the surface slick and the subsurface plumes to break up the oil droplets to speed microbial decomposition.

By late 2010, about a quarter of the oil remained as slicks on the water or as oil and tar on the shores of the Gulf; a quarter had been burned or collected; a quarter had been dispersed as droplets; and a quarter had evaporated or dissolved. Questions remain about potential toxic effects on marine organisms and people of the residual oil and dispersants, and whether microbial decomposition of the oil causes harm to animal life in the Gulf by depleting dissolved oxygen in the water. (See chapter 17 for discussion of dissolved oxygen and water pollution.)

By April 2011, all fisheries in federal waters that had been closed due to the spill had been reopened. Economic losses to commercial and recreational fishing were estimated in the billions of dollars. A number of federal agencies, including the EPA, NOAA, and the U.S. Geological Survey, continue to study the aftereffects of the spill. Shortly after the spill, the government instituted a six-month moratorium on issuance of new permits for offshore drilling in the Gulf. Concerned that the *Deepwater Horizon* accident might, in part, have been due to lax oversight, the Obama administration also created a new agency to regulate offshore drilling, the Bureau of Ocean Energy Management, Regulation, and Enforcement. Permitting has now resumed, but slowly, with new, more stringent regulations in place, aimed at averting another accident like this one. In 2016, BP reached a settlement under which it will pay up to $8.8 billion for restoration of natural resources and to address the loss of recreational uses.

There is no simple, effective, and quick solution to a major oil spill. Taking every precaution to prevent an oil

A

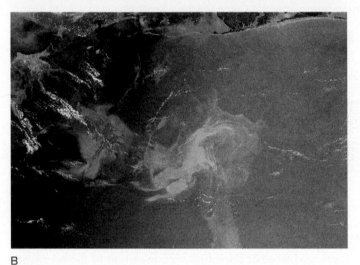

B

Figure 14.23

(A) The *Deepwater Horizon* fire took days to get under control. (B) The resulting oil slick, almost a month after the accident. The rig was near the center of the brightest-grey part of the slick. West of the slick is the Mississippi Delta.

(A) U.S. Coast Guard; (B) NASA image by Jeff Schmaltz, MODIS Rapid Response Team

A

B

Figure 14.24

(A) The Coast Guard conducted a controlled burn on 6 May 2010 to destroy some of the oil from the *Deepwater Horizon* rig before it could reach land. (B) Sediment samples from the beach face at East Ship Island, Mississippi, show alternating layers of oil-rich and cleaner sand.

(A) U.S. Navy photo by Mass Communication Specialist 2nd Class Justin Stumberg; (B) U.S. Geological Survey Mississippi Water Science Center/Photograph by Shane Stocks

spill is in everyone's best interest. And when spills do occur, being able to quickly identify them is critical so that action may be taken before the oil disperses. The U.S. Coast Guard is primarily responsible for oil spill cleanups in U.S. waters. NOAA's Marine Pollution Surveillance Report program, first tested with the Deepwater Horizon incident, uses machine learning and special types of radar to detect anomalies in the smoothness of ocean surface. Reports of detected spills are available at https://www.ospo.noaa.gov/Products/ocean/marinepollution/.

14.4 Coal

Learning Objectives

- Describe the conditions in which coal is formed
- Compare different types of coal
- Recognize the location of proven world reserves of coal
- Describe coal gasification and liquefaction, and explain why they are not likely to expand in the United States in the near future

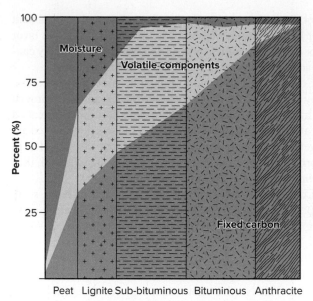

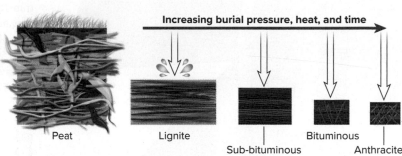

A

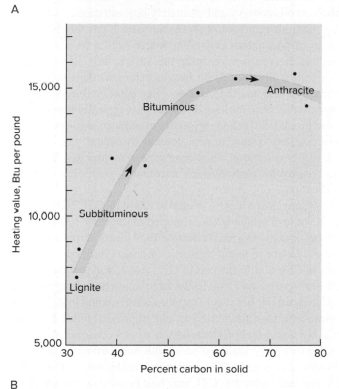

B

Figure 14.25

Change in character of coal with increasing application of heat and pressure. There is a general trend toward higher carbon content and higher heat value with increasing grade, though heat value declines somewhat as coal tends toward graphite.

(A) McGraw Hill

Formation of Coal Deposits

Coal is formed not from marine organisms, but from the remains of land plants. A swampy setting, in which plant growth is lush and where there is water to cover fallen trees, dead leaves, and other debris, is especially favorable to the initial stages of coal formation. The process requires *anaerobic* conditions, in which oxygen is absent or nearly so, since reaction with oxygen destroys the organic matter.

The first combustible product formed under suitable conditions is *peat,* which is still being formed on Earth today. Further burial, with more heat, pressure, and time, gradually dehydrates the organic matter and transforms the spongy peat into soft brown coal (**lignite**) and then to the harder coals (**bituminous** and **anthracite**); see **figure 14.25**. As the coals become harder, their carbon content increases, and so does the amount of heat released by burning a given weight of coal. The hardest, high-carbon coals are the most desirable as fuels because of their potential energy yield. Overly high temperatures lead to metamorphism of coal into graphite.

The higher-grade coals, like oil, apparently require formation periods that are long compared to the rate at which coal is being used. Consequently, coal is a nonrenewable resource. The world supply of coal represents an energy resource far larger than that of petroleum, and this is also true with regard to the U.S. coal supply. In terms of energy equivalence, U.S. coal resources exceed the energy in the remaining U.S. oil and natural gas reserves combined.

Prior to the discovery and widespread exploitation of oil and natural gas, wood was the most commonly used fossil fuel, followed by coal as the industrial age began in earnest in the nineteenth century. Coal is bulky, cumbersome, and dirty to handle and to burn, so it fell somewhat out of favor, particularly for home use, when the liquid and gaseous fossil fuels became available, and again along with carbon dioxide emission concerns. Currently, it supplies only about 11% of total U.S. energy needs. Coal-fired power plants account for 22% of U.S. electric power generation, which, in turn, consumes over 90% of U.S. coal produced.

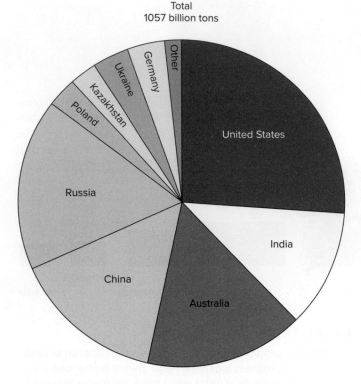

Total
1057 billion tons

United States

India

Australia

China

Russia

Poland

Kazakhstan

Ukraine

Germany

Other

Figure 14.26

World coal reserves, 2020.

Source: bp Statistical Review of World Energy 2021

Coal Reserves and Resources

Coal resource estimates are subject to fewer uncertainties than corresponding estimates for oil and gas. Coal is a solid, so it does not migrate. It is found in the sedimentary rocks in which it formed. It occurs in well-defined beds that are easier to map than underground oil and gas concentrations. And because it formed from land plants, which did not become widespread until 400 million years ago, we don't need to look for coal in more-ancient rocks.

The estimated world reserve of coal is about 1 trillion tons. The United States is particularly well supplied with coal, possessing about 23% of the world's reserves, about 248 billion tons of recoverable coal (**figure 14.26**). In 1974, the USGS assessed total U.S. coal resources at 4 trillion tons, but a newer assessment has not been completed. Even if only the reserves are counted, the U.S. coal supply could satisfy U.S. energy needs for more than 450 years at current levels of energy use, assuming coal could be used for all purposes. As a supplement to other energy sources, coal could last many centuries. Unfortunately, coal use has some serious drawbacks.

Limitations on Coal Use

A primary limitation of coal use is that solid coal is simply not as versatile a substance as petroleum or natural gas. It cannot be used directly in most forms of modern transportation, such as automobiles and airplanes. Coal is also a dirty and inconvenient fuel for home heating, which is why it was abandoned in favor of oil or natural gas. Given present technology, coal simply cannot be adopted as a substitute for oil in all applications.

Coal can be converted to liquid or gaseous hydrocarbon fuels by causing the coal to react with steam or with hydrogen gas at high temperatures. The conversion processes are called **gasification** and **liquefaction,** and the products they created are often referred to as *synfuels.* Both processes are intended to transform coal into a cleaner-burning, more versatile fuel, thus expanding its range of possible applications.

Commercial coal gasification has existed on some scale for 150 years. Many U.S. cities used the method before inexpensive natural gas became widely available following World War II. Europeans continued to develop the technologies into the 1950s. At present, there is only one commercial coal-gasification plant operating in the United States that manufactures natural gas. This North Dakota plant captures about 50% of the CO_2 produced, which is then piped to Canada for use in enhanced oil recovery and ultimately sequestration.

Current gasification processes yield a gas that is a mixture of carbon monoxide and hydrogen with a little methane. The heat derived from burning this mix is only 15 to 30% of what can be obtained from burning an equivalent volume of natural gas. Because the low heat value makes it uneconomic to transport this gas over long distances, it is typically burned only where produced. Technology exists to produce high-quality gas from coal, equivalent in heat content to natural gas, but it is presently uneconomic when compared to natural gas. Research into improved technologies continues, including *in situ* gasification.

There exist a variety of coal to liquid fuels (CTL) technologies, some of which were developed in the early 1900s. There are less than twenty plants outside the United States manufacturing a variety of liquid petroleum products from coal, most of them in China. U.S. interest in CTL was stimulated during the 1970s oil crisis and again in the late twentieth century when it appeared that our petroleum supplies were running low. Technological advances slashed costs and improved yields to about 70%, but liquid fuel products have not been economically competitive on a large scale with conventional petroleum. The most obvious advantage of CTL is that it would allow us to use our plentiful coal resources. CTL may best be used in small-scale or speciality or strategically located applications. The U.S. military is interested in using CTL to create synthetic jet fuels. Some CTL projects have been proposed in the last twenty years, but most have been delayed or canceled.

The CTL process occurs in stages. First the coal is gasified, and some impurities, such as sulfur gases, removed from the resulting gas mix. The gas is then converted to liquid fuels such as gasoline, diesel, and jet fuel. It is a water-intensive process, in which an estimated six to ten barrels of water are consumed for every barrel of liquid fuel produced. It also generates more than twice the greenhouse-gas emissions of producing gasoline from petroleum, so current and future expansion would need to include capturing these gases for use or storage. Other environmental concerns include the obvious expansion of coal mining and its associated issues described in more detail in section 14.5. It is likely that CTL will not be embraced on a commercial scale in the United States within the near future, given the recent expanded reserves and production of natural gas and oil.

14.5 Environmental Impacts of Coal

Learning Objectives

- List the gases emitted from coal and their associated environmental issues
- Describe the problems associated with coal ash
- Recall the hazards of coal mining
- Describe mine reclamation

Why would the United States not use its considerable domestic reserves of coal? One answer is that currently they don't need to because other fossil fuels are available. The other, and perhaps more pertinent answer is that the environmental, and resulting economic consequences of its use are too great. In this section, we will review the numerous environmental impacts of coal.

Gases

A major problem posed by coal is the pollution associated with its mining and use. Like all fossil fuels, coal produces carbon dioxide (CO_2) when burned. In fact, it produces significantly more carbon dioxide per unit energy released than oil or natural gas. The relationship of carbon dioxide to greenhouse-effect heating was discussed in chapter 10. The additional pollutant that is of special concern with coal is sulfur.

The sulfur content of coal can be more than 3%. When the sulfur is burned along with the coal, sulfur gases, notably sulfur dioxide (SO_2), are produced. These gases are poisonous and are extremely irritating to eyes and lungs. The gases also react with water in the atmosphere to produce sulfuric acid which then creates acidic precipitation. Acid rain falling into streams and lakes can kill fish and other aquatic life. It can acidify soil, stunting plant growth. It can dissolve rock; in areas where acid rainfall is a severe problem, buildings and monuments are visibly corroding away because of the acid. The geology of acid rain is explored further in chapter 18.

Oil can contain appreciable sulfur derived from organic matter, too, but most of that sulfur can be removed during the refining process, so that burning oil releases only about one-tenth the sulfur gases of burning coal. Some of the sulfur can be removed from coal prior to burning, but the process is expensive and only partially effective, especially with organic sulfur. Sulfur gases can be trapped by scrubbers in exhaust stacks; this is also expensive, and not perfectly efficient. From the standpoint of environmental quality, low-sulfur coal (1% sulfur or less) is more desirable than high-sulfur coal because it poses less of a threat to air quality, and there is less of sulfur to remove, so stricter emissions standards can be met more cheaply. However, much of the low-sulfur coal in the United States, especially western coal, is also lower-grade coal (see **figure 14.27**), which means that more of it must be burned to yield the same amount of energy.

Another toxic substance of growing concern with increasing coal use is mercury. A common trace element in coal, mercury is also a very volatile (easily vaporized) metal, so it can be released with the waste gases. The consequences are examined in chapter 17.

Ash

Coal use produces a lot of solid waste. The ash residue left after coal is burned typically ranges from 5 to 20% of the original volume. The ash consists mostly of noncombustible silicate minerals but also contains toxic metals. If released with waste gases, the ash fouls the air. If captured by scrubbers or otherwise confined within the combustion chamber, this ash still must be disposed of. If exposed at the surface, the fine-grained ash may weather very rapidly, and the toxic metals such as selenium and uranium can be leached from it, posing a water-pollution threat. Uncontrolled erosion of the ash could likewise cause sediment pollution. The magnitude of this waste-disposal problem should not be underestimated. A single coal-fired electric power plant can produce over a million tons of solid waste a year. There is no obvious safe place to dump all that waste. Over 50% of coal ash is recycled, going into concrete, cement, roadbeds, and fill, and it has been suggested that the ash could actually serve as a useful source of some of the rarer metals concentrated in it. Most coal ash that is not recycled is deposited in landfills.

Between combustion and disposal, the ash may be stored wet, in containment ponds which can pose their own hazard. In December 2008, a containment pond along the Emory River in Tennessee failed, releasing an estimated 5.4 million cubic meters of saturated ash from the Kingston Fossil Plant into the river (**figure 14.28**). The sludge released such pollutants as arsenic, chromium, and lead into the water. Cleanup of the site lasted about six years and cost nearly $1.2 billion. Monitoring of the surface water and groundwater continues. Of the 900 people who worked on the site remediation, about 200 are involved in litigation claiming that exposure to the coal ash left them seriously ill.

Coal-Mining Hazards and Environmental Impacts

Coal mining poses further problems. Underground mining of coal is notoriously dangerous, as well as expensive. Mines can collapse; miners may contract black lung disease from breathing the dust; there is always danger of explosion from pockets of natural gas that occur in many coal seams. A tragic coal-mine fire in Utah in December 1984, a methane explosion in a Chinese coal mine in 2004 that killed over 100 people, and recent

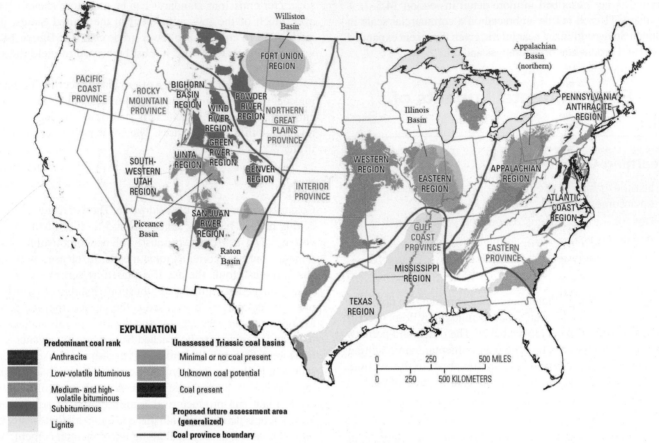

Coal fields of the conterminous United States—National Coal Resource Assessment updated version

EXPLANATION

Predominant coal rank

- Anthracite
- Low-volatile bituminous
- Medium- and high-volatile bituminous
- Subbituminous
- Lignite

Unassessed Triassic coal basins

- Minimal or no coal present
- Unknown coal potential
- Coal present
- Proposed future assessment area (generalized)
- Coal province boundary

0   250   500 MILES

0   250   500 KILOMETERS

Figure 14.27

Distribution of U.S. coal fields.

U.S. Geological Survey

A

B

Figure 14.28

(A) The Kingston Fossil Plant coal ash slurry spill buried nearly half a square mile of land, damaging homes and releasing pollutants.
(B) Lakeshore Park in Kingston, TN, was created as part of the ash spill remediation.

(A-B) United States Environmental Protection Agency

Figure 14.29

Fire out of control in abandoned underground coal mine in Sheridan County, Wyoming.

U.S. Geological Survey Photo Library, Denver, CO/Photograph by C. R. Dunrud

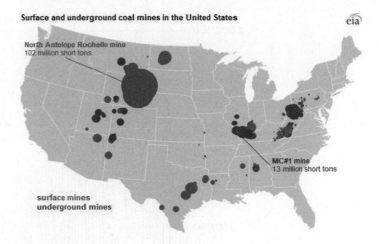

Surface and underground coal mines in the United States   eia

North Antelope Rochelle mine
102 million short tons

MC#1 mine
13 million short tons

surface mines
underground mines

Figure 14.30

Until 1970, virtually all U.S. coal mining occurred east of the Mississippi. Now most mining occurs west of the Mississippi, and the majority of that mining is strip mining.

U.S. Energy Information Administration, Form EIA-7A, Annual Survey of Coal Production and Preparation, and U.S. Department of labor, Mine Safety and Health Administration Form 700-2, Quarterly Mine Employment and coal Production Report.

deadly mine collapses in Siberia, India, and Serbia are reminders of the seriousness of the hazards. Coal miners are also likely exposed to increased cancer risks from breathing the radioactive gas radon, which is produced by natural decay of uranium in rocks surrounding the coal seam.

Even when mining has ceased, fires may start in underground mines (**figure 14.29**). Coal mine fires can start from lightning strikes, humans, or just from high temperatures and are notoriously difficult if not impossible to put out as they provide their own fuel and oxygen to burn. An underground fire has been burning under Centralia, Pennsylvania, since 1962; over 1000 residents have been relocated, at a cost of more than $40 million, and collapse and toxic fumes threaten the handful of remaining residents resisting evacuation. And while Centralia's fire may be among the most famous, it is by no means the only such fire in the United States. At the end of 2021, there were at least 250 coal fires in twelve states and possibly hundreds to thousands more that are undocumented. Given the concerns with wildfires in the dry western states, it isn't surprising that Colorado was recently granted $2 million from the federal government to fight coal mine fires.

Coal-mine fires are a problem in other nations as well. China relies heavily on coal for its energy; it has been estimated that uncontrolled coal-mine fires there consume up to 20% of their coal production, the carbon-dioxide emissions from them are estimated to account for several percent of global CO_2 emissions, and the associated pollution is believed to be aggravating respiratory problems in China and beyond. Mining of India's largest coal field began in 1894; the first fires there started in 1916, and fifty years later, the fires could be found across the whole coal field. They burn on today.

The rising costs of providing a safer working environment for underground coal miners, together with the greater technical difficulty of underground mines, are largely responsible for a steady shift in coal-mining methods in the United States, from about 20% surface mining in 1950 to about 63% surface mining now. As the demand for low-sulfur western coal has increased, so has the location of mining, with Wyoming accounting for 41% of production in 2020 as compared to West Virginia (13%) and Pennsylvania (7%) (**figure 14.30**). Because strip-mining is safer for miners, there has been an accompanying decline in U.S. coal-mine fatalities (**figure 14.31**).

A significant portion of U.S. coal, particularly in the west, occurs in beds very close to the surface. It is relatively cheap to strip off the vegetation and soil and dig out the coal at the surface, thus making strip-minable coal very attractive economically in spite of land-reclamation costs (**figure 14.32**). Strip mining does present its own problems, so of which were discussed in chapter 13. A particular problem with strip mining

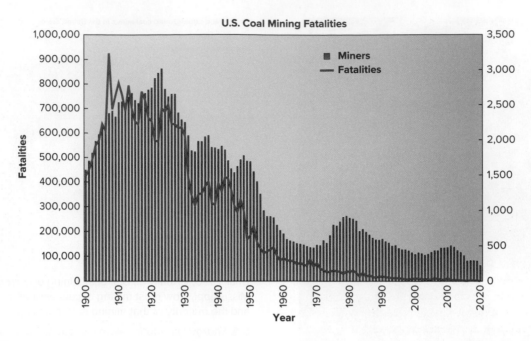

Figure 14.31

Drop in coal-mine fatalities over the last century reflects partly safer mining practices, partly the shift from underground mining to strip mining.

Data from https://arlweb.msha.gov/stats/centurystats/coalstats.asp

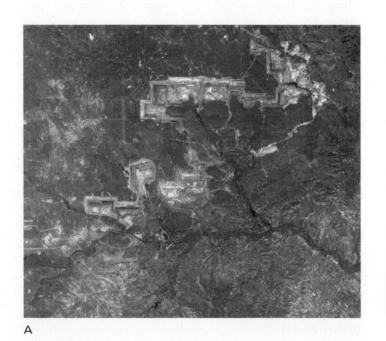

A

B

Figure 14.32

(A) The North Rochelle Antelope Complex in the Powder River Basin of Wyoming is the largest surface coal mine in the world. This image covers approximately 27 km (17 miles) in width. (B) The Eagle Butte Mine is another strip coal mine in Wyoming.

(A) U.S. Geological Survey; (B) Gates Frontiers Fund Wyoming Collection within the Carol M. Highsmith Archive, Library of Congress, Prints & Photographs Division

coal involves, again, the sulfur in the coal. Not every bit of coal is extracted from the surrounding rock, and waste rock is left behind in spoil banks. The sulfur in the spoils can react with water and air to produce runoff water containing sulfuric acid, as was shown in table 12.1.

This same reaction is responsible for most of the acid mine drainage discussed in chapter 13. Because plants grow poorly in very acid conditions, this acid slows revegetation of the area by stunting plant growth or even preventing it altogether. The acid runoff can also pollute area groundwater and surface water, killing aquatic plants and animals in lakes and streams and contaminating the water supply. Very acidic water also can be particularly effective at leaching or dissolving some toxic elements from soils, further contributing to water pollution. Coal strip mines can be reclaimed, but in addition to regrading and replanting the land, it is frequently necessary to replace the original topsoil so that sulfur-rich rocks are not left exposed at the surface, where weathering is especially rapid. Even underground coal mines may have associated acid drainage, but water circulation there is generally more restricted. The question of water availability to support plant regrowth is of special concern in the western United States. Over 50% of U.S. coal reserves are found there, 91% of mine production in the western region is from surface mines, and most western regions underlain by coal are very dry.

When coal surface mines are reclaimed, efforts are commonly made to restore original topography although this may not be easy. The artificial slopes of the restored landscape may not be altogether stable, and as natural erosional processes begin to sculpt the land surface, natural slope adjustments in the form of slumps and slides may occur. In dry areas, thought must be given to how the reclamation will modify drainage patterns. In wetter areas, the runoff may itself be a problem, gullying slopes and contributing to sediment pollution of surface waters until new vegetation is established. In short, reclamation is a multifaceted challenge.

This has proved especially true with the mountaintop-removal style of surface mining that has become common since the 1990s in the Appalachian coal country. With mountaintop removal, vegetation is cleared and then, as the name suggests, explosives and earthmoving equipment are used to break up and remove the entire hilltop. Coal is sorted out from waste rock and soil, and then rinsed, producing a sediment-laden sludgy wastewater that must be contained in ponds to prevent surface-water pollution. The combination of hilltop removals and valley fills produces a much different, and flatter, topography than the original (**figure 14.33**). Yet the 1977 Surface Mining Control and Reclamation Act, discussed further in chapter 19, mandates that the mined land be restored to its "approximate original contour," which is virtually impossible given the scale and complexity. Other issues include ecosystem destruction and water pollution. Even though coal production in the Appalachians has decreased dramatically, mountaintop removal mining continues, but so does the opposition to and litigation against the practice.

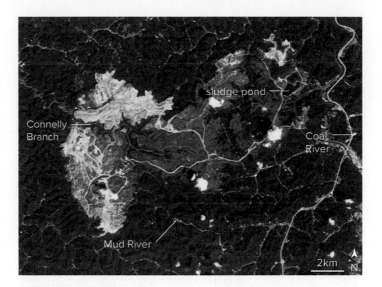

Figure 14.33

Hobet Mine, West Virginia. Some of the surplus spoils have been used to fill in Connelly Branch valley and other streams. Note the scale of this mountaintop mine complex. (2 km = 1¼ mi)

NASA Earth Observatory image created by Robert Simmon, using Landsat data provided by the U.S. Geological Survey

14.6 Oil Shale

Learning Objectives

- Describe oil shale
- Identify reasons the United States is not currently using its large resource of oil shale

Oil shale is a very poorly named source of hydrocarbons. This sedimentary rock doesn't necessarily need to be shaler and the hydrocarbon in it is not oil. The potential fuel in oil shale is a complex, waxy solid called **kerogen,** which is formed from the remains of plants, algae, and bacteria. Kerogen is not a single compound; the term is a general one describing organic matter that is not readily dissolved either in water or in organic solvents, and the term can also be applied to solid precursor organic compounds of oil and natural gas in other types of rocks. The physical properties of kerogen dictate that the oil shale must be crushed and heated to distill out the hydrocarbon as shale oil, which then is refined to produce various liquid petroleum products.

The United States has nearly three-fourths of the world's known oil shale resources. About 70% is in the Green River Formation (**figure 14.34**), which holds enough hydrocarbon to yield an estimated 1½ trillion barrels of shale oil. For a number of reasons, the United States is not yet using this apparently vast

Figure 14.34

Oil shale of the Green River Formation, which is found in parts of Colorado, Utah, and Wyoming.

John A. Karachewski

resource to any significant extent, and is unlikely to do so in the near future.

One reason for this is that much of the kerogen is so widely dispersed through the oil shale that huge volumes of rock must be processed to obtain moderate amounts of shale oil. Even the richest oil shale yields only about three barrels of shale oil per ton of rock processed. The cost has not been competitive with that of conventional petroleum, except when oil prices are extremely high, and to date they have not remained high enough long enough. Nor have large-scale processing facilities yet been built—the few plants built so far are strictly experimental. More efficient and cheaper processing technologies are needed.

Another problem is that a large part of the oil shale is at or near the surface. At present, the economical way to mine it appears to be surface or strip mining with its attendant land disturbance. The Green River Formation is located in areas of the western United States that are already chronically short of water. The dry conditions would make revegetation of the land after strip mining especially difficult. Some consideration has been given to *in situ* extraction of the kerogen, warming the rock in place and using pressure to force out the resulting fluid hydrocarbon, without mining the rock, to minimize land disruption. But the necessary technology is poorly developed.

The water shortage presents a further problem. Current processing technologies require large amounts of water—on the order of three barrels of water per barrel of shale oil produced. Just as water for reclamation is in short supply in the west, so too is the water to process the oil shale.

Finally, since the volume of the rock actually increases during processing, it is possible to end up with a 20 to 30% larger volume of waste rock to dispose of than the original volume of rock mined. Aside from any problems of accelerated weathering of the crushed material, there remains the basic question of where to put it. Because it will not all fit back into the space mined, the topography will inevitably be altered.

Scientists and economists differ widely in the extent to which they see oil shale as a promising alternative to conventional oil and gas. Certainly, the water-shortage, waste-disposal, and land-reclamation problems will have to be solved before shale oil can be used on a large scale. It is unlikely that those problems can be solved within the next few decades, although over the longer term, oil shale may become an important resource, and the higher conventional oil prices rise, the more competitive shale oil becomes, at least in that respect. For now, however, while we have large oil shale *resources,* they cannot be considered fuel *reserves.*

The Arctic National Wildlife Refuge: To Drill or Not to Drill?

The Prudhoe Bay field on the North Slope of Alaska is the largest U.S. oil field ever found. It has yielded over 12.5 billion barrels of oil so far. Production was 271,000 barrels per day in 2019. Its richness justified the $8 billion cost of building the 800-mile Trans-Alaska pipeline. West of Prudhoe Bay lies the 23-million-acre tract now known as the National Petroleum Reserve–Alaska (NPRA), first established as Naval Petroleum Reserve No. 4 by President Warren Harding in 1923. East of Prudhoe Bay is the 19-million-acre Arctic National Wildlife Refuge (ANWR), established in 1980 (**figure CS 14.1.1**).

Geologic characteristics of, and similarities among, the three areas suggested that oil and gas might very well be found in both NPRA and ANWR. When establishing ANWR, Congress deferred a final decision on the future of the coastal portion, some 1.5 million acres known as the "1002 Area" after the section of the act that addressed it.

Little petroleum exploration occurred in NPRA until the late 1990s. In 1996, oil was found in the area between NPRA and Prudhoe Bay. Then, beginning in 2000, exploration wells began to find oil and gas in NPRA.

Estimating potential oil and gas reserves in a new region is a challenge. Drilling and seismic and other data constrain the types and distribution of subsurface rocks. Samples from drill cores, and comparisons with geologically similar areas, allow projection of quantities of oil and gas that may be present. Only a portion of the in-place resources may be technologically recoverable—perhaps 30 to 50% of the oil, 60 to 70% of the gas. And, of course, the market price of each commodity will determine how much is economically recoverable. Combining all of these considerations resulted in a range of estimates of oil and gas reserves in NPRA and the ANWR 1002 Area (**figure CS 14.1.2**). At crude-oil prices of $40 per barrel, the analysis suggested a mean probability of close to 7 billion barrels of economically recoverable oil each in NPRA and in the ANWR 1002 Area. Moreover, it is believed that the oil accumulations in the ANWR 1002 Area are, on average, larger, meaning that they could be developed more efficiently. (Development of the gas reserves would require building another pipeline, something that is particularly unlikely given the success of fracking in developing natural gas deposits in the "lower 48.")

A few years after this USGS analysis was done, oil prices began to rise sharply. One would naturally expect that that would increase the projected volume of reserves (though note that the difference between estimated reserves at $30 per barrel and at $40 per barrel in the original analysis was rather modest). In mid-2008, in fact, oil prices briefly exceeded $140 per barrel—and

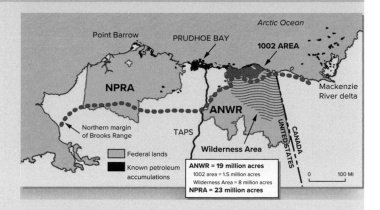

Figure CS 14.1.1

The NPRA and ANWR flank the petroleum-rich Prudhoe Bay area. The future of the "1002 Area" of the ANWR is currently being hotly debated.

After USGS Fact Sheet 040-98.

then dropped precipitously, falling below $40 per barrel by early 2009. Since then, it has been over $100/barrel (2011–2012) and under $30/barrel (2016). As this is written, oil is back around $100/barrel, but there is no assurance of how long it will stay there. Such volatility dictates caution in making reserve projections based on assumptions of very high oil prices, and it certainly prompts oil companies to be cautious in moving forward to develop the more marginal deposits.

For one thing, drilling costs continue to rise, particularly in challenging areas like Alaska. Permafrost is a delicate foundation for construction, as will be explored further in chapter 20, and has become more so with Arctic warming. For this and other reasons, costs of drilling onshore in Alaska rose from $283 per foot drilled in 2000 to $1880 per foot in 2005, and offshore costs were higher.

Furthermore, development of new oil fields is never quick, and this is likely to be especially true in ANWR. A 2008 analysis by the U.S. Energy Information Administration (EIA) conservatively suggested a lag of *ten years* from the time oil leases would be authorized by the federal government to the first oil production. This breaks down as follows:

Two to three years for writing and acceptance of an Environmental Impact Statement, preliminary analysis of exploration data by interested companies, and a lease auction

Two to three years for the drilling of *one* exploratory well (a slow process partly because of the location, and partly because exploratory wells involve extensive data collection, not just drilling)

(Continued)

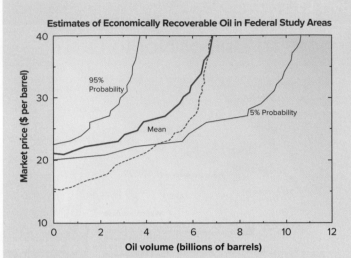

Estimates of Economically Recoverable Oil in Federal Study Areas

(Graph: Market price ($ per barrel) on y-axis from 10 to 40; Oil volume (billions of barrels) on x-axis from 0 to 12. Curves labeled "95% Probability", "Mean", "5% Probability".)

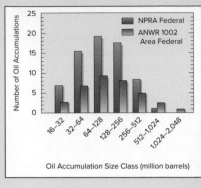

(Bar chart: Number of Oil Accumulations on y-axis from 0 to 25; Oil Accumulation Size Class (million barrels) on x-axis: 16–32, 32–64, 64–128, 128–256, 256–512, 512–1,024, 1,024–2,048. Legend: NPRA Federal; ANWR 1002 Area Federal.)

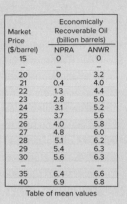

Market Price ($/barrel)	Economically Recoverable Oil (billion barrels) NPRA	ANWR
15	0	0
–	–	–
20	0	3.2
21	0.4	4.0
22	1.3	4.4
23	2.8	5.0
24	3.1	5.2
25	3.7	5.6
26	4.0	5.8
27	4.8	6.0
28	5.1	6.2
29	5.4	6.3
30	5.6	6.3
–	–	–
35	6.4	6.6
40	6.9	6.8

Table of mean values

Figure CS 14.1.2

U.S. Geological Survey estimates of economically recoverable oil in the federally owned portions of NPRA and the ANWR 1002 Area (other areas are Native lands), and size distribution of individual oil accumulations. Red lines are for NPRA; dashed blue line is mean for ANWR 1002 Area.

After USGS Fact Sheet 045-02.

One to two years for the company to formulate a plan for developing its oil field, and for the Bureau of Land Management (the responsible federal agency here) to approve that plan

Three to four years to build the necessary oil-processing facilities and feeder pipelines to connect the new oil to the Trans-Alaska Pipeline, and to drill development wells to extract the oil

A company would need confidence in the price of oil ten or more years in the future before being certain of the economic viability of a particular deposit. The EIA concluded that even prices prevailing at the time of its analysis (over $120 per barrel) would not motivate development of smaller or more technically difficult oil fields in ANWR until after 2030; the richer and more accessible deposits already identified as reserves by USGS would certainly be developed first. As oil prices are lower than at the time of the EIA assessment, speculation about reserves beyond those identified in the USGS study is more complicated. A further complicating factor is the rapid rise in natural gas reserves in the last few years due to new discoveries and the development of shale gas, discussed later in the chapter, which has driven down natural gas prices and made gas a more attractive fuel than oil for many applications.

Those who favor drilling in the ANWR 1002 Area have pointed to the U.S. dependence on imported oil, the financial benefits to Alaskans of additional oil production, the fact that there are tens of millions of acres of other wilderness. They also note that oil production technology has become more sophisticated so as to leave less of a footprint—for example, one can drill one well at the surface that splays out in multiple directions at depth rather than having to drill multiple wells from the surface to tap the same deposits. Those opposed to drilling note the extreme environmental fragility of alpine ecosystems in general, the biological richness of the 1002 Area in particular, and the fact that the United States consumes over 7 billion barrels of crude oil in just a year (so, why risk the environment for so relatively little oil?). New USGS estimates of technically recoverable petroleum resources on the Alaska North Slope are 8.7 billion barrels of oil and 25 trillion cubic feet of natural gas, information which could be used to argue both for and against drilling in ANWR.

The debate continues. At the end of the Trump administration in early 2021, oil leases for drilling in ANWR were put up for auction, although no major companies purchased them, no major banks were involved, and only half of the tracts offered were sold, mostly to a company owned by the state of Alaska. The Biden administration has put a halt on any activities associated with leases, for now.

Energy Prices, Energy Choices

In recent years, consumers have been surprised by rapid swings in fossil-fuel prices. Many factors influence both energy prices and energy consumption. On the price side, the underlying cost of the commodity may vary little except when political events or natural disasters constrain availability (**figure CS 14.2.1**); speculation in the financial markets may exacerbate the effects of these factors, as in 2007–2008. Short-term supply-and-demand issues are often more significant influences on retail prices of petroleum products. As noted in the chapter, crude oil must be refined into a range of petroleum products, and refiners make projections about the relative amounts of gasoline, heating oil, etc. that will be needed at different times of year. Unexpected developments (such as a fierce, prolonged cold spell in heating-oil country) can drive up prices until more of the needed fuel can be produced. Likewise with natural gas; although it's readily delivered by pipeline, it can only be extracted and transported so fast, and if use is unexpectedly heavy for some time, prices will again rise until demand slackens or supply catches up.

Consumers can moderate their own costs through conservation but often feel little incentive to do so if financial pressures are low. Gasoline prices generally stayed relatively low in the 1990s, and consumption slowly rose in complementary fashion

(**figure CS 14.2.2**). In part, this can be related to the rise in popularity of trucks, sport-utility vehicles, and other vehicles with notably lower mileage than is mandated for passenger cars by the Environmental Protection Agency (**figure CS 14.2.3**). The sharp price rise in gasoline in 2008 depressed consumption until 2013, when prices again went down and consumption rose correspondingly. The sharp drop in consumption in 2020 was a reflection of the COVID-19 pandemic, when many of us stayed home rather than driving to work every day, also traveling much less for non-work-related reasons.

And what about the electric vehicles? About 2.3 million electric cars have been sold in the United States since 2010. The electric car may operate cleanly and efficiently, but making that electricity still consumes energy and creates pollution somewhere; currently, about 60% of our electricity is generated using fossil fuels. Sales of both electric and hybrid light vehicles jumped by over 75% in the United States in 2021, but both still make a very small portion of the market. Still, improving the fuel economy of motor vehicles, of whatever sort, will substantially reduce oil consumption, considering that motor gasoline accounts for about 44% of the petroleum consumption in the United States. The newest government mileage standard will be 49 mpg fleet-wide average for model year 2026 cars and light trucks.

Cushing, OK WTI Spot Price FOB

Figure CS 14.2.1

The Arab oil embargo and other events in the Middle East led to a spike in crude-oil prices in the 1980s. Prices then subsided for nearly two decades, but damage to U.S. production by Hurricane Katrina, and wars in the Middle East, combined to push prices higher in the mid-2000s. Speculation drove them sharply higher in 2008, since which time they have fluctuated considerably. (Prices not adjusted for inflation.)

U.S. Energy Information Administration

(Continued)

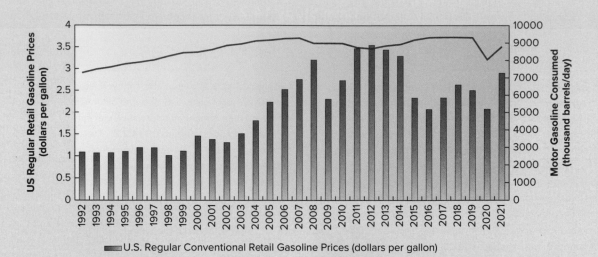

U.S. Regular Conventional Retail Gasoline Prices (dollars per gallon)

Figure CS 14.2.2

U.S. regular retail gasoline prices in comparison to supplied finished motor gasoline (which approximately equals consumption as it represents the disappearance of the product from primary sources).

Source: U.S. Energy Information Administration.

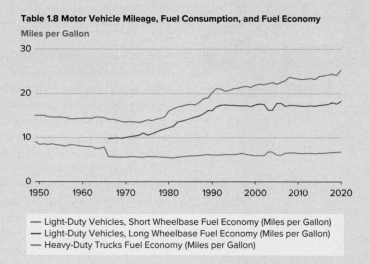

Table 1.8 Motor Vehicle Mileage, Fuel Consumption, and Fuel Economy

— Light-Duty Vehicles, Short Wheelbase Fuel Economy (Miles per Gallon)
— Light-Duty Vehicles, Long Wheelbase Fuel Economy (Miles per Gallon)
— Heavy-Duty Trucks Fuel Economy (Miles per Gallon)

Figure CS 14.2.3

Mileage of passenger cars (light-duty vehicles with a short wheelbase) has risen largely due to EPA mandates; mileage of vans, SUVs, and pickup trucks (light-duty vehicles with a long wheelbase) have lagged consistently behind. Mileage of heavy-duty trucks has barely changed in fifty years.

U.S. Energy Information Administration

Summary

The United States now relies on fossil fuels for about 79% of its energy: about 36% comes from oil, 32% from natural gas, and 11% from coal. All fossil fuels are nonrenewable energy sources. Improvements in drilling technologies have increased U.S. oil and gas reserves, but these still may only last decades given current consumption rates. Oil spills are a subject of concern, especially as much of the remaining U.S. petroleum is offshore or in the Arctic. Direct coal use is rapidly falling in response to its serious environmental consequences even though the United States has large reserves. Coal gasification and liquefaction projects are experimenting with ways to make coal into a more useful fuel. Methane hydrates represent a large resource, but the problem of how to extract their fuel efficiently has not yet been solved, although some countries have engaged in experimental recovery of seafloor methane hydrate. Development of this resource also carries with it the risk of adding more methane to the atmosphere. U.S. oil shale represents a significant hydrocarbon resource but it also presents a number of technical and environmental challenges, so it is likely to remain uneconomic and unused for some time.

Key Terms and Concepts

anaerobic	fracking	liquefaction	oil shale
anthracite	gasification	methane hydrate	petroleum
bitumen	geopressurized zones	natural gas	shale gas
bituminous	hydraulic fracturing	nonrenewable	
coal-bed methane	kerogen	oil	
fossil fuel	lignite	oil sand	

Test Your Learning

1. A society's level of technological development strongly influences its per-capita energy consumption. Explain.

2. Define *fossil fuels,* and list the three that we use most.

3. Briefly describe how conventional oil and gas deposits form and mature.

4. Compare past and projected U.S. consumption of petroleum and coal.

5. Explain the concept of enhanced oil recovery. Give two examples of the method.

6. Describe fracking and how it has allowed shale gas to be added to U.S. natural gas reserves. Note a significant concern that has emerged in connection with it.

7. Outline the advantages that coal-bed methane has over other unconventional gas sources. Describe one problem associated with its extraction.

8. Describe methane hydrates, and indicate where they are found. Explain why they are both a substantial potential energy resource and a concern with respect to global climate change.

9. Explain how, and from what, coal forms.

10. Describe the concept of coal liquefaction and why it is of interest. Suggest one reason it is not widely practiced in the United States.

11. Note what air-pollution problems are associated particularly with coal, relative to other fossil fuels.

12. List and describe at least three potential negative environmental impacts of coal mining. Explain why mountaintop removal is particularly controversial.

13. Describe the nature of oil shale, explaining why it is misnamed. Outline the steps needed to extract the oil, and list three practical and/or environmental problems with its development.

Exploring Further

1. Select a particular region or major city and investigate its energy consumption. Identify the principal energy sources and the proportion that each contributes; see whether these proportions are significantly different now from what they were ten and twenty-five years ago. To what extent have the types and quantities of energy used been sensitive to economic factors?

2. If possible, visit an area that has been or is being drilled for oil or mined for coal. Observe any visible signs of negative effects on the environment, and note any efforts being made to minimize such impacts.

3. Use the most recent **USEIA Total Energy Monthly Energy Review** to update a consumption, production, or reserves trend discussed in this chapter. Note how the trend has changed, and tie the change to technological, economic, and/or societal factors (if applicable).

4. Shale oil and oil shale are two different fossil fuels. Recalling the differences between them, come up with new names for these resources that better represent them. Explain how the new name will help people understand the resource.

CHAPTER 15

Energy Resources—Alternative Sources

Finding energy sources that fulfill our needs without adding carbon dioxide to the atmosphere is now the primary reason for using alternatives to fossil fuels. The United States currently has enough oil and gas to last for decades, and coal for much longer, but we, along with the rest of the world, will need to find and use replacements to avoid the worsening consequences of a warming climate. Alternative energy sources are needed now and for the foreseeable future to

The Nesjavellir Geothermal Heat and Power Plant in Iceland provides electricity and hot water for heating to the people of Reykjavik. All of Iceland's electricity is generated with renewable energy sources.
©Gina S. Szablewski

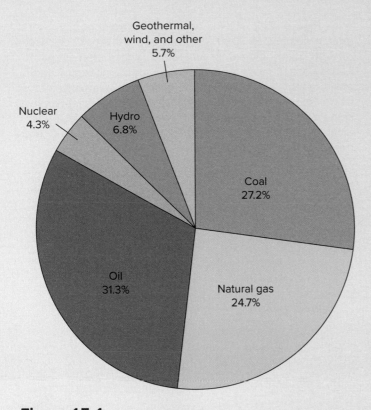

Figure 15.1

World energy consumption by source, 2020.

Source: Data from BP Statistical Review of World Energy 2021 page 11

supply essential energy and to avoid as much disruption as possible (**figure 15.1**). Most of these alternative energy sources are also renewable ones, an important consideration for the long term.

The extent to which alternative energy sources are required and how soon they will be needed is directly related to future world energy demand, which is difficult to predict precisely, especially immediately following a pandemic. In general, consumption can be expected to rise as population increases and as standards of living improve. However, the correlation between energy consumption and standard of living is perhaps even less direct than the correlation between mineral consumption and living standards.

By way of a simplified example, consider the addition of central heating to a home. The furnace is likely to contain about the same quantity of material, regardless of its type or efficiency; the drain on mineral reserves is approximately fixed. The very addition of that furnace represents a certain jump in both mineral and energy consumption. However, depending on the efficiency of the unit, the tightness of the home's insulation, and the climate, the amount of fuel the furnace must consume to maintain a certain level of heat in the home will vary enormously. The best-insulated modern homes may require as little as one-tenth the energy for heating as the average U.S. home. Also, the quantity of material used to manufacture the furnace will not change whether the homeowner maintains the temperature at 60°F or 80°F, but the amount of energy consumed plainly will. And, of course, the climate in which the home is built influences its heating costs. Extreme heat and extreme cold both cause increased energy demand. There are considerable regional variations in energy consumption in the United States because of varying population density, degree of industrialization, and climatic variations.

In the area of transportation, similar variability arises. As standards of living rise to the point that motorized transportation becomes commonplace, consumption of both materials and energy rises. However, the amount of energy used depends heavily on the mode of transportation chosen and its fuel efficiency. It is more fuel-efficient to transport fifty people in a bus than in a dozen automobiles, but the distribution of people and their travel patterns must be clustered appropriately in order for mass transportation to work in a practical sense. Just among passenger cars, fuel efficiency varies greatly. When projecting future energy use in developing countries, the underlying population base is important, too. An increase of 50 vehicles per thousand persons in China, with population over 1.4 billion, means far more added automobiles than an increase of 200 vehicles per thousand in South Korea, with a population of only about 51.8 million, less than one-twentieth of China's.

In deciding just how much energy various alternative sources must supply, assumptions must be made not only about the rates of increase in standards of living (growth of GNP) but also about the degree of energy efficiency with which the growth is achieved. Other factors that influence both demand and projections include prices of various forms of energy (which recent history suggests is difficult to assess years, let alone decades, into the future) and public policy decisions (to pursue or not to pursue nuclear power, for example) in those nations that are major energy consumers.

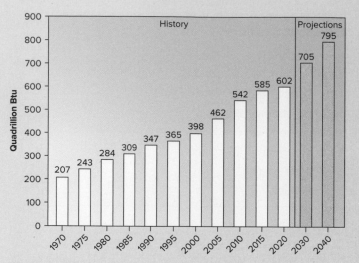

Figure 15.2

World energy consumption, historic and projected to 2040. Projected increase from 2020 to 2040 is about 32%.

Source: International Energy Outlook 2021, U.S. Energy Information Administration.

Figure 15.2 shows one recent projection of world energy demand to 2040. Whether or not it is perfectly accurate, it clearly indicates substantial anticipated growth in energy use. If fossil fuels are inadequate in quantity and/or environmentally problematic, one or more alternatives must be far more extensively developed. This chapter surveys some possibilities. Evaluation of each involves issues of technical practicality, environmental consequences, and economic competitiveness.

15.1 Nuclear Power—Fission

Learning Objectives

- Describe the process of nuclear fission used to make electricity
- Recall how uranium becomes concentrated in sandstone
- Discuss current world estimate of uranium resources
- List ways to extend the nuclear fuel supply
- Identify ways in which nuclear reactors could be damaged and release radiation
- Recall the problems related to the handling of nuclear fuel and waste
- Describe nuclear power plants in terms of location, perceived risk, and future implementation

The phrase *nuclear power* comprises two distinct processes with different advantages and limitations. **Fission** is the splitting apart of atomic nuclei into smaller ones, with the release of energy. **Fusion** is the combining of smaller nuclei into larger ones, also releasing energy. Currently, only fission is commercially feasible. In this section, we will review how we use fission to generate electricity, describe the geology and resource outlook for uranium, and address the risks associated with nuclear power. Fusion is covered in section 15.2. If you need a review of the terminology related to atoms, revisit chapter 2.

Fission—Basic Principles

The basics of fission are outlined in **figure 15.3**. Only about 20 out of more than 250 naturally occurring isotopes undergo fission spontaneously. We can induce some nuclei to split

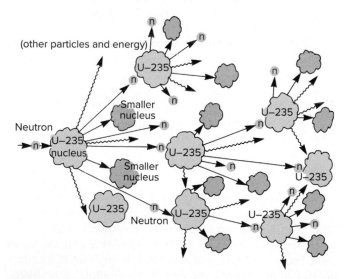

Figure 15.3

Nuclear fission and chain reaction involving ^{235}U (schematic). Neutron capture by ^{235}U causes fission into two smaller nuclei plus additional neutrons, other subatomic particles, and energy. Released neutrons, in turn, cause fission in other ^{235}U nuclei. As ^{235}U nuclei are used up, reaction rate slows; eventually fresh fuel must replace spent fuel.

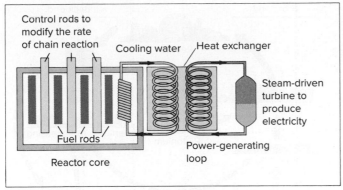

Control rods to modify the rate of chain reaction

Cooling water

Heat exchanger

Steam-driven turbine to produce electricity

Fuel rods

Reactor core

Power-generating loop

Containment building

Figure 15.4

Schematic diagram of conventional nuclear fission reactor. Heat is generated by chain reaction; withdrawing or inserting control rods between fuel elements varies rate of reaction, and thus rate of release of heat energy. Cooling water also serves to extract heat for use. Heat is transferred to power loop via heat exchanger, so the cooling water, which may contain radioactive contaminants, is isolated from the power-generating equipment.

apart and/or split apart more rapidly in order to increase the rate of energy release. Uranium-235 (^{235}U) is the fissionable nucleus most commonly used in modern nuclear power reactors; it is the isotope of uranium with 92 protons and 143 neutrons.

By firing another neutron into its nucleus, we induce ^{235}U to undergo fission. The nucleus splits into two lighter nuclei (not always the same two) and releases additional neutrons as well as energy. Some of the newly released neutrons can induce fission in other nearby ^{235}U nuclei, which, in turn, release more neutrons and more energy in breaking up, and so the process continues in a **chain reaction.** A controlled chain reaction, with a continuous, moderate release of energy, is the basis for fission-powered reactors (**figure 15.4**). The energy released heats cooling water that circulates through the reactor's core. The heat removed from the core is transferred through a heat exchanger to a second water loop in which steam is produced. The steam, in turn, is used to run turbines to produce electricity.

This scheme is somewhat complicated by the fact that a chain reaction is not sustained by ordinary uranium. Only 0.7% of natural uranium is ^{235}U. We must process it to increase the concentration of this isotope to several percent of the total to produce reactor-grade uranium. As the reactor operates, the ^{235}U atoms are split and destroyed so that, in time, the fuel is so depleted in this isotope that it must be replaced with fresh fuel enriched in ^{235}U.

The Geology of Uranium Deposits

Worldwide, 95% of known uranium reserves are found in sedimentary or metasedimentary rocks. In the United States, the great majority of deposits are found in sandstone. They were

Figure 15.5

This uranium roll front deposit in the Rocky Mountains near Denver is located in sandstone, with layers of shale (not shown) above. The geologist is holding a Geiger counter to measure the ionizing radiation from the uranium oxide deposit.

©Gina S. Szablewski

formed by weathering of uranium source rocks, followed by uranium migration in and deposition by groundwater.

Minor amounts of uranium are present in many crustal rocks. Granitic rocks and carbonates may be particularly rich in uranium at concentrations in the range of ppm to tens of ppm. Uranium is concentrated in carbonate rocks during precipitation of the rocks from a solution. After those uranium-bearing carbonates are uplifted to the surface and subjected to weathering, that uranium goes back into solution readily because uranium is particularly soluble in an oxygen-rich environment. The uranium-bearing solutions then infiltrate and join the groundwater system. As they percolate through permeable rocks, such as sandstone, they may encounter chemically reducing conditions, created by some factor such as an abundance of carbon-rich organic matter or of sulfide minerals, in shales bounding the sandstone. Under these conditions, the solubility of uranium is much lower. The dissolved uranium is then precipitated and concentrated in these reducing zones (**figure 15.5**). Over time, as great quantities of uranium-bearing groundwater percolate slowly through such a zone, a large deposit of precipitated uranium ore may form.

World estimates of available uranium are somewhat difficult to obtain, partly because the strategic importance of uranium leads to some secrecy. The World Nuclear Association's most recent estimate for identified recoverable uranium resources at $59/pound ($130/kg) was 13.6 billion pounds (6.2 million tons). Even for the United States, the reserve estimates are strongly sensitive to price, as one would expect. **Table 15.1** summarizes present reserve and resource estimates at different price levels. U.S. nuclear power plant operators purchased the equivalent of about 49 million pounds of uranium in 2020 with the average cost at just under $36/pound. With the type of nuclear reactor currently in commercial operation in the United States, nuclear-generating capacity could not be increased

Table 15.1	U.S. Uranium Reserve Estimates
Recoverable Costs	**Reserves (million pounds of U_3O_8)**
$30/lb U_3O_8	31.2
$50/lb U_3O_8	206
$100/lb U_3O_8	389

Source: Data from Energy Information Administration, U.S. Department of Energy, 2020.

without increased dependence on foreign supplies or the distinct possibility of fuel shortages. The rare isotope uranium-235 is in such short supply that the United States could use up our reserves within a decade, assuming no improvements in reactor technology. Interest in nuclear power has been rising again as more emphasis is placed on using energy sources that don't increase greenhouse-gas emissions. Worldwide, new fission plants continue to be built. So, means of increasing the fuel supply for fission reactors are an issue.

Extending the Nuclear Fuel Supply

^{235}U is not the only possible fission-reactor fuel, although it is the most plentiful naturally occurring one. When an atom of the far more abundant ^{238}U absorbs a neutron, it is converted into plutonium-239 (^{239}U), which is in turn fissionable. ^{238}U makes up 99.3% of natural uranium and over 90% of reactor-grade enriched uranium. During the chain reaction inside the reactor, as freed neutrons move about, some are captured by ^{238}U atoms, making plutonium. Spent fuel can be *reprocessed* to extract this plutonium, which then is purified into fuel for future reactors, as well as to re-enrich the remaining uranium in ^{235}U. How reprocessing alters the nuclear fuel cycle is shown in **figure 15.6**. Fuel reprocessing with recovery of both plutonium and uranium could reduce the demand for new enriched uranium by an estimated 15%.

A **breeder reactor** can maximize the production of new fuel. Breeder reactors produce useful energy during operation, just as conventional burners using up ^{235}U do, by fission in a sustained chain reaction within the reactor core. In addition, they are designed so that surplus neutrons not required to sustain the chain reaction are used to produce more fissionable fuels from suitable materials, like ^{239}Pu from ^{238}U, or thorium-233, which is derived from the common isotope thorium-232. Breeder reactors could synthesize more nuclear fuel for future use than they would actively consume while generating power.

Breeder-reactor technology is more complex than that of conventional water-cooled burners. The core coolant is liquid metallic sodium; the reactor operates at much higher temperatures. The costs would be substantially higher than for burner reactors, perhaps close to $10 billion per reactor by the time any breeders could be completed in the United States. The breeding process is slow: the break-even point after which fuel produced would exceed fuel consumed might be several decades after

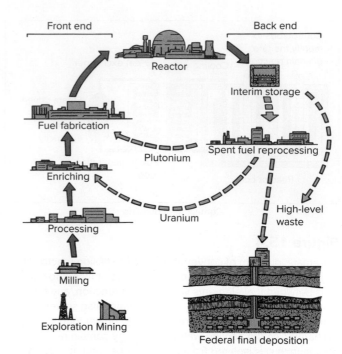

Fuel cycle as it operates currently

Fuel cycle as it would operate with spent fuel reprocessing and federal waste storage

Figure 15.6

The U.S. nuclear fuel cycle, as it currently operates and as it would function with fuel reprocessing.

Source: U.S. Energy Information Administration, Department of Energy.

initial operation. If the nuclear-fission option is to be pursued vigorously in the twenty-first century, the reprocessing of spent fuels and the use of breeder reactors are essential. Yet at present, reprocessing is minimal in the United States, and in early 1985, Congress canceled funding for the experimental Clinch River breeder reactor, in part because of high estimated costs. No commercial breeders currently exist in the United States, and only a handful of breeder reactors are operating in the world, though some—notably the Superphénix reactor in France—are functioning successfully. Even conventional reactors fell out of favor in the United States for some years, for several reasons.

Concerns Related to Nuclear Reactor Safety

A major concern regarding the use of fission power is reactor safety. In normal operation, nuclear power plants release very minor amounts of radiation, which are believed to be harmless. (A general discussion of radiation and its hazards is presented in chapter 16.) People are generally more worried about the small but finite risk of damage to nuclear reactors through accident or deliberate sabotage.

One of the most serious possibilities is a so-called loss-of-coolant event, in which the flow of cooling water to the

Figure 15.7

Three Mile Island near Harrisburg, Pennsylvania.

Doug Sherman/Geofile

reactor core would be interrupted. Resultant overheating of the core might lead to **core meltdown,** in which the fuel and core materials would deteriorate into a molten mass that might or might not melt its way out of the containment building and thus release high levels of radiation into the environment, depending upon the design of the reactor and containment building. A partial loss of coolant, with 35 to 45% meltdown, occurred in one of the two reactors at Three Mile Island in 1979 (**figure 15.7**). That reactor was never brought back online, and the remaining one was shut down in 2019. Multiple independent health studies have found no adverse health effects related to the small amount of radiation released during the accident.

No matter how far awry the operation of a commercial power plant might go, and even if there were a complete loss of coolant, the reactor could not explode like an atomic bomb. Bomb-grade fuels must be much more highly enriched in the fissionable isotope ^{235}U for the reaction to be that intensive and rapid. Also, the newest reactors have additional safety features designed to reduce the risk of accident. However, an ordinary explosion within the reactor could, if large enough, rupture both the containment building and reactor core, and thus release large amounts of radioactive material. The serious accidents at Chernobyl and, more recently, Fukushima, reinforced reservations about reactor safety in many people's minds; see Case Study 15.1.

Plant siting is another problem, as the perceived risk of nuclear plants is most often higher than actual risk. Siting them close to urban areas puts more people potentially at risk in case of accident; placing the plants far from population centers where energy is needed means more transmission loss of electricity. Proximity to water is important for cooling purposes but makes water pollution in case of mishap more likely. There are also concerns about the structural integrity of nuclear plants located close to fault zones—the Diablo Canyon station in California, the Humboldt Bay nuclear plant (decommissioned in 2021), and others; several proposed reactor sites have been rejected when investigation revealed nearby faults. This concern, too, was reinforced in the aftermath of the March 2011 earthquake in Japan.

Concerns Related to Fuel and Waste Handling

The mining and processing of uranium ore occurred in just two states, Wyoming and Utah, from 2018 to 2020, at a total of four locations and including 225 full-time person-years. Nevertheless, these locations pose hazards because of uranium's natural radioactivity. Miners exposed to the higher radiation levels in uranium mines have experienced higher occurrence rates of some types of cancer. Carelessly handled tailings from processing plants have exposed others to radiation hazards; recall the case of Grand Junction, Colorado, mentioned in chapter 13, where radioactive tailings were unknowingly mixed into concrete used in construction.

The use of reprocessing or breeder reactors to produce and recover plutonium to extend the supply of fissionable fuel poses special problems. Plutonium itself is both radioactive and chemically toxic. Many people are particularly concerned that, as a readily fissionable material, it can also be used to make nuclear weapons. Extensive handling, transport, and use of plutonium would pose a significant security problem and would therefore require very tight inventory control to prevent the material from falling into hostile hands.

The radioactive wastes from the production of fission power are another concern, even though the total physical volume created in the United States is relatively small at 83,000 metric tons. Radiation hazards and radioactive-waste disposal are considered within the broader general context of waste disposal in chapter 16. Here, two aspects of the problem are highlighted. First, radioactive materials cannot be treated by chemical reaction, heating, and so on to make them nonradioactive. In this respect, they differ from many toxic chemical wastes that can be broken down by appropriate treatment. Second, there has been sufficient indecision about the best method of radioactive-waste disposal and the appropriate site(s) for it that the great majority of the radioactive wastes generated around the world have yet to be placed in permanent disposal sites. The bulk of the wastes is in temporary storage while various disposal methods are being explored, and many of the temporary waste-holding sites are filled almost to capacity. Acceptable waste-disposal methods will need to be identified and adopted soon, if only to dispose of wastes already accumulated.

Nuclear plants have a volumetrically larger waste problem. The bombardment of the reactor core and structure by neutrons and other atomic debris from the fission process converts some of the structural materials to radioactive ones and changes the physical properties of others, weakening the structure. At some point, each plant will face **decommissioning:** it must be taken out of operation, broken down, and the most radioactive parts delivered to radioactive-waste disposal sites. In late 1982, an electricity-generating plant at Shippingport,

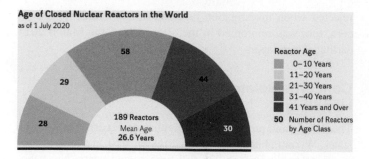

Figure 15.8

Age distribution of nuclear reactors worldwide, July 2020. Many reactors may face decommissioning in the next decade or two.

Sources: WNISR, with IAEA-PRIS, 2020

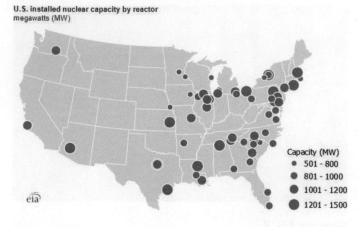

Figure 15.9

Distribution of U.S. nuclear power plants. At the end of 2021, there were ninety-three operating commercial reactors at fifty-five nuclear power plants. In addition, there were two new reactors under construction, and license applications for a few more projects were pending with the Nuclear Regulatory Commission, reflecting renewed interest in the use of nuclear power for generating electricity.

U.S. Energy Information Administration, https://www.eia.gov/todayinenergy/detail.php?id=43256

Pennsylvania, became the first commercial U.S. fission power plant to face decommissioning, after twenty-five years of operation. Costs of demolition and disposal typically approach $1 billion for a single power plant in the United States, and the process may take a decade or more. Dozens of reactors are being, or have been, decommissioned worldwide. A total of thirty-eight U.S. reactors, including those at Shippingport, had been retired by the close of April 2022. Historically, the nominal lifetime allowed by regulatory agencies was about thirty years; currently, it is forty to fifty years. With rigorous plant monitoring and replacement of some key components, it may be possible to operate a fission-powered electricity-generating plant for seventy to eighty years. Sooner or later, decommissioning will take its fiscal and waste-disposal toll. In the United States and worldwide, reactors are aging (**figure 15.8**).

Risk Assessment and Projection

Much of the debate about energy sources focuses on hazards: what kinds, how many, how large. Many of the risks associated with various energy sources may not be immediately obvious. No energy source is risk-free. What constitutes acceptable risk and who assesses it? Underground miners of both coal and uranium face increased health risks, but the power consumer doesn't see those risks directly. Those living downstream from a dam are at risk; so are those downwind from a nuclear or fossil-fuel plant.

A power plant of 1-billion-watt capacity serves the electricity needs of about 1 million people. One study in the mid-1970s projected accidental deaths at 0.2 per year for a uranium-fueled plant this size (virtually all associated with mining), compared to an estimated 2.6 to 4 deaths per year for a comparable coal-fired plant, depending upon mining method, including significant risks associated with processing and transporting the very much larger volume of coal involved. By way of comparison, in the United States in 2020, per million people, there were 124 deaths from motor-vehicle accidents, 265 from accidental poisoning, and 128 from falls.

A further consideration when comparing risks may be how well-defined the risks are. That is, estimates of risks from fires, automobile accidents, and so on are actuarial, based on considerable actual past experience and hard data. By contrast, projections of risks associated with extensive use of fission power are based on limited past experience, plus educated guesses about future events based, in part, on hard-to-quantify parameters like probability of human errors. In 1975, a report prepared for the U.S. Nuclear Regulatory Commission estimated the risk of core meltdown at one per million reactor-years of operation. To date, total global fission-reactor operations add up to just over 18,300 reactor-years, so statistically the accident rate is running far ahead of that estimate. Each such event leads to improvements in design and operation of reactors, but the history has given many people pause.

In the early 1970s, it was predicted that 25% of U.S. energy would be supplied by nuclear power by the year 2000. At the end of 2021, there were fifty-five nuclear power plants (**figure 15.9**) operating in the United States and they accounted for only 8.3% of energy production.

Nuclear plants do have lower fueling and operating costs than gas-fired plants. The small quantity of fuel required for their operation can also be more easily stockpiled against interruption by strikes or transportation delays. However, nuclear plants are more costly to build and take longer to plan, construct, license, and bring online. These economic and time factors, together with increasingly vocal public opposition to nuclear power plants (itself a contributing factor in many licensing delays), made coal, and then natural gas, more appealing to many utilities when considering what type of plant to build.

Worldwide, reliance on nuclear power varies widely. Some nations have no nuclear power plants. They account for only 2 to 5% of electricity production in such countries as Iran, India, China, and Brazil, but over 30% in much of Europe, and in some countries (Slovakia, Ukraine, and France), over 50% (**figure 15.10**). Some nations have turned to nuclear fission for lack of fossil-fuel resources. With the twenty-first century's increasing focus on limiting CO_2 emissions, the clean operation of nuclear fission plants has enhanced their appeal to many. As of April 2022, there were 448 operable nuclear fission plants in the world, and 52 more under construction, including 16 in China, 6 in India, 4 each in Republic of Korea and Russia, and 3 in Turkey.

Different people, and nations, weigh the pros and cons of nuclear fission power in different ways. For many, the uncertainties and potential risks, including the dangers of reactor accident, outweigh the benefits. For others, the problems associated with using fossil fuels appear at least as great. Immediately after the Fukushima disaster, Japan idled its nuclear plants, and the German government declared a commitment to shutting down all German reactors by 2022. A later Japanese government changed course, and energy-hungry Japan continues to operate fission reactors and even plan for more. Germany has rapidly expanded its use of renewable energy sources

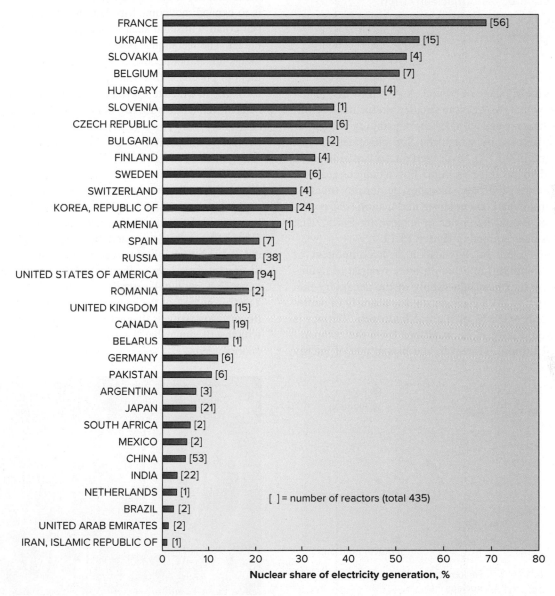

Figure 15.10

Percentage of electricity generated by nuclear fission varies greatly by country, and not simply in proportion to numbers of reactors; where electricity consumption is moderate, a few reactors can account for a large share, while conversely, where electricity demand is large, a great many reactors may meet only a small part of that demand. Above reactor total includes four reactors in Taiwan, which supply 12.7% of electricity there.

Source: Data for April 2022; from International Atomic Energy Agency.

generating 41% of their electricity, and the three remaining nuclear plants are on schedule to be shuttered by the end of 2022, even as some are calling to reverse the decision based upon continued use of coal plants, which will be phased out by 2030. In the balance of this chapter, we explore a variety of nonfossil alternatives to nuclear fission.

15.2 Nuclear Power—Fusion

Learning Objectives

- Describe the process of nuclear fusion
- Summarize how a tokamak will create nuclear fusion
- List the pros and cons of nuclear fusion energy

Nuclear *fusion* is the opposite of fission. As noted earlier, fusion is the process by which two or more smaller atomic nuclei combine to form a larger one, with an accompanying release of energy. It is the process by which the Sun generates its vast amounts of energy. In the Sun, hydrogen nuclei containing one proton are fused to produce helium. For technical reasons, fusion of the heavier hydrogen isotopes deuterium (nucleus containing one proton and one neutron) and tritium (one proton and two neutrons) would be easier to achieve on Earth. That fusion reaction is diagrammed in **figure 15.11**.

Hydrogen is plentiful because it is a component of water; the oceans contain, in effect, a huge reserve of hydrogen, an essentially inexhaustible supply of the necessary fuel for fusion. This is true even considering the scarcity of deuterium, which is only 0.015% of natural hydrogen. Tritium is rarer still and would have to be produced from the comparatively rare metal lithium. However, the magnitude of energy

release from fusion reactions is such that this is not a serious constraint. Helium is the principal product of the projected fusion reactions; it is a nontoxic, chemically inert, harmless gas, and one that is in short supply. There could be some mildly radioactive light-isotope by-products of fusion reactors, but they would be much less hazardous than many of the products of fission reactors.

Since fusion is a far cleaner form of nuclear power than fission, why not use it? The principal reason is technology, or perhaps the expense of the technology needed. To bring about a fusion reaction, the reacting nuclei must be brought very close together at extremely high temperatures and pressures, such as those that occur on the Sun. The natural tendency of hot gases is to expand, not come together, and no known physical material could withstand such temperatures to contain the reacting nuclei. The techniques being used in laboratory fusion experiments are elaborate and complex, primarily involving containment of plasma with strong magnetic fields in a device called a *tokamak,* in which temperatures exceeding those on the Sun can be achieved (**figure 15.12**). In February 2022, researchers at the Joint European Torus (JET) facility were able to sustain the necessary temperatures for about five seconds, creating 59 megajoules of energy.

Scientists have been working on fusion energy for decades spending billions of dollars, and so far haven't been able to generate a net surplus of energy. The ITER project is being funded by an international consortium that includes the United States. It is larger and more advanced than the JET project and is set for initial power production in 2025. The goal is to create more energy than is input. Fusion may seem very expensive, but it will provide a nearly endless source of energy. There is no risk of the equivalent fission meltdown. It is a stationary power plant that will create steam-turbine electricity, which will somewhat limit its contribution to satisfying total energy needs.

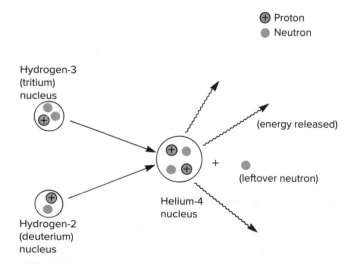

Figure 15.11

Schematic diagram of one nuclear fusion reaction. Other variants are possible.

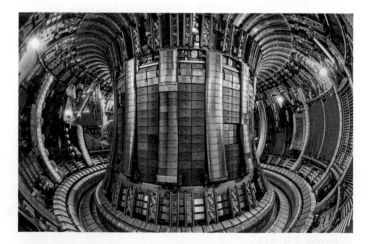

Figure 15.12

The Joint European Torus tokamak is located in Oxfordshire, United Kingdom. It has a major radius of 3 meters; the vacuum chamber is 2.5 meters wide and 4.2 meters high.

ZUMA Press, Inc./Alamy Stock Photo

15.3 Solar Energy

Earth intercepts only a small fraction of the energy radiated by the Sun. Even though much of the energy is reflected or dissipated in the atmosphere, that reaching the surface far exceeds the world's energy needs. The resource is effectively inexhaustible, which contrasts sharply with nonrenewable sources like uranium or fossil fuels. Sunlight falls on Earth without any mining, drilling, pumping, or disruption of the land. Sunshine is free; it is not under the control of any company or cartel, and it is not subject to embargo or other political disruption of supply. The *use* of solar energy is essentially pollution-free, at least in the sense that the absorption of sunlight for heat and the operation of a solar cell for electricity are very clean processes, though solar electricity has other environmental costs. It produces no hazardous solid wastes, air or water pollution, or noise. For most current applications, solar energy is used where it falls. All these features make solar energy an attractive option for the future. Several practical limitations on its use do exist, particularly concerning location, storage, and transmission.

What happens to the solar energy that reaches Earth? Some is reflected and some is absorbed by the atmosphere, land, oceans, and plants. The resultant heat drives ocean currents, winds, the hydrologic cycle, and photosynthesis. After all this, there is still enough to satisfy all human energy needs, but in a very disseminated form. Sunlight varies in intensity not only from region to region (**figure 15.13**) primarily by latitude but also from season to season and day to day as weather conditions change. The two areas in which solar energy can make the greatest immediate contribution are in space heating and in the generation of electricity—uses that together account for about two-thirds of U.S. energy consumption, and which will be discussed in this section.

Solar Heating

Solar space heating typically combines direct use of sunlight for warmth with some provision for collecting and storing additional heat to draw on when the Sun is not shining, and is employed on the scale of an individual building. The simplest approach is *passive-solar heating,* which doesn't require mechanical assistance. The building design should allow the maximum amount of light to stream in through south and west windows (in the Northern Hemisphere) during the cooler months. This heats the materials inside the house, including the structure itself, and the radiating heat warms indoor air. Mediums used specifically for storing heat include water—in barrels, tanks, and even indoor swimming pools—and the rock, brick, concrete, or other dense solids used in the building's construction (**figure 15.14A**). These supply a thermal mass that radiates heat back when needed, in times of less or no sunshine. Additional common features of passive-solar design (**figure 15.14B**) include broad eaves to block sunshine during hotter months and drapes or shutters to help insulate window areas during long winter nights. Variations in passive-solar heating system design are numerous.

Active-solar heating systems usually involve the mechanical circulation of solar-heated water (**figure 15.14C**). The flat solar collectors contain dark-colored tubing of various types that are filled with water and warmed by the Sun; some are contained within glass-covered boxes that decrease heat loss. The warmed water is circulated either directly into a storage tank or into a heat exchanger through which a tank of water is heated. The solar-heated water can provide both space heat and a hot-water supply. If a building already uses conventional hot-water heat, incorporating solar collectors is not necessarily expensive. With the solar collectors mounted on the roof, an active-solar system doesn't require the commitment of any additional land to the heating system. Transpired air systems work by pulling in air through warmed metal cladding, and then the warmed air is pulled into buildings.

While solar heating may be adequate by itself in mild, sunny climates, in areas subject to prolonged spells of cloudiness or extreme cold, a backup heating system is almost always needed, and this is often still one that relies on the consumption of fossil fuels. It has been estimated that, in the United States, 40 to 90% of most homes' heating requirements could be supplied by passive-solar heating systems, depending on location.

Solar Electricity

Direct production of electricity using sunlight is accomplished through **photovoltaic cells,** also called simply *solar cells* (see **figure 15.15**). In simplest form, they consist of two layers of semiconductor material sandwiched together, with a barrier between that allows electrons to flow predominantly in one direction only. Sunlight striking the exposed side dislodges some electrons, which flow as electric current through a circuit, to return to the cell and continue the cycle.

Solar cells have no moving parts and do not emit pollutants during operation. They are the principal power source for satellites and other remote applications and areas difficult to reach with power lines (**figure 15.16**). A major limitation on solar-cell use has historically been cost. As people have adopted the technology in the United States, the costs have dropped steeply in just a decade, decreasing to $37/MWh in 2021, making it the cheapest type of electricity. At the same time, technology has improved and the industry has expanded. Government

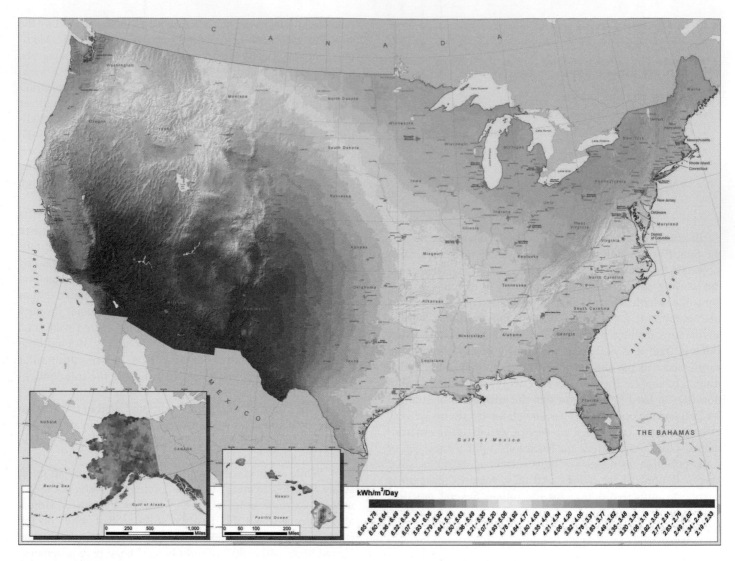

Figure 15.13

Solar-energy potential for electricity generation by photovoltaic (solar) cells in $kWh/m^2/day$. Not all areas are well suited for large-scale solar energy for the simple reason they don't receive enough sunshine.

Source: Map by B. Roberts, courtesy National Renewable Energy Laboratory, U.S. Department of Energy.

tax credits have helped. Total photovoltaic capacity at all scales more than tripled in the United States from 2015 to 2020, and it appears the expansion will continue (**figure 15.17**).

Low solar-cell efficiency and the diffuse character of sunlight continue to make photovoltaic conversion an inadequate option for energy-intensive applications, such as many industrial and manufacturing operations. Even in the areas of strongest sunlight in the United States, incident radiation is of the order of 250 watts per square meter. Operating with commercial solar cells of about 20% efficiency means power generation of only 50 watts per square meter. In other words, to keep one 100-watt lightbulb burning would require at least 2 square meters of collectors in constant sunlight. A 100-megawatt (MW) power plant would require 2 square kilometers of collectors (nearly 1

square mile), and many nuclear generating plants have more than ten times that capacity. Using solar cells at that scale represents a large commitment of both land and the mineral resources from which the collectors are made. The collector array would require tons of steel, glass, concrete, and materials such as gallium, arsenic, selenium, indium, and tellurium.

Siting a sizable array requires a substantial commitment and disturbance of land. Its presence could alter patterns of evaporation and surface runoff. These considerations could be especially critical in desert areas, which are the most favorable sites for such facilities. There are even concerns about how covering up the light-colored, reflective sand with sunlight-absorbing solar panels could increase Earth's net heating, exacerbating global warming.

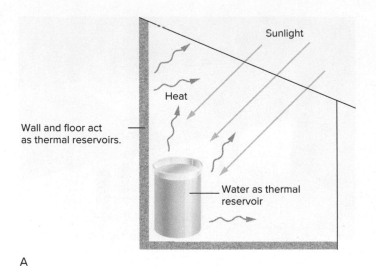

A

In summer, leaves and wide eaves together shade windows, reducing incoming sunlight.

Lower-angle winter sun streams through tree branches and under eaves.

B

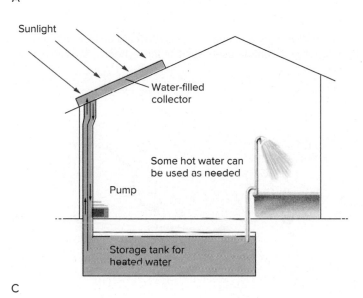

C

Figure 15.14

(A) Basics of passive-solar heating with water or structural materials as thermal reservoir: sunlight streams into greenhouse with glass roof and walls, heat is stored for nights and cloudy days. (B) Design features of home and landscaping can optimize use of Sun in colder weather, provide protection from it in summer. (C) A common type of active-solar heating system with a pump to circulate the water between the collector and the heat exchanger/storage tank.

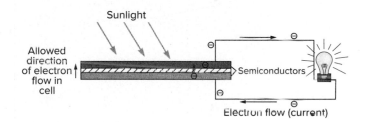

Figure 15.15

In a photovoltaic cell, incident sunlight dislodges electrons in top layer, which flow through wire as electric current, to return to the other side of the cell. Accumulated electrons move back to upper layer of cell and the cycle continues.

For large-scale applications, we concentrate solar to generate electricity. This is done with arrays of parabolic mirrors, which focus sunlight onto a small expanse of solar cells. Alternatively, the solar energy is concentrated with large mirror arrays to heat water or another medium to ultimately generate electricity with steam-driven turbines (**figure 15.18**). The mirrors are typically coated with silver or made from aluminum, for which we are, respectively, 80% and 100% dependent on imports.

Storing solar electricity can be accomplished thermally, electrochemically (batteries), and mechanically. Thermal storage occurs in mediums such as salt, as described above, but other salts that can reach higher temperatures and other materials like sand, are being tried in order to improve storage efficiency. Improvements in lithium-ion battery technology have carried over from vehicle and electronic applications, dropping costs and improving performance so quickly that batteries are now being used for utility-scale electricity applications rather than just for individual homes. The largest battery energy

storage system (associated with both solar and wind energy) is at Moss Landing in California, with a current capacity 400 MW to provide electricity to 300,000 homes for four hours. Despite some setbacks, an additional 350 MW in supersized battery storage is planned; large-scale battery storage systems are planned in other states and countries, too. Research continues, attempting to make batteries safer and hold longer charges, with some researchers recently showing that liquid isomers could store solar energy for eighteen years. For now, lithium-ion batteries used in home systems can hold a charge for one to five days, and excess electricity generated can usually be sold back to utilities if they are connected to the power grid.

Mechanically storing solar energy is accomplished through pumped-storage hydropower. Solar energy is used to

A

B

Figure 15.16

Solar electricity is very useful in remote areas. (A) The solar-powered Sofar spotter buoy deployed by the USGS will collect data on wind, waves, and water data off the coast of Florida. (B) Solar panels bring energy to remote users such as this school in the Himalayas.

(A) BJ Reynolds/USGS; (B) Ashley Cooper/Alamy Stock Photo

Figure 15.17

Utility-scale plants to come online March 2022–February 2023. In the short term, the number of new plants will be dominated by solar electricity.

U.S. Energy Information Administration, https://www.eia.gov/electricity/monthly/

pump water to a higher elevation; when the energy is needed, the water falls back down and generates hydropower (**figure 15.19**). Global expansion of this technology is increasing but slowly due to siting, licensing, and economics. Other mechanical storage systems employ flywheels or compressed air. Solar-generated electricity could also be used to produce renewable fuels such as hydrogen, which now is created primarily from fossil fuels.

About 3% of electricity generated in the United States in 2021 came from solar applications; the US Energy Information Administration (USEIA) predicts this will be 20 to 25% in 2050.

A

B

Figure 15.18

To make solar electricity without photovoltaic cells, solar heat can be used to make steam to power turbines, but it must first be concentrated. (A) Here in the California desert, parabolic mirrors focus sunlight on tubes of water. (B) In the Mojave Desert, a mirror array focuses on a tower in which molten salt is heated; the heat from the salt, in turn, heats water to make steam to run turbines.

(A) Doug Sherman/Geofile; (B) Glow Images

Worldwide, the improvements and cost reductions in solar technologies will allow the widespread expansion of solar power to economically disadvantaged regions, especially those with a lot of sunshine. The Desert to Power initiative plans to deploy solar power plants, water pumps for irrigation, mini-grid electrical systems, and home solar energy systems in the Sahel, helping 250 million people gain access to 10,000 MW of energy that is necessary for both economic and social development in the region.

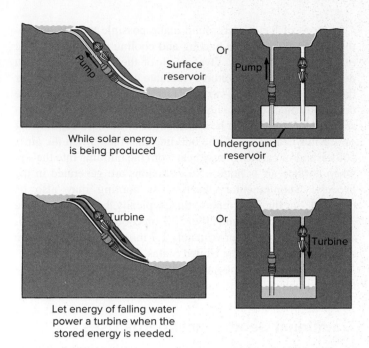

Figure 15.19

Solar energy is used to pump water to a high elevation. When the energy is needed, the water falls and turns a turbine to generate electricity.

15.4 Geothermal Energy

Learning Objectives

- Explain how a geoexchange system is used to heat or cool a home
- Describe how geothermal energy is used to generate electricity or provide heat
- List the limitations of geothermal power
- Describe the potential of hot-dry rock geothermal energy

Earth contains a great deal of heat, some of it left over from its early history, some continually generated by decay of radioactive elements. Slowly, this heat is radiating outward toward the surface. Under normal circumstances, the rate of heat escape is so slow that we don't even notice it and certainly cannot use it. If the heat escaping at the surface were collected over an average square meter for a year, it would be sufficient to heat about 2 gallons of water to the boiling point. Local heat flow can be substantially higher, for example, in young volcanic areas.

Seasonal variations in surface temperatures do not significantly affect this outward heat flow, and don't penetrate very deeply underground, because rocks and soils conduct heat poorly. The temperatures that tend to remain nearly constant for a few tens of feet underground make possible the use of geo-thermal pumps for both heating and cooling needs at the scale of individual buildings. Over much of the United States, at a depth of 20 feet, temperatures fluctuate only a few degrees over the year, and average 50° to 60°F (10° to 15°C). Pipes laid in the deep soil can circulate a liquid between this zone of fairly con-stant temperature and a heat exchanger at the surface, drawing on Earth's heat to warm the building in winter when the air is colder, and/or carrying away heat from the building into the soil when surface air is hotter. No emissions are generated in the process. Supplementary heating or cooling may also be required, but the EPA reports that, typically, homeowners save 30 to 70% on heating and 20 to 50% on cooling using such geo-exchange systems. Approximately 1.7 million such systems are currently installed in the United States.

Larger-scale applications of geothermal energy require more-specialized conditions; these are described below.

Traditional Geothermal Energy Uses

Magma rising into the crust from the mantle brings unusually hot material nearer the surface. Heat from the cooling magma heats any groundwater circulating nearby (**figure 15.20**). This is the basis for extracting **geothermal energy** on a commer-cial scale. The magma-warmed waters may escape at the sur-face in geysers and hot springs, signaling the existence of the shallow heat source below. More subtle evidence of hot rock at depth comes from sensitive measurements of heat flow at the surface: high heat flow signals unusually high

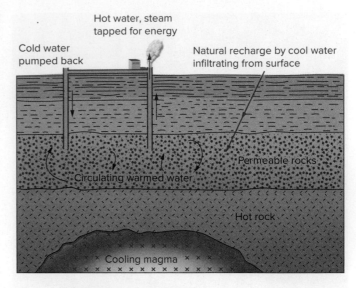

Figure 15.20

Geothermal energy is utilized by tapping circulating warmed groundwater.

temperatures at shallow depths. High heat flow and recent magmatic activity go together and are most often associated with plate boundaries or hot spots, so naturally, most areas in which geothermal energy is being tapped extensively are in these locations (**figure 15.21**).

Exactly how the geothermal energy is used depends largely on how hot the system is. In some places, the

Geothermal Plant Locations

Capacity Installed (Megawatts)

- ○ 0.0 - 13.5
- ◔ 13.6 - 40.2
- ◑ 40.3 - 97.0
- ◕ 97.1 - 303.0
- ● 303.1 - 720.0

Figure 15.21

Geothermal power plants worldwide.

Brett Ketter

Figure 15.22

The Geysers geothermal power complex, in California, is the largest such facility in the world.

U.S. Geological Survey/Photograph by Julie Donnelly-Nolan

Figure 15.23

Mammoth Terraces in Yellowstone National Park have been built from from calcium carbonate minerals dissolved in the park's geothermal waters, a process that continues today.

©Carla Montgomery

groundwater is warmed to temperatures that are comparable to what a home heating unit produces (50° to 90°C, or about 120° to 180°F). Such warm waters can be circulated directly through homes to heat them. Not only is hot groundwater used to heat the vast majority of Icelandic homes, but it is also used to warm sidewalks in urban areas and keep them free of ice and snow.

Other geothermal areas may be so hot that the water is turned to steam. The steam is used to run electric generators. The largest U.S. geothermal-electricity operation is The Geysers Geothermal Field in California (**figure 15.22**), which consists of 322 steam wells and 13 power plants. Millions of gallons of treated wastewater generated nearby is delivered to the site every day and used to replenish the warmed aquifer. Altogether, there are about fifty sites worldwide where geothermal electricity is actively being developed.

Where most feasible, geothermal power is competitive economically with conventional methods of generating electricity. The use of geothermal steam is also largely pollution-free. Some sulfur gases derived from the magmatic heat source may be mixed with the steam. Moreover, there are no ash, radioactive waste, or carbon-dioxide problems as with other fuels. Warm geothermal waters may be a somewhat larger problem. They frequently contain large quantities of dissolved minerals that not only build interesting structures as in **figure 15.23** but also can clog or corrode pipes or may pollute local groundwater or surface water if allowed to run off freely. Sometimes, there are surface subsidence problems, as at Wairakei, New Zealand, where subsidence of up to 0.4 meter per year has been measured; the natural geysers have also stopped erupting. Subsidence problems may be addressed by reinjecting water, but the long-term success of this strategy is unproven. The fluid injection may cause its own problems; studies at The Geysers indicate increased seismicity associated with fluid injection, though most of the earthquakes are below magnitude 4.

While the environmental difficulties associated with geothermal power are relatively small, three other limitations severely restrict its potential. First, each geothermal field can only be used for a few decades, on average, before the rate of heat extraction is seriously reduced. This is a negative consequence of the fact that rocks conduct heat very poorly. As hot water or steam is withdrawn from a geothermal field, it is replaced by cooler water that must be heated before use. Initially, the heating can be rapid, but in time, the permeable rocks become chilled to such an extent that water circulating through them heats too slowly or too little to be useful. The heat of the magma has not been exhausted, but its transmittal into the permeable rocks is slow. Some time must then elapse before the permeable rocks are sufficiently reheated to resume normal operations. Steam pressure at The Geysers has declined since it was first produced in 1960. Despite a capacity of 2 billion watts, by 1991 electricity production was only 1.5 billion watts; power generation had declined to half its 1987 peak within a decade, and by 2019 had decreased to about 750 million watts. There are plans to expand the system in order to bring the power closer to its demand.

A second limitation of geothermal power is that not only are geothermal power plants stationary, but so is the resource itself. Oil and gas can be moved to power-hungry population centers. Geothermal power plants must be put where the hot rocks are, and long-distance transmission of the power they generate is either technically impractical or inefficient. Most large cities are far removed from major geothermal resources.

The total number of sites suitable for geothermal power generation is the third limitation. Plate boundaries and hot spots cover only a small part of Earth's surface, and many of them, like seafloor spreading ridges, are inaccessible. Not all have abundant circulating subsurface water in the area, either. Even accessible regions that do have adequate subsurface water may not be exploited. Yellowstone National Park has the highest concentration of thermal features of any single geothermal area in

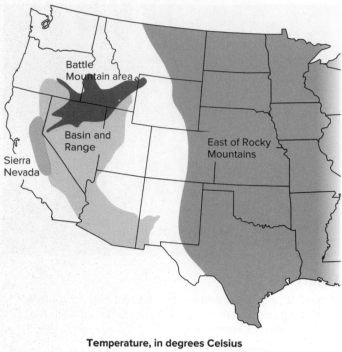

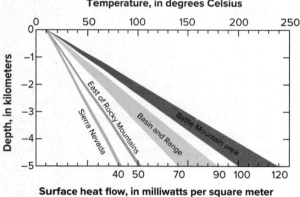

Temperature, in degrees Celsius

Surface heat flow, in milliwatts per square meter

Figure 15.24

The area east of the Rocky Mountains has a geothermal gradient and surface heat flow typical of world average continental crust; selected areas west of the Rockies have more energy potential. The faster temperature increases with depth, the closer to the surface are usefully warm rocks, and—all else being equal—the greater the geothermal-energy potential.

Source: USGS Circular 1249.

the world and estimates are that it has enough power for the entire country, but because of its scenic value and uniqueness, the decision was made years ago not to build geothermal power plants there. Geothermal plants are prohibited in all U.S. national parks. The Puna geothermal plant in Hawaii has continued to be controversial with native Hawaiians who oppose the development of energy associated with the volcano on spiritual and cultural terms, even as they rely heavily on imported oil.

Alternative Geothermal Sources

Many areas away from plate boundaries have heat flow somewhat above the normal level, and rocks in which temperatures

increase with depth more rapidly than in the average continental crust. The **geothermal gradient** is the rate of increase of temperature with increasing depth. Even in ordinary crust, the geothermal gradient is about 30°C/kilometer (about 85°F/mile). Where geothermal gradients are at least 40°C/kilometer, even in the absence of much subsurface water, the region can be regarded as a potential geothermal resource of the **hot-dry-rock** type. Deep drilling to reach usefully high temperatures must be combined with induced circulation of water pumped in from the surface, and perhaps artificial fracturing to increase permeability, to make use of these hot rocks. The amount of heat extractable from hot-dry-rock fields is estimated at more than ten times that of natural hot-water and steam geothermal fields, just because the former are much more extensive. Most of the regions identified as possible hot-dry-rock geothermal fields in the United States are in less populated western states with restricted water supplies (**figure 15.24**). There is a large degree of uncertainty about how much of an energy contribution they could ultimately make. Certainly, hot-dry-rock geothermal energy will be less economical than that of the hot-water or steam fields where circulating water is already present. Although some experimentation with hot-dry-rock fields is underway, commercial development in such areas is not expected in the immediate future.

The geopressurized natural gas zones described in chapter 14 represent a third possible type of geothermal resource. As hot, gas-filled fluids are extracted for the gas, the heat from the hot water might, hypothetically, also be used to generate electricity. Substantial technical problems probably will be associated with developing this resource. It also would likely be economically infeasible to pump the spent water back to such depths for recycling.

15.5 Hydropower

Learning Objectives

- Recall U.S. regional variations in hydropower
- List reasons hydropower won't expand in the United States
- Explain the conflict between the use of reservoirs for flood control, hydropower, and water supply
- Recall concerns with further development of hydropower globally

The energy of falling or flowing water has been used for centuries. It is now used primarily to generate electricity. Hydroelectric power has consistently supplied a small percentage of U.S. energy needs for several decades; it currently provides about 6% of U.S. electricity. The principal requirements for the generation of substantial amounts of hydroelectric power are a large volume of water and the rapid movement of that water. Commercial generation of hydropower typically involves damming

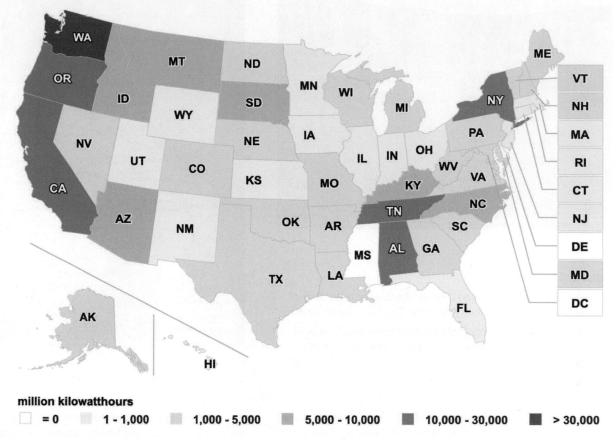

Hydroelectricity generation by state in 2021

million kilowatthours

☐ = 0 | ■ 1 - 1,000 | ■ 1,000 - 5,000 | ■ 5,000 - 10,000 | ■ 10,000 - 30,000 | ■ > 30,000

Figure 15.25

2021 hydroelectricity generation by state.

U.S. Energy Information Administration, https://www.eia.gov/energyexplained/hydropower/where-hydropower-is-generated.php

up a high-discharge stream, impounding a large volume of water, and releasing it as desired, rather than operating subject to great seasonal variations in discharge. The requirement of plentiful surface water is reflected in large regional variations in water use for hydropower generation (**figure 15.25**).

Hydropower is a very clean energy source in that the water is not polluted as it flows through the generating equipment. No chemicals are added to it, nor are any dissolved or airborne pollutants produced. The water itself is not consumed during power generation; it merely passes through the generating equipment. In fact, water use for hydropower in the United States is estimated to be more than 2½ times the average annual surface-water runoff of the nation, which is possible because the same water can pass through multiple hydropower dams along a stream. Hydropower is renewable as long as the streams continue to flow. Its economic competitiveness with other sources is demonstrated by the fact that over one-quarter of U.S. electricity-generating plants are hydropower plants; worldwide, about 16% of all electricity consumed is hydropower.

The Federal Power Commission has estimated that the potential energy to be derived from hydropower in the United States, if it were tapped in every possible location, is about triple current hydropower use. In principle, then, hydropower could supply one-fifth of U.S. electricity if consumption remained near present levels. However, development is unlikely to occur on such a scale.

Limitations on Hydropower Development

We have already considered, in chapters 6 and 11, some of the problems posed by dam construction, including silting-up of reservoirs, habitat destruction, water loss by evaporation, and even earthquakes. Evaluation of the risks of various energy sources must also consider the possibility of dam failure. There are over 92,000 dams in the United States of all sizes and purposes with an average age of 61 years; 2140 are for generating hydroelectric power. According to the National Inventory of Dams, over 16,000 dams are considered to be high hazard potential, meaning that their failure would result in loss of life and significant economic damages. Aside from age and poor design or construction, a dam might fail because it is located on an active fault zone, or one that might be reactivated by filling the reservoir. Not all otherwise-suitable sites are safe for hydropower dams.

A

Figure 15.26

(A) The Lower Granite Lock and Dam in Washington is just one of many dams on the Snake River and in the Columbia River Basin (B).

(A) USACE—Walla Walla District; (B) United States Army Corps of Engineers, https://www.nwd.usace.army.mil/Media/Fact-Sheets/Fact-Sheet-Article-View/Article/475820/columbia-river-basin-dams/

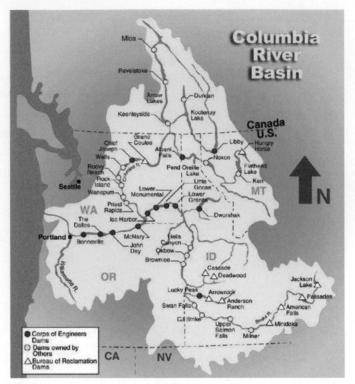

B

The late 1990s saw some reversals in U.S. hydropower development. For the first time, the Federal Energy Regulatory Commission ordered the dismantling of a dam used for hydropower generation, over the owner's objections: the Edwards Dam in Maine, the removal of which began in July of 1999. The move was a response to concerns over the cumulative environmental impact of 162 years of the dam's existence, and as it was a small facility (accounting for less than 1% of Maine's electric power), opposition to the dismantling was limited. Hundreds of dams in the United States have been dismantled over the past two decades, but most were not used for hydropower, and many were simply obsolete. More-vigorous objections are being made to the proposed removal of four hydropower dams on the Snake River in Washington State (**figure 15.26**). Concern for local salmon populations is among the reasons for the removal, but as these dams and reservoirs are more extensively used both for power generation and to supply irrigation water, their destruction would have more negative impact on local people than did the Edwards Dam removal. As of this writing, these dams' ultimate fate is uncertain.

Other sites with considerable power potential may not be available or appropriate for development. Construction might destroy a unique wildlife habitat or threaten an endangered species. It might deface a scenic or historic area or alter its natural character; suggestions for additional power dams along the Colorado River that would have involved backup of reservoir water into the Grand Canyon were met by vigorous protests. Many potential sites are just too remote from population centers to be practical unless power transmission efficiency is improved.

An alternative to development of many new hydropower sites would be to add hydroelectric-generating facilities to dams already in place for flood control, recreational purposes, and so on. Although the release of impounded water for power generation alters streamflow patterns, it is likely to have far less negative impact than either the original dam construction or the creation of new dam/reservoir complexes. Still, it is clear that flood control and power generation are somewhat conflicting aims. One requires considerable reserve storage capacity, while the other is enhanced by impounding the maximum volume of water.

Conventional hydropower is limited by the stationary nature of the resource and is more susceptible to natural disruptions than other sources considered so far. Torrential precipitation, 100-year floods, and prolonged droughts are rare, but they do happen. The ongoing western drought that started in 2000 has dropped water levels in Lake Powell, the reservoir behind Glen Canyon Dam (**figure 15.27**), to the lowest levels since the reservoir was filled. Water storage is at about 23% of its maximum reached in 1983. Hydropower generation has declined irregularly from a 1997 high of 6.7 billion kWh to 4 billion kWh in 2016. As of May 2022, an additional water level drop of about 30 feet will result in no electricity generation whatsoever. Emergency plans to hold back water that was set to be released, and to add water released from an upstream reservoir, will at least for the short-term allow the turbines to spin. It is not implausible, and probably more than likely, that further restrictions and emergency measures will need to be implemented to keep the electricity flowing to over 5.5 million customers in the

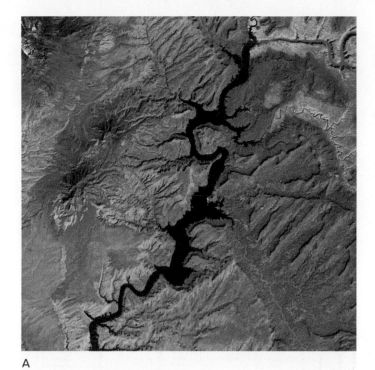

A

B

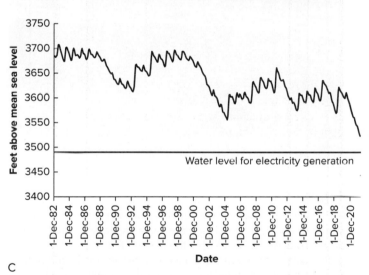

C

Figure 15.27

Lake Powell in 2001 (A) and 2021 (B). As of May 2022, the water level in Lake Powell was 3522' (C); the dam can only produce electricity if the water level is above 3490'. The water level dropped 39' between May 2020 and May 2021. Check out what it is now at the Bureau of Reclamation's site.

(A-B) NASA; (C) Source: Data for graph from https://www.usbr. gov/uc/water/crsp/cs/gcd.html

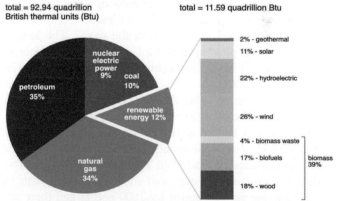

Figure 15.28

Hydropower has now fallen to the second most-used, single renewable energy source in the United States. Renewable sources still provide a relatively small proportion of energy used.

Annual Energy Review 2021, U.S. Energy Information Administration.

coming years. And difficult decisions about this reservoir as a source of water for millions of people will also need to be considered.

For various reasons, it is unlikely that numerous additional hydroelectric power plants will be developed in the United States. This clean, cheap, renewable energy source can continue indefinitely to make a modest contribution to energy use, but it cannot be expected to supply much more energy in the future than it does now. Still, hydropower is an important renewable energy source in the United States (**figure 15.28**).

Worldwide, future hydropower development is expected to be rapid in Asia and South America, but opposition exists there, too. The largest such project ever, the Three Gorges Dam project in China, went forward despite international protests from environmental groups. A more recent project in Tanzania is nearing completion after critics were told by their government they would be jailed for resisting the building of the dam. Supply disruptions are not unique to the United States. (See further discussion of dams in chapter 20.) Brazil is also dealing with drought problems. There, 61% of electricity is created with hydropower, so they are looking to use natural gas plants to make up for the drop in water flow energy.

15.6 Energy from the Oceans

Learning Objectives

- Describe the ways in which tides are used to make power
- Describe ocean thermal energy conversion
- Compare the benefits and drawbacks of extracting energy from the oceans

Three different approaches exist for extracting energy from the oceans: harnessing the energy of waves or tides, or making use of temperature differences between deep and shallow waters.

Tides represent a great volume of shifting water. In fact, all large bodies of standing water on Earth, including the oceans and large lakes like the Great Lakes, show tides. Why not also harness this moving water as an energy source? Unfortunately, the energy represented by tides is too dispersed in most places to be useful. Average beach tides reflect a difference between high-tide and low-tide water levels of about 1 meter. A commercial tidal-power electricity-generating plant using a barrage construction requires at least 5 meters of difference between high and low tides for efficient generation of electricity and a bay or inlet with a narrow opening that could be dammed to regulate the water flow in and out. The proper conditions exist in very few places in the world. The largest tidal barrage power plant is in South Korea and has a generating capacity of 254 MW. The second largest, and oldest, is in France (**figure 15.29A**).

Newer designs to generate tidal power involve turbines that are placed in the tidal flow (**figure 15.29B**) on or near the sea floor. They can be like wind turbines and placed individually in a row, or can be designed with horizontally-oriented barrel-shaped blades connected in a fence-like structure. The four tidal-stream turbines deployed in the Pentland Firth in Scotland generate enough power for 2600 homes and there are plans to install at least 50 more; similar projects are underway in Wales. Demonstration projects are ongoing in the northeastern United States, and in other countries including Canada, Japan, and China.

Tidal power shares the environmental benefits of conventional hydropower. Limitations include the cost to build turbines that must withstand strong forces in salty water, limited locations that have strong enough tides and are near enough to population centers, and concerns about altering sea bed

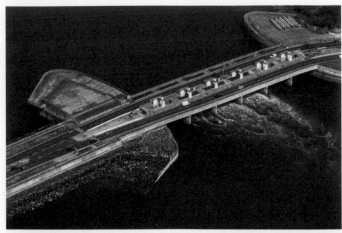

A

B

Figure 15.29

Tidal power can be acquired in different ways. (A) The La Rance, France, tidal barrage power plant has a capacity of 240 MW. It works by allowing water from the rising tide to flow through sluice gates, which are then closed; water returns on the outgoing tide by passing through turbines and thereby generating electricity. (B) Tidal turbines act much in the same way as those used for wind power, except the fluid here is water.

(A) Francois BOIZOT/Shutterstock; (B) Alex Mit/Shutterstock

ecosystems and disturbing marine creatures with turbine noise and possible collisions.

Ocean thermal energy conversion (OTEC) is another clean, renewable technology that is not yet widely deployed. It exploits the temperature difference between warm surface water and the cold water at depth. Either the warm water is vaporized and used directly to run a turbine, or its heat is used to vaporize a working fluid to do so; the vapor is recondensed by chilling with the cold water. No fuel is burned, no emissions released, and the vast scale of the oceans ensures a long

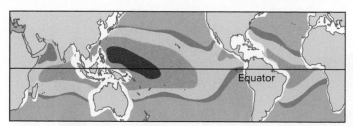

Temperature difference between surface and depth of 1000 m

- Less than 18°C
- 18° to 20°C
- 20° to 22°C
- 22° to 24°C
- More than 24°C
- Depth less than 1000 m

Figure 15.30

Given the thermal and other requirements of OTEC, tropical islands are likely to be the first sites for its development.

After National Renewable Energy Laboratory, U.S. Department of Energy.

life for such facilities. Where the water is vaporized, the vapor is pure water, which, when recondensed, can be used for water supply—this is essentially a distillation process. The cold seawater can be used for other purposes, such as air conditioning or aquaculture.

Specialized conditions are needed for OTEC to be a productive energy source. The deep cold water must be accessible near shore; coastlines with broad shelves are unsuitable. The temperature difference between warm and cold seawater must be at least 40°F (22°C) year-round, which is true only near the equator (**figure 15.30**) making this technology most suitable for select tropical islands. And in such settings, there is concern about possible damage to reefs and coastal ecosystems from the pipes and other equipment. In 2015, Hawaii became the first state to have an operational OTEC facility. Facilities in other countries are planned or in development, mostly to provide electricity and desalinized water for islanders.

Over much of the oceans, waves ripple incessantly over the surface. The up-and-down motion of the water can be harnessed in various ways to generate electricity. The bobbing water can drive a pump to push water through a turbine, for example; or, in an enclosed chamber that is partially submerged, the rise and fall of the water surface can produce pulses of compression in the air above, which can drive an air-powered turbine. These systems too are clean and renewable, and have the potential to generate electricity for millions of homes, even if we used just a small portion of total resources available. The U.S. has just one active wave energy project in Hawaii, and the Department of Energy has just given $25 million for eight pre-permitted projects to be deployed off the coast of Oregon by 2023. Europe has 17 current projects with a goal of deploying 100 MW of wave and tidal energy by 2025. There has only been one small operational wave-energy system that is now defunct and many other never got off the ground over the past 30 years, but the renewed interest means that wave energy will likely be a

small and necessary part of our energy portfolio in the near future. Concerns with these systems include the visual impact of the equipment on coastal areas, and possible disruption of natural sediment-transport patterns.

15.7 Wind Energy

Learning Objectives

- Recall the windiest locations in the United States
- List the benefits and drawbacks of wind-generated electricity
- Describe the recent trends in wind power
- Explain why growth in wind power will likely be concentrated offshore

As a variant of solar energy, wind energy is clean and renewable. People have utilized wind power to some extent for more than 2000 years; the windmills of the Netherlands are probably the best-known historic example. Recently, the generation of electricity from wind power has expanded rapidly and there is considerable interest in making more extensive use of wind power.

Wind energy shares certain limitations with solar energy. It is dispersed, not only in two dimensions but in three. Wind is also erratic, highly variable in speed both regionally and locally. The regional variations in potential power supply are even more significant than they may appear from average wind velocities because windmill power generation increases as the cube of wind speed. So if wind velocity doubles, power output increases by a factor of 8.

Figure 15.31 shows that most of the windiest places on land in the United States are rather far removed physically from most of the heavily populated and heavily industrialized areas. As with other physically localized power sources, the technological difficulty of long-distance transmission of electricity will limit wind power's near-term contribution in land-locked areas of high electricity consumption until transmission efficiencies improve.

Even where average wind velocities are great, strong winds do not always blow. This presents the same storage problem as does solar electricity. At present, wind-generated electricity is used most commonly in conjunction with conventionally generated power when wind conditions are favorable. Design improvements have steadily increased the reliability of wind generators; the newest wind turbines are available for some power generation at least 95% of the time.

Winds blow both more strongly and more consistently at high altitudes. Tall wind turbines, perhaps a mile high, may provide more energy with less of a storage problem. However, such tall structures are technically more difficult to build and require substantial quantities of material. They may also be deemed more objectionable visually.

Blowing wind represents far more total energy than we can use, but most of it cannot be harnessed. Most commercial

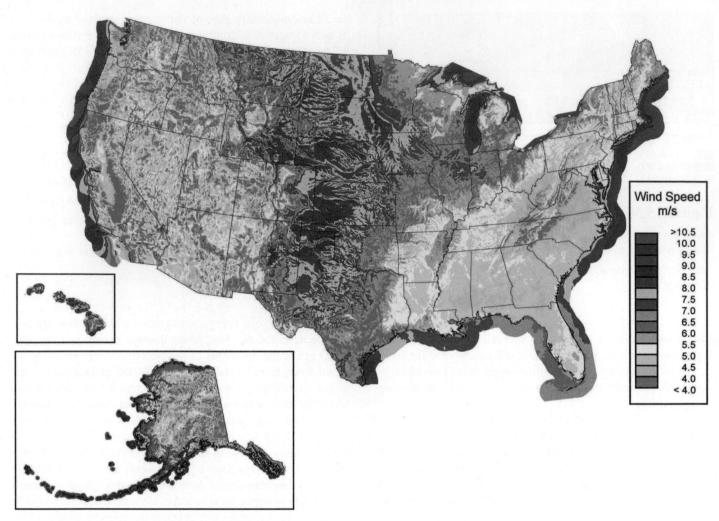

Figure 15.31

Average annual wind speed 80 meters (about 250 ft) above the surface, a common height for wind turbines. Topographic irregularities, including tall vegetation, disrupt wind flow, so the strongest consistent winds are found in the Central Plains and offshore.

Source: National Renewable Energy Laboratory, U.S. Department of Energy.

electricity wind-power generation involves wind farms, concentrations of many generators in a few especially favorable windy sites (**figure 15.32**). The limits to wind-power use include the area that can be committed to wind-generator arrays, as well as the distance the electricity can be transmitted without excessive loss in the power grid. About 500 2-MW wind generators are required to generate as much power as a sizeable conventional fossil-fuel or nuclear-powered electricity-generating plant. The units have to be spread out, or they block each other's wind flow. Spacing them at four windmills per square kilometer requires about 125 square kilometers (48 square miles) of land to produce energy equivalent to a 1000-MW power plant. The land below the turbines can be used for other purposes such as agriculture, grazing, and even for solar panel installation.

In recent years, interest in generating wind power using offshore turbines has grown. This is not only because these locations are windy, but are also are more accessible to large cities located at or near coastlines, as is evident from **figure 15.31**. There are both technical issues and questions of jurisdiction to be addressed; a utility cannot own or lease an area of the ocean as it can a plot of land, and even offshore, a wind-turbine array covers a substantial area. Aesthetic and practical objections to the presence of offshore turbine arrays exist also. Still, the fact that transmission losses of wind-generated electricity could be greatly reduced through offshore generation makes this a potentially attractive option. In Europe, several offshore arrays have already been built in the North Sea. Recent additions of offshore capacity were concentrated almost entirely in China and Europe. In 2017, Rhode Island became the first state to have a commercial-scale offshore wind-turbine array. Another project is under construction off the coast of Virginia Beach and will be fully operational in 2026 with 8.8 million MW, enough to power over 650,000 homes.

A

B

Figure 15.32

(A) A wind turbine farm along the Texas Gulf coast is also used for agriculture. (B) A drone is used to inspect a wind turbine. Wind energy projects in 2020 employed nearly 117,000 full-time workers in the U.S. Land lease payments to individuals and tax revenues to communities are other benefits.

(A) Roschetzky Photography/Shutterstock (B) TimSiegert-batcam/Shutterstock

Other concerns relating to wind farms, on land or offshore, include the aesthetic impact; interference with and deaths of migrating birds and bats; the noise associated with a large number of windmills operating; electromagnetic interference that may disrupt communications systems; and disruptions to animals and ecosystems during construction and ongoing operation of the turbines. While the number of estimated birds killed by wind turbines each year is tiny compared to deaths by window strikes and cats, the placement of new installations are considering migration routes and employing new technologies and design to decrease mortalities. Used wind turbines do have a limited lifetime of twenty-five to thirty years, so they can pose a waste disposal problem. Besides being landfilled, people are finding unique ways to use them in various types of structures.

Soon they will also be shredded and then used as a raw material in cement manufacturing. The very newest blades made of thermoplastic resin will be fully recyclable.

Addition of wind capacity in the United States during the 2000s varied from 1 to over 14 gigawatts (GW) yearly, reaching the highest amounts in 2012 and 2020 in response to the upcoming phaseout of tax credits; total capacity is now over 118 GW. The USEIA expects another 14 GW to be added by the end of 2023. Texas generates the most wind energy, and Iowa generates the largest percent of its energy (58%) from wind. As with solar, technology improvements and increased production have made wind the second cheapest source of electricity in the United States. In 2021, wind-powered electricity-generating facilities in the United States accounted for about 9.2% of U.S. electricity generation; on March 29, 2022, wind-generated electricity reached an all-time daily high of 19%. Wind-power electricity generation has been rising sharply in the United States in recent years (**figure 15.33A**), and with the federal push to generate 30 GW of offshore wind energy by 2030, this trend will most likely continue.

Globally, electricity generated from wind grew 11% from 2019 to 2020. China had the largest increases. The International Energy Agency estimates that global growth will need to be 18% every year from 2020 to 2030 in order to reach the goal for wind's part in reaching net zero carbon output by that time. Most of the increased capacity would need to come from offshore locations, which still have a number of technical and political hurdles to overcome before more turbines can be installed.

15.8 Biofuels

Learning Objectives

- Compare biomass to biofuels
- Explain the problems in describing wood as a carbon-neutral fuel
- Describe the source and use of biogas
- Compare past and current reasons for using/not using ethanol as a gasoline substitute

The term *biomass* technically refers to the total mass of all the organisms living on Earth. In an energy context, the term **biofuels** has become a catchall for various ways of deriving energy from biomass, from organisms, or from their remains. Biomass-derived energy is ultimately solar energy, since most biofuels come from plant materials, and plants need sunlight to grow. We can also think of biofuels as unfossilized fuels because they represent fuels derived from living or recent organisms rather than ancient ones; they have not been modified extensively by geologic processes acting over long periods of time.

There are many types of biofuels, and the terms we use to describe them can overlap a bit. Most of these fuels fall into

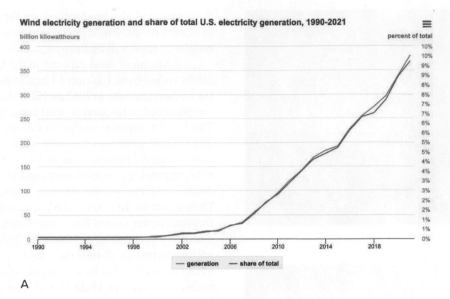

Wind electricity generation and share of total U.S. electricity generation, 1990-2021

billion kilowatthours / percent of total

— generation   — share of total

A

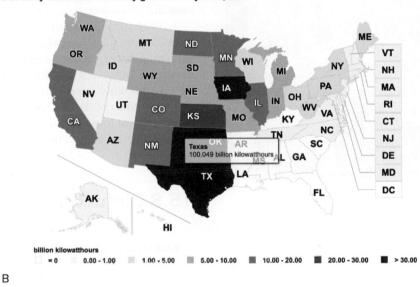

U.S. utility-scale wind electricity generation by state, 2021

Texas
100.049 billion kilowatthours

billion kilowatthours
□ = 0   0.00 - 1.00   1.00 - 5.00   5.00 - 10.00   10.00 - 20.00   20.00 - 30.00   ■ > 30.00

B

Figure 15.33

(A) Wind electricity generation in billion kilowatt-hours and share of total U.S. electricity generation. (B) U.S. utility-scale wind electricity generation by state, 2021. Visit the U.S. Wind Turbine Database for more detailed information about wind turbine placement and capacities.

(A) U.S. Energy Information Administration, https://www.eia.gov/energyexplained/wind/electricity-generation-from-wind.php; (B) U.S. Energy Information Administration, https://www.eia.gov/energyexplained/wind/where-wind-power-is-harnessed.php

three broad categories that we will cover briefly in this section: wood, waste, and alcohol fuels. Some are used alone, and others are burned in combination with conventional fuels (co-fired or co-generation). All biofuels are burned to release their energy, so they share the carbon-dioxide-pollution problems of fossil fuels. Some, like wood, also contribute to particulate air pollutants. Their appeal is that they are renewable, we can produce them domestically and even locally, and they are often a leftover waste product from another industry. They can be used to heat a home, a business, a college; to power a vehicle; or to generate electricity for a utility.

Wood

More than 2 billion people, and many of these on the African continent, still use wood as their primary source of heating and cooking fuel. In the United States, over 12 million households use wood for some type of space heating. The largest U.S. consumer of wood is the industrial sector, particularly manufacturers of paper and wood products who use wood waste to produce steam and electricity (**figure 15.34**). The most serious environmental concern related to wood fuel is deforestation and its role in climate change. The U.S. Environmental Protection Agency

Figure 15.34

This power plant in Anderson, California, uses chipped forest debris as a fuel.

Warren Gretz/National Renewable Energy Laboratory

and the European Union, along with other countries and governmental agencies, have declared that wood is a carbon-neutral fuel. While the argument that burning trees for fuel results in a net zero addition of carbon to the atmosphere is technically true, many scientists are concerned about the rate at which that interchange is occurring, and equate the burning of wood mass for energy on a large scale to that of burning coal.

The U.S. exported nearly 7 million tons of wood pellets in 2021, most of the material coming from forest lands in the Southeast. There are over 20 facilities in this region that process wood pellets that are then exported to Europe. A lot of these pellets are going to the United Kingdom, the world's largest consumer, where they are being used to power former coal plants to generate electricity. Japan and South Korea are also starting to import wood pellets for the same purpose. Given that researchers estimate it can take at least forty years for the carbon emitted from burning trees to work itself through the carbon cycle, we may want to reconsider using trees on a large-scale to generate power.

Waste-Derived Fuels

Biofuels include agricultural or other wastes that would historically have been burned in a field or dumped in a landfill. They also include the combustible portion of urban refuse which can be burned to provide heat for electricity-generating plants. About 25 million tons of municipal solid waste was burned in 65 power plants in the United States in 2020, and over 60% of the waste was biomass that created 45% of the electricity generated. More extensive use of incineration as a means of solid-waste disposal, both in the United States and abroad, would further increase the use of biomass for energy worldwide. Countries like Japan burn a lot of their waste because they don't have the landfill space, and the power generated is a benefit.

Some waste-derived fuels are liquids. Research is ongoing on ways to derive inoffensive liquid fuels from animal manures, which are rich in organic matter. Vehicles have been designed to run on used vegetable oil from food-frying operations. Some diesel-powered vehicles can run on *biodiesel,* a general term for fuels derived from vegetable oil or animal fats, or a blend of such oil with petroleum diesel fuel. The alcohol fuels discussed below are now mainly produced from corn, but processes to derive alcohol from plant wastes rather than potential food are being developed.

Another waste-derived biomass fuel growing in use is *biogas,* also called landfill gas. When broken down in the absence of oxygen, organic wastes yield a variety of gaseous products. Some of these are useless, or smelly, or toxic, but among them is methane (CH_4), the same compound that predominates in natural gas. Sanitary-landfill operations, described in chapter 16, are suitable sites for methane production as organic wastes in the refuse decay. Landfill gas can be burned directly or treated to remove CO_2 and other gases, or blended with purer, pipelined natural gas to extend the gas supply. Methane can also be produced from decaying manures. Where large quantities of animal waste are available, biogas is created by microbial digestion in enclosed tanks, reducing waste-disposal and water pollution problems while generating useful fuel (**figure 15.35**). Large sewage districts use methane gas created through their treatment process (which includes anaerobic digestion of organics) to supply a significant portion of the electricity to run their operations or heat buildings. Excess biogas can also be purified and sold back to local utilities.

Alcohol Fuels

One biofuel that has received special attention is alcohol. Initially, it was extensively developed mainly for incorporation into *gasohol.* Gasohol as originally created was a blend of 90% gasoline and 10% alcohol, the proportions reflecting a mix on which conventional gasoline engines could run. Gasohol came into vogue following the OPEC oil embargo of the 1970s. Its popularity waned when oil prices subsequently declined, though most gasoline still contains some *ethanol* (the most common type of fuel alcohol).

The higher the proportion of alcohol in the mix, the further the gasoline can be stretched. There is, in principle, no reason why engines cannot be designed to run on 20%, 50%, or 100% alcohol, and, in fact, vehicles have been made to run solely on alcohol. Recently, increasing numbers of vehicles have been designed to run on *E85,* a blend of 85% ethanol and 15% gasoline. Such vehicles, most common in the Midwest and South, are now the majority of alcohol-fueled vehicles. Some can run either on E85 or on regular gasoline (flex fuel vehicles), though they must be specifically designed to do so.

Government policy can encourage expanded use of alcohol and alcohol/gasoline blends. The Clean Air Act

A

B

Figure 15.35

Anaerobic digesters are used to turn organic waste into usable natural gas. (A) Natural gas made from manure can be used on the farm or sold back to a utility. The process also creates a solid material commonly used as animal bedding. (B) Digesters are used on a larger scale in the sewage treatment process.

(A) Rudmer Zwerver/Shutterstock; (B) Teresatky/Shutterstock

Amendments of 1990 included provisions for reduced vehicle emissions, and alcohol is somewhat cleaner-burning. The Energy Policy Act of 1992 required gradual replacement of some government, utility-company, and other fleet vehicles with alternative-fuel vehicles, which would include vehicles powered by straight alcohol or E85. The same act extended an excise-tax break on ethanol-blend fuels. There are about 22 million flex-fuel vehicles on the road but this number is still small compared to the 253 million light-duty vehicles. Ethanol consumption appears to have plateaued; the highest consumption was in 2019 at 14.55 billion gallons, and was a bit lower in 2021 (**figure 15.36**). Global trends are similar to those in the United

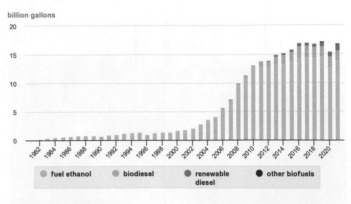

Figure 15.36

U.S. biofuel consumption by type, 1981 to 2021. Renewable diesel is made from the same product as biodiesel but through a different process.

Source: U.S. Energy Information Administration, Monthly Energy Review, March 2022, preliminary data for 2021

States. The USEIA suggests that global demand will grow by over 25% by 2026 but that stronger policies and improved technologies are needed to bring cost down that will stimulate demand.

Substantial concerns about ethanol fuels were raised as world production rose sharply in the early 2000s. It takes energy to grow, harvest, and derive ethanol from corn or other starting plant material. Depending on the process, it can take *more* energy than is released burning the ethanol. Alcohol fuels weren't so much a new source that would add to our energy reserves but rather a way to reduce dependence on imported oil. With the United States currently energy independent and climate concerns looming in the forefront, any biofuel has the same concerns as discussed above with wood: that their production and use releases more carbon that is taken up when they are growing.

The other concern about biofuels is the diversion of food crops to create fuel rather than to feed people and animals. Not only does this result in less food but also in higher prices for that food. In the United States, the push to grow more corn for ethanol production created pressure to grow crops on marginal cropland previously set aside for erosion-control purposes under the Conservation Reserve Program, discussed in chapter 12. And studies have suggested that, to the extent that more forests and grasslands are cleared to provide additional cropland for biofuel production, that land-use change will increase net greenhouse-gas emissions by decreasing carbon sinks (see chapter 18). Research is underway to develop efficient means to produce ethanol from cellulosic biomass such as native prairie grasses or fast-growing trees, but any process involving growing crops specifically for biofuel production will involve some of the foregoing issues as well. Over the long term, concentrating on producing ethanol from waste organic matter may be the best option.

A Tale of Two Disasters: Chernobyl and Fukushima

On 26 April 1986, an accident occurred at the nuclear power plant at Chernobyl, Ukraine, reinforcing many fears about the safety of the nuclear industry. It resulted from a combination of design deficiencies, equipment problems, and human error, as plant operators conducted an experiment to test system response to unusual conditions. The results were disastrous. A power surge caused a steam explosion; the reactor vessel ruptured causing more steam eruptions that destroyed the reactor core and damaged the building it was in. A graphite fire burned for ten days, releasing radioactive materials that drifted with atmospheric circulation over Scandinavia and eastern Europe.

According to the United Nations report, 31 people directly because of the accident; 134 of 600 people at the site on the day of the accident got acute radiation sickness; and 115,000 people were immediately evacuated. The republics most severely affected were Ukraine and Belarus. Radioactive fallout settled on the surrounding land (**figure CS 15.1.1**).

The clearest health effect related to Chernobyl was an increased rate of thyroid cancer in children, particularly those who drank milk with high levels of radioactive iodine. Some of the long-range health effects of the accident may never be known because the effects of radiation are not always distinguishable from the effects of other agents (see chapter 16), although there was a slight increase in leukemia to emergency and recovery workers (some 200,000 people). The World Health Organization estimated a total of 4000 deaths attributable to the accident over the lifetimes of emergency workers and local residents in the most contaminated areas; other estimates were as high as 30,000. Other long-term exposure concerns for people were from radionuclides in the soil that would get incorporated into plants and animals. Within twenty to thirty kilometers of the site, there was increased mortality of plants and animals due to acute exposure. A total of 350,000 people were permanently relocated.

Can a Chernobyl-style accident happen at a commercial nuclear reactor in the United States? Basically, no, because the design of these reactors is fundamentally different from the design of the Chernobyl reactor. In any reactor, some material must serve as a *moderator*, to slow the neutrons streaming through the core enough that they can interact with nuclei to sustain the chain reaction. In the Chernobyl reactor, the moderator was graphite. In commercial U.S. reactors, the moderator is water. Should the core of a water-moderated reactor overheat, the water would vaporize. Steam is a poor moderator, so the rate of chain reaction would slow down. However, graphite remains solid up to extremely high temperatures, and hot graphite and cold graphite are both effective moderators. In a graphite reactor, therefore, a runaway

Figure CS 15.1.1

Distribution of radioactive fallout from Chernobyl can be estimated from surface-ground deposition of cesium-137. The star marks the location of the Chernobyl site.

Source: Data from Mark A. Fischetti, "The Puzzle of Chernobyl" in IEEE Spectrum 23, July 1986, pp. 38–39.

chain reaction is not stopped by changes in the moderator. At Chernobyl, as the graphite continued to overheat, it began to burn, turning the reactor core into the equivalent of a giant block of radioactive charcoal. (To put it out, workers eventually had to smother it in concrete.) The fuel rods in the burning core ruptured and melted from the extreme heat. The core meltdown, together with the explosion, led to the release of high levels of radioactive materials into the environment. The extent of that release was also far greater for the Chernobyl reactor than it would have been for a U.S. reactor suffering a similar accident because the containment structures of U.S. reactors are much more substantial and effective. The accident focused international attention sharply on the issue of reactor safety and also on the reactor-operator interface and operator training. The incident also emphasized that, like air pollution, radioactive fallout ignores international boundaries.

Japan, with a highly technological society and no fossil-fuel resources, has long embraced nuclear power. The Fukushima accident has caused considerable lessening of that enthusiasm.

(Continued)

As noted in chapter 4, the Fukushima Dai-Ishi plant was designed with both earthquakes and tsunamis in mind. When the 11 March 2011 quake hit, all six reactors were shut down within seconds by automated control-rod insertion; when external power supplies were cut off, a backup generating system to circulate cooling water to the still-hot reactor cores was promptly activated. A tsunami warning did not raise undue concern, as the predicted height was 3 meters and the plant was more than 10 meters above sea level. Unfortunately, the actual tsunami was far higher, estimated later at 14 meters or more. It overwhelmed the protective seawall, surged into the plant, and flooded and knocked out eleven of twelve backup generators. Without power, operators had no functioning instruments to monitor the reactors. The company had additional power supplies on trucks that could be brought in in case of emergency—but the scale of this natural disaster was so great that the trucks could not get through the flood of survivors trying to flee the area on badly damaged roads. Plant operators cobbled together car batteries to power their instruments, but the data they obtained were incomplete and, in retrospect, apparently not entirely accurate. Cooling water was boiling to steam and pressure was building inside several reactor buildings; reactor 1 was in the worst condition, with core meltdown beginning. Fire trucks were brought in to pump first fresh water, then (when the fresh ran out) salt water, to try to cool the reactors' cores. Local residents were evacuated to allow workers to vent some of the accumulated (now-radioactive) gas from inside containment vessels to reduce pressure. Plant workers struggled to get control of the situation and bring all reactors to stable shutdown. But on the afternoon of 13 March, a stray spark ignited hydrogen gas that had formed from the hot steam inside reactor 1; the explosion released considerable radiation, and made things both harder and more dangerous for workers by littering the whole complex with rubble and radioactive material. Explosions eventually occurred in units 2, 3, and 4, and meltdowns in units 2 and 3, as well.

Given that Japan is an island nation, and given the prevailing winds and currents, there was no major radiation threat to other countries. The World Health Organization has also confirmed that radiation in fallout and runoff to the Pacific Ocean would have been diluted quickly enough not to contaminate the region's seafood supplies. Over 100,000 people were evacuated in connection with the Fukushima accident. More than eleven years later, people were allowed to return to the last of the evacuated areas, although about 60% say they are not interested in doing so. No incidents of radiation sickness or casualties were recorded for the accident. As with Chernobyl, there remains a risk of radiation-related health problems, particularly among the 160 plant workers and emergency personnel who were exposed to the highest radiation levels, and the full extent of these problems will not be known for decades.

Electricity's Hidden Energy Costs

Have you ever looked closely at your electric bill? Many utilities now provide information on the sources of their power, and the associated pollution (**figure CS 15.2.1**). Such considerations are very much part of current discussions about energy sources. At one time, the main concern was availability of the fossil fuels. Now global focus has shifted to the emission of greenhouse gases like CO_2, and the relative pollutant outputs of different energy sources are a significant part of the debate. With nuclear power, the issue of waste disposal is also a concern.

U.S. households have become more electrified, and that trend will continue as we acquire more power from wind and solar. By 2020, 99.5% of U.S. households had a refrigerator, and 34% had two or more; 89% had a conventional range, the majority of these electric; 83% reported having a clothes dryer, and 73% a dishwasher. From 1978 to 2020, the proportion of households having a microwave oven went from 8% to 96%. Forty percent of U.S. households reported using electricity as their primary source of heat, 88% had some sort of air conditioning, and 66% had central air conditioning. In 2020, 97% of U.S. households had a television, 54% had two or three, and 6% had five or more. From 1990 to 2018, U.S. households with personal computers increased from 16% to 92%. By 2020, 41% had at least one desktop computer, 75% at least one laptop, and 60% at least one tablet or e-reader.

A hidden issue with electricity is efficiency, or lack of it. Only *one-third* of the energy consumed in generating electricity is delivered to the end user as power (**figure CS 15.2.2A**), so as electricity consumption grows, the corresponding energy consumption grows three times as fast (**figure 15.2.2B**). The other two-thirds is lost, most as waste heat during the process of conversion from the energy of the power sources (e.g., the chemical energy of fossil fuels) to electricity, and the rest in transmission and distribution.

Part of the rise in residential electricity consumption is a result of what have been called "vampire appliances," electric devices that for various reasons draw current all the time, not just when actively in use. Avoiding such appliances would help restrain demand. So would conservation through the use of more-efficient devices such as light-emitting diode (LED) bulbs in place of incandescent lightbulbs. The last longer and use much less energy, which means they save money. In 2020, nearly half of all households already used LEDs for most or all of their indoor lighting.

The inefficiency of the ways we now generate our electricity also has a bearing, indirectly, on the utility of hydrogen fuel cells. In a hydrogen fuel cell, which operates somewhat like a battery, hydrogen and oxygen are combined to generate electricity, typically to power a vehicle, and this process is actually quite efficient. The only waste product is water, so fuel-cell-powered

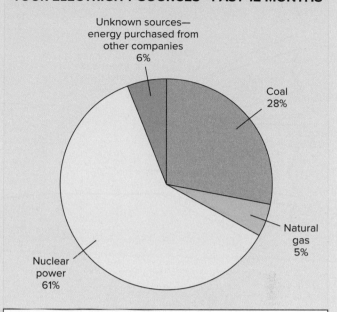

YOUR ELECTRICITY SOURCES—PAST 12 MONTHS

- Unknown sources—energy purchased from other companies 6%
- Coal 28%
- Natural gas 5%
- Nuclear power 61%

Average emissions and nuclear waste per 1000 kilowatt-hours (kWhs) produced from known sources for the past 12 months	
Carbon dioxide	659.84 lb
Nitrogen oxides	1.69 lb
Sulfur dioxide	2.97 lb
High-level nuclear waste	0.0037 lb
Low-level nuclear waste	0.0003 cu. ft.

Figure CS 15.2.1

Typical profile of municipal electricity supply, a mix of sources with varying waste and pollution consequences. The gaseous wastes are associated with the fossil fuels, the sulfur primarily with the coal. Note the relative quantities of wastes from fossil versus nuclear fuel. Bear in mind, too, that amounts of coal ash generated are not tabulated; it is not a waste product monitored and regulated in the same way as the gaseous air pollutants or radioactive waste.

cars are very clean to operate, and the technology already exists. So far, so good. But if we are to run millions of vehicles on fuel cells, where will the hydrogen come from? The obvious answer is water, of which we have a plentiful global supply. However, the usual method for separating water into hydrogen and oxygen rapidly and in quantity uses electricity. Before we can embrace fuel cells on a large scale, we need either to figure out how to generate still *more* electricity cleanly and sustainably, or to develop alternate ways to produce the necessary large quantities of hydrogen.

(Continued)

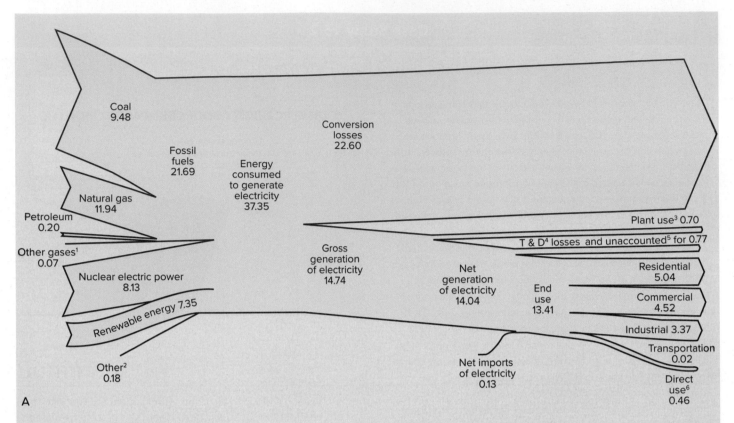

A

(A) Electricity flow 2021: Energy sources on the left, outputs on the right. Units are quadrillion Btu. Note the huge conversion losses. "T & D losses" are additional losses during electricity transmission and distribution.

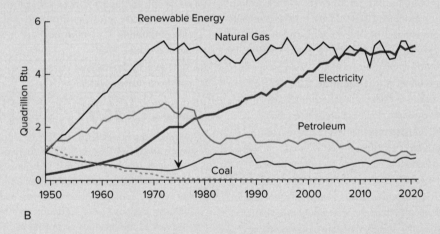

B

(B) Residential and commercial energy consumption in the United States is increasingly in the form of electricity, with correspondingly larger losses. This graph is for residential energy consumption.

Figure CS 15.2.2

Source: Annual Energy Review, U.S. Energy Information Administration.

Summary

As we contemplate the limited nature and environmental impacts of fossil fuels, we find an almost bewildering variety of alternatives available. None is as versatile as liquid and gaseous petroleum fuels, or as immediately and abundantly available as coal. Some of the already-viable, clean, renewable alternatives are place-bound and have limited ultimate potential (e.g., hydropower, geothermal power). Nuclear fission produces minimal emissions but entails waste-disposal problems and concerns about reactor safety; the fuel reprocessing necessary if fission-power use is to be greatly expanded raises security concerns. Solar and wind energy are expanding rapidly with recent technological improvements but are impractical for certain applications. Biofuels, like fossil fuels, yield carbon dioxide (and perhaps other pollutants also) and may require substantial expansion of cropland, potentially increasing soil-erosion problems.

This incomplete sampling of alternatives and their pros and cons illustrates some of the complexity about the future of our energy sources.

The energy-use picture in the future will not be dominated by a single source as has been the case in the fossil-fuel era. Rather, a blend of sources with different strengths, weaknesses, and specialized applications is likely. Different nations, too, will make different choices, for reasons of geology or geography, economics, or differing environmental priorities (**figure 15.37**).

In the meantime, as the alternatives are explored and developed, vigorous efforts to conserve energy would yield much-needed time to make a smooth transition from present to future energy sources. However, considerable pressures toward increasing energy consumption exist in countries at all levels of technological development.

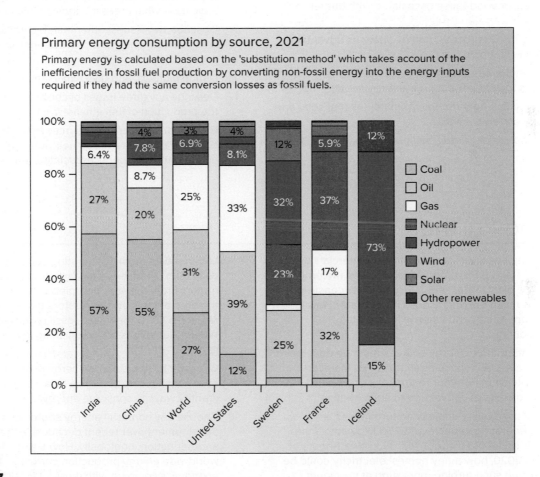

Figure 15.37

Shares of fossil and nonfossil fuels in energy consumption for selected countries and the world, 2020.

Source: Statistical Review of World Energy - BP (2022)

Key Terms and Concepts

biofuels
breeder reactor
chain reaction
core meltdown

decommissioning
fission
fusion
geothermal energy

geothermal gradient
hot-dry-rock
ocean thermal energy
  conversion

photovoltaic cells

Test Your Learning

1. Briefly describe the nature of the fission chain reaction used to generate power in commercial nuclear power plants, and how the energy released is utilized.

2. If nuclear power is to be expanded, breeder reactors and fuel reprocessing will be necessary. Explain why, and indicate what additional safety and security concerns are then involved, beyond those associated with burner reactors.

3. Describe the fusion process, and evaluate its advantages and present limitations.

4. Indicate in what areas solar energy can make the greatest contributions toward meeting our energy needs.

5. Cite three practical/technological limitations that solar and wind energy currently share.

6. Explain the nature of geothermal energy and how it is extracted.

7. Identify factors that restrict the use of geothermal energy in time and in space, and indicate how hot-dry-rock geothermal areas expand their potential.

8. Compare the potential of (a) conventional hydropower and (b) tidal power to supply needed energy.

9. Explain the basis of ocean thermal energy conversion; indicate in what areas it is most viable and why.

10. Describe two issues, other than technological ones, that limit the locations in which wave energy might be harnessed.

11. Define *biofuels,* and give two examples.

12. Describe any three issues of concern with rapid growth in the production of fuel ethanol from corn or other foodstuffs.

13. Choose any two energy sources from this chapter and compare them in terms of their potential and the negative environmental impacts associated with each.

Exploring Further

1. Investigate the history of a completed commercial nuclear power plant project. How long a history does the project have? Did the plant suffer regulatory or other construction delays? How much electricity does it generate? How long has it been operating, and when is it projected for decommissioning?

2. While most alternative energy sources must be developed on a large scale, solar space heating can be installed building by building. For your region, explore the feasibility and costs of conversion to solar heating.

3. A commonly used utility-scale wind turbine has an electricity-generating capacity of 2750 kilowatts. If the average U.S. home used about 11,000 kilowatt-hours of electricity in 2020, how many homes' electricity could be supplied by one such turbine, operating at maximum capacity? Why is this number really an upper limit?

4. Look up data on the generating capacity of a commercial wind farm or solar-electric facility, preferably close to your region. From the area occupied by the facility, calculate the area of the same generators that would be required to replace a 1000-megawatt fossil or nuclear power plant. What assumptions are you making in calculating your result?

5. Consider having a home energy audit (often available free or at nominal cost through a local utility company) to identify ways to conserve energy.

6. Choose any nonfossil energy source and investigate its development over recent decades, in the United States, in another nation, or globally. How significant a factor is it in worldwide energy production currently? How do its economics compare with those of fossil fuels? What are its future prospects?

CHAPTER 16

Waste Disposal

Thane Creek

Deonar
—dumping
ground

Smoke

High-consumption technological societies tend to generate copious quantities of wastes. In the United States alone in 2018, each person generated an average of 5 pounds (about 2.3 kilograms) of garbage every day. That represents an increase of nearly 60% over 1970 per-capita waste production. Add in industrial, agricultural, and mineral wastes, and that makes an estimated total of over 4 billion tons of solid waste produced in the United States each year. Most of the 39 billion gallons of water withdrawn daily by public water departments end up as sewage-tainted

Chapter Outline

16.1 Solid Wastes—General

16.2 Municipal Waste Disposal and Ocean Dumping

16.3 Reducing Solid-Waste Volume

16.4 Toxic Waste Disposal

16.5 Sewage Treatment

16.6 Radioactive Waste

For much of the world, municipal waste disposal means an open dump. This site in Mumbai, India, receives more than 8 million pounds of trash a day—which is only about one-third of the city's waste. It covers more than half a square mile and some piles of trash are over 100 feet high. When vandals started fires there in 2016, choking smoke spread over the city; schools were closed for four days.

NASA Earth Observatory image by Joshua Stevens, using Landsat data provided by the U.S. Geological Survey.

wastewater, and more-concentrated liquid wastes are generated by industry. Each day, the question of where to put the growing accumulations of radioactive waste materials becomes more pressing. Proper, secure disposal of all these varied wastes is critical to minimizing environmental pollution. Americans make up 4.25% of the population, consume one-third of the world's resources, and produce about half the world's solid waste. In this chapter, we will survey various waste-disposal strategies and examine their pros and cons.

16.1 Solid Wastes—General

Learning Objectives

- List sources of solid waste

There are many sources of solid waste in the United States. More than half of them are linked to agricultural activities, with the dominant component being waste from livestock, particularly on feedlots or other places where large numbers of animals are concentrated. Most of this waste is not highly toxic except when contaminated with agricultural chemicals, nor is it collected for systematic disposal, so it is difficult to determine its volume; an estimate from the early 2000s by the U.S. Environmental Protection Agency was over 1 billion tons of waste per year.

The other major waste source is the mineral industry, which generates immense quantities of spoils, tailings, slag, and other rock and mineral wastes. Typically, more than 90% of the material handled during mining for metals is discarded. Tailings and spoils are generally handled on-site—as, for example, when surface mines are reclaimed, or as shown in **figure 16.1**. The amount of waste involved makes long-distance transportation or sophisticated treatment of the wastes uneconomical. The weathering of mining wastes can be a significant water-pollution hazard, depending on the nature of the rocks, with metals and sulfuric acid among the principal pollutants. Shielding the pulverized rocks from rapid weathering with a soil cover is a common control/disposal strategy. In addition, certain chemicals used to extract metals during processing are toxic and require special handling in disposal.

Much of the attention in solid-waste disposal is devoted to the comparatively small amount of municipal waste, an estimated 2 to 5% of the solid-waste stream. It is concentrated in cities, is highly visible, and must be collected, transported, and disposed of at some cost. Section 16.2 explores historical and present strategies for dealing with municipal wastes.

Non-mining industrial wastes likewise command a relatively large share of attention, despite their relatively small volume, because many industrial wastes are highly toxic. Most of the highly publicized unsafe hazardous-waste disposal sites involve improper disposal of industrial chemical wastes. Some of the principal industrial solid-waste sources are shown in **figure 16.2**.

Figure 16.1

Spoils from the Escondida copper mine in Chile's Atacama Desert are simply impounded on-site. The spoils are pumped in solution (green) into the basin, in which they then dry out. For scale, the retaining dam (straight line at lower left) is about 1 km (0.6 mi) long.

Image Science and Analysis Laboratory, NASA Johnson Space Center

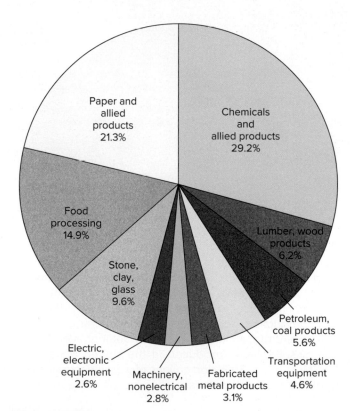

Figure 16.2

Principal industrial solid-waste sources. All these wastes together amount to only a few percent of the quantities of mineral and agricultural wastes.

Data from U.S. Environmental Protection Agency.

While industrial wastes may supply the largest amounts of toxic materials, municipal waste is far from harmless. Aside from organic materials, such as food waste and paper, a wide variety of poisons are used in households: corrosive cleaning agents, disinfectants, solvents such as paint thinner and dry-cleaning fluids, insecticides and insect repellents, and so on. These toxic chemicals together represent a substantial, if more dilute, potential source of pollution if carelessly handled and not disposed of properly.

16.2 Municipal Waste Disposal and Ocean Dumping

Learning Objectives

- Recall the composition of municipal solid waste
- List the problems with open dumps
- Summarize the goal and design of a sanitary landfill
- Describe the current need for and problems with new landfill siting
- List the benefits and problems with waste incineration
- Summarize the current state of ocean dumping

A great variety of materials collectively make up the solid-waste disposal problem that costs municipalities several billion dollars each year (**figure 16.3**). The complexity of the waste-disposal problem is compounded by the mix of different materials to be dealt with. The best disposal method for one kind of waste may not be appropriate for another. In this section, we will review some of these disposal methods. We will also briefly summarize the practice of dumping waste in the oceans.

Open Dumps

A long-established method for solid-waste disposal that demands a minimum of effort and expense has been the open dump. Drawbacks to such facilities are fairly obvious, especially to those having the misfortune to live nearby. Open dumps are unsightly, unsanitary, and generally smelly; they attract scavenging birds, rats, insects, and other pests; they are fire hazards. Surface water percolating through the trash can dissolve out, or leach, harmful chemicals that are then carried away from the dump site in surface runoff or through percolation into groundwater. Trash may be scattered by wind or water. Some of the gases rising from the dump may be toxic.

All in all, open dumps are an unsatisfactory means of solid-waste disposal. Since 1976, they have been illegal in the United States. Yet even now, after decades of concerted efforts to find and close all open-dump sites, the Environmental Protection Agency has estimated that there may still be several hundred thousand sites of illicit open dumping, large and small, in the United States. Open dumps remain the primary method of municipal solid-waste disposal in many countries where they

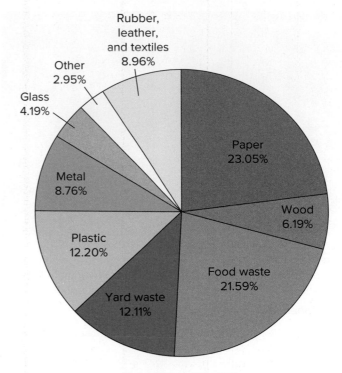

Total: 292.4 million tons

Figure 16.3

Typical composition of municipal solid waste in 2018 before recycling. Individual municipalities' waste composition may vary considerably.

Data from U.S. Environmental Protection Agency.

Figure 16.4

This waste picker is collecting reusable and recyclable materials to sell. They are often responsible for the high recycling rates in some cities.

Mohamed Abdulraheem/Shutterstock

are often the only source of income for waste pickers and recyclers (**figure 16.4**). Closing these dumps would remove the serious health risks to these people but also their livelihoods.

Figure 16.5

The basic principle of a sanitary landfill is to cover the wastes with soil each day. There are other standards to follow.

Doug Sherman/Geofile

Sanitary Landfills

The major share of municipal solid waste in the United States ends up in **sanitary landfills** (**figure 16.5**). The method has been in use since the early twentieth century. In a basic sanitary-landfill operation, a layer of compacted trash is covered with a layer of earth at least once a day. The earth cover keeps out vermin and helps to confine the refuse. Landfills have generally been sited in low places such as natural valleys, old abandoned gravel pits, or surface mines. When the site is full, a thicker layer of earth is placed on top, and the land can be used for other purposes, provided that the nature of the wastes and the design of the landfill are such that leakage of noxious gases or toxins is minimal. The most suitable uses include parks, golf courses, pastureland, parking lots, and other facilities not requiring much excavation. The city of Evanston, Illinois, built a landfill up into a hill, and the now-complete Mount Trashmore is a ski area. Closed landfills are attractive locations for solar array installations. Attempts to construct buildings on old landfill sites are likely to be less successful. They may be complicated by the limited excavation possible because of refuse near the land surface, by the settling of trash as it later decomposes, and by the possibility of pollutants escaping from the landfill site.

Pollutants can escape from improperly designed landfills in a variety of ways. Gases are produced in decomposing refuse in a landfill just as in an open dump, though the particular gases differ somewhat as a result of the exclusion of air from landfill trash. Initially, decomposition proceeds aerobically, consuming oxygen and producing such products as carbon dioxide (CO_2) and sulfur dioxide (SO_2). When oxygen in the covered landfill is used up, anaerobic decomposition yields such gases as methane (CH_4) and hydrogen sulfide (H_2S). If the surface soil is

permeable, these gases may escape through it. Sealing the landfill to prevent the free escape of gases can serve a twofold purpose: reduction of pollution, and retention of useful methane as described in chapter 15. The potential buildup of excess gas pressure requires some venting of gas, whether deliberate or otherwise. Where the quantity of useable methane is insufficient to be recoverable economically, the vented gases may be burned directly at the vent over the landfill site to break down noxious compounds.

If soil above or below a landfill is permeable, **leachate** can escape to contaminate surface water or groundwater (**figure 16.6**). This is a particular problem with landfills so poorly sited that the regional water table reaches the base of the

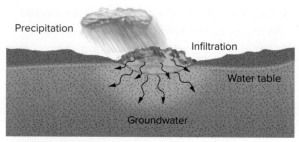

A. Direct contamination of groundwater: Water table intersects landfill.

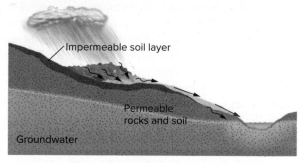

B. Leachate runs off over sloping land to pollute lake or stream below, despite impermeable material directly under landfill site.

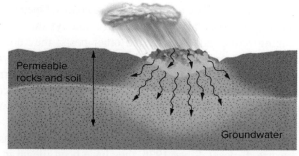

C. Lack of impermeable liner below allows leachate to infiltrate to groundwater.

Figure 16.6

Leachate can escape from a poorly designed or poorly sited landfill to contaminate surface water or groundwater.

landfill during part or all of the year. It is also a problem with older landfills that were constructed without impermeable liners beneath. Increasing awareness of the danger of groundwater pollution has led to improvements in the location and design of sanitary landfills. They are now ideally placed over rock or soil of limited permeability (commonly clay-rich soils, sediments, or sedimentary rocks, or unfractured bedrock) well above the water table. Where the soils are too permeable, liners of plastic or other waterproof material may be used to contain percolating leachate, or thick layers of low-permeability clay may be placed beneath the site before infilling begins.

There is another potential problem with landfills sealed below with low-permeability materials, if the layers above are fairly permeable. Infiltrating water from the surface will accumulate in the landfill as in a giant bathtub, and the leachate may eventually spill out and pollute the surroundings (**figure 16.7A**). Use of low-permeability materials above as well as below the refuse minimizes this possibility (**figure 16.7B**). Leachate problems are lessened in arid climates where precipitation is light, but most large U.S. cities are not located in such places. Modern landfills are carefully monitored to detect developing leachate-buildup problems. Leachate can be pumped out to prevent leakage but then itself has to be properly disposed of.

A subtle pathway for the escape of toxic chemicals is provided by plants growing on a finished landfill site that is not covered by an impermeable layer. As plant roots take up water, they also take up chemicals dissolved in the water. While these locations are generally not well suited for cropland or pastureland, plants can be used specifically to take up toxins from contaminated sites in the process of *phytoremediation*.

A well-designed modern municipal landfill is a complex creation involving a large piece of land (**figure 16.8**). A rule of thumb for a sanitary landfill for municipal wastes is that one acre is needed for every 10,000 people each year, if the landfill is filled to a depth of about 3 meters per year. For a large city, that represents considerable real estate. True, land used for a sanitary landfill can later be used for something else when the site is full, but then another site must be found, and another, and another. All the time, the population grows and spreads and also competes for land. More than half the cities in the United States are pressed for landfill facilities, with municipal solid waste piling up at a rate approaching 300 million tons per year.

Local residents typically resist when new landfill sites are proposed, a phenomenon sometimes described by acronyms such as NIMBY—Not In My Back Yard. Between the political, social, and geological constraints on landfill siting that have evolved as public understanding of pollutants has increased, the number of potential sites for new landfill operations is limited. As existing landfills fill, the total number of landfills has declined, from nearly 8000 in 1988 to about 1250 in 2018. In states with high population density and limited landfill sites, the ratio of people to landfills may be 80,000 to 1 or higher, and obviously, the greater this ratio, the more rapidly the landfills fill. Small states like New Jersey, ship some of their trash for disposal in neighboring states. Elsewhere, remaining active landfills have

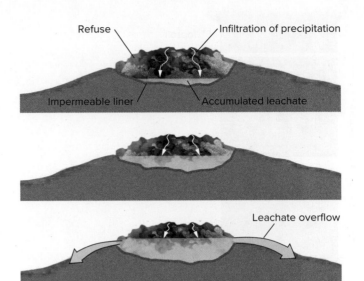

B

Figure 16.7

(A) The bathtub effect caused by the accumulation of infiltrated leachate above impermeable liner in unmonitored landfill. Overflow may not occur for months or years. (B) Installation of impermeable plastic membrane and gas vents, K. I. Sawyer Air Force Base, Michigan.

(B) U.S. Army Corps of Engineers/Photograph by Harry Weddington

been enlarged to accept more trash. In efforts to conserve dwindling landfill volume, at least half of the states have decreed that residents can no longer include landscape waste with trash collected for landfill disposal; the landscape waste is collected separately for composting. The landfill squeeze is especially acute in the Northeastern states that are relatively small and have large populations. Here regulations such as single-use plastic bans are necessarily employed in order to create less trash.

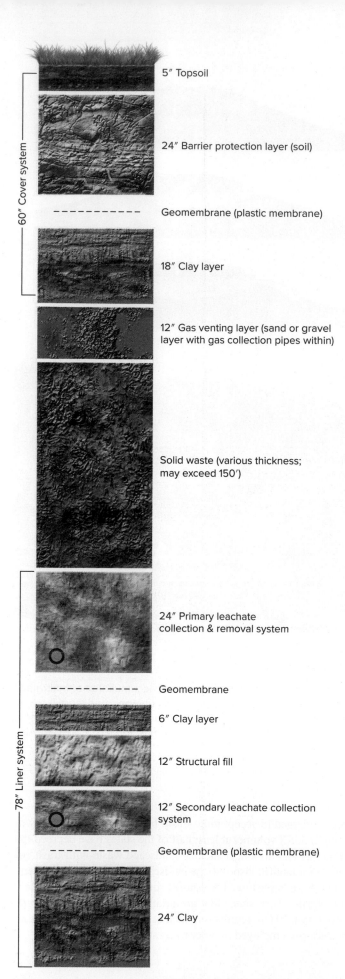

5″ Topsoil

24″ Barrier protection layer (soil)

60″ Cover system

Geomembrane (plastic membrane)

18″ Clay layer

12″ Gas venting layer (sand or gravel layer with gas collection pipes within)

Solid waste (various thickness; may exceed 150′)

24″ Primary leachate collection & removal system

78″ Liner system

Geomembrane

6″ Clay layer

12″ Structural fill

12″ Secondary leachate collection system

Geomembrane (plastic membrane)

24″ Clay

Incineration

Incineration as a means of waste disposal provides a partial solution to the space requirements of landfills. It is an imperfect solution, since burning wastes contributes to air pollution, adding CO_2 if nothing else. At moderate temperatures, incineration may also produce a variety of toxic gases, depending on what is burned. For instance, plastics when burned can release chlorine gas and hydrochloric acid, both of which are toxic and corrosive, or deadly hydrogen cyanide; combustion of sulfur-bearing organic matter SO_2; and so on.

The technology of incineration has been improved in recent years. Modern very-high-temperature (up to 1700°C, or 3000°F) incinerators break down hazardous compounds, such as the complex organic compounds, into much less dangerous ones, especially CO_2 and water. However, individual chemical *elements* are not destroyed. Volatile toxic elements, mercury and lead among others, may escape with the waste gases. Much of the volume of urban waste is noncombustible, even at high temperatures. Refuse that will not burn must be separated before incineration or removed afterward and disposed of in some other way. Harmful nonvolatile, noncombustible substances in the wastes will be concentrated in the residue. If the potential toxicity of the residue is high, it may require handling comparable to that for toxic industrial wastes, raising the net cost of waste disposal considerably. Still, general-purpose municipal incinerators can be operated quite cheaply and are useful in reducing the total volume of solid wastes by burning the paper, wood, and other combustibles before landfill disposal of the rest.

A further benefit of incineration can be realized if the heat generated is recovered. The combined benefits of land conservation and energy production have led to extensive adoption of incineration in a number of European nations and Japan (**figure 16.9**). The United States has been slower to adopt this practice, probably because of more abundant supplies of other energy sources and land.

A growing number of U.S. cities have put the considerable quantity of heat energy released by an incinerator to good use. New York City burns about 30% of its trash to generate enough power to run the incineration plant and for 46,000 homes. Scrap metal is recovered and recycled, leftover ash is used as cover in nearby landfills, and resulting gases are treated before being released. Since New York ships some of the remaining trash to different states for landfilling, the 90% reduction in volume from local incineration also saves money in shipping costs. A facility in downtown Minneapolis also burns trash to generate enough electricity for 25,000 homes.

Figure 16.8

A properly designed modern municipal landfill provides for sealing below and (when completed) above, venting of gas, and collection of leachate. Monitoring wells may also be placed around the site.

After New York State Department of Environmental Conservation.

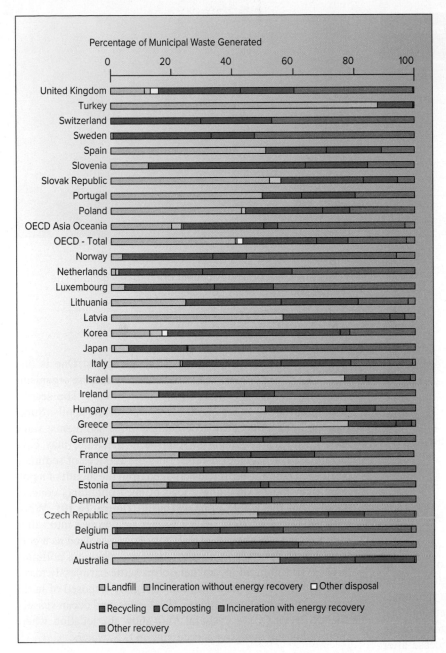

Percentage of Municipal Waste Generated

Legend:
- □ Landfill
- □ Incineration without energy recovery
- □ Other disposal
- ■ Recycling
- ■ Composting
- ■ Incineration with energy recovery
- ■ Other recovery

Countries listed (top to bottom): United Kingdom, Turkey, Switzerland, Sweden, Spain, Slovenia, Slovak Republic, Portugal, Poland, OECD Asia Oceania, OECD - Total, Norway, Netherlands, Luxembourg, Lithuania, Latvia, Korea, Japan, Italy, Israel, Ireland, Hungary, Greece, Germany, France, Finland, Estonia, Denmark, Czech Republic, Belgium, Austria, Australia

Figure 16.9

Municipal-waste handling methods of choice vary widely around the world.

Data from OECD, Environment at a Glance, 2019.

Disposing of garbage in a way that results in less CO_2 and methane emissions compared to it ending up in a landfill and in useful electricity both contribute to the acceptance of incineration as a waste-disposal strategy, though as a heat source its efficiency is modest (**table 16.1**). Considerable public resistance to new incinerators exists, however, which is part of the reason that it may take three to five years from inception to opening of a new incinerator. Still, as landfills fill, something has to be done to reduce waste volume, and by 2018, incineration with energy recovery accounted for about 12% of municipal solid-waste disposal in the United States.

Ocean Dumping

Solid and liquid wastes of all kinds end up in the ocean either through intentional dumping directly into the water or through improper land disposal. Historically, people put their garbage in the oceans because the oceans were big and the garbage seemed to disappear in their vastness. As populations grew and society became more technologically advanced, the problems of dumping materials in the ocean soon became apparent. Most ocean dumping is banned through the regulations set forth in the 1972 London Convention and the later 1996 London Protocol, but not all countries have ratified these treaties, so ocean dumping bans are applied rather unevenly. The large islands of plastic garbage that have concentrated in the middle of ocean gyres, and the accumulations of microplastics on the ocean floor, are likely the most obvious of current ocean pollution problems.

Ocean dumping has been used for chemical wastes, municipal garbage, and other refuse. In some cases, shifting currents can bring the waste back to shore rather than

Table 16.1	Comparative Heat Values of Fuels and Wastes	
Fuel		**Btu/lb**
Coal (anthracite)		13,500
Coal (bituminous)		14,000
Peat		3600
#2 fuel oil		18,000
Natural gas (Btu/cu ft)		1116
Mixed municipal solid waste		**4800**
Mixed paper		6800
Newsprint		7950
Corrugated		7043
Junk mail		6088
Magazines		5250
Mixed food waste		2370
Wax milk cartons		11,325
Polyethylene		18,687
Polystyrene		16,419
Mixed plastic		14,100
Tires		13,800
Leaves (10% moisture)		7984
Cured lumber		7300

While average municipal waste yields much less energy per pound than most fuels, incineration also reduces remaining waste volume.

Source: U.S. Environmental Protection Agency.

Figure 16.10

Rivers are dredged to keep them open to navigation. The dredge spoils are dumped in oceans.

Gira/Getty Images

dispersing it in the oceans as intended. Increasing recognition of the dangers of dumping untreated wastes in the sea led to drastic curtailment of the practice by the Environmental Protection Agency in most areas. In the late 1980s, public outcry over incidents of wastes washing up onto beaches, high bacterial counts in nearshore waters, and deaths of marine mammals combined to add pressure on Congress to do something. The Ocean Dumping Ban Act of 1988 decreed that ocean dumping of U.S. sewage sludge and industrial waste would cease after 1991. Britain, the other nation dumping large quantities of sewage sludge, phased out the practice by 1998; it had ended ocean dumping of its industrial waste by 1993.

Shipboard incineration in the open ocean was once thought of as an innocuous way to dispose of waste far away from the public view. Stockpiles of particularly hazardous chemical waste were burned on specially designed ships at sea during the 1970s and 1980s, and the U.S. began to develop regulations to support the practice. Ocean incineration was abandoned along with other ocean dumping as described above, as the emissions from any burning would be deposited in the open ocean.

The oceans do remain a dumping site for one very-high-volume waste product: *dredge spoils.* Dredge spoils are sediments dredged from reservoirs and waterways to enlarge capacity or improve navigation. They are dumped in the oceans at rates up to 550 million tons a year. There are two primary

problems with ocean dumping of dredge spoils. One is that fine-grained sediments may harm or destroy marine organisms that cannot survive in sediment-clouded waters. The second potential problem relates to the chemistry of the spoils. Some solid pollutants may have accumulated in the sediments; some toxic chemicals originally in solution in the water may have precipitated or become adsorbed onto the surfaces of sediment particles. Dumping sediments dredged from the mouth of a polluted river means dumping a load of such pollutants concentrated from the whole drainage basin (**figure 16.10**). When sediments dredged from fresher waters are dumped into saline waters, some adsorbed chemicals may be redissolved as a consequence of the change in water chemistry, posing a pollution threat even if the sediments themselves settle harmlessly to the bottom. A few other materials can be legally disposed of in the ocean including human remains, fish wastes, and ocean vessels. The USEPA's Ocean Disposal Map show the location where approved ocean dumping is occurring in U.S. waters.

16.3 Reducing Solid-Waste Volume

Learning Objectives

- List strategies to minimize waste production
- Describe the benefits of composting
- Compare the successes and limitations of recycling glass, paper, plastic, and e-waste
- Exemplify reuse of materials
- List issues associated with e-cycling

The sheer volume of the solid-waste-disposal problem has led to a variety of attempts to reduce it. In industrialized countries,

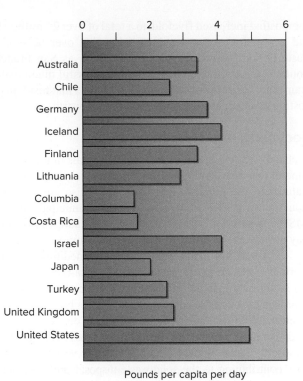

Figure 16.11

Relative amounts of municipal solid waste generated per capita in selected countries.

Data from OECD Statistical Country Profiles 2020.

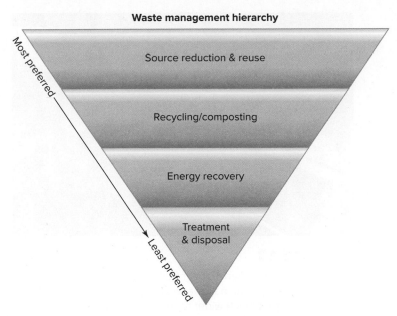

Waste management hierarchy

Most preferred

Source reduction & reuse

Recycling/composting

Energy recovery

Treatment & disposal

Least preferred

Figure 16.12

The U.S. Environmental Protection Agency's preferred strategy is to minimize production of waste in the first place. Next best is recycling or composting. Energy recovery includes incineration with recovery of the heat energy and extracting methane from a landfill. Least desirable are incineration and landfilling with no recovery of useful heat or methane, respectively.

Source: U.S. Environmental Protection Agency

the problem can be especially acute (**figure 16.11**); high gross national product (GNP) tends to be correlated not only with material and energy consumption but also with waste production. Indeed, the U.S. Environmental Protection Agency (EPA) now formally advocates a waste management hierarchy that puts a premium on reducing the amount of waste to be managed (**figure 16.12**). At the disposal stage, another volume-reduction strategy is compaction, either in individual homes with trash compactors or at large municipal compaction facilities. Less volume means less landfill space used or the slower filling of available sites. However, it also means no reuse of any potentially recoverable material, and the slower decay of organic material.

Studies of landfill sites over the last several decades have revealed that decomposition in landfills takes place much more slowly than once believed, especially if the landfill is relatively dry. The University of Arizona's Garbage Project found well-preserved food wastes two decades old, and readable newspapers three or four decades old. Decomposition rates seem to be faster in wet waste but modern landfills are most often designed to minimize water inflow. In the future, careful manipulation of water content and perhaps judicious addition of suitable microorganisms can maximize biodegradation while minimizing the risks from toxic leachate.

In this section, we will discuss some ways to recycle and reuse materials that are better left out of landfills. We will also introduce some of the benefits, limitations, and roadblocks to recycling and reusing food and yard waste, glass, paper, plastic, and electronics, and suggest ways in which recycling rates could be improved.

Handling (Nontoxic) Organic Matter

Putting your organic matter in the garbage disposal rather than the garbage can is really not a way to reduce the volume of waste. The practice merely diverts some organic matter to become part of the water-pollution problem. The organic-matter content of the water is increased (see chapter 17 for the consequences of this), and more of a load is placed on municipal sewage-treatment plants.

Organic matter can be turned to good use through composting, a practice long familiar to gardeners and farmers. Many kinds of plant wastes and animal manures can be handled this way. Partial decomposition of the organic matter by microorganisms produces a crumbly, brown material rich in plant nutrients. Finished compost is a very useful soil additive, improving soil structure and water-holding capacity, as well as adding nutrients. Compositing can be practiced by individuals with enough outdoor space for a composter, or as part of a curbside program that collects organic materials to be made into compost which is later resold in the community (**figure 16.13**). In just seven years, one commercial collection program in the Midwest collected 6 million pounds of compostable materials from individuals, offices, schools, florists, restaurants, and breweries, waste that instead would have been sent to landfills.

Figure 16.13

Home composters often just put fruit, vegetable, and egg shells in their bins, along with paper and yard waste. Curbside collectors may accept a much wider range of food products.

ChameleonsEye/Shutterstock

In parts of Europe and Asia, demand for finished compost has long made it economically practical to establish municipal composting facilities. The city of Auckland, New Zealand, began such a facility as early as 1960 after it became evident that remaining landfill sites were severely limited. The process is not without complications, however. Organic matter must be separated from glass, metal, and other noncompostables. Occasionally, chemical contamination of the refuse with weed-killers or other toxic substances makes some batches of compost unusable for agriculture. Still, sale of the finished compost helps to pay the cost of the composting operation, and the volume of waste going to the landfill is reduced.

As landfills filled, the United States increasingly turned to composting, too. By 1997, nearly half of the states had banned yard waste from landfills, and yard-waste composting programs had been established in nearly every state. From 1990 to 2007, the quantity of yard trimmings collected for composting increased fivefold, to a total of over 21 million tons, and it remains at that level, meaning that over 60% of yard waste is composted. American cities such as San Francisco, Boulder, and Portland, Oregon, have successful municipal programs that collect yard waste and food scraps that result in composting as much as 50% of their municipal waste.

Recycling and Reusing

In chapter 13, we considered recycling metals in the context of mineral resources and noted that an energy saving can often be realized in the process. Recycling and reuse are also waste-reduction strategies. **Figure 16.14** shows municipal solid waste (MSW) trends for various products, indicating how well we may or may not be doing in recycling or composting them.

Glass is not made from scarce commodities, but just as quartz is a weathering-resistant mineral, silica-rich glass is virtually indestructible in dumps, landfills, and along roadsides. It can be broken up, but it does not readily dissolve or break down. Reuse of returnable glass bottles requires only about one-third the energy needed to make new bottles. Recycling all glass beverage containers would reduce by over 5 million tons per year the amount of glass contributing to the solid-waste-disposal problem. Glass is 100% recyclable. Glass containers vary in composition and color, and need to be sorted when used to make more glass. Recycled glass is also used in insulation, abrasives, countertops, road base aggregate, and even replacement sand for beach nourishment.

Imposing deposits on beverage containers provides a financial incentive not to throw them in the trash or out the door. In Oregon, passage of a mandatory-deposit law for beverage containers reduced roadside litter by up to 84%. Since Oregon pioneered the idea in 1971, nine other states have followed suit. Currently, the United States recycles about 31% of its glass altogether; in countries such as Sweden and Belgium, the proportion recycled approaches 95%.

Paper might also be recycled more extensively. In the United States, about 68% of the paper and 89% of the cardboard we discard is recycled; about 35% of new paper has recycled content and nearly all cardboard does. Paper recycling is easiest

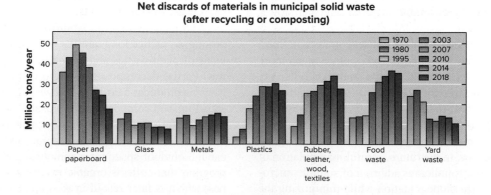

Figure 16.14

Despite recycling and composting, municipal solid waste in the United States includes large quantities of plastic and food waste. Upward trends with time on this graph indicate worsening rates of recycling or composting.

Source: U.S. Environmental Protection Agency.

and most effective when a single type of paper is collected in quantity. That limits the variety of inks and other chemicals that must be handled during reprocessing. Mixtures of printed, waxed, and plasticized papers are somewhat harder to handle economically but can nevertheless be recycled. The Recycling Council of Ontario estimates that every ton of paper recycled means about eighteen trees, and 3 cubic meters of landfill space, saved. Moreover, making paper from the recycled fibers requires 60% less energy than does manufacturing paper from newly cut trees. The U.S. EPA reports that every ton of mixed paper recycled saves energy equivalent to 185 gallons of gasoline.

Plastics continue to be something of a disposal problem. The same durability that makes them useful also makes them difficult to break down when no longer needed, except by high-temperature combustion. Some degradable plastics have been developed to break down in the environment after a period of exposure to sunlight, weather, and microbial activity, but these plastics are suitable only for applications where they need only hold together for a short time—for example, fast-food containers, which are increasingly being made of paper products. Another difficulty in recycling plastics is similar to the problem with different steels. A mix of plastics, when reprocessed, is unlikely to have quite the right properties for any of the applications from which the various scrap plastics were derived. Still, the blend may be suitable for other uses, such as plastic piping, plastic lumber, or a shredded-plastic stuffing for upholstery.

One approach to facilitating plastic recycling is to mark those plastics that can be more easily recycled with the triangular symbol of three arrows head-to-tail that is widely used to represent recycling and identify the basic type of plastic by a number within the triangle (see **figure 16.15** for examples and key). A remaining obstacle is that there must exist an identifiable market or demand for a particular plastic in order for its recycling to be economically feasible; so a given municipal waste hauler might be able to collect soda bottles and milk jugs for recycling, but not foam packing materials. Markets vary by region. It should also be realized that plastic is typically not recycled into the same form—old soda bottle to new soda bottle, for instance. In this respect, plastic recycling differs from that of other materials. A recycled aluminum can is likely to come back as a new aluminum can, a recycled glass bottle as a refilled or remanufactured bottle. A plastic's properties tend to change during recycling, and recycled plastic may be less strong than new plastic. So a soda bottle may be recycled, but it will be transformed into something else such as fiber for carpeting, plastic trash bags, or plastic lumber for park benches.

The rate of plastic recycling compared to other products is 9% (or lower), in the United States, and that hasn't changed much in recent years. At the same time, global plastic production continues to increase by about 15 million tons per year, and Americans use a lot of plastic. Why are plastic recycling rates so low? The primary reason is economics: it is cheaper to throw away plastic than it is to recycle it. China banned all plastic imports in 2018 and India followed with their own ban in 2019; and in 2020, 180 countries agreed to not ship plastic waste from richer to poorer countries under the Basel Convention. Because the United States did

Symbol (letters may be omitted)	Plastic type	Examples
1 PETE	Polyethylene terephthalate	Soda bottles, salad-dressing bottles
2 HDPE	High-density polyethylene	Milk jugs, motor-oil bottles, detergent bottles
3 V	Polyvinyl chloride	Shampoo bottles, wrapping film
4 LDPE	Low-density polyethylene	Grocery and other shopping bags
5 PP	Polypropylene	Cereal-box liners, dairy-product containers, prescription bottles
6 PS	Polystyrene	Foam cups, packing materials

Figure 16.15

Use of standard symbols facilitates separation of distinct types of plastics for recycling. Sometimes the same symbol with a numeral "7" is used for "other."

not ratify this convention, it is effectively banned from trading most plastic waste with other nations. In other words, there is too much plastic waste and nowhere to recycle it. Low plastic recycling rates are also the result of confusion as to what can and cannot be recycled, and questions if the plastic put in the bin will end up being recycled in the end. A significant portion of exported plastic does end up in landfills (and through improper disposal, in rivers and the oceans) because facilities in those countries are overwhelmed with not just the volume, but also the low-quality of the plastic received. There is much room for improvement in plastic recycling in the United States that will likely need to rely on a circular approach from initial product design to recycled material markets. Using alternative products for packaging will also help to reduce the amount of plastic to be disposed of.

There are additional general obstacles to recycling, some of which were mentioned earlier. Where recovery of materials from municipal refuse is desired, **source separation** is generally necessary. This means that individual homeowners, businesses, and other trash generators must sort that trash into categories prior to collection (**figure 16.16**). This has been successful in some communities and where legally mandated, it can work well once residents adjust to the new realities. Source separation creates less material contamination and results in higher revenues for recyclers, but it may discourage participation because of its logistics.

A

B

Figure 16.16

Source separation of recyclable materials. (A) This public recycling bin makes it fairly simple to sort glass by color. (B) Individual bins for each type of material can make source separation of recyclables more appealing.

(A) robeo/iStock/Getty Images; (B) Jacobs Stock Photography Ltd/DigitalVision/Getty Images

Recycling may conflict in some measure with other waste-disposal objectives. For example, recycling combustible materials reduces the energy output of municipal incinerators used to generate power. Paper recyclers are already encountering this problem, and as uses are found for recycled plastics, the same difficulty may arise with those materials.

In the United States, waste recovery for recycling or composting increased dramatically from 10% of municipal wastes in 1985 until 2010 when it leveled out around 35%; it fell to 32% in 2018 (**figure 16.17A**). The recycling rates of individual materials vary considerably, in part reflecting differences in economics, or technical or practical constraints described previously (**figure 16.17B**).

Whether or not source reduction in the volume of municipal solid wastes has been successful, there has been a clear shift in the handling of municipal wastes collected in the United States,

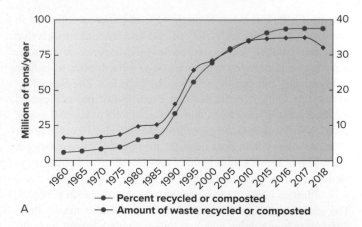

A

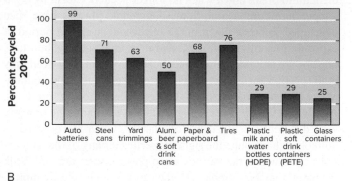

B

Figure 16.17

(A) After rising sharply since 1980, the amount and percentage of municipal solid waste recycled/composted and the percentage, have plateaued. (B) Recycling rates vary greatly by material. Note that although the two types of plastic containers highlighted here are recycled at rates close to 30%, only 9% of plastics overall are recycled.

Source: U.S. Environmental Protection Agency.

toward recovery and away from land disposal (**figure 16.18**). Still, some materials are too difficult to reuse efficiently, for reasons already outlined, and these will continue to require ultimate disposal. Beyond municipal wastes are toxic by-products of industrial processes that are not themselves useful, or are too toxic for safe handling during extensive reprocessing. These require more-specialized and careful disposal, as many are also liquids. We will address toxic wastes in the next section of this chapter.

The amount of construction and demolition (C&D) waste generated each year is approximately twice that of municipal solid waste. Recycling rates for these materials are high, primarily because reusing them or selling them saves the money for shipping and disposal. In 2018, 95% of concrete and asphalt from C&D were reused to make gravel, paving materials, aggregates, and other construction and infrastructure materials. General project recycling rates are around 76%, and recent carefully managed building projects in New York saved 96% of their C&D waste from going into landfills. Waste exchanges are also successful in recycling and reusing materials, whereby one industry's discarded waste becomes another's raw materials. Ideally, this both reduces waste disposal and saves the waste

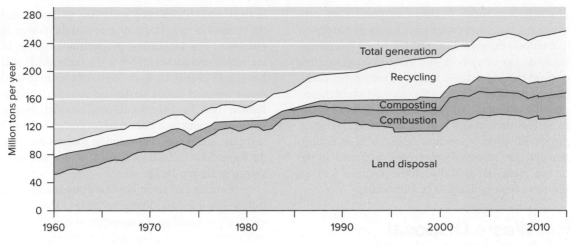

Figure 16.18

The trend in municipal waste is away from traditional disposal, but there are limits to the feasibility of the alternatives.

Source: U.S. Environmental Protection Agency.

generator money. One company's scrap wood becomes another's air freshener; waste isopropyl alcohol becomes a cleaning solvent. Exchange listings for wanted products include used shoes, peanut shells, wood ash, and electronic motors. Listings for products being sold include sulfuric acid, PET soda bottles, salt brine water, and bales of nylon carpet.

E-waste

One somewhat new area of concern is the recycling of electronics waste (e-waste), or e-cycling. The problem has been growing rapidly in recent years as the quantity of consumer electronics has soared. U.S. sales of such devices—computers, TVs, mobile devices, and so on—grew from about 80 million in 1990 to over 870 million in 2020. Sooner or later every such device reaches the end of its useful life and becomes a candidate for disposal. The concern relates particularly to toxic elements in electronics: lead in cathode-ray tubes and circuit boards; cadmium in semiconductors; mercury in switches, circuit boards, lamps, and batteries; and more. Regulations regarding e-waste in the United States are inconsistent: only twenty-five states have regulations, proposed federal legislation is sitting in Congress, USEPA regulations do not apply to households and small businesses, and the United States has not ratified the Basel Convention (mentioned above) to better track hazardous waste. Most U.S. e-waste, which is 14% of the global total, is exported to South and East Asia because the labor needed to handle it is cheaper, regulations are generally less strict, and laws protecting workers may be limited or nonexistent (**figure 16.19**).

Similar to the problems with plastic recycling in the United States, there is a great need for a much more comprehensive and circular infrastructure to handle electronics: more processing centers; more opportunities for scrap materials to be worked

Figure 16.19

This e-waste recycling and reclamation company in the Philippines salvages gold and other precious metals from microprocessors of discarded electronics. About 4000 discarded mobile phones can produce about 100 grams of gold.

REUTERS/Alamy Stock Photo

back into manufacturing; more producer responsibility programs; better waste tracking; and easier ways for people to repair, refurbish, sell, or recycle their used electronics. Many electronics companies already do have buy-back and trade-in programs that help to keep e-waste out of landfills. Apple's Daisy robot disassembles 200 phones per hour to access fourteen different minerals including cobalt and lithium for recycling. Recycling e-waste not only results in less toxic releases to air and water, but saves money and energy in mining and processing new materials. Overall, the demand for accessible recycling programs in the United States is not being met even though the potential for economic growth in this sector is likely to be substantial.

16.4 Toxic-Waste Disposal

Learning Objectives

- List ways to dispose of toxic waste
- Describe a secure landfill
- Describe the geologic conditions necessary for deep well injection of liquid waste

Toxic-waste problems come in many forms. A notable everyday example is the problem of used oil. Presently, over 1.4 billion gallons of used lubricants derived from petroleum are generated in the United States each year; 40% of this waste is poured into the ground or into storm drains, and the fate of another 20% is unknown. An increasing proportion of toxic waste is being reclaimed and recycled, but many individuals still discard toxic household chemicals in the trash to end up in landfills. The two major types of liquid wastes are sewage, which is discussed in section 16.5, and the more-concentrated, highly toxic, liquid waste by-products of industrial processes—acids, bases, organic solvents, and so on. This section focuses on disposal strategies for hazardous industrial wastes, major sources of which are shown in **figure 16.20**.

Handling of toxic liquid wastes has historically tended to follow one of two divergent paths. The *dilute-and-disperse* approach is based on the assumption that, if toxic substances are sufficiently diluted, they will be rendered harmless. That had been the rationale behind much dumping into oceans and large lakes and rivers. With the increasing recognition of substances that are toxic even at levels below 1 ppb in water, including many of the complex organic solvents, agricultural chemicals, and others, the basic premise has been brought into question. Also, certain pollutants can accumulate in organisms and become more concentrated up a food chain (see chapter 17), which means that the chemicals can return to hazardous concentrations.

The opposite approach is the *concentrate-and-contain* alternative. Thoughtless disposal of concentrated wastes followed by inadequate containment has led to disasters like

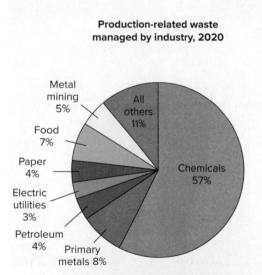

A. Seven industrial sectors account for 89% of the toxic wastes generated in the United States. The good news is that most of these wastes are now recycled (55%), treated (24%), or incinerated for energy recovery (10%), so that only about 11% of the original total is released into air or water, or disposed of.

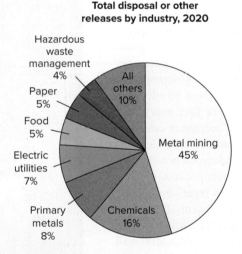

B. The major industry source of this remaining 11% is mining, and the bulk of these wastes are kept on-site, as described earlier.

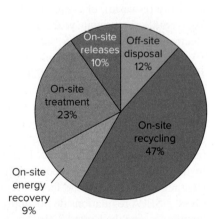

C. Overall disposition of the 11%. The volume of toxic wastes ultimately requiring off-site disposal is thus less than 1.5% of the total toxics generated.

Figure 16.20

Source: Toxics Release Inventory 2020, U.S. Environmental Protection Agency.

Love Canal, New York, or Woburn, Massachusetts (see also **figure 16.21**). In the past, some concentrated liquid industrial wastes have been dumped in trenches or pits directly and buried, while other wastes have been placed in metal or plastic containers and consigned to dumps or landfills. The disposal sites frequently were not evaluated with respect to their suitability as toxic-waste disposal sites and, over the longer term, the wastes were not contained. Metal drums rusted, plastic cracked and leaked, and wastes seeped out to contaminate groundwater and soil. Superfund, discussed in Case Study 16.2, was created in large part to address such issues.

A

B

Figure 16.21

Careless toxic-waste disposal leads to pollution. (A) Trinitrotoluene (TNT)-contaminated water seeps to the surface during excavation at Weldon Springs Ordnance Works, St. Louis, Missouri. (B) Once soils are contaminated, cleanup usually means digging out that soil, then disposing of it elsewhere. Thorium-contaminated soil, Wayne, New Jersey.

(A) Photograph by Bill Empson, U.S. Army Corps of Engineers; (B) Bill Empson/U.S. Army Corps of Engineers

Secure Landfills

A **secure landfill** is designed to store toxic solid and liquid wastes. An example of one type is shown in **figure 16.22**. The wastes are put in sealed drums before disposal. Beneath the drums are layers of plastic and/or compacted clay to contain any unexpected leaks. Wells and piping are installed so that the groundwater below and around the site can be checked periodically for any sign of leakage of the waste chemicals. Excess accumulating leachate can be pumped out before it leaks out. Such a system provides multiple safeguards against accidental environmental contamination, and the monitoring wells allow prompt detection of any leaks. The design shares many features with modern municipal landfills, but with still more provisions for waste containment and site monitoring.

Unfortunately, a growing body of evidence indicates that no site is truly secure, even if conscientiously designed. Carefully compacted clay may be very low in permeability but is probably never completely impermeable, especially over long time intervals. Chemical and biological reactions in the wastes and leachate can rupture or decompose plastic, and the stress caused by the weight of wastes and cover can fracture a clay liner. Even relatively innocuous municipal waste can prove hard to contain. When built, the Mount Trashmore landfill in Evanston, Illinois, was hailed for its state-of-the-art design. But monitoring wells later revealed detectable leakage of at least a dozen volatile organic compounds deemed high-priority toxic pollutants by the EPA, including benzene, toluene, vinyl chloride, and chloroform. Leakage from secure toxic-waste dumps, in which hundreds or thousands of barrels of

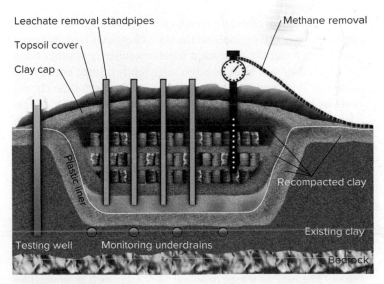

Figure 16.22

A secure landfill design for toxic-waste disposal, including provisions for leachate containment and for monitoring the chemistry of subsurface water nearby.

concentrated toxic liquid chemicals are stored, has far more potential for harm. The historical response to detection of leakage from one toxic-waste dump has been to dig up as much of the hazardous material and contaminated soil as possible and transfer it to a more secure landfill. There is now some question as to whether a wholly different disposal method might be preferable.

Deep-Well Disposal

Another alternative for disposal of liquid industrial waste is injection into deep wells (**figure 16.23**). The rock unit selected to receive the wastes must be relatively porous and permeable (commonly, sandstone or fractured limestone), and it must be isolated by low-permeability layers (e.g., shale) above and below. The subsurface geology must be known in sufficient detail that there is reasonable confidence that the disposal stratum remains isolated for some distance from the well site in all directions. Information about that geology may be derived from many sources: direct drilling to obtain core samples that provide a vertical section of the rock units present; geophysical studies that provide data on depths to and thicknesses of different rock layers, and on the distribution of groundwater; geologic mapping on the basis of cores, surface outcrops, and geophysical data to interpolate between points sampled directly.

These disposal wells are hundreds to thousands of meters deep, far removed from the surface, and below the regional water table. The pore water in the disposal stratum should be brackish or saline water not suitable for a water supply. Where the well intersects any shallower aquifers that are or might be used for water supply, it must be snugly lined (cased) to prevent leakage of the wastes into those aquifers. Local well water is monitored to detect any accidental leaks promptly.

Movement of deep groundwater is generally slow, and the assumption is that by the time the toxic chemicals have migrated far enough laterally to reach a useable aquifer or another body of water, they will have become sufficiently diluted not to pose a threat. This presumes knowledge of the toxicity of the chemicals in low concentrations. When the wastes are more or less dense than the groundwater they displace and not miscible with it, folds or other geologic structures may help to contain them and slow their spread, as oil traps contain petroleum (**figure 16.23B**). The behavior of chemicals that dissolve in the pore water is much less well understood. They can diffuse through the water more rapidly than the water itself moves, so that even if deep groundwater transport is slow, contaminant migration may not be.

Costs for deep-well disposal are comparable to or somewhat less than those associated with secure landfill sites. The rate of waste disposal in a deep well is limited by the permeability of the rocks of the disposal stratum, while landfills have no equivalent limitation. The region's geology must be such that suitable strata exist for disposal by injection, while landfills can be constructed in a much greater

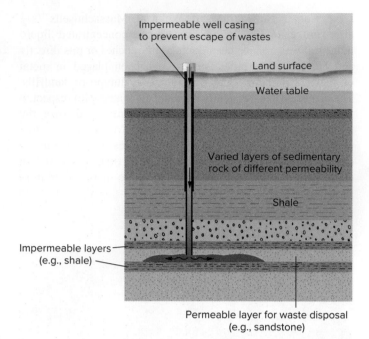

A

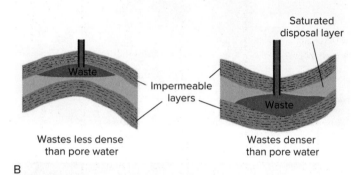

B

Figure 16.23

Deep-well disposal for liquid wastes. (A) Basic design: Wastes are placed in a deep permeable layer that is geographically and geologically isolated by low-permeability strata. (B) Containment of wastes assisted by geologic structures, much as petroleum is trapped.

variety of settings. Like landfills, deep injection wells may leak. Finally, as noted in chapter 4, deep-well waste injection can trigger earthquakes in faulted rocks, and such induced seismicity has been associated with injection of wastewater from natural-gas production.

Other Strategies

Currently there are twenty-five commercial hazardous waste combustor facilities in the United States. These are furnaces designed to incinerate fine streams of liquid at very high temperatures, which destroys toxic liquid organic chemicals. The by-product is often just carbon dioxide. Other liquid wastes can be neutralized or broken down by chemical treatment, which may avoid the necessity for ultra-secure disposal altogether.

As with solid wastes, it may even be possible to use certain liquid waste via waste exchanges. The nitric acid used in quantity by the electronics industry to etch silicon wafers can be neutralized to produce calcium nitrate and then incorporated in high-grade fertilizers. Spent acid used in the steel industry is rich in dissolved iron and can be used at geothermal power plants to control hydrogen sulfide gas emissions, which react with the iron in solution to precipitate iron sulfides.

These and other means of handling toxic liquid industrial wastes will continue to be developed. What these methods generally have in common is that they are designed to deal either with specialized types of waste or with limited quantities. Volumetrically, at least, a far larger liquid-waste disposal problem is posed by that very commonplace material, sewage.

16.5 Sewage Treatment

Learning Objectives

- Describe the design and operation of a septic system
- Describe the design and operation of a municipal sewage treatment system

Treatment of sewage in wastewater is necessary to avoid the problems arising from excess organic matter in water that include oxygen depletion and algal bloom, as described in chapter 17, and to address the concern about the spread of disease through biological contamination of drinking-water supplies by *pathogenic* organisms. Most of the approximately 39 billion gallons of water withdrawn for public water supplies in the United States each day winds up as wastewater. So does urban surface runoff water collected in storm drains, along with a portion of the water used in rural areas. Two primary treatment strategies, which vary based on population density and local geology, are addressed below.

Septic Systems

On an individual-user level, modern sewage treatment typically involves a septic system of some kind (**figure 16.24**). Wastes are first transferred to a settling tank in which solids settle out, to be broken down slowly through bacterial action. The remaining liquid carries a load of dissolved organic matter and of microorganisms—some pathogenic—whose metabolism requires little or no oxygen. The dissolved organic matter represents food for those microorganisms. The liquid is allowed to seep out through porous pipes into the soil of the **leaching field** or **absorption field**. There, oxygen is available in the pore spaces, and aerobic soil microorganisms that can use that oxygen in metabolizing the organic matter compete for the nutrients with the microorganisms in the sewage, breaking down the organic matter and destroying some pathogens. Passage through the soil, especially if it is fine-grained, also filters the liquid, removing remaining fine suspended solids and even the larger

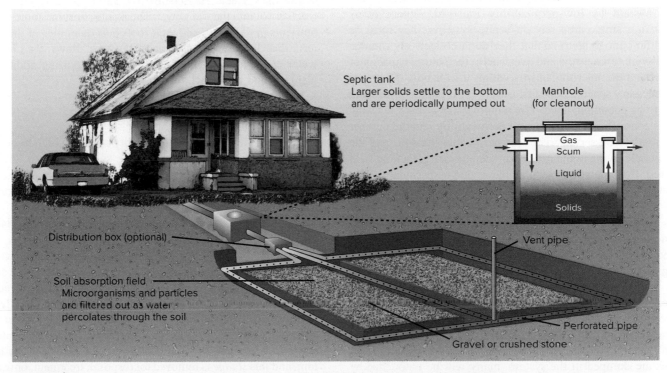

Figure 16.24

Basics of septic tank system. A settling tank for solids and the slow release of liquids into the soil of the leaching field for natural decomposition.

pathogenic organisms. Inorganic reactions in the soil can also break down undesirable compounds in the sewage. Ideally, by the time any of the liquid reaches either the surface or groundwater supplies, it has been purified of biological and many chemical contaminants. Some compounds do remain in solution, however; nitrate is ordinarily the most significant potential pollutant among them. Septage must be collected from the tank every few years and then can be taken to a municipal system or independent company for treatment, dumped in approved landfills, used on a farm for fertilizer, or used to create methane for the generation of electricity.

There are several geologic requirements for a properly functioning septic system. The soil must be sufficiently permeable that the fluids will flow through it rather than merely backing up in the septic tank, but not so permeable that the flow into water supplies or out at the surface occurs before the wastes have been sufficiently purified. Ordinarily, the water table should be well below the level of the septic system: first, to avoid immediate groundwater contamination by raw sewage; second, because oxygen levels in saturated soil are often too low to permit rapid aerobic breakdown of the organic matter. There must be sufficient soil depth so that the wastewater will be adequately filtered by the time it reaches either surface or bedrock; this filtration usually requires at least 60 centimeters of soil above the pipes and 150 centimeters below. The leaching field should not extend to within about 15 meters of any body of surface water, for similar reasons.

If a well is used to obtain drinking water on the same site, it must be upgradient of or far enough removed from the septic system that partially decomposed sewage does not reach the well, or else it should tap an aquifer that is isolated from the septic system by low-permeability material. Where many houses in a single area rely on septic systems, they must be spaced far enough apart so that they don't collectively saturate the soil with raw sewage and overwhelm the natural capacity of the soil and the microorganisms within it to handle the waste. The necessary spacing depends, in part, on the sizes of leaching fields to be accommodated. The required size of each leaching field is controlled, in turn, by soil permeability and the number of persons to be served. The less permeable the soil, and the more people living in the residence, the larger the leaching field required. Considerations such as these, and assessments of the potential impact of the nitrate released from these systems, lead local authorities to stipulate minimum lot sizes where septic tanks are to be used. Typical lot sizes might be one-half acre to one acre per dwelling, but again, appropriate limits are controlled largely by local geology. In some cases, geology and/or topography may preclude use of a septic system altogether.

The sewage treatment associated with a septic system is entirely natural, and its thoroughness is highly variable from site to site. One can enhance the suitability of the site but the natural soil chemistry, biology, and physical properties largely control the effectiveness of the system. Also, if toxic household wastes are dumped in the system, many will be untouched by the chemical processes and microbial activity that attack and decompose the organic wastes, and some might even kill those helpful microorganisms.

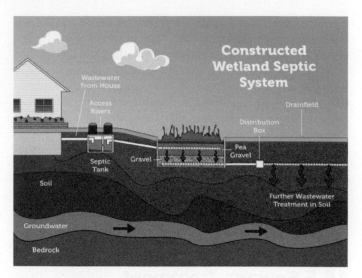

Figure 16.25

There are many designs for septic systems. This one with a constructed wetland employs plants to remove pathogens and nutrients.

Source: U.S. Environmental Protection Agency

Just as natural wetlands can serve a water-cleaning role, so artificial wetlands and ponds can enhance a septic system and increase its capacity, with the aid of plants. The conventional subsurface leaching field can be replaced by one or more gravel-filled basins planted with wetland plants (**figure 16.25**). While the effluent stays below the surface, the plants' roots reach down into the wet gravel. The plants consume some of the nutrients in the water, and their roots host quantities of the bacteria that decompose the sewage. If space allows, the cleaned water could be channeled into a pond supporting additional plant life and waterfowl.

Municipal Sewage Treatment

In urbanized areas, population density is far too high to permit effective sewage treatment by septic systems. About 84% of U.S. households are now served instead by sewer systems. Municipal sewage treatment does vary; the basic steps involved are summarized in **figure 16.26** and below.

Primary treatment first involves the use of screens (#4) to physically remove larger solids including trash, rocks, toys, and other objects that have been carried along with the wastewater. Then the wastewater moves to the grit chamber (#5) where sand and gravel settle out to the bottom. The materials removed in these first two stages are most commonly transported to a landfill for disposal. The remaining liquid and organic matter are moved to a clarifier or sedimentation tank (#7) where the rate of water flow is carefully controlled to let solids sink to the bottom, and this *sludge* is removed for disposal, treatment, or use as a fertilizer. Oil and grease are skimmed off the surface. The remaining dissolved organic matter and other dissolved chemicals remain in solution and move to the next stage of treatment.

SEWAGE TREATMENT PLANT

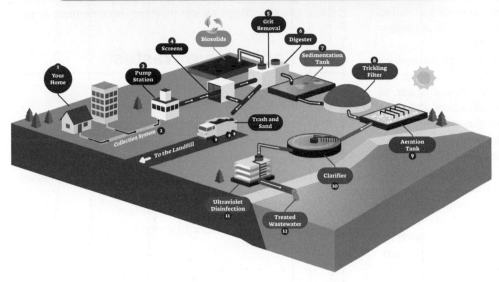

Figure 16.26

Schematic diagram of primary, secondary, and tertiary stages of municipal sewage treatment. Step #1 is where the wastewater is pumped or otherwise brought to the surface for treatment. See text for fuller description of the role of each stage.

VectorMine/Shutterstock

Secondary treatment of the remaining liquid is mainly biological. The effluent from primary treatment is aerated, and bacteria act on the dissolved and suspended organic matter to break it down (#9). The treated sewage passes through another clarifier (#10) to remove solids, which now contain the useful microorganisms, too. This activated sludge can be cycled back to mix with fresh input water and air to continue the breakdown process. At this point, the water way also be disinfected with chlorine, ozone, or UV light (#11). Primary plus secondary treatment together can reduce suspended solids and oxygen demand (see chapter 17) by about 90%, nitrogen by one-half, and phosphorus by one-third. Other dissolved chemicals remain. Virtually all municipal wastewater treatment plants in the United States provide secondary treatment.

After secondary treatment, wastewater can also be subjected to various kinds of tertiary treatment before it is discharged, generally to a nearby surface water body. Fine filtration, passage through activated charcoal, distillation, further chlorination, and various chemical treatments to remove dissolved minerals and other chemicals are some possible forms of tertiary treatment. Such thorough tertiary wastewater treatment can add significant costs to the treatment process, which is why tertiary treatment occurs in only about one-third of sewage treatment plants. It is likely that we will increasingly use tertiary treatment as a means of extending dwindling water supplies and in response to the challenges of making our water sources safe to drink. More diverse compounds are finding their way into municipal sewage that are not being removed through the usual treatment process, and that scientists are yet unsure how they affect both ecosystems and humans. These include herbicides and many compounds associated with pharmaceuticals and personal care products such as antibiotics, hormones, microplastics, and steroids. While there may exist ways to remove these contaminants of concern from the waste stream, it may not be economical to do so.

A weakness in some municipal treatment systems is that where flow from storm and sanitary sewers is combined, large amounts of rain or meltwater runoff in a short time can overload the capacity of treatment plants. At such times, some raw sewage may be released to streams and lakes untreated. There are currently 860 municipal systems that experience these combined sewer overflows. Remedies are generally expensive and include separating the storm and sanitary sewers, and adding storage to the system. In response to numerous overflows into Lake Michigan and streams that drain into the lake, the Milwaukee Metropolitan Sewage District added a Deep Tunnel system with a storage capacity of 521 million gallons (**figure 16.27**).

Figure 16.27

This deep tunnel from Chicago's TARP system is very similar to the one in Milwaukee's sewage system, both serving to reduce the effects of untreated waste reaching Lake Michigan during large storm events.

Chicago Tribune/Getty Imges

Since its completion in 1994, 141 billion gallons of untreated waters have been saved from flowing directly into surface water bodies. The nearly 30-mile long, up to 300-feet deep tunnel did come at a cost of $98.5 million. Even so, the general population in Milwaukee County is still asked to refrain from using water-draining appliances during large storm events.

Wastewater treatment can yield other benefits besides cleaned water. The city of Arcata, California, uses sewage that has passed through primary treatment to supply water to marshes and ponds, where natural microbial action and plant growth purify the water further and remove nutrients from it. Ultimately, the water purity is such that the output water can be used at a fish hatchery. The Tapia Water Reclamation Facility in northwestern Los Angeles County takes 10 million gallons of wastewater per day and transforms it into water of high enough quality to use for irrigation. And Calpine, the company that currently operates The Geysers geothermal-power complex, is transporting close to 20 million gallons a day by pipelines from Santa Rosa and Lake County, California, to recharge the geothermal reservoir and increase power production.

A by-product of wastewater treatment is large quantities of sludge, which can present a solid-waste-disposal problem. Recall that the sludge generated in primary treatment is made of organic waste and microbes, whereas activated sludge is mostly the deceased microbes and bacteria that consumed organics in the secondary stage of treatment. Turning either kind of sludge (or both combined) into fertilizer, usually after it has been composted in a digester or kiln-dried at high temperatures, is a way to recover some money rather than paying for disposal. Chicago's municipal sewage-treatment facilities process 1.3 billion gallons of sewage and generate approximately 600 tons of sludge (on a dry weight basis) every day. In the early 1970s, Chicago developed a plan for using some of its sludge to fertilize 11,000 acres of formerly strip-mined land during reclamation. Austin, Texas, composts and transforms sludge into "Dillo Dirt" with which parks and athletic fields are fertilized; it is also sold to gardeners. The city of Milwaukee has been selling the sludge as the fertilizer "Milorganite" for decades. Urban sludge can contain high concentrations of heavy metals or other toxic industrial chemicals. This may limit the use of such sludge fertilizer to land not to be used for food crops or pastureland (parks, golf courses, and so on). Currently, over half the sewage sludge produced in the United States is used for fertilizing/soil conditioning; the rest is either incinerated or landfilled.

In speaking of sewage treatment, it is worth keeping in mind the disparities between the industrialized and other nations. According to the United Nations, over 90% of the sewage in developing countries is discharged into surface waters with no treatment whatsoever, and that in many parts of the world, about 70% of the wastewater reaching fresh or coastal waters is untreated; areas where this is true include the Caribbean, West and Central Africa, and South and East Asia. It should also be noted that while the focus here has been on U.S.-style sewage transport and treatment, which involve a great deal of water, treatment of human wastes does not necessarily require those water-intensive approaches. In this country, composting toilets are perhaps the most familiar example of a non-aqueous system. Other nations, where water is in shorter supply, have made larger-scale use of such approaches, and they become increasingly appealing as water shortages become more acute.

16.6 Radioactive Wastes

Learning Objectives

- Describe radioactive decay
- List the possible effects of radiation on humans
- Compare low-level and high-level radioactive wastes
- List possible locations for the storage of high-level radioactive waste
- Describe Waste Isolation Pilot Plant
- Summarize the history of the proposed Yucca Mountain storage facility

Radioactive wastes differ somewhat in character from chemical wastes. Proposals for the disposal of solid and liquid radioactive wastes usually differ somewhat from the methods used for other wastes, although some radioactive materials are chemical toxins as well. Here we will review what radioactive decay is and the effects of radiation. We will also look at the nature of radioactive wastes and the proposed ways in which they may be stored or disposed of.

Radioactive Decay

Some isotopes, because of the combination of protons and neutrons in their atomic nuclei, are basically unstable, and sooner or later they change into other elements through a process called *radioactive decay*. In doing so, they release radiation: alpha particles (nuclei of helium-4, with two protons, two neutrons, and a +2 charge), beta particles (electrons, with a −1 charge), or gamma rays (electromagnetic radiation, similar to X rays or microwaves but shorter in wavelength and more penetrating). A given decay may emit more than one type of radiation.

Radioactive decay is a statistical phenomenon. It is impossible to predict the instant at which an atom of a radioactive element will decay. Over a period of time, a fixed percentage of a group of atoms of a given radioactive isotope (radioisotope) will decay. The idea is somewhat like flipping a coin. One cannot know beforehand whether a given flip will come up heads or tails, but over a great many flips, one can expect that half will come up heads, half tails.

Each different radioisotope has its own characteristic rate of decay, often described in terms of a parameter called **half-life**. The half-life of a radioisotope is the length of time required for half the atoms of that isotope initially present in a system to decay. The concept is illustrated in **figure 16.28**. Suppose, for example, that the half-life of a particular radioactive isotope is ten years and that, initially, there are 2000 atoms of it present. After ten years, there will be about 1000 atoms left;

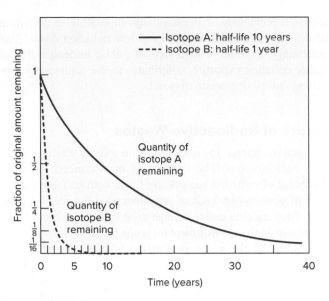

Figure 16.28

The phenomenon of radioactive decay: Each radioisotope decays at its own characteristic rate, defined by its half-life.

after ten more years, 500 atoms; after ten more years, 250; and so on. After five half-lives, only a few percent of the original amount is left; after ten half-lives or more, the fraction left is vanishingly small.

Half-lives of different radioactive elements vary from fractions of a second to billions of years, but for any specific radioactive isotope, the half-life is constant. Uranium-238 has a half-life of about 4.5 billion years; carbon-14, 5730 years; radon-222, 3.82 days. The half-life of a given isotope is constant regardless of the physical or chemical state in which it exists, whether it is in a mineral or dissolved in water, in the atmosphere or deep in the crust at high temperature and pressure. Spontaneous radioactive decay cannot be accelerated or slowed down by chemical or physical means. Because this is so, naturally occurring radioisotopes are powerful tools for studying Earth's history, as described in appendix A. Unfortunately, it also means that radioactive wastes will continue to be radioactive, and each isotope will decay at its own rate, regardless of how the wastes are handled. The wastes cannot be made to decay faster to get rid of the radioactive material more quickly, and they cannot be treated to make them nonradioactive. This is a key difference between radioactive wastes and many chemical toxins; many of the latter can, through proper treatment, be broken down or neutralized, which increases the disposal options available for dealing with them.

Effects of Radiation

Alpha and beta particles and gamma rays all are types of ionizing radiation, meaning they can strip off electrons from atoms or split molecules into pieces. Depending on which particular atoms or molecules in an organism are affected and the intensity of the radiation dose, the results could include genetic mutations, cancer, tissue burns, or nothing significant at all. Alpha particles are more massive but not very energetic.

They can be very damaging but cannot travel far through matter; they can be stopped by a sheet of paper. Alpha-emitting radioisotopes are dangerous only when inhaled or ingested. Beta particles are lighter and less damaging but travel farther; they can be stopped by wood. Gamma rays are the most energetic and penetrating, but can be stopped by concrete.

How much damage is done by a given dose of radiation? How small a dose is harmless? Scientists don't know precisely. Most controlled radiation studies have been done on animals, not humans. The most definitive data on radiation effects on people come from accidental exposures of scientists or technicians to sizeable doses of radiation or from observing the after-effects of the explosion of atomic weapons in Japan during World War II. In those cases, the doses received by the victims are generally very high and very imprecisely known, and their relevance to concerns about exposure to low levels of radiation is not clear in any case. Data from Chernobyl and Fukushima may, in time, provide some clarification.

The effects of low doses of radiation on humans are particularly hard to quantify for several reasons. One is that some of the consequences, such as slowly developing cancers, don't appear for many years after exposure. By that time, it is much harder to link cause and effect clearly than it is, for example, in the case of severe radiation burns resulting immediately from a massive dose. A related problem is that many of the results of radiation exposure—like cancer or mutations—can have other causes. There are hosts of known or suspected chemical carcinogens and mutagens; how does one uniquely identify the cause of each particular cancer or mutation, years after exposure to whatever caused it?

Another uncertainty arises from the question of linearity in cause and effect (**figure 16.29**). If a population is exposed to a certain dose of a certain type of radiation and one hundred cases of cancer result, does that mean that one-tenth the exposure would induce ten cases of cancer, and so on? Or is there a minimum harmful level of radiation exposure, below which we are not at risk?

We are, after all, surrounded by radiation. Cosmic rays bombard us from space; naturally occurring radioactive elements decay in the soil beneath our feet and in the wood and brick and rock and concrete in our homes. Perhaps, in consequence, humans have developed a tolerance of radiation doses that are not significantly above natural background levels, as with line B in **figure 16.29**. Nobody can avoid some natural exposure to radiation (**table 16.2**). We increase our exposure with every medical X ray, every airplane ride, every journey to a high mountaintop, and every time we use a ceramic cup or dish. Each of us is radioactive, a consequence of naturally occurring carbon-14 and other radioisotopes in the body. Any additional exposure from radioactive waste is just that, an incremental increase over our inevitable, constant exposure to radiation. Does such a small exposure to low-level radiation represent a significantly increased risk? How small is small? Frustratingly, perhaps, there are no simple, direct answers to questions such as these. Individuals' background exposure to radiation is universal, yet quite variable; for this and other reasons

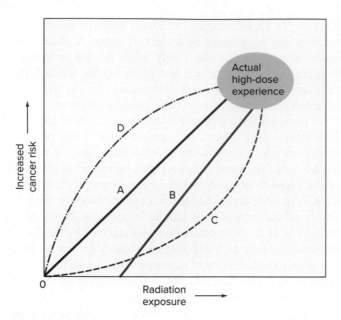

Figure 16.29

Current U.S. radiation-exposure limits are based on line A, linear—no threshold: if a massive dose of radiation causes a certain increase in cancer, we can extrapolate back on a straight line to zero dose, zero increased risk. Line B is linear with threshold: it assumes we can tolerate small doses with no ill effects but that beyond some threshold amount, risk increases in direct proportion to radiation exposure. Or maybe low doses add some risk but large doses are proportionately much more hazardous (line C); or maybe low doses increase risk a lot, and beyond that, higher doses don't increase risk much more (line D). None of us has zero exposure, and cancer has many causes, so we simply can't pin down the shape of the curve, for lack of clear data.

discussed previously, it is essentially impossible to determine definitively the effects of additional low radiation doses. That uncertainty, together with a lack of public understanding of natural radiation exposure, contributes to the controversy concerning radioactive-waste disposal.

Nature of Radioactive Wastes

As noted in chapter 15, nuclear fission reactor wastes include many radioisotopes. The isotopes of most concern from the standpoint of radiation hazards are those with half-lives of the order of years to hundreds of years that pose the greatest problems. They are radioactive enough to be significant hazards, yet will persist in the environment for some time.

Some of the waste radioactive isotopes are also toxic chemical poisons, so they are dangerous independently of their radioactivity, even if their half-lives are long. Plutonium-239, with a half-life of 24,000 years, is one example of such an isotope. Other isotopes pose special hazards because the corresponding elements are biologically concentrated, often in one particular organ, leading to the possibility of a concentrated radiation source in the body. Included in this category are such products as iodine-131 (half-life, eight days), which would be concentrated, like any iodine, in the human thyroid gland; iron-59 (forty-five days), which would be concentrated in the iron-rich hemoglobin in blood; cesium-137 (thirty years), which is concentrated in the nervous system; and strontium-90 (twenty-nine years), which tends to be concentrated with calcium in bones, teeth, and milk.

Radioactive wastes are often classified collectively as **low-level** or **high-level**. The division is an imprecise one. The low-level wastes are relatively low-radioactivity wastes, not

Table 16.2	Typical Natural Radiation Exposure
Source	**Dosage, mrem/yr***
radon	200
external (cosmic rays, rocks, soil, masonry building materials)	55
medical (mainly X rays)	53
internal (in the body)—includes doses from food, water, air	39
consumer products	10
other, including normal operation of nuclear power plants	3
Where and how you live can affect your exposure	
worldwide radioactive fallout	4 mrem/yr
chest X ray	10–39 mrem
lower-gastrointestinal-tract X ray	500 mrem
jet travel—for each 2500 miles, add	1 mrem
TV viewing, 3 hours/day	1 mrem/yr
cooking with natural gas, add	6 mrem/yr

*Radiation dosages are expressed in units of *rems* or, for very small doses, *millirems* (mrem). A *rem* ("radiation equivalent man") is a unit that takes into account the type of radiation involved, the energy of that radiation, the amount of it to which one has been exposed, and the effect of that type of radiation on the body. The U.S. average annual dosage is 360 mrem.

Source: U.S. Department of Energy.

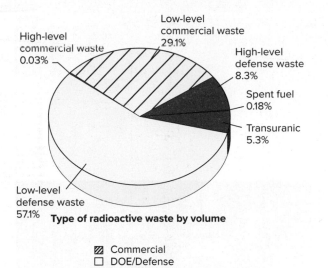

Type of radioactive waste by volume

☒ Commercial
☐ DOE/Defense

Type of radioactive waste by level of radioactivity

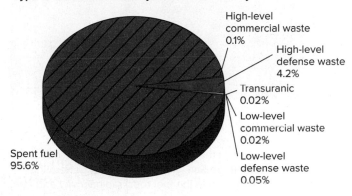

Figure 16.30

By volume, high-level waste is a small fraction of the radioactive-waste-disposal problem—but it accounts for nearly 100% of the radioactivity, and thus the radiation hazard.

Data from U.S. Department of Energy report DOE/RW-0006.

requiring extraordinary disposal precautions. Volumetrically, they account for over 90% of radioactive wastes generated (**figure 16.30**). Some low-level wastes are believed to be harmless even if released directly into the environment. The routine small releases of radioactive gas from operating fission reactors are in this category. Liquid low-level wastes—from laundering of protective clothing, from decontamination processes, from floor drains—are often released also, diluted as necessary so that the concentration of radioactivity is suitably low. Solid low-level wastes, such as filters, protective clothing, and laboratory materials from medical and research laboratories, are commonly disposed of in landfills. Some solid low-level wastes from commercial reactor operations are held on-site until their radioactivity has decreased essentially to background levels, at which point they are treated like ordinary trash. Other low-level wastes are collected on site until enough has accumulated to warrant shipment for disposal.

Federal legislation required states to take the responsibility for their own low-level wastes in the early 1990s (though they

may enter into multi-state compacts, and several of these exist). States that benefit most from nuclear power, and research and other facilities using nuclear materials, bear more of the burden of the disposal of the corresponding low-level wastes. There is often concern over local disposal of low-level wastes. As a result, no states have yet developed the mandated disposal sites to the point of actual disposal, and in some cases, planning is not far advanced. However, there are four sites currently licensed by the Nuclear Regulatory Commission (NRC) to receive low-level wastes for disposal, one each in South Carolina, Texas, Utah, and Washington State, and states may contract with these facilities for disposal of their low-level nuclear wastes.

Spent reactor fuel rods and by-products from their fabrication and reprocessing are examples of high-level wastes. Given that they represent much more concentrated and intense sources of radiation, there is correspondingly greater concern about their proper disposal. Over the years, various more-or-less elaborate disposal schemes have been proposed. Currently, these materials sit in temporary storage—often on-site at nuclear reactors and other nuclear facilities—awaiting permanent disposal (**figure 16.31**). The basic disposal problem with the chemically varied high-level wastes is how to isolate them from the biosphere with some confidence that they will stay isolated for thousands of years or longer. In 1985, the EPA specified that high-level wastes be isolated in such a way as to cause fewer than 1000 deaths in 10,000 years (the expected fatality rate associated with unmined uranium ore); in 2008, they increased the required time period of waste isolation to 1 million years. The sections that follow survey some of the many strategies proposed for dealing with these high-level and other concentrated radioactive wastes.

Possibilities for High-Level Waste Disposal

Some of the more exotic schemes proposed in the past for disposal of high-level radioactive wastes include: putting wastes in space, perhaps rocketing them into the Sun; encapsulating them under thick ice sheets in Antarctica; burying them in sediments near ocean trenches so that they would be pulled into subduction zones; and burying them in the thick, clay-rich sediments of the deep, isolated ocean floor. The logistics and costs of these proposals, even if technically feasible, coupled with serious questions regarding risk, make these schemes impractical (at least for now).

Disposal of both liquid and solid high-level radioactive wastes could, more practically, occur in specific bedrock conditions. Some, such as by-products of fuel reprocessing, are presently held in liquid form rather than solid, in part to keep the heat-producing radioactive elements dilute enough that they don't melt their storage containers. These liquid wastes are currently stored in cooled underground tanks, holding up to 1 million gallons apiece. Instead, they could be stored in caverns hollowed out of low-permeability, unfractured rocks such as basalt or granite. Geologists would need to thoroughly investigate the rocks' physical properties beforehand to ensure that the caverns could contain the liquid effectively, and that the rocks

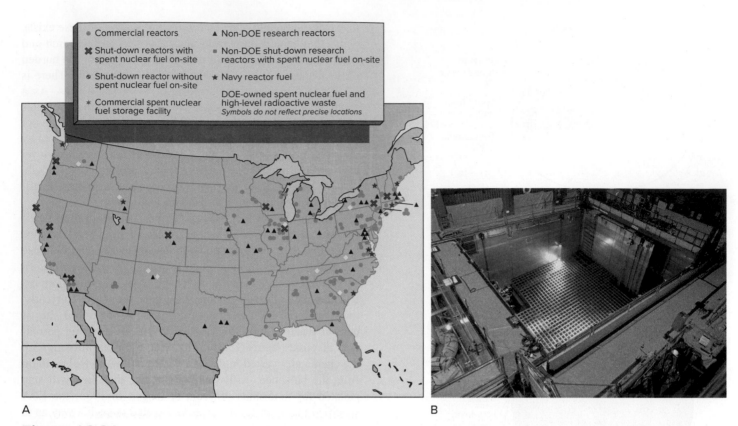

- Commercial reactors
- ✖ Shut-down reactors with spent nuclear fuel on-site
- ⊙ Shut-down reactor without spent nuclear fuel on-site
- ✳ Commercial spent nuclear fuel storage facility
- ▲ Non-DOE research reactors
- ■ Non-DOE shut-down research reactors with spent nuclear fuel on-site
- ★ Navy reactor fuel

DOE-owned spent nuclear fuel and high-level radioactive waste
Symbols do not reflect precise locations

A

B

Figure 16.31

(A) Locations (approximate) of spent nuclear fuel elements and high-level radioactive wastes in the United States awaiting disposal. Currently, about 90,000 metric tons of such wastes have accumulated. (B) Spent-fuel pool at the San Onofre, California, generating station. Spent reactor fuel is often stored on-site at the reactor facility, in deep pools of water, while the fuel rods' radioactivity and heat production decrease.

Sources: (A) Map from U.S. Department of Energy, Office of Civilian Radioactive Waste Management; (B) United States Nuclear Regulatory Commission

were geologically stable. Faulting or fracturing of the rocks would create conduits through which the highly radioactive liquid could escape rapidly, perhaps to the surface or into nearby aquifers. The wells leading down to the caverns, through which wastes would be pumped, would have to be tightly cased to prevent shallower leaks.

The Hanford, Washington, federal radioactive-waste repository has over 50 million gallons of high-level liquid wastes in storage; an estimated 1 million gallons have leaked, some into the Columbia River, and leakage continues (**figure 16.32**). Remediation efforts are in process there including bioremediation of toxic metals, but these methods can do nothing to reduce the radioactivity. Full cleanup at the Hanford Superfund site will involve containing and chemically stabilizing the wastes, moving them to a new storage site, and dealing with the contaminated soil. The Department of Energy (DOE) currently spends about $2.5 billion a year on the Hanford cleanup employing over 8000 people, now projected to take up to 75 years and ultimately cost between $323 and $677 billion.

Solidifying, or vitrifying, liquid high-level wastes prior to disposal decreases the mobility of the waste. Certain minerals, ceramics, and glasses have proven to be relatively resistant to

Figure 16.32

The Hanford Site was part of the Manhattan project and produced plutonium weapons during World War II and the Cold War. It is now the largest environmental cleanup in the United States.

Department of Energy/NRC

leaching over centuries or even millennia in the natural environment. Conversion of the wastes into a relatively stable solid form would be a first step in many high-level-waste disposal schemes. The solid waste would then be sealed in canisters and the canisters placed in some kind of bedrock cavern, old mine, or trench. Solidification would render the wastes less susceptible to rapid escape if the storage facility were breached. Operational vitrification plants already exist in Europe and Japan. Currently under construction, Hanford's Waste Treatment and Immobilization Plant will be the largest of its kind, turning it liquid radioactive waste into vitrified low-level solid waste to be store on-site and high-level waste to be stored at an off-site facility (yet to be determined.)

A variety of bedrock types are being investigated worldwide as host rocks for solid high-level-waste disposal. The general design with any bedrock disposal site or repository would involve the **multiple barrier concept:** surrounding solid waste with several different types of materials to create multiple obstructions to waste leakage or invasion by groundwater. A major variable is the nature of the host rock. Each of the rock types introduced below has some particular characteristics that make it promising as a potential host rock, as well as some potential drawbacks.

Granite is plentiful, strong, and stable, but it commonly contains fractures. It also has a low porosity and minerals that are quite insoluble in a temperate climate. Fresh, un-fractured, thick *basalt* is another option. It is strong, has high-temperature minerals and sometimes glass, and a fairly high thermal conductivity. Weaknesses include possible fractures, porous gas-bubble zones, and that it may easily weather. Massive deposits of *tuff* are another possible hot rock for solid high-level wastes. Tuff has some similar physical properties to basalt, but is very brittle and easily fractured. Some tuffs have abundant zeolites, which are potentially valuable for their ion-exchange capacity, but those that do are weaker, more porous, and more permeable than welded tuffs. Thick *clays* would both slow down and adsorb migrating elements but might dehydrate at high temperatures. *Shale* is low in permeability and somewhat plastic under stress but is also weak, fractures readily, and is commonly interbedded with permeable sedimentary rocks.

Salt, in thickly bedded deposits or domes, has several special properties that may make it a particularly suitable repository. It has a relatively high melting point, so can withstand considerable heating by radioactive wastes without melting. Although soluble under wet conditions, salt typically has low porosity and permeability. Under dry conditions, it can provide very tight storage with minimal leakage potential. Salt flows plastically under pressure while remaining solid, making it somewhat self-sealing. If it were fractured by seismic activity or by accumulated stress in the rocks, flow in the salt could seal the cracks. The oil from Strategic Petroleum Reserve is successively stored in salt caverns in the southeastern United States.

Waste Isolation Pilot Plant (WIPP)

While commercial high-level wastes await permanent disposal, the U.S. government has inaugurated a disposal site in southeastern New Mexico for some of its own high-level wastes, **transuranic wastes** created during the production of nuclear weapons. The Waste Isolation Pilot Plant, or WIPP, officially became the first U.S. underground repository for such wastes on 26 March 1999, when it received its first waste shipment. This event was the culmination of twenty years of site evaluation, planning, and construction following congressional authorization of the site in 1979.

The disposal unit at WIPP is bedded salt, underlain by more evaporites and overlain by mudstones. The disposal layer is 2150 feet below the surface. The region is now dry, so the water table is low, and examination of the geology indicates that it has been stable for over 200 million years, offering reassurance with respect to future stability and waste containment. The facility is shown in **figure 16.33**.

So far, WIPP has received more than 3.5 million cubic feet of transuranic wastes (a bit over half of its capacity), from 14 sites around the country. To address public concerns about the transport of these highly radioactive materials, they are shipped in extremely sturdy containers (**figure 16.34**), and all the WIPP trucks, driven by carefully screened drivers, are tracked by satellite. Collectively the wastes have traveled over 15 million miles from their sources to WIPP for disposal, without incident. This should reassure those who fear accidents associated with transportation of spent fuel from its many current locations to any centralized disposal facility.

WIPP is ultimately expected to receive transuranic wastes from twenty-some sites across the nation. WIPP's authorization extends only to defense-related transuranic waste, not to commercial or other high-level radioactive wastes. For these, site analysis and development are still in progress.

The Long Road to Yucca Mountain: A Dead End?

Much of the delay in the selection of a site for disposal of high-level radioactive waste from commercial nuclear reactors has resulted from the need for thorough investigation of the geology of any proposed site. However, in the United States at least, there is also a political component to the delay. Simply put, nobody wants to host a high-level-waste disposal site.

The Nuclear Waste Policy Act of 1982 provided for the establishment of two high-level-waste disposal sites for commercial reactor waste, one in the western United States, one in the East, both to be identified by the DOE. The same Act created the Office of Civilian Radioactive Waste Management and established the Nuclear Waste Fund (NWF), supported by spent-fuel fee charges on waste generators and a fee of 0.1 cent per kilowatt-hour on nuclear electricity. This sounds small, but to date, over $44 billion has been paid into NWF. The NWF was to finance disposal-site selection and investigation; over $10 billion has been spent so far.

In 1986, DOE announced three candidates for the western site: Hanford, Washington (where disposal would be in basalt); Yucca Mountain, Nevada (ash-flow tuff); and Deaf Smith County, Texas (salt). All three states protested. So did

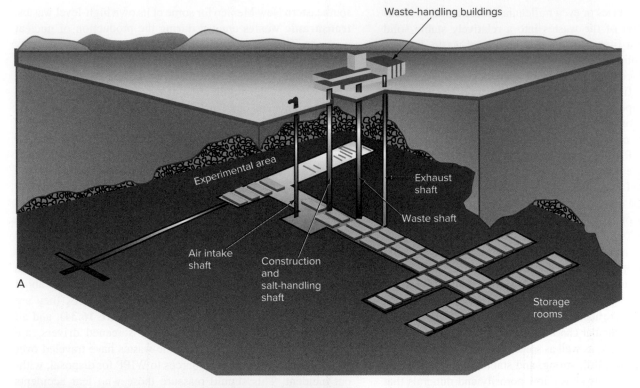

Waste-handling buildings

Experimental area

Exhaust shaft

Waste shaft

Air intake shaft

Construction and salt-handling shaft

Storage rooms

A

B

Figure 16.33

Waste Isolation Pilot Plant (WIPP) in New Mexico. (A) Plan of storage facility. (B) Waste canisters are emplaced, by remotely controlled machinery, into holes bored into salt walls.

Sources: (A) After Office of Environmental Restoration and Waste Management, Environmental Restoration and Waste Management: An Introduction, U.S. Department of Energy; (B) DOE Photo

Tennessee, where—at Oak Ridge—the waste-packaging site was to be built to prepare the wastes for disposal. Original plans would next have called for a billion-dollar testing program to be undertaken at each of the three proposed disposal sites, to be followed by selection of one for *the* western site. Congress subsequently called a year-long halt to further action.

Then, in late 1987, Congress suddenly moved decisively to end the uncertainties of site choice. The Yucca Mountain site (**figure 16.35**) would be developed, and investigation of the other sites halted, at a cost savings estimated at $4 billion. The state of Nevada would receive $20 million per year and special consideration of certain research grants by way of compensation.

Properties of the site that contribute to its attractiveness include (1) the tuff host rock; (2) generally arid climate, with no perennial streams and limited subsurface water percolation in the area; (3) low population density; (4) low regional water table, so the repository would be 200 to 400 meters above the water table; and (5) relative geologic stability. There are no known useful mineral or energy resources in the vicinity. The site is also located on lands already owned by the federal government, and overlaps the Nevada Test Site previously used for nuclear-weapons testing.

There are several geologic issues to be considered. There are a number of faults in the area, some related to the collapse of ancient volcanic calderas, some related to the formation of the mountains, some related to plate motions. Current seismicity is low, and evidence suggests limited local fault displacement for the last 10 million years. Still, repository design would provide for earthquakes up to magnitude 6.8 or so, the largest plausible on the known faults. The most recent volcanism in the region is a cinder cone that may be as young as 20,000 years old, and basaltic volcanism is possible within the next 10,000 years or so. There are also questions about how likely the region is to stay as dry as it currently is, particularly in light of anticipated global climate change, so the rocks' response to increased water may be relevant.

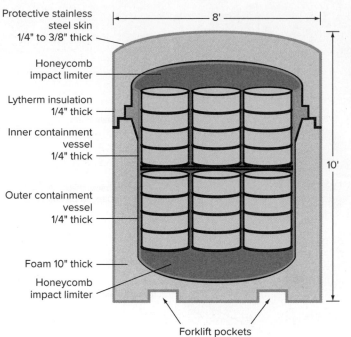

Trupact-II

Protective stainless steel skin 1/4" to 3/8" thick

Honeycomb impact limiter

Lytherm insulation 1/4" thick

Inner containment vessel 1/4" thick

Outer containment vessel 1/4" thick

Foam 10" thick

Honeycomb impact limiter

Forklift pockets

8'

10'

Weight
12,705 lb empty
19,250 lb loaded

Material
Stainless steel
Polyurethane foam
Ceramic fiber insulation

Figure 16.34

Wastes destined for WIPP are contained in specially designed vessels such as this even before transport to the site. Similar vessels would be used for waste headed for Yucca Mountain or an alternate repository. Trupact-II was the first canister design; two similar, newer designs are now also in use.

Source: U.S. Department of Energy.

A

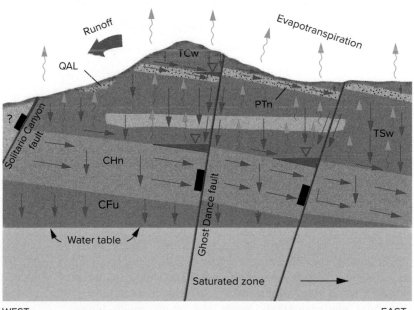

Runoff

Evapotranspiration

QAL

TCw

PTn

TSw

Solitario Canyon fault

?

CHn

Ghost Dance fault

CFu

Water table

Saturated zone

WEST   EAST

QAL	Alluvium (stream deposits)	Liquid-water flow
TCw	Tiva Canyon welded unit	Water-vapor flow
PTn	Paintbrush nonwelded unit	Normal fault
TSw	Topopah Spring welded unit	Possible perched-water zone
CHn	Calico Hills nonwelded unit	
CFu	Crater Flat (undifferentiated) unit	

B

Figure 16.35

(A) View to the south of Yucca Mountain; coring activities are visible along the ridge. The desert site is 90 miles from Las Vegas, the closest sizeable population center. (B) Cross section of proposed waste-disposal site at Yucca Mountain, Nevada. Generalized east-west section, showing potential water and water-vapor flow paths; the "welded" and "unwelded" units are tuffs. Repository level highlighted (yellow).

Sources: (A) Office of Civilian Radioactive Waste Management/DOE; (B) After Site Characterization Plan Overview, U.S. Department of Energy, 1988

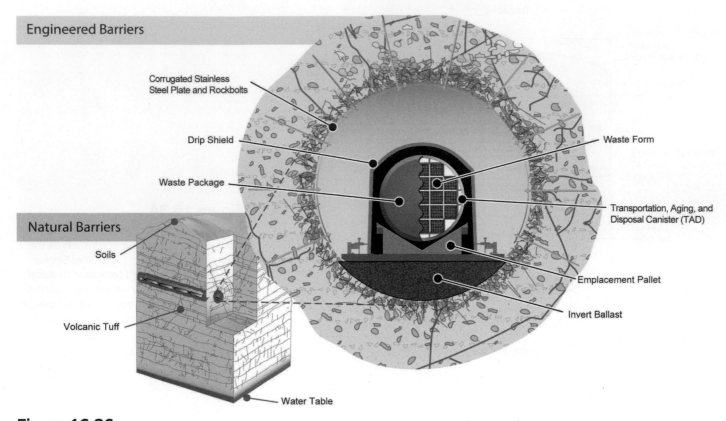

Engineered Barriers

Corrugated Stainless
Steel Plate and Rockbolts

Drip Shield

Waste Package

Waste Form

Transportation, Aging, and
Disposal Canister (TAD)

Natural Barriers

Soils

Volcanic Tuff

Emplacement Pallet

Invert Ballast

Water Table

Figure 16.36

Both natural and engineered barriers contribute to secure containment of wastes in the Yucca Mountain design.

Diagram courtesy Office of Civilian Radioactive Waste Management/DOE.

However positive the geologic factors, waste containment would not rely solely on geology (**figure 16.36**). In addition to the natural barriers are such engineered barriers as a stable waste form, durable container, drip shield over the waste to deflect downward-percolating water, steel-plate liner for the tunnels, and more. This illustrates the multibarrier concept previously mentioned.

Site-investigation activities have included very detailed structural, hydrologic, and geochemical studies, and more detailed analysis of geologic-stability issues (**figure 16.37**). The evaluation has been comprehensive, with laboratory studies supplementing research in the field. The site characterization plan was a nine-volume, 6000+ -page document, the draft and final environmental impact statements much longer. Meanwhile, the high-level wastes remain on hold, in temporary storage, as the Yucca Mountain timetable slips. Site characterization began in 1986. At one time, the projection was for the start of construction in 1998 and first waste disposal in 2003. In fact, the final site analysis was not completed until late in fiscal year 2001, the final environmental impact statement was issued by the DOE in February 2002, and the site was officially designated by Congress and the president for a high-level waste repository in July 2003. NRC review of

DOE's 2008 license application was expected to take three to four years; officials hoped for construction to begin in time for waste disposal to start in 2017, their estimate of the best achievable timetable.

In 2009, then-President Obama suspended all funding for the Yucca Mountain project, after over 20 years and $9 billion spent studying Yucca Mountain and planning for the waste-disposal site. This did not halt NRC review of the license application, which continued until late 2011, when the matter became mired in legal action. Review of the license application resumed in 2013, and the NRC issued a safety-evaluation report in 2015 and a final supplemental report in 2016. The Trump Administration requested funds to restart licensing of the project in three of four of their budgets, but none were enacted. The Biden administration has no yet sought funding for the project.

Throughout this, high-level waste continues to accumulate in temporary storage, and the federal government has had to pay utility companies $9 billion to store it themselves. The amount of high-level waste accumulated so far exceeds the Yucca Mountain repository design to accommodate 70,000 metric tons. In the end of 2021, two additional private-sector temporary storage sites were proposed, immediately facing opposition from the two host states. The relicensing of the

A

B

C

Figure 16.37

Tunnels were bored into Yucca Mountain using alpine mining equipment (A) to provide access to study the rocks in place, including their response to heat (B) and their hydrologic properties (C).

(A-C) Office of Civilian Radioactive Waste Management/DOE

Yucca Mountain project, with an expanded acceptable volume of waste, is necessary to fully decommission nuclear sites that are already shutdown but are still storing waste, to license new nuclear power plants, to halt enormous federal payments to utility companies seeking damages, and to ultimately make storage of high-level waste in the United States much safer than it currently is.

No High-Level Radioactive Waste Disposal Yet

At least two dozen nations are making plans to dispose of high-level nuclear wastes; but so far, no high-level radioactive wastes from power-reactor operation have been consigned to permanent disposal sites elsewhere in the world by any nation. All this waste is in temporary storage awaiting selection and construction of disposal sites. Geologists can identify sites that have a very high probability of remaining geologically quiet and secure for many generations to come, but absolute guarantees are not possible. Uncertainties about the long-term integrity and geologic stability of any given site, coupled with increasing public resistance in regions with sites under consideration, have held up the disposal efforts. Sweden's planned facility was approved by their government in early 2021. Finland is constructing a bedrock disposal site in granitic gneiss that should start receiving waste in 2024.

Radioactive-waste disposal remains an issue even if the expanded nuclear-fission power option is not pursued. Currently, France, Russia, Japan, India, and China reprocess spent reactor fuel elements, which reduces the volume of high-level waste but does not eliminate the problem. Like the United States, Canada, Finland, and Sweden do not reprocess, but plan simple disposal of spent fuel. Worldwide, spent fuel is accumulating at over 11,300 metric tons a year. In 60 years of nuclear-power generation, over 400,000 metric tons of spent fuel have been produced, about a third of which have been reprocessed. A growing number of nations have accumulated high-level wastes, not only from past power generation but also from nuclear-weapons production and radiochemical research, including medical applications. All of that waste must ultimately be put somewhere in permanent disposal. The only alternative is to keep the wastes at the surface, closely monitored and guarded to protect against accident or sabotage, assuming that long-term political stability is more likely than geologic stability. It is expected that Finland, Sweden, and France will have operational geologic repositories for disposal of high-level wastes by the mid to late 2020s. Elsewhere, including in the United States, it remains an open question where the wastes will be put, and how soon.

Decisions, Decisions . . .

"Well, *I'm* not wasting a foam cup for my coffee!" declared Bonnie firmly. "Paper makes a lot less waste. Besides, these cups are partly recycled paper. I get them from the same place I get my recycled-paper coffee filters."

"But when you've used up that paper cup, it's all trash," retorted Rod. "You can't recycle paper that's dirty with food waste. I, on the other hand, simply rinse my plastic foam cup, and voilà— I'm ready to recycle, having carefully observed the '6-PS' on the bottom, meaning that it's polystyrene, which our garbage-disposal company does take. I checked."

"Of course, you've just used—and dirtied—some water rinsing out the cup, Mr. Greener-than-thou. And my paper cup will biodegrade in the landfill and make methane! Your plastic cup is just going to take more energy to recycle—probably to make more trash-can liners like the one in this wastebasket."

"Well, it beats just making more trash we didn't have room for. And speaking of trash-can liners, your paper cup isn't going to do much of anything for years and years, neatly sealed in plastic in the landfill. Don't hold your breath waiting for that methane."

"Okay, but how about where the cups come from? Yours took oil to make; mine just took trees, and they're renewable."

Bonnie and Rod had progressed to the point of arguing the relative energy and materials efficiency of paper versus polystyrene cups when Jason sauntered in, waving his mug.

"I do hate to interrupt a good argument," he commented, "but may I suggest a superior alternative? What we have here is the basic china mug—made from abundant natural material, clay, and infinitely reuseable, unless I get clumsy."

Figure CS 16.1.1

Paper, plastic, ceramic, glass? The choice with the least negative environmental impact is not always obvious.

©*Carla Montgomery*

There was a pause, and then Bonnie muttered, "It still needs washing, using water and detergent. So it's not perfect. At least I *hope* you wash that thing once in a while!"

This is a first look at the kind of analysis that can go into trying to choose the most environmentally benign of two or more alternatives. A complete comparison of paper versus polystyrene cups would require a good deal of research on the resource demands of, and pollution and waste produced by, the cutting of logs and milling and production of the paper cup, the drilling for oil to make and the process for the manufacturing of the polystyrene, and so on. An analysis considering just resource use for a single complete place setting is shown in **figure CS 16.1.2**.

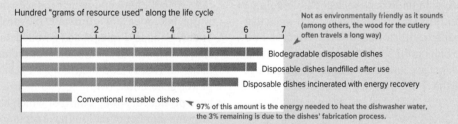

Everyday alternatives: biodegradable, disposable or conventional tableware?

Hundred "grams of resource used" along the life cycle

Not as environmentally friendly as it sounds (among others, the wood for the cutlery often travels a long way)

Biodegradable disposable dishes
Disposable dishes landfilled after use
Disposable dishes incinerated with energy recovery
Conventional reusable dishes

97% of this amount is the energy needed to heat the dishwasher water, the 3% remaining is due to the dishes' fabrication process.

The calculations consider all resources necessary to support the life cycle of a single table setting (plate, glass, knife, fork, spoon, and coffee cup).

Source: Friedrich Schmidt-Bleek et al., *Der ökologische rucksack, Wirtschaft für eine zukunft mit zukunft*, Hirzel Editions, Stuttgart, 2004.

Figure CS 16.1.2

This resource analysis of tableware alternatives considers both energy and materials involved in calculating resources used, but not such factors as the pollution associated with mining or drilling for raw materials (minerals, oil) or the manufacturing processes. Assumptions must always be made—for example, about efficiency of manufacturing processes or material transport. One could go still further and consider greenhouse-gas production from the energy used, or methane that might be derived from the biodegradable set, or . . .?

Graphic by Emmanuelle Bournay, UNEP/GRID-Arendal Maps and Graphics Library. Reprinted with permission UNEP. old.grida.no/graphicslib/ detail/everyday-alternatives-biodegradable-disposable-or-conventional-tableware_a4ed.

Source: Friedrich Schmidt-Bleek et al., Der ökologische rucksack, Wirtschaft für eine zukunft mit zukunft, Hirzel Editions, Stuttgart, 2004.

Another common debate of a similar kind involves disposable paper versus reuseable cloth diapers. Both alternatives, in this case, use renewable raw materials, but the energy required for, and the environmental consequences of, their manufacture will differ; the disposables commonly end up in landfills, while the reuseables require water, detergents, and energy to clean, and the wash water in turn becomes polluted. Again, a thorough, cradle-to-grave analysis of each alternative may be a good deal more complex, and the choices not as straightforward as they may initially appear. For some consumers and product types, too, economic considerations may enter into the choices made.

Consider other everyday products that offer a choice of materials. That choice probably has waste-disposal and pollution, as well as resource-use, consequences. Think about what those might be and how you might decide which choice to make.

Case Study 16.2

The Ghost of Toxins Past: Superfund

Disposal of identifiably toxic wastes in the United States is carefully controlled today. It has not always been so. In some cases, disposal companies hired by other firms disposed of toxic wastes illicitly and inappropriately. In other instances, the waste-generating companies dumped or buried their own wastes on-site. Decades later, when the firm responsible is long gone, pollutants may ooze from the land surface or may suddenly be detected in groundwater. Sometimes, the waste disposer can be found but denies that the problem is serious and/or declines to spend the money to clean up the site. Sometimes, the responsible company is no longer in business.

In response to these and other similar situations, in late 1980, Congress passed the Comprehensive Environmental Response, Compensation, and Liability Act which, among other provisions, created what is popularly known as Superfund. The original $1.6 billion trust fund was generated primarily through taxes on oil and on producers of forty-two specific hazardous chemicals. Superfund monies are intended to pay for immediate emergency cleanup of abandoned toxic-waste sites and of sites whose owners refuse to clean them up. Superfund was reauthorized several times, its trust fund increasing to $13.6 billion before its taxing authority expired in 1995. Since that time, its operations have been supported by federal appropriations. The American Recovery and Reinvestment Act of 2009 provided an infusion of $600 million to Superfund to help speed up cleanup activities. The new Bipartisan Infrastructure Law will add another $3.5 billion and reinstate the taxing authority.

The size of the toxic waste problem was and is still far greater than originally estimated. At the time Superfund was created, Congress envisioned cleanup of about 400 toxic-waste sites. As of March 2022, 1333 were on the National Priorities List (NPL) (**figure CS 16.2.1**), 43 more listings were pending, and 431 had been deleted. Many other sites posing less-acute risks have also been identified. Another 3781 hazardous waste sites are on the 2019 list for corrective action under the Resource Conservation and Recovery Act (RCRA). And there are an estimated 450,000 brownfields, sites where redevelopment may be complicated by a hazardous substance, contaminant, or pollutant.

Some Superfund NPL sites have been well publicized and notorious. One famous example is Love Canal, New York, where Hooker Chemical dumped 21,000 tons of waste chemicals from 1942 through 1952. The wastes were subsequently covered up, and homes and a school built nearby. Unusually high rates of cancer and other health problems among residents eventually led to recognition of the pollution, and Love Canal was placed on the NPL in 1983. Other sites may be smaller or less well-known, but all NPL sites are considered to pose major hazards.

Cleanup of NPL sites is meticulously planned and executed, so is generally both expensive and slow. Even after construction is completed, it may be several more years before treatment of the site is sufficient for EPA to delete it from the NPL. Love Canal was not deleted from the NPL until 2004.

The average cost of cleaning up a Superfund site is over $40 million, not counting any associated litigation, and some projections now put eventual total cleanup costs of Superfund sites at around $1 trillion. Where responsible parties can be identified, they do contribute to the cleanup costs but massive cleanup bills remain.

In the meantime, every bit of hazardous waste cleaned up and removed from one spot still has to be disposed of in another. Waste minimization becomes increasingly appealing as these problematic NPL, RCRA, and brownfield sites multiply.

Curious about what is going on around you? Visit the Cleanup in My Community map that contains information not just about hazardous waste but other environmental risks in your area.

(Continued)

Figure CS 16.2.1

NPL sites in the fifty states, as of March 2022. Every state has at least one, some more than a hundred. Dozens more sites are currently being evaluated for possible addition to the NPL.

U.S. Environmental Protection Agency

Summary

Methods of solid-waste disposal include open dumping, use of landfills, incineration, and disposal in the oceans. Incineration contributes to air pollution but can also provide energy. Landfills require a commitment of space, although the land may be used for other purposes when filled, and useful methane produced by waste breakdown can be recovered from some landfills. Composting is the best option for yard waste and other nontoxic organic matter. Secure landfills for toxic solid or liquid industrial wastes should isolate the hazardous wastes with low-permeability materials and include provisions for monitoring groundwater chemistry to detect any waste leakage. Increasing evidence suggests that construction of wholly secure landfill sites is unlikely. Deep-well disposal and incineration are alternative methods for handling toxic liquid wastes. Reduction in initial waste volume and recycling are both necessary to minimize the quantity of waste requiring disposal.

The principal liquid-waste problem in terms of volume is sewage. Areas of low population density can rely on the natural sewage treatment associated with septic tanks, provided that soil and site characteristics are suitable. Municipal

sewage-treatment plants in densely populated areas are constructed to purify water so well that it can be safely returned to rivers and streams or recycled for irrigation, and even used to supplement other sources of drinking water, although most municipalities do not go to such expensive lengths.

A complicating feature of radioactive wastes is that they cannot be made nonradioactive; each radioisotope decays at its own characteristic rate regardless of how the waste is handled. Given the long half-lives of many isotopes in high-level wastes and the fact that some are chemical poisons as well as radiation hazards, long-term waste isolation is required. Disposal in granitic bedrock, tuff, shale, or salt deposits is among the methods under particularly serious consideration worldwide. In the United States, the Waste Isolation Pilot Plant has become the first operational geologic repository for high-level waste disposal, receiving the government's transuranic waste. Yucca Mountain, Nevada, had been designated for development of a permanent disposal site for commercial high-level wastes, but actual disposal is still years away even if a decision to proceed here is ultimately made.

Key Terms and Concepts

absorption field	leachate	multiple barrier concept	source separation
half-life	leaching field	sanitary landfill	transuranic wastes
high-level waste	low-level waste	secure landfill	

Test Your Learning

1. Identify the two kinds of activities that generate the most solid wastes.

2. Describe the advantages that a sanitary landfill has over an open dump. Outline three pathways through which pollutants may escape from an improperly designed landfill site.

3. Describe any three features of a modern municipal landfill that reduce the risk of pollution, and explain their function.

4. Landfills and incinerators have in common the fact that both can serve as energy sources. Explain.

5. Compare the relative ease of recycling of (a) glass bottles, (b) paper, (c) plastics, (d) copper, and (e) steel, noting what factors make the practice more or less feasible in each case.

6. Describe how electronics recycling could be improved.

7. Compare the dilute-and-disperse and concentrate-and-contain philosophies of liquid-waste disposal.

8. Outline the relative merits and drawbacks of deep-well disposal and of incineration as disposal strategies for toxic liquid wastes.

9. Describe three kinds of limitations that would restrict the use of septic systems.

10. Municipal sewage treatment produces large volumes of sludge as a by-product. Suggest possible uses for this material, and note any factors that might restrict its use.

11. Describe two reasons that it is difficult to assess the hazards of low-level exposure to radiation.

12. Identify characteristics of the Yucca Mountain site that made it attractive as a possible disposal site for high-level wastes. Outline three geologic concerns relative to the long-term security of waste containment there.

Exploring Further

1. Where does your garbage go? How is it disposed of? Is any portion of it utilized in some useful way? If your community has a recycling program, find out what is recycled, in what quantities, and what changes have occurred over the last five or ten years in materials recycled and levels of participation.

2. For a period of one or several weeks, weigh all the trash you discard. If every one of the 330 million people in the United States or the 38 million in Canada discarded the same amount of trash, how much would be generated in a year? Suppose that each of the 7.9 billion people on Earth did the same. How much garbage would that make? Is this likely to be a realistic estimate of actual world trash generation? Why or why not?

3. Investigate your community's sewage treatment. What level is it? What is the quality of the outflow? How does the storm-sewer system connect with the sewage-treatment system?

4. Choose a single Superfund NPL site, or several sites within a state or region, and check the history and current status of cleanup activities there. A website including all the NPL (National Priorities List) sites by region, which would be a place to start, is at **www.epa.gov/superfund/superfund-national-priorities-list-npl**.

5. Check on the current status of the Yucca Mountain license application at **www.nrc.gov/waste/hlw-disposal/yucca-lic-app.html**, or read the Congressional Research Service's Civilian Nuclear Waste Disposal report from 2021 at **https://sgp.fas.org/crs/misc/RL33461.pdf**.

6. Evaluate the advantages, disadvantages, technical issues, and possible safety concerns relating to disposal of high-level radioactive wastes in (a) subduction zones, (b) sediments on the deep-sea floor, (c) space, and (d) in Antarctic ice.

CHAPTER 17

Water Pollution

In chapter 11, we reviewed the issues surrounding the availability and volume of freshwater on Earth, which constitutes less than 1% of the water in the hydrosphere. Not only are we facing dwindling supplies of water, but most of the accessible water near major population centers contains chemical compounds that degrade the water, making it unsafe to drink. The number and variety of potentially harmful water pollutants are staggering. The Flint, Michigan, disaster mentioned below shows us that not all of the problems are related to new chemicals but rather ones in which the risks are relatively well understood.

Chapter Outline

17.1 General Principles

17.2 Organic Matter

17.3 Agricultural Pollution

17.4 Industrial Pollution

17.5 Surface Water Remediation

17.6 Groundwater Pollution

17.7 Groundwater Remediation

Flint, Michigan. The current advice to Flint residents from the USEPA includes the following: DO NOT drink unfiltered water. It is not safe. DO NOT cook or brush teeth with unfiltered water. DO NOT allow babies and children to drink bathwater. What led to severe degradation of water quality in Flint?

Jim West/Alamy Stock Photo

A *pollutant* is sometimes defined simply as a chemical out of place, a substance found in a high enough concentration in some setting that it creates a nuisance or hazard. Some pollution is physical and some thermal, but most of the pollutants with which we are concerned are dissolved chemicals in the water. Some are naturally occurring; others are synthetic or are added to natural systems by human activity. In this chapter, we review some important categories of water pollutants, their common sources, and the nature of the problems each presents. We also look at some possible solutions to existing water-pollution problems. While it is true that substantial quantities of surface water and groundwater are unpolluted by human activities, much of this water is far removed from population centers that require it, and the quality of much of the most accessible water has indeed been degraded. Addressing water pollution is necessary for enhancing water supplies.

17.1 General Principles

Learning Objectives

- Recognize geochemical cycles
- Know how to calculate residence time
- Explain residence time of a pollutant in a water reservoir
- Summarize the difficulty of determining the health effects of a trace element
- Exemplify point and nonpoint pollution sources

Natural sources of water contain more than just H_2O molecules. As saltwater, clearly the oceans contain all sorts of dissolved compounds, but so do lakes, streams, and groundwater in lesser amounts. The very definition of *freshwater*, according to the U.S. Geological Survey, is water containing less than 1000 mg/liter of dissolved solids. It isn't necessarily the salts themselves that make drinking seawater a poor choice, but the concentration of them dissolved in the water. Some of the other substances that do find their way naturally into water are unhealthy for us or other life forms, as well as are pollutants introduced by modern industry, agriculture, and people themselves.

In this section, we will review some general principles and terminology to help us better understand the water pollution topics discussed in the rest of the chapter.

Geochemical Cycles

Earth processes are involved in many types of natural cycles. We have already introduced the rock cycle and the hydrologic cycle earlier in chapters 2 and 6, respectively. All of the chemicals in the environment participate in geochemical cycles of some kind. You have likely heard about the carbon cycle for it important role in biological processes and climate change, but there are others that are not as well understood.

For example, a very simplified natural cycle for calcium is shown in **figure 17.1**. Calcium is a major constituent of most common rock types. When weathered out of rocks, usually by solution of calcium carbonate minerals or by the breakdown of calcium-bearing silicates, calcium goes into solution in surface water or groundwater. From there it may be carried into the oceans. Some is taken up in the calcite of marine organisms' shells, in which form it may later be deposited on the sea floor after the organisms die. Some is precipitated directly out of solution into limestone deposits, especially in warm, shallow water; other calcium-bearing sediments may also be deposited. Over

Calcium (derived from weathering of rocks) dissolved in runoff from continent

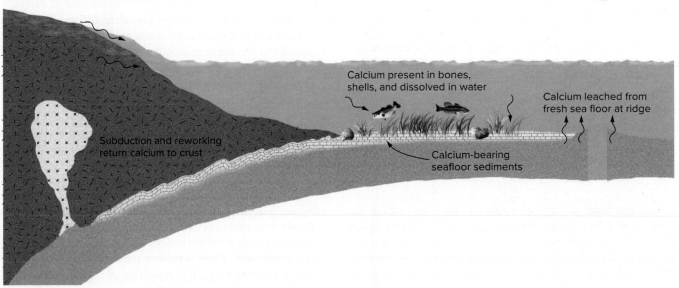

Figure 17.1

Simplified calcium cycle. Calcium weathered out of rocks is washed into bodies of water. Some remains in solution, some is taken up by organisms, and some is deposited in sediments. The deposited calcium is, in time, reworked through the rock cycle into new rock.

geologic time, the sediments are lithified, then metamorphosed, or subducted and mixed with magma from the mantle, or otherwise reincorporated into the continental crust as the cycle continues. Subcycles also exist. For instance, some mantle-derived calcium is leached directly into the oceans by the warmed seawater or other hydrothermal fluids interacting with the fresh seafloor basalts at spreading ridges; after precipitation, it may be reintroduced into the mantle without cycling through the continental crust. Some dissolved weathering products are carried into lakes from which they may be directly redeposited in continental sediments, and so on.

Residence Time

Within the calcium cycle outlined, there are several important calcium reservoirs, including the mantle, continental crustal rocks, submarine limestone deposits, and the ocean, in which large quantities of calcium are dissolved. A measure of how rapidly the calcium cycles through each of these reservoirs is a parameter known as **residence time.** We can define this in a variety of ways. For a substance that is at saturation level in a reservoir, so that the reservoir can hold no more of it, residence time is given by the following relation:

$$\text{Residence time} \atop \begin{array}{c}\text{(of a substance}\\ \text{in a reservoir)}\end{array} = \frac{\begin{array}{c}\text{Capacity}\\ \text{(of the reservoir to hold that substance)}\end{array}}{\begin{array}{c}\text{Rate of influx}\\ \text{(of the substance into the reservoir)}\end{array}}$$

Let's illustrate this with a non-geologic example. Consider a hotel with 100 single rooms. Some lodgers stay only one night; others stay for weeks. On average, ten people check out each day, so (assuming this is a popular hotel that can always fill its rooms) ten new people check in each day. The residence time of lodgers in the hotel, then, is

$$\text{Residence time} = \frac{100 \text{ persons}}{10 \text{ persons/day}} = 10 \text{ days}$$

In other words, people stay an average of ten days each at this hotel.

There are many other situations in which the quantity of a substance in a reservoir stays approximately constant (a *steady-state* situation) though the reservoir is not at capacity. In such a case, residence time could be defined as

$$\frac{\text{Amount of the substance in the reservoir}}{\text{Rate of input (or outflow) of the substance}}$$

To pursue our hotel analogy, consider a competing hotel in the next block that has similar turnover of clientele but a much lower occupancy rate—only 50 of its 100 rooms are typically occupied. Then if ten people check in and out each day, the residence time of patrons in the second hotel is only (50 persons)/(10 persons/day) = 5 days; each patron stays only half as long, on average, in the second hotel. Or, assuming for the moment that world population is a constant 7.5 billion, then if 150 million people are born and 150 million die each year, the residence time of people on Earth would be (7½ billion) ÷ (150 million) = 50 years. That would not mean that everyone would live for 50 years, but that would be the *average* life span.

Which brings us to a more qualitative way of viewing residence time: as the average length of time a substance remains in a system or reservoir. This may be a more useful working definition for substances whose capacity is not limited in a given reservoir (the continental crust, for instance) or for substances that don't achieve saturation but are removed from the reservoir by decomposition after a time, as is the case with many synthetic chemicals that cycle through water or air.

The reservoir in **figure 17.1** in which the residence time of calcium can be calculated most directly is the ocean. The capacity of the ocean to hold calcium is controlled by the solubility of calcite; if too much dissolved calcium and carbonate have poured into the ocean, some calcite is removed by precipitation. We can estimate the rate of influx by measuring the amount of dissolved calcium in all major rivers draining into the ocean. The resulting calculated residence time of calcium in the oceans works out to about 1 million years. In other words, the average calcium atom normally floats around in the ocean for 1 million years from the time it is flushed or leached into the ocean until it is removed by precipitation in sediments or in shells.

Oceanic residence times for different elements vary widely. Sodium, for instance, is put into the oceans less rapidly than calcium, and the capacity of the ocean for sodium is much larger. The residence time of sodium in the oceans, then, is longer than that of calcium: 100 million years. By contrast, iron compounds are not very soluble in the modern ocean. The residence time of iron in the ocean is only 200 years; iron precipitates out rather quickly once it is introduced into the sea. Residence times will differ for any given element in different reservoirs with different capacities and rates of influx.

Human activities may alter the natural figures somewhat, chiefly by altering the rate of influx. In the case of calcium, for instance, extensive mining of limestone for use in agriculture and construction increases the amount of calcium-rich material exposed to weathering; calcium salts are also used for salting roads, and these salts gradually dissolve away into runoff water. These additions are probably insufficient to alter appreciably the residence times of materials in very large reservoirs like the ocean. Locally, however, as in the case of a single lake, stream, or aquifer, they may change the rate of cycling and/or increase a given material's concentration throughout a cycle.

Residence Time and Pollution

In principle, one can speak of residence times of more complex chemicals in natural systems, with the added complication that we don't fully understand the behavior of many synthetic chemicals in the environment. There is no long natural history to use for reference when estimating the capacities of different reservoirs for recently created compounds. Some compounds' residence times are limited by rapid breakdown into other chemicals, but breakdown rates for many are slow or unknown. So is the ultimate fate of many synthetic chemicals. Uncertainties of this sort make evaluation of the seriousness of industrial pollution very difficult.

Where residence times of pollutants are known in particular systems, they indicate the length of time over which those pollutants will remain a problem in those reservoirs, and also the rapidity with which the pollutants will be removed from that system. Suppose, for example, that a factory has been discharging a particular toxic compound into a lake. Some of the compound is drained out of the lake by an outflowing stream. The residence time of the water, and of the toxic chemical in the lake, has been determined to be twenty years. If that is the average length of time the water and synthetic chemical remain in the lake, and if input of the toxin ceases but flushing of water through the lake continues, then after one residence time, half of the chemical should have been removed from the lake; after a second twenty years, another half; and so on. Mathematically, the decrease in concentration follows the same pattern as radioactive decay. One can then project how long it would take for a safely low concentration in the water to be reached. The calculations are more complex if additional processes are involved (if, for instance, some of the toxin decomposes and some is removed into lake-bottom sediment), but the principles remain the same. However, substances removed from one system, unless broken down, will only be moved into another, where they may continue to pose a threat. If our hypothetical toxic chemical is at saturation concentration in a lake and is removed only by precipitation into or adsorption onto the lake-bottom sediments, then if the rate of input into the lake doubles, so will the rate of output, and the concentration in the sediments will be twice as high. And just because a toxin is removed into sediments, it isn't really gone; if the sediments are disturbed or the water chemistry changes, the toxin may go back into solution again.

Trace Elements, Health, and Pollution

Much of the concern about pollution is associated with the possible negative health effects, whether of humans or other organisms. Small quantities of chemicals can have profound impacts, good or bad. Centuries ago, scientists began to observe geographic patterns in the occurrence of certain diseases. For example, until the twentieth century, the northern half of the United States was known as the "goiter belt" for the high frequency of the iodine-deficiency disease goiter. Subsequent analysis revealed that the soils in that area were relatively low in iodine. Consequently, the crops grown and the animals grazing on those lands constituted an iodine-deficient diet for people living there. When their diets were supplemented with iodine, the occurrence of goiter decreased.

Such observations provided early evidence of the importance to health of *trace elements,* those elements that occur in very low concentrations in a system. Trace elements' concentrations are typically a few hundred parts per million (ppm) or less. Their effects may be beneficial or toxic, depending on the element and the organism. To complicate matters further, the same element may be both beneficial and harmful, depending upon the dosage; fluorine is an example. Teeth are composed of apatite, a calcium phosphate mineral. Incorporation of some fluoride into its crystal structure makes the apatite harder, more resistant to decay. Persons whose drinking water contains fluoride concentrations even as low as 1 ppm show a reduction in tooth-decay incidence. This corresponds to a fluoride intake of 1 milligram per liter of water consumed. Normal total daily fluoride intake from food and water ranges from 0.3 to 5 milligrams.

Fluoride is not an unmixed blessing. Where natural water supplies are unusually high in fluoride, so that two to eight times the normal dose of fluoride is consumed, teeth may become mottled with dark spots. The teeth are still quite decay-resistant, however; the spots are only a cosmetic problem. Of more concern is the effect of very high fluoride consumption (twenty to forty times the normal dose), which may trigger abnormal, excess bone development and calcification of ligaments. A **dose-response curve** is a graph illustrating the relative benefit or harm of a trace element or other substance as a function of the dosage. Fluoride in humans is characterized by the dose-response curve shown in **figure 17.2**.

The prospect of reduced tooth decay is the principal motive behind fluoridation of public water supplies where

Figure 17.2

Generalized dose-response curve for fluoride in humans.

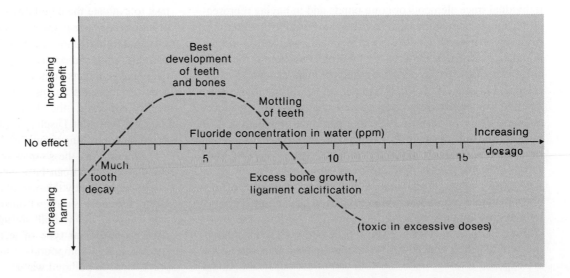

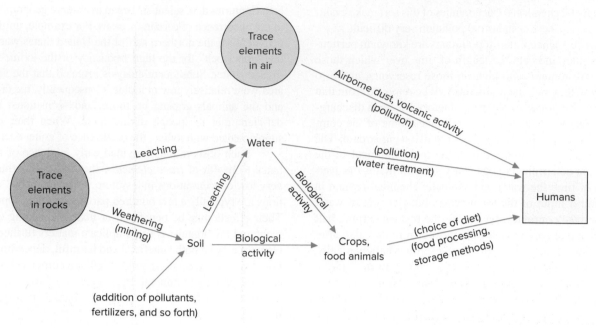

Figure 17.3

Pathways through which trace elements enter the body, including processes through which trace-element concentrations may be modified and other substances added to our intake. Human influences are in parentheses.

natural fluoride concentrations are low. Because extremely high doses of fluoride is toxic to humans, there has been some localized opposition to fluoridation programs. The concern is a misplaced one, for the difference between a beneficial and toxic dose, in this case, is enormous. The usual concentration achieved through purposeful fluoridation is 1 ppm (1 milligram per liter). A lethal dose of fluoride would be about 4 grams (4000 milligrams). To take in that much fluoride through fluoridated drinking water, one would not only have to consume 4000 liters of water but also would have to do so within a day or so, because fluoride is not accumulative in the body; it would be readily excreted along with all that water.

As a practical matter, there are limits on the extent to which detailed dose-response curves can be constructed for humans. The limits are related to the difficulties in isolating the effects of individual trace elements or compounds and in having representative human populations that have been exposed to the full range of doses of interest. Controlled experiments on plant or animal populations or tissue cultures are possible, but there is the difficulty of trying to extrapolate the results to humans with any confidence. Different individuals and persons of different ages may respond somewhat differently to any chemical, so any dose-response curve must be specific to a particular population, or it must be only an approximation. Such considerations make it difficult to establish safe exposure limits for hazardous materials and permissible levels of pollutant emissions.

The challenge of isolating the effects on health of any one substance is also illustrated, in part, by the complex paths by which elements move from rocks or air into the body (**figure 17.3**). A variety of natural processes and human activities modify their concentrations along the way. In addition, we are exposed to other materials that can affect our health, and a huge and growing variety of synthetic chemicals as well as naturally occurring elements and chemical compounds.

Point and Nonpoint Pollution Sources

Sources of pollution may be subdivided into point sources and nonpoint sources (**figure 17.4**). **Point sources** are those from which pollutants are released at one readily identifiable spot: a sewer outlet, a factory, a septic tank, and so forth. **Nonpoint sources** are more diffuse; examples would include fertilizer runoff from farmland, acid drainage from an abandoned strip mine, or runoff of sodium or calcium chloride from road salts, either directly off roadways or via storm drains. The point sources are easier to identify as potential pollution problems, and easier to monitor systematically. It is a comparatively straightforward task to evaluate the quality of water from a single sewer outfall or industrial waste pipe, and also to control and treat it, if necessary. Sampling runoff from a field in a representative way is more difficult, and it may also be more difficult to assess potential negative effects on human health. So, both for practical reasons and because its historic focus has been particularly on health concerns, the Environmental Protection Agency's National Pollutant Discharge Elimination System has concentrated on controlling point sources of water pollution.

Identifying the sources of water pollutants once a problem is recognized can be a challenge. Sometimes, the source can be identified by the overall chemical fingerprint of the pollution, for rarely is a single substance involved. Different pollutant sources—municipal sewage, a factory's wastewater, a farmer's particular type of fertilizer—are characterized by a specific mix of compounds, which may be identified in similar proportions in polluted water.

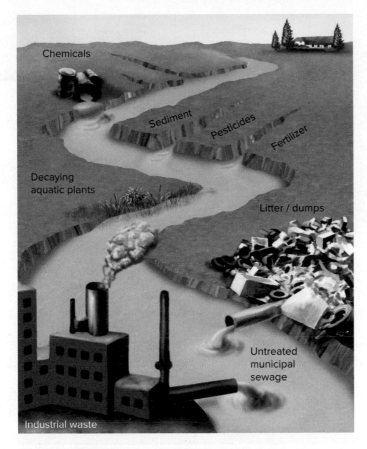

Figure 17.4

Point sources and nonpoint sources of water pollution in a stream. Downward infiltration of contaminant-laden water from the nonpoint sources eventually also pollutes groundwater.

Source: After USDA Soil Conservation Service.

17.2 Organic Matter

Learning Objectives

- List sources of organic matter, and associated nutrients, in wastewater
- Compare aerobic and anaerobic decomposition
- Define BOD as a measure of organic matter in water
- Summarize eutrophication

This section lists the primary sources of organic matter in wastewater, and describes how, as the organic matter breaks down, water quality is affected. This discussion also includes how excess nutrients can affect bodies of water and the organisms that live within them. The subject matter of this section is distinguished from the smaller subset of toxic organic compounds included later in section 17.4.

Sources and Decomposition of Organic Matter

Organic matter in general is the substance of living and dead organisms and their by-products. It includes a variety of

Figure 17.5

Uncontrolled runoff from this livestock yard may be a major contributor of organic waste to local streams or shallow groundwater.

Photo by Tim McCabe, U.S. Department of Agriculture Natural Resources Conservation Service.

materials, ranging from dead leaves settling in a stream to algae on a pond. Its most abundant and problematic form in the context of water pollution is human and animal wastes. Feedlots and other animal-husbandry activities create large concentrations of animal wastes (**figure 17.5**), and the trend in U.S. agriculture is toward ever-denser concentrations of cattle, or pigs, or poultry in such operations. Food-processing plants are other sources of large quantities of organic matter discharged in wastewater.

Sewage becomes mixed with a large quantity of wastewater not originally contaminated with it. Domestic wastewater is typically all channeled into one outlet, and in municipalities, domestic wastes, industrial wastes, and frequently even storm runoff collected in street drains all ultimately come together in the municipal sewer system. Moreover, wastewater treatment plants are commonly located beside a body of surface water, into which their treated water is discharged. Because wastewater is incompletely treated, the pollution of the surface water is increased.

Aside from its potential for spreading disease, organic matter creates another kind of water-pollution problem. In time, organic matter in water is broken down by microorganisms, especially bacteria. If there is ample oxygen in the water, **aerobic decomposition** occurs, consuming oxygen, facilitated by aerobic organisms. This depletes the dissolved oxygen supply in the water. Eventually, so much of the oxygen may be used up that further breakdown must proceed by **anaerobic decomposition.** Anaerobic decay produces a variety of noxious gases, including hydrogen sulfide (H_2S), the toxic gas that smells like rotten eggs, and methane (CH_4), which is unhealthful and of no practical use in a body of water. More importantly, anaerobic decay signals oxygen depletion. Animal life, including fish, requires oxygen to survive. When the dissolved oxygen is used up, the water is sometimes described as dead, though anaerobic bacteria and some plant life may continue to live and even to thrive.

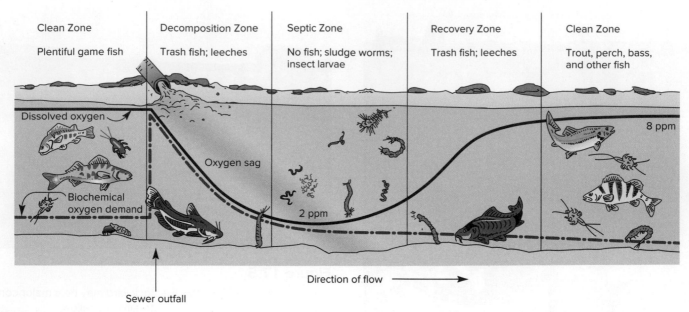

Clean Zone — Plentiful game fish

Decomposition Zone — Trash fish; leeches

Septic Zone — No fish; sludge worms; insect larvae

Recovery Zone — Trash fish; leeches

Clean Zone — Trout, perch, bass, and other fish

Dissolved oxygen

Oxygen sag

Biochemical oxygen demand

8 ppm

2 ppm

Sewer outfall

Direction of flow ⟶

Figure 17.6

Typical oxygen sag curve downstream from an organic-matter input, and its effect on both dissolved oxygen and the fauna in the stream. The dissolved-oxygen content of this stream is normally 8 ppm; it drops to 2 ppm below the sewer outfall where BOD spikes abruptly.

Biochemical Oxygen Demand

The organic-matter load in a body of water is described by a parameter known as **biochemical oxygen demand, or BOD** for short. The BOD of a system is a measure of the amount of oxygen required to break down the organic matter aerobically; the more organic matter, the higher the BOD. The BOD may equal or exceed the actual amount of dissolved oxygen in the water. As oxygen is depleted, more tends to dissolve in the water to restore chemical equilibrium, but, initially, this gradual reoxygenation lags behind oxygen consumption. Measurements of water chemistry along a stream below a sizeable source of organic waste matter characteristically define an **oxygen sag curve** (**figure 17.6**), a graph of dissolved-oxygen content as a function of distance from the waste source that shows sharp depletion near the source, recovering downstream. Persistent oxygen depletion is more likely to occur in a body of standing water, such as a lake or reservoir, than in a fast-flowing stream. Flowing water is better mixed and circulated, and more of it is regularly brought into contact with the air, from which it can take up more oxygen. Climate also plays a role because oxygen dissolves more readily in cold water than in warm (**table 17.1**). Removal of organic matter to reduce BOD is a key function of municipal sewage-treatment plants.

Eutrophication

The breakdown of excess organic matter not only consumes oxygen but also releases a variety of compounds into the water, among them nitrates, phosphates, and sulfates. The nitrates and phosphates, in particular, are critical plant nutrients, and an abundance of them in the water strongly encourages the growth of plants, including algae. Excess fertilizer

Table 17.1	Dissolved Oxygen as a Function of Temperature
Temperature, °C (°F)	**Dissolved oxygen, mg/L**
0 (32)	14.6
5 (41)	12.8
10 (50)	11.3
15 (59)	10.2
20 (68)	9.2
25 (77)	8.6

As temperature goes up, dissolved oxygen concentrations decline. (1 mg/L is approximately 1 ppm.)

Source: U.S. Geological Survey Water Resources Division.

runoff from farmland puts excessive nutrients into water. This development is known as **eutrophication** of the water; the water body itself is then described as *eutrophic*. The exuberant algal growth that often accompanies or results from eutrophication is also called an algal bloom and may appear as slimy green scum on the water (**figure 17.7**). Once such a condition develops, it acts to continue to worsen water quality (**figure 17.8**). Algal growth proceeds vigorously in the photic zone near the water surface until the plants are killed by cold or crowded out by other plants. The dead plants sink to the bottom, where they, in turn, become part of the organic-matter load on the water, increasing the BOD and, as they decay, re-releasing nutrients into the water. The consequences are most acute in very still water, where bottom waters don't readily circulate to the surface to take up oxygen. It should be noted that eutrophication can be a normal process in the

A

B

Figure 17.7

(A) Typical algal bloom on a pond in summer. In this case, the source of the nutrients is waterfowl. The seemingly benign action of feeding flocks of ducks and geese around a suburban pond like this one not only alters the birds' behavior and teaches them to depend on handouts; it also leads to an increased pollutant load of organic fertilizer in the pond, and resultant degradation in water quality. (B) In extreme cases, algae may cover the whole water surface, as on this Wisconsin lake.

(A-B) ©Carla Montgomery

development of a natural lake. Pollution, however, may accelerate or enhance it to the point that it becomes problematic. Aside from the aesthetic impact of algal bloom, oxygen depletion in eutrophic waters can kill aquatic organisms such as fish that require oxygen to survive.

Thorough waste treatment, as described in chapter 16, reduces the impact of organic matter, at least in terms of oxygen demand. However, breakdown of the complex organic molecules releases simpler compounds into the water that may serve as nutrients. Human or animal wastes are not the only agents in sewage that contribute to eutrophication, either. Fertilizer runoff from farmland is a factor, as noted later in the chapter. Phosphates from detergents are potential nutrients, too. The phosphates are added to detergents to enhance their cleaning ability, in part by softening the water. Realization of the harmful environmental effects of phosphates in sewage has led to restrictions on the amounts used in commercial detergents, but some phosphate component is still allowed in most places. And most sewage treatment does not remove all of the dissolved nutrient load.

17.3 Agricultural Pollution

Learning Objectives

- Relate fertilizer use to water quality problems
- Recall ways to reduce water pollution from fertilizers and organic waste
- List ways excess sediment load causes water quality problems
- Describe problems with, and alternatives to, chemical pesticide use

The same kinds of water-pollution problems that result from agricultural activities also occur in urban and suburban areas. We apply fertilizers, herbicides, and insecticides to yards, parks, and gardens just as we do to farmland, and their residues correspondingly appear in water just about everywhere (**figure 17.9**). Just as we noted that soil erosion during urbanization can be much more intense than erosion during cultivation of crops, so pesticide accumulations may be more notable in urban areas, as seen in **figure 17.10**. This section focuses on water pollution associated with agriculture because the much greater land area involved means much larger quantities of potential pollutants applied and greater impact of agricultural activities on pollution extent overall.

Fertilizers and Organic Waste

The three principal constituents of commercial fertilizers are nitrates, phosphates, and potash. When applied to or incorporated in the soil, they aren't immediately taken up by plants. The compounds must be soluble in order for plants to use them, which means that they can also dissolve in surface-water and groundwater bodies. These plant foods then contribute to eutrophication problems. There is typically a clear correlation with applications of fertilizer (**figure 17.11**), so the concerns are greatest in such areas.

Reduction in fertilizer applications to the minimum needed, perhaps coupled with the use of slow-release fertilizers, would minimize the harmful effects. Nitrates can alternatively be supplied by the periodic planting of legume, on the roots of which grow bacteria that fix nitrogen in the soil. This practice reduces the need to apply soluble synthetic fertilizers.

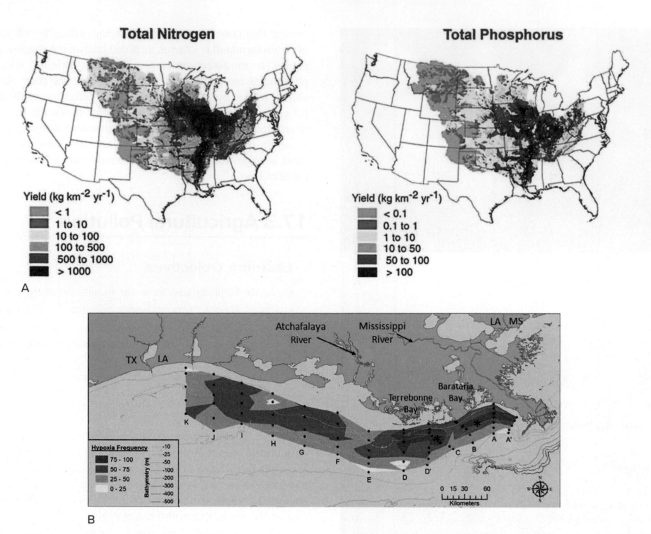

Figure 17.8

Nutrients supplied to the Gulf of Mexico by the Mississippi River (A) primarily through runoff from agricultural land support abundant growth of plankton near the water surface in the Gulf. When the algae die and settle toward the bottom, they increase BOD, deplete oxygen, and create a dead zone, a condition known as *hypoxia,* when oxygen concentrations in the near-bottom waters are so low (below 2 mg/L) that marine life cannot survive. The dead zone as measured July 25–31, 2021 (B) covered an area of 6,330 square miles. High temperatures in the Gulf area exacerbated the problem; recall table 17.1.

(A) Maps Courtesy U.S. Geological Survey National Water Quality Assessment Program; (B) Source: Image courtesy NOAA Hypoxia Watch.

Unfortunately, the use of other natural fertilizers, like animal manures, fails to eliminate the water-quality problems associated with fertilizer runoff. Manures are organic wastes, so they also contribute to the BOD of runoff waters in addition to adding nutrients that lead to eutrophication.

Runoff of wastes from domestic animals, especially from commercial feedlots, is a potential problem. One concern that arises in connection with some such operations is not unlike the problem of municipal sewage-treatment plants overwhelmed by the added volume of runoff water from major storms. It is increasingly common for the wastes to be diverted into shallow basins where they may be partially decomposed and/or dried and concentrated for use as fertilizer. If poorly designed, the surrounding walls may fail, releasing the contents. Even if the structure holds,

in an intense storm, the added water volume may cause overflowing of the wastes. One approach is to build broad channels downslope of the waste source, planted with grass or other vegetation, to act as filter strips for the runoff. The filter strip catches suspended solids and slows fluid runoff, allowing time for decomposition of some dissolved constituents, and nutrient uptake by the vegetation (depending on season), plus increased infiltration to reduce the volume of runoff leaving the site. Studies have shown reductions of up to 80% in dissolved phosphorus, nitrogen, and other constituents in the surface runoff water leaving the filter strip, so the practice clearly helps to protect surface-water quality. The primary potential disadvantage is that, by increasing infiltration, the filter strip may increase the dissolved-nutrient load in groundwater below the site.

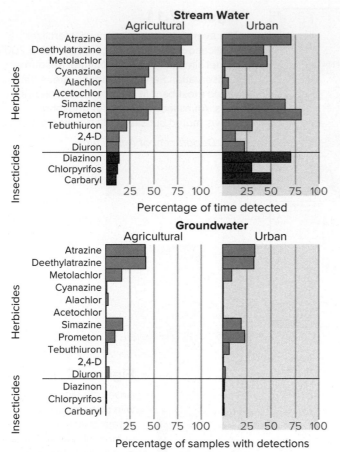

Figure 17.9

Herbicides and insecticides are everywhere in our water and as common in urban as agricultural areas. Pesticides were detectable in nearly all stream samples in this national study.

Source: USGS Circular 1291, as revised online 2007.

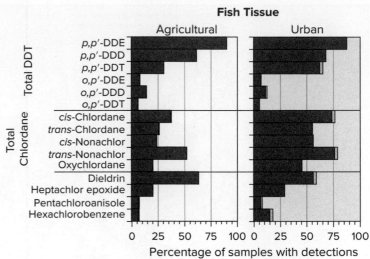

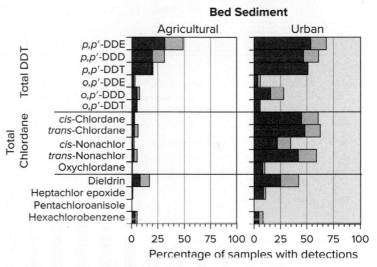

Figure 17.10

This 2007 national study showed continued presence of pesticides in sediments and fish, even decades after the pesticides' last use: dichlorodiphenyl trichloroethane (DDT) was banned for most applications in 1972, dieldrin in 1974, and chlordane in 1988. Darker portion of each bar represents higher detection levels, above 5 parts per billion (ppb) for fish tissue and 2 ppb in dry sediment.

Source: U.S. Geological Survey Circular 1291, as revised online 2007.

An alternative, productive solution to the excess-manure problem is more extensive use of these wastes for the production of methane for fuel. Anaerobic digesters dramatically reduce water pollution by killing pathogens. The liquid by-product still contains nitrogen and phosphorus, but the compounds are made more readily available for plants so it can be used as a free fertilizer. Digesters also reduce greenhouse gas emissions and create an energy source that can be used on site or sold back to a utility. Energy output from these systems has tripled in the last decade, with significant growth in the use of compressed natural gas as a vehicle fuel (**figure 17.12**). There are thousands more dairy and hog operations in the United States where such biogas recovery systems would be technically feasible.

Sediment Pollution

Some cases of high suspended-sediment load in water occur naturally (**figure 17.13**). These are commonly localized situations, and may be temporary as well. In many agricultural areas, sediment pollution of lakes and streams is the most serious water-quality problem and is ongoing. Farmland and forest land together are believed to account for about 75% of the 3 billion tons of sediment supplied annually to U.S. waterways. Of this total, the bulk of sediment is derived from farmland; sediment yields from forest land are typically low.

Sediment pollution not only causes water to be murky and unpleasant to look at, swim in, or drink, it also reduces the light available to underwater plants and blankets food supplies and the nests of fish, reducing fish and shellfish populations. Channels and reservoirs may be filled in. The sediment also clogs water filters and damages power-generating equipment. Some sediment transport is a perfectly natural consequence of stream erosion, but agricultural development typically increases erosion rates by four to nine times unless agricultural practices are chosen carefully.

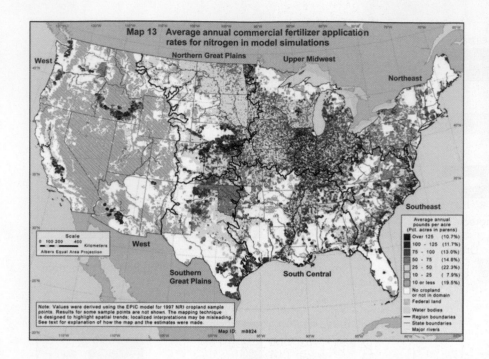

Figure 17.11

The average annual commercial fertilizer application rates for nitrogen, as used in 2006 model simulations by the Natural Resource Conservation Service to identify areas that would most benefit from conservation practices. An associated map for phosphorus shows muted but similar distribution patterns.

U.S. Department of Agriculture

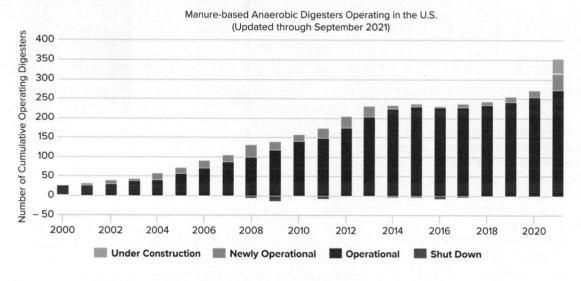

Figure 17.12

Anaerobic digesters help with numerous environmental issues. The number of new and under-construction digesters increased in 2021 by 30% from the previous year, spurred on by California's Low Carbon Fuel Standard's need for renewable transportation fuels.

U.S. Environmental Protection Agency , https://www.epa.gov/agstar/agstar-data-and-trends

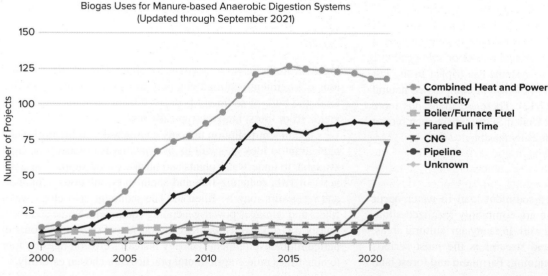

Figure 17.13

Sediment clouds stream waters. The stream at right is murky with suspended sediment, probably because of bank collapse. Note the contrast with the clear tributary at left. Junction of Soda Butte Creek and the Lamar River, Yellowstone National Park.

©Carla Montgomery

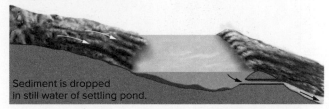

Surface runoff carries
a suspended sediment load.

Sediment is dropped
in still water of settling pond.

Outflow water is clear
of suspended sediment.

A

B

Figure 17.14

(A) Settling ponds enhance water quality and reduce soil loss, reducing sediment pollution and also trapping pollutants adsorbed onto the sediment particles. Removal of sediment from the water may increase erosion below the settling pond. (B) Heavy surface runoff may contribute to both flooding and soil erosion, muddying floodwaters with sediment pollution. This watershed dam in eastern Iowa traps the sediment-laden waters, reducing flooding downstream and containing the sediment, which will settle out of the basin's still water.

(B) Photo by Lynn Betts, USDA Natural Resources Conservation Service.

Chapter 12 examined some strategies for limiting soil loss from farmland. From a water-quality standpoint, an additional approach is the use of settling ponds below fields subject to serious erosion by surface runoff (**figure 17.14**). These ponds don't stop erosion from the fields, but by impounding the water, they cause the suspended soil to be dropped before the water is released to a lake or stream. Settling ponds can also be helpful below large construction projects, logging operations, or anywhere ground disturbance is accelerating erosion.

Pesticides

Herbicides and insecticides are a significant source of pollution in agriculture. Most such compounds now in use are complex organic compounds of the kinds described in section 17.4 later in the chapter, and are in some measure toxic to humans or other life-forms. U.S. farmers use more than a billion pounds of pesticides each year, of which about 90% are synthetic organic compounds. The use of pesticides increased dramatically from 1960 to 1981, leveled off, and then declined by 40% over the last three decades. While the decreased volume may seem like good news, many of these compounds are extremely long-lasting in the environment (recall **figure 17.10**) and recent studies show that the toxicity of the compounds used has increased, especially to insects and aquatic invertebrates.

The agricultural products themselves are not the sole problem. Many of the chemicals used in producing them are also toxic and persistent (dioxin, for instance); so are some of the early decomposition products of herbicides and pesticides, when and if they do break down.

Commonly, spring applications of agricultural chemicals are followed by concentration spikes in runoff water (**figure 17.15**); clearly, a great deal of what is applied is wasted. The potential risks could be reduced by eliminating routine applications of such chemicals as a sort of preventive medicine. Applying a remedy only when a problem is demonstrated to exist may initially reduce synthetic chemical use at the cost of crop yield. Over the longer term, the results may be more economical, in part because the bugs or weeds will not as readily develop resistance to the substance being used.

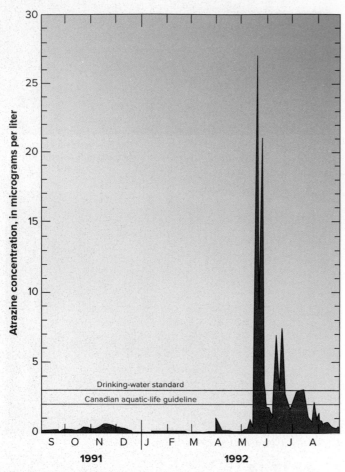

Figure 17.15

A study in the central Nebraska corn belt showed a sharp rise in pesticide runoff (here illustrated by atrazine) following early-spring application to fields. Spring and summer rains continued to wash off significant, though declining, amounts. While the average annual concentration in runoff did not exceed drinking-water standards, peak concentrations were up to ten times that standard.

Source: After USGS Circular 1225.

A growing variety of nonchemical insect-reduction strategies are finding increasing acceptance. Some are specific to a single pest. For instance, the larvae of a small wasp that is harmless to people and animals prey parasitically on tomato hornworms and kill them. A particular bacterium *Bacillus thuringiensis (Bt)* attacks and destroys several types of borer insects but again produces no known ill effects in humans, mammals, birds, or fish, and leaves no persistent chemical residues on food. Unfortunately, sometimes the introduced species itself becomes a problem, as in the case of the Asian Lady Beetle, originally used to combat aphids.

Other remedies may be more broadly applicable. An example is the practice of sterilizing a large population of a particular insect pest by irradiation and then releasing the sterilized insects into the fields. Assuming the usual number of matings, the next generation of pests should be smaller in number because some of the matings will have involved infertile insects.

Other growers bait simple insect traps with scents or hormones carefully synthesized to simulate female insects, thereby luring the males to their doom. Procedures such as these offer promise for reducing the need for synthetic chemicals of uncertain or damaging environmental impact.

In the 1990s, breakthroughs in genetic engineering created new possibilities. For example, scientists modified corn plants to produce their own insecticidal defense against corn borers, and *Bt* genes have been transferred into some crops. There are several factors that constrain genetically modifying plants. One is concern for possible unintended negative consequences, such as the apparent toxicity of the above-mentioned corn's pollen to monarch-butterfly larvae that eat milkweed leaves on which stray wind-borne pollen has fallen. Another is regulatory-agency concern about the safety of these crops for human consumption. And there continues to be public resistance to the general concept of genetically engineered foods, independent of any specific health or safety concerns.

17.4 Industrial Pollution

Learning Objectives

- Explain the difficulties in demonstrating the safety of new chemicals
- Describe how a compound bioaccumulates and biomagnifies in a food chain
- Exemplify metals, other inorganics, and organic compounds associated with industrial pollution
- List the factors involved in controlling industrial water pollution
- Describe thermal water pollution

Hundreds of new chemicals are created by industrial scientists each year. The rate at which new chemicals are developed makes it impossible to demonstrate the safety of the new chemicals as fast as they are invented. To prove a chemical safe, it would be necessary to test a wide range of doses on every major category of organism at several stages in the life cycle. That simply is impossible in terms of time, money, laboratory space, and ethics. Tests have historically been carried out on a few representative populations of laboratory animals to which the chemical is believed most likely to be harmful. New alternative test methods include computer models and those based on biochemical or cell responses. Independent of the test method, whether a specific organism will turn out to be unexpectedly sensitive to a chemical won't be known until after they have been exposed to it in the environment. The completely safe alternative would be to stop developing new chemicals, of course, but that would deprive society of many important medicinal drugs, as well as materials to make life safer, more comfortable, or more pleasant. The magnitude of the mystery about the safety of industrial chemicals is emphasized by a National Research Council report. A committee compiled a list of nearly 66,000 drugs, pesticides, and other industrial chemicals and discovered that no toxicity data at all were available for 70% of them; a complete health hazard evaluation was possible for only 2%. There are over 194 million organic and inorganic substances in the Chemical Abstract Service registry, a record kept by the American Chemical Society of new chemicals created or identified since 1957. Complete toxicity assessments for that many substances would be a formidable task.

In the following, we will review the primary categories on industrial chemical pollutants, the major concerns with them in our water supplies, and very briefly, the issues associated with controlling their use and release. Issues with thermal pollution are also covered. What industries are most responsible for the release of harmful chemicals? **Figure 17.16** shows the sectors that reported the largest releases of Toxic Release Inventory chemicals into water bodies in 2020.

Inorganic Pollutants—Metals

Many of the inorganic industrial pollutants of particular concern are potentially toxic metals. Manufacturing, mining, and mineral-processing activities can all increase the influxes of these naturally occurring substances into the environment. The four nonfood industry sectors identified in **figure 17.16** commonly release such metals as arsenic, cadmium, chromium, copper, lead, mercury, nickel, uranium, and zinc. If their releases are not carefully controlled, their wastewaters can locally increase concentrations from harmless to toxic levels. Sometimes, too, natural processes can increase the concentration of a toxic metal after release.

The element mercury is a case in point. In nature, the small amounts of mercury in rocks and soil are weathered out into waters, or into the air from which it is quickly removed with precipitation. Mercury in water may be taken up by plants and animals; eventually it is removed into sediments. Its residence time in the oceans is about 80,000 years.

Mercury is one of the **heavy metals,** a group that includes lead, cadmium, plutonium, and others. A feature that the heavy metals have in common is that they tend to accumulate in the bodies of organisms that ingest them (*bioaccumulation*). Their concentrations also increase up a food chain (**figure 17.17**) in a phenomenon called **biomagnification.** Some marine algae may contain heavy metals at concentrations more than 1000 times that of the water in which they are living. Small fish eating the algae develop higher concentrations of heavy metals in their flesh; larger fish who eat the smaller fish concentrate the metals still further. The flesh of predator fish high up a multistage food chain may show mercury concentrations 100,000 to 1 million times that of the water in which they swim. Birds and mammals who eat those fish may concentrate it further, though the diversity of most people's diets is greater than that of most animals and will moderate the effects of consuming mercury-rich fish. Almost all fish has some mercury in it. Still, different fish and shellfish present significantly different degrees of risk, especially to children, pregnant people, and people who are breastfeeding. Fish eating advice and guidelines were updated by the EPA in 2021.

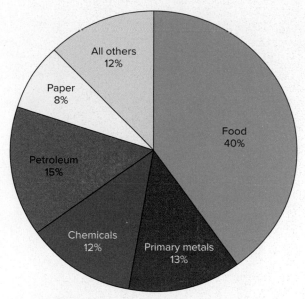

Figure 17.16

Wastewater releases by industry are subject to EPA permits and regulations. Note that the food industry accounts for the largest single share of water releases, and 98% of its waste consists of nitrates.

Data from U.S. Environmental Protection Agency.

In most natural settings, heavy-metal accumulations in organisms are not very serious because the natural concentrations of these metals are low in waters and soils to begin with. The problems develop when human activities locally upset the natural cycle. Mining and processing heavy metals in particular can increase the rate at which the heavy metals weather out of rock into the environment, and these industries have also discharged concentrated doses of heavy metals into some bodies of water.

Mercury is a very volatile metal. When mercury vapor is emitted into the atmosphere, it disperses widely before being deposited on land and water, either directly (dry deposition) or with precipitation (wet deposition). The United Nations estimates that atmospheric mercury concentrations have increased 300 to 500% in the last century. Current activities that release the most mercury into the environment are artisanal and small-scale mining, combustion of coal, and metals and cement production. What begins as a potential air-pollution issue becomes a water-pollution problem. In fact, because mercury in air is typically elemental mercury, it becomes a health threat only after it enters water bodies where it can be methylated. The increasing use of coal for energy led to sharply increasing number of mercury-related fish consumption advisories in the late 1990s and early 2000s. Now that coal burning has decreased dramatically in the United States, will our mercury concerns also disappear? Unfortunately, no, because metals such as mercury are generally persistent and mobile in the environment, and many are nondegradable.

Methylmercury is the most common form of organic mercury in the environment, is highly toxic, and is created from inorganic mercury by microorganisms (**figure 17.17**). It acts on the central nervous system and the brain in particular. It can

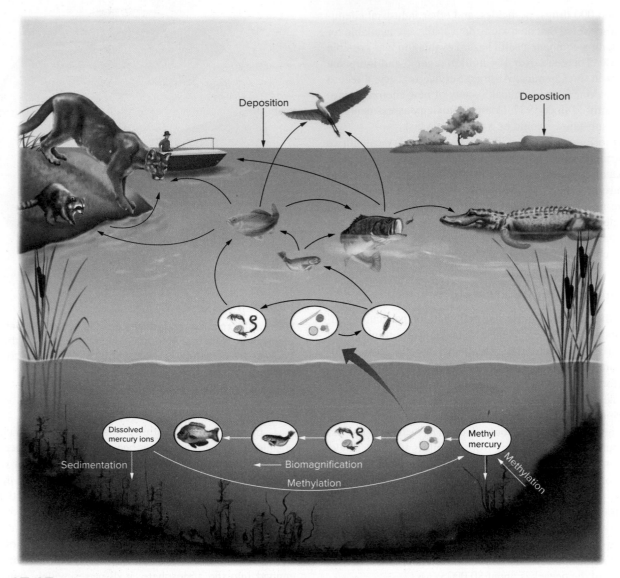

Figure 17.17

Mercury behavior in a body of water. Mercury is deposited from the atmosphere into water, where it is converted to methylmercury by microorganisms. Those microorganisms are consumed by small fish, who are in turn consumed by larger fish, and so on up the food chain to birds and mammals. Toxins, such as mercury and Polychlorinated biphenyls (PCBs), can biomagnify enough to become a human health threat.

cause loss of sight, feeling, and hearing, as well as nervousness, shakiness, and death. By the time the symptoms become apparent, the damage is irreversible. Mercury nitrate is a toxic, inorganic form of mercury that was commonly used in the past in the making of felt; the phrase "mad as a hatter" comes from hat makers suffering from mercury poisoning. Mercury has also been used as a fungicide, and in the manufacture of plastics.

Minimata disease is the name given to the severe methylmercury poisoning suffered by the people of Minimata, Japan. In the early 1950s, industrial wastewaters full of mercury used in the production of acetaldehyde were pumped into the ocean near this small fishing village, where it magnified up the aquatic food chain and ultimately to the fish, which was the primary diet of the coastal people living there. Babies were born with severe deformities; some victims were paralyzed; others slurred their speech and suffered convulsions. Nearly 3000 people contracted Minimata disease and over 1700 died of it. Thousands more had less severe symptoms cause by consumption of mercury-laden seafood. The investigation of Minimata disease alerted other nations to the toxicity of mercury and people were advised to limit their intake of tuna and certain other fish with particularly high mercury contents. After some years of enhanced pollution control, the mercury levels associated with industrial wastewater decreased.

Mercury is not the only toxic heavy metal. Cadmium poisoning was most dramatically demonstrated in Japan when cadmium-rich mine wastes were dumped into the upper Zintsu

River. The people used the water for irrigation of the rice fields and for domestic use. Many developed *itai-itai* (literally, "ouch-ouch") disease, cadmium poisoning characterized by abdominal pain, vomiting, and other unpleasant symptoms. Fear of the toxic-metal effects of plutonium is one concern of those worried about radioactive-waste disposal.

For some industrial applications, there are substitutes for toxic metals. Mercury is no longer used in the manufacture of felt, for instance. High concentrations of toxic metals also may exist in natural geological materials. As a volatile metal, mercury may be emitted during explosive volcanic eruptions, but these are short-lived, and the metal is quickly removed from the atmosphere by rain or snow. Human activities have had a more lasting impact (**figure 17.18**).

Arsenic is an example of potentially toxic metals that can occur naturally in lakes, streams, and groundwater at concentrations above U.S. drinking-water standards. The toxic effects of large doses of arsenic are well documented. It can cause skin, bladder, and other cancers; in large doses, it is lethal. The effects of low doses are much less clear. In some cultures, arsenic is deliberately consumed to promote a clear complexion and glossy hair. It is accumulative in the body, and over time, some tolerance to its toxicity may be developed. Arsenic is not known to be an essential nutrient, but whether it is harmful even in the smallest quantities is not known. Drinking-water limits for such elements are established by extrapolation from high doses known to be toxic to a dose low enough to correspond to a perceived acceptable level of risk (e.g., one death per million persons, or serious illness per 100,000, or whatever) coupled with statistical studies of the incidence of toxic effects in populations naturally exposed to various doses of the substance, often with an additional safety margin thrown in.

For many years, the drinking-water standard for arsenic was set at 50 parts per billion (ppb). A new standard of 10 ppb was introduced in 1999 but did not become effective until 2006 due to objections based, in part, on the fact that natural arsenic levels in groundwater in much of the country already exceeded that level (**figure 17.19**).

Arsenic in groundwater is not a problem unique to the United States. The World Health Organization estimates that over 100 million people in South and Southeast Asia drink groundwater containing harmful concentrations of arsenic. The problem is particularly severe in Bangladesh. Ironically, the issue has developed there only since the 1970s, when the people were encouraged to switch from using surface water

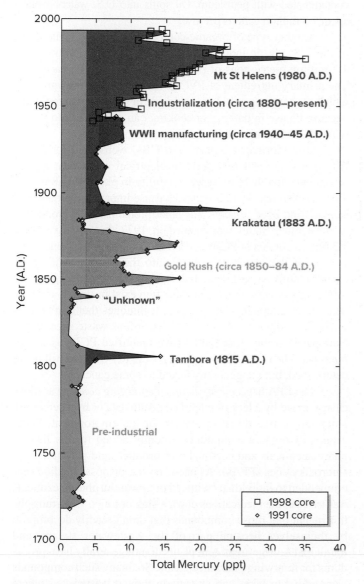

Figure 17.18

A USGS study of nearly 100 ice cores from a Wyoming alpine glacier shows clear evidence of both natural and human influence on mercury.

After USGS Fact Sheet 051-02.

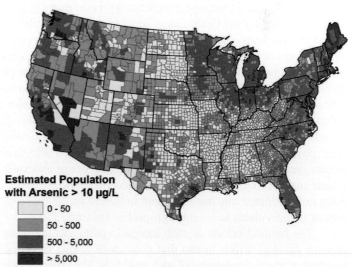

Figure 17.19

Sampling of over 20,000 wells indicates that arsenic in groundwater in many parts of the United States naturally exceeds the EPA drinking-water standard of a maximum of 10 micrograms per liter (µg/L); 2.1 million people may be at risk (2017).

U.S. Geological Survey

contaminated with organic waste to using well water. Now, 90% of the population drinks well water, despite the fact that potentially dangerous arsenic levels in the groundwater have been recognized since the 1990s, and by now many cases of arsenic poisoning have been documented. Solutions that include treating the water or drilling deeper wells to cleaner aquifers are being developed, but the problem will not be solved quickly. Researchers have recently found strong evidence of correlation between widespread arsenic poisoning in Bangladesh and a decline in productivity and cognition.

Other Inorganic Pollutants

Some nonmetallic elements commonly used in industry are also potentially toxic to aquatic life, if not to humans. For example, chlorine is widely used to kill bacteria in municipal water- and sewage-treatment plants and to destroy various microorganisms that might otherwise foul the plumbing in power stations. Released in wastewater, it can also kill algae and harm fish populations.

Acids from industrial operations used to be a considerably greater pollution problem before stringent controls on their release were imposed. Acid mine drainage remains a serious source of surface-water and groundwater pollution in coal- and sulfide-mining areas, as described in chapters 13 and 14. The acids, in turn, may leach additional toxic metals from rocks, tailings, or soil.

The toxic effects of certain asbestos minerals were not manifested or well-defined until long after their initial release into the environment by human activities. Asbestos minerals have been prized for decades for their fire-retardant properties and have been used in ceiling tiles and other building materials for years. Asbestos was also mixed in cement and used extensively in North America to create water pipes. Wastes from asbestos mining and processing were dumped into many bodies of water, including the Great Lakes because they were believed to be inert, harmless mineral materials. By the time the carcinogenic effects of some asbestos minerals were realized, asbestos workers had been exposed and segments of the public had been drinking asbestos-bearing waters for twenty years or more. Sustained exposure to asbestos minerals, whether through inhalation or ingestion, is linked to a rare form of cancer that develops decades after the exposure. In Japan, many prefer rice coated with powder, and the powder is mainly pulverized talc that occurs geologically in association with some types of asbestos, which may account for higher incidence of stomach cancer. Such contaminated talc has also been indicated as the cause of cancer in individuals in recent high-profile lawsuits.

The harmful effects of such inorganic pollutants are often slow to develop, which means that decades of testing might be required to reveal the dangers of such materials. The costs of such long-term testing are extremely high and generally prohibitively so.

Organic Compounds

The majority of new chemical compounds created each year are organic (carbon-containing) compounds. Many thousands of these compounds, naturally occurring and synthetic, are widely used as herbicides and pesticides, as well as being used in a variety of industrial processes. Examples include dichlorodiphenyl trichloroethane (DDT) and dioxin. Their negative effects in organisms vary with the particular type of compound. Some are carcinogenic, some are directly toxic to humans or other organisms, and others make water unpalatable. Some also accumulate in organisms as the heavy metals do.

Oil spills, discussed in chapter 14, are another kind of organic-compound pollution. At least as much additional oil pollution occurs each year from the careless disposal of used crankcase oil, dumping of bilge from ships, and the runoff of oil from city streets during rainstorms. Underground tanks and pipelines may also leak, and drilling muds and waste brines discarded in oil fields may be contaminated with petroleum. Oil spills into U.S. waters average about 1.3 million gallons per year.

Another type of organic-compound pollution is the result of the U.S. plastic industry's demand for production of nearly 7.2 million metric tons of polyvinyl chloride (PVC) each year. The primary ingredient of PVC is vinyl chloride, the vapors of which are carcinogenic. Vinyl chloride in water is also regulated because its use in bathing or cooking can release it into the air where it can be inhaled.

Polychlorinated biphenyls (PCBs) were used for nearly fifty years as insulating fluid in electrical equipment and as compounds that help preserve flexibility in plastics. Laboratory tests revealed that PCBs in animals cause impaired reproduction, stomach and liver ailments, and other problems. PCB production in the United States was banned in 1977, but approximately 1.3 billion pounds of PCBs had already been produced and some portion of that quantity released into the environment. PCBs adsorb readily to sediment, and do bioaccumulate and biomagnify in food chains. Contaminants often exist in stream and lake beds, so cleanups involve dredging techniques that don't remobilize the sediments. At least 500 hazardous waste sites on the National Priorities List (NPL) have identified PCBs as a contaminant. The risks of low-level exposure to PCBs are still being determined, but are generally linked to some cancers.

The EPA lists *contaminants of emerging concern* as those characterized by a lack of published health data or by a perceived, potential, or real threat to humans or the environment. Some groups of organic compounds recently added to this list are pharmaceuticals and personal care products and per- and polyfluoroalkylenes (PFAs). At one time the disposal method recommended to individuals with old or unwanted pharmaceuticals was to flush the medications down a sink or toilet. Unfortunately, this meant that those compounds that didn't readily decompose by themselves became part of the wastewater stream, and municipal sewage-treatment plants do not generally have procedures for removing pharmaceuticals. So, many such compounds were added to lakes and streams in treated wastewater, where some cause adverse health effects in aquatic animals. Traces of common medications have been found in some drinking-water supplies as well. The potential negative effects on humans are still being evaluated but seem to be concentrated on them being *endocrine disrupters,* compounds that alter the normal function of hormones and that result in a variety of health effects. Meanwhile, disposal of pharmaceuticals by the health-care industry is now regulated, with these compounds classified as hazardous waste. For individuals, the EPA recommends taking advantage of pharmaceutical take-back or household hazardous-waste collection programs.

PFAs have been in use since the 1940s, and there are thousands of types of them. Many of them break down slowly and they bioaccumulate in the environment and within organisms. They are used in a large variety of products and exposure can occur through breathing, drinking, eating, or just touching them. PFAs are everywhere. Exposure may lead to a wide range of health effects including obesity, hormone disruption, developmental delays, cancers, and reduced immune system responses. Because PFAs are both ubiquitous and variable in type, and people can be exposed to them in many ways, scientists will likely be researching them for a long time to specify their adverse health effects.

Problems of Control

Cost and efficiency are factors in pollution control. Those substances not to be released with effluent water (or air) must be retained. This becomes progressively more difficult as complete cleaning of the effluent is approached. First, there is no wholly effective way to remove pollutants from wastewater before discharge. No chemical process is 100% efficient. Second, as cleaner output is approached, costs skyrocket. If it costs $1 million to remove 90% of some pollutant from industrial wastewater, then removal of the next 9% (90% of the remaining 10% of the pollutant) costs an additional million dollars. To get the water 99.9% clean costs a total of $3 million, and so on. Some toxin always remains, and the costs escalate rapidly as purer water is approached. At what point is it no longer cost-effective to keep cleaning up? That depends on the toxicity of the pollutant and on the importance of the process of which it is the product. And, the toxins retained as a result of the cleaning process don't just disappear. They become part of the growing mass of industrial toxic waste requiring careful disposal.

The very nature of the ways in which hazardous compounds are used is an obvious problem. Herbicides and pesticides are commonly spread over broad areas, where it is typically impossible to confine them. The result, not surprisingly, is that they become nearly ubiquitous in the environment, as was illustrated in **figure 17.9**.

Even when the threat posed by a toxic agent is recognized and its production actually ceases, it may prove long-lived in the environment. Once released and dispersed, it may be impossible to clean up. The only course is to wait for its natural destruction, which may take unexpectedly long. DDT is a case in point; see Case Study 17.2.

DDT was undoubtedly a uniquely powerful tool for fighting insect pests transmitting serious diseases. It was neither the first nor the last new compound to be found to have significant toxic effects long after it was put into general use. This may be inevitable, given the rate of synthesis of new compounds and the complexity of testing their safety, but one unfortunate result has been the long-term persistence of toxins in water, sediments, and organisms.

Not all of the synthetic organic chemicals last long in the environment. Some (e.g., vinyl chloride) evaporate rapidly, even if released into water; some (e.g., benzene) evaporate or break down in a matter of days. Even among the pesticides, behavior varies widely. One can speak of the half-life of such a compound in the environment, as a measure of rate of breakdown for those that decompose in soil or water. While some, such as the insecticide malathion, have half-lives of a few days, the majority of common pesticides have half-lives on the order of a year. Among the longest-lasting are dieldrin (half-life about 10 years), chlordane (over 30 years), and DDT (15 to over 120 years depending on the exact compound; recall **figure 17.10**).

Thermal Pollution

Thermal pollution, the release of excess or waste heat, is a by-product of the generation of power. The waste heat from automobile exhaust or heating systems contributes to thermal pollution of the atmosphere, but its magnitude is generally believed to be insignificant. Potentially more serious is the local thermal pollution of water by electricity-generating plants and other industries using cooling water.

Only part of the heat absorbed by cooling water can be extracted effectively and used constructively. The still-warm cooling water is returned to its source and replaced by a fresh supply of cooler water. The resulting temperature increase usually exceeds normal seasonal fluctuations in the stream or lake and results in a consistently higher temperature regime near the effluent source.

Fish and other cold blooded organisms, including a variety of microorganisms, can survive only within certain temperature ranges. Excessively high temperatures may result in the wholesale destruction of organisms. More-moderate increases can change the balance of organisms present. For example, green algae grow best at 30° to 35°C (86° to 95°F); blue-green algae, at 35° to 40°C (95° to 104°F). Higher temperatures thus favor the blue-green algae, which are a poorer food source for fish and may be toxic to some. Many fish spawn in waters somewhat cooler than they can live in; a temperature rise of a few degrees might have no effect on existing adult fish populations but could seriously impair breeding and thus decrease future numbers. Furthermore, changes in water temperature change the rates of chemical reactions and critical chemical properties of the water, such as concentrations of dissolved gases, including the oxygen that is vital to fish; recall table 17.1.

Thermal pollution can be greatly reduced by holding water in cooling towers before release, but it usually is still at a somewhat elevated temperature. The extent of thermal pollution is restricted in both space and time. It adds no substances to the water that persist for long periods or that can be transported over long distances, and reduction in the output of excess waste heat results in an immediate reduction in the magnitude of the continuing pollution problems.

17.5 Surface Water Remediation

Learning Objectives

- Lists ways to remediate surface-water contamination

All of the pollutants already mentioned in this chapter together have significantly affected surface waters over broad areas of the United States. Reduction in pollutant input is one strategy for addressing such problems, but it requires time to succeed, especially with relatively stagnant waters such as lakes. A variety of more aggressive treatment methods are available.

Many water pollutants, including phosphates, toxic organics, and heavy metals, can become attached to the surfaces of fine sediment particles on a lake or stream bottom. After a long period of pollutant accumulation, the bottom sediments may contain a large reserve of undesirable or toxic chemicals, which could, in principle, continue to be re-released into the water long after input of new pollutants ceased. Wholesale removal of the contaminated sediments by dredging takes the pollutants permanently out of the system. However, they still need to be disposed of, commonly in landfills.

Dredging operations must be done carefully to minimize the amount of very fine-grained material churned back into suspension in the water. Fine resuspended sediment, with its high surface-to-volume ratio, tends to contain the highest concentrations of pollutants adsorbed onto grain surfaces. Also, if fine resuspended sediments remain suspended for some time, the increased water turbidity may be harmful to aquatic life. The process can also be expensive.

Dredging is used in the United States for both emergency cleanup of toxic wastes and as a final remediation solution. When 250 gallons of PCBs spilled into the Duwamish Waterway near Seattle in 1974, the EPA chose to dredge nearly 4 meters' thickness of contaminated sediment from the waterway, recovering 220 to 240 gallons of the spilled chemicals. While that action may have resulted in decreasing acute pollution of the river, the Lower Duwamish Waterway was declared a Superfund site in 2001 (**Figure 17.20A**). PCBs, arsenic, and other contaminants from industrial wastewater sources adhered to the sediments on the river bed, 50% of which were removed by the end of 2015. Preparation for the cleanup of the remaining sediments is ongoing and estimated to cost $342 million. Other dredging projects have been much larger. The Hudson River PCBs Superfund Site remediation has resulted in the removal of 2.75 million cubic yards of contaminated river-bottom sediments, which appears to have successfully reduced pollution in the river overall, including in the water and in fish. Cleanup has already cost billions of dollars but there are calls from environmental groups to extend the PCB cleanup to other portions of the river (**Figure 17.20B**).

Another way to reduce the escape of contaminants from bottom sediments is to leave the sediments in place but to isolate them wholly or partially with a physical barrier. Over limited areas, like lagoons and small reservoirs, impermeable plastic liners have been placed on top of the sediments and held in place by an overlying layer of sand. The permanence of the treatment is questionable, and it has not been used on a large scale. Compacted clay layers of low permeability might, in principle, serve a similar purpose, but this treatment has not been attempted on any significant scale.

The addition of salts of aluminum, calcium, and iron to sediment changes the sediment chemistry, fixing phosphorus in the sediment and thereby reducing eutrophication. The technique has been used in small lakes in Wisconsin, Ohio, Minnesota, Washington, Oregon, and elsewhere. While not always successful, the method has often reduced phosphate levels in the water and corresponding algal growth. The cost compares favorably with

A

B

Figure 17.20

PCB Superfund sites. (A) A recent legal settlement requires those industries responsible for contamination in the Lower Duwamish Waterway to implement habitat restoration valued at more than $47 million. (B) PCB-laden sediment from the Hudson River was emptied onto barges, and then moved to an upstream processing facility where it was dewatered. Drained water was treated on-site and the dried PCB-contaminated sediments were sent to waste-disposal facilities in six different states.

(a) National Oceanic and Atmospheric Administration; (b) U.S. Environmental Protection Agency

the use of synthetic algicides to destroy the algae. However, careless over-treatment may prove toxic to fish, and the treatment must be repeated every two to three years. An alternative, biological means of addressing eutrophication in ponds and lakes might be the introduction of algae-eating fish that have been sterilized to prevent their overwhelming native species. Can algal blooms themselves be a way to clean up a body of water? There is some evidence from a severely polluted lake in China that algal blooms scavenge mercury and arsenic from the water, and then can be collected and disposed of later, leaving behind cleaner water.

Active decontamination is most often used in response to toxic-waste spills. This usually includes containing and removing as much of the contaminant as possible through absorbent booms, pigs, or granular materials, and sucking up the waste if it's floating on the water's surface. The specific treatment methods depend on the toxin involved. In 1974, a toxic organic-chemical herbicide washed into Clarksburg Pond in New Jersey from an adjacent parking lot. Many fish were killed, and local wildlife dependent on the pond for water were endangered. The contamination also threatened to spread to groundwater by infiltration and to the Delaware River to which the pond water flows overland. The EPA set up an emergency water-cleaning operation. The herbicide was removed principally by activated charcoal through which the water was filtered. Subsequent tests showed the groundwater to be uncontaminated, and within two years, fish were again plentiful in the pond.

Artificial aeration addresses oxygen depletion in a lake. There are several possible procedures, including bubbling air or oxygen up through the waters, thereby providing more oxygen to oxygen-starved deep waters, or simply circulating the water mechanically, cycling up deep waters to the surface where they can dissolve oxygen directly from the atmosphere. Many small artificial ponds are now designed with fountains whose function is not purely decorative. Spraying water into the air enhances aeration through the droplets' considerable surface area. When successful, aeration can transform water from an anaerobic to an aerobic condition, to the great benefit of fish populations.

From a public-health perspective, the principal concern with respect to quality of surface water relates to its use for drinking water. Here the news is generally good, as the most common contaminants in most areas, especially agricultural regions, are excess nutrients—and in those areas, most public supplies are drawn from groundwater, not surface water. However, the groundwater in these areas is often contaminated too.

17.6 Groundwater Pollution

Learning Objectives

- List reasons why groundwater pollution is misperceived
- Explain the connection between groundwater pollution and surficial contamination
- Describe a contaminant plume

Groundwater pollution, whether from point or nonpoint sources, is especially insidious because it isn't visible and often goes undetected for some time. Municipalities using well water must routinely test its quality. Homeowners relying on wells may be less anxious to go to the trouble or expense of testing, particularly if they are unaware of any potential danger. Yet comprehensive analyses of water quality in both domestic (private) and municipal wells have suggested potential health concerns in a significant fraction of cases (**figure 17.21**). In most instances, the passage of pollutants from their source into an aquifer used for drinking water is slow because it occurs by percolation of water through rock and soil, not by overland flow. There may be a significant time lapse between the introduction of a pollutant into the system in one spot and its appearance in groundwater or surface water elsewhere. Conversely, groundwater pollution in karst areas, with their rapid drainage, may spread unexpectedly swiftly. The "out of sight, out of mind" aspect of groundwater contributes to the problem. So does imperfect understanding of the interrelationships between surface water and groundwater in specific cases.

Below we will review how contaminants get into groundwater and then how they move within an aquifer, also including how geologists go about determining those interactions so that the risk of polluted groundwater can be assessed.

The Surface Water–Groundwater Connection

Given the nature of groundwater recharge, the process can readily introduce soluble pollutants along with recharge water. For example, agricultural pesticides are applied at the surface. Those pesticides that are neither used up nor broken down may remain dissolved in infiltrating water as well as runoff, resulting in pesticide contamination of groundwater. Acid mine drainage, storm-sewer runoff, and just about any liquid chemical product spilled on the surface are potential groundwater pollutants. The susceptibility of the groundwater to pollution from such sources is a function of both inputs and local geology (**figure 17.22**).

Acidic recharge can be a particular problem because many toxic metals and other substances are more soluble in more acidic water. How severely groundwater chemistry is affected is related, in part, to the residence time of the groundwater in its aquifer before it is extracted through a well or spring. That is, if water residence time is short, as with the shallow aquifers in **figure 17.23**, there is less opportunity for chemical reactions in the aquifer to buffer or moderate the impacts of the acid. Longer residence times may allow more moderation of the chemistry of the acid waters as well as the concentrations of various dissolved constituents, providing more opportunity for breakdown of potentially toxic compounds. With dramatic reductions in air emissions that cause it, acid rain and its recharge into aquifers is no longer the widespread problem that it was just thirty years ago.

Altogether, substantial groundwater-quality problems exist in the United States. Most are natural; for example, high dissolved mineral content, arsenic (**figure 17.19**), or radium (**figure 11.22**). Some are due to human activities (**figure 17.24**). The nature of the problems and their causes can vary greatly from place to

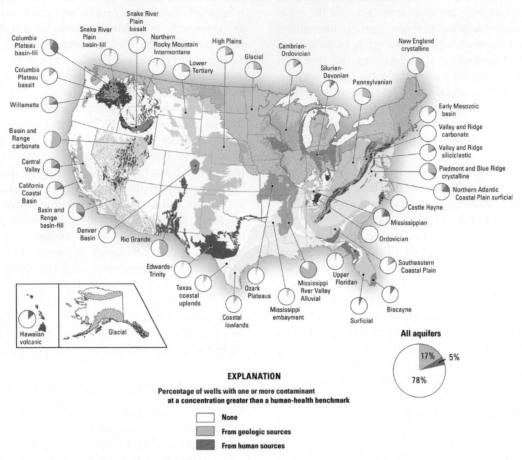

Exceedances of human-health benchmarks by one or more contaminants

EXPLANATION

Percentage of wells with one or more contaminant
at a concentration greater than a human-health benchmark

☐ None
☐ From geologic sources
■ From human sources

All aquifers

17% — 5%

78%

Figure 17.21

In a recent study of aquifer water quality, 22% of the 6600 municipal and domestic wells sampled contained a chemical constituent that exceeded a human health benchmark. About 3/4 of the exceedances were from geologic sources; the most commonly found constituents were manganese, arsenic, and radon. The most common contaminants from human sources were nitrate and dieldrin.

U.S. Geological Survey

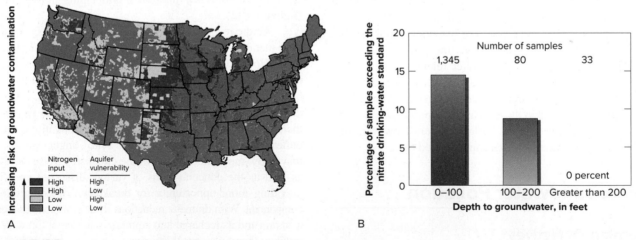

A

Increasing risk of groundwater contamination

Nitrogen input	Aquifer vulnerability
High	High
High	Low
Low	High
Low	Low

B

Number of samples

1,345   80   33

0 percent

Depth to groundwater, in feet

Percentage of samples exceeding the nitrate drinking-water standard

Figure 17.22

(A) Susceptibility of groundwater to nitrate pollution is related not only to nitrogen applied in fertilizer (as shown in **figure 17.11**) but to speed of drainage; well-drained soils are permeable soils through which water infiltrates readily. (B) In a national study of major aquifers, nitrate concentrations in groundwater decreased as depth to aquifer increased; most samples exceeding the drinking-water standard of 10 milligrams per liter were from shallow aquifers.

Source: U.S. Geological Survey National Water Quality Assessment Program.

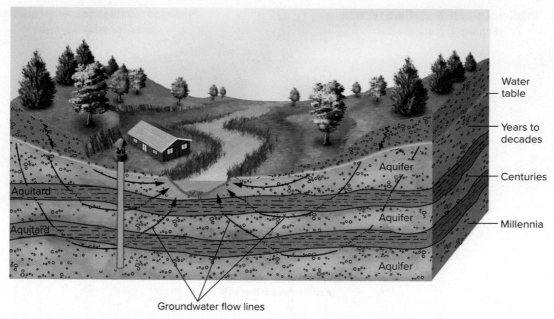

Figure 17.23

Groundwater flow rates may vary widely; in deep aquifers beneath multiple aquitards, centuries or more may elapse between infiltration in the recharge area and appearance in surface water.

place. Not surprisingly, pesticides or their breakdown products are common in groundwater beneath agricultural areas. Nitrate concentrations tend to be higher in groundwater below agricultural areas than in groundwater under undeveloped areas. On the other hand, as noted, fertilizers, insecticides, and herbicides are not unique to agricultural regions; they may be used abundantly in urban/residential areas, too. And where improvements in municipal sewage treatment have been made to reduce ammonia concentrations, this has often been accomplished by converting the nitrogen in the ammonia to nitrate—thereby increasing nitrate concentrations in the water system. In virtually all cases involving human activity, a groundwater-pollution problem has evolved from surface-water pollution.

Groundwater pollution has sometimes appeared decades after the industry or activity responsible for it has ceased to operate and has disappeared from sight and memory. Chemicals dumped or spilled into the soil long ago might not reach an aquifer for years. Even after the source has been realized, so large an area may have been contaminated that cleanup is impractical and/or prohibitively expensive. This is a problem with many old, abandoned toxic-waste dump sites such as Superfund sites discussed in chapter 16. Groundwater pollution from nonpoint sources, like farmland, may also be so widespread that cleanup is not feasible.

Some strategies for addressing groundwater contamination are explored later in section 17.7. With groundwater, it is especially important to avoid or limit pollution initially, because it is typically more difficult to clean up afterward than is polluted surface water. For example, underground storage tanks, such as the ones used to store gasoline at a service station, used to be made of bare steel, which corroded readily

and was easy to puncture. Since 1988, tanks have to be either constructed of materials that don't rust or be lined with rust-proof materials. Tanks constructed with double-walls or with plastic and fiberglass might be used to prevent corrosion and leakage and the soil and groundwater contamination that would ensue.

Bearing in mind the potential of groundwater pollution from surface-water infiltration, one can also assess the vulnerability of a groundwater system to such pollution in terms of how well protected it is. Such an assessment may, in turn, suggest how closely the groundwater system should be monitored. There are many sources of surface-water pollution. In addition to those already cited, there are a variety of casual pollution sources related to improper disposal: crankcase oil dumped into a storm sewer by a home mechanic, solvents and cleaning fluids poured down the drain by homeowners or dry cleaners or painters or furniture refinishers, oil that was deposited on roads by traffic flushed off and running into storm sewers in a heavy rain, even chemicals from a college science laboratory thoughtlessly flushed down a laboratory drain—all these, many simply reflecting individual carelessness, may ultimately pollute groundwater.

Tracing Pollution's Path

Groundwater pollutants migrate from a point source in two ways. As groundwater flows, it carries pollutants along. Dissolved constituents can also diffuse through the water, even when it is barely moving, from areas of high concentrations toward areas of lower concentration. Depending on the nature of the pollutants, they may gradually break down as they move away from their

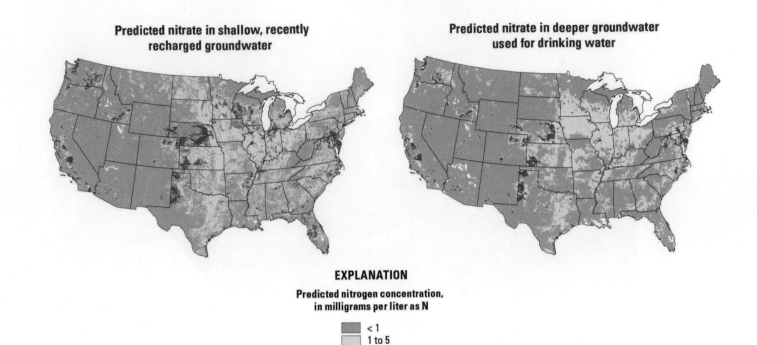

Predicted nitrate in shallow, recently recharged groundwater

Predicted nitrate in deeper groundwater used for drinking water

EXPLANATION

Predicted nitrogen concentration, in milligrams per liter as N

- < 1
- 1 to 5
- > 5 to 10
- > 10

A

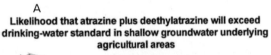

Likelihood that atrazine plus deethylatrazine will exceed drinking-water standard in shallow groundwater underlying agricultural areas

Probability of exceeding 3.0 µg/L

- 1.0 - 10.0%
- > 10.0%
- < 0.1%
- 0.1 - 1.0%

B

Figure 17.24

Groundwater contamination due to human activities. (A) Predicted nitrate concentrations are higher in areas of intense agriculture than in undeveloped regions. (B) Where atrazine use is high, the differences in predicted concentrations are due to geologic factors such as soil permeability and groundwater residence time.

(A-B) U.S. Geological Survey

17.7 Groundwater Remediation

Learning Objectives

- Describe how contaminated soil can be a continuing source of groundwater pollution
- List methods to remediate groundwater after it has been extracted
- Describe *in situ* groundwater remediation methods
- Summarize the ongoing cleanup at the Rocky Mountain Arsenal site

source, as by reaction with aquifer rocks or through the action of microorganisms.

Where pollutant migration is primarily by groundwater flow, a **contaminant plume** develops down-gradient from the point source. This is a tongue of polluted groundwater that may extend for hundreds of meters (**figure 17.25**). Leachate plumes seep from leaky landfills, ill-designed toxic-waste sites, and other pollutant sources.

Geologists determine the extent of each plume and concentration of each contaminant by sampling via wells, knowing that a plume's flow direction is constrained by groundwater flow. Some pollutants float or sink, depending on their density, and don't readily mix with water. Some dissolve and diffuse very rapidly. Understanding the behavior of the plumes and the fate of the various pollutants is important to containing and cleaning up the contaminants.

Earlier sections of this chapter noted various means of reducing the input of pollutants into water. The approach of reducing or stopping the infiltration of further pollutants, then waiting for natural processes to remove or destroy the pollutants already in the system (monitored natural attenuation), is sometimes the only approach technically or economically possible in the case of contaminated groundwater, given its relative inaccessibility. Systematic monitoring of groundwater can be difficult without an extensive (and usually expensive) network of wells because, by the time the pollution problems are recognized, the contaminants have typically spread widely in the aquifer system. Even after the original

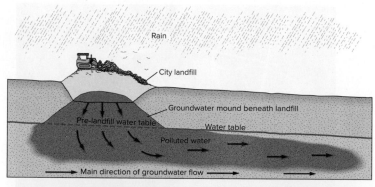

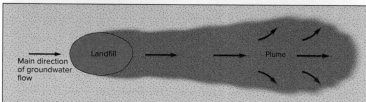

A. Schematic view, in cross-section (top) and map view (bottom). Note that the plume tends to spread as it flows. When it reaches a well or body of surface water, serious pollution may appear very suddenly.

Figure 17.25

Groundwater contaminant plumes.

(B) After U.S. Geological Survey Fact Sheet 040-03.

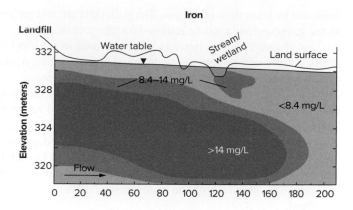

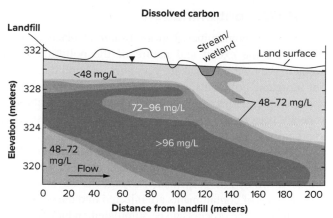

B. A groundwater contaminant plume from the Norman (Oklahoma) Landfill Environmental Research Site.

source of the pollutants is removed or contained, further groundwater contamination may occur intermittently for some time, too. If pollutants have adsorbed onto soil in the unsaturated zone, some may be dissolved or dislodged each time precipitation percolates through that soil, adding a new pulse of pollution to the groundwater system. In other cases, pollutants may have diffused into fine sediments or become associated with organic matter in an aquifer, and may seep back out into the groundwater long after introduction of new contaminants into the system has been stopped. Removing the worst of the contaminated soil, and preventing the infiltration of water through any remaining tainted soil, is often an important step in cleaning up groundwater contamination. The slow migration and limited mixing of most groundwaters complicate *in situ* treatment of contamination; bear in mind the way groundwater occurs, in cracks and pores dispersed throughout a large volume of rock. Often, then, polluted groundwater is only treated after it is extracted for use.

In the following, we review and exemplify groundwater remediation methods.

Decontamination After Extraction

The **pump-and-treat** method of groundwater remediation involves removing the polluted water from the aquifer and applying some sort of method to remove the contaminant of

concern. It is most often used to limit the spread of contamination, as described in the Rocky Mountain Arsenal (RMA) situation later in this section, but may also be employed when impure groundwater must be used for some purpose such as a municipal water supply. The type of treatment depends on the intended water use and the particular pollutants involved.

Inorganic compounds can be removed from extracted groundwater by adjusting its acidity (pH). Addition of alkalies may result in the precipitation of many heavy or toxic metals as hydroxides. Many of the same metals may be precipitated as sulfides, which are relatively insoluble, or as carbonates. All of these strategies yield a solid sludge (the precipitate) that contains the toxic metals and must still be disposed of.

Just as microorganisms can be used to decompose organic matter in sewage, microbial activity can break down a variety of organic compounds in groundwater after it is extracted. The organisms can be mixed in bulk with the water; after treatment, the biomass must then be settled or filtered out. Alternatively, the organisms can be fixed on a solid substrate, and the contaminated water passed over them.

Air stripping encompasses a set of methods by which volatile organic pollutants are transferred from water into air and thus removed from the water. The mechanical details of the process vary, but each case involves some form of aeration of the water, followed by separation of the gas. The pollutants are still present in the gas phase and must be removed for alternate disposal.

Activated charcoal (activated carbon), used in a variety of filters, adsorbs organic compounds dissolved in groundwater. Again, the compounds themselves remain intact and require disposal.

In short, treatment methods available *after* groundwater is extracted are many and span the same broad range as methods of treating surface-water pollution or purifying a municipal water supply. But as long as the polluted groundwater remains in the ground, unless it is contained (which is not common because of costs), the contaminants can go right on spreading, carried with the water flow or diffusing through the fluid.

In Situ Decontamination

Treatment of contaminated groundwater in place is possible only when the extent and nature of the pollution are well defined; the methods used are very site- and pollutant-specific. For inorganic pollutants, such as heavy metals, immobilization is one strategy. Injection wells placed within the contaminated zone are used to add chemicals that cause the toxic substances to precipitate and thus to become stationary.

Biological decomposition, or **bioremediation,** is effective on a broad range of organic compounds, many of which can be broken down by microorganisms. Organisms that will consume certain types of contaminants are injected into the contaminated groundwater. The process can be stimulated, in individual cases, by the addition of oxygen or nutrients to accelerate growth of the microorganisms; additional microorganisms can be introduced to attack the particular compound(s). Geologists often use this approach to accelerate cleanup in cases where monitored natural attenuation is the primary strategy. Occasionally, objectionable residues affect water taste and odor. Not all organics can be eliminated in this way, but most of the organics on the EPA's Priority Pollutant List can be successfully attacked by biological means. Genetic engineering may also allow the development of strains of microorganisms targeted to specific pollutants, increasing the efficiency of bioremediation efforts.

With petroleum products that don't mix well with water but instead tend to float on it, wells that reach barely to water-table level may selectively pump out contaminant-rich fluid, leaving much purer water behind.

The *permeable reactive barrier* is a remediation method in which pollutants are broken down as the groundwater flows through the barrier. Such an approach has successfully dealt with toxic organic compounds that previously escaped from the site of the Denver Federal Center (**figure 17.26**).

Containment at the Rocky Mountain Arsenal

One human activity can have multiple environmental impacts. Deep-well disposal of liquid waste at the RMA, near Denver, Colorado, was identified as the cause of a series of small earthquakes in the area. Careless surface disposal of toxic wastes at the same site also resulted in contamination of

A

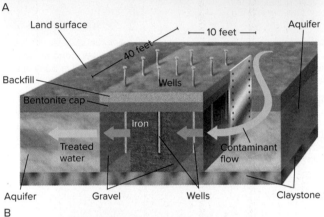

B

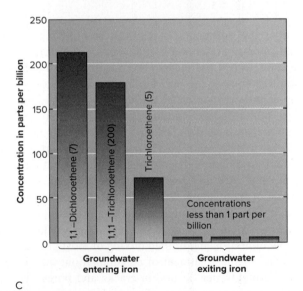

C

Figure 17.26

(A) A trench is dug across the groundwater flow path, and the reactive material installed. (B) Groundwater flows through gravel into permeable barrier containing metallic iron, which reacts with and destroys the organic solvents, leaving clean water to flow on through the aquifer (C).

(A) U.S. Geological Survey/Photograph courtesy David Naftz; (B-C) After U.S. Geological Survey, National Wetlands Research Center.

local groundwater. Beginning in 1943, wastewater containing a variety of organic and inorganic chemicals was discharged into unlined surface ponds. The principal contaminants were the organics diisomethylphosphonate and dichloropentadiene.

The water table in the shallow aquifer is locally only 2 to 5 meters below the ground surface. Water from this aquifer is widely used for irrigation and for watering livestock. Incidents of severe crop damage were reported in the early 1950s. To limit further contamination, an asphalt-lined disposal pond was constructed in 1956, but in time, the liner leaked. Contaminants already in the groundwater system continued to spread. When new complaints of damage to crops and livestock were made in the early 1970s, the Colorado Department of Health investigated. They detected a variety of toxic organic compounds in water and soil on the arsenal property; some of these compounds were present, though in much lower concentrations, in water drawn from off-site municipal supply wells. The Department of Health ordered a cessation of the waste discharges and cleanup of the existing pollution, with provision to prevent further discharge of pollutants from the arsenal property. The site was placed on the Superfund NPL in 1987.

Because groundwater contamination was already so widespread, the only feasible approach to containing the further spread of pollutants was to halt them at the arsenal boundaries. This effort was combined with cleanup activity (**figure 17.27**). A physical barrier of clay-rich material was placed along the arsenal side of the boundary at several locations on the north and west sides, where the natural groundwater flow was outward from the arsenal. On the arsenal side of the barrier, wells extract the contaminated groundwater. It is treated by a variety of processes, including filtration, chemical oxidation, air stripping, and passage through activated charcoal. The cleaned water is then reintroduced into the aquifer on the outward side of the barrier. The net groundwater flow regime outside of the arsenal has been only minimally disrupted, and pollutants in nearby water supplies have been greatly reduced. This, then, is a version of pump-and-treat remediation: the water is not immediately used after treatment, but is returned, cleaned, to the aquifer system.

The efforts have not been inexpensive. Over $2 billion has been spent so far, and the boundary barrier system will have to continue in operation indefinitely as the slowly migrating contaminants in the groundwater continue to move toward the limits of the site. Cleanup of one section of the site was completed in 2010 and is now the home of RMA National Wildlife Refuge, 15,000 acres of urban land with over 330 species of birds, mammals, fish, and other creatures including bald eagles, bison, burrowing owls, and black-tailed prairie dogs (**figure 17.28**). The activities described for the RMA site illustrate the decontamination and control success that can be achieved in selected cases, the practical and economic value of

appropriate planning to prevent contamination in the first place, and the community and ecological benefits that can result from remediation.

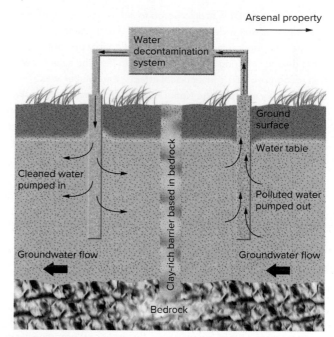

Figure 17.27

Sketch of boundary control system for contaminated groundwater at Rocky Mountain Arsenal. Again, the barrier—in this case, an impermeable one—is placed across the path of groundwater flow.

Figure 17.28

Bison are just one of the many species that make their home at the Rocky Mountain Arsenal National Wildlife Refuge just east of Denver, CO.

U.S. Fish and Wildlife Service

The Long Shadow of DDT

The compound DDT was discovered to act as an insecticide during the 1930s. Its first wide use during World War II, killing lice, ticks, and malaria-bearing mosquitoes, unquestionably prevented a great deal of suffering and many deaths. In fact, the scientist who first recognized DDT's insecticidal effects, Paul Muller, was subsequently awarded the Nobel Prize in Medicine for his work. Farmers undertook wholesale sprayings with DDT to control pests in food crops. The chemical was regarded as a panacea for insect problems; cheery advertisements in the popular press promoted its use in the home.

Then the complications arose. One was that whole insect populations began to develop some resistance to DDT. Each time DDT was used, those individual insects with more natural resistance to its effects would survive in greater proportions than the population as a whole. The next generation would contain a higher proportion of insects with some inherited resistance, who would, in turn, survive the next spraying in greater numbers, and so on. This development of resistance resulted in the need for stronger and stronger applications of DDT, which ultimately favored the development of ever-tougher insect populations that could withstand these higher doses. Insects' short breeding cycles made it possible for all of this to occur within a very few years.

Early on, DDT was also found to be quite toxic to tropical fish, which wasn't considered a big problem. Moreover, it was believed that DDT would break down fairly quickly in the environment. In fact, it did not. It is a persistent chemical in the natural environment. Also, like the heavy metals, DDT is a bioaccumulative chemical. It is fat-soluble and builds up in the fatty tissues of humans and animals. Fish not killed by DDT nevertheless accumulated concentrated doses of it, which were passed on to fish-eating birds (**figure CS 17.1.1**). Then another deadly effect of DDT was realized: it impairs calcium metabolism. In birds, this effect was manifested in the laying of eggs with very thin and fragile shells. Whole colonies of birds were wiped out, not because the adult birds died, but because not a single egg survived long enough to hatch. The eggs might be crushed just by the weight of the adults sitting on the nest. Robins picked up the toxin from DDT-bearing worms. Whole species were put at risk. All of this prompted Rachel Carson to write *Silent Spring* in 1962, warning of the insecticide's long-term threat.

Finally, the volume of data demonstrating the toxicity of DDT to fish, birds, and valuable insects such as bees became so large that, in 1972, the U.S. EPA banned its use, except on a few minor crops and in medical emergencies (to fight infestations of disease-carrying insects). So persistent is it in the environment that more than forty years later, fish often had detectable (though much

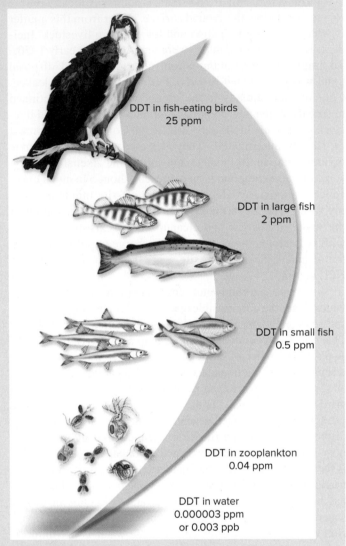

DDT in fish-eating birds
25 ppm

DDT in large fish
2 ppm

DDT in small fish
0.5 ppm

DDT in zooplankton
0.04 ppm

DDT in water
0.000003 ppm
or 0.003 ppb

Figure CS 17.1.1

Biomagnification can increase DDT concentration a million times or more from water to predators high up the food chain.

lower) levels of DDT in their tissues. Encouragingly, many wildlife populations have recovered since DDT use in the United States was sharply curtailed. Bald eagles, ospreys, and California condors are among the predator bird species that have rebounded in numbers across the United States. Still, DDT continues to be used in other countries to control mosquitoes that spread malaria.

Acid mine drainage is a widespread problem. Many metal ore deposits contain sulfide minerals, and other sulfur compounds such as pyrite are associated with coal. Sulfides weather to produce sulfuric-acid-rich drainage. The acid drainage, in turn, leaches a number of metals, some of them toxic, including mercury, lead, cadmium, and arsenic. The hazard is obvious where polluted surface drainage is visible (**figure CS 17.2.1**). Where it is not, visual inspection doesn't help much in assessing degrees of risk from acid drainage. Recently, the use of high-resolution airborne spectrometers has provided a powerful tool for assessing such sites.

Mining near Leadville began in 1859. The sulfide-rich ores were mined for gold, silver, lead, and zinc. In the process, tailings piles were scattered over about 30 square kilometers (12 square miles) in and around Leadville. Predictably, sulfide mineral residues in the tailings weathered and metal-laden acid drainage resulted. In 1983, 18 square miles of the area were placed on the Superfund National Priorities List.

Investigators involved in helping to plan the cleanup selected a single tailings pile and made an aerial traverse across it (**figure CS 17.2.2A**). Comparing spectral data to analysis of surface samples, they could identify zones of different mineralogy across the pile by the distinctive spectrographic signatures of the minerals: different minerals reflect different wavelengths of radiation, just as they reflect different wavelengths of visible light, resulting in different colors. These mineral zones corresponded to varying degrees of acidity of drainage (**figure CS 17.2.2B**). In general, increasing acidity (lower pH) correlated with higher

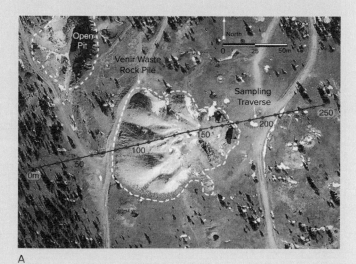

A

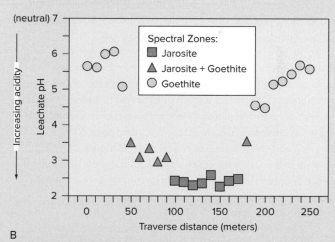

B

Figure CS 17.2.2

(A) Traverse track across a tailings pile at California Gulch. Red dots are sampling sites. (B) Different minerals found across the tailings pile correspond to different acidity (pH) of leachate; goethite is an iron hydroxide; jarosite, a hydrous iron sulfate formed by weathering of sulfides in acid conditions. Many toxic metals at this site are much more leachable in acid waters.

(A) Image courtesy U.S. Geological Survey Spectroscopy Lab

(Continued)

Figure CS 17.2.1

Acid runoff from the Lead Queen mine after a monsoon storm in 2014.

Glen E. "Gooch" Goodwin/USGS

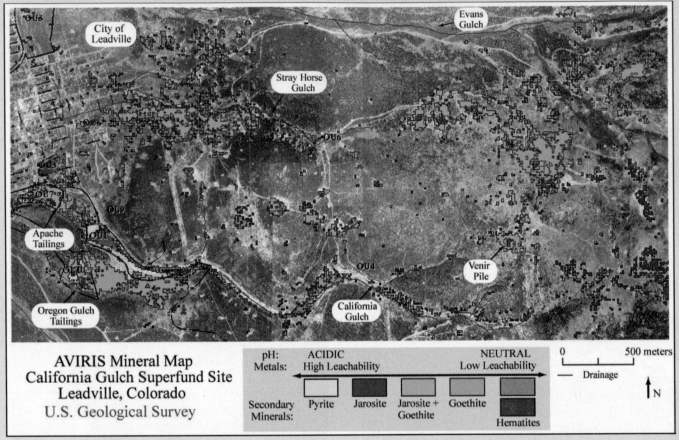

Figure CS 17.2.3

Airborne-spectroscopic scan shows distribution of minerals, which correspond to varying degrees of acidity of water and thus of leachability of metals; combined with drainage information, the image helps target areas of greatest water-pollution danger.

Source: U.S. Geological Survey Spectroscopy Lab.

solubility/leachability of metals. In short, the mineralogy—which could be determined by overflight—was an indicator of drainage geochemistry. This offers a much quicker way to map relative degrees of hazard from dissolved toxic metals over a broad area, and focus cleanup activities accordingly, than does surface sampling and conventional chemical analysis (**figure CS 17.2.3**). Cleanup of the California Gulch site is 90% complete (in nine of the twelve operational units) after being on the NPL for nearly forty years, and many areas have been redeveloped for public recreational purposes.

There are close to 50,000 sites in the United States alone, now inactive, where metals were once mined. Each of these represents a potential acid-drainage hazard, and only a small fraction of them are receiving the close scrutiny of Superfund NPL locations. Having efficient ways to quickly identify those sites, or parts of sites, posing the greatest risk allows prioritization of cleanup efforts. The result of inaction will ultimately lead to more pollution, as exemplified by the 2015 accidental spill of 3 million gallons of wastewater containing high levels of arsenic and mercury into a tributary of the Animas River at the Gold King Mine site near Silverton, Colorado (**figure CS 17.2.4**).

Figure CS 17.2.4

The orange-colored water in the Animas River resulted from the accidental spill of mine wastewaters near Silverton, CO.

KaraGrubis/iStock/Getty Images

Summary

Much of the concern about pollution relates to the effects of trace elements or other chemicals on health. A dose-response curve describes the degree of benefit or harm from a particular substance in varying doses. The same substance may be beneficial or harmful depending upon the dose, and different organisms may respond quite differently to the same chemical.

A variety of substances, some naturally occurring and some added by human activities, cycle through the hydrosphere. How long each substance spends in a particular reservoir is described by the residence time of that substance in the reservoir. The materials of greatest concern in the context of pollution are usually toxic ones with long residence times in the environment. Little is known about many synthetic chemicals—their movements, longevity in natural systems, method of breakdown, and often even the extent of their toxicity.

Sediment pollution is a problem, particularly in agricultural areas, where better erosion control could greatly reduce it. Organic matter, whether contributed by domestic sewage, agriculture, or industry, increases the BOD of the water, making it inhospitable to oxygen-breathing organisms. Where the water is rich in plant nutrients from organic wastes, fertilizer runoff, or other sources, such as phosphate from detergents, a eutrophic condition develops that encourages undesirably vigorous growth of algae and other plant life. The toxic organic compounds in herbicides and insecticides are widely used in both agricultural and urban areas, and their residues are often persistent in both groundwater and surface water. Some of these compounds also show biomagnification in organisms.

In addition to adding many new and sometimes-toxic chemicals to the environment, industrial activities can unbalance the natural cycles of other harmful substances, such as the heavy metals, by increasing the rate at which they are introduced into the environment. These elements share with some other chemicals the tendency to accumulate in organisms and increase in concentration up the food chain, becoming a greater hazard higher up in the chain. Power plants may present a thermal, rather than a chemical, pollution threat.

Reduction in pollutant release into aqueous systems is the most common strategy for reducing pollution problems. As a general rule, pollutants from point sources are more easily controlled or confined than are pollutants from nonpoint sources. More-aggressive physical and/or chemical treatments are sometimes used in water-quality restoration efforts, particularly in the case of small, still water bodies, or to combat toxic-chemical spills. However, even if harmful materials are contained rather than released into the environment or are removed from water or underlying sediment, they still pose a waste-disposal problem. Groundwater pollution is a particular challenge, not only because it is often less readily detected, but also because the nature of groundwater makes it harder to treat *in situ*. Bioremediation and the use of permeable reactive barriers are among techniques that can successfully address groundwater pollution without disrupting flow patterns. New technological tools can assist in focusing efforts to remediate water-pollution problems on areas of most acute need.

Key Terms and Concepts

aerobic decomposition	biomagnification	heavy metals	pump-and-treat
anaerobic decomposition	contaminant plume	nonpoint source	residence time
biochemical oxygen demand (BOD)	dose-response curve	oxygen sag curve	
	eutrophication	point source	

Test Your Learning

1. Describe the purpose of a dose-response curve. Sketch an example for (a) a substance that is essential in small doses but harmful at high doses and (b) a substance that is essential to health in low doses and harmless at higher doses.

2. Explain the concept of residence time; illustrate it with respect to some dissolved constituent in seawater or another water reservoir.

3. Describe how human activities most commonly alter the cycles of naturally occurring elements.

4. Define *biomagnification,* and explain why it is a particular concern of humans. Give an example of a chemical that is subject to biomagnification.

5. Define *BOD.* Sketch an oxygen sag curve for a stream below a source of organic-waste matter, and explain its shape and significance.

6. Define *thermal pollution,* and explain why it is, in a sense, a less-worrisome kind of pollution than most types of chemical pollution.

7. Cite at least three possible sources of the nutrients that contribute to eutrophication of water; explain the concept, the problem that it creates, and why, in general, nutrient-rich waters are often considered undesirable.

8. Describe the kinds of pollution that can be reduced by the use of settling ponds.

9. Briefly describe two after-the-fact approaches to reducing water pollution, and note their limitations.

Exploring Further

1. Investigate your own water quality. If your water comes from a well that taps a major aquifer, you may want to look at the USGS's recent **Regional Assessments of Groundwater Quality**. How is the type and amount of pollution related to the aquifer characteristics and the land use? If you use a municipal water supply, look into its water quality by finding your Consumer Confidence Report from the EPA at **www.epa.gov/ccr**. What, if any, constituents exceed the EPA drinking-water standards?

2. As the largest freshwater lakes in the United States, the Great Lakes receive considerable attention and study; in 2004, the president declared them a "national treasure" by executive order. Investigate the current status of pollution and remediation efforts, in any of the Great Lakes, at **www.epa.gov/greatlakes/**.

3. States, territories, and tribes put out their own fish-consumption advisories. The link to these contacts, as well as information from the historical National Listing of Fish Advisories, can be found at **https://www.epa.gov/fish-tech/state-territorial-and-tribal-fish-consumption-advisories**.

4. A lake contains 200 million m^3 of water. It is fed and drained by streams; the average inflow (and outflow) is 15 m^3/sec. (a) What is the residence time of water in the lake, in days? (b) A pollutant in the lake has reached a concentration of 75 ppm before its input is stopped. Thereafter, it is removed in solution in the outflow stream. Given the meaning of residence time, half the lake water is replaced in a period equal to the residence time, so in this case the quantity of the pollutant is reduced by half in that time. Approximately how long will it take for the concentration of the pollutant to reach 10 ppm or less?

5. Investigate the Flint, Michigan, water crisis. Describe the socioeconomic factors involved, and show how citizen science helped in identifying the water pollution problem. What could scientists and engineers have done better to avoid the crisis?

CHAPTER 18

Air Pollution

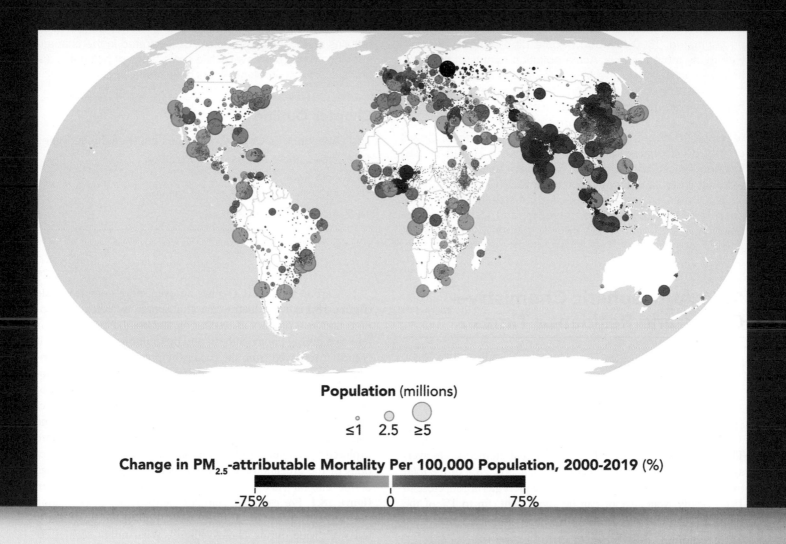

Population (millions)

≤1   2.5   ≥5

Change in PM₂.₅-attributable Mortality Per 100,000 Population, 2000-2019 (%)

-75%   0   75%

Donora, Pennsylvania, 1948: "The fog closed over Donora on the morning of Tuesday, October 26th. . . . By Thursday, it had stiffened . . . into a motionless clot of smoke. That afternoon, it was just possible to see across the street . . . the air began to have a sickening smell, almost a taste . . ." (Roueché, p. 175). By Friday, residents with asthma and other lung disorders began to find it difficult to breathe. Then more of the residents took sick, becoming nauseous, coughing and choking, suffering from headaches and abdominal pains. The fog persisted for five days. Altogether, nearly 6000 people were stricken by the polluted fog; 20 of them died.

Air pollution can affect visibility, climate, and health. Researchers have modeled the increase in mortality due to fine particulate air pollution, considering changes in air quality from 2000–2019. City dwellers are the most at risk; in 2019, 86% lived in areas where PM₂.₅ concentrations exceed World Health Organization guidelines. That pollution resulted in 1.8 million premature deaths.

NASA Earth Observatory images by Joshua Stevens

Donora was a mill town with a steel plant, a wire plant, and a zinc and sulfuric acid plant among its industries. The litany of toxic chemicals in its air identified during later investigations included fluoride, chloride, hydrogen sulfide, sulfur dioxide, and cadmium oxide, along with soot and ash. These materials were all routinely released into the air, but not generally with such devastating effects.

The events of late October 1948 in Donora might be called an acute air-pollution episode—sudden, obvious, and dramatic. Other still-more-serious single episodes are known. For four days in early December 1952, weather conditions trapped high concentrations of smoke and sulfur gases from coal burning in homes and factories over London, England; an estimated 3500 to 4000 people died in consequence, with many more made ill. Acute pollution events of this kind are, fortunately, rare. The health impact of moderately elevated levels of pollutants found widely over Earth, especially near urban areas, is more difficult to assess.

The costs of air pollution are staggering. The most recent estimate from the World Bank puts the annual global cost of health damages at $8.1 trillion. The Health Effects Institute indicates that in 2019, 6.6 million deaths worldwide were attributable to air pollution, over 60,000 of those in the United States. In addition, there are the sizeable costs in illness, medical expenses, absenteeism, and loss of production, which are somewhat harder to quantify. Other annual economic costs in the billions include cleaning dirty items of particulates, damage to crops and livestock, and reduction in forest-product harvests.

One estimate suggests that reduction of air-pollution levels by 50% in major urban areas of the United States would save more than $2 billion per year in health costs. Implicit in the decreased health costs is an increase in longevity, decrease in illness, and improvement in quality of life among those now adversely affected by air pollution. The financial and human considerations together are powerful incentives for trying to limit air pollution. In the United States, great progress has been made in reducing most air pollutants, as we will see in this chapter. Elsewhere in the world, air pollution remains a very serious problem, as illustrated in the chapter-opening image. Moreover, while some types of pollutants are primarily a local problem, others have global impacts.

Chapter Outline

18.1 Atmospheric Chemistry— Cycles and Residence Times

Learning Objectives

- Recall the composition of the atmosphere
- Recognize the reservoirs and fluxes involved in the global carbon cycle

The atmosphere consists of three principal elements. On average, nitrogen (mostly as N_2) comprises nearly 78% of the total by weight; oxygen (O_2), about 21%; the inert gas argon, close to 1%. Locally, water vapor can be significant, up to 1% of the total. Everything else in the atmosphere together makes up much less than 1% of it.

Materials cycle through the atmosphere as they do through other natural reservoirs. One can speak of the residence times of gases or particles in the atmosphere just as one can discuss residence times of chemicals in the ocean. As with dissolved substances, residence times of gases are influenced by the amounts present in a given reservoir such as the atmosphere, and the rates of addition and removal. Oxygen, for example, is added to the atmosphere during photosynthesis by plants; it is removed by oxygen-breathing organisms, by solution in the oceans, by reaction with rocks during weathering, and by combustion. Its residence time in the atmosphere is estimated at 5,000 years.

Carbon dioxide (CO_2) has an even more complex cycle (**figure 18.1**). It is added to the atmosphere by volcanic eruptions and as a product of respiration and combustion, and it is removed during photosynthesis and by solution in the oceans. It is further removed from the oceans by precipitation in carbonate sediments. The concentration of carbon dioxide in the atmosphere is about 420 parts per million (ppm), far less than that of oxygen, and carbon dioxide has a correspondingly much shorter residence time, estimated at less than 100 years. Some of the consequences of increased CO_2 content in the atmosphere were explored in chapter 10. One of the factors complicating projections of the greenhouse effect relates to imprecise knowledge of the carbon fluxes between pairs of reservoirs shown in **figure 18.1**. For example, consider the difficulty of estimating precisely the amount of CO_2 consumed by plants and phytoplankton worldwide during photosynthesis. What might seem a far simpler problem, estimating net CO_2 flux between atmosphere and oceans, is not so simple after all, for the extent to which CO_2 dissolves in the oceans varies with temperature, and air and water temperatures, in turn, vary both regionally and seasonally. Still, the broad outline of the global carbon cycle, in terms of reservoirs and processes, is reasonably well understood.

The chemically inert gases, including argon and trace amounts of helium and neon, have virtually infinite residence times. They do not, by definition, react chemically with other substances, so they are not readily removed by natural processes. The geochemical cycle of nitrogen is so complex that its overall residence time in the atmosphere is not known, although residence times of

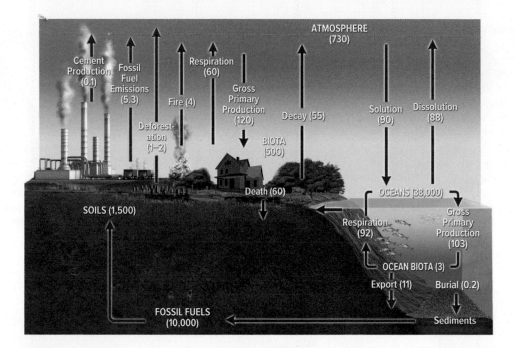

Figure 18.1

Reservoirs and fluxes in the global carbon cycle. Reservoirs are indicated in capital letters, fluxes from one reservoir to another by arrows. Numbers are billions of tons of carbon. While flux changes due to human activities (red arrows) seem relatively small, they can have a perceptible effect on atmospheric CO_2 concentrations.

individual compounds have been estimated. They vary considerably, from about 44 million years for N_2 to a matter of weeks for the trace amounts of nitric acid (HNO_3). As with many synthetic water pollutants, the fate and residence times of air pollutants added by human activity often are also poorly known.

18.2 Types and Sources of Air Pollution

Learning Objectives

- Describe the types, sources, trends, and health concerns of particulate pollution
- List the sources, trends, and health concerns with carbon monoxide and sulfur dioxide emissions
- Describe the relationship between nitrogen dioxide emissions and ground-level ozone pollution
- Summarize the formation of ozone and how it varies by latitude and season
- Explain how CFCs lead to the destruction of ozone
- Describe how leaded gasoline resulted in lead poisoning

Most air pollutants are either gases or **particulates.** The principal gaseous pollutants are oxides of carbon, nitrogen, and sulfur. They share some common sources but create distinctly different kinds of problems. The principal sources for each of the major pollutants and emissions are indicated for the United States in **figure 18.2**.

The amount of data regarding the types and sources of global air pollution is vast, and has increased drastically with the expanding use of satellites to analyze processes occurring in Earth's atmosphere. In this section, we will review a variety of pollutant groups, including their sources, the problems they cause, and emission trends in the United States. Overall, emissions have decreased significantly in the past 30 years in response to regulation; some pollutant concentrations have increased recently, but still remain below the national air quality standards.

Particulates

Particulate air pollutants generated by human activities are largely derived from point sources. The particulates include soot, smoke, and ash from fuel (mainly coal) combustion; dust released during industrial processes; and other solids from accidental and deliberate burning of vegetation. Estimates of the magnitude of global anthropogenic contributions vary widely, from about 35 million tons/year (two-thirds from combustion) to 180 million tons/year (mostly industrial). Additional particulates are added by many natural processes, including volcanic eruptions, natural forest fires, erosion of dust by wind, and blowing salt spray off the sea surface.

Typically neither natural nor anthropogenic particulates have long residence times in the atmosphere. They are quickly removed by precipitation, usually within days or weeks. In rare cases, such as fine volcanic ash shot high into the atmosphere by violent eruptions, the finest, lightest material may stay in the air for as long as several years. Particulate pollution is usually a local problem, being most severe close to its source and lasting only a short time. There are

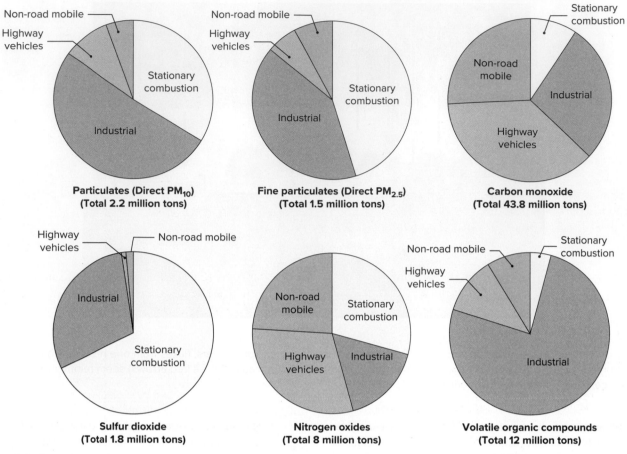

Particulates (Direct PM₁₀)
(Total 2.2 million tons)

Fine particulates (Direct PM₂.₅)
(Total 1.5 million tons)

Carbon monoxide
(Total 43.8 million tons)

Sulfur dioxide
(Total 1.8 million tons)

Nitrogen oxides
(Total 8 million tons)

Volatile organic compounds
(Total 12 million tons)

Figure 18.2

Emission sources of some major air pollutants. Stationary combustion is primarily utilities; non-road mobile includes such forms of transportation as airplanes, trains, and ships. Particulates, carbon monoxide, sulfur dioxide, and nitrogen dioxide are four of the six criteria air pollutants regulated by the EPA. The others are lead and ground-level ozone.

Source: Data from EPA, Our Nation's Air 2021.

incidences of particulates traveling long distances and causing severe problems thousands of kilometers from their source. It isn't unusual for wind to carry dust across the Atlantic from the Sahara, and in the summer of 2020, a plume thousands of kilometers long reached the Caribbean and Southeastern United States, resulting in poor air quality and the closure of beaches.

The nature of the problem(s) posed by particulate pollution depends somewhat on the nature of the particulates. Layers of dense smoke are ugly, whatever their makeup. In areas showered with ash from a power station or industrial plant, cleaning expenses increase. Fine rock and mineral dust of many kinds has been shown to be carcinogenic when inhaled; the finest particulates are especially unhealthy and are linked to cardiovascular and respiratory diseases. Dust can be a vector for viral and bacterial diseases. Many particulates are also chemically toxic. Coal ash may contain both heavy metals and uranium stuck to the particles, for example. The dust was a significant concern after the 2001 destruction of the World Trade Center. It was a complex mixture of materials, including glass fibers, concrete fragments, gypsum wallboard, (chrysotile) asbestos insulation, paper, and metal-rich particles. Care had to be

taken during cleanup because some of this dust could be harmful if inhaled. The size of particulates is important in determining their health and environmental effects, so beginning in 1999, the Environmental Protection Agency (EPA) began tracking emissions of particulates with diameters less than 10 microns (about 0.0004 inch), designated PM₁₀, and fine particulates less than 2.5 microns across, designated PM₂.₅.

In the United States, particulate emissions from many industrial and commercial sources have been reduced (**figure 18.3**); dust raised by such activities as agriculture, construction, and driving on unpaved roads, all remaining significant sources, is harder to control, so complete elimination of particulate pollution is not feasible. Recent exceedances of the PM₂.₅ concentration standard (35μg/m³) have occurred nearly exclusively in western states experiencing large and long-lasting wildfires. Globally, the control of particulate pollution is a mostly a matter of health rather than climate or aesthetics. Over the past 30 years, 53 of every 100,000 deaths are attributed to exposure to particulates, and the most intense exposures occur in North Africa, the Middle East, and Asia. Satellites are increasingly being used to track particulates and other

Direct PM$_{2.5}$ Emissions

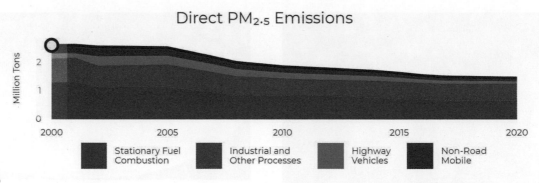

Stationary Fuel Combustion — Industrial and Other Processes — Highway Vehicles — Non-Road Mobile

Figure 18.3

Direct emissions of PM$_{2.5}$ have decreased by over 40% in the last 20 years; the national average 24-hour concentration (how much a person is exposed to in $\mu g/m^3$) also decreased, by 30%.

Source: U.S. Environmental Protection Agency, Our Nation's Air 2021.

pollutants, and even to distinguish between pollutants from natural processes and those resulting from human activities (**figure 18.4**).

Worldwide, particulate pollution in the form of soot is most intense in the air above China and India, where cooking and heating are done with wood, coal, and other biofuels at low temperatures, resulting in incomplete combustion (**figure 18.5**). Soot is also intense wherever fires are used in slash-and-burn farming. Recent severe soot pollution in Nigeria has been blamed on both the operation and destruction of illegal oil refineries there. Soot in the atmosphere absorbs sunlight and heats the air above it, causing convection and affecting the hydrologic cycle. Soot also drifts widely through the atmosphere, even reaching the Arctic.

As it settles out of the air, it darkens the surface, decreasing albedo but causing the underlying materials to warm; soot has been implicated in helping to melt both Arctic ice and glaciers in the Himalayas.

Carbon Gases

The principal anthropogenic carbon gases are carbon monoxide and carbon dioxide. Carbon dioxide is naturally present in the atmosphere and is essential to the life cycles of plants. It is a natural end product of the complete combustion of carbon-bearing fuels:

$$\underset{\text{carbon}}{C} + \underset{\text{oxygen}}{O_2} = \underset{\text{carbon dioxide}}{CO_2}$$

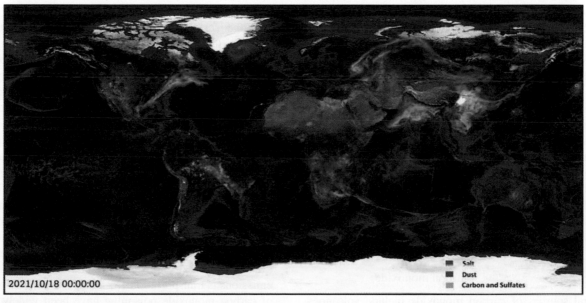

Figure 18.4

Aerosol optical thickness is determined by satellites which measure how much light can pass through the aerosols. The bright colors of each type represent thicker layers. NASA's GEOS model provides a ten-day aerosol thickness forecast. Notice the dust from the Sahara moving west out over the Atlantic.

NASA's Scientific Visualization Studio/NASA Center for Climate Simulation.

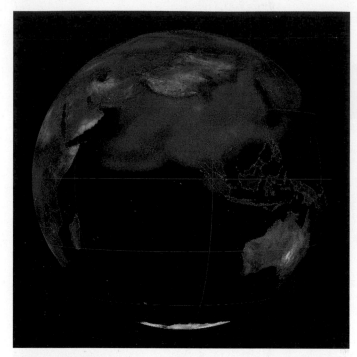

Figure 18.5

Soot, or black carbon, in the atmosphere on January 21, 2007.

NOAA/Global Systems Division (GSD) David Himes

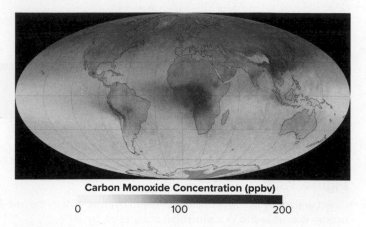

Carbon Monoxide Concentration (ppbv)

0   100   200

Figure 18.6

Fires in Africa and vehicle and industrial combustion in China produce not only visible particulates, but invisible carbon monoxide as well, which can travel far from its source. It can still be detected by satellite, as shown in this 2014 image. Globally, sources may change with the seasons (ppmv = ppm by volume).

NASA Earth Observatory image created by Jesse Allen and Joshua Stevens using data from the National Center for Atmospheric Research and the University of Toronto MOPITT teams.

As such, carbon dioxide is continually added to the atmosphere through the combustion of fossil fuels, as well as by natural processes, including respiration by all oxygen-breathing organisms. As noted in chapter 10, atmospheric carbon dioxide levels have increased over 50% in the last 160 years because of heavy fossil-fuel use, and concentrations continue to climb, spurring concerns about resultant climate change. An estimated 36 billion metric tons of anthropogenic carbon dioxide are added to the atmosphere each year. The United States accounts for approximately 15% of this total; the primary sources of our CO_2 emissions are electric utilities and transportation, each of which produces about a third of them. The U.S.EPA does consider CO_2 as a pollutant that can be regulated because of its indirect yet substantial impact on human health through climate change.

Carbon monoxide (CO), while volumetrically far less important than carbon dioxide, is more immediately deadly. It is produced during the incomplete combustion of carbon-bearing materials, when less oxygen is available:

$$2C + O_2 = 2CO$$
$$\text{carbon} \quad \text{oxygen} \quad \text{carbon monoxide}$$

Natural sources of carbon monoxide include volcanoes, fires, lightning, and the oxidation of coal deposits. Most anthropogenic CO comes from fossil-fuel burning. The residence time of CO in the atmosphere is a few months, where it reacts with oxygen in the air to create CO_2. It can constitute a serious local health hazard, and occasionally a regional one as well (**figure 18.6**).

Carbon monoxide is an invisible, odorless, colorless, tasteless gas. Its toxicity to animals arises from the fact that it replaces oxygen in the hemoglobin in blood. The vital function of hemoglobin is to transport oxygen through the bloodstream. Carbon monoxide molecules can attach themselves to the hemoglobin molecule in the site that oxygen would normally occupy; moreover, CO bonds there more strongly. As CO builds up in the bloodstream, cells (especially brain cells) begin to die from the lack of oxygen, and, eventually, enough cells may fail to cause the death of the whole organism. The difficulty of detecting carbon monoxide, together with the sleepiness that is an early consequence of a reduced supply of oxygen to the brain, helps to explain the number of accidental deaths from carbon monoxide poisoning in inadequately ventilated spaces.

Carbon monoxide doesn't remain in the blood indefinitely. If someone in the early stages of carbon monoxide poisoning is removed to fresh air or, if possible, given concentrated medical oxygen, the carbon monoxide is gradually released (though some irrevocable brain damage may have occurred). The problem is to recognize in time what is taking place. Carbon monoxide can build up to dangerous levels wherever combustion is concentrated and air circulation is poor.

In the United States, as in most industrialized countries, the single largest anthropogenic source of carbon monoxide emissions is transportation (recall **figure 18.2**). Nonfatal cases of carbon monoxide poisoning have been widely reported in congested urban areas with high traffic density. This is one argument in favor of cleaning up auto emissions and making engines burn fuel as efficiently as possible. Another is simply that, if an engine is producing carbon monoxide, it is wasting energy. Burning carbon completely to make carbon dioxide releases three times the energy that is produced from incomplete combustion to carbon monoxide. Substantial progress in reducing CO emissions has been made, particularly with respect to CO from highway vehicles. Federal and state regulations together have resulted in reduction of U.S. carbon monoxide emissions by 81% between 1980 and 2020.

Sulfur Gases

The principal sulfur gas produced through human activities is sulfur dioxide, SO_2. More than 97 million metric tons are emitted worldwide each year, the vast majority associated with the burning of fossil fuels for power generation by utilities and industry. Within a few days of its release into the atmosphere, sulfur dioxide reacts with water vapor and oxygen in the atmosphere to form sulfuric acid (H_2SO_4), which is a strong and highly corrosive acid:

$$SO_2 \quad + \quad H_2O \quad + \tfrac{1}{2} \ O_2 = \quad H_2SO_4$$

sulfur dioxide \quad water vapor \quad oxygen \quad sulfuric acid

Much of this is scavenged out of the atmosphere in the form of acid rain (discussed later in the chapter) to contribute to acid runoff. This, then, is another example of an air-pollution problem that becomes a water-pollution problem. As long as it remains in the air, sulfuric acid is severely irritating to lungs and eyes, mostly causing respiratory problem in humans. Controls on power-plant emissions, as implemented in response to regulations of the Clean Air Act and its amendments, have been the major factor in reducing SO_2 emissions in the United States (**figure 18.7**). In 2020, sulfur emissions were just 8% of what they were in 1990.

Nitrogen Gases and Ground-Level Ozone

The geochemistry of nitrogen oxides (NO_x) in the atmosphere is complex. Since nitrogen and oxygen are by far the most abundant elements in air, it isn't surprising that, at the high temperatures found in engines and furnaces, they react to form nitrogen oxide compounds. Nitrogen monoxide (NO) can act somewhat like carbon monoxide in the bloodstream, though it rarely reaches toxic levels. In time, it reacts with oxygen to make nitrogen dioxide (NO_2). Nitrogen dioxide reacts with water vapor in air to make nitric acid (HNO_3), which is both an irritant and corrosive:

$$2\,NO_2 \quad + \quad H_2O \quad + \tfrac{1}{2}\,O_2 = \quad 2\,HNO_3$$

nitrogen dioxide \quad water vapor \quad oxygen \quad nitric acid

Combustion of various kinds adds approximately 24 million tons of nitrogen oxides to the air each year, compared to the estimated 20 to 90 million tons produced by natural biological action, volcanoes, oceans, and lightning. Most anthropogenic NO_2 production is strongly concentrated in urban and industrialized areas and may create serious problems in those areas (**figure 18.8**). In fact, as engines and furnaces are made more efficient to reduce emission of CO and unburned hydrocarbons, nitrogen-oxide emissions may increase.

The most obvious harmful effect of nitrogen dioxide is its role in the production of photochemical smog. Key factors in the formation of photochemical smog are high concentrations of nitrogen oxides and strong sunlight. Dozens of chemical reactions may be involved, but the critical one involving sunlight is the breakup of NO_2 to produce NO and a free oxygen atom, which reacts with the common oxygen molecule (O_2) to make *ozone* (O_3). Ozone is a strong irritant to the lungs, especially dangerous to those with lung ailments or those who are exercising and breathing hard in the polluted air. Significant adverse medical effects can result from ozone concentrations below 1 ppm. Ozone also inhibits photosynthesis in plants. The dual requirement of nitrogen dioxide plus sunlight to produce ozone at ground level explains why ozone alerts are more often broadcast in cities with heavy traffic, and during the summertime, when sunshine is abundant and strong. Areas in which air quality is problematic with respect to ozone are, for the most part, urbanized areas (**figure 18.9**). Globally, most of this near-ground ozone is concentrated in the Northern Hemisphere.

The Ozone Layer and Chlorofluorocarbons

If ozone is so harmful, why has so much concern been expressed over the possible destruction of the ozone layer? The answer is that ozone at ground level, where it can interact with animals and plants,

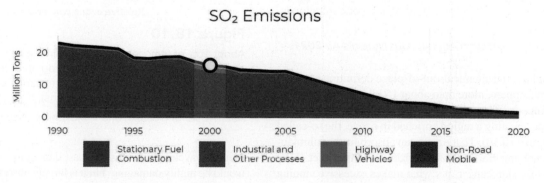

Figure 18.7

Sulfur dioxide emissions have been dramatically reduced, primarily by regulation of coal-fired power plants.

Source: U.S. Environmental Protection Agency, Our Nation's Air 2021.

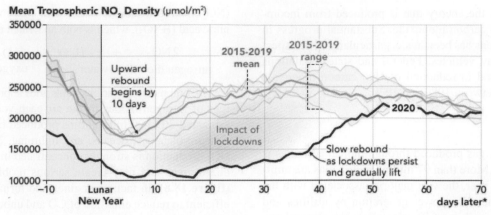

Mean Tropospheric NO₂ Density (µmol/m²)

Figure 18.8

Tropospheric NO_2 levels in China fell dramatically in February at the beginning of the pandemic, as the country shut down transportation and most of its economy. Three months later, levels returned to near normal. It is usual for NO_2 to fall during Lunar New Year celebrations when people are working and driving less.

Source: National Aeronautics and Space Administration

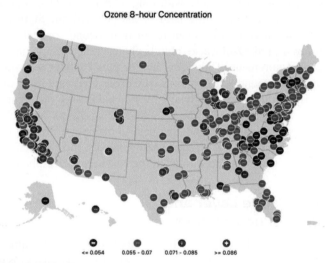

Ozone 8-hour Concentration

Figure 18.9

Snapshot of maximum ozone levels in 2020. For each site, the value is the fourth-highest 8-hour ozone concentration, in ppm. The EPA's air-quality standard for 8-hour ozone concentration is 0.070 ppm.

U.S. Environmental Protection Agency, Our Nation's Air 2021.

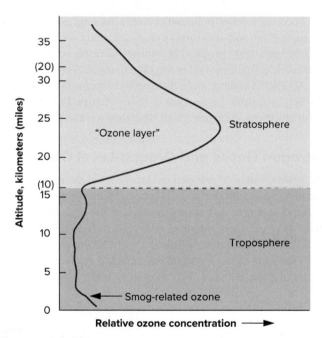

Figure 18.10

About 10% of atmospheric ozone is in the troposphere. The other 90% is in the ozone-enriched region of the stratosphere known as the *ozone layer*. Even there, ozone concentration is less than half a part per million, at most.

After National Oceanic and Atmospheric Administration.

is a good example of the chemical-out-of-place definition of a pollutant. In the stratosphere, more than about 15 kilometers above the surface, ultraviolet rays from the Sun interact with ordinary oxygen to produce ozone, creating a region enriched in ozone, the so-called **ozone layer** (**figure 18.10**). That ozone can absorb further ultraviolet radiation, shielding Earth's surface from it. **Ultraviolet** (UV) radiation can cause skin cancer; it's what makes excessive tanning and Sun exposure unhealthy. The presence of the ozone layer decreases that risk. UV radiation spans a range of wavelengths shorter than those of visible light (**figure 18.11**). UVA, the

longest-wavelength UV, is the least damaging. UVC, the shortest, would be highly damaging, but it is largely absorbed by oxygen and water vapor in the atmosphere. It is the intermediate-wavelength UVB, also potentially damaging, that is particularly absorbed by ozone. Measurements worldwide document that the amount of

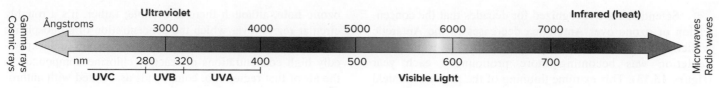

Figure 18.11

Ultraviolet radiation is to the shorter-wavelength, higher-energy side of the visible-light spectrum. 1 nm (nanometer) = 10^{-9} meters = one-billionth of a meter = 10 Å (Ångstroms).

damaging UV reaching the surface increases as the ozone in the ozone layer decreases. Scientists of the U.N. Environment Programme estimate that each 1% reduction in stratospheric ozone could result in a 3% increase in non-melanoma skin cancers in light-skinned people, in addition to increased occurrence of melanoma, blindness related to development of cataracts, gene mutations, and immune-system damage.

Normally, O_2 molecules in the stratosphere absorb UVC radiation, which splits them into two oxygen atoms, each of which can attach to another O_2 molecule to make O_3. The O_3 molecule, in turn, absorbs UVB radiation, which splits it into O_2 plus a free oxygen atom, and the latter, colliding with another O_3, causes regrouping back into two O_2 molecules. Over time, a balance between ozone creation and destruction develops.

The concentration of ozone in the ozone layer varies seasonally and with latitude (**figure 18.12**). Ozone production rates are most rapid near the equator, as a consequence of the strong sunlight there, but generally, the total ozone in a vertical column of air increases with latitude as a result of the balance between natural ozone production and destruction rates, and atmospheric circulation patterns. UVB exposure is typically higher at low latitudes and lower near the poles.

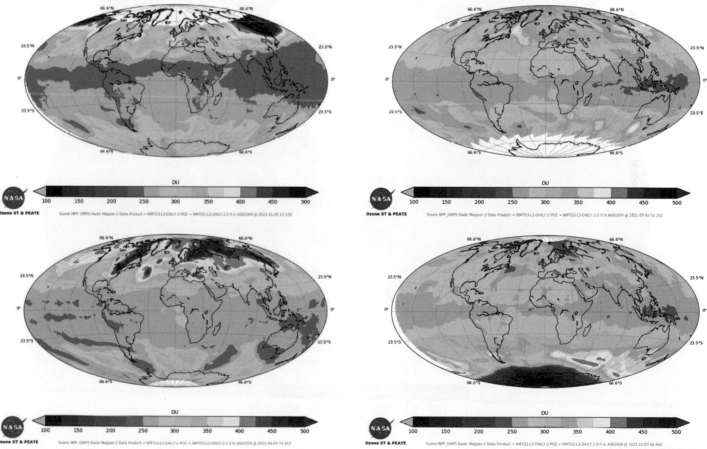

Figure 18.12

The seasonal variation in global ozone distribution on the firsts of January, April, July, and October 2021 as measured by backscattering UV satellite sensors in **Dobson units.**

National Aeronautics and Space Administration

Scientists have recognized for decades that the concentration of ozone over Antarctica decreases in the Antarctic winter. In the 1980s, it became apparent that the extent of the depletion was becoming more pronounced each year (**figure 18.13**). This extreme thinning of the protective shield of stratospheric ozone came to be described inaccurately as an **ozone hole,** although there is no hole; rather, it's a roughly circular region over which the concentration of stratospheric ozone is relatively depleted. Measurements showed unexpectedly high concentrations of reactive chlorine compounds in the air of that region too, compounds associated with anthropogenic chlorofluorocarbons (CFCs).

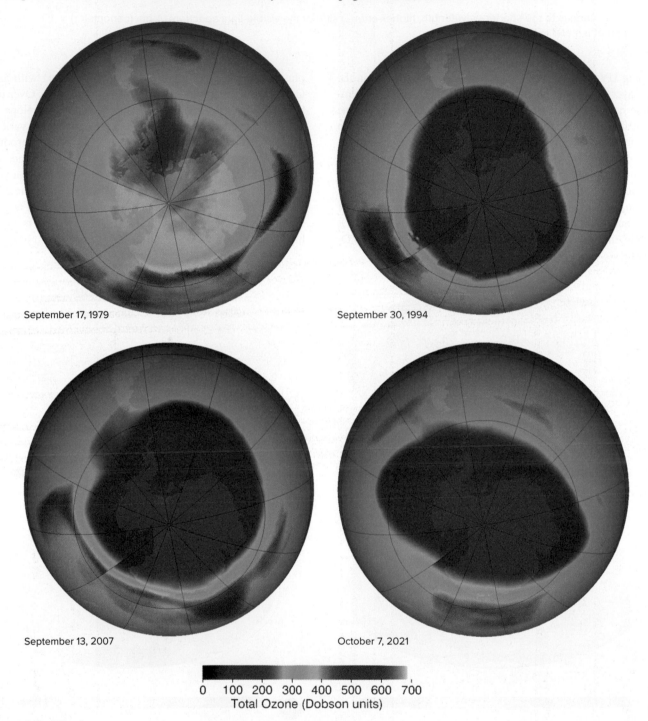

September 17, 1979

September 30, 1994

September 13, 2007

October 7, 2021

0 100 200 300 400 500 600 700
Total Ozone (Dobson units)

Figure 18.13

Measurements of ozone over the Southern Hemisphere showed a deepening "hole" beginning in 1979. The seasonal ozone depletion continued to intensify for years after the banning of CFCs; 2006 showed the deepest hole yet. By now, measurements suggest, the situation has begun to improve, but full recovery of the ozone layer is still decades away.

NASA images courtesy NASA Ozone Hole Watch.

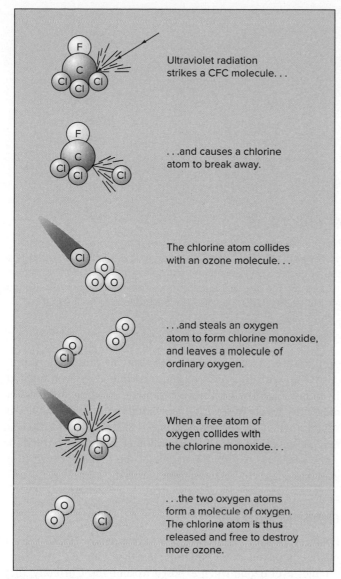

Figure 18.14

Schematic of the role CFCs play in ozone destruction. This involves destruction of ozone beyond what normally occurs by interaction with ultraviolet light. Not shown here, bromine also causes the destruction of ozone.

Images courtesy NASA.

The chemistry of ozone destruction is complex (**figure 18.14**). Most CFCs remain in the atmosphere for fifty to several hundred years, slowly migrating upward and breaking down over time. Chlorine atoms are freed by this CFC decomposition, which occurs especially readily in polar clouds during the winter, when ice crystals provide a surface on which the key chemical reactions can take place. Moreover, a single chlorine atom can destroy many ozone molecules.

Recognizing the evident threat of ozone destruction by CFCs, the global community banded together to support the Montreal Protocol on Substances that Deplete the Ozone Layer. First adopted in 1987, strengthened in 1990, and signed by 175 nations, the Montreal Protocol stipulated phaseout of production of CFCs and other ozone-depleting compounds by the year 2000. One early concern was that some populous developing countries were reluctant to sign, uncertain about their ability to find technically and economically workable substitutes. Even in developed nations, there was some delay in the CFC phaseout. Ultimately, the very real concerns about ozone depletion, and inclusion of provisions designed to assist developing countries to meet its objectives, encouraged additional nations to sign the Montreal Protocol. It has, in fact, become the first treaty in the history of the United Nations to achieve universal ratification—all 197 members have signed it.

The original Protocol was not an immediate or perfect solution. Some of the compounds being substituted for CFCs can themselves contribute to ozone depletion or are potent greenhouse gases (GHGs). The 2016 Kigali amendment to the Protocol, outlining the 80% cut in production and consumption of hydrofluorocarbons by 2046, was signed again by all 197 countries. Also, the residence times of CFCs and their breakdown products are years to decades, so their effect on the atmosphere continues long after their production ceases. And old air conditioners and refrigerators continue to leak CFCs. Nevertheless, by 2017 researchers could see evidence that recovery of the ozone in the stratosphere was beginning. However, recent projections suggest that the ozone layer may not recover fully until about 2065 in the mid-latitudes, later in Antarctica. A complicating factor, only realized relatively recently, is that nitrous oxide also contributes to ozone destruction in the stratosphere. Nitrous oxide is a GHG, is not one of the gases covered by the Montreal Protocol, and global emissions are increasing.

Reductions in ozone have also occurred in the Northern Hemisphere, as it did in March 2020 when scientists detected a pronounced ozone "hole" over the North Pole (**figure 18.15**). The development of holes such as this, which also occurred in March 2011, is attributed to prolonged, unusual conditions in which frigid air is trapped by strong polar westerly winds. As we discussed in chapter 10, extreme weather conditions like the increased intensity of polar vortexes may become more common in the near future as the climate system becomes imbalanced by the rapid buildup of heat energy.

Lead

One air pollutant that has by now been virtually eliminated is lead. Levels of lead in U.S. air decreased by 98% between 1980 and 2014 as result of the removal of lead from gasoline. Lead is not naturally a significant component of petroleum, but one particular lead compound, tetraethyl lead, had been a gasoline additive since the 1940s as an antiknock agent to improve engine performance.

Lead became and still remains a pollutant of great concern because of its serious and numerous health effects. It is one of the heavy metals that accumulate in the body.

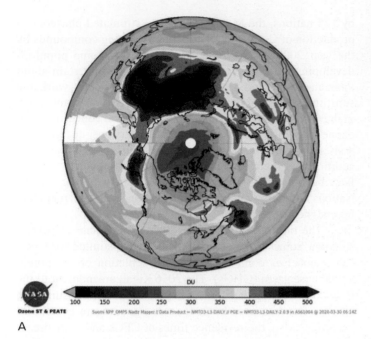

A

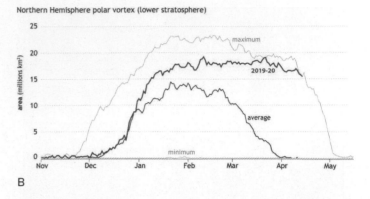

Northern Hemisphere polar vortex (lower stratosphere)

B

Figure 18.15

North pole ozone depletion. (A) The strong winds of the polar vortex kept frigid air from dispersing to lower latitudes, setting up the necessary conditions for destruction of ozone in the stratosphere. (B) The size of the polar vortex remained higher than normal for in the 2019–2020 winter, and lasted longer than usual.

National Aeronautics and Space Administration

Exposure to lead in high concentrations causes brain damage. Mild lead poisoning in the nervous system can cause depression, nervousness, apathy, and other psychological disorders, as well as learning difficulties and deficits. Other symptoms include infertility and gastrointestinal disorders. Children in urban areas who breathe lead-laden air and also consume chips of old lead-based paint often develop high and harmful lead levels in their blood. As many as 2.5% of U.S. children may have blood level levels above levels that cause adverse effects

Lead levels in most paints have been greatly reduced, and lead is no longer added to gasoline. Beginning in the early 1970s, the EPA began to mandate reductions in the lead content of gasoline, with lead phased out almost completely from automotive fuel by 1987. Lead consumed in gasoline dropped 70% from 1975 to 1982. Concentrations of atmospheric lead decreased correspondingly, by about 65% over the same period, and this, in turn, was reflected in greatly reduced lead levels in blood (**figure 18.16A**). For a time, a limited quantity of leaded fuel remained available, primarily for the benefit of old engines in farm equipment that, for technical reasons, need some lead to protect their valves. By 1996, a complete ban on leaded gasoline took effect, and U.S. lead emissions have been virtually eliminated (**figure 18.16B**); 664 tons were emitted in 2020. Leaded gasoline was globally phased out in 2021, although it is still used in some aircraft, race cars, and boats.

Unfortunately, this is not yet the end of the lead-pollution story. Lead from gasoline emitted in vehicle exhaust was commonly deposited in and on soils adjacent to highways, reaching concentrations of hundreds of ppm along busy urban streets. From there it may be leached into surface water or groundwater, transported with the soil in runoff, or simply stirred up with fine dust, to be redistributed or even inhaled. Even now, significant numbers of urban children still develop unhealthy levels of lead in their blood, with values higher in summer when they are more exposed to soil outdoors. So lead may no longer be a notable air pollutant, but clearly, it will be some years yet before the residue of the tetraethyl lead from gasoline ceases to be a pollution concern.

Other Pollutants

Easily vaporized volatile organic compounds (VOCs) are a major component of air pollution; together with NO_x, they form ground-level ozone when exposed to sunlight. Currently, they come primarily from industrial processes. U.S. emissions of VOCs in 2020 were about half they were thirty years previous, with the largest decrease (81%) occurring in the transportation sector, which now is responsible for about one-fifth the yearly amount. Emissions of some VOCs are declining, while other are increasing (**figure 18.17**).

Other air pollutants are of more localized significance. Heavy metals other than lead, for example, can be a severe problem close to mineral-smelting operations. Such elements as mercury, lead, cadmium, zinc, and arsenic may be emitted either as vapors or attached to particulate emissions from smelters. While they are quickly removed from the atmosphere by precipitation, the metals may then accumulate in local soils, from which they are concentrated in organic matter, including growing plants. They can constitute a serious health hazard if accumulated in food or forage crops.

Finally, there are air pollutants of particular concern only indoors; see Case Study 18.1.

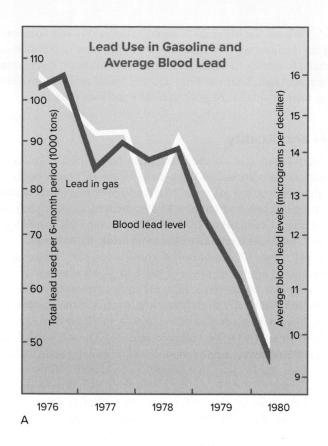

A

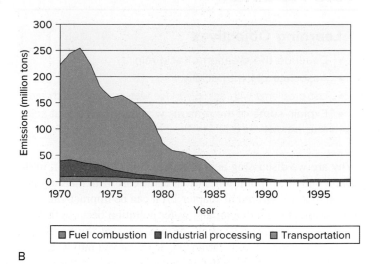

B

Figure 18.16

(A) Declines in lead used in gasoline were directly reflected in lower lead levels in blood. (B) By now, lead emissions are negligible.

(A) After L. Whiteman, "Trends to Remember: The Lead Phase Down," EPA Journal, May/June 1992, p. 38; (B) From U.S. Environmental Protection Agency, National Emissions Trends 1990–1998.

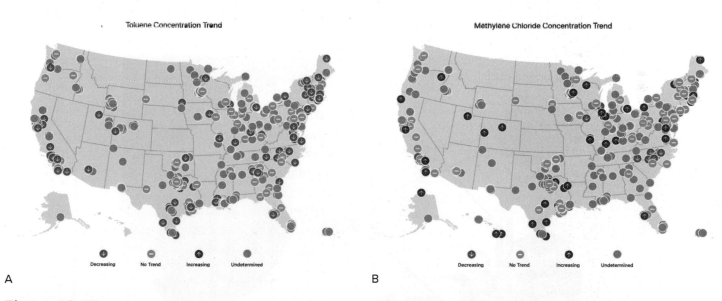

A   B

Figure 18.17

Trends in VOC emissions; most are decreasing, but some are not. (A) The national median toluene concentration, from 146 sites, has decreased by 57% in fifteen years. (B) The national median methylene chloride concentration hasn't changed but individual sites show increasing measured concentrations. To assess these and other VOC trends, go to Our Nation's Air 2021, and scroll down to the Air Toxics section.

(A-B) United States Environmental Protection Agency

18.3 Acid Rain

Learning Objectives

- Describe the creation of acid rain
- Associate pH with acidity
- List problems with acidic surface water and groundwater
- Explain some of the reasons why concerns about acid rain persist

Why are we discussing acid rain, which is clearly water, in the chapter about air pollution? Just as is the case with mercury pollution that we covered in chapter 17, it can be impractical, if not impossible, to separate air and water pollution because they are intimately connected through the processes of the hydrologic cycle. Emissions into the atmosphere get moved into water, and aerosols from surface water bodies move up into the air. Air pollutants can be carried long distances by wind, and after combining with other compounds in the atmosphere to make acids, they eventually make their way to Earth's surface along with precipitation in wet deposition (**figure 18.18**). All precipitation is naturally acidic, but **acid rain** is precipitation that is more acidic than normal, and the term is usually tied to anthropogenic emissions of sulfur dioxide (SO_2) and nitrogen oxides (NO_x). There are natural sources of SO_2 and NO_x, but currently most comes from the combustion of fossil fuels to create electricity and power vehicles. Acid-forming particles can also be dry-deposited on the surface, which then form acids when combined with surface water, runoff, or groundwater.

We use the pH scale to describe the acidity of a solution. In this section, we will first review the pH scale and how it is related to acidity to better understand the following discussion about acid rain. Then our brief analysis covers reasons why acid rain intensity may vary geographically and over time.

pH and Acidity

Acidity is reported on the **pH scale.** The pH of a solution is inversely proportional to the hydrogen-ion (H^+) concentration in the solution; the pH scale, like the Richter magnitude scale, is a logarithmic one (**figure 18.19**). Neutral liquids, such as pure water, have a pH of 7. Acid substances have pH values less than 7; the lower the number, the more acidic the solution. Typical pH values of common household acids like vinegar and lemon juice are in the range of pH 2 to 3. Alkaline solutions, like solutions of ammonia, have pH values greater than 7. Precipitation is naturally slightly acidic as a result of the solution of gases that make acids. For example, CO_2 in the air dissolves in rain to form H_2CO_3, or carbonic acid. H_2CO_3 is a weak acid because a relatively small proportion of its molecules dissociate to release the H^+ ions that contribute to acidity. While many gases in the air contribute to the acidity of rain, discussions of acid rain focus on the sulfur gases that react to form atmospheric sulfuric acid (H_2SO_4), a strong acid that dissociates extensively.

Acid rain has a variety of detrimental effects. It can damage structures and outdoor sculptures through its corrosive effects. Plants may suffer defoliation, or their growth may be stunted as changes in soil-water chemistry reduce their ability to

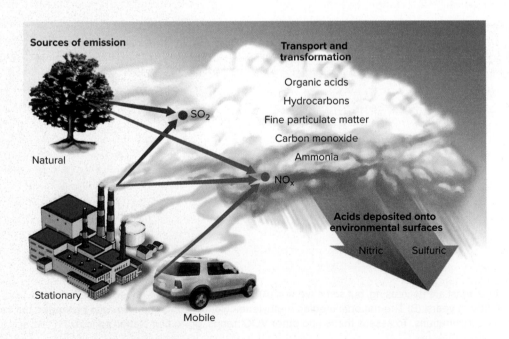

Figure 18.18

Acid-forming molecules emitted from natural and anthropologic sources are deposited directly or along with rain, and are the source of acid rain. A more encompassing term for these processes is *acid deposition.*

Source: Enger, Environmental Science, 2022, 16e—MHE.

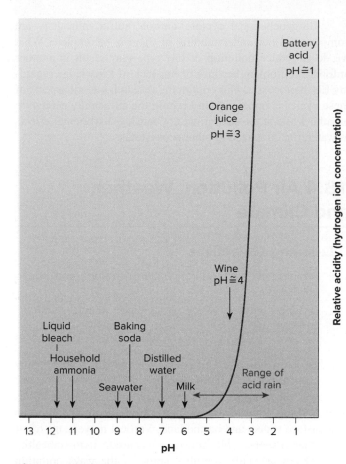

Figure 18.19

The pH scale is a logarithmic one; small differences in pH value translate into large differences in acidity. Range of acid precipitation is compared with typical pH values of common household substances.

take up key nutrients such as calcium (**figure 18.20**). Fish may be killed in acidified streams; or the adult fish may survive, but their eggs may not hatch in the acidified waters. Acidic water more readily leaches potentially toxic metals from soils, transferring them to surface-water or groundwater supplies, and reduces the capacity of soils to neutralize further additions of acid. Acid water leaching nutrients from lake-bottom sediments may contribute to algal bloom. Impacts of acid rain on groundwater were also explored in chapter 17. Studies have found a correlation between rainfall acidity and potentially toxic levels of mercury in lakes in northern North America. Other metals that may be linked to acid rain include lead, zinc, selenium, copper, and aluminum; recall that acid mine drainage is correlated with a higher concentration of leached metals. The accumulative properties of the heavy metals, in turn, raise concerns about toxic levels in fish and consequent risks to humans and other fish-eaters.

Regional Variations and Impacts

Rainfall is demonstrably more acidic in regions downwind from industrial areas releasing significant amounts of sulfur among their emissions. These observations formed the basis for decisions

Figure 18.20

These trees in the Appalachian Mountains of Virginia have been damaged by acid rain.

Mary Terriberry/Shutterstock

to control sulfur pollution. Acid rain and its associated problems have declined significantly in the United States because sulfur emissions have dramatically decreased. However, our concerns about acid rain have not disappeared for a few reasons: our understanding of the sulfur cycle is incomplete, nitrogen oxides can also create acid rain, soils are still affected by past interactions with acidic waters, and the pollutants that cause acid rain can travel long distances, namely from places that don't have such stringent regulations to control them. There are also questions about the nature of the local geology and how it affects the resulting acidity of groundwater

Scientists are still working out the specifics about the biogeochemical cycling of sulfur, including the role that anthropogenic sulfur plays. There is little information about the natural background levels of SO_2 and related sulfur species. And because all air is now polluted to some extent by human activities, it is difficult to determine the precise anthropogenic impacts on the chemistry of precipitation. Scientists also don't yet understand the role of marine algae and microbes in creating the 300 million tons of dimethyl sulfide that are released from the oceans every year, which makes up an estimated one-third of atmospheric sulfur.

It may be the case that an increase in rainfall acidity due to the formation of nitric acid from nitrogen oxides formed in internal-combustion engines is a much more significant problem in and near urban areas. The initial focus on sulfuric acid in rain partially reflects the fact that sulfur gas pollution from stationary sources is more readily controllable. Vehicles emission standards for NO_x for light duty vehicles in 2025 will be a 98% improvement over those from 1975. But as stricter standards were/are phased in over that time period, they only apply to new cars, so there are plenty of older cars that don't meet the newest standards. Still, NO_x emissions from highway vehicles has decreased 75% in the past 30 years.

Some research suggests that natural background pH of rainwater may be somewhat lower than the historically assumed

5.6 as a result of other natural chemical reactions in the atmosphere, and that pH may vary from place to place as a consequence of localized conditions. There also may be natural variety in water acidity due to the nature of the local geology through which water moves.

Chemical reactions between water and rock are complex. Farmers and gardeners have known for years that certain kinds of rock and soil react to form acidic pore waters, while others yield alkaline water. For instance, because limestone can reacts to make water more alkaline, it serves, to some extent, to neutralize acid waters. The effects of acid rain falling on carbonate-rich rocks and soils are therefore moderated. Conversely, granitic rocks and the soils derived from them are commonly more acidic in character. They cannot buffer or moderate the effects of acid rain; the acid rain just makes the surface and subsurface waters more acidic in such regions. Some operators of coal-fired plants argued that the predominantly granitic soils of the Northeast were principally at fault for the very acid lakes and streams in that region. Certainly, the geology doesn't reduce the problem. However, the rainfall in the Northeast has historically been distinctly more acidic, and only after regulations sharply reduced sulfur emissions from the power plants and other industries has this situation changed (**figure 18.21**).

The interaction of acid rain with underlying geology can cause problems, both in the short and longterms. Recently published research shows that the depletion of metals such as calcium, in soils in the Northeast United States is still affecting the recovery of surface water bodies and ecosystems suffering from the effects of acid rain, even while the quality of many streams are improving. In China, infiltration of acid rain from coal-fired power plants has been linked to massive landslides, caused when calcite in the rocks was dissolved and the rocks weakened.

The northeastern United States hasn't been the only area concerned about sulfuric acid generated in this country. The Canadian government expressed great concern over the effects of acid rain caused by pollutants generated in the United States that subsequently drift northeast over eastern Canada. Much of that area, too, is underlain by rocks and soils highly sensitive to acid precipitation and is already subject to acid precipitation associated with Canadian sulfate sources. In Europe, acid rain produced by industrial activity in central Europe and the United Kingdom falls on Scandinavia.

Control of SO_2 and NO_x emissions is plainly desirable for reducing acid rain and air pollution, and our efforts have been successful in doing both. The negative impacts of acid rain are well documented. Where sulfur gas release has been reduced, recovery of water quality in lakes and streams has occurred. However, the reduction in acid rain has made one type of air pollution worse. Ammonia gas (NH_3), linked to widespread use of agricultural fertilizer, is effectively scrubbed from the atmosphere by acid rain; less acid rain, more ammonia gas, which contributes to particulate pollution and high levels of nitrates in surface waters. Hotspots of ammonia gas have recently been identified by scientists using satellite data, indicating areas where acid rain has decreased and fertilizer use has increased simultaneously (**figure 18.22**).

Biogeochemical cycles—carbon, sulfur, nitrogen—are complex and our understanding of them is incomplete. What we do to reduce pollution of one type may result in another unforeseen problem because of our lack of knowledge regarding Earth processes. Our continuing research into all aspects of these cycles is imperative in keeping the air healthy enough to breathe while at the same time not causing detrimental effects to water, the climate, and other ecosystems.

18.4 Air Pollution, Weather, and Climate

Learning Objectives

- Explain how a thermal inversion develops and traps pollutants
- Describe ways in which air pollution affects weather and climate

Weather doesn't just affect air pollution by moving it horizontally from place to place. High temperatures and sunshine can cause and even speed up chemical reactions that create conditions such as smog. Rain washes pollution from the air and deposits it in surface waters and on the land. Vertical movements of air can either disperse pollution or concentrate it to unhealthy levels. Here we will exemplify some of the ways pollution interacts with weather and climate.

Thermal Inversion

Air-pollution problems are naturally more severe when air is stagnant and pollutants are confined. Particular atmospheric conditions can contribute to acute air-pollution episodes of the kinds mentioned at the beginning of this chapter. A frequent culprit is the condition known as a **thermal inversion.** Within the lower atmosphere, air temperature normally is warmer near the surface and gets cooler at higher altitudes (**figure 18.23a**). In a thermal inversion, there is a zone of relatively warmer air at some distance above the ground. That is, going upward from Earth's surface, temperatures decrease for a time, then increase in the warmer layer (inversion of the normal pattern), then ultimately continue to decrease at still higher altitudes. Inversions may become established in a variety of ways, including cold air flowing under and pushing up warm air (**figures 18.23b** and **c**) or warmer air moving in over colder air close to the ground. Rapid cooling of near-surface air on a clear, calm night may also lead to a thermal inversion.

Most air pollutants, as they are released, are warmer than the surrounding air. Warm air is less dense than colder air, so ordinarily, warm pollutant gases rise and keep rising and dispersing through progressively cooler air above. When a thermal inversion exists, a warm-air layer becomes settled over a cooler layer. The warm pollutant gases rise only until they reach the warm-air layer. At that point, the pollutants are no longer less dense than the air above, so they stop rising. They are effectively

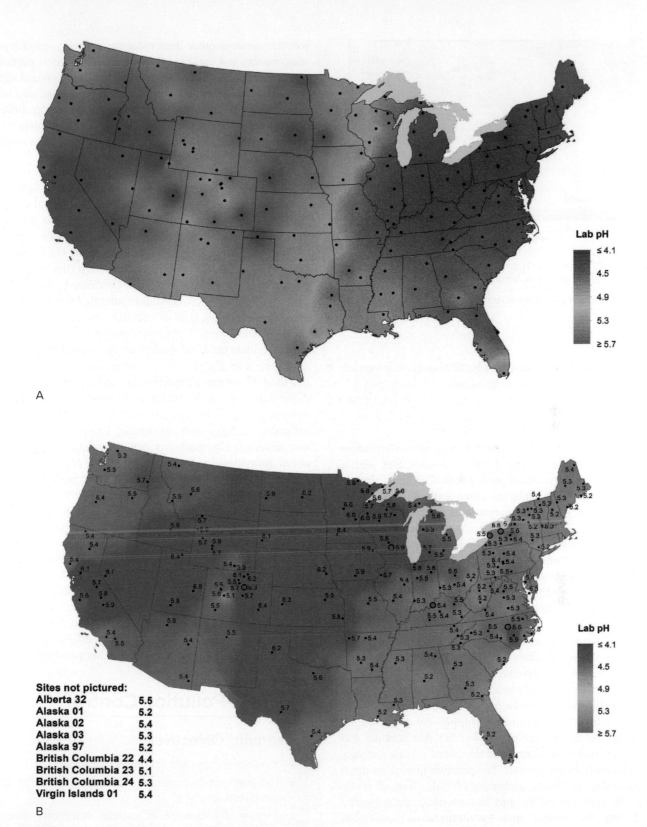

Figure 18.21

Hydrogen ion concentrations as pH, 1990 (A) compared to 2020 (B). The acidity of rainfall is decreasing just about everywhere in the United States; the Northeast has improved the most dramatically.

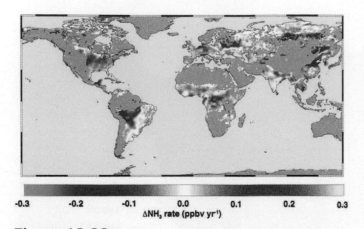

Figure 18.22

Change in global atmospheric ammonia from 2002 to 2016.

National Aeronautics and Space Administration

trapped close to the ground and simply build up in the near-surface air. Sometimes, the cold/warm air boundary is so sharp that it is visible as a planar surface above the polluted zone (**figure 18.24**). The Donora, Pennsylvania, episode mentioned at the beginning of the chapter was triggered by an inversion.

Every one of the half-dozen major acute air-pollution episodes of the twentieth century has been associated with a thermal inversion, as have many milder ones. Certainly, air pollution can be a health hazard in the absence of a thermal inversion. The inversion, however, concentrates the pollutants more strongly. Moreover, inversions can persist for a week or longer, once established, because the denser, cooler air near the ground does not tend to rise through the lighter, warmer air above.

Certain geographic or topographic settings are particularly prone to the formation of thermal inversions. Donora, Pennsylvania, lies in a valley. When a warm front moves across the hilly terrain, some pockets of cold air may remain in the valleys, establishing a thermal inversion (**figure 18.23b**). Los Angeles has the misfortune to be located in such a spot. Cool air blowing across the Pacific Ocean moves over the land and is trapped by the mountains to the east beneath a layer of warmed air over the continent (**figure 18.23c**). Germany's coal-rich Ruhr Valley suffered so badly from pollution trapped by a prolonged inversion during the winter of 1984–1985 that schools and many factories had to be closed and the use of private automobiles banned until the air cleared. Temperature inversions often occur seasonally. In China, outbreaks of $PM_{2.5}$ haze at levels between the very unhealthy and hazardous range generally occur during the winter due to temperature inversions (**figure 18.25**). As with most weather conditions, nothing can be done about a thermal inversion except wait it out.

Impact on Weather and Climate

While weather conditions such as thermal inversions can affect the degree of air pollution, air pollution, in turn, can affect the weather in ways other than reducing visibility, modifying air temperature, and making rain more acidic. This is particularly true when particulates are part of the pollution. Water vapor in the air condenses most readily when it has something to condense on. This is the principle behind cloud seeding: fine, solid crystals are spread through wet air, and water droplets form on and around these seed crystals. Particulate pollutants can perform a similar function.

This is illustrated in **figure 18.26**. Southwest of Lake Michigan lies the industrialized area of Chicago, Illinois, and Gary and Hammond, Indiana. The Gary/Hammond area, especially, is a major steel milling region. At the time this satellite photograph was taken, waste gases from the mills were being blown northeast across the lake (note faint smoke plumes at bottom end of the lake, just left of center). At first, they remained almost invisible, but as water vapor condensed, clouds appeared and, in this scene, snow crystals eventually formed. A large area of snow in Michigan is on a direct line with the pollution plumes from the southwest. The particulate pollutants in the exhaust gases facilitate the condensation and precipitation.

The role of aerosols in weather, and climate, is complex, and their effects can vary depending on their type and size. Dark particulates can absorb sunlight and warm an area, while light-colored particulates and droplets may promote cooling by reflecting sunlight back into space. These temperature effects can influence air-circulation patterns, water evaporation, and cloud formation, and thus precipitation patterns. Studies in China have implicated sooty particulate pollution in more-frequent flooding in south China and droughts in north China. NASA researchers have also proposed that dust from Saharan dust storms, carried across the Atlantic, was partly responsible for a relatively quiet hurricane season in 2006. The dust may have reduced the warming of surface water that supplies energy to storm systems, and suppressed convection of water vapor upward in the atmosphere to precipitate as rain. Our near future concerns regarding the interaction of climate and weather will not only include research into how changes in climate will affect pollution levels, but also how mitigation efforts to reduce input of CO_2 could affect the emissions of other pollutants.

18.5 Air-Pollution Control

Learning Objectives

- Summarize the AQI and how it is used
- List methods to control different types of air pollutant emissions
- Explain the change in vehicle emissions and fuel economy with time
- Name ways to sequester and store carbon dioxide

The costs of air pollution control may seem high, but the benefits of doing so are more than worth the amount spent. Rough economic estimates of just the direct health benefits to those living in the United States are in the hundreds of

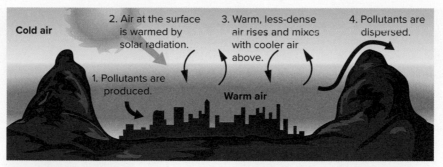

A. Normal situation

1. Pollutants are produced.
2. Air at the surface is warmed by solar radiation.
3. Warm, less-dense air rises and mixes with cooler air above.
4. Pollutants are dispersed.

Cold air

Warm air

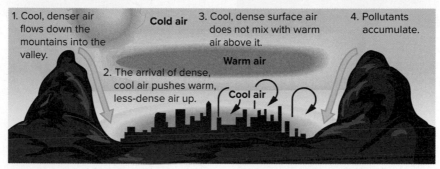

B. Thermal inversion

1. Cool, denser air flows down the mountains into the valley.
2. The arrival of dense, cool air pushes warm, less-dense air up.
3. Cool, dense surface air does not mix with warm air above it.
4. Pollutants accumulate.

Cold air

Warm air

Cool air

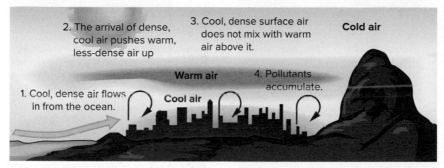

C. Thermal inversion

2. The arrival of dense, cool air pushes warm, less-dense air up
3. Cool, dense surface air does not mix with warm air above it.
4. Pollutants accumulate.
1. Cool, dense air flows in from the ocean.

Cold air

Warm air

Cool air

Figure 18.23

In a thermal inversion, the usual pattern of warm air underneath cold air (A) is reversed, trapping pollutants near the surface, sometimes for days; (B) Inversions are common in valleys; (C) and where inflowing cold ocean air is trapped by geographic features.

Source: Enger, Environmental Science, 2022, 16e—MHE.

billions of dollars annually. In this section, we will review how the EPA translates measured pollution levels into usable information; ways in which pollutant emissions, particularly from vehicles, have been controlled; and ways in which we are addressing the necessary capture and sequestration of carbon dioxide in order to avoid the worst of predicted climate outcomes.

Air-Quality Standards

Growing dissatisfaction with air quality in the United States led, in 1970, to the Clean Air Act Amendments, which empowered the Environmental Protection Agency to establish and enforce air-quality standards (see also chapter 19). The standards developed are designed to protect human health, clear the visible pollution from the air, and prevent crop and structural damage and other adverse effects. When meteorologists and others report on local air quality, they are referring to the Air Quality Index (AQI), developed by the EPA to translate pollutant concentrations into descriptors that the general public can readily understand. The AQI is illustrated in **figure 18.27**. On any given day, the AQI may be different for different pollutants. Overall air quality is assigned the descriptors and color applicable to the pollutant with the highest (worst) AQI that day. The AQI provides a quick way to assess air quality and improvements over time (**figure 18.28**).

Figure 18.24

"The Brown Cloud" is created along the Rocky Mountain Front Range when air pollution, much of it nitrogen dioxide from vehicle emissions, is trapped underneath warm air in a temperature inversion.

WeatherVideoHD.TV

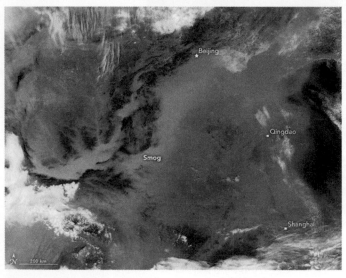

Figure 18.25

Nitrate is the principal component of the winter haze pollution in Chinese cities such as Beijing, made worse by seasonal temperature inversions. Notice how pollution is trapped in the valleys to the west.

NASA Earth Observatory images by Jeff Schmaltz

Control Methods

Air-pollutant emissions can be reduced either by trapping the pollutants at the source or by converting dangerous compounds to less harmful ones prior to effluent release. A variety of technologies for cleaning air exist.

Where particulates are the major concern, filters can be used to clear the air, with the fineness of the filters adjusted to

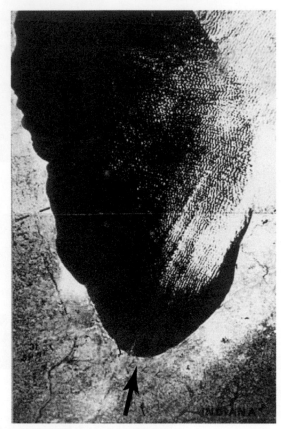

Figure 18.26

Air-pollution plumes from the Chicago, Illinois, and Gary/Hammond, Indiana, area bring clouds and snow to the area to the Northeast. Black area is Lake Michigan; arrow points to source of pollution and initial trend.

NASA

the size range of particles being produced. Filters may be 99.9% efficient or better, but they are commonly made of paper, fiber, or other combustible material. These cannot be used with very high-temperature gases.

An alternate means of removing particulates is electrostatic precipitators or scrubbers. In these, high voltages charge the particulates, which are then attracted to and caught on charged plates. Efficiencies are usually between 98 and 99.5%. These systems are sufficiently costly that they are practical only for larger operations. They are widely used at present in coal-fired electricity-generating plants.

Wet scrubbers are another possibility. In these scrubbers, the gas is passed through a stream or mist of clean water, which removes particulates and also dissolves out some gases. Wet chemical scrubbing, in which the gas is passed through a water/chemical slurry rather than pure water, is used particularly to clean sulfur gases out of coal-fired plant emissions. Currently, the most widely used designs use a slurry of either lime (CaO) or limestone ($CaCO_3$), which reacts with and removes the sulfur gases. Wet scrubbing

Air Quality Index (AQI) Value	Descriptor	Meaning
0–50	Good	Air quality is satisfactory and air pollution poses little or no risk.
51–100	Moderate	Air quality is acceptable, but a few persons unusually sensitive to the pollutant may experience negative health effects.
101–150	Unhealthy for sensitive groups	While the general public is not likely to be affected, members of sensitive groups are more likely to experience health effects. (For example, ozone at this level may be problematic for those with lung disease; those with lung or heart disease are at risk from particulates.)
151–200	Unhealthy	The population at large may experience negative health effects, and impact on sensitive groups will be more severe.
201–300	Very Unhealthy	Everyone may experience more-serious health effects.
> 300	Hazardous	This AQI level triggers health warnings of emergency conditions. The whole population is more likely to be affected.

Figure 18.27

For each of six major air pollutants regulated by the Clean Air Act Amendments (ground-level ozone, particulates, carbon monoxide, sulfur dioxide, nitrogen dioxide, and lead), an air-quality standard (concentration or exposure limit) has been set. An AQI = 100 corresponds to the standard level for the particular pollutant. Various ranges of AQI are assigned different descriptors based on likely health effects and color codes that allow the general public to understand easily the corresponding degree of health risk.

Adapted from U.S. Environmental Protection Agency.

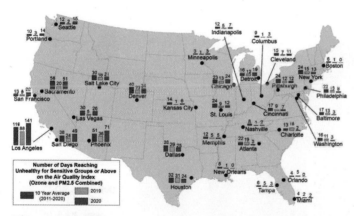

Figure 18.28

Number of days per year on which overall AQI was over 100, for selected cities. For most, air quality continues to improve as standards are tightened and new pollution-control techniques are implemented; in 2020, there were 641 unhealthy air days for these thirty-five cities combined.

U.S. Environmental Protection Agency, Our Nation's Air 2021

produces contaminated fluids, which present equipment-cleaning and/or waste-disposal problems.

Both carbon monoxide and nitrogen oxides are best controlled by modifying the combustion process. Lower combustion temperatures, for example, minimize the production of nitrogen oxides and moderate the production of carbon monoxide.

The use of afterburners to complete combustion is another way to reduce carbon monoxide emission, by converting it to carbon dioxide.

Combustion is an effective way to destroy organic compounds, including hydrocarbons, producing carbon dioxide and water. More-complex scrubbing or collection procedures for organics also exist but are not discussed in detail here. Such measures are necessary for low-temperature municipal incinerators burning materials that might include toxic organic chemicals that can only be thoroughly decomposed in the high-temperature incinerators designed for toxic waste disposal.

Altogether, the United States spends about $65 billion a year on air-pollution abatement. We have reduced emissions and markedly improved air quality in the past thirty years. Efforts at reducing acid deposition have also been successful. Phase I of the 1990 reauthorization of the Clean Air Act Amendments targeted a number of sources of sulfur pollution, with a view to reducing acid rain. Sulfate and acidity both showed substantial reductions downwind of Phase I targets. Tighter regulations imposed in 2005 resulted in further reductions in sulfur emissions and rainfall acidity (recall **figures 18.7** and **18.21**). Reductions in acid precipitation, in turn, have resulted in reduced acidity of surface waters in the northeastern United States. Although concentrations of both nitrate and sulfate in precipitation may remain somewhat elevated locally, both have shown notable declines across the United States (**figure 18.29**).

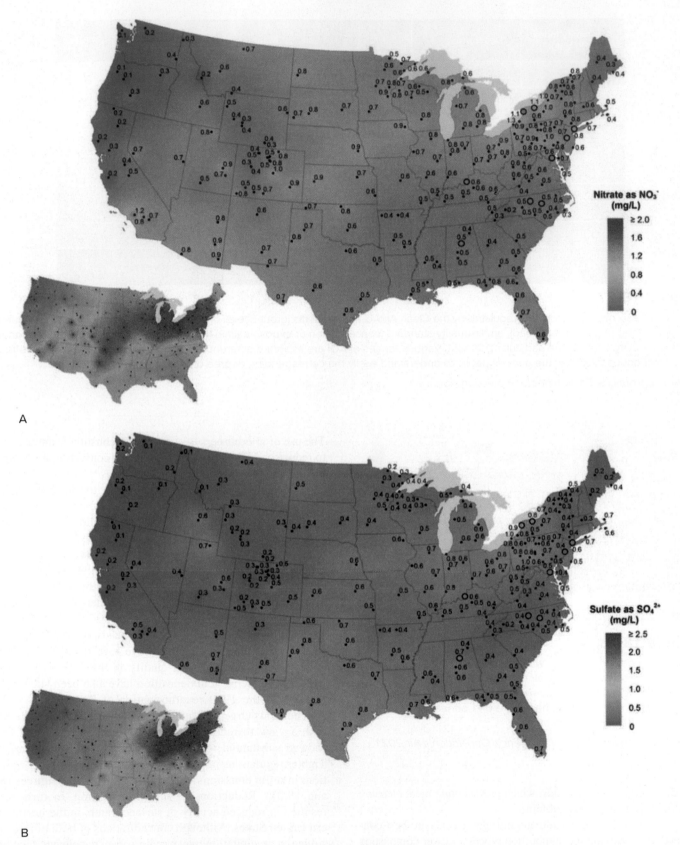

Figure 18.29

(A) Nitrate (NO_3^-) concentration in precipitation (mg/L) reflects particularly nitrogen-oxide emissions from urban and industrial areas upwind; (B) sulfate (SO_4^{2-}) forms from sulfur dioxide emissions. Large maps show 2020 data; inset maps, 1994. Dots represent sampling locations.

Maps from National Atmospheric Deposition Program.

Automobile Emissions

As part of the efforts to reduce air pollution, the EPA has imposed limits on permissible levels of carbon monoxide, nitrogen oxide, hydrocarbon, and particulate emissions from vehicles. There are also sulfur standards for gasoline and diesel fuels to control emissions of sulfur dioxide. Manufacturers must show through testing that their vehicles and engines comply with emission standards. They have continued to do so through technology improvements in combustion systems, the use of computers to monitor performance, and the use of catalytic converters and particulate filters to remove pollutants before they are emitted. In a catalytic converter, a catalyst (commonly platinum, palladium, or rhodium) enhances the oxidation, or combustion, of hydrocarbons, NO_X and CO with the resulting end products being water, nitrogen gas, and carbon dioxide.

Carbon dioxide became a pollutant that the EPA could regulate in 2007. The first phase of standards for cars and light trucks was promulgated in 2010, and the latest set of GHG emissions standards were finalized in December 2021. Compliance with the GHG standards isn't just a matter of combustion engine technology decreasing the amount of GHG coming out of individual tailpipes. Rather, this averaging, banking, and trading program "means that the standards may be met on a fleet average basis, manufacturers may earn and bank credits to use later, and manufacturers may trade credits with other manufacturers." How is the United States doing at reducing CO_2 vehicle emissions? Since 2002, they have decreased by 24%. All vehicle types are emitting less CO_2, but the market shift toward larger vehicles has offset some of the improvements, and not all car makers have improved emissions over the last five years. The EPA's Annual Automotive Trends Report contains more details.

Improvements in fuel economy reduce all emissions simultaneously, of course, as less fuel is burned for a given distance traveled. The National Highway Transportation Safety Administration has been setting fuel economy standards for new cars and light trucks (pickups, vans, and SUVs) since the late 1970s. They are known as the Corporate Average Fuel Economy (CAFE) standards (**figure 18.30A**). Raising fuel efficiency of new vehicles gradually raises the average fuel efficiency of all vehicles on the roads, but the latter always lags behind the new-vehicle standards. For model years 1990 to 2010, the CAFE standard for new passenger cars was 27.5 mpg, but the average fuel economy for all cars in 1990 was only 20.2 mpg, rising to 23.5 mpg by 2010. The CAFE standards rose again in 2011; by 2017 they were 38.5 to 39.6 for cars and 29.4 for light trucks. The Federal Highway Administration now subdivides all the light-duty vehicles by wheelbase rather than strictly by vehicle type, but in any case, it is noteworthy that by 2020, the fuel efficiency of the vehicles on the highway in the short-wheelbase group ranged from 28.4 to 31.7 mpg, and for the longer-wheelbase group, 19.2 to 23.8 mpg (**figure 18.30B**). In short, the fuel-efficiency lag persists and indeed seems to be increasing. Still, the stricter fuel-economy

A

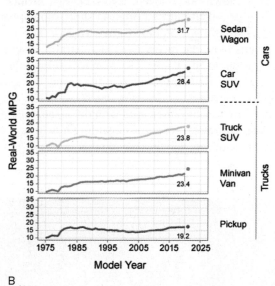

B

Figure 18.30

The success at improving vehicle efficiency can be estimated by comparing (A) The CAFE standards for fuel economy and the actual mileage achieved (B). Manufacturers who do not meet the standards are fined; in 2017, the NHTSA collected over $79 million.

Source: (a) National Highway Transportation Safety Administration (NHTSA) (b) Source: https://www.epa.gov/automotive-trends/highlights-automotive-trends-report

requirements are slowly improving overall average vehicle mileage, and limits for both cars and trucks have helped bring emissions down.

Carbon Capture and Storage

Historically, CO_2 had not been a pollutant monitored for public-health reasons by the EPA. However, given the serious negative health effects associated with global warming, the EPA does have an obligation to treat CO_2 pollution as a health issue. For various reasons, then, there has been increased interest in methods of **carbon capture and storage** (CCS), also known as *carbon sequestration:* storing carbon in some reservoir to remove it from atmospheric circulation, using physical or biological means. We've already discussed some of the ways to reduce carbon emissions and enhance removal of CO_2 from the atmosphere in chapter 10, and we introduced CCS in chapter 14 discussions on the continued use of fossil fuels as an energy source. We'll expand those ideas here.

Enhancement of natural carbon sinks is one approach in sequestering carbon. Reforestation is one example; a lush rainforest's vegetation contains much more carbon than a typical grassland or field of cultivated crops. Another strategy is ocean fertilization, a form of geoengineering; this promotes the growth of algae, which die and sink deep into the ocean, taking their carbon with them. Small experiments and natural events have shown that, indeed, fertilizing sea surface water with iron, a key algal nutrient often in short supply, can promote the desired algal bloom (**figure 18.31**). The iron-rich smoke clouds from the recent fires in Australia, discussed in the opening photo of chapter 10, were a natural experiment confirming this. Remaining uncertainties and risks include the potential impacts on broader ocean ecosystems, the long-term fate of the carbon submerged with the algae, the possible creation of dead zones, questions regarding how iron is cycled naturally in the ocean, and the role of other nutrients in marine productivity. A recent study in the equatorial Pacific suggests that more iron would decrease productivity because the amount needed to stimulate algal growth there is already supplied by upwelling currents. Another iron fertilization experiment resulted in other creatures consuming the iron and passing the CO_2 along the food chain instead of sinking it to the deep ocean.

To address the voluminous CO_2 output of industrial processes, a variety of physical methods are being considered, tested, and used to capture and store carbon dioxide before it is ever released into the atmosphere in the first place. Carbon dioxide could be pumped into the oceans, to dissolve in the water (at depths below about 1 km), or deposited by pipeline on the sea floor (below 3 km), where it would be expected to form a denser-than-water "lake" of liquid CO_2. Either approach raises questions about the long-term stability of the CO_2 storage and, in the case of dissolved CO_2, the effect on seawater chemistry. A number of geological reservoirs have also been considered (**figure 18.32A**). One might pump CO_2 into coal seams too thin or low-quality to mine, perhaps extracting useable coal-bed methane in the process. The CO_2 could be pumped into depleted petroleum reservoirs and used to enhance oil recovery, or into groundwater too high in dissolved minerals to be of interest as a water source. A test site in Iceland has been successful in injecting CO_2 into porous basalt and transforming 90% of it into geologically stable carbonate minerals. Some researchers have cautioned that rapid pumping of large quantities of CO_2 into geologic reservoirs may trigger earthquakes, just as pumping of liquid wastes for deep-well disposal can do, but this hasn't been a problem at the present scale of CCS activities.

Some commercial carbon-sequestration facilities are already in operation. For example, natural gas from the Sleipner field offshore from Norway is high in associated CO_2. But Norway has a high carbon-emission tax. So it is economically advantageous for Statoil (now Equinor), the company operating the field, to separate the CO_2 and inject it back into saline pore fluid in a permeable sandstone below the sea floor, rather than releasing it into the atmosphere. Equinor is currently expanding the CCS infrastructure to accept CO_2 from other sources for sequestration for an initial annual storage capacity of 1.5 million tons in their Northern Lights project.

In 2017, the first commercial post-combustion CCS facility in the United States became operational. At the Petra Nova plant near Houston, Texas (**figure 18.32B**), flue gases are passed through a separate, natural-gas-powered plant that scrubs the gases chemically and concentrates the CO_2. The carbon dioxide is then sent by pipeline 82 miles to the West Ranch oil field, where it is used for enhanced oil recovery. The estimated total cost of the CCS retrofit was about $1 billion, and the energy consumed in the CO_2 capture is about 15% of the total energy produced by the coal- and gas-fired plants together. Whether the enhanced-recovery operation will ultimately

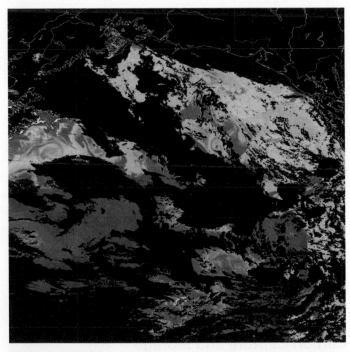

Figure 18.31

In this NASA satellite image of part of the Gulf of Alaska, brighter colors indicate higher chlorophyll levels, in turn reflecting abundance of phytoplankton (algae). Small bright patch at bottom center is algal bloom resulting from a 2002 iron-fertilization experiment.

Jim Gower, Bill Crawford, and Frank Whitney of Instituteof Ocean Sciences, Sidney BC; the IOS SERIES team; and NASA

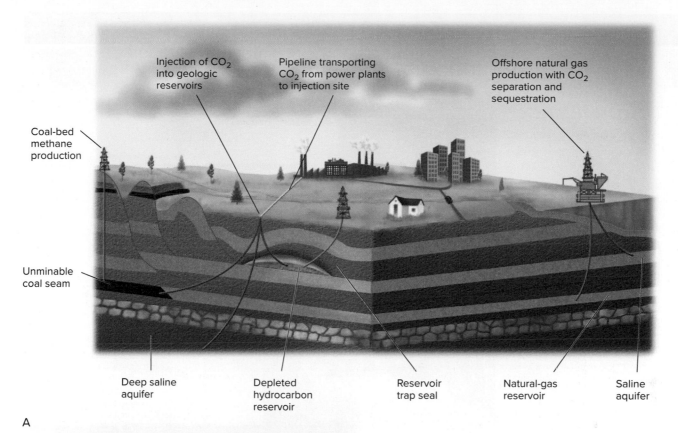

Coal-bed methane production

Injection of CO_2 into geologic reservoirs

Pipeline transporting CO_2 from power plants to injection site

Offshore natural gas production with CO_2 separation and sequestration

Unminable coal seam

Deep saline aquifer

Depleted hydrocarbon reservoir

Reservoir trap seal

Natural-gas reservoir

Saline aquifer

A

B

Figure 18.32

(A) Examples of geological methods of carbon sequestration. (B) At the Petra Nova coal-fired electricity-generating plant, CO_2 emissions are captured for use in enhanced oil recovery.

Sources: (A) After USGS Fact Sheet FS-026-03; (B) National Energy Technology Laboratory, U.S. Department of Energy

provide stable, permanent storage of the CO_2 remains to be seen. Nevertheless, Petra Nova captured approximately 1.9 million tons of CO_2 emissions per year. The facility was placed in reserve shutdown in May 2020 in response to the worldwide economic downturn.

Such sequestration activities are becoming more common, in part because the economics support them. Currently, there are approximately eighteen major CCS projects in operation worldwide. As noted, some countries, such as Norway, have used carbon taxes to encourage industries to pursue CCS. The Petra Nova project was partially funded by a grant from the U.S. Department of Energy. The world has more than enough geological capacity to store the amount of CO_2 necessary to

keep global temperature rise to 2°C by mid-century, and much of that is in the United States, close to where the CO_2 is produced. Estimates do range quite a bit for the amount of CO_2 storage needed to succeed, but some researchers suggest that a goal of 2700 gigatons is reachable with the current CCS development trajectory. Capturing industrial emissions is just one way to control CO_2. Another is direct capture; there are nineteen operational plants worldwide, mostly small, and that sell the captured gas for uses such as beverage carbonation. Other strategies include energy conservation, improvements in energy efficiency, development of low-emission sources, and the widespread electrification of our vehicles, power systems, and appliances.

Indoor Air Pollution?

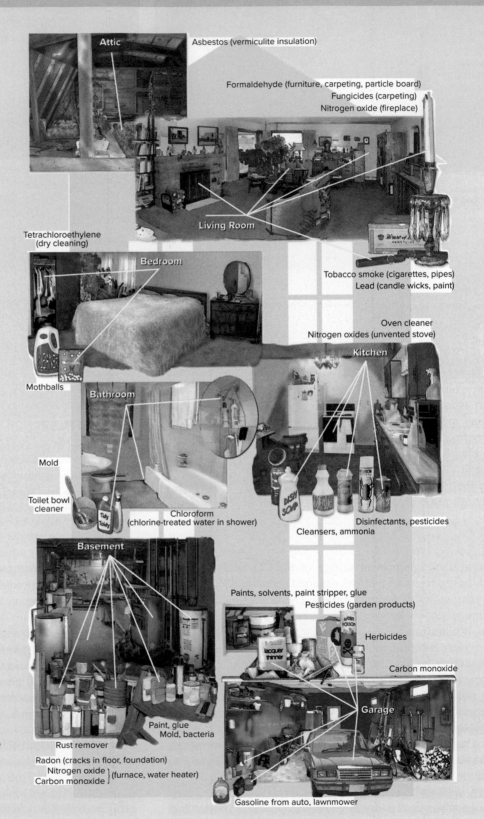

Asbestos (vermiculite insulation)

Formaldehyde (furniture, carpeting, particle board)
Fungicides (carpeting)
Nitrogen oxide (fireplace)

Attic

Living Room

Tetrachloroethylene
(dry cleaning)

Tobacco smoke (cigarettes, pipes)
Lead (candle wicks, paint)

Bedroom

Oven cleaner
Nitrogen oxides (unvented stove)

Kitchen

Mothballs

Bathroom

Mold

Toilet bowl
cleaner

Chloroform
(chlorine-treated water in shower)

Disinfectants, pesticides

Cleansers, ammonia

Basement

Paints, solvents, paint stripper, glue
Pesticides (garden products)

Herbicides

Carbon monoxide

Garage

Figure CS 18.1.1

Sources of indoor air pollution are many.
Sources of radon in the home include the
soil, walls, and water.

*Source: After U.S. Environmental
Protection Agency, EPA Journal, Sept./
Oct. 1990.*

Paint, glue
Mold, bacteria

Rust remover

Radon (cracks in floor, foundation)
Nitrogen oxide ⎫
Carbon monoxide ⎭ (furnace, water heater)

Gasoline from auto, lawnmower

Many potential air-pollution hazards have been created or intensified by our search for energy efficiency in response to dwindling fuel supplies and rising costs. For example, it was learned after the fact that blown-in foam insulation releases dangerous quantities of the volatile gas formaldehyde as it cures. (Formaldehyde is the pungent liquid in which creatures are commonly preserved for dissection by biology classes.) Many homes in which this insulation had been installed had to be torn apart later to remove it. The very act of insulating tightly has meant that toxic gases that might once have escaped harmlessly through cracks and small air leaks are now being sealed in, and their concentrations are building up. Among them are carbon monoxide (from furnaces, heaters, gas appliances, and cigarettes), VOCs (such as paint and solvent fumes), and radon (**figure CS 18.1.1**).

Radon is a colorless, odorless, tasteless gas that also happens to be radioactive. It is produced in small quantities in nature by the decay of the natural trace elements uranium and thorium. When radon is free to escape into the open atmosphere, its concentration remains low. When it is confined inside a building, the radioactivity level increases (**figure CS 18.1.2**). Tightly sealed homes may have radon levels above the levels found inside uranium mines.

How does all this radon get in? Uranium and thorium occur in most rocks and soils and in building materials made from them, like concrete or brick. Being a gas, radon can diffuse into the house through unsealed foundation walls or directly into unfinished crawl spaces from the soil. It can emanate from masonry walls. Radon seeps into groundwater from aquifer rocks and then enters the house via the plumbing. The potential radon input from rock and soil depends on local geology (**figure CS 18.1.3**).

In late 2008, there was even a flurry of concern raised in the media about granite countertops as a radiation hazard and radon source. But the radioactivity in granite is very low, as is the amount of radon it produces. Furthermore, granite is not very permeable, and granite used in countertops is typically sealed, too, so radon would not readily escape from it. Considering all the environmental sources of radiation noted in chapter 16, a granite countertop would be a negligible addition.

As radon is a chemically inert gas, it will not remain in the lungs; one breathes it in, then out. If an atom of radon happens to decay while in the lungs, it decays to radioactive isotopes of lead, bismuth, and other metals that might lodge in the lungs and later decay there, posing a cancer risk. The U.S. EPA estimates that radon is the second leading cause of lung cancer overall and responsible for 21,000 lung cancer deaths every year.

Given all the air pollutants that can accumulate inside the home, many experts recommend that snugly insulated homes being built or remodeled with emphasis on insulation for energy efficiency also include an air exchanger designed to provide adequate ventilation without loss of appreciable heat. Home carbon monoxide detectors are highly recommended and readily available. Radon test kits are also available (even through checkout from local libraries), or you can hire a professional to test your home; more info is available at https://www.epa.gov/radon.

Figure CS 18.1.2

Ranges of radioactivity due to radon in natural settings—groundwater, air in soil pores, outdoor air—and in the home. Note that the radioactivity scale is exponential, so radiation levels in high-radon homes may be hundreds of times what is typical in outdoor air.

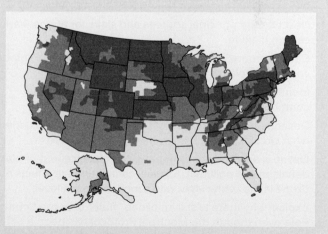

Figure CS 18.1.3

Geology influences the likelihood of elevated radon levels. On the EPA's Radon Zones map, areas shaded in red have predicted radon levels in homes above 4 picocuries/liter (pCi/L), the EPA's action level; areas shaded in orange have predicted radon of 2–4 pCi/L; areas in yellow, less than 2 pCi/L. Note that these zones are generalizations; there can be significant small-scale local variations in geology, and thus in radon production.

Summary

The principal air pollutants produced by human activities are particulates; gaseous oxides of carbon, sulfur, or nitrogen; and VOCs. The largest source of this pollution is combustion, and especially transportation. Most pollutants have natural contributions too, but human inputs tend to be spatially concentrated in cities and in areas of high agricultural output. Control of air pollutants is complicated because so many of them are gaseous substances generated in immense volumes that are difficult to contain; once released, they may disperse widely and rapidly in three dimensions in the atmosphere, thwarting efforts to recover or remove them. Climatic conditions can influence the severity of air pollution. Strong sunlight in urban areas produces photochemical smog, and temperature inversions in the atmosphere trap and concentrate pollutants. Air pollutants, in turn, may influence weather conditions by contributing to acid rain, by providing particles on which water or ice can condense, and by blocking or absorbing solar radiation. Ozone in photochemical smog can be a serious health threat, especially to those with respiratory problems. Yet in the ozone layer in the stratosphere, it provides important protection against damaging UV radiation. Enforcement of air-quality and emissions standards by the Environmental Protection Agency for pollutants with potential health consequences has led to dramatic improvements in air quality over the last four decades, particularly with respect to pollutants for which anthropogenic sources are relatively important, such as lead. However, even when pollutants are controlled, the environment may be slow to recover. For example, despite international agreements curtailing the release of the CFCs that contribute to the destruction of stratospheric ozone, CFC levels remain high as a consequence of the compounds' decades-long residence times, so ozone concentrations have not yet recovered proportionately. CCS will help to reduce CO_2 pollution. Given that the world produces a far greater volume of CO_2 than of any other pollutant gas, such techniques will need to be adopted on a large scale in order to play a significant role in moderating the effects of CO_2 in global warming and climate change.

Key Terms and Concepts

acid rain	**carbon capture and storage**	**ozone layer**	**pH scale**
aerosol	**ozone hole**	**particulates**	**thermal inversion**
			ultraviolet radiation

Test Your Learning

1. Describe the principal sources and sinks for atmospheric carbon dioxide.

2. Carbon monoxide is a pollutant of local, rather than global, concern. Explain.

3. Identify the origin of the various nitrogen oxides that contribute to photochemical smog.

4. Describe how photochemical smog develops, and under what circumstances this problem is most severe.

5. Ozone is an excellent example of the chemical-out-of-place definition of a pollutant. Explain, contrasting the effects of ozone in the ozone layer with ozone at ground level.

6. Explain why there is still an ozone "hole" over the South Pole even though the Montreal Protocol has been a success.

7. Explain why UVC radiation is more damaging than UVB, and if that is so, why we are so concerned with ozone, which principally absorbs UVB rather than UVC.

8. Lead additives in gasoline were effectively eliminated decades ago in the United States, but lead from vehicle exhaust is still a pollution concern in urban areas; explain why.

9. Identify two pollutant species responsible for acid rain; indicate the principal source of each, and compare how each has been reduced as an air-pollution problem in the United States.

10. Briefly summarize why the seriousness of the problems posed by acid rain may vary geographically.

11. Describe the phenomenon of a thermal inversion. Give an example of a geographic setting that might be especially conducive to the development of an inversion. Outline the role of thermal inversion in intensifying air-pollution episodes.

12. Summarize how federal regulations for automobile emissions control have resulted in improvements in air quality with respect to different pollutants.

13. Describe a geological and biological approach to carbon capture and sequestration, noting a possible concern or problem with each, and the overall challenge of using carbon sequestration to moderate global CO_2.

14. Identify the radiation hazard associated with indoor air pollution; describe how it develops, how it is related to one's local geology, and why it has become a subject of increasing concern in recent years.

Exploring Further

1. EPA monitors air quality and issues air-pollution hazard warnings. The daily air-quality forecast for your area may be available at **https://airnow.gov/**. If so, keep track, over a period of weeks, of the weather conditions on days when warnings are issued, and see what patterns emerge.

2. Examine air-quality trends for your local area at **www.epa.gov/air-trends**. What pollutants have declined, and by how much? Have any increased? If so, consider why.

3. Check the latest information on stratospheric ozone depletion. A good place to start is at NASA's Ozone Watch at **ozonewatch.gsfc.nasa.gov/**.

4. Take a close look at **figure 18.4**. Hypothesize, and then investigate, the sources of aerosols shown in Mexico, South America, and Europe. The image is created from data collected in September 2021.

5. Visit the **National Atmospheric Deposition Program's** site, specifically their annual national trends network database. Choose a parameter (such as NH_4 or Cl) and then analyze changes in its concentration or deposition over a specific time frame. Try to explain the changes. How is this parameter tied to air (and/or water) pollution?

CHAPTER 19

Environmental Law and Policy

Environmental laws of one sort or another have a long history. Many centuries ago, English kings restricted the burning of coal in London because the air pollution had become choking, and the penalty for violators could be execution. Penalties for breaking environmental laws in the United States today are more often financial than physical. The number of such laws and the areas they regulate continue to increase. In addition to environmental laws, which are regulations enforceable through the judicial system, there are policies designed to achieve particular environmental goals, which may influence the actions of individuals, companies, organizations, and governments. Through time, the philosophies behind resource development, pollution control, and land use have changed, and the focus of environmental laws and policies has shifted in response. A comprehensive treatment of environmental legislation is well beyond the scope of this book, but in this chapter, we will look at common themes and problems with some environmental laws related to geologic matters.

Chapter Outline

19.1 Resource Law: Water

19.2 Resource Law: Minerals and Fuels

19.3 International Resource Disputes

19.4 Pollution and Its Control

19.5 Cost-Benefit Analysis

19.6 Law Relating to Geologic Hazards

19.7 The National Environmental Policy Act (1969)

Among the national parks and monuments protected by U.S. environmental laws are some geologically special and spectacular places. Acadia National Park, Maine.

MIHAI ANDRITOIU/Alamy Stock Photo

19.1 Resource Law: Water

Learning Objectives

- Compare Riparian Doctrine to Prior Appropriation
- Compare the Rule of Capture to the American Rule

Given the importance of water as a resource and the increasing scarcity of high-quality water, it is not surprising that laws specifying water rights are necessary. What is perhaps unexpected is that the basic principles governing water rights vary not only from nation to nation but also from place to place within a single country. Also, quite different principles may be applied to surface-water rights and groundwater rights, even though surface and subsurface waters are inevitably linked through the hydrologic cycle.

Surface-Water Law

Navigable surface water bodies are treated as community property, accessible to all, resources that cannot be appropriated or assigned to individuals (**figure 19.1**). To the extent that they can be viewed as being owned, ownership lies with the state, on behalf of the public. However, while an individual cannot own a stream or a lake, one may have rights to use its water.

The two principal approaches to surface-water rights in the United States are the *Riparian Doctrine* and the *Doctrine of Prior Appropriation*. The term *riparian* is derived from the Latin word for "bank," as in riverbank, and sums up the essence of the **Riparian Doctrine**. Whoever owns land adjacent to a body of surface water has a right to use that water, and all individuals with land bordering a given body of water have an equal right to that water. Provision is also generally made to the effect that the water must be used for *natural purposes* or *beneficial uses* and returned to the body of water from which it came in

Figure 19.1

Anyone can paddle or boat on, or even swim in, the Chicago River, but access is restricted by the property owners along its banks. Most rivers and lakes have public access points.

Chris Tobin/Photodisc/Getty Images

essentially the same amount and quality as when it was removed. The Riparian Doctrine, long the basis for English surface-water law, became the basis for assigning surface-water rights in the eastern United States. Often implicit or explicit in surface-water laws based on the Riparian Doctrine is the concept that no one individual's water use should substantially interfere with others' use—but where water is relatively plentiful, as in the eastern United States, this isn't typically a problem. On the other hand, strictly speaking, only landowners bordering the water body have any rights to its water.

In the western United States, where surface water is in shorter supply, the prevailing doctrine is that of **Prior Appropriation**. Under this scheme, whoever is first, historically, to use water from a given surface-water source has top-priority rights to that water. Users who begin to draw on the same water source later have subordinate rights, in the order in which they begin to use the water. One need not own any land at all to have water rights under this scheme. One's water rights under this doctrine are established and maintained by continuing to divert the water for beneficial use. Future use is to be proportional to the quantity used when the rights were established. An advantage of this system is that it makes clearer who is entitled to what in times of shortage. Ideally, it ought also to discourage excessive settlement or development in arid regions, because latecomers have no guarantee of surface-water rights if and when demand exceeds supply. In practice, it doesn't appear to have served as such a deterrent; in some cases, development has simply proceeded using groundwater instead. Moreover, because failure to use the water for a period of time causes one to lose water rights under the appropriation doctrine, it may unintentionally lead to waste of water, to water use simply for the sake of preserving those rights rather than out of genuine need—not a wise arrangement in regions where water supplies are inadequate.

In states operating under prior-appropriation principles, special provisions apply to Native American tribes living on reservations. Formal establishment of the reservation has been judged to imply establishment of water rights at the same time. Users with water rights predating the reservation retain priority rights, but Native American rights take precedence over rights of users established after creation of the reservation, even if the Native American rights were not immediately exercised. Moreover, the quantity of water involved is to be sufficient to irrigate all arable land on the reservation, and those rights are not lost due to non-use.

What constitutes beneficial use of water also varies regionally. Domestic water use is commonly accepted as beneficial. So is hydropower generation, which doesn't appreciably consume water, though some may be lost to evaporation. Irrigation that is critical to agriculture may be regarded as a legitimate beneficial use, but watering one's lawn might not be. Industrial water use is often low-priority. In some areas, a hierarchy of beneficial uses has been established. A typical sequence, from highest to lowest priority, might be: domestic water use, municipal supply, irrigation, watering livestock, diversion of water for power generation, use for mining or drilling for oil or gas, navigation, and recreation. These sort of rankings are increasingly coming into play as water shortages in the western United States get worse and emergency use restrictions are emplaced; watering the lawn is usually the first activity prohibited.

Such prioritizing of usage complicates water-rights questions. An additional complication is that several states, usually those with both water-rich and water-poor areas, have tried to combine both the riparian and prior-appropriation principles (see **figure 19.2**); each state's particular scheme is different. Some states—for example, California—have extended the principle of appropriation to give the state the right to appropriate surface water and to transfer it from one region to another or otherwise redistribute it to maximize its beneficial use.

Groundwater Law

Groundwater law can be an even more confusing problem, in part because groundwater law was, to a great extent, developed before there was general appreciation of the hydrologic cycle, the movement of groundwater, or its relationship to surface water. As with surface-water rights, different principles may govern groundwater rights in different jurisdictions, producing a similar patchwork distribution of water-rights laws across the country (**figure 19.3**). In most states, one of two distinctly different principles has been applied. The so-called English Rule, or *Rule of Capture,* gives property owners the right to all the groundwater they can extract from under their own land. This is sometimes considered analogous to riparian surface-water rights. It does ignore the reality of lateral movement of groundwater. Groundwater doesn't recognize property lines, so heavy use by one landowner may deprive adjacent property owners of water as groundwater flows toward actively pumping wells. The American Rule, or *Rule of Reasonable Use,* limits a property owner's groundwater use to beneficial use in connection with the land above, and it includes the idea that one person's water use shouldn't be so great that it deprives neighboring property owners of water. Unfortunately, as with surface-water laws, there is no consistent definition of what constitutes beneficial use. These rules have then been variously combined with appropriation or riparian philosophies of water rights.

Some states, notably California, have specifically addressed the problem of several landowners whose properties overlie the same body of groundwater (aquifer system), specifying that each is entitled to a share of the water that is proportional to his or her share of the overlying land *(correlative rights).* Many other states have introduced a prior-appropriation principle into the determination of groundwater rights, as indicated in **figure 19.3**.

One factor that complicates the allocation of groundwater is its comparative invisibility. The water in a lake or stream can be seen and measured. The extent, distribution, movement, quantity, and quality of water in a given aquifer system may be so imperfectly known that it is difficult to recognize at what level one individual's water use may deprive others who draw on the same water source.

As an aside, note that similar problems arise with other reservoirs of fluids underground—specifically, oil and gas. Rights to petroleum from an oil field underlying the Iraq–Kuwait border were a point of contention between those nations prior to the 1990 invasion of Kuwait by Iraq.

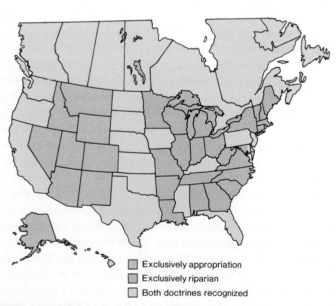

Figure 19.2

Distribution of principles of surface-water law applied in the United States.

Source: Data from New Mexico Bureau of Mines Circular 95.

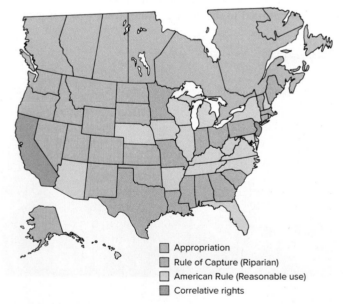

Figure 19.3

Distribution of principles of groundwater law applied in the United States.

Source: Data from New Mexico Bureau of Mines Circular 95.

19.2 Resource Law: Minerals and Fuels

Learning Objectives

- Describe the 1872 Mining Law
- List reasons for modernizing the 1872 Mining Law
- Summarize the stipulations of the Surface Mining Control and Reclamation Act of 1977

Federal regulations regarding mineral exploitation on U.S. federal lands were formalized 150 years ago. Despite multiple attempts to do so, the provisions in the 1872 Mining Law haven't changed much since then. Below we will review this law and its history.

Mineral Rights

For more than a century, U.S. federal laws have included provisions for the development of mineral and fuel resources. The fundamental underlying legislation for the development of mineral and fuel resources is the 1872 Mining Law. Its intent was to encourage exploitation of mineral resources by granting mineral rights and often land title to anyone who located a mineral deposit on federal land and invested some time and money in developing that deposit. Federal land could be converted to private ownership for $2.50 to $5 an acre—and as those prices were set in the legislation, they would remain in effect for any minerals still subject to the 1872 law, even decades later when inflation had raised land values. The individual (or company) did not necessarily even have to notify the government of the exact location of the deposit, nor was the developer required to pay royalties on the materials mined. The style of mining was not regulated, and there was no provision for reclamation or restoration of the land when mining operations were completed. What this amounted to was a colossal giveaway of minerals on public lands (and land too) in the interest of resource development.

Subsequent laws, particularly the Mineral Leasing Act of 1920 and the Mineral Leasing Act for Acquired Lands of 1947, began to restrict this giveaway. Certain materials, such as oil, gas, coal, potash, and phosphate deposits, were singled out for different treatment. Identification of such deposits on federally controlled land would not carry with it unlimited rights of exploitation. Instead, the extraction rights would be leased from the government for a finite period, and royalties had to be paid, compensating the public in some measure for the mineral or fuel wealth extracted from the public lands. Still, no requirements concerning the mining or any land reclamation were imposed, and the so-called hard-rock minerals (including iron, copper, gold, silver, and others) were exempted. The Outer Continental Shelf Lands Act of 1953 extended similar leasing provisions to offshore regions under U.S. jurisdiction.

Sparked by a 1989 General Accounting Office report recommending reform of some provisions of the 1872 Mining Law, intense debate has continued to take place in Congress about possible reform of that law. Under reform proposals in a 1995 bill passed by the House of Representatives, rights to ore deposits still governed by the 1872 law would not be so freely granted, but would be treated more like the rights to oil, coal, and other materials already explicitly removed from the jurisdiction of the 1872 law. Fairer (nearer market) prices for land would have to be paid, as would royalties, and strict mine-reclamation standards would have to be met. An obvious motive is to assure the public, in whose behalf federal lands are presumably held, fair value for mineral rights granted, a return on minerals extracted, and environmental protection. Industry advocated lower royalty rates, deduction of certain mineral-extraction expenses from the amount on which royalties would be paid, and other differences from the House-passed bill, backing a less radical reform passed by the Senate in 1996. Both bills stalled in committee. Subsequent attempts to revive the reform efforts failed. From time to time during the late 1990s, Congress did pass moratoria on the issuing of new mineral rights.

Other attempts to modernize the mining law came in 2007, 2015, and 2019 with the introduction of a Hard Rock Mining (Leasing) and Reclamation Act. These bills would add protections for special lands, establish environmental standards, implement fiscal reform through royalties and more, and establish a fund to clean up abandoned mines, among others. These, too, were not passed, primarily due to opposition from lobbyist and special interest groups associated with large mining companies operating in mineral-rich states such as Nevada. Until a revision is passed, the 1872 Mining Law, for all its perceived flaws, remains in effect.

Mine Reclamation

Growing dissatisfaction with the condition of many mined lands eventually led to the imposition of some restrictions on mining and post-mining activities in the form of the Surface Mining Control and Reclamation Act of 1977. This act applies only to coal, whether surface-mined or deep-mined, and attempts to minimize the long-term impact of coal mining on the land surface. Where farmland is disturbed by mining, for example, the land's productivity should be restored afterward. In principle, the approximate surface topography should also be restored. However, in practice, where a thick coal bed has been removed over a large area, there may be no nearby source of sufficient fill material to make this possible, so this provision may be disputed in specific areas. (This is one of the fiercely contested aspects of mountaintop-removal coal mining, discussed in chapter 14: shearing off the top of a mountain and putting the tailings in a valley profoundly alters the topography.) Similarly, the act stipulates that groundwater recharge capacity be restored, but particularly if an aquifer has been removed during mining, perfect restoration may not be possible. Still, even when a complete return to pre-mining conditions is not possible, this legislation and other laws patterned after it go a long way toward preserving the long-term quality of public lands and preventing them from becoming barren wastes. Unfortunately, it doesn't require reclamation of lands mined before it was enacted, which may be continuing sources of acid runoff. The National Environmental Policy Act (NEPA) (discussed later in the chapter) has provided a vehicle for controlling mining practices for other kinds of mineral and fuel deposits; and Superfund helps deal with legacy unreclaimed sites that present acute pollution problems.

Many states impose laws similar to or stricter than the federal resource and/or environmental laws on lands under state control. However, the impact of the federal laws alone should not be underestimated. Nearly one-third of the land in the United States, most of it in the West, is under federal control (**figure 19.4**). Indeed, part of the pressure for reforming the 1872 Mining Law comes from state land-management agents in the western states, where the great disparities between mineral-rights and mineral-development laws on federal and on state land create some tensions. For example, it may be economically advantageous under current laws for a company to seek mineral rights on, and mine, federal rather than adjacent state-controlled lands; but the states (like the federal government) receive no returns from mineral extraction on federal lands within their borders, while they do have to deal with problems such as air and water pollution associated with the mining and mineral processing.

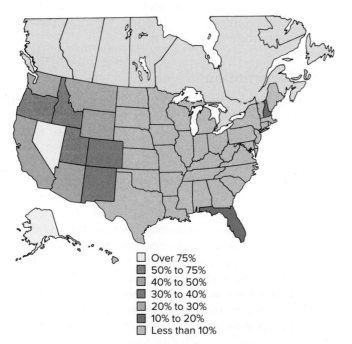

- ☐ Over 75%
- ■ 50% to 75%
- ■ 40% to 50%
- ■ 30% to 40%
- ☐ 20% to 30%
- ■ 10% to 20%
- ☐ Less than 10%

Figure 19.4

Federal land ownership in the United States. Of the federal lands, about 40% is under the control of the Bureau of Land Management, 30% is the national forest system, and about 15% each, the national park system and the national wildlife refuge system.

Source: Data from 2007 National Resources Inventory, *2009.*

19.3 International Resource Disputes

Learning Objectives

- Describe an Exclusive Economic Zone
- List seafloor mineral resources likely to be developed
- Recall the jurisdiction of the International Seabed Authority
- Describe the duties of the Commission on the Limits of the Continental Shelf
- Summarize the jurisdiction of activities in Antarctica

As noted in chapter 13, additional legal problems develop when it is unclear in whose jurisdiction valuable resources fall. International disputes over marine mineral rights have intensified as land-based reserves have been depleted and as improved technology has made it possible to investigate and plan the development of underwater mineral deposits.

Traditionally, nations bordering the sea claimed territorial limits of 3 miles (5 kilometers) outward from their coastlines. Realization that valuable minerals might be found on the continental shelves led some individual nations to assert their rights to the whole extent of the continental shelf. That turns out to be a highly variable extent, even for individual countries (**figure 19.5**) Off the east coast of the United States, for example, the continental shelf may extend beyond 150 kilometers (about 100 miles) offshore. The continental shelf off the west coast is less than one-tenth as wide. Nations bordered by narrow shelves quite naturally found this approach unfair. Some consideration was then given to equalizing claims by extending territorial limits out to 200 miles offshore. That would generally include all of the continental shelves and a considerable area of sea floor as well. The biggest beneficiaries of such a scheme would be countries that were well separated from other nations and had long expanses of coastline. Closely packed island nations, at the other extreme, might not realize any effective increase in jurisdiction over the old 3-mile limit. Landlocked nations, of course, would continue to have no seabed territorial claims at all.

Law of the Sea and Exclusive Economic Zones

In 1982, after eight years of intermittent negotiations collectively described as the "Third United Nations Conference on the Law of the Sea," an elaborate convention on the Law of the Sea emerged that attempted to bring some order out of the chaos. Resources are only one area addressed by the treaty. Territorial waters are extended to a 12-mile limit, and various navigational, fishing, and other rights are defined. **Exclusive Economic Zones** (EEZs) extending up to 200 nautical miles from the shorelines of coastal nations are established, within which those nations have exclusive rights to mineral-resource exploitation. These areas aren't restricted to continental shelves only. Part of the idea behind the 200-mile limit was to help geologically disadvantaged nations lacking broad continental shelves. For example, Mexico can claim

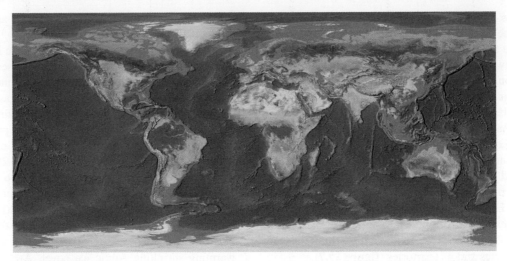

Figure 19.5

World map with continental shelves designated in lighter blue (shallower water).

Jan Rysavy/E+/Getty Images

resource rights to some deep-sea manganese nodules within their EEZ; Saudi Arabia and Sudan have the rights to the metal-rich muds on the floor of the Red Sea. Included in the EEZ off the west coast of the United States is a part of the East Pacific Rise spreading-ridge system, the Juan de Fuca rise, along which sulfide mineral deposits are actively forming. Areas of the ocean basins outside EEZs (which is 2/3 of the ocean) are under the jurisdiction of an International Seabed Authority (ISA). Exploitation of

mineral resources in areas under the ISA's control requires payment of royalties to it, and the treaty includes some provision for financial and technological assistance to developing countries wishing to share in these resources. While such deep-sea resources have yet to be commercially exploited, the ISA has already issued 31 exploratory contracts for mining deep-sea deposits (**figure 19.6**). They are currently working to finalize regulations for mining in international waters that could start as soon as 2026.

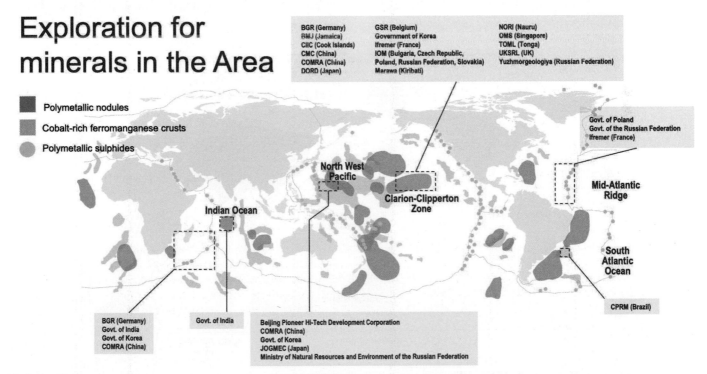

Figure 19.6

Most of the seafloor, and its mineral resources, are outside the reach of EEZs. The International Seabed Authority regulates mining in those areas. Given the large potential, there is a lot of interest.

IUCN World Conservation Union, https://www.iucn.org/resources/issues-briefs/deep-sea-mining

The ISA also adjudicates EEZ disputes between nations, some of which can be unusual. China has built a number of islands on coral reefs in the South China Sea, and then claimed rights to EEZs around the new islands. The ISA rejected those claims, noting that the island-building damaged the reef systems. China has so far declined to recognize the ISA's ruling, and has since fully militarized three of the islands.

The United States was one of only four nations (out of 141) to vote against the Law of the Sea convention, at the time objecting principally to some of the provisions concerning deep-seabed mining. The United States has tended to abide by most of the treaty's provisions. In March 1983, President Reagan unilaterally proclaimed a U.S. EEZ that extends to the 200-mile limit offshore. This move expanded the undersea territory under U.S. jurisdiction to some 3.9 billion acres, more than 1½ times the land area of the United States and its territories (**figure 19.7**).

Exploration, characterization, and mapping of the U.S. EEZ are ongoing. Seafloor resources with the greatest potential of development include cobalt-rich ferromanganese crusts, manganese (polymetallic) nodules, massive (polymetallic) sulfides, and phosphorites (**figure 19.8A**). Areas of greatest interest include old seafloor in the northwest Pacific Basin and along the Atlantic Coast, and hydrothermal vent activity on the Pacific continental margin and near volcanic islands. There is also potential for finding oil and gas deposits in the thick sediments off the Atlantic coast, although exploratory wells drilled there in the 1970s and 1980s were determined to be uneconomical and were abandoned; there are no current oil and gas leases. The western and northern margins of Alaska also are thickly blanketed with sediment hosting known and potential petroleum deposits (**figure 19.8B**).

A provision of the Convention on the Law of the Sea that has prompted a flurry of seafloor mapping seems likely to lead to some conflicts. The 200-mile limit does encompass the whole width of the continental margin in many cases, but not all. The provision in question allows a nation to claim rights beyond the 200-mile limit, extending its EEZ, if it can prove that the area in question represents a natural prolongation of its landmass. If a nation's continental margin extends beyond 200 miles offshore, it may claim jurisdiction over resources out to 350 miles offshore, or as much as 100 miles out from the point on the continental slope corresponding to a water depth of 2500 meters (the conventional base of the continental slope beyond the shelf), depending on the thickness of sediments and other criteria. With global warming shrinking Arctic ice cover, Arctic resources become more accessible. Of course, there is growing interest among all five Arctic nations (the United States, Canada, Denmark, Norway, and the Russian Federation) in staking their maximum possible jurisdictional claims in the Arctic. Some indication of the possible magnitude of such claims can be seen in **figure 19.5**. Note how far north of Alaska the continental slope extends in some areas, and realize that the Convention provisions could extend jurisdiction up to 100 miles beyond that.

The Convention has a Commission on the Limits of the Continental Shelf that reviews such claims, which must be supported by detailed mapping of seafloor topography and sediment thickness. Of the ninety-two submissions so far, the Commission has reviewed and recommended twenty-five, ten

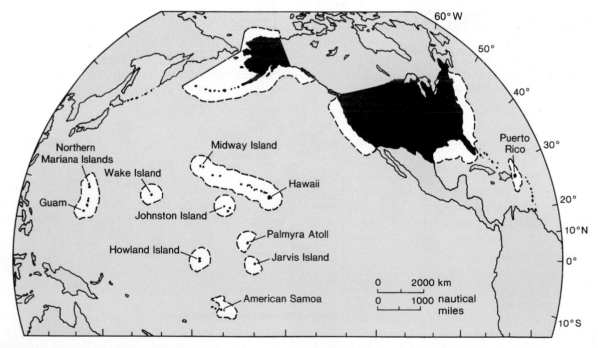

Figure 19.7

Exclusive Economic Zone of the United States, the largest of any country in the world.

Source: U.S. Geological Survey Annual Report 1983.

A

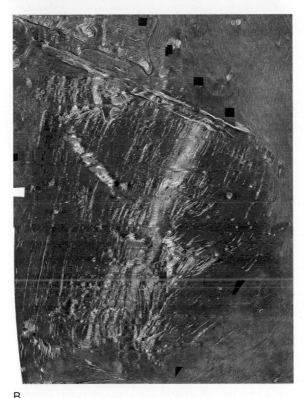

B

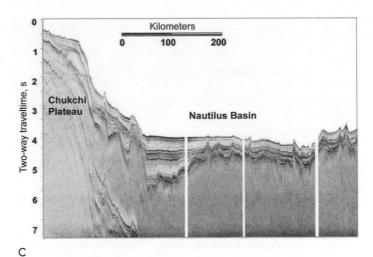

C

Figure 19.8

(A) This phosphorite rocked formed on the seafloor off the coast of Southern California. (B) Detailed sonar map of a portion of the Juan de Fuca ridge off the northwestern United States. (C) Seismic reflection profile from the Chukchi Plateau northwards off the north coast of Alaska reveals sediment thicknesses. (Two-way traveltime is the time taken for a pulse to travel from the surface to a reflecting layer and back again; one second equals about 750 m of water or 1500 m of sediment.)

(A) Mosaic of images from USGS Open-File Report 2010-1332; (B) USGS publication Sound Waves, 2012 Source: from USGS publication Sound Waves, 2012.

of which have been adopted by the United Nations. As with the 200-mile limit, there is plenty of potential for overlapping or conflicting claims by different nations bordering the same shelf. Because the United States is still not a party to the Convention, it cannot participate in this claims process, although it might choose to claim such extended jurisdiction by itself.

Antarctica

Antarctica in some ways presented a foretaste of the jurisdictional problems we now see with respect to the ocean basins. Prior to 1960, seven countries—Argentina, Australia, Chile, France, New Zealand, Norway, and the United Kingdom—had laid claim to portions of Antarctica, either on the grounds of their exploration efforts there or on the basis of geographic proximity (see **figure 19.9**). Several of these claims overlapped. None was recognized by most other nations, including five

(Belgium, Japan, South Africa, the United States, and the former Soviet Union) that had themselves conducted considerable scientific research in Antarctica.

After a history of disagreement punctuated by occasional violence, these twelve nations signed a treaty in 1961 that set aside all territorial claims to Antarctica. Various additional points were agreed upon: (1) the whole continent should remain open; (2) military activities, weapons testing, and nuclear-waste disposal should be banned there; and (3) every effort should be made to preserve the distinctive Antarctic flora and fauna. Other nations subsequently signed; there are now twenty-nine nations that are consultative parties to the Antarctic treaty, and another twenty-five non-consultive parties.

The treaty didn't address the question of mineral resources, in part because the extent to which minerals and petroleum might occur there wasn't realized, and in part because their exploitation in Antarctica was then economically quite impractical. Not until 1972 was the question of minerals even raised. The issue continued to be a source of contention for some years. Finally, in 1988, after six years of negotiations, the consultative parties (including the United States) produced a convention to regulate mineral-resource development in Antarctica: the Convention on the Regulation of Antarctic Mineral Resource Activities (CRAMRA). Under this convention, no such activity—prospecting, exploration, or development— could occur without prior environmental-impact assessment; significant adverse impacts would not be permitted. Actual mineral-resource development would require the unanimous

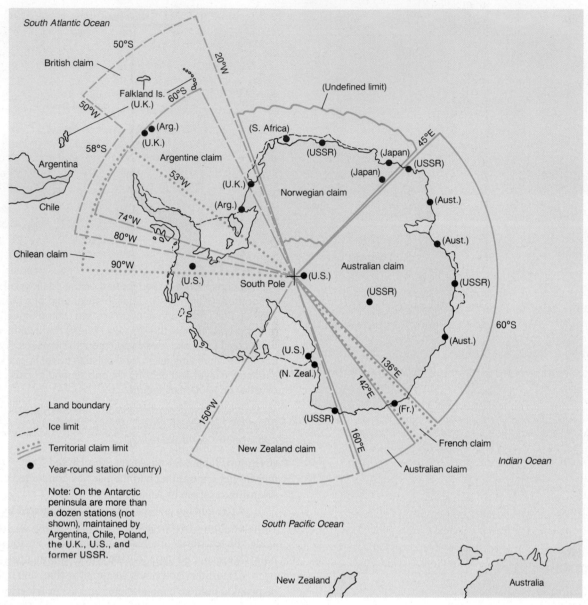

Figure 19.9

Conflicting land claims in Antarctica prior to the 1961 treaty. Note especially the overlapping claims. Find more info about the Antarctic Treaty at https://www.ats.aq/index_e.html.

Sources: Data from C. Craddock, et al., Geological Maps of Antarctica, Antarctic Map Folio Series, Folio 12, copyright 1970 by the American Geographical Society; and J. H. Zumberge, "Mineral Resources and Geopolitics in Antarctica," American Scientist 67, 1989, pp. 68–79.

consent of a supervisory commission composed mainly of representatives of the consultative parties.

In the years after CRAMRA's formulation, debate has continued. Sixteen of the twenty consultative parties must approve the treaty in order for it to take effect. Part of the reason for resistance to ratification was a concern that CRAMRA, by establishing guidelines for mineral-resource development, was effectively opening up the continent to such development. In the United States, this negative reaction to CRAMRA led to the passage of the Antarctic Protection Act of 1990. Briefly, the Act prohibits Antarctic mineral prospecting and development by U.S. citizens and companies and urges other nations to join

in an indefinite ban on such activities and to consider banning them permanently. The 1991 Protocol on Environmental Protection to the Antarctic Treaty, which ultimately replaced CRAMRA, involves an indefinite prohibition on activities related to development of geologic resources. Designating Antarctica as a "natural reserve, devoted to peace and science," it entered into force in 1998. In 2001–2002, the U.S. Environmental Protection Agency (EPA) issued a rule on environmental impact statements (EISs) (described in a later section) for other nongovernmental activities—including tourism and scientific study—to be undertaken in Antarctica, again with the intent of protecting the unique Antarctic environment.

19.4 Pollution and Its Control

Learning Objectives

- Recall U.S. laws regarding water pollution
- Recall U.S. laws regarding air pollution
- Describe Resource Conservation and Recovery Act of 1976
- Describe the EPA and its authority in pollution control
- List problems in defining the limits of pollution
- Describe international agreements used to control pollution and greenhouse gas emissions

At the outset of a discussion of laws regulating pollution, one might legitimately ask what legal basis or justification exists in the United States for antipollution legislation. Have we, in fact, any legal right to a clean environment? The primary constitutional justification depends on an interpretation of the Ninth Amendment: "The enumeration in the Constitution of certain rights shall not be construed to deny or disparage others retained by the people." The argument made is that the right to a clean, healthful environment is one of those individual rights so basic as not to have required explicit protection under the Constitution.

This interpretation generally hasn't been upheld by the courts. However, those harmed or threatened by pollution have often sought relief through lawsuits alleging nuisance or negligence on the part of the polluter. Nuisance and negligence are varieties of *torts* that are violations of individual personal or property rights, punishable under civil law, and distinct from violations of basic public rights. The government has also exercised its authority to promote the general welfare by enacting legislation to restrict the spread of potentially toxic or harmful substances, including pollutants.

Water Pollution

The first legislative attempt to control water pollution in the United States was the Refuse Act of 1899, which prohibited dumping or discharging refuse into any body of navigable water. Because most streams drain into larger streams, which ultimately become navigable, or into large lakes, this act presumably banned the pollution of most lakes and streams. Enactment of a law is different that its enforcement, and this one did little to prevent water pollution.

In succeeding decades, stricter and more specific anti-water-pollution laws were enacted. The Federal Water Pollution Control Act of 1956 focused particularly on municipal sewage-treatment facilities. The much broader Water Quality Improvement Act of 1970 and the subsequent amendments in the Clean Water Act of 1977 also address oil spills and various chemical pollutants, from both point and nonpoint sources. One of its objectives was to eliminate the discharge of pollutants into navigable waters of the United States by 1985—something that was, in principle, mandated eighty-six years earlier. The Clean Water Act and amendments nominally required all municipal sewage-treatment facilities to undertake secondary treatment by 1983, although practice lagged somewhat behind mandate; only about two-thirds of municipalities actually met that deadline. Another provision was that *best available technologies* to control pollutant releases by twenty-nine categories of industries were to be in place by July 1984. This concept of best available treatment is discussed further later in the chapter.

The fundamental underlying objective is "to restore and maintain the chemical, physical, and biological integrity of the Nation's waters." To this end, the EPA is directed to establish water-quality criteria for various types of water uses and to monitor compliance with water-quality standards (**table 19.1**). This is an increasingly complex task as the number of substances of concern grows, as noted in chapter 17.

Locally, significant improvements in surface-water quality resulting from improved pollution control have certainly occurred, but as is evident from chapter 17, more remains to be done. A major limitation of water-pollution legislation generally is that it treats surface waters only, although there is growing realization of the urgency of groundwater protection as well as of the links between surface and groundwater. Polluters have been prosecuted successfully, if indirectly, for groundwater pollution by toxic wastes after the groundwater seeped into and contaminated associated surface waters.

The fate of water-quality legislation in the near future has become increasingly cloudy. Since 1991, authorization and funding for many programs of the Clean Water Act and its later reauthorizations have lapsed; for example, continuing funding for municipal wastewater-treatment assistance lapsed in 1994. Repeated efforts since 1992 to enact further reauthorizations have failed. Versions of such bills, introduced in 1995 and 1996, included controversial wetlands-protection provisions that were opposed by some private-property owners. What funding is provided now is on an annual, ad-hoc basis. Reauthorization with assured multiyear funding continues to be problematic, and enforcement of environmental-quality standards requires funds.

Air Pollution

In the United States, air pollution was not addressed at all on the federal level until 1955, when Congress allocated $5 million to study the problem. The Clean Air Act of 1963 first empowered agencies to undertake air-pollution-control efforts. It was amended in 1965 to allow national regulation of motor-vehicle emissions. The Air Quality Act of 1965 required the establishment of air-quality standards to be based on the known harmful effects of various air pollutants. An overriding objective of this act and a series of 1970 amendments was to protect and improve air quality in the United States, particularly from the perspective of public health. Substantial progress has been made with respect to many pollutants, as noted in chapter 18.

The Clean Air Act was reauthorized and amended in 1990. Various provisions of these Clean Air Act Amendments (CAAA) deal with many areas of air quality. The CAAA goes beyond the Montreal Protocol (chapter 18) in reducing production of

Table 19.1 The EPA's Primary Safe-Drinking-Water Standards for Selected Chemical Components

Contaminants	Health Effects	MCL* (ppm)	Significant Sources
Inorganic Chemicals (Total of 16 regulated)			
arsenic	skin, circulatory-system damage; possible increased cancer risk	0.01	geological; pesticide runoff from orchards; glass and electronics production waste
barium	increased blood pressure	2	petroleum drilling waste; geological
cadmium	kidney damage	0.005	corrosion of galvanized pipes; geological; mining and smelting; runoff from waste batteries and paints
fluoride	skeletal damage	4	geological; additive to drinking water; toothpaste; discharge from fertilizer, aluminum processing
lead	developmental delays in infants and children; kidney problems, high blood pressure in adults	0.015**	leaching from lead pipes and lead-based solder pipe joints; geological
mercury (inorganic)	kidney damage	0.002	geological; refineries; runoff from landfills, croplands
nitrite	"blue-baby syndrome"	1	fertilizer, feedlot runoff; sewage; geological
selenium	gastrointestinal effects; circulatory problems; hair or fingernail loss	0.05	geological; mining; petroleum refineries
Organic Chemicals (Total of 53 regulated)			
alachlor	eye, liver, kidney, spleen problems; anemia; increased cancer risk	0.002	herbicide runoff from row crops
benzene	anemia; increased cancer risk	0.005	factory discharge; leaching from gas storage tanks and landfills
2,4-D	liver, kidney, or adrenal-gland problems	0.07	herbicide runoff from row crops
dichloromethane	liver problems; increased cancer risk	0.005	discharge from drug, chemical factories
dioxin	reproductive difficulties; increased cancer risk	0.00000003	emission from waste incineration, combustion; discharge from chemical factories
diquat	cataracts	0.02	herbicide runoff
ethylbenzene	liver or kidney problems	0.7	discharge from petroleum refineries
methoxychlor	reproductive difficulties	0.04	runoff/leaching from insecticide use on fruits, vegetables, alfalfa, livestock
polychlorinated biphenyls (PCBs)	skin changes; thymus-gland problems; immune deficiencies; reproductive or nervous-system difficulties; increased cancer risk	0.0005	runoff from landfills; discharge of waste chemicals
styrene	liver, kidney, circulatory-system problems	0.1	discharge from rubber or plastic factories; leaching from landfills
1,2,4-trichlorobenzene	changes in adrenal glands	0.07	discharge from textile-finishing factories
trichloroethylene	liver problems; increased cancer risk	0.005	discharge from metal degreasing sites, factories
vinyl chloride	increased cancer risk	0.002	leaching from PVC pipes; discharge from plastic factories

Source: U.S. Environmental Protection Agency.

*MCL (Maximum Contaminant Level): Maximum concentration allowed in drinking water. These levels are set based on the public-health risks of particular contaminants (note how different the values are for different substances). They are designed to minimize the projected risks while taking into account available water-treatment methods and costs.

**Not an MCL, but an "action level" that requires municipalities to control the corrosiveness of their water; see lead sources for the reason.

chlorofluorocarbons (CFCs) and other compounds believed harmful to the ozone layer. It provides for reductions in industrial emissions of nearly 200 airborne pollutants; requires further reduction of vehicle emissions; sets standards for improved air quality with respect to ozone, carbon monoxide, and particulates in urban areas; and provides for reduction of SO_2 emissions from power plants.

The SO_2/acid-rain provisions have an interesting economic component. Beginning in 1995, electricity-generating utility plants were assigned SO_2 allowances (one allowance = 1 ton SO_2 emission per year) based on power output, and allowances can be traded between plants or utility companies for cash. Utilities that reduce their emissions below their allowed output can sell their excess SO_2 allowances, which is an economic incentive to reduce pollution. In the early years, millions of allowances were traded, at prices that occasionally exceeded $1500 each. This **cap-and-trade** approach has proven successful in achieving the desired emissions reductions. Other air pollutants addressed by emissions trading programs include nitrogen oxides, ozone, and particulates.

The EPA holds an auction of a portion of the allowances. Each year, the 8.95 million allowances (a limit fixed in 2010) are reassigned among the various existing SO_2 generators, but some are held back for the auction, in part to ensure that new electricity-generating plants can acquire the allowances necessary to operate. Anyone can bid on any amount of allowances, and some individuals and environmental groups have bought them simply to take them off the market and permanently reduce total SO_2 emissions. This has become quite affordable, as improvements in sulfate-control technologies have reduced the prices utilities are willing to pay for allowances at the auction: the year-2022 allowances were auctioned at an average of about 4 cents each for a total amount paid of $4810; compare that to 2000, when the average allowance cost was $130 and $8.5 million was the total amount paid for them. It's apparent that fewer and fewer utilities need to buy additional allowances; the most recent auction included just one coal power plant of the four entities involved.

The impact of the regulations on sulfate and acid rain, and certain other pollutants, was noted in chapter 18. Significant air-quality improvements have clearly occurred with respect to many pollutants. As will be seen later in the chapter, however, a few air-pollutant emissions continue to rise, despite international agreements in principle to reduce them. There are also trade-offs. For example, successful efforts to reduce acid rain are linked to an increase in atmospheric ammonia, which is itself associated with both particulate pollution and surface deposition of nitrogen.

Waste Disposal

The Solid Waste Disposal Act of 1965 was intended mainly to help state and local governments to dispose of municipal solid wastes. Gradually, the emphasis shifted as it was recognized that the most serious problem was hazardous or toxic wastes. The Resource Conservation and Recovery Act of 1976 (RCRA) includes provisions for assisting state and local governments in developing solid-waste management plans, but it also regulates waste-disposal facilities and the handling of hazardous waste and establishes cooperative efforts among governments at all levels to recover valuable materials and energy from solid waste. Hazardous wastes are defined and guidelines issued for their transportation and disposal.

The EPA monitors compliance with the act's provisions. Development of waste-handling standards, issuance of permits to approved disposal facilities, and inspections and prosecutions for noncompliance with regulations are ongoing activities. However, there are tens of thousands of identified producers and transporters of hazardous waste in the United States subject to RCRA. In part because of this, the EPA has delegated many of its responsibilities under RCRA to the states.

The 1984 Hazardous and Solid Waste Amendments Amendments to RCRA considerably expanded the scope of the original legislation. Approximately 100,000 firms generating only small amounts of hazardous waste (less than 1000 kilograms per month) are now subject to regulation under the act; originally, these small generators were exempt. Before 1984, a waste producer could request *delisting* as a generator of hazardous waste by demonstrating that its wastes no longer contained whatever hazardous chemicals (as identified by the EPA) had originally caused the operation to be listed; now, delisting requires that the waste producer demonstrate essentially that its wastes contain no identified hazardous chemicals at all. A whole new set of regulations has been imposed on landfills and underground storage tanks containing hazardous substances, which in many cases require expensive retrofitting of the disposal site to meet higher standards; landfill disposal of toxic liquid waste is banned. Additional amendments in 1992 and 1996 reflected the growing recognition of the seriousness of the toxic-waste problem.

The U.S. EPA

In 1970, the U.S. EPA was established in conjunction with the reorganization of many federal agencies. Among its primary responsibilities was the establishment and enforcement of air- and water-quality standards. Under the Toxic Substances Control Act of 1976, the EPA was given the authority to require toxicity testing of all chemical substances entering the environment and to regulate them as necessary. The EPA also has a responsibility to encourage research into environmental quality and to develop a body of personnel to carry out environmental monitoring and improvement. As the largest and best-funded of the pollution-control agencies, it tends to have the most clout in the areas of regulation and enforcement. However, the scope of its responsibilities is broad, and expanding. The number and variety of situations with which the EPA is expected to deal, restrictions imposed by limited budgets, and the varying emphases placed on different aspects of environmental quality by different EPA administrators are all factors that have contributed to limiting the agency's effectiveness. Legislation in the late 1980s and early 1990s tended to strengthen the EPA's enforcement ability by establishing clear penalties for noncompliance with regulations. The CAAA, for example, allows the EPA to impose penalties of up to $200,000 for violations and even to pursue criminal sanctions against serious, knowing violations of CAAA. It is also true that efforts of the U.S. EPA are often supplemented by activities of state EPAs. But enforcement also requires staff and budget for investigations, legal actions, and so on. Sharp budget cuts for the EPA in the early twenty-first century, reflecting reduced priority in Congress for environmental-protection activities in general, certainly reduced the EPA's ability to carry out its various mandates and monitoring and enforcement activities.

A Supreme Court decision in early 2007 expanded the EPA's responsibilities under the CAAA still further. Since 1999, a dozen states and many cities had been urging the EPA to add carbon dioxide to the list of air pollutants that it regulates, particularly in the context of automobile emissions. The EPA's authority to regulate the six air pollutants discussed in chapter

18 derives from a mandate to protect public health, and CO_2 is not toxic or, by itself, considered a threat to health. On the other hand, to the extent that CO_2 causes global warming and climate change and that negative health effects of global warming are increasingly being recognized, one could argue that CO_2 indeed poses a risk to public health. Initially, the EPA declined to take on CO_2 regulation, claiming both lack of legal authority and lack of a definitive connection between greenhouse gases (GHGs) and global warming. The states, cities, and other concerned groups sued, and the issue made its way to the Supreme Court. In a split decision, the Court ruled that the EPA does have statutory authority to regulate GHG emissions such as CO_2 from automobiles—that given the harm these gases (and resultant climate change) cause, they are clearly pollutants under the CAAA, and thus subject to such regulation. The ruling did not mean that the EPA must immediately impose CO_2 emissions limits, but the pressure to begin moving in that direction sharply increased, so the EPA began plans to do so.

The regulation of CO_2 continues to advance since the seminal ruling in 2007, beginning in 2010 with vehicle emission standards as discussed in chapter 18. The EPA also imposed rules requiring reporting of GHG emissions from large sources in 2010. Recently, multiple plans have been proposed limiting GHG emissions at new and existing power plants and other stationary sources. Final rules have not yet been adopted.

The EPA has considered a federal cap-and-trade scheme for CO_2 emissions, given the success of the SO_2-allowance scheme. However, the number and variety of significant sources of CO_2 emissions are far greater, making emissions allocations correspondingly more difficult. The transportation sector presents an extra challenge, for while it represents the largest volume of CO_2 emissions, they come in small quantities from millions of individual vehicles. At least for now; the electrification of the transportation sector will remove this barrier. Meanwhile, a cooperative between eleven northeastern states called the Regional Greenhouse Gas Initiative has been in operation since 2008. Since that time, emissions from power plants in those states have dropped dramatically because of a switch to natural gas from coal. Currently, there are cost-benefit concerns (see section 19.5) of reducing emissions further, and states are considering leaving the co-op. The European Union does have a market for carbon emission allowances that are auctioned among more than 11,000 companies in twenty five of the twenty-seven member nations and three non-member nations. The recent economic crisis hasn't been good for the market, though. Emissions have been reduced more than expected, and the surplus of allowances has resulted in a decrease in carbon prices and less incentive to reduce emissions. Over forty countries have some type of carbon emissions trading market.

Defining Limits of Pollution

Enforcement of antipollution regulations remains a major challenge, in part, because of the terms of antipollution laws themselves. Consider, for example, an objective of *zero pollutant discharge* by a certain year. At first inspection, zero might seem a well-defined quantity. However, from the standpoint of practical science, it is not. Water quality must be measured by instruments, and each instrument has its own detection limits for each chemical. No commonly used chemical-analysis techniques for air or water can detect the occasional atom or molecule of a pollutant. A given instrument might, for example, be able to detect manganese down to a concentration of 1 parts per million (ppm), lead down to 500 parts per billion (ppb) (0.5 ppm), mercury to 150 ppb. Discharges of these elements below the instrument's detection limit will register as zero when the actual concentrations in the wastewater might be hundreds of ppb. How close one must actually come to zero discharge, then, depends strongly on the sensitivity of the particular analytical techniques and instruments used.

The alternative approach is to specify a (detectable) maximum permissible concentration for pollutants discharged in wastewater or exhaust gases. But pollutants vary so widely in toxicity that one may be harmless at a concentration that for another is fatal. That means a need to specify *different* permissible maxima for individual chemicals, as in **table 19.1**. Given the number and variety of potentially harmful chemicals, a staggering amount of research would be needed to determine scientifically the appropriate safe upper limit for each. Thus, regulations are limited to a moderate number of toxins known to pose health risks. Even so, extensive data are needed to set safe exposure limits. With less-comprehensive data, the EPA and other agencies may set standards that later may be found to be too liberal or unnecessarily conservative. For some naturally occurring toxic elements, they have occasionally set standards stricter than the natural ambient air or water quality in an area. To meet the discharge standards, then, a company might be required to release wastewater cleaner than the water it took in. For instance, this might be the case with arsenic, as suggested in figure 17.19.

No treatment process is perfectly efficient, and emissions-control standards must necessarily recognize this fact. For instance, 1982 incinerator standards established under the Resource Conservation and Recovery Act specify 99.99% destruction and/or removal efficiency for certain specified toxic organic compounds. While this is a high level of destruction, if 1 million gallons of some such material is incinerated in a given facility in a year, the permitted 0.01% would amount to a total of 100 gallons of the toxic waste released.

Imprecision in terminology remains a concern. The CAAA, for example, includes such expressions as *lowest achievable level* and *nonessential applications* (uses) of CFCs and other controlled ozone-threatening substances. The exact definition and interpretation of terms such as this often become the subject matter of litigation as they certainly change with technological improvements and time. In some instances, the introduction of economic considerations further muddies the regulatory waters, both in the framing of laws and regulations and in their implementation; see "Cost-Benefit Analysis" later in the chapter.

International Initiatives

The latter half of the twentieth century saw a rapid rise in international treaties relating to the environment (**figure 19.10**).

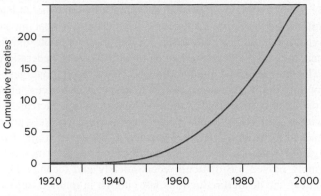

Figure 19.10

Increasing environmental awareness has led to increasing numbers of international environmental treaties.

Source: Data from U.N. Environment Programme.

Some especially notable steps were taken toward international cooperation and collective action to address problems of global impact and concern during the 1990s. Some of these initiatives relate to problems that affect localized areas, widely distributed around the world—for example, the U.N. Convention to Combat Desertification, which came into force in late 1996. Others recognize that all nations share some resources—the atmosphere especially—and no nation individually causes or can solve problems associated with pollution of such a common global resource (though to be fair, it is certainly true that some nations have historically contributed a disproportionately large share to that pollution).

One important outgrowth of the Earth Summit in Rio de Janeiro in 1992 was the United Nations Framework Convention on Climate Change (UNFCCC). This Convention recognizes the role of atmospheric CO_2 and other GHGs in global warming and other possible climatic consequences. It calls for nations to report on their releases of GHGs and efforts to reduce these and to strive for sound management and conservation of GHG sinks (plants, oceans), and it calls upon economically developed nations to assist developing nations in such efforts. This Convention came into force, following its ratification by over fifty nations, in March 1994. The stated goal is "stabilization of greenhouse-gas concentrations in the atmosphere at a level that would prevent dangerous anthropogenic interference with the climate system." Its very existence is significant, as is the promptness of ratification. The Montreal Protocol on Substances that Deplete the Ozone Layer, discussed in chapter 18, is another example of such international commitments to pollution control for the common good.

These accords reflect a relatively new concept in international law and diplomacy, the **precautionary principle**. Historically, nations' activities were not restricted or prohibited unless a direct causal link to some specific damage had been established. However, there are cases in which, by the time scientific certainty has been achieved, serious and perhaps irreversible damage may have been done. Under the precautionary principle, if there is reasonable likelihood that certain actions may result in serious harm, they may be restrained before great damage is demonstrated. So, given the known heat-trapping behavior of the GHGs and documented increases in their atmospheric concentrations, nations may agree to limit releases of GHGs before a specific amount of global warming or climate change has occurred; given the links between ozone depletion and increased UVB transmission, and between UV radiation and certain health risks such as skin cancer, they agree to reduce or stop releasing ozone-depleting compounds before significant negative health impacts have been documented.

This can be important to minimizing harm in cases in which there may be a significant time lag between exposure to a hazard and manifestation of the adverse health effects or those in which natural systems' response to changes in human inputs are slow. The CFCs are an example of the latter. Global emissions have been sharply reduced under the Montreal Protocol, and later amendments strengthened it and expanded its scope, so that ninety-six substances are now covered by it. Given the residence times of CFCs in the atmosphere, and that economically disadvantaged nations were given longer to phase out their production, atmospheric concentrations have only recently leveled off, and the ozone layer is not expected to recover fully for some decades. Still, the longer nations waited to take action, the further into the future the problem would have persisted.

Of course, sometimes solving one problem can create another. Initial substitutes for the CFCs included HFCs (hydrofluorocarbons) and HCFCs (hydrochlorofluorocarbons), and their atmospheric concentrations began to rise as CFCs were phased out (**figure 19.11**). They do contain some chlorine; moreover, HFCs and HCFCs are powerful GHGs. They too now fall under the Montreal Protocol as amended, their use to be phased out by 2030 in developed countries (2040 in developing countries). As their atmospheric residence times are of the order of ten years (versus about 100 for CFCs), their effects will be less lasting. But as they are phased out in turn, it remains to be seen how environmentally benign the successors to HFCs and HCFCs will be.

It should also be noted that agreement in principle is one thing, implementation another, and enforcement yet another. Following the Rio Summit, years of negotiations on specific targets and strategies produced the Kyoto Protocol of 1997. A major component of the Kyoto Protocol was a commitment by industrialized nations to reduce their GHG emissions collectively to 5.2% below 1990 levels by 2008–2012. In practice, the focus was on CO_2, overwhelmingly the principal GHG emitted in terms of quantity. But global CO_2 emissions, meanwhile, had been continuing to climb since Rio, so reductions relative to then-current CO_2 emissions would have had to be much more than 5% in many cases. For the United States, CO_2 emissions in 2007 were 20% above 1990 levels; to get to 5% below 1990 levels would have meant reducing CO_2 emissions by about 21% from 2007 levels. Concerns over the economic impact of making such sharp reductions in CO_2 emissions led to the inclusion of certain flexibility provisions in the Kyoto Protocol: allowing nations to meet a portion of their commitment by enhancing CO_2 sinks (such as forests) rather than actually reducing emissions; permitting industrialized nations some credits for providing developing nations with low-emissions technologies; and providing for the

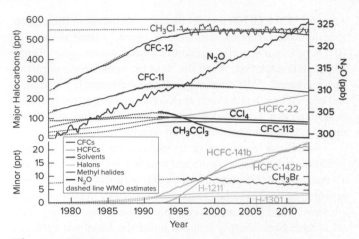

Figure 19.11

CFCs have leveled off or begun to decline, but HCFCs are on the rise. Other compounds listed are also among the nearly 100 ozone-depleting substances covered by the amended Montreal Protocol, with the exception of N_2O, a growing concern yet to be addressed. (ppt = parts per trillion)

Sources: Data from World Meteorological Organization and National Oceanic and Atmospheric Administration; graph courtesy NOAA Earth System Research Laboratory.

trading of emissions credits so that one nation could acquire the right to the unused portion of the emissions allowance of another.

Another principle of international environmental laws reflected in the Kyoto Protocol (as well as the Montreal Protocol and others) is the **common but differentiated responsibility of states**. (*States,* in U.N. usage, essentially means nations.) The idea is that we all share the global environment, and the responsibility to protect its quality, but some nations contribute more to pollution and other negative impacts (see, for example, **figure 19.12**), while some have more financial resources to develop less-harmful alternatives than other nations have. Those nations that create more of the problem and can better afford the costs of addressing it are expected to do more toward a solution. So, under the Montreal Protocol, developing nations have longer to phase out HFCs and HCFCs. Likewise, in principle, industrialized nations that produce more CO_2 and other GHGs (or CFCs, or other pollutants) and can more readily invest in developing less-polluting alternatives are expected to reduce their emissions more rapidly, and sooner. (Indeed, developing nations such as India and China were not subject to emissions limits under the Kyoto Protocol, though their GHG emissions had been rising rapidly. India's CO_2 emissions from fossil-fuel burning increased by over 150% from 1990 to 2008, and China's increased by 185%, so that China ranked #1 and India #3 in such emissions worldwide, with the United States and Russian Federation at #2 and #4, respectively.) Consequently, those developed nations on whom the greater financial and compliance burdens fall are not always eager to embrace what other nations see as their fair share of responsibility. This contributes to the difficulty of achieving consensus on the details of such international agreements.

Not until late 2004, with acceptance by the Russian Federation, did the Kyoto Protocol even achieve the critical mass of signatories to come into force. (Some observers note that by then, thanks to a depressed economy, the Russian Federation's GHG emissions had already dropped below its Kyoto target, so not only did it have no sacrifices to make; it had excess emissions credits to sell!) By then, it was becoming increasingly evident that the overall Kyoto targets were unlikely to be reached by around 2010 as originally hoped, even if all of the signatory nations held to their commitments in principle. It simply proved too difficult, economically and technically, to achieve the desired reductions fast enough in many developed nations. The United States opted out of Kyoto altogether, and this further limited achievement of reduced GHG levels globally, though U.S. companies operating in countries bound by the Kyoto Protocol were affected there. The unrestrained growth in emissions in the developing countries, especially China, was a growing concern. Finally, it should be noted that, like many international environmental accords, the Kyoto Protocol did not actually include enforcement provisions or penalties for failing to meet one's obligations, which certainly reduced its practical effectiveness.

The international community then began looking beyond Kyoto and its limitations to the challenge of crafting future strategies and agreements for reining in GHGs over the longer term, beyond the end of the Kyoto commitment period in 2012. For more than three decades, there have been international treaties limiting/prohibiting waste disposal in the oceans; the current agreement on the subject is the London Protocol of 1996. In 2007, the London Protocol was amended specifically to allow carbon dioxide from capture and storage operations to be handled by sub-seabed geological disposal in the ocean basins. The U.N. Climate Conference in Copenhagen in late 2009 was specifically planned as the successor to Kyoto, where the crafting of a new, major international treaty on climate would begin. That conference produced only a nonbinding *accord* among the 193 participating nations, including a target maximum global temperature rise of 2°C (about 4°F), and a commitment by industrialized nations to set voluntary emissions targets to be reached by 2020. A conference in Cancún in 2010 yielded a framework agreement to reduce deforestation (thereby increasing CO_2 sinks) by paying nations to preserve their forests, and another to create a $100-billion-per-year green fund to help developing nations produce clean energy and adjust to the impacts of climate change. The next convention of the delegates in Durban, South Africa, in 2011, resulted in a firm commitment to negotiate a successor treaty to the Kyoto Protocol by 2015, and to extend Kyoto while those negotiations proceeded.

The result was the Paris Agreement on climate change, adopted in December 2015 and signed by all 197 parties to the original Rio framework convention (the UNFCCC). It entered into force late in 2016, when the condition was met that it was ratified by at least fifty-five nations representing at least 55% of global GHG emissions. (Signing the Agreement signified a country's intention to pursue formal ratification through that

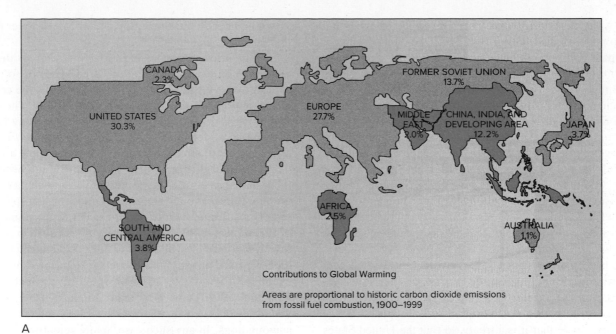

A

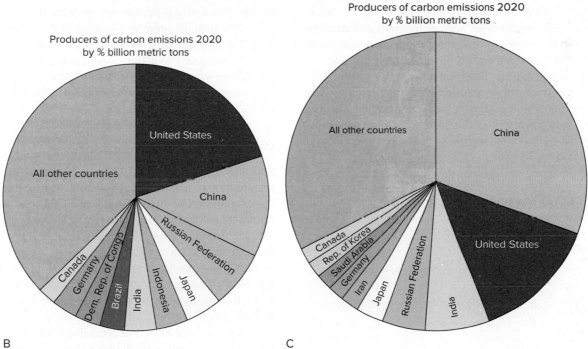

B   C

Figure 19.12

(A) This map shows sizes of countries/regions in proportion to their contributions to global CO_2 emissions over the twentieth century. (Pink areas are industrialized countries; green, developing countries.) In the twenty-first century, the pattern has become quite different. (B) In 2000, the United States was still the primary producer of CO_2 emissions, accounting for about 20% of the world total. By 2020 (C), China's emissions had more than tripled, accounting for most of the approximately 30% increase in global CO_2 emissions over that interval that is reflected in the larger size of the circle in (C).

(A) After World Resources Institute "Contributions to Global Warming" map by Equator Graphics, Inc.; (B) Data from United Nations Environment Programme Environmental Data Explorer as compiled from UNFCCC; (C) Data from Carbon Dioxide Information Analysis Center, U.S. Department of Energy.

country's appropriate official process, which might, for example, involve approval by a head of state, or by a legislative body or bodies.) The Paris Agreement, informed by the growing body of scientific evidence on climate change and reports from the United Nations' Intergovernmental Panel on Climate Change, sets a target of keeping global temperature rise to not more than 2°C (about 4°F) above pre-industrial levels, and to make efforts to keep the warming below 1.5°C. Unlike Kyoto, the Paris Agreement does not set a specific target for individual nations to meet. Instead, each of the Parties to the Agreement must prepare and publish an *intended nationally determined contribution* (INDC), its plan to work toward the overall global goal, and must report on its own GHG emissions. Every five years, beginning in 2023, each Party must submit a new INDC, which should be more ambitious than the last in moving toward the goal, and there will be a *global stocktake* assessing the Parties' collective progress toward it. The previously established green fund, supported by contributions from the developed countries, would be used to help developing countries reduce their emissions and grow their economies with greater reliance on clean energy sources, and also to assist poorer countries adversely affected by such consequences of climate change as sea-level rise and extreme storms.

The history of the United States' involvement with respect to the Paris Agreement has been somewhat unusual. President Obama adopted the Agreement by executive order, rather than handling it as a treaty that would require Senate approval. Some legal scholars argue that it is a treaty, so that the United States could not be committed to it by executive order; but U.S. presidents can enter into various other sorts of international arrangements by executive order, so that order was accepted internationally as U.S. ratification of the Agreement, and the country was counted among those that had ratified it. President Trump withdrew from the Agreement in November 2020 after a one-year waiting period notifying the United Nations that the United States was leaving. President Biden rejoined the agreement in February 2021 after indicating that the country would do so on his first day in office.

19.5 Cost-Benefit Analysis

Learning Objectives

- Describe how the costs of ecosystem services are calculated
- List problems in environmental cost-benefit analysis
- Illustrate how the federal government does or does not take cost into consideration in pollution control regulation

In 1997, a working group with the National Center of Ecological Analysis and Synthesis published a paper that included calculations and estimates on the value of natural capital. They analyzed seventeen different **ecosystem services,** which included the regulation of gas, climate, and water; waste treatment; food production; and nutrient cycling. This formative assessment put a $33 trillion price tag on natural processes that contribute to human health and welfare. One of the goals was to give policymakers and resource managers the data they needed to make decisions on the cost-benefit analysis of pollution control regulations. How did they, and others that preceded and then followed, go about putting an actual number on the worth, say, of the cost of water filtration that occurs in wetlands?

Numerical estimates of ecosystem services depend on what type of service is being assessed. The cost of a provisioning service, such as food production, may be calculated based on real market values for the commodity. The cost of a regulating service, such as water regulation, may be calculated based on technically replacing that process. The recreational value of a resource might be what people are willing to pay to access it or to keep it clean. Researchers making these assessments are increasingly taking advantage of Geographic Information System (GIS), models, and specific ecosystem studies in order to provide accurate assessments. One of the biggest problems in quantifying ecosystem services is that we don't fully understand biogeochemical cycles, an idea we have already approached in multiple chapters. For example, it is difficult to value fungi in nutrient cycling if you don't even know how many kinds of fungi there are, what they do, their numbers, or how they interact with their surroundings. In any such case, many scientists believe that we are drastically underestimating the value of ecosystems.

The quantification of pollution control measures (which in a way, are technical substitutes for natural ecosystem services) is necessary to making sound decisions that benefit people and are economically feasible. In the rest of this section, we will address some of the other issues associated with environmental cost-benefit analysis.

Problems of Quantification

Antipollution laws that specify using the "best available technology economically achievable" for treatment, or contain some similar clause that introduces the idea of economic feasibility, often result in additional conflicts and enforcement problems. As noted in earlier chapters, costs for exhaust or wastewater treatment rise exponentially as cleaner and cleaner air and water are achieved. These costs can be defined and calculated with reasonable precision for a given mine, manufacturing plant, automobile, or whatever. Balanced against these cleanup costs are the adverse health effects or general environmental degradation caused by the pollutants. The latter are harder to quantify and, sometimes, even to prove, given the number and variety of pollutant types and sources.

There are human costs outside of health concerns. The closing of a coal-fired power plant will certainly result in less sulfur and CO_2 being released in the atmosphere, but there is also the financial loss to employees and the surrounding community that supports them. Assuming that each can be quantified, the writing and interpretation of regulations need to consider the balance.

Similar debates arise in many areas. Allowing development on a barrier island may increase tourism and expand the local tax base, helping schools and allowing municipal improvements—but at what risk to the lives and property of those who would live on the vulnerable land, and cost to taxpayers providing disaster assistance and rebuilding costly infrastructure? The dangers of increased atmospheric CO_2 are widely accepted, and fossil-fuel extraction may cause environmental disturbances of various sorts—but our global energy consumption continues to grow; clearly we tend to decide, implicitly at least, that the benefits outweigh the risks, perhaps because the benefits are more readily perceived than the risks.

In the realm of pollution control, companies frequently contest strict regulation of their pollutant discharges on the grounds that meeting the standards is economically impossible and/or unreasonable. Each individual, each judge, and each jury has somewhat different ideas of what is reasonable; of the value of clean air and water; of the importance of preserving wildlife, vegetation, natural topography, or geology; and of what constitutes an acceptably low health risk. Enforcement is very uneven and inconsistent. Virtually everyone agrees that it is desirable to maintain a high-quality environment. But just what does that mean? What should it cost? And who should bare those costs?

Cost-Benefit Analysis and the Federal Government

A month after assuming office in 1981, President Reagan issued Executive Order 12291, which decreed, in part, that a (pollution-control) regulation may be put forth (by an agency such as EPA) only if the potential benefits to society outweigh the potential costs. A Regulatory Impact Analysis must be performed on any proposed new major regulation, one that is likely to result in either (1) an impact on the economy of $100 million or more per year; (2) a major cost increase to consumers, individual industries, governmental agencies, or geographic regions; or (3) significant adverse effects on competition, employment, investment, productivity, or the ability of U.S.-based corporations to compete with foreign-based concerns.

This represented a major departure from previous policy. The Clean Water Act ordered the EPA to be sensitive to costs in establishing standards, but it didn't require cost-benefit analysis; the Clean Air Act actually prohibited considering costs in setting health standards. Now, for example, for the contaminants regulated in drinking water, the EPA defines not only the Maximum Contaminant Levels (MCLs) as shown in table 19.1 but also identifies MCLGs—Maximum Contaminant Level Goals, concentrations below which there are no known negative health effects in humans. For some of these substances (e.g., barium, cadmium, mercury, nitrate), the MCLGs are the same as the MCLs. For others (e.g., arsenic, lead, benzene, dioxin, polychlorinated biphenyls), the MCLG is actually zero. In such cases, to quote the EPA, "MCLs are set as close to MCLGs as feasible using the best available treatment technology and taking cost into consideration." Cost-benefit analysis is still not the decisive factor in regulation, but it is at least taken into account.

Cost-benefit analysis was a key part of the debate surrounding the Clear Skies Act of 2003, proposed as part of President Bush's Clear Skies initiative, which would have capped emissions of sulfur and nitrogen oxides and of mercury by electricity-generating plants. The EPA, asked to do the cost-benefit analysis, projected the costs of compliance as rising to $6.3 billion per year. Set against those costs was an impressive array of benefits, including reduction in acidification of lakes and streams in the United States from acid rain; reduced mercury in the environment; over $3 billion per year worth of benefit from improved visibility in national parks and wilderness areas; an estimated 30,000 fewer visits to hospitals and emergency rooms, 12.5 million fewer days with respiratory symptoms (meaning fewer days of school or work missed) per year by 2020; and at least 8400 premature deaths prevented, with $21 billion per year in health benefits. In this case, the projected benefits far outweighed the anticipated costs, but the cost-benefit argument was not enough, and the measure ultimately failed to get out of committee after years of congressional debate. Since then, some limits of the kind envisioned in the Clear Skies Act have been implemented through EPA rulemaking.

19.6 Laws Relating to Geologic Hazards

Learning Objectives

- List reasons for and against regulations restricting construction in areas of known geologic hazards
- Explain why restrictive building codes don't succeed in protecting all buildings from hazards
- Describe ways in which California (CA) earthquake regulations allow construction in at-risk areas
- Summarize how the National Flood Insurance Program works
- List some common problems with laws designed to reduce geologic hazards risk

Laws or zoning ordinances restricting construction or establishing standards for construction in areas of known geologic hazards are a more modern development than antipollution legislation. Their typical objective is to protect persons from injury or loss through geologic processes, even when the individuals themselves may be unaware of the existence or magnitude of the risk. While such an objective sounds benign and practical enough, such laws may be vigorously opposed (see especially chapters 4, 6, and 7). Opposition doesn't come only from real-estate investors who wish to maximize profits from land development in areas of questionable geologic safety. Individual homeowners also often oppose these laws designed to protect them, perhaps fearing the costs imposed by compliance with the laws. For example, engineering surveys may be required before construction can begin, or special building codes may have to be followed in places vulnerable to earthquakes or floods. People may also feel that it is their right to live where they like, accepting any natural risks. And some, of course, simply do not believe that whatever-it-is could possibly happen to them.

Occasionally, lawsuits have been filed in opposition to zoning or land-use restrictions, on the grounds that such restrictions amount to a *taking*—depriving the property owner of the right to use that property freely—without compensation. Such action is prohibited by the Fifth Amendment to the Constitution. However, the courts have generally held that such restrictions are, instead, simply the government's exercise of its police power on behalf of the public good and that compensation is thus unwarranted.

Construction Controls

Restrictive building codes can be strikingly successful in reducing damage or loss of life. After a major earthquake near Long Beach, California, in 1933, the California legislature passed a regulation known as the Field Act, which created improved construction standards for school buildings. Many of those standards are still being followed. Their effectiveness was demonstrated in subsequent earthquakes, notably the major one in San Fernando in 1971. It was noted afterward that damage to school buildings from the earthquake was almost wholly limited either to schools built prior to 1933 or to those later buildings whose engineers had been permitted to deviate from strict compliance with the Field Act. The 1995 Kobe earthquake further illustrated the value of earthquake-resistant design (**figure 19.13**). The problem of preexisting structures remains a major problem rarely addressed even by new building codes aimed at hazard reduction, as was

illustrated yet again in the 1989 Loma Prieta earthquake and the 1994 Northridge quake. Before these earthquakes occurred, there had been talk of retrofitting some older buildings, freeways, and bridges to improve safety in accordance with newer design specifications, but due to cost and inertia, many modifications went unmade, and structural failures resulted. Retrofitting of privately owned buildings tends to be even rarer (**figure 19.14**).

Perhaps because the state is subject to so many geologic hazards, California and its municipalities have been among the leaders in passage of legislation designed to

A

B

Figure 19.14

Hazard-mitigation legislation rarely requires privately owned buildings to be modified for greater resistance. (A) The first story of this two-story house completely collapsed in the Loma Prieta earthquake. (B) The lowest floor of this three-story apartment building, built with garages at each end and apartments in between, collapsed in the Northridge earthquake, and sixteen people were killed.

(A) U.S. Geological Survey/Photograph by D. Schwartz; (B) Photograph by J. Dewey, U.S. Geological Survey, courtesy NOAA/NGDC

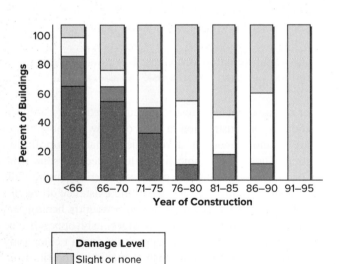

Figure 19.13

Affects of the 1995 Kobe earthquake on buildings according to age. Newer buildings with improved building codes suffered far less damage than older ones.

Table 19.2	Comparison of Landslide Damage With and Without Restrictive Construction Guidelines in Los Angeles, California		
	Prior to 1952	**1952–1962**	**1962–1967**
controls pertaining to landslide hazards	none	limited grading codes in place	much stricter requirements for engineering and soil studies; modern grading procedures required
sites constructed (approx.)	10,000	27,000	11,000
sites damaged (%)	1040 (10.4%)	350 (1.3%)	17 (0.2%)
total damage to residences (approx.)	$3,300,000	$2,800,000	$80,000
average damage per site built	$330	$100	$7
average cost per site damaged	$3200	$7900	$4700

Source: Data from J. F. Slosson, The Role of Engineering Geology in Urban Planning, Colorado Geological Survey Special Publication 1, 1969.

minimize the damage from geologic hazards. One successful effort in this regard has been the imposition of much-more-stringent regulations governing construction on potentially unstable, landslide-prone hillsides near Los Angeles. Over a ten-year period of rapid development (1952–1962), increasingly strict requirements for slope and soil stability studies, site grading, and construction engineering were developed. The resulting reduction in the rate of landslide damage was dramatic (**table 19.2**). Although the costs of structural damage to those homes actually damaged by landslides or soil failure continued to be high (as the result of escalating property values), the number and percentage of individual units damaged both dropped sharply, especially after 1963. Plainly, at least where the geologic hazards are moderate, the risk of damage from the hazards can be minimized by taking them into account and designing around them. This is a large part of engineering geology (see chapter 20).

Other Responses to Earthquake Hazards

Not all hazard-mitigation legislation is equally effective, as evidenced by a few California examples prompted by the 1971 San Fernando earthquake, described briefly below. Each of these bills is well-intentioned and represents a positive step toward reducing earthquake damage. Collectively, they also illustrate a number of weaknesses that limit their effectiveness.

The Seismic Safety Element Bill (1971) requires that communities take seismic and related hazards (ground shaking, tsunamis, and so forth) into account in planning and regulating development. This is plainly sensible in a high-seismic-risk area. But the bill does not prohibit or regulate construction explicitly; public officials are left to decide what constitutes acceptable risk in siting and building. The qualifications of the person(s) evaluating the seismic risks are not stipulated either, and there is no penalty if the community actually ignores the bill altogether.

The Dam Safety Bill (1972), despite its name, has nothing directly to do with the engineering adequacy of dams. It requires dam owners to prepare maps showing the areas that would be flooded in the event of dam failure, and it requires local authorities to prepare evacuation procedures for those areas threatened. Building in the areas at risk is not prohibited, and those living there may not necessarily be made aware of the dangers.

The Alquist-Priolo Geologic Hazards Act (1972; amended 1975) is a more specific and detailed bill. Under its provisions, the state geologist identifies "Special Studies Zones" along active faults. Proposed construction in these zones requires review by local authorities. The principal objective is to prevent damage through fault offset. Local officials are expected to consider the size of earthquakes and the extent of damage anticipated along each particular fault in deciding where or whether to allow building. Engineering and site-evaluation studies can be costly, and this prompted the 1975 amendment that exempted individual single-family homes; housing developments, apartments, and commercial buildings are still covered. Once again, economic considerations vie with scientific ones in selecting a course of action—another sort of cost-benefit analysis.

Flood Hazards, Flood Insurance

When geologic catastrophes occur, federal disaster-relief funds are often part of the rescue/recovery/rebuilding operations. In other words, all taxpayers pay for the damage suffered by those who, through ignorance or by choice, have been living in areas of high geologic risks. This has struck some as inequitable and was part of the motivation behind the Flood Insurance Act (1968), its successor, the Federal Flood Disaster Protection Act (1973), and five subsequent laws meant to strengthen the National Flood Insurance Program (NFIP). These measures provide for federally subsidized flood insurance for property owners in identified flood-hazard areas, whether in stream floodplains or in flood-prone coastal regions. Those most at risk, then, pay for insurance against their possible flood losses. The idea is that the flood insurance will replace after-the-fact disaster assistance in flood-prone areas. To encourage property owners to purchase the insurance, the legislation provides that it be required of those obtaining federally insured mortgages or mortgages through federally affiliated banks and lending institutions. The objective isn't simply to compel homeowners to protect their own property or to make them aware that they are buying in a floodplain, but also to broaden the premium base that provides funds to pay off claims when flooding does occur.

For a community to remain eligible for the insurance program, it must enact strict floodplain-zoning regulations. Initially, all proposed new construction in the floodplain must be carefully evaluated and structures designed so as to minimize potential flood damage. The community works with the Federal Emergency Management Agency to develop and adopt flood insurance control maps that precisely delineate areas that will be affected by 50-year, 100-year, and 500-year floods (**figure 19.15**). Then stricter provisions are enacted that require floodproofing or elevation above the 100-year flood level of new structures in the floodplain. The community may wish to include additional provisions, such as one requiring the replacement of water-storage volume lost in the floodplain due to construction. The laws don't completely ban building in flood-prone areas but there is significant impetus not to do so.

An unintended and unfortunate side effect of the availability of subsidized flood insurance has been to encourage people to rebuild, sometimes several times, in severely flood-prone areas and, in a sense, the insurance has encouraged continued development in such areas. The costs to the government can then far exceed the premiums collected (**figure 19.16**). In some unstable coastal zones, this has happened to such an extent that Congress has voted to remove some of the most

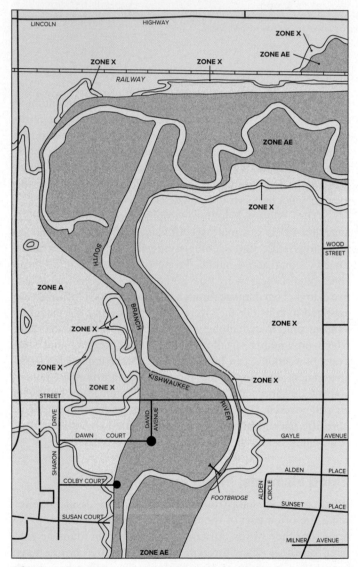

Figure 19.15

Sample section of a flood insurance rate map. Zone A is the estimated 100-year floodplain. In zone AE, height of 100-year flood is determined by detailed analysis of records. Zone X includes a variety of lower-risk situations: areas where the 100-year flood depth is estimated at less than 1 foot (narrow zone adjacent to zone A), areas protected by levees, and areas outside the 100-year but in the 500-year floodplain. View this map and other at FEMA's Flood Map Service Center.

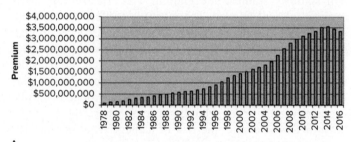

A

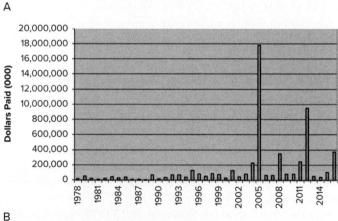

B

Figure 19.16

Premiums paid for federal flood insurance (A) kept pace with payouts for insured losses (B) until Hurricane Katrina in 2005. Since then, the program has operated in the red, in part because it was never designed to deal with disasters of that scale. Note that Hurricane Sandy occurred in 2012.

Source: FEMA

Graphs from Federal Emergency Management Agency.

Figure 19.17

Barrier islands regularly take a beating from coastal storms; most have now been removed from the flood-insurance program. The Outer Banks of North Carolina after Hurricane Joaquin in 2015.

U.S. Geological Survey Coastal and Marine Geology Program

Figure 19.18

Landslide in Potrero Canyon, Pacific Palisades area, California, after 7 inches of rain in January 1956. Later building codes made newer structures safer but left old ones in areas of known risk.

U.S. Geological Survey/Photograph by J. T. McGill

severely threatened areas from the program (**figure 19.17**; see also chapter 7). The losses from Hurricane Katrina put the program deeply in debt, where it remains; the catastrophic 2017 hurricane season with just about $12 billion in claims payouts didn't help. The large debt of the program, the increasing number of flooding events causing extensive damage, and a goal of providing more equitable risk assessment has resulted in a new NFIP pricing policy implemented in two stages in late 2021 and early 2022. Risk Rating 2.0 takes advantage of new mapping technologies and uses more variables to provide house-specific insurance rates.

Problems with Geologic-Hazard Mitigation Laws

A pervasive weakness in all kinds of geologic-hazard mitigation laws, including all of the examples discussed in this section, is the fact that they typically apply only to new structures. Only new schools in California need conform to Field Act building codes; only new hillside homes need to be built using modern slope-stabilization methods; only new federally insured mortgages on flood-prone properties mandate that the property owner carry flood insurance; and so on. Where considerable development has preceded the legislation, those affected by the new law in the near term may be a very small percentage of those at risk. In the 1993 Mississippi River basin floods, many thousands of victims were uninsured, aware of neither the need for nor the availability of flood insurance. Where new construction is banned in very-high-risk areas, older structures persist. A small move toward eliminating past poor practice would be a regulation requiring that, if an older structure in a geologically high-risk area were destroyed—whether by geologic processes or another means, such as fire—it could be rebuilt only if the new structure conformed to the newer laws for floodproofing, fault-trace setback, slope grading, and so forth. Even modest proposals of this kind, however, commonly meet with vocal citizen opposition.

Laws designed to reduce the risk of damage from geologic hazards have several problems in common. The basic scientific information on which to base sensible regulations may be lacking. Laws may be poorly written, failing to specify fully what must be done and by whom, or they may be weakened by exceptions or omissions that exempt many from their provisions. A major omission, and a very common one, is that existing structures in areas at risk may continue unaffected (**figure 19.18**; recall also **figure 19.14**).

19.7 The National Environmental Policy Act (1969)

Learning Objectives

- Describe the purpose of an EIS
- List the NEPA requirements for an EIS
- Name some perceived shortcomings of EISs

The (U.S.) NEPA established environmental protection as an important national priority and provided for the creation of the Council on Environmental Quality in the Executive Office of the President. The **environmental impact statement** (EIS) is the most visible outgrowth of the NEPA. In this section, we will consider briefly what goes into an EIS, what the point is, and how well the legislation's objectives are being met.

The NEPA actually pertains only to federal agencies and their actions. Whenever a federal agency proposes legislation or "other major Federal action" (which can include anything from policy changes to construction projects that are wholly or partially supported by federal funds) that can significantly affect the quality of the human environment, an EIS must be prepared. Many states have adopted similar legislation related to projects involving funding or approval by state agencies, so the scope of the EIS process has been considerably broadened. The preparation of EISs should ensure that, before an agency acts, it considers as fully as possible the potential environmental ramifications of its action, in order to make the best possible decisions. The discussion that follows is restricted to federally mandated EISs.

The NEPA specifies that an EIS should include

1. A description of the proposed action, its purpose, and why it is needed

2. A discussion of various alternatives (including the proposed action)

3. An indication of the environment to be affected and the environmental consequences anticipated

4. Lists of preparers of the statement and those agencies, organizations, or persons to whom copies of the statement are being sent

Additional supplementary material may also be included. The statement is expected to "rigorously explore and objectively evaluate all reasonable alternatives" (an analysis that presumably will ultimately supply appropriate justification for the particular course of action preferred by the agency). Possible environmental consequences might include not only those that are in some sense geologic (e.g., altering natural processes like stream flow or runoff, causing air or water pollution, inevitable consumption of energy or other resources) but also biological ones (loss of habitat, destruction of organisms) and social ones (affecting quality of life, demands on infrastructure, employment, historic or cultural resources). Cost-benefit analyses of various alternatives may be included if the agency is taking them into account.

After a draft EIS is prepared, a variety of others, ranging from other federal agencies to the general public, are invited to comment. In fact, comments should be actively solicited from all persons or organizations that are particularly likely to be interested or affected. The agency preparing the EIS is then expected to respond to those comments, which might mean modifying the proposed action or the analysis in the EIS or considering additional alternatives.

Over the first decade following passage of the NEPA, an average of more than 1000 EISs each year were prepared by federal agencies. Lawsuits were filed contesting about 10% of this number of projects. In many such cases, it was alleged that the EIS was inadequate, incomplete, or inaccurate. In other cases, the charge was that an EIS ought to have been prepared but was not (the responsible agency felt that no significant environmental impact was involved). Close to half the lawsuits involved citizen or environmental groups as plaintiffs. The number of EISs filed annually began to decline in the late 1970s (**figure 19.19**), but the number of lawsuits rose, peaking at 157 in 1982; by the mid-1980s, the number of lawsuits began to drop off sharply in turn.

In recent years, about 330 EISs have been produced each year. As might be expected, the number of suits filed against a particular agency is broadly proportional to the number of EISs that it prepares. A small number of agencies account for most of the EISs. In 2019, for example, over 75% of EISs were filed by just four agencies (**figure 19.20**): the Department of Agriculture; Department of the Interior; Department of Transportation; and Department of Defense.

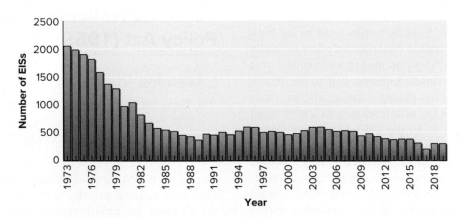

Figure 19.19

Number of EISs published by all federal agencies, 1973–2019.

Source: https://naep.memberclicks.net/assets/annual-report/2019_NEPA_Annual_Report/NEPA_Annual_Report_2019.pdf

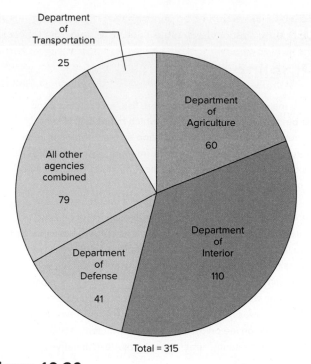

Figure 19.20

Distribution of EIS filings with the U.S. EPA in 2019, by agency.

Source: https://naep.memberclicks.net/assets/annual-report/2019_ NEPA_Annual_Report/NEPA_Annual_Report_2019.pdf

In that same year, twenty-one lawsuits were filed in connection with the NEPA and seven of those were with the Department of Agriculture.

The usefulness of EISs has been limited by a number of factors. We have already noted in the context of antipollution laws the difficulty of applying cost-benefit analysis to situations where not all factors have well-defined costs associated with them. Another potential problem is the impartiality (or lack of it) of EIS preparers or proposers. An agency preparing an EIS for a project of its own has already selected a preferred course of action, and it might, consciously or otherwise, be inclined to present that action in an unduly favorable light. Where the proposed action is to be carried out by an individual or corporation—such as a company needing a federal permit to mine on federal lands—the EIS may, in part, draw on information supplied by the proposer, which could again be biased. There are also only general

statutory requirements that the analysis in an EIS be thorough, which is no guarantee that all the appropriate kinds of specialists will be involved in the preparation of a given EIS. Audits of the accuracy with which EIS preparers predicted environmental impacts have often shown, in retrospect, rather low accuracy.

A major purpose of the NEPA is to allow the public to be informed about and involved in decision making related to the environment. To prevent EISs from becoming prohibitively long and verbose, the NEPA stipulates that the length should normally be 150 pages or less, or up to 300 pages for especially complex projects. However, tiering is allowed, whereby one EIS may be written for each level of decision making of a large or multistage project. One might then have to read through several statements of 150–300 pages each to comment intelligently on a particular project. Many people don't have the time to do that much reading, so they may not become involved at all. (For that matter, not all the relevant employees of the federal agencies involved may do their EIS homework adequately either, for the same reason.) An analysis of EISs filed in 1996 showed that the 243 draft EISs ranged in length from 55 to 1622 pages, with 20% over 300 pages of text; the 270 final EISs ranged from 12 to 1638 pages, averaging 204 pages, with 24% over 300 pages and 6% over 500 pages of text.

Nor can one always project the length of the EIS from the size of the geographic area affected or the apparent complexity of the project. For example, a proposal to designate a portion of the Bering Land Bridge National Preserve as a wilderness area resulted in a 300-page final EIS. This was longer than the final EIS for a project in Wyoming that involved building a phosphate fertilizer plant, phosphate slurry pipeline, railroad spur, microwave communication system, and electrical power transmission system, and relocating a road! A project of the scale and complexity of the Trans-Alaska Pipeline, however, would be expected to involve a lengthy EIS, and it did; see Case Study 19.

A major loophole of the NEPA, in the eyes of many environmental groups, is that statements must be prepared only when the anticipated environmental impact of a proposed project is *significant,* where significance is evaluated by the federal agency involved in proposing or permitting the action in question. Clearly, significant is an inexact term. For each project for which an EIS is filed, many more have been deemed by the agency not to warrant one, and others may dispute that conclusion. As with other kinds of environmental laws, the NEPA has some very desirable objectives, but practice has often fallen short of the ideal.

A Tale of Two Pipelines

The discovery of economically important reserves of oil along Alaska's North Slope raised the immediate question of how to transport that oil to refineries in the lower forty-eight states. It was proposed that a pipeline be constructed to carry the oil south to the port of Valdez, from which tankers could move the oil farther south. The planned route (**figure CS 19.1.1A**) crossed over 1000 kilometers of federal lands, which necessitated a permit from the government; hence the requirement of an EIS under the then newly passed NEPA.

Various possible alternatives to the proposed route included alternate pipeline routes across Alaska; transportation by some combination of pipeline, rail, and highway across both Alaska and Canada; and tanker transport directly from the North Slope (**figure CS 19.1.1B**). Transport exclusively by tanker was ultimately deemed impractical, for the harbors and surrounding seas of northern Alaska are blocked by ice much of the year. Schemes involving pipelines across Alaska and subsequent tanker transport posed threats to both terrestrial and marine environments; routes involving only pipelines through Alaska and Canada would leave marine life intact but would disturb far more land. A pipeline through Canada would also necessarily be under the jurisdiction of a foreign government.

The possible environmental impacts were many and varied. Construction and operation of overland pipelines could disturb the ground, water, vegetation, fish, and wildlife. Land must be committed to the project, which might cost wildlife habitats or inhibit the migration of land animals. Warm oil in an underground pipeline could melt permafrost, with possible subsidence of and damage to the pipeline in the resulting mud. The pipeline would cross major fault zones; southern Alaska, in particular, is subject to relatively frequent earthquakes, and severe earthquakes also occur on the Denali Fault; pipeline rupture from any cause raised concerns about oil spills onto land or into lakes and streams. The use of tankers would mean possible marine oil spills also, and heavy tanker traffic could affect commercial fishing operations. On the plus side, pipeline construction would create jobs, besides providing access to major oil reserves. The influx of money and work crews and the impacts of a pipeline on the physical environment, however, would also significantly affect the lives of Alaskan natives, perhaps negatively.

On balance, the original proposed route was deemed likely to have the least overall adverse impact. Principal unavoidable impacts were disturbances of terrain, fish, wildlife, and some human environments; some pollution at the port of Valdez from oil discharge from treatment of tanker ballast water; and secondary effects from the influx of more people into the region. (We now know that tanker accidents can also be a significant problem.)

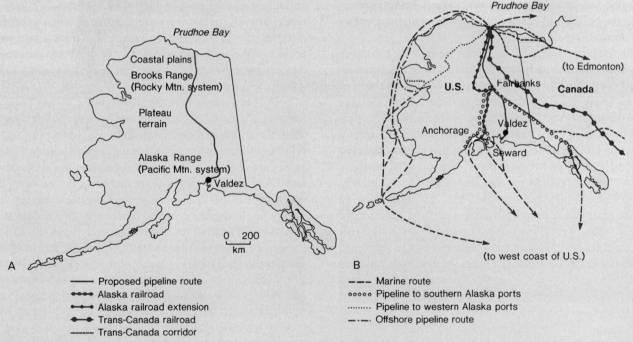

Figure CS 19.1.1

(A) Route of the Trans-Alaska pipeline. (B) Proposed alternative means for transport of the oil.

Source: After D. A. Brew, Environmental Impact Analysis: *The Example of the Trans-Alaska Pipeline, U.S. Geological Survey Circular 695, 1974.*

The analysis was a large and complex task, reflected in the six-volume final EIS. However, such detailed analysis allowed the anticipation and minimization or elimination of a variety of negative environmental effects and problems, in part through thoughtful engineering, as explored further in chapter 20. The finished pipeline (**figure CS 19.1.2**) has functioned effectively for over four decades, during which time over 18 billion barrels of oil have moved through it. The largest spill occurred in 2006 when a transit pipeline in Prudhoe Bay ruptured, releasing over 250,000 gallons.

In 2008, TransCanada Corporation proposed construction of a pipeline ("Keystone XL") to carry oil from the Canadian Athabasca oil sands to Oklahoma for distribution to refineries in the Gulf Coast region. The 1700-mile-long pipeline would cross parts of seven states. Two EISs were filed on the project. The EPA commented on both draft EISs, citing a range of concerns that it felt were not fully addressed, including air quality; pollution; impacts on water resources, wetlands, and migratory birds; possible subsidence in karst terranes; impacts on native tribes living in areas to be crossed; and adequacy of emergency-response plans. The final EISs were filed in 2008 and 2010.

Because the Keystone XL pipeline would cross an international boundary, a presidential permit was required. Congress, anxious for action, stipulated a two-month time limit for consideration of the permit application after filing the second final EIS. President Obama declined to be rushed into approval on a project of this magnitude and denied the permit. (Interestingly, this was also the recommendation of the State Department, lead agency for the EIS process on Keystone XL.) In May 2012, a new application for the presidential permit was submitted, involving a modified and shorter route—just under 1200 miles—ending in Nebraska. Citing environmental concerns, the president denied the permit again in 2015.

Shortly after taking office, President Trump approved the Keystone XL permit. Each state through which the pipeline would pass also had to approve. Montana and South Dakota

Figure CS 19.1.2

The view at mile 562 along the pipeline. Note that in many places, the pipeline was elevated, in part to facilitate wildlife migration across the route.

©Carla Montgomery

did so first. Nebraska's Public Service Commission did not grant approval until late 2017—and, instead of approving TransCanada's preferred route, the Nebraska commission approved a different, somewhat longer (and thus more costly) alternative route. This brought into play a new set of potentially affected landowners and other stakeholders. Opponents stalled the project with lawsuits until President Biden canceled the cross-border permit soon after taking office in 2021, and the project was canceled outright by it primary corporate backer in June of that year.

Summary

Environmental laws and policies relating to geologic matters include some relating to the right to exploit certain resources, some designed to limit pollution and other kinds of environmental deterioration, and some intended to force individuals and agencies to take geologic hazards into account in development and construction. They may fall short of their goals for a variety of reasons. Terms may be inadequately defined or scientifically meaningless, making consistent enforcement difficult. Economic pressure may force pursuit of a course that may be less desirable in purely scientific terms, or a rigorous cost-benefit analysis may be impossible. There may be individual or institutional resistance or bias to combat, or a particular law may recognize so many exceptions that the majority of cases ultimately end up unregulated. These and other difficulties, many without straightforward solutions, will continue to complicate efforts to develop environmental legislation. The future will likely see increasing development of intergovernmental agreements on critical global environmental issues, but as those agreements commonly lack strong enforcement provisions, their effectiveness also may be limited. The NEPA paved the way for systematic analysis of the environmental consequences of various actions through the Environmental Impact Statement process and for citizen input, but not every project gives rise to an EIS, or to an appropriately thorough one.

Key Terms and Concepts

common but differentiated responsibility of states
ecosystem services

environmental impact statement
Exclusive Economic Zone

precautionary principle
Prior Appropriation
Riparian Doctrine

Test Your Learning

1. Compare the basic concepts of the Riparian and Appropriation Doctrines underlying much surface-water law.

2. Explain why groundwater rights are inherently somewhat more difficult to define than surface-water rights.

3. Identify the principal objective behind early (nineteenth-century) federal mineral-resource laws, and describe how the emphasis shifted over the last century.

4. Explain the concept of EEZs, and give two examples of types of mineral resources they might encompass.

5. Discuss some of the difficulties of defining and achieving "zero pollutant discharge."

6. Changes in federal policy have changed the degree of emphasis put on cost-benefit analysis in setting pollution-control standards. Explain briefly.

7. Cite two common problems with construction or zoning restrictions that limit their effectiveness, particularly in densely populated areas.

8. Briefly explain the *precautionary principle* and the concept of the *common but differentiated responsibility of states* and how they relate to the development of international environmental-protection agreements.

9. Summarize the kinds of information included in an EIS.

10. Cite and explain at least two features of the EIS process that may limit its effectiveness.

Exploring Further

1. Consider what restrictions, if any, you as a legislator might place on development in a floodplain, on landslide-prone hills, or in active fault zones. If your community is subject to such geologic hazards, look up existing ordinances to see how extensive their restrictions are, and inquire about how routinely variances and exceptions are granted.

2. Choose a state and investigate its groundwater laws. Compare what you learn to the information shared in figure 19.3. Is your interpretation of the laws different than that shown on the map? Hypothesize why.

3. Identify the geologic (plate tectonic) settings in U.S. EEZs, and investigate the types of mineral resources forming in each of those settings.

4. Pick an item of environmental legislation, and investigate its current status. You might choose the reauthorization of the Clean Water Act, Clean Air Act Amendments, or CERCLA, or revisions of the 1872 Mining Law; the review

of the Canadian Environmental Protection Act; or the ratification status of one of the U.N. climate conventions.

5. Interested in bidding on SO_2 allowances? Find out more about EPA's annual auction at **https://www.epa.gov/airmarkets**.

6. Seek out the EIS for a local project or other project of interest. Note what kinds of impacts are projected and how their importance is evaluated. Is the EIS comprehensible to you as an interested individual? Would you suggest any areas in which more thorough analysis seems to be needed? (A listing of EIS filings with EPA is found at **https://www.epa.gov/nepa**.)

7. Another controversial pipeline project has been the Dakota Access Pipeline. Although it has now been built and is operational, legal challenges are still being mounted by the Standing Rock Sioux Tribe and others, in part on the grounds that the EIS process was not properly carried out. Explore the current status of the project.

CHAPTER 20

Land-Use Planning and Engineering Geology

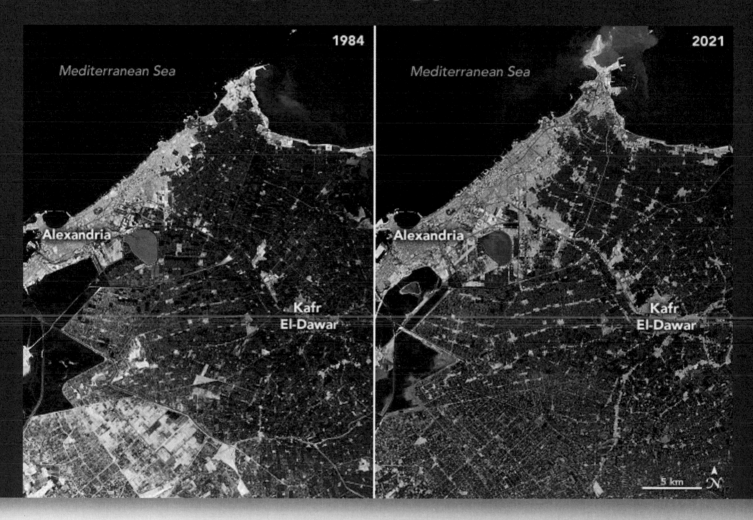

Most of Egypt's 100 million people live within the Nile River's floodplain. Only 4% of the country's land is arable, and it is being lost to urbanization at a rate of 2% a year. Fertile farmland is also being lost to sea-level rise and saltwater intrusion. In hindsight, planning the use of the land before it was developed may have preserved it for agriculture.

NASA Earth Observatory images by Lauren Dauphin

The purpose of land-use planning is to make the best, most sensible, practical, safe, efficient use of each parcel of land. Because such decisions are based, in part, on geologic considerations, land-use planning and engineering geology overlap. In general, however, land-use planning involves a much larger body of geologic and nongeologic facts and concerns. These may, for instance, include potential economic or practical benefits from a given use of the land and possible negative environmental or aesthetic impacts. Often, land-use planning takes the form of assessing the suitability of a particular parcel of land for a particular purpose and proceeds somewhat like an environmental impact assessment.

A frequent problem in land-use planning, as in other areas such as cost-benefit analysis in pollution control or evaluation of energy options, is that individual judgments about the relative importance or value of different considerations are involved. A strip mine may be beautiful to an unemployed miner, an uninterrupted river may be more precious than a dam and hydroelectric plant to an angler, and so on. This chapter does not attempt to make such judgments, but instead reviews briefly some principles and practices of land-use planning.

Civil engineering, in some sense, goes back to the oldest human communities, to the irrigation canals of Egypt dug in 2400 BCE, to the pyramids, to the Great Wall of China, and to the aqueducts and roads of ancient Rome. The ancestral Puebloans at Mesa Verde built not just dwellings, but also water reservoirs to collect and store that precious resource.

The building of many ancient structures required some understanding of earth materials. Mistakes resulted from gaps in that understanding. The Leaning Tower of Pisa leans, not by the design of its builders and not because of any structural flaw within the edifice proper, but because it was built on unstable soils.

Over the last 200 years, it has become increasingly possible to incorporate geologic principles and considerations into construction plans. The paramount concern of the modern engineering geologist is to take a site's geology fully into account in designing a structure so that the structure will be safe and stable. Increasingly, engineering geologists also consider the impact that the building or structure, in turn, will have on the geologic environment.

Chapter Outline

20.1 Land-Use Planning

20.2 Land-Use Options

20.3 The Federal Government and Land-Use Planning

20.4 Maps and geographic information system (GIS) as Planning Tools

20.5 Engineering Geology

20.6 Case Histories, Old and New

20.7 Dams, Failures, and Consequences

20.1 Land-Use Planning

Learning Objectives

- List reasons for and considerations taken in land-use planning
- Recall recent trends in U.S. land use

Why take the time, money, and effort to plan the use of land? Given that about half of land area in the United States is not developed or being used for agriculture, it may seem like just about any open piece of land would be good enough for any type of project. One motive behind land-use planning is safety. Some land is geologically unstable and unsuitable for certain kinds of structures. Another motive is where to put the large number of people that encompass our ever-growing population. When human communities were small, if they determined a piece of land was unsuitable for their purpose or it became unusable due to poor management, they could just move to a new spot and begin again.

Centuries ago, there was ample space for all. That is becoming less true all the time. From 1982 to 2017, an average of 1.26 million acres of rural land were converted to developed uses each year just in the United States. Admittedly, 1.26 million acres may not seem like a lot compared to the 2300 million acres

of land in the United States. Year after year, however, it adds up (**figure 20.1**). From 1982 to 2017, the total area developed—44.2 million acres—was altogether nearly the size of the state of Washington state. Excluding the amount of land devoted to roads, railroads, and other transportation corridors, which has stayed relatively constant, the amount of developed land in the United States increased by about 61% over that period.

The nature of the land being developed (**figure 20.1B**) may have special consequences. Over the same period, the amount of prime farmland decreased by about 15 million acres, and total cropland by nearly 53 million acres. Loss of cropland means less land available for producing food or biofuels. This is a particular concern in parts of the Midwest, where urban sprawl has been overrunning prime farmland developed on nutrient-rich loess. Clearing forest land for development means reducing a CO_2 sink.

It is becoming increasingly important to consider not only what we can safely do with a given piece of land but also how we can put the land to optimal use. Many of the considerations involved are geologic ones—presence of hazards such as flooding, landslides, or earthquakes; availability of water; depth to water table; runoff/drainage patterns; soil character; resource potential; susceptibility of the land to pollution, erosion, or other disruption if certain kinds of land uses are permitted; and so on. Again, biological/ecological, economic, and political factors also enter into the decisions we ultimately make. So may aesthetic factors. For example, as noted in chapter 15, Yellowstone National

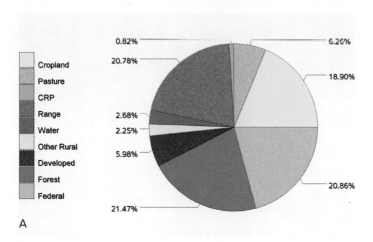

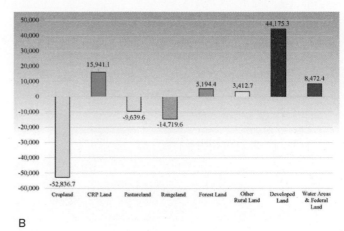

A

B

Figure 20.1

(A) U.S. surface area by land cover/use in 2017. (B) Net changes in land use, 1982–2017 in thousands of acres.

(A) Source: United States Department of Agriculture,https://www.nrcs.usda.gov/Internet/NRCS_RCA/reports/nri_nat.html
(B) Source: United States Department of Agriculture

Park represents a tremendous geothermal resource. However, it is also a great scenic resource and important wildlife habitat, and these considerations are deemed important enough to override the potential gain from exploiting the geothermal energy. In early oil-boom days, on the other hand, oil was so sought-after that considerations of whether or not one might want a derrick in one's front yard were often overlooked (**figure 20.2**).

Figure 20.2

Drilling for oil, Long Beach, California, 1901.

Source: U.S. Geological Survey Photo Library, Denver, CO/ Photograph by C. W. Hayes.

The Bingham Canyon copper mine in Utah, one of the world's largest open-pit mines as noted in chapter 13, has irrevocably altered some 7500 acres of undeveloped western land, but it has yielded, on average, over $9 million per acre in metallic minerals. As with pollution controls, costs and benefits of particular land-use choices are weighed in making land-use decisions.

20.2 Land-Use Options

Learning Objectives

- Compare multiple use and sequential use of land
- Illustrate multiple uses and sequential uses of land

Particularly in densely populated areas, there is growing interest not only in making the best use of each piece of land but, if possible, in using the same parcel of land for several different purposes. The strategies involved are called *multiple use* and *sequential use*.

Multiple use is using the same land for two or more purposes at the same time. Two examples are shown in **figure 20.3**. At first glance, the area in **figure 20.3A** appears just to be a ball field. However, note that it is built into a basin. When it rains— when the field wouldn't be used as a ball field anyway—the field acts as a recharge basin, catching fresh rainwater and allowing it to infiltrate slowly down to the seriously depleted aquifers under Long Island. So the same land serves two purposes: one recreational, the other directed toward conserving/ increasing groundwater resources. Other examples of multiple land use would be the generation of wind power from wind-generator arrays in fields used simultaneously for crops (**figure 20.3B**) or grazing land (or even residential areas, as is done in Denmark), or underground mining deep below the surface while urban development proceeds above.

A

B

Figure 20.3

Multiple land uses. (A) This ball field doubles as a recharge basin for Long Island aquifers. (B) Near Snyder, Texas, arrays of wind generators cross fields of cotton (note contour planting).

(A) U.S. Geological Survey Photo Library, Denver, CO; (B) NASA Earth Observatory image by Joshua Stevens, using Landsat data from USGS

The alternative approach of **sequential use** involves using land for two or more different purposes, one after another. Because the land uses can be incompatible, a greater variety of combinations is possible. Abandoned underground mines, if dry and adequately ventilated, can be used for warehouse space, manufacturing, or even office space. This has been done, for example, in the Kansas City area, in abandoned underground limestone quarries. The old quarries now include nearly 1 million cubic meters of frozen-food storage space (about one-tenth of the U.S. total), and the rock's insulating properties make the practice very energy-efficient. The great strength of rock floors makes possible the use of heavy manufacturing equipment, and the controlled humidity is an advantage to print shop. Rock walls are also fireproof and contain noise well. On a smaller scale, abandoned mine space at Wampum, Pennsylvania, has been converted to office and laboratory space for the mining company. A portion of the nearly mile-deep Homestake gold mine in South Dakota is home to the Sanford Underground Research Facility where scientists conduct physics experiments; another portion is being excavated to house other experiments into the nature of the universe.

Not all abandoned underground mines are equally suitable for subsequent occupation. The rock structure must be sufficiently strong for safety and not prone to deterioration through slow weathering with time. Even if above the water table, the space must be protected by overlying impermeable rock from infiltration of subsurface water from above. Flat-lying rock strata facilitate conversion to occupied space. There must be no likelihood of quantities of dangerous gases present; old limestone mines would be safe in this respect, while old coal mines, with possible associated methane present, probably would not be.

Alternatively, abandoned mines could be used for waste disposal (recall the proposal in chapter 16 to place high-level radioactive wastes in abandoned salt mines). Strip mines have been suggested as possible landfill sites when mining is completed; this would follow resource extraction with waste disposal, after which the land could be covered, regraded, and put to another use. Denver, Colorado, has carried out a scheme of this sort. Old gravel pits were used for sanitary landfill, and then, after they had been filled, the Denver Coliseum and its parking lots were constructed over the top. (Using such permeable materials for landfill may or may not be advisable in a given situation.)

An alternative post-landfill use might be the extraction of methane gas for energy. An abandoned quarry can be flooded to make a recreational lake. Many such sequences of activities can be imagined; see, for examples, **figure 20.4**. For that matter, each time a Superfund site is reclaimed and converted to a productive purpose, that, too, is sequential land use. The advantage all sequential-use arrangements share is land conservation. If one piece of land can be made to do double or triple duty, that much less land must be used for or disturbed by human activities overall. Sequential use also lessens the likelihood of being forced to use land that is unstable or vulnerable to degradation through human use.

20.3 The Federal Government and Land-Use Planning

Learning Objectives

- Describe the two primary types of federal lands
- Exemplify controversy regarding the preservation of public lands

A

B

Figure 20.4

(A) Limestone, marble, and dozens of useful minerals were extracted from this quarry in Riverside, California; then the pits were allowed to fill, making decorative lakes, and the quarry was converted to a golf course. (B) The limestone mined here was used to build the city of St. Emilion above. The space is now an underground pottery museum; tool marks from the quarrying are still visible on walls and ceiling. (C) Osaka's Kansai Airport is built on landfill. Though it does settle 2–4 cm/yr, it suffered little damage in the 1995 Kobe earthquake and essentially none in the 2011 Tohoku quake.

(A, B) ©Carla Montgomery; (C) NASA Image Collection/Alamy Stock Photo

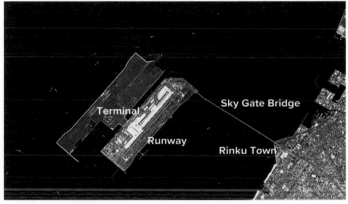

C

The federal government controls approximately one-third of the land in the United States. The proportion does vary widely among states, from less than 1% in many northeastern and mid-continent states to over 60% of Utah, Idaho, and Alaska, and over 80% of Nevada (recall **figure 19.4**). The impact of federal land-use policies, like other federal resource-related laws, is felt in very different degrees in different states.

Federal land-use policies have changed through time. As with mineral-resource laws (chapter 19), federal emphasis was initially on resource development in preference to preservation. Early preservation efforts were limited. The first of the national parks, Yellowstone, was established in 1872; the first national forests were created two decades later. Still, resource development continued to be a high priority until about the middle of the twentieth century.

Federal lands can be broadly divided into two types: those intended primarily for preservation (including national parks and wilderness areas) and those on which multiple and compatible land uses can be allowed (e.g., national forests). On the latter lands, additional uses beyond recreation or habitat preservation might include livestock grazing, mining, logging, and exploration and drilling for petroleum. Problems arise when the multiple uses allowed turn out not to be compatible after all. Livestock may outcompete wildlife for limited food;

overgrazing, careless timbering, or even just too many tourists passing through may accelerate soil erosion and loss. Because the preservation function has sometimes been less successful when multiple land uses are allowed, many groups have tried to pressure the federal government to put more of its lands into the highly protected categories.

In 1978, President Carter did that with 107 million acres of Alaska. Under the Alaska Lands Act, that land—almost one-third of the state—was designated as national monuments, wilderness areas, or similarly protected land. Partially as a result of that action, mining and petroleum development now are prohibited or severely restricted on 40% of Alaska's total land. Some hailed the decision as a forward-looking move to preserve dwindling wilderness. Oil and mining interests were very unhappy at the reduction in land available for exploration and warned of possible future resource shortages. State residents who had hoped to benefit from more jobs or from taxes on those companies' profits from resource exploitation were likewise not pleased. The state, in turn, wanted more control over what was to be done with its lands. Recently, proposals to open parts of the Arctic National Wildlife Refuge to petroleum exploration have intensified the debate.

Figure 20.5

Bears Ears National Monument is managed by the U.S. Forest Service, the Bureau of Land Management, the Hopi Nation, the Navajo Nation, the Ute Tribe, the Ute Mountain Ute Tribe, the Ute Indian Tribe of the Uintah Ouray, and the Zuni Tribe.

Source: U.S Department of Interior Bureau of Land Management

In late 2017, President Trump declared an intention to shrink two large national monuments in Utah: Bears Ears (**figure 20.5**), created by President Obama, and Grand Staircase–Escalante, created by President Clinton, by 85% and 45%, respectively. This would open up about 2 million acres of previously protected land to a variety of uses, including mining and oil and gas leasing. Conservation groups, anglers, hunters, Native American tribes, and others filed suit to block the reductions. Both monuments were restored by President Biden in October 2021.

Often, controversies arising in connection with the environmental impact statement process reflect significantly different opinions about appropriate or optimal land uses for the area in question. There are also conflicting opinions about the appropriateness of resource development by private corporations on public lands, even when that is allowed by the land's classification, and whether or not royalties are paid. Is it possible for a government to manage so much public land in everyone's best interests? Which interests take precedence?

20.4 Maps and GIS as Planning Tools

Learning Objectives

- List land-use planning information that can be stored in a map
- Summarize how information is digitized to create a map
- Describe the benefits of GIS over traditional maps
- List some limitations of maps

Many kinds of information, geologic and otherwise, go into comprehensive land-use planning. Historically, much of this information has been most quickly and clearly examined in map form, first in paper and then in a digital context. Now most of this information is contained within and viewed from a **geographic information system** (GIS), examples of which have been included throughout this text. Any geologic property or process that varies from place to place, including topography or steepness of slopes, bedrock geology, surficial materials or soil types, depth to water table, rates of cliff erosion, and so on, can be represented in a map. We can also use maps to show locations of hazards past or present, such as fault zones, floodplains, and landslides. **Figure 20.6** illustrates some simple examples; recall also figure 19.15. Nongeologic factors—vegetation, population density, or land use, for instance—may also lend themselves to representation in map form (**figure 20.7**).

Using maps such as these, a land-use planner can then seek sites for particular land uses based on whatever set of criteria seems most appropriate. Maps make it possible to see quickly where several different conditions are satisfied. In siting a major interstate highway, for example, a planner might seek gently sloping terrain not underlain by expansive clay soils or active fault zones, where the present population density is low. A survey of an appropriate set of maps permits the planner to find potentially suitable sites swiftly. Conversely, maps can aid in long-term land-use planning even when conversion of rural land is not immediately contemplated. If the location of a stream's 100-year floodplain or a major fault zone is known, restrictive zoning ordinances can prevent unwise development before it occurs, or special construction requirements for those areas can be imposed. The U.S. Geological Survey Coastal and Marine Geology Program has maps reflecting long-term coastal change but also maps showing short-term vulnerability to storms.

Of course, the suitability of a particular area for a specific land use cannot always be determined in a clear-cut, yes-or-no fashion. A land-use planner may instead need to consider a number of alternative sites for a project, each of which is less than fully desirable in some different way. Or there may be degrees of suitability for some purpose, such as housing developments, considering a variable like slope stability. In the latter case, the planner might stipulate different levels of intensity of the same land use in different areas. A higher density of homes might be permitted on gently sloping terrain than on steeper hillsides, for instance.

In recent decades, computers have played an increasingly important role in the land-use planning process because

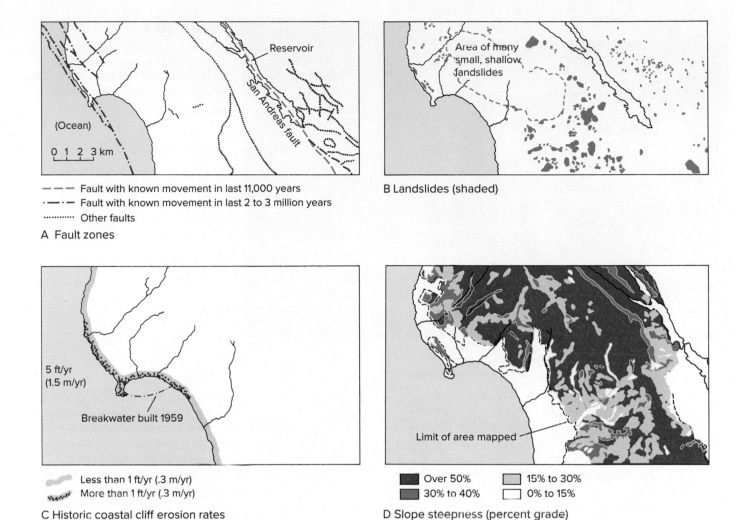

Figure 20.6

Map representation of several kinds of geologic considerations, each map focusing on a single issue: note that the scale of the map requires averaging over significant areas, resulting in some loss of detail.

Source: Modified from B. Atwater, "Land-Use Controls Arising from Erosion of Seacliffs, Landsliding, and Fault Movement," in U.S. Geological Survey Professional Paper 950.

of their capacity to manipulate large volumes of quantitative information. A map can be broken down into a grid of numbers, each data point representing a particular property (slope, soil type, and so forth) over some area (1 square kilometer, 10 square kilometers, or whatever) (**figure 20.8**). The computer can then combine as many kinds of information as desired, each represented on a separate array, and produce a composite measure for each point of the grid to indicate the overall suitability of that area for the land use under consideration, which is most often urbanization/housing development (**figure 20.9**). The same basic geologic or nongeologic data can be combined in different ways for different purposes by weighting various factors differently. In mapping the suitability of land for wildlife habitat or refuge, for

example, abundant surface and near-surface water might be a positive factor and the presence of expansive clay soils irrelevant. A land-use planner looking at the same region with high-density housing in mind might well regard both factors as negative. The data-processing programs can be adjusted accordingly to produce summary data tailored to particular objectives.

As computers have become more powerful and their programming more sophisticated, they have increasingly been used for GIS, a computer-based system for storing and manipulating mappable data—in other words, data that can be associated with, or plotted on, points on a map. The data are often stored and displayed in *layers,* where each layer is a distinct type of information (**figure 20.10**). The results can be examined

Figure 20.7

U.S. land use/cover map created from Landsat satellite images collected in 2016. From 2001 to 2016, 7.6% of the land cover changed at least once.

Source: USGS

NLCD 2016 Landcover

Key to Land Cover Types
- Open Water
- Perennial Ice and Snow
- Developed, Open Space
- Developed, Low Intensity
- Developed, Medium Intensity
- Developed, High Intensity
- Barren Land
- Deciduous Forest
- Evergreen Forest
- Mixed Forest
- Dwarf/Scrub
- Shrub/Scrub
- Grassland/Herbaceous
- Sedge/Herbaceous
- Moss
- Pasture/Hay
- Cultivated Crops
- Woody Wetlands
- Emergent Herbaceous Wetlands

NLCD 2016 Land Cover for the conterminous United States represented as 16 land cover classes.

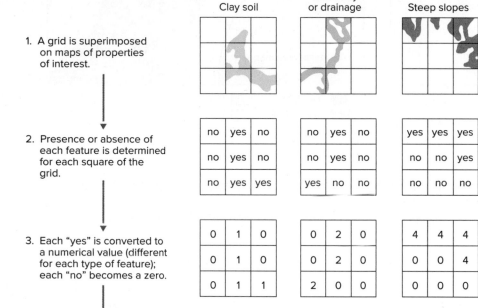

1. A grid is superimposed on maps of properties of interest.

2. Presence or absence of each feature is determined for each square of the grid.

3. Each "yes" is converted to a numerical value (different for each type of feature); each "no" becomes a zero.

4. Sum of numerical values for each grid square indicates number and kind of problem(s) present. A "7" means all three factors present; "0," none of them; "3," clay soil plus drainage, but not steep slopes; and so on.

5. Numbers are converted to symbols for easier scanning when suitable sites for various activities are to be chosen.

Clay soil

no	yes	no
no	yes	no
no	yes	yes

0	1	0
0	1	0
0	1	1

Waterway or drainage

no	yes	no
no	yes	no
yes	no	no

0	2	0
0	2	0
2	0	0

Steep slopes

yes	yes	yes
no	no	yes
no	no	no

4	4	4
0	0	4
0	0	0

4	7	4
0	3	4
2	1	1

0: No problems

1: Clay soil only

2: Drainage only

3: Clay soil and drainage

4: Steep slopes only

5: Steep slopes and clay soil

6: Steep slopes and drainage

7: All features

Figure 20.8

Digitized maps can represent data in a form that both human planners and computers can use. In this simplified example, for each parameter (clay soil, water drainage, steep slopes), each point (area) has one of two values: either the area is underlain by clay or it is not; either the slopes are steeper than a certain grade or they are not; and so on. In a more complex scheme, there might be several possible values for each parameter.

Source: A. J. Froelich et al., "Planning a New Community in an Urban Setting: Lehigh," U.S. Geological Survey Professional Paper 950.

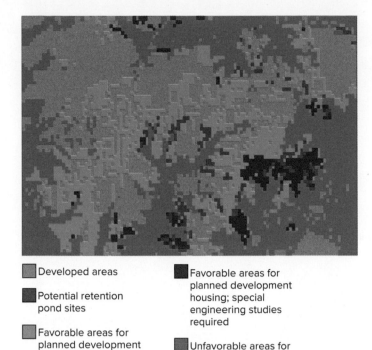

Developed areas

Potential retention pond sites

Favorable areas for planned development housing

Favorable areas for planned development housing; special engineering studies required

Unfavorable areas for planned development housing

Figure 20.9

The computer can rapidly combine digitized data such as that of **figure 20.8** into a composite map for land-use planning, showing more- and less-suitable areas for development or any other intended purpose.

Source: A. J. Froelich et al., "Planning a New Community in an Urban Setting: Lehigh," U.S. Geological Survey Professional Paper 950.

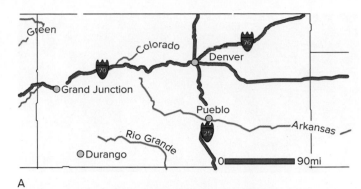

A

B

C

Figure 20.10

Sample GIS representations from USGS National Map Viewer. A roadmap-style map of this region near Denver (A) shows only major rivers, interstate highways, and cities. Adding the topography layer (B) or land cover (C) adds much more information.

Images generated by ArcIMS Viewer, courtesy U.S. Geological Survey.

visually and/or analyzed by computer, which can combine data from multiple layers.

Maps as tools have limitations, too. One is the matter of scale: The information presented must be given in sufficient detail so that features of interest will, in fact, show up in the data. On a map on which 1 centimeter represents 1 kilometer of distance on the surface (1:100,000 scale), all kinds of relatively small features—cliffs, sinkholes, small streams, and so on—might not appear at all; yet these features could be of great concern to a prospective home owner. At the same time, superimposing too much detail can obscure information (**figure 20.11**).

Sometimes, the difficulty is that the information itself is available on only a gross scale. Maybe the only maps of vegetation or surficial geology were made from satellite images, for instance. Sometimes, the data are simply unavailable. Producing many kinds of topographic, geologic, or other earth-science maps means compiling many observations or measurements, which takes time, personnel, and funds. Even in the United States, which is rather thoroughly mapped by world standards, not all types of information have been obtained for all areas. In these cases, land-use decisions

may have to be based on incomplete data, perhaps supplemented by broad, reconnaissance-type surveys of the area under study. Sometimes, too, the problem is not so much lack of data as conflict among competing options. The pressure to develop areas where risks are high or incompletely assessed is particularly acute where development is rapid, especially if population density is already high so land available for new development is scarce (compare **figure 20.12** with **figure 20.7**).

A

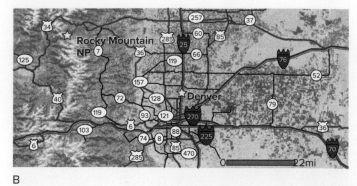

B

Figure 20.11

When topography and land cover are mapped on a finer scale, more information can be displayed; compare (A) with **figure 20.10C**. But adding the highways layer (B) obliterates some detail.

Images generated by ArcIMS Viewer, courtesy U.S. Geological Survey.

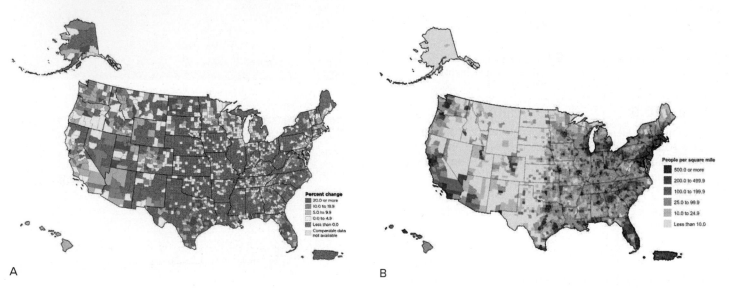

A

B

Figure 20.12

(A) Population change by county, 2010–2020. (B) Population density, 2020. Despite rapid population growth, much of the West remains sparsely settled, in part due to limited water resources (recall chapter 11).

(A-B) Source: Maps from U.S. Census Bureau, 2020 census

Satellite imagery can both assist with and supplement conventional mapping, and supply key data for GIS analysis. As we have seen in the context of resource exploration, satellites can efficiently survey broad areas with specialized sensors to distinguish different types of rocks, vegetation, etc., and resolution is improving all the time. The information on topography and geology that they provide is often directly relevant to land-use planning. Satellite images made during flood events facilitate accurate flood-hazard mapping. Current land-use patterns can be identified, urbanization monitored, and land-use changes tracked (**figure 20.13**). Satellite data are integral to both local and global land-use discussions, as the impact of land-use decisions is recognized in new contexts, notably that of global climate change.

20.5 Engineering Geology

Learning Objectives

- List issues that engineering geologists address
- Describe the particular challenges of building an oil pipeline in the tundra
- Recall the types of models engineers use to understand natural systems

In this section, we will introduce some of the issues engineering geologists need to consider when approaching a project, give a specific example of an especially challenging geologic

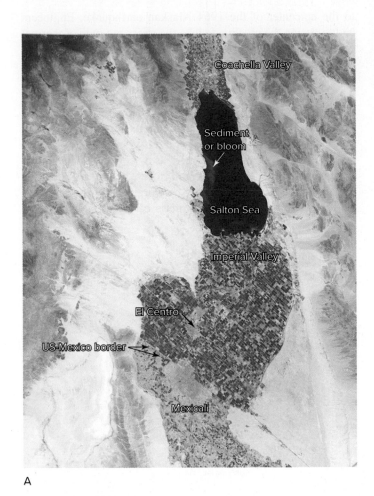

A

B

C

Figure 20.13

(A) The extent of agricultural development supported by water from the Salton Sea is evident; the U.S.–Mexico border stands out sharply, as agricultural land use is much less developed on the drier Mexican side. (B) The vulnerability of the densely urbanized island city of Venice, Italy, to rising sea level is strikingly evident in this photograph. It sits in a shallow lagoon, a narrow barrier island between it and the Adriatic Sea. The Grand Canal snakes across the city. Satellites can monitor incursions of water into St. Mark's Square (small white area near the southeastern end of the Grand Canal) and other landmarks. (C) Land-use change in progress: Deforestation in eastern Bolivia reduces a carbon sink, as agricultural development radiates outward from small communities established by resettlement for the purpose of growing soybeans.

(A) Image courtesy of Earth Sciences and Image Analysis Laboratory, NASA Johnson Space Center; (B, C) Earth Sciences and Image Analysis Laboratory, NASA Johnson Space Center.

engineering situation, and then look at how engineers use models to better understand the natural systems they need to work in.

Engineering Issues to Consider

Engineering geologists address a broad range of issues and problems related to the topics of previous chapters—slope stability, earthquake-resistant design, groundwater and surface-water resources, mine construction and development, coastal erosion, stability of foundations, building of highways and tunnels, and more. Obviously, not every geologic factor or process is equally important to every project, but a major construction project—for example, a housing development—might require

consideration of a very broad range of geologic matters. Some are outlined below.

What rock types are present? Are they strong enough to support the proposed structure? Do they fracture easily? Are there structural features within the rocks—folds, faults, bedding planes—that make the rocks' properties nonuniform and that should be taken into account? How porous and permeable are the rocks? Over the longer term, are they unusually prone to erosion or weathering?

Many of the same questions about porosity, permeability, strength, and stability might be asked about the soils. Are any slopes likely to give way to landslides? Does the construction itself have the potential to trigger slides? How cohesive is the soil?

How compressible and prone to settling? How elastic? Do the soils tend to expand and contract as moisture content varies? Are they subject to hydrocompaction? This problem of expansive clay soils as a construction hazard is not a trivial one (**figure 20.14**). Their failures take few, if any, lives, by contrast with more obvious hazards like floods and earthquakes, but the total structural damage to homes, commercial buildings, pavements, and utilities in the United States each year has been estimated as approximately equal to the total costs of all other geologic hazards combined. Just one example is the new, luxury skyscraper in San Francisco that is sinking and tilting, mentioned in chapter 8; it appears the planned $100 million to fix it may not work afterall.

What about water? What quality and quantity of surface water or groundwater is available? What are the surface runoff patterns? Does part of the site now serve as a recharge area for an aquifer system? If septic tanks are planned, are soil properties and topography appropriate for them?

Then there are the possible catastrophic hazards. Earthquakes and volcanoes are significant hazards in relatively few places, but where they are, they need to be taken into account in building siting and/or design. Landslides and floods potentially affect many more areas. Do sinkholes occur in the area? Has there ever been underground mining in the area?

Long as this list of geologic concerns is, it does not exhaust the possibilities. Consider the case of the Trans-Alaska Pipeline, discussed in chapter 19. At 1300 kilometers long, it spans a variety of geologic settings and potential problems. Much of the terrain it crosses is rugged, including two major mountain ranges, the Brooks Range and the Alaska Range. It also crosses about 800 streams, the largest of which is the Yukon River. All of those streams freeze in winter, and some thaw partially in summer, complicating the engineering (**figure 20.15A**). A number of faults crisscross the area, and the southern end of the pipeline, the city of Valdez, is close to the epicenter of the 1964 Alaskan earthquake. Engineers had to design the pipeline to take the stress not only of thermal expansion and contraction but also of earthquakes and ground displacement (**figure 20.15B**); near the Valdez end and along the Denali Fault, it was built to withstand earthquakes up to Richter magnitude 7.5.

A further construction headache peculiar to very cold climates is permafrost, described in chapter 10. As long as the permafrost stays frozen, it makes a fairly solid base for structures. If it is disturbed and warmed, as by construction, some of the ice melts. The deeper soil is still frozen, however, so depending on the topography, the water may drain slowly or not at all. The result is a mucky, sodden mass of saturated soil that is

A

B

Figure 20.15

(A) The Trans-Alaska Pipeline crosses the Tanana River via suspension bridge to avoid the problem of freezing/thawing and possible ice jams on the river. (B) Pipeline design takes into account thermal expansion and contraction as temperatures rise and fall, and possible displacement resulting from earthquakes; recall **figure 4.11**. Kinks in the pipeline, and loose coupling to supports (see also **figure 20.1**), help accomplish this. So the pipeline zigzags across the land, as shown in **figure CS 19.1.2** of Case Study 19.1; here (B), it also dives under the shallow river in the foreground.

(A, B) ©Carla Montgomery

Figure 20.14

Failure of masonry structure on unstable soil.

Source: Photograph courtesy of USDA Soil Conservation Service.

difficult to work in or with and that is structurally weak (**figure 20.16**). The work crews on the Trans-Alaska Pipeline were faced with trying to install, under very unstable conditions, structural members that would be solidly supported after construction was done and the melted permafrost had refrozen. Even the oil in the pipeline posed a problem. The oil is warm (about 60°C, or 140°F) as it comes up from deep in the earth, and its temperature stays well above freezing as it flows through the pipeline. The warm oil could thaw a zone within the permafrost immediately around the pipeline, and the sagging of the pipeline in the resultant mud could lead to pipeline rupture.

Ultimately, buried sections of the Trans-Alaska Pipeline were refrigerated in places to keep the permafrost frozen. Elsewhere, cooling systems were installed to cool the supports of aboveground sections of pipeline, to keep them firmly in place (**figure 20.17**).

Of course, the pipeline was engineered for the conditions at the time it was built. We have already noted (chapter 10) that warming seems to be particularly pronounced in the Arctic, and this includes much of Alaska. One consequence is extensive melting of permafrost, including that in which pipeline supports had been anchored. A recent study has suggested that up to one-third of these supports may be becoming unstable as a result, jeopardizing the integrity of the pipeline. Replacing

A

B

Figure 20.16

Permafrost affects construction work and transportation systems. (A) Melted permafrost produces a sodden mass of waterlogged soil like that surrounding this tractor. (B) Differential subsidence of railroad tracks due to partial thawing of permafrost, Copper River region, Alaska. This track was abandoned in 1938.

(A) USGS Photo Library, Denver, CO/Photograph by T. L. Péwé; (B) USGS Photo Library, Denver, CO/Photograph by L. A. Yehle, both courtesy of O. J. Ferrians.

Figure 20.17

Close-up of pipeline support system shows how pipeline can shift from side to side between support posts, protected by bumpers on the pipeline. Thermosyphons on top of posts are part of a system designed to cool the underground sections of the posts. Coolant fluid circulates vertically, and the fins expose maximum surface area to the cool air to counteract the warming effects of oil or sunlight heating the ground.

©Carla Montgomery

these supports, some of which extend as much as 70 feet below the surface, could cost an estimated $100,000 each.

Even where some permafrost remains at depth, warmer temperatures mean later freezing at the surface in the fall, and earlier melting in the spring. But the heavy equipment used in oil exploration and drilling cannot navigate the soft muck of the melted permafrost; that equipment can only be used in permafrost areas while the ground is solidly frozen. In areas of the Alaskan tundra, the number of days that this equipment can be deployed has shrunk dramatically over the past three decades, from 200 or more per year to less than 100. To the extent that the dwindling of the permafrost is related to warming caused by CO_2 from the burning of fossil fuels, some see real irony here.

Testing and Scale Modeling

Relatively modern additions to the arsenal of tools available to the geological engineer are the use of scale models of natural systems, and testing—actual or theoretical—of the behavior of natural and construction materials.

Many scale-model experiments take place in the laboratory. Nowadays, much testing is done using sophisticated computer models rather than by building physical models or working with actual samples of materials. However, it is often still useful to conduct physical experiments and observe the results. We have already noted this in the context of earthquake-resistant design, but scale models have been applied to other geologic hazards as well.

The photographs in **figure 20.18** show a 1:500-scale model of the Swift Dam and Reservoir on the south flank of Mount St. Helens. The model was built by Western Canada Hydraulic Laboratories for the dam's owners, Pacific Power and Light Company of Portland, Oregon.

High-density muds like those used in drilling oil wells simulated volcanic mudflows entering the reservoir from the creeks that feed it. **Figures 20.18B** and **20.18C** show the progress of one such flow, from the model Swift Creek, into the reservoir. With each test, maximum wave height at the model dam and overall reservoir response to mudflow input were closely observed. At the request of the state of Washington, maximum mudflow volumes corresponding to about half the reservoir volume were tested. Under these conditions, waves from single flows, and the final reservoir height, nearly reached the dam crest.

Experiments such as these can play a key role in engineering safer structures in high-risk areas.

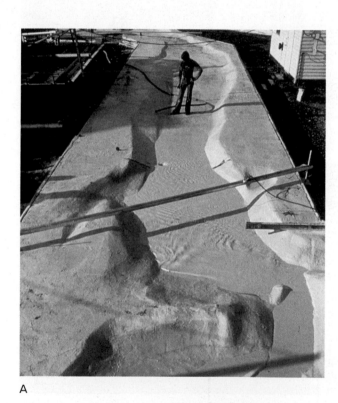

A

B

C

Figure 20.18

Completed model reservoir being filled with water to normal depth prior to test runs (A). Mudflow scaled to correspond to 44,000 acre-feet is released from model Swift Creek into reservoir (B); as flow progresses into reservoir (C), water height increases and waves lap at the dam.

(A, B, C) USGS Photo Library, Denver, CO/Photographs by J. E. Peterson

20.6 Case Histories, Old and New

Learning Objectives

- Describe why the Leaning Tower of Pisa leans, and how it was stabilized
- Summarize the construction history of the Panama Canal
- Describe Boston's Big Dig

This section briefly outlines several case histories, in partial illustration of the range of problems encountered in engineering geology and some approaches to their solution.

The Leaning Tower of Pisa

The Leaning Tower of Pisa is not Pisa's only bell tower, nor is it the only one that leans. At least two other bell towers built during the same period have also tilted, by as much as 5 degrees from the vertical. The underlying reason is the same in all cases: soft, unstable ground. But it is the elegant, freestanding bell tower of the Pisa Cathedral (**figure 20.19**) that is by far the greatest tourist attraction among these towers, so it has drawn the most vigorous efforts to stabilize it.

Pisa is underlain by several layers of sediment. The upper 10 meters consist mainly of silt, with about 30 meters of soft marine clay below that. These two layers are water-rich and compressible under load, and can also flow under stress. Below them are much more rigid sand layers.

The Leaning Tower was built in several phases between 1173 and 1370. It began to tilt even before it was completed. The weight of the tower forced water out of the silt and clay layers, compacting them, and they may also have flowed somewhat. Once the tilt was well established, it seemed to be self-reinforcing— that is, as the structure began to lean, more and more pressure was concentrated on the lower side, causing more flow and compaction of the unstable clay and silt layers and more tilt. The 58-meter-high tower has also sunk about 2½ meters into the soft ground. By 1911, the top had tilted more than 5 meters from vertical, and careful measurements thereafter showed that the tilt was continuing to increase. When a similarly built bell tower at the Cathedral of Pavia collapsed in 1989, the Leaning Tower of Pisa was closed to tourists, and what would become an eleven-year stabilization effort was begun.

Initial suggestions included physical methods, such as boring and selectively removing some material from the north (high) side of the foundation to reduce the tilt, and chemical methods, such as treating the clays to stiffen them and prevent further sinking of the settled south side. In 1993, 900 tons of lead were placed on the high side of the Tower's base to anchor and counterweight it, halting further tilting. In 1999, a careful three-year program of selective soil extraction from under the north side was implemented to reduce the tilt a bit and stabilize the tower. By 2002, the Leaning Tower, though still leaning, was pronounced stabilized, and was opened to visitors again.

Figure 20.19

The Leaning Tower of the Pisa Cathedral has been leaning to some extent since the twelfth century. The leaning began during construction, and builders' attempts to compensate resulted in some curvature in the tower itself.

Javier Larrea/AGEfotostock.

The Panama Canal

A French company began excavation of a canal across the Isthmus of Panama in 1880. The initial design, patterned after the Suez Canal, was for a sea-level cut from Atlantic to Pacific, with no locks—this being the preference not of the engineers, but of those financing and directing the project. The Suez Canal was excavated mainly through sand and soil, not rock, and in a climate with much less rainfall to promote sliding. Also, because of the higher elevation of the Isthmus of Panama, a much larger volume of material would have had to be excavated for a sea-level canal across Panama.

Little or no investigation of the geology of the Panama Canal site was made before excavation began. Such investigation would have revealed layers of young volcanic rocks, lava flows, and pyroclastic deposits, interbedded with some shale and sandstone. Because the rocks dip in many places toward the canal, excavation removed the support from some of these rock layers, which then tended to slide (**figure 20.20A**). Sliding was facilitated by very high rainfall, which averages 215 centimeters (85 inches) a year. Elsewhere, as the canal was dug, the weight of the rocks on the side caused flow and buckling of the rocks at the bottom of the excavation, which might rise 10 meters or more, requiring redigging to open the canal and removal of material from the sides to relieve some of the pressure, as well as repair of railway lines (**figures 20.20B** and **C**).

Added to the geologic obstacles were heat, humidity, and rampant diseases, such as the mosquito-spread yellow fever, which caused many deaths among the workers. After seven frustrating years, the French company was bankrupt. A five-year lull in excavation was followed by renewed efforts by a second French firm, but it later abandoned the attempt.

The United States acquired the rights to take over the project in 1903. At that point, the whole project design was reconsidered, and instead of a sea-level canal, a more complex design was chosen. A river was dammed to create a lake in mid-isthmus, and locks were to be built on either side to carry ships, in several stages, from sea level up to lake level and back down to sea level on the other side. One technical advantage of the lock design was that it would reduce the total volume of excavation required. An ecological benefit was that the lock design would better separate the Atlantic and Pacific oceans; a concern with a sea-level canal and associated free flow of water

and organisms between oceans was the possible impact on various ecosystems that had previously been separated by land.

Even so, finishing the canal wasn't easy (**figure 20.21**). A geologist was not invited to examine the excavation and associated landslides until 1910, by which time many large slides were already beyond control. Sliding and excavation, in fact, continued long past the time of the canal's nominal completion and substantially increased its costs. Early consideration of the geology of the canal zone might not have eliminated the sliding problems, especially considering the lesser understanding of rock mechanics at the time of construction. However, some of the instability problems could have been anticipated and reduced, and the costs could certainly have been projected more accurately. As it was, the canal was not completed until 1914, more than thirty years after it was begun.

At that, it has continued to evolve. Parts were widened in the mid-1900s, and a new phase of widening was begun in 1992. A new set of locks was completed in 2016. The geology remains a challenge, as seen in **figure 20.21B**; better understanding of it, coupled with better equipment, makes modern excavation more manageable.

Boston's Big Dig

Like many cities, Boston experienced increasing traffic congestion over the last half of the twentieth century. The traffic load on the six-lane, elevated Central Artery nearly tripled from 1959 to the end of the century, creating estimated costs of half a billion dollars a year in wasted fuel in idling gridlocked cars, late delivery charges, and elevated accident rates. In a bold move, the city embarked on a massive transportation overhaul. The two

Bedding planes are possible zones of weakness, sliding.

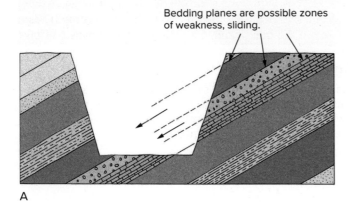

A

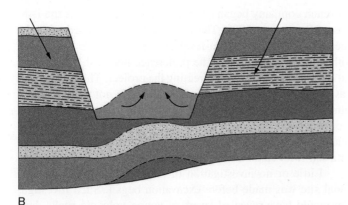

B

C

Figure 20.20

Geologic factors complicated construction of the Panama Canal. (A) Dipping beds sliding into a canal reduce the canal's capacity, requiring more excavation. (B) Removing the weight of the rocks in the canal may allow buckling of rocks below, which are under pressure from the rocks on the canal's sides. (C) Buckling of the floor of the canal was severe enough to disrupt railroad tracks essential to the project.

(C) Library of Congress Prints & Photographs Division [LC-DIG-det-4a24696]

Figure 20.21

(A) Excavation of the Panama Canal, 1913. Note standing water under/around the train tracks. (B) Landslides remain a threat today.

(A) Library of Congress Prints & Photographs Division [LC-DIG-hec-03152]; (B) USGS Open-File Report OFR 01-0276

major components of this Central Artery/ Tunnel Project (quickly christened the Big Dig) were an eight -to-ten-lane underground expressway to replace the Central Artery, and an extension of the Massachusetts Turnpike across Boston Harbor to Logan Airport via a four-lane tunnel under the harbor. The Central Artery, slated for demolition once the new expressway opened, had to be kept open throughout construction, and various underground structures pass beneath trains and subways that had to remain operational. Logistics aside, the local geology was a challenge. The city is underlain by a complex assortment of glacial sediments, volcanic rocks, and weakly metamorphosed mudstone; the depth to reach strong bedrock is over 120 feet in places. It is also densely built, with many historic buildings whose foundations could be destabilized by careless construction. And the water table is shallow; but pumping water out from active excavations could destabilize adjacent rock and soil.

To keep the excavation walls stable while digging down to bedrock, the project relied on extensive use of *slurry walls.*

As trenches were dug, a clay-water slurry was poured in to replace the earth removed. Slurry is dense and stable enough to hold up excavation walls until steel reinforcing beams can be sunk into place and concrete poured in—displacing the slurry—to form the permanent wall. In other places, the ground was purposely frozen to keep it firm during construction.

A different approach was taken for the tunnel under Boston Harbor. Tunnel segments were completed on land, towed to their intended positions, sunk into place, joined, and then pumped dry.

The Big Dig wasn't cheap. The size and complexity of the project resulted in a cost of over $14 billion. But the objectives were achieved, with minimal disruption of the city's operation along the way, and the project, begun in 1991, was essentially complete in the spring of 2006, with all roads and tunnels open.

20.7 Dams, Failures, and Consequences

Learning Objectives

- Explain the St. Francis Dam failure
- Describe the Three Gorges Dam project

When a building cracks and crumbles due to settling or subsidence, the repairs may be costly, but the toll rarely includes lives. Even when a bridge or tunnel collapses, only those few persons on or in it at the time are directly affected, although indirect consequences can last for months to years. A catastrophic dam failure, on the other hand, can destroy whole towns and take thousands of lives in a matter of minutes (**figure 20.22**). The motivation for intelligent and careful design and siting of

Figure 20.22

Hoover Dam can impound up to 28.9 million acre-feet of water in Lake Mead; it's a lot less than that now due to drought. Lake Mead stretches into the distance behind the dam.

PBH/Alamy Stock Photo.

dams is thus particularly strong. Unfortunately, past practice has frequently fallen short of the ideal. In chapter 8, we examined the Vaiont Reservoir disaster. In this section, we will look briefly at another well-documented dam disaster. Neither need have occurred; careful geologic investigation beforehand would have shown that neither site was suitable.

The St. Francis Dam

The St. Francis Dam was built in California, about 70 kilometers north of Los Angeles. The reservoir it impounded was principally intended for a water supply. The dam, 60 meters high and over 150 meters wide, was completed in 1926. The walls on one side of the valley are coarse sandstones and other sedimentary rocks. On the other side, the rocks are schists, mica-rich metamorphic rocks that tend to break along parallel planes and slope toward the valley floor. The contact between the two rock types, over which the dam was built, is a fault. Geologists had identified and mapped the fault before the dam was built. The construction itself omitted several features that would have minimized water seepage under the dam. Cracks in the structure were noted as the reservoir began to fill.

On 12 March 1928, the dam abruptly failed. The contents of the reservoir swept down San Francisquito Canyon. Huge chunks of the concrete dam were carried up to a mile downstream. Damages were estimated at $10 million; the official death toll was 430.

Possible reasons for the failure became clear from subsequent laboratory tests; the fault zone didn't appear to have been a significant factor. The rocks of the valley walls had seemed to be sufficiently strong when dry. But the sedimentary rocks on one side of the dam are rich in gypsum, which is very soluble. When a fist-sized sample of one of the sedimentary rocks was placed in water, it bubbled out air, soaked up water, and disintegrated, in less than an hour, into a heap of sand and clay. Reinvestigation of the site later also revealed an old landslide complex, a mile and a half long, within the schists. Recent studies and modeling now suggest that the filling of the reservoir reactivated the slide which put pressure on the dam. Hydrostatic pressure from water seepage underneath it may have tended to buoy up the dam, stressing it further. Apparently, the west side of the dam, destabilized and damaged by softening of the gypsum-rich rocks, failed first, and a rapid, muddy washout began. This reduced the support for the rest of the system. As the water began to pour out, it further eroded the base of the dam. The east side of the dam then collapsed. Somewhat astonishingly, the central section of the dam remained standing (**figure 20.23**).

As with the Vjont (Vaiont) Dam case, a little more careful investigation in advance could have averted disaster. In the case of the St. Francis Dam, it seems that testing the reaction of the wallrocks to immersion in water and looking hard enough at the slopes to recognize the old landslide complex should have told the engineers everything they needed to know about site suitability.

Figure 20.23

Failure of the St. Francis Dam in California: the remains of the dam after its collapse. Note the waterline on the surrounding hills, marking the water level in the reservoir before failure.

USGS Photo Library, Denver, CO/Photograph by H. T. Stearns

Three Gorges Dam

As long ago as 1919, officials proposed damming the Yangtze River as a means to control flooding and enhance navigation; an estimated half-million people died in floods on the Yangtze in the twentieth century alone. However, not until 1993 did construction begin on a dam in the Three Gorges area, at a site that offered solid granite bedrock and soaring cliffs along the river to contain a huge reservoir.

The Three Gorges Dam (**figure 20.24**) is 185 meters tall, and nearly 2.3 kilometers (about a mile and a half) wide. With thirty-four hydropower generators, the complex has a collective generating capacity of about 22,500 megawatts, making it the world's largest hydropower project. The complex reduces both China's burning of coal and damage and loss of life from flooding. Impoundment of water upriver makes the Yangtze navigable by much larger ships farther upstream than before, and a system of locks to get the ships past the dam was part of the project.

As with all large projects of this kind, however, there were negative consequences, and there was vocal opposition to the Three Gorges Dam long before construction began. The reservoir behind the dam stretches over 600 kilometers (370 miles) long. It submerged approximately 125,000 acres of prime farmland, covering 1000 square kilometers (440 square miles) overall. Cities along the banks in the area the reservoir occupies had to be abandoned and their residents moved elsewhere; an estimated 1.3 to 1.9 million people were ultimately displaced. Cultural and archaeological features and sites that could not be moved—1300 excavated sites, and an estimated 8000 unexcavated sites—were lost. The scenic qualities of the Three Gorges area have been significantly and permanently

A

B

Figure 20.24

(A) Three Gorges Dam site. The dam was completed in May 2006. Most of the sediment carried by the Yangtze is now trapped far upstream where the river enters the still-growing reservoir. (B) Reinforcing slide-prone slopes along the Yangtze is one strategy for reducing ongoing landslide risks.

(A) NASA Image Collection/Alamy Stock Photo; (B) U.S. Geological Survey/Photograph by Lynn Highland

altered. Ecologists fear the impact of habitat loss, especially on rare or endangered species that live in the Yangtze—for example, the Chinese river dolphin and Chinese paddlefish, which may already have vanished; the Yangtze soft-shell turtle, Chinese sturgeon, Siberian crane, and others.

Basic safety concerns are natural with a project of this scale, beyond the structural integrity of the construction. Reservoir-induced seismicity has already been observed, whether due to reactivation of old faults or collapse of karst caves in limestone wallrocks of the reservoir. Thousands of small quakes, with magnitudes up to 4.1, have been noted since the reservoir began to fill in 2003. A larger concern is the possible effect on two major fault zones in the region, which historically have experienced quakes over magnitude 6, and the dam's ability to withstand such quakes: It is built to withstand earthquakes as large as magnitude 7, but China has certainly experienced larger quakes than that.

Landslides are a major issue. Historic slides in weak, dipping sedimentary rocks are common along the Yangtze. Even the relocation of villages has triggered some new sliding. Within a month of the start of reservoir filling, old landslides at Shuping and Qianjiangping were reactivated. Many known old slides are being monitored, which enabled scientists to warn residents as slip of the Qianjiangping slide began, but even so, fourteen lives were lost, with ten people missing, and 4 factories and over 100 homes were destroyed by the huge, 24-million-cubic-meter slide. In 2007, reactivation of the Xemaomian slide forced evacuation of a village. Monitoring and bank-strengthening efforts (**figure 20.24B**) continue. The filling of the reservoir is not the only potential trigger, either. For flood-control purposes, the level of the reservoir, filled to 175 meters above sea level during the winter, is lowered to 145 meters during the rainy spring/summer season. Geoscientists are concerned that such an annual wetting/drying cycle will further destabilize slopes in the affected zone; studies indicate that this is already occurring to some extent.

The annual fluctuation in reservoir level may have further consequences. The land exposed when the level is lowered is attractive as farmland. However, when the level is raised and the land again submerged in the winter, fertilizers, herbicides, and insecticides will be dissolved and washed into the reservoir, polluting the water.

Preliminary studies indicate additional early environmental impacts of the Three Gorges Dam project. The Yangtze River Basin is the largest in South Asia, and thousands of dams within it had already considerably reduced the sediment load reaching the East China Sea; the Baihetan dam (fourth largest in the world), completed in 2021, is located on a tributary of the Yangtze and is holding back even more sediment. Within two years after the Three Gorges Dam first began impounding water (2003), that sediment load was cut again by almost half. That which reaches the delta is now not much over 100 million tons per year, compared to nearly 500 million tons per year in the mid-twentieth century. Observed results include increased channel erosion and erosion of the Yangtze delta. Future coastal erosion may threaten such important areas as Shanghai. The large reservoir behind the dam, with its broad surface subject to evaporation, is even affecting local weather—the evaporation process cooling the air and altering circulation/wind patterns, with demonstrable changes in the distribution of rainfall not just locally but regionally. Monitoring the longer-term impacts should prove interesting.

Other Examples and Construction Issues

Dams, more dramatically than most other structures, underscore the need for careful application of the principles of engineering geology before and during construction. They may also underscore the limits of our understanding. The Baldwin Hills reservoir was built near Los Angeles, California, between 1947 and 1951. The geologic setting was thoroughly studied and (so it was believed) adequately taken into account. The reservoir area is underlain by an active fault zone; the design incorporated a seismic safety factor double that used by engineers in other earthquake-prone areas. Drainage was provided to prevent saturation of the foundations. The design was conservative, and provision was made for monitoring after construction. Nevertheless, on 14 December 1963, a differential slip of more than 10 centimeters occurred along a fault, water began to scour its way out, and 2 hours later, the dam was abruptly breached. Clearly, the designers did not understand the geology as well as they thought.

Geology and geography may, in fact, lead to the building of dams on faults. The zone of weakness created by a major fault zone may become a topographic low along which streams flow and lakes accumulate. This is a natural site, then, for a dam and reservoir, but damming such a stream necessarily involves placing a dam on or very close to the fault. The San Andreas is only one of many faults marked, in part, by the presence of natural sag ponds and artificial reservoirs (**figure 20.25**).

The majority of the approximately 50,000 dams built in the United States over the last hundred years were not built on faults, but they may nevertheless suffer from inadequate design. All dams are designed to survive any imaginable flood load, typically through the provision of a *spillway* that allows water to flow over or around the dam unimpeded when the water level in the reservoir becomes too high (**figure 20.26**). The worst-case scenario for which a spillway is constructed is commonly a 1000-year flood. Still, as noted in chapter 6, it may be difficult to project the volume of water any large, infrequent flood involves, and changing land-use patterns and climate change may be altering the flood-frequency curves. Dam failures have occurred, too, from dam washout as the sides or face of the spillway were scoured away by floodwaters. The U.S. Army Corps of Engineers National Inventory of Dams currently identifies nearly 16,500 dams as ones whose failure would case serious property damage or loss of life. Of that number, a recent review indicated that about 2200 of those are unsafe to human life because of disrepair. Many other are in the highly hazardous category because of inadequate spillways. The Oroville Dam has two spillways, of different designs, and in 2017 experienced problems with both at once (Case Study 20.2).

Dam construction has become increasingly complex and sophisticated, aided by computers, laboratory tests, field studies, and scale-model experiments; closer monitoring also increases safety. Still, failures continue. A recent one occurred at the Taum Sauk hydroelectric project in Missouri. This project, completed in 1963, is a pumped storage system, with an upper and lower reservoir. During peak-demand times, turbines generate electricity as water flows from the upper to the lower reservoir. At night or when demand is light, water is pumped back up from the lower to the upper reservoir. On the night of 14 December 2005, the dam on the upper reservoir abruptly failed (**figure 20.27**). Investigation suggested that faulty instrumentation/monitoring of the water level in the upper reservoir caused it to be overfilled, overtopping and then washing out the dam (which had no spillway). Approximately a billion gallons of water swept downslope. Fortunately, the lower reservoir is designed to hold most of the volume of the upper; there was limited overflow from the lower reservoir, its dam held, and there were no fatalities. The upper reservoir has since been rebuilt, its design improved, and the project is back in operation.

Global concern now is focusing increasingly on the rapid proliferation of dams in some developing countries. The World Commission on Dams monitors the progress of large dam projects worldwide. For this purpose, large dams are defined as

Figure 20.25

Sag ponds and reservoirs abound along the San Andreas Fault; here, San Andreas and Crystal Springs reservoir.

Photograph by R.E. Wallace, USGS Photo Library, Denver, CO.

Figure 20.26

Awoonga High Dam in Australia illustrates modern construction methods. View of dam and reservoir; spillway in foreground.

©Carla Montgomery

A

B

Figure 20.27

(A) Cross-sectional view of failed earthen dam at Taum Sauk upper reservoir. Earthen dams without spillways are especially prone to washout. (B) Rushing floodwater from the upper reservoir scoured down to bedrock, carrying away forest and soil.

(A, B) Federal Energy Regulatory Commission

those over 15 meters high and those 5 to 15 meters high with reservoirs over 3 million cubic meters. Altogether, there are more than 57,000 such dams. Nearly 40% are in China, and most of China's large dams have been built since 1950, which means an average of about six dams per week built since that time. The reasons historically have been primarily flood control and irrigation, with increased hydropower-generating capacity a growing motive. In addition to such issues as displacement of people and flooding of habitat and cultural resources (**figure 20.28**), the World Commission noted serious safety concerns that have emerged as so many large dams have been built so quickly.

Figure 20.28

Benefits and issues involved in dam construction are many. Safety is an additional, very basic one, which can arise even with new dams, especially if design and construction are hurried.

Source: World Commission on Dams report, 2000.

How Green Is My Golf Course?

Historically, golf courses (of which the United States has more than 15,000) have often been roundly criticized as environmentally unfriendly land uses, except perhaps in floodplains (**figure CS 20.1.1**). By nature, they sprawl over a good deal of real estate; a typical new golf course occupies about 150 acres. Natural habitat has often been cleared to make way for vast expanses of tidy close-mown grass, kept lush by lavish applications of water and fertilizer and protected by herbicides and insecticides. The water use has been especially criticized in deserts and other dry settings where a golf course can consume a million gallons of water in a day; chemical runoff could pose pollution problems. However, things are changing for the better as a result of increased environmental sensitivity, reinforced by laws requiring environmental impact assessment of new construction.

Intensive research is producing turf grasses that require minimal use of pesticides and other chemical additives and need less water. Course designers are reducing the amount of area planted to short turf, leaving cactus-studded waste areas in desert courses in places the golfers shouldn't be hitting anyway, or letting much of the deep rough become very deep indeed (**figure CS 20.1.2**), perhaps even engaging in prairie restoration. Preservation of wetland or wooded habitat around the course can actually provide sanctuary for wildlife in an area otherwise being overrun by urban sprawl, while enhancing the course visually.

Building courses on landfills is an increasingly common practice (**figure CS 20.1.3**). The first such proposal was, in fact, put forth in 1930 in New York; now there are several dozen examples nationwide, a third of them in California, and more under development. The newest twist on this sequential-use concept is a true ugly-duckling transformation, beginning with a Superfund site near Butte, Montana, site of accumulated toxic wastes from an old copper-smelting operation in the town of Anaconda. The hazardous material has been carefully covered up, sealed with clay and plastic, and—with the help of Jack Nicklaus as designer, more than $40 million from ARCO (the site's last owner when the smelting ceased, faced with even higher potential costs with conventional site cleanup), and the collaboration and blessing of the EPA—the site has been transformed into the Old Works, an attractive, wildflower-decked golf course (**figure CS 20.1.4**). Once completed, it was deeded to the local residents, so the operation not only eliminated a hazard but also now offers a new source of revenue to the region, which suffered economically from the cessation of mining and smelting activities.

Figure CS 20.1.1

This course in DeKalb, Illinois, is unusable when the river floods, but there is minimal damage cost.

©Carla Montgomery

Figure CS 20.1.2

It adds to the challenge to have long expanses of unmanicured land on each hole, but is generally kinder to the environment. (Recall also **figure 20.4A.**)

©Carla Montgomery

Figure CS 20.1.3

This golf course west of Chicago, Illinois, is built on a finished section of landfill; ongoing landfill operation continues on adjacent land (in distance and at right).

©Carla Montgomery

Figure CS 20.1.4

The Old Works golf course in Anaconda, Montana, was a Superfund project founded on a sprawling abandoned smelting operation. To preserve some of the historic flavor of the site, designer Jack Nicklaus kept rows of brick smelters along one side of the course, and filled the bunkers with crushed black slag rather than sand.

©Carla Montgomery

The Oroville Dam

The Oroville Dam, on the Feather River in the foothills of the Sierra Nevada in northern California, was built in the 1960s to be a cornerstone of the California State Water Project. Like many dams, it has several functions, including hydropower generation, flood control, and water storage for water supply. The highest dam in the United States at 770 feet (235 meters), this earthfill dam impounds the second-largest reservoir in California, Lake Oroville, with a capacity of 3.5 million acre-feet. It was designed with two spillways beside the dam. Outflow from the concrete-lined primary spillway is controlled by gates; the emergency spillway, which comes into play only when the reservoir is at capacity and inflow exceeds what can be released via the primary spillway, had a concrete lip with a slope of exposed rock and soil below.

During construction, questions were raised about the solidity of the rocky slope beneath the primary spillway, but the designer was assured that it was adequate. Cracks appeared in it in 1968 soon after construction was completed, and water was seen seeping beneath it as drainage pipes underneath it cracked also. This was accepted as normal. A question of possible reservoir-induced seismicity arose in 1975 when a magnitude-5.7 earthquake occurred in the area, but this has not been a significant ongoing issue. When the Federal Energy Regulatory Commission reauthorized dam operation in 2005, it was suggested by some groups that concrete facing be put on the hillside below the emergency spillway to protect it from erosion during use, but this was not deemed necessary.

As we have noted, managers of dam/reservoir systems such as this face a challenge from competing priorities. In particular, maximizing water storage for future supply reduces capacity available for flood control if needed. Northern California experienced severe drought in 2012–2015; during the winter of 2015–2016, the reservoir was filled to less than 50% of its capacity. It was not surprising, then, that by 2016, the California Department of Water Resources was inclined to favor increasing storage when water was available, in anticipation of another possible summer drought the next year. Unfortunately, the winter of 2016–2017 was unusually wet, including some warm rains that melted snowpack in the mountains, further increasing runoff to the Feather River.

In early February 2017, outflow from the primary spillway was held at 30,000 cu ft/sec (cfs), approximately balancing inflow to the reservoir. Reservoir level was being kept at 850 feet, fifty feet below the level at which the emergency spillway might be needed. With more heavy rain predicted, releases were increased to 54,000 cfs, when operators noted something odd along the primary spillway. They halted outflow on 7 February to inspect it, and found a substantial hole in the concrete, with erosion below and alongside it (**figure CS 20.2.1**). But the rains kept coming—nearly 13 inches altogether from 6 to 10 February. As the water level in Lake Oroville rose, releases from the primary spillway were resumed, at about 55,000 cfs. Runoff into the lake soared with the heavy rains, to a peak of 190,000 cfs on 9 February, much higher than forecast. Thus, the lake level rose with the higher inflow, until on 11 February it reached 901 feet, still rising. At that point, water began to flow over the emergency spillway, for the first time since the dam was built.

Some washout of the hillside below the concrete lip was expected, but erosion was much more serious than anticipated (**figure CS 20.2.2**). Headward erosion of the gully toward the lip

Figure CS 20.2.1

Damage to the primary spillway as of 9 February 2017.

Kelly M. Grow, California DWR.

Figure CS 20.2.2

Water rushing downslope from the emergency spillway on 12 February severely eroded the slope and even took out a road.

National Oceanic and Atmospheric Administration

was so rapid that it was feared that the concrete at the top of the emergency spillway might fail, at which point a major washout could result in uncontrolled drainage of a great deal of water from the lake, flooding the valley below and endangering many lives. On 12 February, a mandatory evacuation order was issued to over 180,000 people living below the dam. The lake level rose to 902.6 feet. So, despite the obvious damage to the primary spillway, release through it was increased to 100,000 cfs to bring the lake level down as rapidly as possible to take pressure off the emergency spillway. Fortunately, by then the rains had stopped, so the lake level dropped fairly quickly to the point that water was no longer pouring down via the emergency spillway. On 14 February, the evacuation order was modified to an evacuation warning. The high releases via the primary spillway continued, to bring the lake level down further (**figure CS 20.2.3**). By 20 February, it was back down to 850 feet, and it was possible to halt releases to examine the damage and undertake emergency repairs to both spillways.

Through all of this, the dam itself remained perfectly sound (**figure CS 20.2.4**). However, the damage to the primary spillway was extensive (**figure CS 20.2.5**). Repairing it involved digging deeper to stronger, more coherent rock to serve as a foundation, and concrete panels were placed near the top of the

emergency spillway to protect the upper slope from erosion if and when that spillway is next needed. Total costs of the repairs have exceeded $500 million.

Figure CS 20.2.4

Astronaut photograph of the Oroville Dam on 22 February—intact dam at right, primary spillway in the center, and emergency spillway at left.

ISS Crew Earth Observations Facility and the Earth Science and RemoteSensing Unit, Johnson Space Center, courtesy NASA.

Figure CS 20.2.3

By 15 February, the lake level had come down, but water was still being released from the primary spillway at 100,000 cfs. Note the gully carved below the emergency spillway to the left of the primary spillway.

Dale Kolke, California DWR

Figure CS 20.2.5

Damage to and around the primary spillway when outflow was cut off on 27 February.

Historic Collection/Alamy Stock Photo

Summary

Engineering geology is concerned with making structures as safe and stable as possible, given various kinds of geologic hazards and potential problems. Land-use planning encompasses the same concerns as engineering geology, combined with additional geologic and nongeologic considerations. The aim of the land-use planner is to make the best possible use of limited land, taking all these factors into account. Using the same land for several purposes, either simultaneously or sequentially, is one approach to conserving the land resource. The success of both geological-engineering and land-use-planning efforts is heavily dependent on the accuracy and completeness of the data available to the individuals carrying out these tasks. As with pollution control, relative costs and benefits may have to be weighed in deciding whether, and how, to undertake a particular construction project.

Key Terms and Concepts

geographic information system

multiple use

sequential use

Test Your Learning

1. Describe the concepts of *multiple use* and *sequential use,* and give two examples of each.

2. Explain how preservation of federal land can be contentious.

3. Maps of geologic or other factors are useful tools in land-use planning, but their usefulness can be limited by practical problems. Describe two such possible limitations.

4. Briefly explain how GIS can assist in the land-use planning process, and how it can be adjusted for different planning objectives.

5. Cite at least ten geologic considerations that might be important in siting an apartment building, and indicate how many of the same considerations would be relevant to siting its parking lot.

6. Define *permafrost,* and explain how it makes construction more difficult.

7. Describe how engineers may be able to minimize problems posed by unstable clays in near-surface rock and soil layers.

8. Dams are not infrequently built over faults; explain why this is so.

9. Identify and explain any five concerns or negative consequences associated with large dam projects, such as Three Gorges.

10. Describe the function of a dam's spillway and what can happen if the face of the slope below is too readily eroded.

Exploring Further

1. Most city or county planning offices have long-range plans for development in undeveloped areas. Seek out such plans for your area, if available, and investigate what kinds of considerations (geologic or otherwise) went into those plans.

2. Make a walking tour of your neighborhood or of another area to look for signs of careless or thoughtless engineering practice. Is any particular problem especially common?

3. Go to NASA's Earth Observatory "Image of the Day" site, **earthobservatory.nasa.gov/IOTD/**. What image is featured, and how was it acquired? Were any special remote-sensing techniques involved? Explore other images from the archive to see how satellites with specialized detectors help scientists investigate and monitor Earth.

4. Check the history/progress/current status of a major project such as the Big Dig, Three Gorges Dam, Keystone XL pipeline, Oroville Dam, or the Qinghai-Tibet Railway. Or explore in more detail a particular aspect of Three Gorges impacts, such as the history of landslides or seismicity since reservoir filling began in 2003.

5. The National Map is a program of the U.S. Geological Survey that allows users to create and examine all types of map products; found at **nationalmap.gov/**.

6. Revisit a GIS tool introduced in an earlier chapter to see what geological information it contains that would be useful for a geological engineer or land-use planner.

Introduction

Much of our understanding of geologic processes, including the rates at which they occur and the kinds of impacts they may have on human activities, is made possible through the development of various means of telling time in geologic systems. In this appendix, we will explore several of these methods. A final section examines some applications of geologic age determinations to the study of process rates.

Relative Dating

Arranging Events in Order

Before scientists understood how to establish numerical ages of rocks or geologic events, people were working to at least place a sequence of events in the proper order. Nicolaus Steno was among the first to publish his efforts in this direction. In 1669, he set forth two very basic principles that could be applied to sedimentary rocks. The first, the **Principle of Superposition** (**figure A.1**), points out that, in an undisturbed sequence of sediments or sedimentary rocks, those on the bottom were deposited first, followed in succession by the layers above them, ending with the youngest on top. Today, this idea may seem so obvious as not to be worth stating, but at the time it represented a real step forward in thinking logically about rocks. The second, the **Principle of Original Horizontality,** is based on the observation that sediments are deposited in approximately horizontal, flat-lying layers. If one comes upon sedimentary rocks in which the layers are folded or dipping steeply, they most

likely have been displaced or deformed after deposition and solidification into rock (**figure A.2**).

In later centuries, scientists extended these principles to work igneous rocks into such sequences (**figure A.3**). If an igneous rock cuts across layers of sedimentary rocks, the sedimentary rocks must have been there first, the igneous rock introduced later. This is called the *Principle of Crosscutting Relationships,* and it applies whether the rocks that are crosscut are

A

B

Figure A.2

(A) A tilted sequence of sedimentary rocks in the Canadian Rocky Mountains. (B) These folded marbles and slates in Kings Canyon National Park, California, were originally flat-lying beds of limestone and shale
(A-B) ©Carla Montgomery.

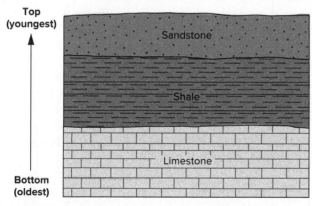

Figure A.1

The Principle of Superposition: the rocks on the bottom of a sequence of undisturbed sedimentary rocks were deposited first, and the depositional ages become younger higher in the sequence.

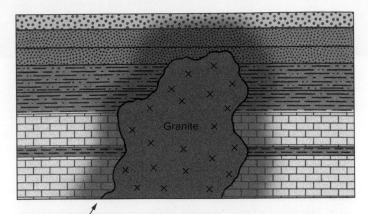

Rocks adjacent to intruding magma may also be metamorphosed by its heat.

A

B

Figure A.3

(A) An igneous rock crosscutting a sedimentary sequence or other rock must postdate the rocks it cuts across. (B) Here, veins of granite cut through a schist; the finer-grained granite at bottom partially crosscuts, and is thus younger than, the coarser.

©Carla Montgomery

sedimentary or not. Often, there is a further clue to the correct sequence; the hot magma that created the igneous intrusion may have baked, or metamorphosed, preexisting rocks immediately adjacent to it, so again the igneous rock must have come second. If an igneous rock contains pieces of other rock types (**figure A.4**), the invading magma must have picked up those pieces as solid chunks, so the rocks making up the inclusions must predate the host rock (*Principle of Inclusion*).

Such geologic common sense can be applied in many ways. If a strongly metamorphosed sedimentary rock is overlain by a completely unmetamorphosed one, for instance, the metamorphism must have occurred after the first sedimentary rock formed but before the second. We can unravel quite complex sequences of geologic events by taking into account such principles. See **figure A.5** for examples.

Figure A.4

The darker rock is a schist that was invaded by the magma that formed the (lighter-colored) granite, which picked up pieces of schist. The schist is thus older than the granite.
©Carla Montgomery

Correlation

Fossils play an integral role in the determination of relative ages around the year 1800, William Smith put forth the **Law of Faunal Succession.** The basic principle is that, through time, life-forms change (**figure A.6**); old ones disappear from and new ones appear in the fossil record, but the same form is never exactly duplicated at two different times in history. This principle implies that when one finds exactly the same type of fossil preserved in two rocks, even if the rocks are quite different compositionally and/or geographically widely separated, they are the same age. Smith's law allowed age *correlation* between rock units exposed in different places (**figure A.7**). A limitation on its usefulness is that it can be applied only to rocks in which fossils are preserved, which are almost exclusively sedimentary rocks.

The foregoing ideas were all useful in clarifying age relationships among rock units. They did not, however, help answer questions like: How old is this granite? How long did it take to deposit this limestone? How recently, and over how long a period, did this apparently extinct volcano erupt? Has this fault been active in modern times?

Uniformitarianism

James Hutton is another famous early geologist who was intrigued with both the natural processes and the rock sequences he observed. He is widely credited with developing and popularizing the concept of **uniformitarianism,** which is often summarized by the phrase "the present is the key to the past." It assumes

A

B

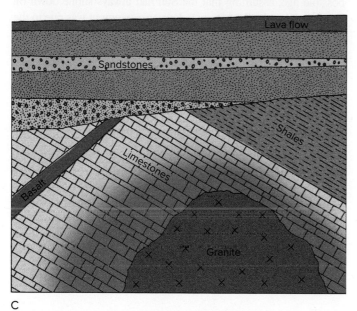

C

Figure A.5

(A) The Principle of Superposition and Principle of Original Horizontality together indicate that the sequence of events shown here in sedimentary rocks of the Grand Canyon was: deposition of the lower layers (dark red and gray); tilting and erosion; deposition of the still-flat-lying upper layers (green, cream, and light red). (B) The coarse gray granite is younger than the darker metasedimentary rocks (Principle of Inclusion) but older than the light quartz/feldspar vein (Crosscutting Relationships). (C) Sorting out relative ages of rock in an outcrop: Of the sedimentary rocks, the limestones must be the oldest (Principle of Superposition), followed by the shales. The granite intrusion and basalt must both be younger than the limestone they crosscut; the granite has also metamorphosed the surrounding limestone. It is not possible to tell whether the igneous rocks predate or postdate the shales (because they are not crosscut or metamorphosed by the igneous rocks), or to determine whether the sedimentary rocks were tilted before or after the igneous rocks were emplaced. After the limestones and shales were tilted and the basalt injected, they were eroded, and then the sandstones were deposited on top. Finally, the lava flow covered the entire sequence.

(A) Stephen Reynolds; (B) ©Carla Montgomery

that the same physical laws have operated throughout Earth's history, so by studying current, active, geologic processes and their products, geologists can infer much about how ancient rocks formed and changed. For example, an active volcanic eruption of flowing magma that hardens into basaltic rock in front of our eyes is a strong indicator of the general source of an ancient basalt flow even if geologists cannot now find the volcano from which it erupted. Studying the processes that occur on a sandy beach or those that create dunes in the desert provides geologists useful insights into the origin of some sandstones. Receding glaciers in Iceland leave moraines behind, which help researchers to understand the source and significance of the moraines in the midwestern United States, Canada, and elsewhere.

Some people have misinterpreted uniformitarianism to mean that the *rates* of geologic processes have been the same through time as well, that by observing processes like erosion, sedimentation, and seafloor spreading as they now occur, one

can know the rate at which they occurred in the past. Unfortunately, this isn't always the case, and wasn't really part of uniformitarianism as originally conceived. Volcanism may have involved the same types of physical and chemical processes throughout geologic time, but it likely was much more active early in Earth's history, before the planet cooled appreciably; rates of chemical sedimentation, as well as the proportions of different minerals in the sediments, could have been very different in the past, when the temperature and chemistry of the atmosphere and oceans were also different; and so on. In short, even uniformitarianism doesn't allow numerical answers to questions about geologic process rates in the past.

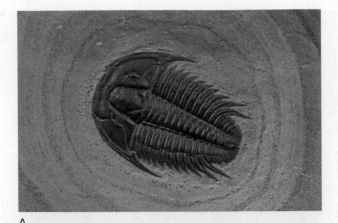

A

B

Figure A.6

Trilobites are an extinct class of marine arthropods found only in rocks from the Paleozoic Era. Paleontologists have described over 20,000 species of them (so far).
(A) Scott Orr/iStock/360/Getty Images (b) Aneese/iStock/Getty Images

How Old Is the Earth?

Geologists and nongeologists alike have been fascinated for centuries with the very basic question of Earth's age. Many have attempted to answer it, but with little success until the twentieth century.

Early Efforts

One of the earliest widely publicized determinations of the age of Earth was the seventeenth-century work of Archbishop Ussher of Ireland. He painstakingly and literally counted up the generations in the Bible and arrived at a date of 4004 BCE for the formation of Earth. This very young age was impossible for most geologists to accept; the complex geology of the modern Earth seemed to require far longer to develop.

Even less satisfactory from that point of view was the estimate of the philosopher Immanuel Kant. He tried to find a maximum possible age, assuming that the Sun had always shone down on Earth and that the Sun's tremendous energy output was due to the burning of some sort of conventional fuel. But a mass of fuel the size of the Sun would burn up in only 1000 years, given the rate at which the energy is released, plainly an impossible result in view of several millennia of recorded human history. Kant, of course, knew nothing of the nuclear fusion that actually powers the Sun.

Nineteenth-Century Views

About 1800, Georges L. L. de Buffon attacked the problem from another angle. He assumed that Earth was initially molten, modeled it as a ball of iron, and calculated how long it would take this quantity of iron to cool to present surface temperatures: 75,000 years. To most geologists, this still was not nearly long enough.

Around 1850, physicist Hermann L. F. von Helmholtz took the approach of supposing that the Sun's luminosity was due to

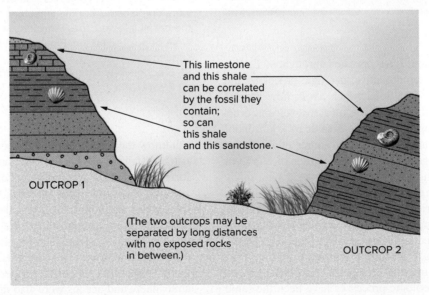

This limestone and this shale can be correlated by the fossil they contain; so can this shale and this sandstone.

OUTCROP 1

(The two outcrops may be separated by long distances with no exposed rocks in between.)

OUTCROP 2

Figure A.7

Similarity of fossils suggests similarity of ages even in different and quite widely separated rocks.

infall of particles into its center, converting gravitational potential energy to heat and light. This gave a maximum age for the Sun (and presumably Earth) of 20 to 40 million years. Again, his assumptions were wrong, so his answer was also.

In the late 1800s, Lord William T. Kelvin reworked Buffon's calculations, modeling Earth more realistically in terms of rock properties rather than those of metallic iron. Interestingly, he, too, arrived at an age of 20 to 40 million years. What he did not take into account, because natural radioactivity was then unknown, was that some heat is continually being *added* to Earth's interior through radioactive decay, so it has taken longer to cool down to its present temperature regime.

The calculations went on, and something was wrong with each. In 1893, U.S. Geological Survey geologist Charles D. Walcott tried to compute the total thickness of the sedimentary rock record throughout geologic history and, dividing by typical modern sedimentation rates, to estimate how long such a pile of sediments would have taken to accumulate. His answer was 75 million years. Walcott was hampered in his efforts by several factors, including the gaps in many sedimentary rock sequences (periods during which no sedimentation occurred or some sediments were eroded away), and the reality that there is no one spot on Earth where sediments have always been accumulating. He also had no good idea of the total thickness of sediments deposited in the time before organisms capable of preservation as fossils became widespread, which turns out to be most of Earth's history.

In 1899, physicist John Joly published calculations based on the salinity of the oceans. Taking the total dissolved load of salts delivered to the seas by rivers, and assuming that the ocean was initially pure water, he determined that it would take about 100 million years for the present concentrations of salts to be reached. Aside from Joly's assumption that rates of weathering and salt input into the oceans throughout Earth's history were constant, he didn't consider that the buildup of salts is slowed by the removal of some of the dissolved material—for example, as chemical sediments.

So the debate continued, imaginative and frequently heated, until the discovery of radioactivity provided a much more powerful and accurate tool with which to undertake a solution.

Radiometric Dating

Henri Becquerel didn't set out to solve any geologic problems nor even to discover radioactivity. He was interested in a curious property of some uranium salts: if exposed to light, the salts continued to emit light for a while afterward even in a dark room (phosphoresce). He had put a vial of uranium salts away in a drawer on top of some photographic plates well wrapped in black paper. When he next examined the photographic plates, they were fogged with a faint image of the vial, as if they had somehow been exposed to light right through the paper. The uranium was emitting something that could pass through light-opaque materials. We now know that uranium is one of several substances that are naturally radioactive, that undergo spontaneous decay. Once the phenomenon of radioactive decay was reasonably well understood, physicists and geoscientists began to realize that it could be a very useful tool for investigating Earth's history.

Radioactive Decay and Dating

As noted in chapter 16, one key property of any particular radioisotope is that it decays at a constant, characteristic rate, with a distinct half-life, which can be determined in the laboratory. One can then, in principle, use the relative amounts of a decaying isotope (conventionally called the *parent*) and the product isotope into which it decays (*daughter*) to find the age of the sample; see **figure A.8**. Suppose that parent isotope A decays to daughter B with a half-life of 1 million years. In a rock that contained some A and no B when it formed, A will gradually decay and B will accumulate. After 1 million years (one half-life of A), half of the atoms of A initially present will have decayed to yield an equal number of atoms of B.

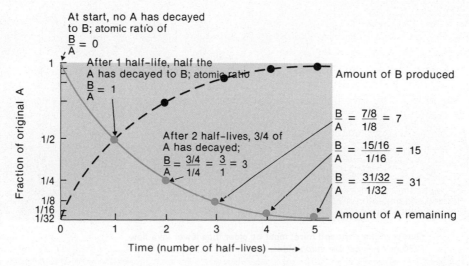

Figure A.8

Decay of radioactive parent A proceeds with complementary accumulation of daughter B. All ratios represent relative numbers of atoms.

After another half-life, half of the remaining half will have decayed, so three-fourths of the initial number of atoms of A will have been converted to B and one-fourth will remain; the ratio of B to A will be 3:1. After another million years, there will be seven atoms of B for every atom of A, and so on. This is **exponential decay,** decreasing abundance over time in a pattern that is just the opposite of the exponential growth of populations discussed in chapter 1. If a sample contains 24,000 atoms of B and 8000 atoms of A, then, assuming no B was present when the rock formed and noting the 3:1 ratio of B to A, we would conclude that the sample was two half-lives of A (in this case, 2 million years old).

This description is an oversimplification of the process we use to date natural geologic samples. Many samples contain some of the daughter isotope as well as the parent at the time of formation, and correction for this must be made. Many other samples haven't remained chemically closed throughout their histories; they have gained or lost atoms of the parent or daughter isotope of interest, which would make the apparent date incorrect. In the case of samples with a complex history—an igneous rock that has been metamorphosed once or twice and also weathered, for example—it can be difficult both to obtain a date and to decide which event, if any, is being dated. Fortunately, there are ways to recognize, and in many cases correct for, disturbances in isotopic systems.

Choice of an Isotopic System

Several conditions must be satisfied by an isotopic system to be used for dating geologic materials. The parent isotope chosen must be abundant enough in the sample for its quantity to be measurable. Either the daughter isotope must not normally be incorporated into the sample initially, or there must be a means of correcting for the amount initially present. The half-life of the parent must be appropriate to the age of the event being dated: long enough that some parent atoms are still present but short enough that some appreciable decay and accumulation of daughter atoms have occurred. A radioisotope with a half-life of ten days would be of no use in dating a million-year-old rock; the parent isotope would have decayed away completely long ago, and there would be no way to tell how long ago. Conversely, a radioisotope with a half-life of 10 billion years would be useless for dating a fresh lava flow because the atoms of the daughter isotope that would have accumulated in the rock would be too few to be measurable. Only about half a dozen isotopes are widely used in dating geologic samples. They include both uranium-235 and uranium-238 (with half-lives of 700 million years and 4.5 billion years, respectively), potassium-40 (1.3 billion years), rubidium-87 (49 billion years), and carbon-14 (5730 years).

Radiometric and Relative Ages Combined

Radiometric dates (dates determined using radioisotopic methods) are sometimes imprecisely called *absolute* ages, or numeric ages, to distinguish them from the relative ages determined as described earlier. For various technical reasons, we cannot determine accurate radiometric dates for many rocks and fossils, so we often use numeric and relative dating methods in conjunction (**figure A.9**). Undateable sedimentary rocks crosscut by an igneous intrusion must at least be older than the igneous rock, which may be dateable; a fossil found in a rock

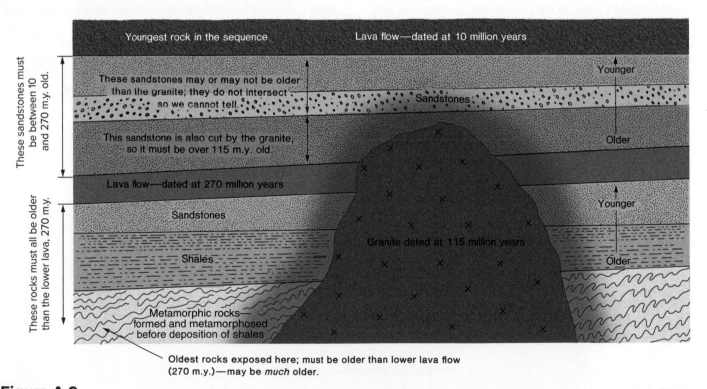

Figure A.9
Radiometric and relative dating together can put constraints on ages of units not dateable directly.

sandwiched between two dateable lava flows has its age bracketed by the ages of the flows; and so on.

Radiometric dating has proven an extremely powerful tool for studying our planet's history. Because Earth is geologically still very active, no rocks have been preserved unchanged since its formation. However, the dating of meteorites and Moon rocks, coupled with geochemical evidence that the whole solar system formed at the same time, has led to the determination of an age of about 4.55 billion years for Earth. The oldest rock samples from the continents are close to 4 billion years old.

Geologists have found evidence that at least some igneous rock formed well before 4 billion years ago. Grains of the mineral *zircon* collected in the Jack Hills of Australia have been dated at 4.4 billion years. The rocks of the Jack Hills are sedimentary and metasedimentary; but zircon crystallizes from magma. The zircon must have weathered out of its original igneous-rock host to be deposited in a sediment that was later lithified and became a part of the Jack Hills. That means that at least some solid crust must have existed 4.4 billion years ago, geologically very soon after the formation of Earth.

The Geologic Time Scale

Particularly in the days before radiometric dating, it was necessary to establish some subdivisions of Earth's history by which particular intervals of time could be indicated. This initially was done using those rocks with abundant fossil remains. The resultant time scale was subsequently refined with the aid of radiometric dates.

Early geologists used the Law of Faunal Succession and the appearance and disappearance of particular fossils in the sedimentary rock record to mark the boundaries of the time units. The principal divisions are the *eras*—**Paleozoic, Mesozoic,** and **Cenozoic.** The eras are subdivided into *periods* and the periods into *epochs.* The names of these smaller time units were assigned in various ways. Often, they were named for the location of the *type section,* the place where rocks of that age are well exposed and that time unit was first defined. For example, the Jurassic period is named for the Jura Mountains of France. Other periods were named on the basis of some characteristics of the rocks of the type section. The Cretaceous period derives its name from the Latin *creta,* for the chalky strata of that age in southern England and northern Europe; rocks of Carboniferous age commonly include coal beds. All of the relatively unfossiliferous rocks that seemed to be older (lower in the sequence) than the Cambrian period were lumped together as pre-Cambrian, later formally named **Precambrian.**

With the advent of radiometric dating, researchers could attach numbers to the units and boundaries of the time scale. It became apparent that, indeed, geologic history spanned long periods of time and that the most detailed part of the scale, the **Phanerozoic** (Cambrian Quaternary), was by far the shortest, comprising less than 15% of Earth's history. The approximate time framework of the Phanerozoic is shown in **table A.1**.

The Precambrian is divided into three eras: the *Hadean* from Earth's formation to 4000 million years ago, the *Archean* from 4000 to 2500 million years ago, and the *Proterozoic* from 2500 million years ago to the start of the Cambrian. The farther we go back in time, the more sparse the fossils are and the harder it is to interpret any evidence left behind in the rock record, but geologists have divided the Proterozoic and Archean based on global tectonic activity (or lulls therein) rather than

Table A.1	The Phanerozoic Time Scale			
Era	**Period**	**Epoch**	**Start of Interval***	**Distinctive Life-Forms**
		Holocene	0.12	modern humans
	Quaternary	Pleistocene	2.6	Stone-Age humans
		Pliocene	5.3	
		Miocene	23	flowering plants common
		Oligocene	34	ancestral pigs, apes
		Eocene	56	ancestral horses, cattle
Cenozoic	Tertiary	Paleocene	66	
	Cretaceous		145	dinosaurs become extinct; flowering plants appear
	Jurassic		200	birds, mammals appear
Mesozoic	Triassic		251	dinosaurs, first modern corals appear
	Permian		299	rise of reptiles, amphibians
	Carboniferous		359	coal forests; first reptiles, winged insects
	Devonian		416	first amphibians, trees
	Silurian		444	first land plants, coral reefs
	Ordovician		488	first fishlike vertebrates
Paleozoic	Cambrian		541	first widespread fossils

Dates, in millions of years before the present, from compilation by the Geological Society of America for the Decade of North American Geology.

fossils and stratigraphy. It is likely that these divisions will change as more evidence emerges.

At the other end of the time scale, a debate has recently developed about whether human impacts on the planet have become so significant that a new name should be assigned for the time from the present going forward—a new epoch, perhaps. The name *Anthropocene* has been proposed. A key point of disagreement is whether humans' effects will, over the long march of geologic time, really leave a distinct and substantial mark in the geologic record. This discussion is ongoing.

Geologic Process Rates

We can estimate the rates at which geologic processes occur in a variety of ways, many of which rely on radiometric dating techniques. For example, the rate of seafloor spreading away from a particular spreading ridge can be found by dividing the age of a sample of sea floor into its present distance from the ridge. If a 10-million-year-old seafloor sample is now 400 kilometers from the ridge, then, on average, it has been moved away from the ridge at a rate of

$$\frac{400 \text{ km}}{10,000,000 \text{ yr}} \times \frac{1000 \text{ m}}{1 \text{ km}} \times \frac{100 \text{ cm}}{1 \text{ m}} = \frac{4 \text{ cm}}{\text{yr}}$$

The minimum rate of uplift of rocks in a mountain range might be estimated from the age of marine sedimentary rocks in the mountains, once deposited under water, now high above sea level. Such uplift rates are typically 1 centimeter per year or less, often much less. Beaches formed on Scandinavian coastlines during the last Ice Age have been rising since the ice sheets melted and their mass was removed from the land. From the beach deposits' ages and present elevation, uplift rates can be approximated. Typical rates of this postglacial rebound are on the order of 1 centimeter per year. Radiometric dating has shown that a small volcano may stay active for over 100,000 years, a major volcanic center for 1 to 10 million years. To build a large mountain range may take 100 million years.

Rates of continental erosion due to weathering can be deduced from the loads of major rivers draining the continents, dividing the volume of rock those loads represent by the surface area of the corresponding drainage basin(s). The North American continent is being leveled by erosion at an average rate of 0.03 millimeter per year, or about a tenth of an inch per century. The larger the area over which such measurements are made, however, the greater the potential for large local deviations from the average due to special local soil or weather conditions. We must remain cautious about extrapolating present process rates too far into the past or future.

Summary

Before the discovery of natural radioactivity, only relative age determinations for rocks and geologic events were possible. Geologists used field relationships and fossil correlations as the principal methods for this purpose. Starting in the early 1900s, radiometric dating made it possible to measure the quantitative age of many geologic materials and events, to assign dates to the units of the geologic time scale, and to determine the age of Earth. Researchers have also used radiometric dates to explore the rates at which different kinds of geologic processes occur, considerably advancing our understanding of Earth's development. However, while we can use our observations of present geologic processes to understand past geologic history, we cannot assume that all of those processes have proceeded at rates comparable to those presently observed.

Key Terms and Concepts

Cenozoic
exponential decay
Law of Faunal Succession

Mesozoic
Paleozoic
Phanerozoic

Precambrian
Principle of Original
   Horizontality

Principle of Superposition
uniformitarianism

Mineral Identification

Table B.1 lists many of the more common minerals. Representative chemical formulas are provided for reference. Some appear complex because of opportunities for solid solution; some have been simplified by limiting the range of compositions represented, although additional elemental substitutions are possible.

A few general identification guidelines and comments:

Hardness is an approximate measure of how readily a mineral scratches, or is scratched by, other minerals. Values of hardness range from 1 (softest) to 10 (hardest) and are measured against the ten reference minerals of the Mohs hardness scale, shown in table 2.1. For example, a mineral that scratches quartz and is scratched by topaz would have a hardness of 7½.

Luster describes the surface sheen of a mineral sample, and the terms (such as metallic, pearly, earthy) are self-explanatory. *Cleavage* is a mineral's tendency to break preferentially along certain planes of weakness in the crystal structure.

Minerals showing metallic luster are usually sulfides (or native metals, but these are much rarer). Native metals have been omitted from table B.1. Those few that are likely to be encountered, such as native copper or silver, may be identified by their resemblance to household examples of the same metals.

Of the nonmetals, the silicates are generally systematically harder than the nonsilicates. Hardnesses of silicates are typically over 5, with exceptions principally among the sheet silicates; many of the nonsilicates, such as sulfates and carbonates, are much softer.

Distinctive luster, cleavage, or other identifying properties are listed under the column "Other Characteristics." In a few cases, this column notes restrictions on the occurrence of certain minerals as a possible clue to identification; for example, "occurs only in metamorphic rocks," or "usually found in pegmatites."

A Note on Mineral Formulas

Each chemical element is denoted by a one- or two-letter symbol. Many of these symbols make sense in terms of the English name for the element—O for oxygen, He for helium, Si for silicon, and so on. Other symbols reflect the fact that, in earlier centuries, scientists were generally versed in Latin or Greek: The symbols Fe for iron and Pb for lead, for example, are derived from *ferrum* and *plumbum,* respectively, the Latin names for these elements.

The chemical symbols for the elements can express the compositions of substances very precisely and concisely. Subscripts after a symbol indicate the number of atoms/ions of one element present in proportion to the other elements in the formula. For example, the formula $Fe_3Al_2Si_3O_{12}$ represents a compound in which, for every twelve oxygen atoms, there are three iron atoms, two aluminum atoms, and three silicon atoms. (This happens to be a variety of the mineral garnet.) The chemical formula is much briefer than describing the composition in words. It is also more exact than the mineral name "garnet," for there are several compositions of garnets with the same basic kind of formula and crystal structure: Other examples include a magnesium-aluminum garnet with formula $Mg_3Al_2Si_3O_{12}$, a calcium-aluminum garnet with the formula $Ca_3Al_2Si_3O_{12}$, and a calcium-chromium garnet, $Ca_3Cr_2Si_3O_{12}$. Moreover, chemical formulas are understood by all scientists, while mineral names are known primarily to geologists.

Formulas can become very complex, especially when different elements can substitute for each other in the same site in the crystal structure (**figure B.1**). Iron and magnesium often do this in silicates. Biotite, a common, dark-colored mica, may be rich in iron and have a formula of $KFe_3AlSi_4O_{10}(OH)_2$, or it may be rich in magnesium and have a formula of $KMg_3AlSi_4O_{10}(OH)_2$, or, more commonly, it may contain some iron and some magnesium, which together total three atoms per formula. The generalized formula is then $K(Fe,Mg)_3AlSi_4O_{10}(OH)_2$, as it appears in **table B.1**. See also "garnet" in the table.

Rock Identification

One approach to rock identification is to decide whether the sample is igneous, sedimentary, or metamorphic and then look at the detailed descriptions and illustrations in chapter 2. How does one identify the basic rock type? Here are some general guidelines:

1. Glassy rocks or rocks containing bubbles are volcanic.
2. Coarse-grained rocks with tightly interlocking crystals are likely to be plutonic, especially if they lack foliation.
3. Coarse-grained sedimentary rocks differ from plutonic rocks in that the grains in the sedimentary rocks tend to be more rounded and to interlock less closely. A breccia does have angular fragments, but the fragments in a breccia are typically rock fragments, not individual mineral crystals.
4. Rocks that are not very cohesive, that crumble apart easily into individual grains, are generally clastic sedimentary rocks. One other possibility would be a poorly consolidated volcanic ash, but this should be recognizable by the nature of the grains, many of which will be glassy shards. (Note, however, that extensive weathering can make even a granite crumble.)

Mineral	Formula	Color	Hardness	Other Characteristics
amphibole (e.g., hornblende)	$(Na,Ca)_2(Mg,Fe,Al)_5Si_8O_{22}(OH)_2$	green, blue, brown, black	5 to 6	often forms needlelike crystals; two good cleavages forming 120-degree angle
apatite	$Ca_5(PO_4)_3(F,Cl,OH)$	usually yellowish	5	crystals hexagonal in cross-section
azurite	$Cu_3(CO_3)_2(OH)_2$	vivid blue	3½ to 4	often associated with malachite
barite	$BaSO_4$	colorless	3 to 3½	high specific gravity, 4.5 (denser than most silicates)
beryl	$Be_3Al_2Si_6O_{18}$	aqua to green	7½ to 8	usually found in pegmatites
biotite (a mica)	$K(Mg,Fe)_3AlSi_3O_{10}(OH)_2$	black	5½	excellent cleavage into thin sheets
bornite	Cu_5FeS_4	iridescent blue, purple	3	metallic luster
calcite	$CaCO_3$	variable; colorless if pure	3	effervesces (fizzes) in weak acid; cleaves into rhombohedra
chalcopyrite	$CuFeS_2$	brassy yellow	3½ to 4	
chlorite	$(Mg,Fe)_3(Si,Al)_4O_{10}(OH)_2$	light green	2 to 2½	cleaves into small flakes
cinnabar	HgS	red	2½	earthy luster; may show silver flecks
corundum	Al_2O_3	variable; colorless in pure form	9	most readily identified by its hardness
covellite	CuS	blue	1½ to 2	metallic luster
dolomite	$CaMg(CO_3)_2$	white or pink	3½ to 4	powdered mineral effervesces in acid
epidote	$Ca_2FeAl_2Si_3O_{12}(OH)$	green	6 to 7	glassy luster
fluorite	CaF_2	variable; often green or purple	4	cleaves into octahedral fragments; may fluoresce in ultraviolet light
galena	PbS	silver-gray	2½	metallic luster; cleaves into cubes
garnet	$(Ca,Mg,Fe)_3(Fe,Al)_2Si_3O_{12}$	variable; often dark red	7	glassy luster
graphite	C	dark gray	1 to 2	streaks like pencil lead
gypsum	$CaSO_4 \cdot 2H_2O$	colorless	2	
halite	$NaCl$	colorless	2½	salty taste; cleaves into cubes
hematite	Fe_2O_3	red or dark gray	5½ to 6½	red-brown streak regardless of color
kaolinite	$Al_2Si_2O_5(OH)_4$	white	2	earthy luster
kyanite	Al_2SiO_6	blue	5 to 7	found in high-pressure metamorphic rocks; often forms bladelike crystals
limonite	$Fe_4O_3(OH)_6$	yellow-brown	2 to 3	earthy luster; yellow-brown streak
magnetite	Fe_3O_4	black	6	strongly magnetic
malachite	$Cu_2CO_3(OH)_2$	green	3½ to 4	often forms in concentric rings of light and dark green
molybdenite	MoS_2	dark gray	1 to 1½	cleaves into flakes; more metallic luster than graphite
muscovite (a mica)	$KAl_3Si_3O_{10}(OH)_2$	colorless	2 to 2½	excellent cleavage into thin sheets
olivine	$(Fe,Mg)_2SiO_4$	yellow-green	6½ to 7	glassy luster
phlogopite (a mica)	$KMg_3AlSi_3O_{10}(OH)_2$	brown	2½ to 3	closely resembles biotite, but less black
plagioclase feldspar	$(Na,Ca)(Al,Si)_2Si_2O_8$	white to gray	6	may show fine parallel striations on cleavage surfaces
potassium feldspar	$KAlSi_3O_8$	white; often stained pink or aqua	6	two good cleavages forming a 90-degree angle; no striations
pyrite	FeS_2	yellow	6 to 6½	metallic luster; black streak
pyroxene (e.g., augite)	$(Na,Ca,Mg,Fe,Al)_2Si_2O_6$	usually green or black	5 to 7	two good cleavages forming a 90-degree angle

Mineral	Formula	Color	Hardness	Other Characteristics
quartz	SiO_2	variable; commonly colorless or white	7	glassy luster; conchoidal fracture
serpentine	$Mg_3Si_2O_5(OH)_4$	green to yellow	3 to 5	waxy or silky luster; may be fibrous (asbestos)
sillimanite	Al_2SiO_5	white	6 to 7	occurs only in metamorphic rocks; often forms needlelike crystals
sphalerite	ZnS	yellow-brown	3½ to 4	glassy luster
staurolite	$Fe_2Al_9Si_4O_{20}(OH)_2$	brown	7 to 7½	found in metamorphic rocks; elongated crystals may have crosslike form
sulfur	S	yellow	1½ to 2½	sulfurous odor
sylvite	KCl	colorless	2	cleaves into cubes; salty taste, but more bitter than halite
talc	$Mg_3Si_4O_{10}(OH)_2$	white to green	1	greasy or slippery to the touch
tourmaline	$(Na,Ca)-(Li,Mg,Al)$ $(Al,Fe,Mn)_6(BO_3)_3Si_6O_{18}(OH)_4$	black, red, green	7 to 7½	elongated crystals, triangular in cross-section; conchoidal fracture
zircon	Zr_2SiO_4	usually brown	7½	typically forms very small crystals in igneous rocks; durable, so also common in clastic sedimentary rocks

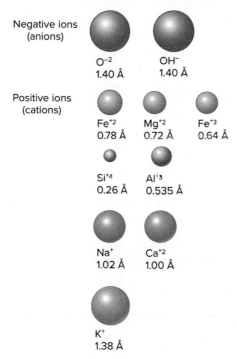

5. More cohesive, fine-grained sedimentary rocks may be distinguished from fine-grained volcanics because sedimentary rocks are generally softer and more likely to show a tendency to break along bedding planes. Phenocrysts, of course, indicate a (porphyritic) volcanic rock.

6. Foliated metamorphic rocks are distinguished by their foliation (schistosity, compositional banding). Also, rocks containing abundant mica, garnet, or amphibole are commonly metamorphic rocks.

7. Nonfoliated metamorphic rocks, like quartzite and marble, resemble their sedimentary parents but are harder, denser, and more compact. They may also have a shiny or glittery appearance on broken surfaces, due to recrystallization during metamorphism.

Once a preliminary determination of category (igneous/sedimentary/metamorphic) has been made, **table B.2** and chapter 2 can be used to identify the rock type. (Keep in mind, however, that the key and text focus on relatively common rock types.)

Figure B.1

Sizes of common ions in the crust with radius of each given in Ångstroms (Å). (1 Å = 10^{-10} meters = 0.000000004 in) Ions of similar size and charge can substitute for each other in minerals' crystal structures. Iron and magnesium frequently do so in the ferromagnesians. Some Al^{+3} substitutes for Si^{+4} in many silicates; Na^+ and Ca^{+2} can replace each other in some feldspars; and so on.

Igneous Rocks

I. Extremely coarse-grained: Rock is a pegmatite. (Most pegmatites are granitic, with or without exotic minerals.)

II. Phaneritic (coarse enough that all grains are visible to the naked eye).

   A. Significant quartz visible; only minor mafic minerals: *granite*.

   B. No obvious quartz; feldspar (light-colored) and mafic minerals (dark) in similar amounts: *diorite*.

   C. No quartz; rock consists mostly of mafic minerals: *gabbro*.

   D. No visible quartz or feldspar: Rock is ultramafic.

III. Porphyritic with fine-grained groundmass: Go to Part IV to describe groundmass (using phenocryst compositions to assist); adjective "porphyritic" will preface rock name.

IV. Aphanitic (grains too fine to distinguish easily with the naked eye).

   A. Quartz is visible or rock is light in color (white, cream, pink): probably *rhyolite*.

   B. No visible quartz; medium tone (commonly grey or green); if phenocrysts are present, commonly plagioclase, pyroxene, or amphibole: probably *andesite*.

   C. Rock is dark, commonly black; any phenocrysts are olivine or pyroxene: *basalt*.

   D. Rock is glassy and massive: *obsidian* (regardless of composition).

   E. Rock consists of gritty mineral grains, ash, and glass shards: *ignimbrite* (welded tuff).

Sedimentary Rocks

I. Rock consists of visible shell fragments or of oolites: *limestone*.

II. Rock consists of interlocking grains with texture somewhat like that of igneous rock and is light in color: probable chemical sedimentary rock.

   A. Tastes like table salt: *halite*.

   B. No marked taste; hardness of 2 (if grains are large enough to scratch); does not effervesce: *gypsum*.

   C. Effervesces in weak HCl: limestone (calcite).

   D. Effervesces weakly in HCl, only if scratched: *dolomite*.

III. Rock consists of grains apparently cemented or compacted together: probable clastic sedimentary rock.

   A. Coarse grains (several millimeters or more in diameter), perhaps with a finer matrix: *conglomerate* if the grains are rounded, *breccia* if they are angular.

   B. Sand-sized grains; gritty feel: sandstone. If predominantly quartz grains, *quartz sandstone;* if roughly equal proportions of quartz and feldspar, *arkose;* if many rock fragments, and perhaps a fine-grained matrix, *greywacke*.

   C. Grains too fine to see readily with the naked eye: *mudstone*. If rock shows lamination, and a tendency to part along parallel planes, *shale*.

IV. Relatively dense; compact, dark, no visible grains; massive texture, conchoidal fracture: *chert* (silica).

Metamorphic Rocks

I. Nonfoliated; compact texture with interlocking grains: identified by predominant mineral(s).

   A. If quartz-rich, perhaps with a sugary appearance: *quartzite*.

   B. If calcite or dolomite (identified by effervescence, hardness): *marble*.

   C. Rock consists predominantly of amphiboles: *amphibolite*.

II. Foliated: classified mainly by texture.

   A. Very fine-grained; pronounced rock cleavage along parallel planes, to resemble flagstones: *slate*.

   B. Fine-grained; slatelike, but with glossy cleavage surfaces: *phyllite*.

   C. Coarser grains; obvious foliation, commonly defined by prominent mica flakes, sometimes by elongated crystals like amphiboles: *schist*.

   D. Compositional or textural banding, especially with alternating light (quartz, feldspar) and dark (ferromagnesian) bands: *gneiss*.

Length

1 cm = 0.394 in.
1 m = 39.37 in. = 1.09 yd.
1 km = 0.621 mi.
1 in. = 2.54 cm
1 yd. = 0.914 m
1 mi. = 1760 yd. = 1.61 km

Area

1 sq. cm = 0.155 sq. in.
1 sq. m = 1.20 sq. yd. = 1550 sq. in.
1 sq. km = 0.386 sq. mi.
1 sq. in. = 6.45 sq. cm
1 sq. yd. = 1296 sq. in. = 0.836 sq. m
1 sq. mi. = 2.59 sq. km
1 acre = 4840 sq. yd. = 4047 sq. m

Volume

1 cu cm = 0.061 cu in.
1 cu m = 1.31 cu yd.
1 cu km = 0.240 cu mi.
1 cu in. = 16.4 cu cm
1 cu yd. = 0.765 cu m
1 cu mi. = 4.17 cu km

Liquid Volume

1 ml = 0.0338 fl. oz.
1 liter = 1.06 qt.
1 fl. oz. = 29.6 ml
1 qt. = 0.946 liter
1 gal. = 4 qt. = 3.78 liter
1 acre-foot = 326,000 gal. = 1220 cu m

Weight/Mass

1 g = 0.0353 oz.
1 kg = 2.20 lb.
1 metric ton (tonne) = 1000 kg = 2200 lb.
1 oz. (avoirdupois) = 28.4 g
1 lb. = 454 g = 0.454 kg
1 ton = 2000 lb. = 909 kg
1 troy oz. = 1.10 oz. avoirdupois = 31.2 g

Energy

1 cal (calorie) = amount of heat required to raise
temperature of 1 ml of water by 1°C
1 Btu (British thermal unit) = amount of heat required
to raise temperature of 1 lb of water by 1°F
1 Btu = 252 cal
1 quad = 1 quadrillion Btu = 1,000,000,000,000,000 Btu

Average Energy Contents
of Various Fossil Fuels

Fuel	Calories	Btu
1 barrel crude oil	1,460,000,000	5,800,000
1 ton coal	5,650,000,000	22,400,000
1 cu ft. natural gas	257,000	1020

Explanation of Prefixes
and Their Values

deci-: one-tenth   1 deciliter (dl) = 0.1 liter
centi-: one-hundredth   1 centimeter (cm) = 0.01 meter
milli-: one-thousandth   1 milliliter (ml) = 0.001 liter
kilo-: one-thousand   1 kilometer (km) = 1000 meters

Glossary

A

A horizon Usually top zone in soil profile, consisting of mix of mineral and organic material.

ablation The loss of glacier ice by melting or evaporation.

abrasion Erosion by wind-transported sediment or by the scraping of rock fragments frozen in glacial ice.

absorption field *See* leaching field.

accreted terrane A terrane that has been moved from somewhere else to its present relative position on a continent.

acid rain Rain that is more acidic (has lower pH) than normal precipitation.

acre-foot The amount of water necessary to cover 1 acre of land (43,560 square feet) by 1 foot of water; equal to 325,851 gallons.

active margin A continental margin at which there is significant volcanic and earthquake activity; commonly, a convergent plate margin.

aerobic decomposition Decomposition using or consuming oxygen.

aerosol A suspension of particulate matter in air.

aftershocks Earthquakes that follow the main shock when a fault has slipped; of magnitude equal to or lower than the main shock.

albedo The fraction of incoming radiation that is reflected back by a surface.

algal bloom An overly exuberant growth of algae in eutrophic water.

alluvial fan A wedge-shaped sediment deposit left where a tributary flows into a more slowly flowing stream, or where a mountain stream flows into a desert.

alpine (valley) glacier A glacier occupying a valley in mountainous terrain.

anaerobic decomposition Decomposition that occurs without using, or in the absence of, oxygen.

andesite Volcanic rock intermediate in composition between basalt and rhyolite.

angle of repose The maximum slope angle at which a given unconsolidated material is stable.

anion An ion with a net negative charge.

anthracite The hardest and most mature of naturally occurring coals.

anthropogenic Produced by human activity.

aquifer Rock that is sufficiently porous and permeable to be useful as a source of water.

aquitard Rock of low permeability, through which water flows very slowly.

arête Sharp-spined ridge created by erosion by valley glaciers flowing along either side of the ridge.

artesian system A confined aquifer system in which groundwater can rise above its aquifer under its own pressure.

B

asthenosphere Partially molten, plastic zone within the upper mantle immediately below the lithosphere.

atom The smallest particle into which a chemical element can be subdivided.

atomic mass number The sum of the number of protons and the number of neutrons in an atomic nucleus.

atomic number The number of protons in an atomic nucleus; characteristic of a particular element.

B horizon Soil layer located below the soil's A and E horizons; also known as the zone of accumulation.

banded iron formation A sedimentary rock consisting of alternating iron-rich and iron-poor layers, found predominantly in Precambrian rocks, that may serve as an ore of iron.

barrier islands Constantly changing, long, low, narrow islands made of sand, formed parallel to a coastline due to wave and current action.

basalt A volcanic rock rich in ferromagnesian minerals; relatively low in silica.

base level The lowest elevation to which a stream can cut down; for most streams, this is the level of the body of water into which they flow, such as another stream, lake, or ocean.

beach Gently sloping shoreline area of sand or other sediment.

beach face That part of the beach that is regularly washed by waves as tides rise and fall.

bed load Material moved along a stream bed by flowing water.

Benioff zone A dipping plane of progressively deeper earthquake foci at a convergent plate boundary associated with a subducting plate.

berm Flatter portion of a beach landward of the beach face.

bioaccumulation Accumulation of a chemical in an organism.

biochemical oxygen demand (BOD) Quantity of oxygen required for aerobic decomposition of organic matter in a system.

biodiesel Fuel derived from vegetable oil or animal fat that can be used in some diesel-powered vehicles, or a blend of regular diesel with such fuels.

biofuels Energy derived from living organisms or from organic matter (biomass).

biogas Methane derived from decaying organic matter.

biomagnification Process through which the concentration of a harmful substance, such as a heavy metal, increases in organisms as it moves up a food chain.

bioremediation Reduction of a pollution hazard by use of organisms.

biosphere The sum of all living things on Earth.

bitumen The dark, heavy, viscous petroleum found in oil sands.

bituminous A form of coal that is softer than anthracite but harder than lignite.

BOD *See* biochemical oxygen demand.

body waves Seismic waves that pass through Earth's interior; include P waves and S waves.

braided stream A stream with multiple channels that divide and rejoin.

breeder reactor A reactor in which new fissionable material is produced in quantity at the same time as energy is generated.

brittle Describes materials that tend to rupture before appreciable plastic deformation has occurred.

C

C horizon Soil layer located directly below the soil's B horizon; consists of coarsely broken bedrock.

caldera A large, bowl-shaped summit depression in a volcano; may be formed by explosion or collapse.

calving The formation of icebergs by the breakup of a glacier flowing out over water.

cap-and-trade A strategy for reducing pollutant emissions by setting an overall limit on a group (as for SO_2 emissions from all electric utilities) but allowing units within the group to buy or sell emissions allowances, providing a financial incentive to units to reduce emissions.

capacity (stream) The load that a stream can carry.

carbon capture and storage (CCS) The process of trapping carbon-containing emissions or filtering CO_2 from the atmosphere and then isolating the carbon in some reservoir, from which it doesn't contribute to atmospheric greenhouse gases.

carbonate Nonsilicate mineral containing carbonate groups (CO_3), carbon and oxygen in the proportions of one atom of carbon to three atoms of oxygen.

carrying capacity The ability of a system (or the whole Earth) to sustain its population in reasonably healthy and comfortable conditions.

cation An ion with a net positive charge.

Cenozoic Geologic era spanning the time from 66 million years ago to the present.

chain reaction (nuclear) The process during which fission of one nucleus triggers fission of others, which, in turn, induces fission in others, and so on.

channelization The modification of a stream channel, such as deepening or straightening of the channel, usually with the objective of reducing flood hazards.

chemical sediment Sediment formed at low temperature by direct precipitation from solution.

chemical weathering The breakdown of minerals by chemical reaction with water, with other chemicals dissolved in water, or with gases in the air.

cinder cone A relatively small volcano built of cinders and other pyroclastics piled up around the volcanic vent; also called a scoria cone.

cirque Bowl-shaped depression formed at the head of an alpine glacier by erosion.

clastic Broken or fragmented; describes sediments or sedimentary rocks that are formed from fragments of preexisting rocks or minerals.

cleavage (mineral) The tendency of a mineral to break preferentially along planes in certain directions in the crystal structure.

coal A solid, carbon-rich fuel formed from the remains of land plants through the effects of heat, pressure, and time.

coal-bed methane Methane associated with, and extracted from, coal deposits.

common but differentiated responsibility of states The concept that all nations share responsibility for protecting the global environment, but individual nations' responsibilities differ with their differing contributions to the problems and resources available to address them.

composite volcano *See* stratovolcano.

compound A chemical combination of two or more elements, in specific proportions, having a distinctive set of resultant physical properties.

compressive stress Occurs when forces press inward on an object.

concentration factor (ore) The concentration of a metal in a given ore deposit divided by its average concentration in the continental crust.

conditional resources *See* subeconomic resources.

cone of depression A broadly conical depression of the water table or potentiometric surface caused by pumped groundwater withdrawal.

confined aquifer An aquifer overlain by an aquitard or aquiclude.

confining pressure Pressure that is uniform in all directions around a rock or other body.

constructive plate boundary A plate boundary at which new lithosphere is created; e.g., a seafloor spreading ridge.

contact metamorphism Local metamorphism adjacent to a cooling magma body.

contaminant plume A tongue of contaminant-rich water extending away from a point source of groundwater pollution in the direction of groundwater flow.

continental drift The concept that the continents have moved about over Earth's surface.

continental glacier A large glacier covering extensive land area; also known as an *ice cap* or *ice sheet;* may be several kilometers thick.

convection cells Circulating masses of material driven by temperature differences (hot material rises, then moves laterally, cools, sinks, and is reheated to rise again).

convergent plate boundary Plate boundary at which lithospheric plates are moving toward each other; for example, a subduction zone or continental collision zone.

coral bleaching Event during which coral expel the symbiotic algae on which they rely for their color and much of their nourishment; commonly caused by too-warm water.

core The innermost zone of Earth; composed largely of iron.

core meltdown A possible nuclear reactor accident resulting from loss of core coolant and subsequent overheating.

covalent bonding Bonding involving sharing of electrons between atoms.

creep (fault) Slow, gradual slip along an "unlocked" fault zone without major, damaging earthquakes; caused by plate motions.

creep (rock or soil) Slow, gradual downslope movement of unstable surficial materials.

crest The maximum stage reached during a flood event.

crust The outermost compositional zone of Earth, made predominantly of relatively low-density silicate minerals.

crystalline Describes materials possessing a regular, repeating internal arrangement of atoms.

cumulative reserves The total reserves, including materials already consumed or exploited.

curie A unit of radioactivity equal to 3.7×10^{10} decays/sec.

Curie temperature The temperature above which a magnetic material loses its magnetism; different for each such material.

cut bank Steep stream bank being eroded by lateral migration of meanders.

D

Darcy's law Relationship of groundwater flow rate between two points to the difference in hydraulic head between them.

debris avalanche A mixed flow of rock, soil, vegetation, or other materials.

debris flow Another term for *debris avalanche;* especially, used for a water-saturated one.

decommissioning (nuclear plant) The shutdown of a nuclear reactor at the end of its safe, useful life; includes the disposal of radioactive parts.

decompression melting Melting due to a decrease in pressure.

deep-focus earthquakes Those with focal depths over 100 kilometers.

deflation The wholesale removal of loose sediment by wind erosion.

delta A fan-shaped deposit of sediment formed at a stream's mouth.

desert An imprecise term that generally describes a region receiving less than 25 cm of precipitation annually and that has limited vegetation.

desert pavement A desert surface produced by the combined effects of wind erosion and overland surface-water runoff, in which the larger rocks are exposed by the selective removal of fine sediment; these rocks, in turn, protect finer material below from erosion.

desertification The process by which marginally habitable arid lands are converted to desert; typically accelerated by human activities.

destructive plate boundary A plate boundary at which lithosphere is consumed; e.g., a subduction zone.

dip-slip fault A fault with predominantly vertical displacement.

discharge (groundwater) The water exiting an aquifer, either by natural means or human causes.

dissolved load Sum of dissolved material transported by a stream.

divergent plate boundary A boundary along which lithospheric plates are moving apart; for example, seafloor spreading ridges and continental rift zones.

divide Topographic high separating drainage basins of different streams; also, plane in a groundwater system from which groundwater flows away in different directions on either side.

Dobson unit 1 Dobson unit = 0.001cm thickness of ozone at 0 degrees Celsius and the pressure of 1 atmosphere, distributed over the whole vertical column of the atmosphere.

dormant volcano A volcano with no recent eruptive history but that still looks relatively fresh and unweathered; may become active again in future.

dose-response curve A graph illustrating the relative benefit or harm from a trace element or other substance as a function of the dosage received or amount consumed by a person or organism.

doubling time The length of time required for a population to double in size.

downstream flood A flood affecting a large area of drainage basin or a large stream system; typically caused by prolonged rain or rapid regional snowmelt.

drainage basin The region from which surface water drains into a particular stream.

dredge spoils Sediment dredged from waterways to improve navigation or to increase water capacity.

drift Sediment transported and deposited by a glacier; *see also* till, outwash.

drowned valley Along a coastline, a stream valley that is partially flooded by seawater as a consequence of land sinking and/or sea level rising.

ductile Describes material that undergoes extensive plastic deformation without rupturing.

dune A low mound or ridge of sediment (usually sand) deposited by wind.

E

E horizon Layer typically located between A and B soil horizons; also known as zone of leaching.

E85 A liquid fuel consisting of 85% ethanol, 15% gasoline.

earthquake Ground displacement and energy release associated with the sudden motion of rocks along a fault.

earthquake cycle The concept that there is a periodic quality about the occurrence of major earthquakes on a given fault zone, with repeated cycles of stress buildup, rupture, and relaxation of stress through smaller aftershocks.

ecosystem services Benefits provided to humans by the natural environment.

EEZ *See* Exclusive Economic Zone.

EIS *See* environmental impact statement.

elastic deformation Deformation proportional to applied stress, from which the affected material will return to its original size and shape when the stress is removed.

elastic limit The stress above which a material will cease to deform elastically.

elastic rebound Phenomenon whereby stressed rocks snap back elastically after an earthquake to their pre-stress condition.

electron A subatomic particle with an electrical charge of −1; generally found orbiting an atomic nucleus.

end moraine A ridge of till accumulated at the end of a glacier.

enhanced recovery Means to recover subsurface petroleum resources that extend beyond pumping (primary) and water flushing (secondary).

environmental geology The study of the interactions between humans and their geologic environment.

environmental impact statement (EIS) An analysis of the environmental impacts to be anticipated from a proposed action and its alternatives; mandated by the National Environmental Policy Act and legislation patterned after it.

eolian Deposited or shaped by wind action.

ephemeral stream A stream that flows only occasionally, in response to a precipitation event.

epicenter The point on Earth's surface directly above the focus of an earthquake.

equilibrium line Boundary on the surface of a glacier at which accumulation just equals ablation.

estuary A body of water along a coastline that contains a mix of fresh and salt water.

eutrophication The development of high nutrient levels (especially high concentrations of nitrates and phosphates) in water; may lead to algal bloom.

evaporite A sedimentary mineral deposit formed when shallow or inland seas dry up; also, the minerals commonly deposited in such an environment.

evapotranspiration Movement of water vapor from Earth's surface into air by a combination of evaporation from rocks and transpiration through plants.

Exclusive Economic Zone (EEZ) A zone extending to 200 miles offshore from a nation's coast, within which the 1982 Law of the Sea Treaty recognizes that nation's exclusive right to resource exploitation.

exponential decay Breakdown of a fixed percentage or fraction of a substance per unit time.

exponential growth Growth characterized by a constant percentage increase per unit time.

extinct volcano A volcano that has no recent eruptive history and appears very weathered in appearance; not expected to erupt again.

F

fall (rock) Mass wasting by free-fall of material not always in contact with the ground underneath.

fault Planar break in rock along which one side has moved relative to the other.

fault scarp A steep slope formed along the fault plane at the ground surface.

ferromagnesian A term describing silicates containing significant amounts of iron and/or magnesium; these minerals are usually dark-colored.

firn Dense, coarsely crystalline snow partially converted to ice.

fission (atomic) The process by which atomic nuclei are split into smaller fragments.

fissure eruption The eruption of lava from a crack in the lithosphere, rather than from a central vent.

flash flood A variety of upstream flood characterized by a very rapid rise in stream stage.

flood Condition in which stream stage is above channel bank height.

flood-frequency curve A graph of stream stage or discharge as a function of recurrence interval (or annual probability of occurrence).

floodplain A flat region or valley floor surrounding a stream channel, formed by meandering and sediment deposition, into which the stream overflows during flooding.

flow Mass wasting in which materials move in chaotic fashion.

focus The point of first break on a fault during an earthquake; also called hypocenter.

foliation Parallel alignment of linear or platy minerals in a rock.

fossil fuels Hydrocarbon fuels formed from organic matter.

fracking *See* hydraulic fracturing.

fractional crystallization Process by which magma composition is changed as minerals crystallize from it and are removed from the residual magma.

fusion (nuclear) The process by which atomic nuclei combine to produce larger nuclei.

G

gasification Any process by which coal is converted to a gaseous hydrocarbon fuel.

geoengineering Any large-scale scheme to influence Earth's climate by modifying some aspect of the climate system—atmosphere, oceans, or land.

geographic information system (GIS) A computer-based system for storing, manipulating, and analyzing data associated with particular geographic locations.

geopressurized zones Deep aquifers under unusually high pressure, exceeding normal hydrostatic (fluid) pressure.

geothermal energy Energy derived from the internal heat of Earth; its use usually requires a near-surface heat source, such as young igneous rock, and nearby circulating subsurface water.

geothermal gradient The rate of increase of temperature with depth in Earth.

GIS *See* geographic information system.

glacier A mass of ice that moves or flows over land under its own weight.

glass Solid (especially silicate) lacking a regular internal crystal structure.

gouge Finely pulverized rock along a fault plane.

gradient The slope (steepness) of a stream channel along its length.

granite Coarse-grained plutonic igneous rock, typically rich in quartz and feldspars.

greenhouse effect The warming of the atmosphere due to trapping of infrared rays by atmospheric gases, especially as due to the increased concentration of carbon dioxide derived from the burning of fossil fuels.

groundwater Water in the zone of saturation, below the water table.

H

half-life The length of time required for half of an initial quantity (of atoms) of a radioactive isotope to decay.

halide Compound of one or more metals plus a halogen element (fluorine, chlorine, iodine, or bromine).

hard water Water containing substantial quantities of dissolved calcium, magnesium, and/or iron.

hardness (mineral) The ability of a mineral to resist scratching.

harmonic tremors A distinctive type of seismic signal, rhythmic and continuous, often associated with magma movement in or beneath a volcano.

heavy metals A group of dense metals, including mercury, lead, cadmium, plutonium, and others, that share the characteristic of being accumulative in organisms and tending to become increasingly concentrated in organisms higher up a food chain.

high-level (radioactive) waste Waste sufficiently radioactive to require special handling in disposal.

horn A peak formed by headwall erosion by several alpine glaciers diverging from the same topographic high.

hot spot An isolated center of volcanic activity; often not associated with a plate boundary.

hot-dry-rock Geothermal resource area in which geothermal gradients are high but indigenous groundwater is lacking.

hydraulic fracturing Method for increasing permeability of rocks by injecting fluid under high pressure, to promote development of fractures, in order to facilitate extraction of oil and natural gas.

hydraulic gradient The difference in hydraulic head between two points, divided by the distance between them.

hydraulic head Potential energy of water above a given point, reflected in the height of the water surface (aboveground), water table in an unconfined aquifer, or potentiometric surface in a confined aquifer.

hydrocarbon Any compound consisting of carbon and hydrogen.

hydrograph A graph of stream stage or discharge against time.

hydrologic cycle The cycle through which water in the hydrosphere moves; includes such processes as evaporation, precipitation, and surface and groundwater runoff.

hydrosphere All water at and near Earth's surface that is not chemically bound in rocks or living creatures.

hydrothermal Relating to warm water; hydrothermal ores are ores deposited by circulating warm fluids in Earth's crust.

hypothesis A conceptual model or explanation for a set of data, measurements, or observations.

hypothetical resources That quantity of a resource material expected to be found in areas in which like deposits are known to exist.

hypoxia Condition in which water contains too little dissolved oxygen for aquatic organisms to survive.

I

ice age Any period of extensive continental glaciation; when capitalized (Ice Age), the phrase refers to the most recent such episode, from 2 million to 11,000 years ago.

igneous rock A rock formed or crystallized from a magma.

induced seismicity Seismicity caused/triggered by human activity.

infiltration The process by which surface water sinks into the ground.

infrared Electromagnetic radiation just to the long-wavelength side of the visible light spectrum; heat radiation.

intensity (earthquake) A measure of the damaging effects of an earthquake at a particular spot; commonly reported on the Modified Mercalli Scale.

ion Atom that has gained or lost electrons, so it has a net electrical charge.

ionic bonding Bonding due to attraction between oppositely charged ions.

island arc Chain of volcanic islands formed parallel to a subduction zone, on the overriding plate.

isotopes Atoms of a given chemical element having the same atomic number but different atomic mass numbers.

K

karst Terrain characterized by abundant formation of underground solution cavities and sinkholes; commonly underlain by limestone or gypsum.

kerogen A waxy solid hydrocarbon in oil shale.

kimberlites Ultramafic igneous rocks that occur as pipelike intrusive bodies that originated in the mantle.

L

lahar A volcanic mudflow deposit formed from hot ash and water, the latter often derived from melting snow on a snow-capped or glaciated volcano.

landslide A general term applied to a rapid mass-wasting event.

laterite An extreme variety of pedalfer soil that is highly leached; common in tropical climates.

lava Magma that flows out at Earth's surface.

lava dome A compact, bulbous, steep-sided structure built of very viscous, silicic lava emitted from a central pipe or vent.

Law of Faunal Succession The concept that life-forms change through time and that therefore each given fossil organism corresponds to only one interval of time.

leachate Water containing dissolved chemicals; applied particularly to fluids escaping from waste disposal sites.

leaching The removal of elements or compounds by dissolution.

leaching field A network of porous pipes and surrounding soil from which septic-tank effluent is slowly released.

levees Raised banks along a stream channel that tend to contain the water during high-discharge events.

lignite The softest of the coals.

liquefaction (coal) Any process by which coal is converted into a liquid hydrocarbon fuel.

liquefaction (soil) A quicksand condition arising in wet soil shaken by seismic waves; soil loses its strength as particles lose contact with each other.

lithification Conversion of unconsolidated sediment into cohesive rock by compaction over time.

lithosphere The solid, outermost zone of Earth, including the crust and a portion of the upper mantle; approximately 50 kilometers thick under the oceans and commonly more than 100 kilometers thick beneath the continents.

littoral drift Sand movement along the length of a beach.

load (stream) The total quantity of material transported by a stream: sum of bed load, suspended load, and dissolved load.

loess Wind-deposited sediment composed of fine grained particles, typically silt.

longitudinal profile Diagram of elevation of a stream bed along its length.

longshore current Net movement of water parallel to a coastline, arising when waves and currents approach the shore at an oblique angle.

low-level (radioactive) waste Wastes that are sufficiently low in radioactivity that they can be released safely into the environment or disposed of with minimal precautions.

M

magma A naturally occurring silicate melt, which may also contain mineral crystals, dissolved water, or gases; molten rock beneath Earth's surface.

magnitude (earthquake) Measure of earthquake size; often reported using the Richter magnitude scale.

manganese nodules Lumps of manganese and iron oxides and hydroxides, with other metals, found on the sea floor.

mantle The zone of the Earth's interior between crust and core; rich in ferromagnesian silicates.

mass movement See mass wasting.

mass wasting The downslope movement of material due to gravity; also known as mass movement.

meanders The curves or bends in a stream channel.

mechanical weathering The physical breakup of rock or mineral grains by surface processes.

Mesozoic Geologic era from 251 to 66 million years ago.

metamorphic rock Rock that is changed in form (deformed and/or recrystallized) through the effects of heat and/or pressure.

methane hydrate A crystalline solid of natural gas and water molecules, found in arctic regions and marine sediments.

Milankovitch cycles Cyclic variations in the amount of sunlight reaching a given latitude, caused by variations in Earth's orbit and tilt on its axis.

milling Erosion by the grinding action of sand-laden waves on a coast.

mineral A naturally occurring, inorganic, solid element or compound with a definite composition or range in composition, usually having a regular internal crystal structure.

moment magnitude A measure of earthquake magnitude directly proportional to energy release, calculated from the area of fault plane broken, amount of fault displacement, and shear strength of the rock; represented by M_w.

moraine Landform made of till.

mouth (stream) The point where a stream ends; where it reaches its base level.

multiple barrier concept Waste-disposal approach that involves several mechanisms or materials for isolating the waste from the environment; often used in the context of high-level radioactive wastes.

multiple use The land-use practice in which a given piece of land is used for two or more purposes simultaneously.

N

native element Nonsilicate mineral consisting of a single chemical element.

natural gas Gaseous hydrocarbons, especially methane (CH_4).

neutron An electrically neutral subatomic particle with a mass approximately equal to one atomic mass unit; generally found within an atomic nucleus.

nonpoint source A diffuse source of pollutants, such as runoff from farmland or drainage from a strip mine.

nonrenewable Not being replenished or formed at any significant rate on a human timescale.

normal fault A dip-slip fault where the block above the fault moves down relative to the block below it; indicates tensional stress.

nucleus The center of an atom, containing protons and neutrons.

O

O horizon Top layer of soil, consisting wholly of organic matter; not always present.

ocean thermal energy conversion (OTEC) Power generation making use of the temperature difference between deep, cold seawater and warmer near-surface water.

oil Any of various liquid hydrocarbon compounds.

oil sand Loose or semi-consolidated mixture of sand, clay, water, and bitumen.

oil shale A sedimentary rock containing the waxy solid hydrocarbon kerogen.

ore A rock in which a valuable or useful mineral occurs at a concentration sufficiently high to make it economically practical to mine.

organic sediments Carbon-rich sediments derived from the remains of living organisms.

orographic effect The influence of mountains on the flow of air currents that affects the distribution of precipitation.

OTEC *See* ocean thermal energy conversion.

outwash Glacial sediment moved and redeposited by meltwater.

oxbows Old meanders now cut off or abandoned by a stream.

oxide Nonsilicate mineral containing oxygen combined with one or more metals.

oxygen sag curve A graph depicting oxygen depletion followed by reoxygenation in a stream system below a source of organic waste matter; caused by aerobic decay of the organic matter.

ozone hole Area over which the ozone concentration in the ozone layer is lower than over surrounding areas; commonly develops annually over Antarctica.

ozone layer An ozone-rich layer within the stratosphere, between about 15 and 35 kilometers above Earth's surface; absorbs potentially harmful ultraviolet radiation.

P

P waves Compressional seismic body waves (primary waves).

paleomagnetism The fossil magnetism preserved in rocks formed in the past.

Paleozoic Geologic era from 541 to 251 million years ago.

Pangaea The name given to the ancient supercontinent that existed some 200 million years ago.

partial melting The process of incomplete melting of a rock due to minerals having different melting temperatures; results in a magma with a composition different from the source rock.

particulates Fine-grained, solid particles suspended in air; includes soot, ash, and dust.

passive margin A geologically quiet continental margin, lacking significant volcanic or seismic activity.

pathogenic Disease-causing; often applied to harmful microorganisms.

pedalfer A moderately leached soil rich in residual iron and aluminum oxide minerals.

pedocal A soil in which calcium carbonate and other readily soluble minerals are retained; characteristic of drier climates.

pegmatite A very coarsely crystalline igneous rock.

percolation Movement of subsurface water through rock or soil under its own pressure.

periodic table A chart in which the chemical elements are arranged in a way that reflects patterns of chemical behavior related to the electronic structure of their atoms.

permafrost A condition found in cold climates, wherein ground remains frozen year-round at some depth below the surface.

permeability A measure of how readily fluid can flow through a rock, sediment, or soil.

petroleum Hydrocarbons derived from organic matter and used as fuel.

pH scale Scale for reporting acidic or alkaline quality of a liquid based on the concentration of hydrogen ions.

Phanerozoic The time from 541 million years ago to the present.

photovoltaic cells (solar cells) Devices that convert solar radiation directly to electricity.

phreatic eruption A violent, explosive volcanic eruption like a steam-boiler explosion, occurring when subsurface water is heated and converted to steam by hot magma underground.

phreatic zone *See* saturated zone.

physical weathering *See* mechanical weathering.

placers The ores concentrated by stream or wave action on the basis of mineral densities and/or resistance to weathering.

plastic deformation Permanent strain in material stressed beyond the elastic limit; the material will not return to its original dimensions when the stress is removed.

plate tectonics The theory that relates large-scale deformation of Earth's crust to the existence and movement of mobile pieces of rigid lithosphere.

plutonic Describes an igneous rock crystallized well below Earth's surface; typically coarse-grained.

point bar A sedimentary feature built in a stream channel, on the inside of a meander, or anywhere the water slows.

point source A single, concentrated, identifiable source of pollutants, such as a sewer outfall or factory smokestack.

polar-wander curve A plot of apparent magnetic pole positions at various times in the past relative to a continent, assuming the continent's position to have been fixed on Earth.

pore pressure Pressure of fluid filling cracks and pores in rock or soil.

porosity Proportion of void space in rock, sediment, or soil.

potentiometric surface A feature analogous to a water table but applied to confined aquifers; indicates the height to which the water's pressure would raise the water if the water were unconfined.

Precambrian The time from the formation of Earth to 541 million years ago (the start of the Cambrian Period).

precautionary principle A concept in international law/diplomacy under which nations' activities may be restricted if there is reasonable likelihood that those activities may be harmful; significant damage (as to the environment) need not already have occurred.

precession Cyclic wobble of Earth's rotational axis, changing its orientation in space.

precursor phenomena Phenomena that precede an earthquake, volcanic eruption, or other natural event, which may be used to predict the upcoming event.

Principle of Original Horizontality The concept that sediments are deposited in approximately horizontal, flat-lying layers.

Principle of Superposition The concept that, in an undisturbed sequence of sedimentary rocks, those on the bottom were deposited first, with successively younger layers deposited atop them.

Prior Appropriation The principle of surface-water law by which users of water from a given source have priority rights to it on the basis of relative time of first use.

proton Subatomic particle with a charge of $+1$ and a mass of approximately one atomic mass unit; generally found within an atomic nucleus.

pump-and-treat Approach to groundwater purification whereby the water is extracted prior to treatment.

pyroclastic flow Denser-than-air flow of hot gas and ash from a volcano.

pyroclastics The hot fragments of rock and magma emitted during an explosive volcanic eruption.

Q

quick clay The sediment formed from glacial rock flour deposited in a marine setting, weakened by subsequent flushing with fresh pore water.

R

rain shadow A dry zone on the downwind side of a mountain range which is caused by loss of moisture from air passing over the mountains.

rare-earth elements Elements with atomic numbers from 57 through 71.

recharge The processes of infiltration and migration by which groundwater is replenished.

recrystallization Atomic rearrangement to form new crystals in a rock while it remains in the solid state; often, the grain sizes in the rock after recrystallization are coarser than in the rock originally.

recurrence interval The average length of time between floods of a given size along a particular stream.

regional metamorphism Metamorphism on a large scale, involving increased heat and pressure; often associated with mountain building.

regolith Surficial sediment deposit formed in place, whether or not capable of supporting life. (*See also* soil.)

remote sensing Investigation without direct contact, as by using aerial or satellite photography, radar, and so on.

reprocessing (nuclear fuel) The treatment of used fuel elements to extract plutonium or other fissionable material to make new fuel elements.

reserves That quantity of a (resource) material that has been found and is recoverable economically with existing technology.

residence time The average length of time a substance persists in a system; may also be defined as (capacity)/(rate of influx).

resources Those things important or necessary to human life and civilization; also, the total quantity of a given (resource) commodity on Earth, discovered and undiscovered, that might ultimately become part of the reserves.

retention pond A large basin designed to catch surface runoff to prevent its flow directly into a stream.

reverse fault A dip-slip fault in which the block above the fault is pushed up and over the lower block; indicates compressional stress.

rhyolite Silica-rich volcanic rock; the volcanic compositional equivalent of granite.

Richter magnitude scale A logarithmic scale used for reporting the size of earthquakes; Richter magnitude (local magnitude, M_L) is proportional to amount of ground shaking (displacement) at the epicenter and is computed on the basis of the amplitude (height) of the largest seismic wave on a seismogram.

Riparian Doctrine The principle of surface-water law by which all landowners bordering a body of water have equal rights to it.

rock cycle The concept that rocks are continually subject to change and that any rock may be transformed into another type of rock through an appropriate geologic process.

rock A solid, cohesive aggregate of one or more types of Earth materials.

rock flour A fine sediment of pulverized rock produced by glacial erosion.

rockfall *See* fall (rock).

rupture Breakage or failure of material under stress.

S

S waves Shear seismic body waves or secondary waves; S waves do not propagate through liquids, and travel more slowly through solids than do P waves.

saltation The process by which particles are moved in short jumps over the ground surface or stream bed by wind or water.

saltwater intrusion A process by which salt water replaces fresh groundwater when the fresh water is being used more rapidly than it is being recharged; especially common in coastal areas.

sanitary landfill A disposal site for solid or contained liquid waste; in simplest form, a dump site at which wastes are covered with layers of earth daily or more often.

saturated zone The region of rock or soil in which pore spaces are completely filled with liquid; also known as the phreatic zone.

scientific method Means of discovering scientific principles by formulating hypotheses, making predictions from them, and testing the predictions.

scrubbers (electrostatic precipitators) Devices that remove pollutant gases and particulates from exhaust gases of power or manufacturing plants.

seafloor spreading The process by which new lithosphere is created at spreading ridges as plates of oceanic lithosphere move apart.

secure landfill A sanitary landfill designed to contain toxic chemical wastes; typically includes one or more impermeable liners and often is monitored by nearby wells.

sediment Surface accumulation of loose, unconsolidated mineral or rock particles.

sedimentary rock A rock formed from sediments at low temperature.

seismic gap A section of an active fault along which few earthquakes are occurring, in contrast to adjacent fault segments; presumably, a locked section of fault.

seismic tomography Technique using velocity variations of seismic body waves to map regions of relatively higher and lower temperature/density within Earth.

seismic waves The form in which most energy is released during earthquakes; divided into body waves and surface waves.

seismograph An instrument used to measure ground motion caused by seismic waves (and other disturbances).

seismology The science of studying the nature, behavior, and timing of seismic waves as they travel through different (Earth) materials.

sensitive clay A material similar in behavior to quick clay, but derived from different materials, such as volcanic ash.

septic system A sewage-disposal method involving the slow dispersal of wastes into soil for natural aerobic breakdown.

sequential use The land-use principle by which the same land is used for two or more different purposes, one after another.

settling pond A pond in which sediment-laden surface runoff is impounded to allow sediment to settle out before the water is released.

shale gas Natural gas occurring in shale typically of low permeability.

shallow-focus earthquakes Those with focal depths less than 100 kilometers.

shear strength The ability of a material to resist shearing stress.

shearing stress Stress that tends to cause different parts of an object to slide past each other across a plane; with respect to mass movements, stress tending to pull material downslope.

shield volcano A volcano with gentle slopes, broad relative to its height, built from many flows of fluid, low-viscosity basaltic lava.

silicate Mineral containing silicon and oxygen and usually one or more additional elements.

sinkhole A circular depression in the ground surface commonly caused by collapse into an underground cavern formed by solution.

slide A form of mass wasting in which a relatively coherent mass of material moves downslope along a well-defined surface.

slip face The downwind side of a dune, on which material tends to assume the slope of the angle of repose of the sediment of the dune.

slow-slip earthquake Displacement along a previously locked section of a fault zone over a period of days to months, far more rapid than creep but gradual enough to cause little or no seismicity while releasing considerable stored energy.

slump A slide moved only a short distance, often with a rotational component to the movement.

soil The accumulation of unconsolidated rock and mineral fragments and organic matter formed in place at Earth's surface; capable of supporting life.

soil moisture The water in the soil in the zone of aeration, or vadose zone.

solifluction Creep or very slow flow in saturated soil.

source separation The sorting of (waste) material by type prior to collection, usually to facilitate the recycling of individual materials or to ready

material for a particular disposal strategy, such as incineration.

speculative resources That quantity of a resource material that may be found in areas where similar deposits are not already known to exist.

spoil banks Piles of waste rock and soil left behind by surface mining, especially strip mining.

stage (stream) The height (elevation) of a stream surface at a given point along the stream's length; usually expressed as elevation above sea level.

storm tide Water level that is the sum of storm surge plus normal tide level.

strain Deformation resulting from the application of stress.

stratovolcano A composite volcanic cone built of interlayered lava flows and pyroclastic materials.

stream A body of flowing water confined within a channel.

stress A force applied to an object.

striations Parallel grooves in a rock surface cut when a glacier containing rock debris flows over that rock.

strike-slip fault A fault with predominantly horizontal displacement.

strip mining The surface-mining technique in which the cover rock or soil is removed from a near-surface mineral or fuel deposit, and the deposit is then extracted by open excavation, in a series of parallel strips.

subduction zone A convergent plate boundary at which oceanic lithosphere is moving beneath another plate (continental or oceanic) into the asthenosphere.

subeconomic resources Those resources already found that cannot be exploited economically with existing technology; also known as conditional resources.

subsurface water The water in the hydrosphere below the ground surface.

sulfate Nonsilicate mineral containing sulfate (SO_4) groups, each made up of one atom of sulfur and four atoms of oxygen.

sulfide Nonsilicate mineral containing sulfur but no oxygen.

surface waves The seismic waves that travel along Earth's surface.

surge (storm) Localized increase in water level of an ocean or large lake; caused by the high winds and extreme low pressure associated with major storms.

suspended load (stream) The material that is light or fine enough to be moved along suspended in the stream, supported by the flowing water.

sustainable development Development that causes neither serious environmental damage nor such acute resource depletion as to imperil future development or quality of life.

T

tailings The piles of crushed waste rock created as a by-product of mineral processing.

talus Accumulated debris from rockfalls and rockslides.

tectonics The study of large-scale movement and deformation of Earth's crust.

tensile stress Stress tending to pull an object apart.

terminal moraine The end moraine marking the farthest advance of a glacier.

terrane A geologically distinct region that can be distinguished from adjacent regions of different geology and/or history; commonly bounded by faults.

theory A generally accepted explanation for a set of data or observations; its validity has been tested by the scientific method.

thermal inversion The condition in which air temperature increases, rather than decreases, with increasing altitude; the overlying layer of warmer air may then trap warm, rising, pollutant-laden gases.

thermohaline circulation Major ocean circulation pattern, driven by winds and by differences in temperature and salinity of water masses.

thrust fault A reverse fault with very shallowly dipping fault plane.

till Poorly sorted sediment deposited by melting glacial ice.

topsoil The topmost portion of soil, richest in organic matter.

trace elements Those elements present in a system at very low concentrations, typically 100 ppm or less.

transform fault A fault between offset segments of a spreading ridge, along which two plates move horizontally in opposite directions.

transuranic Describes elements with atomic numbers higher than that of uranium, none of which occur naturally on Earth, and all of which are radioactive.

triple junction Point at which three plates (and three plate boundaries) meet.

tsunami A seismic sea wave, generated by a major displacement of water, commonly by an earthquake in or near an ocean basin.

turbidity current A denser-than-water suspension of sediment in water that flows rapidly downslope; develops in ocean basins as a result of submarine landslides or earthquakes.

U

ultramafic Extremely rich in ferromagnesian silicates.

ultraviolet Electromagnetic radiation just to the short-wavelength side of the visible light spectrum; biologically hazardous.

unconfined aquifer An aquifer not overlain directly by an aquitard.

uniformitarianism The concept that the same basic physical laws have operated throughout Earth's history, and therefore, by studying present geologic processes and their products, we can infer much about geologic processes operative in the past.

unsaturated zone A partly saturated region of rock or soil, above the water table; also known as the vadose zone.

upstream flood Flood affecting only localized sections of a stream system; caused by such events as an intense, local cloudburst or a dam failure; typically brief in duration.

V

vadose zone *See* unsaturated zone.

ventifact A rock shaped by wind erosion.

viscosity Resistance to flow.

volcanic Describes any igneous rock formed at or near Earth's surface.

Volcanic Explosivity Index A scale for reporting the size of an explosive volcanic eruption.

volcano Structure built of lava and/or pyroclastics erupted from a central pipe or vent.

W

water table The top of the zone of saturation.

watershed *See* drainage basin.

wave base Depth at which the movement of water molecules in waves becomes negligible.

wave refraction The deflection of waves as they approach shore.

wave-cut platform Steplike surface cut in rock by wave action at sea or lake level.

wave refraction The deflection of waves as they approach shore.

weathering The physical and/or chemical breakdown of rocks and minerals.

well-sorted Describes sediments displaying uniform particle size and/or density.

Z

zone of accumulation *See* B horizon.

zone of leaching *See* E horizon.

Index